AF327153

QUATERNARY COASTS OF THE UNITED STATES: MARINE AND LACUSTRINE SYSTEMS

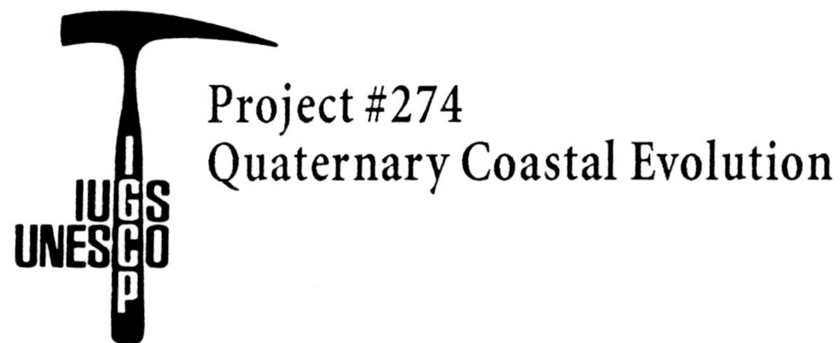

Project #274
Quaternary Coastal Evolution

edited by

Charles H. Fletcher, III

and

John F. Wehmiller

Barbara H. Lidz, Editor of Special Publications
SEPM Special Publication No. 48

Tulsa, Oklahoma, U.S.A. *December, 1992*

A PUBLICATION OF

SEPM (SOCIETY FOR SEDIMENTARY GEOLOGY)

ISBN 0-918985-98-6

P.O. Box 4756
Tulsa, Oklahoma 74131

Printed in the United States of America

PREFACE

IGCP Project #274: Quaternary Coastal Evolution

The International Geological Correlation Programme (IGCP) is a multinational, multidisciplinary effort sponsored jointly by UNESCO and the International Union of Geological Societies (IUGS). Founded in 1973, the aims of IGCP are to foster specific projects that contribute to correlation of stratigraphic sequences and the study of techniques that may ultimately bear on those objectives. IGCP Project 274, "Quaternary Coastal Evolution: Case Studies, Models, and Regional Patterns," conceived as a successor project to the highly successful IGCP Project 200, "Sea-Level Correlations and Applications," was accepted by the IGCP Board in 1988. The primary aims of Project 274 are to: (1) document and explain local to global variations in coastal and continental-shelf evolution, incorporating knowledge of coastal and shelf processes and environment with geodynamic, climatic, oceanographic and other data to produce local and regional models, ranging from descriptive to numerical, leading to a better understanding of interactive forces responsible for past, present, and future changes to the coasts of the world; and (2) promote specified thematic studies, which are necessary to solve problems of coastal change affecting human occupation of the coastal zone. Given the ever-increasing surge of human population to the coastal zone and attendant developmental stresses, not to mention the projections and debates regarding long-term sea-level rise related to climatic change, such endeavors are timely indeed. In addition, the project hopes to develop a globally coherent framework for integrated analysis and prediction of coastal change on different spatio-temporal scales, concentrating on the last 125,000 years.

At present, Project 274 has over 600 corresponding members, representing 69 countries. An important aim of all IGCP projects is to improve international cooperation in science, particularly involving countries in less-developed parts of the world. One of the primary mechanisms for this is international meetings, which incorporate technical sessions and field trips. Including its inaugural meeting in Amsterdam, the Netherlands, in 1988, Project 274 has sponsored global and regional scientific meetings in Malaysia, Thailand, the United States, Argentina, French Guyana, and China. In 1992, sponsored meetings were held in New Zealand and Peru. The Project will culminate in 1993 with meetings in Africa.

Researchers in participating countries are organized into Working Groups. The Working Group of the United States

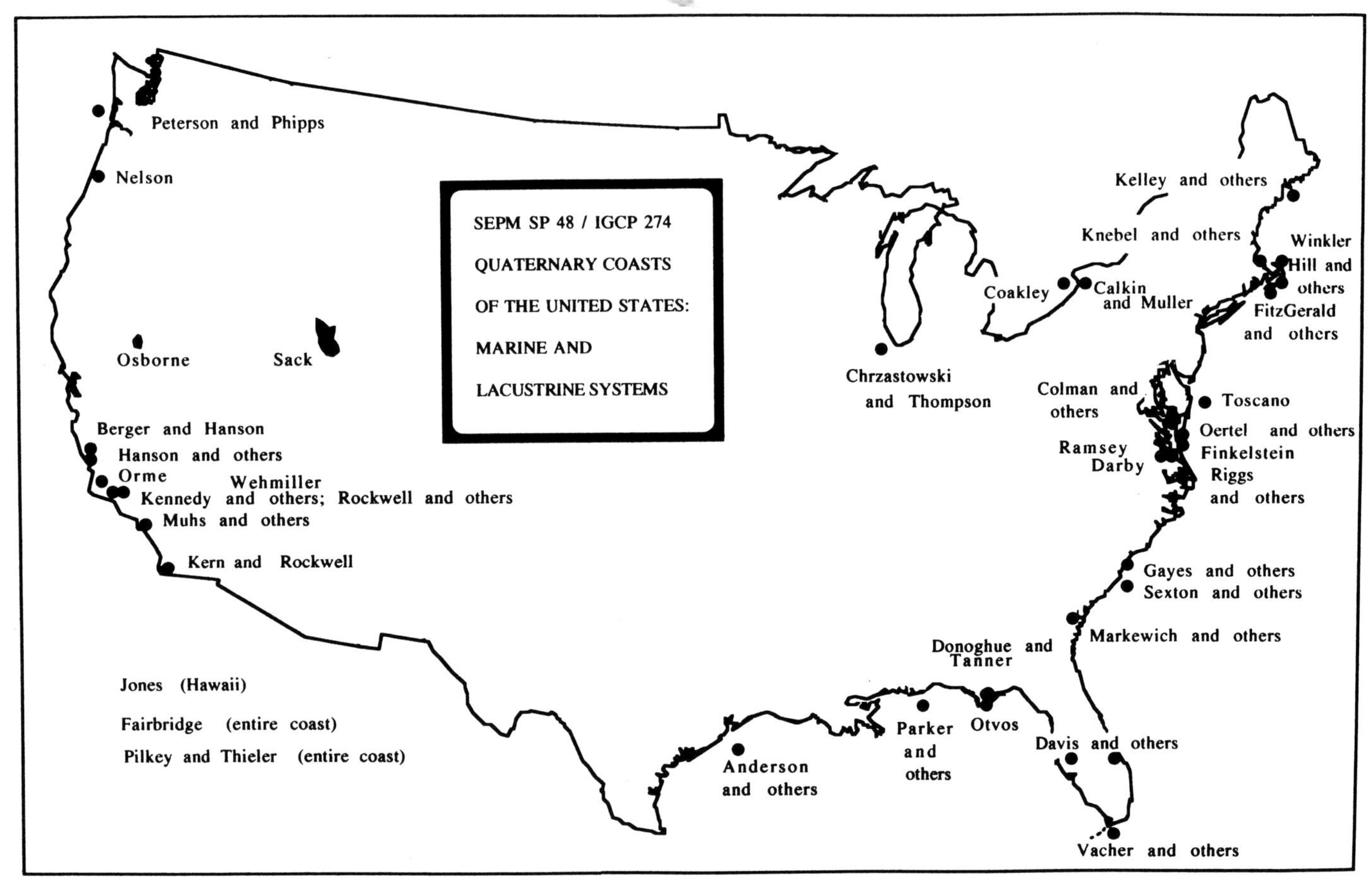

FIG. 1.—Location map of study sites in this volume.

of America has been very active in IGCP Project 274. At this writing, the Group has over 60 members, including two who serve on the Executive Board of the Project. The Radiocarbon Dating Pool, in which several laboratories have generously donated analyses to facilitate dating of coastal sequences from lesser developed countries, is managed from the United States. In 1991, a highly successful research conference, sponsored jointly by SEPM and the U.S. Working Group of IGCP Project 274, was held in Tallahassee, Florida, and attended by over 60 workers from 15 countries.

Formal contributions to IGCP Projects are the publications of individual corresponding members who acknowledge them as such. This SEPM Special Publication, *Quaternary Coasts of the United States: Marine and Lacustrine Systems,* represents the major cumulative contribution of the Working Group of the United States of America to IGCP Project 274. The volume contains 39 papers, and includes sections on Atlantic, Pacific, Gulf, and lacustrine shorelines (Fig. 1), covering both Holocene and Pleistocene deposits. Representing a summary of decades of research into coastal and continental-shelf evolution of North America, this volume should be useful to the international community of coastal researchers.

John R. Suter
National Representative of the United States of America
IGCP Project 274

It is with deep gratitude that we acknowledge the excellent work of the many reviewers of the papers in this volume. Many thanks to: John B. Anderson, Brian Atwater, Abhijit Basu, Daniel F. Belknap, Kelvin Berryman, Harold Borns, W. B. Bryan, David A. Budd, Nicholas K. Coch, Steven M. Colman, Doak C. Cox, T. M. Cronin, Robin Davidson-Arnott, Richard A. Davis, William P. Dillon, Joe Donoghue, Robert H. Dott, Jr., Jules R. DuBar, William Easton, Kenneth Fink, Kenneth Finkelstein, Jane L. Forsyth, Robert W. Frey, Paul T. Gayes, Jerry Glenn, Richard W. Grigg, Gary B. Griggs, Robert B. Halley, P. M. Harris, Sytze van Heteren, Mike C. Hill, Al C. Hine, Paul F. Karrow, Philip J. Kern, Harley J. Knebel, E. D. Koozmin, J. Chris Kraft, LaVerne Kulm, Rufus Le Blanc, Mike Machette, Fred T. Mackenzie, Louis J. Maker, Robert McMaster, Dorothy Merritts, James G. Moore, Randall L. Moory, Daniel R. Muhs, Charles Nittrouer, Robert N. Oldale, Ervin Otvos, Charles G. Oviatt, J. P. Owens, Jeff Paine, Curt Peterson, Orrin Pilkey, James E. Pizzuto, Orson van de Plassche, Kelvin Ramsey, Patricia R. Sanford, E. A. Shinn, Frank W. Stapor, Robert S. Tie, Gerald Weber, S. Jeffress Williams, Linda York, Kimo Zaiger.

Chip Fletcher
John Wehmiller

CONTENTS

PART IV. PACIFIC COASTAL SYSTEMS

PART V. LACUSTRINE COASTAL SYSTEMS

PART I
NATIONAL PERSPECTIVES

EROSION OF THE UNITED STATES SHORELINE

ORRIN H. PILKEY, JR AND E. ROBERT THIELER
Duke University, Department of Geology, Old Chemistry Building, Durham, North Carolina 27706

ABSTRACT: Over 75 percent of the United States ocean shoreline is eroding (retreating landward). Shoreline progradation, where occurring, is generally assumed to be a temporary phenomenon. When affecting a developed area, shoreline retreat is usually termed *erosion*, but considerable confusion remains over the use of this term. *Retreat* and *progradation* refer to a change in shoreline position, whereas *erosion* and *accretion* refer to volumetric changes in the subaerial beach. As used in this paper, however, *erosion* refers to any form of shoreline retreat, consistent with common usage.

Coastal erosion is a fundamental and widespread process on U.S. and world shorelines. Evidence, particularly on barrier-island coasts, indicates that in the past few decades or millenia, erosion may have become a more widespread process. Possible causes of this change include the effect of humans, shoreface steepening, or an increase in the rate of eustatic sea-level rise.

Mechanisms responsible for shoreline erosion are highly variable, both temporally and geographically. In addition, our understanding of shoreline sediment-transport dynamics is incomplete. Consequently, we are presently unable to predict accurately future shoreline-retreat rates related to continued sea-level rise. The Bruun Rule, for example, predicts little shoreline retreat relative to using, as a predictive tool, the slope of the land surface over which sea level is expected to rise.

WHAT IS COASTAL EROSION?

Defining erosion in the coastal environment is difficult. Various terms are used to describe the process, including *drowning, coastal erosion, shoreline erosion, beach erosion, shoreline retreat, beach retreat,* and *shoreline recession*. Strictly speaking, *retreat* and *progradation* refer to a change in shoreline position; *erosion* and *accretion* refer to volumetric changes in the subaerial beach (Wood and others, 1988; Oertel and others, 1989). The term *beach erosion,* however, is deeply ingrained in the public lexicon to refer to any form of shoreline retreat (usually the change in position of the high-tide line); we shall here use it in the same broad sense as does the general public.

When dealing with shorelines and beaches, it is crucial to distinguish between *erosion* and an *erosion problem*. Many kilometers of undeveloped U.S. shoreline are eroding (retreating landward) and, as a rule, such locations are not considered to be societal problems. Erosion and retreat of the shoreline are, in fact, critically important processes in the evolution of Holocene coastal landforms. Estuaries, lagoons, spits, barrier islands, sea cliffs, and many other coastal landforms all owe their shape and depositional patterns to erosion, at least in part. It is only when humans interfere with or get in the way of shoreline erosion that it is recognized as a problem. A broad discussion of the U.S. coastal-erosion problem, along with methodology, political implications, and costs for mitigation is furnished by the National Research Council (1990).

U.S. SHORELINE-EROSION RATES

Erosion rates for the contiguous U.S. shoreline, based on data in the Coastal Erosion Information System (CEIS; May and others, 1982), have been compiled by May and others (1983) and Dolan and others (1985; Tables 1 and 2; Fig. 1). CEIS includes shoreline-change data for the Atlantic, Gulf of Mexico, Pacific and Great Lakes coasts, as well as for major bays and estuaries. The data in CEIS are drawn from a variety of sources, including published reports, historical shoreline-change maps, field surveys, and aerial-photograph analyses. However, the lack of a standard method among coastal scientists for analyzing shoreline changes has resulted in the inclusion of data utilizing a variety of reference features, measurement techniques, and rate-of-change calculations. Thus, whereas CEIS represents the best available data for the U.S. as a whole, much work is needed to document regional and local erosion rates accurately.

When examined in detail, such as on a state-by-state basis, the actual percentage of eroding shoreline is generally much higher than the coastwide averages shown in Figure 1 suggest. This discrepancy is likely due to the large amount of shoreline for which "no data" exist and the coarse resolution of CEIS. Dolan and others (1990b), for example, report that 78 percent of the Delaware shoreline is eroding; Maryland, 94 percent; Virginia, 72 percent; and North Carolina, 72 percent. Griggs (1987a) states that approximately 86 percent of the California shoreline is eroding.

WHY UBIQUITOUS EROSION?

Coastal erosion is a fundamental and widespread process not only on U.S. but also world shorelines. There is some evidence, however, that such has not always been the case. Examination of the U.S. barrier-island shoreline, for example, indicates that fundamental changes in depositional patterns have recently taken place. Many barrier islands are regressive, indicating that a seaward accretion or widening of the island occurred over the last 2 to 4 ka. Within the last several hundred years, however, a profound change has taken place and shoreline retreat has replaced shoreline progradation. In other words, barrier islands are becoming increasingly transgressive. Most regressive barrier islands (e.g., Bogue Banks, North Carolina, and Galveston Island, Texas) are now eroding on both the ocean and lagoon sides.

Possible explanations for increased worldwide erosion include the effects of humans—shoreline stabilization and river damming; sand supply reduction—shoreface steepening; and a recent increase in the rate of eustatic sea-level rise. However appealing, each of these possibilities leaves certain questions unanswered. For example, humans certainly contribute to coastal erosion in some locations, but completely undeveloped areas such as the barrier-island system on the Pacific coast of Colombia are also eroding at high rates.

Quaternary Coasts of the United States: Marine and Lacustrine Systems, SEPM Special Publication No. 48

TABLE 1.—SHORELINE RATES OF CHANGE FOR REGIONS AND STATES

Region	Mean (m/yr)[1]	Standard Deviation	Range[1]
Atlantic Coast	−0.8	3.2	25.5/−24.6
Maine	−0.4	0.6	1.9/−0.5
New Hampshire	−0.5	—	−0.5/−0.5
Massachusetts	−0.9	1.9	4.5/−4.5
Rhode Island	−0.5	0.1	−0.3/−0.7
New York	0.1	3.2	18.8/−2.2
New Jersey	−1.0	5.4	25.5/−15.0
Delaware	0.1	2.4	5.0/−2.3
Maryland	−1.5	3.0	1.3/−8.8
Virginia	−4.2	5.5	0.9/−24.6
North Carolina	−0.6	2.1	9.4/−6.0
South Carolina	−2.0	3.8	5.9/−17.7
Georgia	0.7	2.8	5.0/−4.0
Florida	−0.1	1.2	5.0/−2.9
Gulf of Mexico	−1.8	2.7	8.8/−15.3
Florida	−0.4	1.6	8.8/−4.5
Alabama	−1.1	0.6	0.8/−3.1
Mississippi	−0.6	2.0	0.6/−6.4
Louisiana	−4.2	3.3	3.4/−15.3
Texas	−1.2	1.4	0.8/−5.0
Pacific Coast	0.0	1.5	10.0/−4.2
California	−0.1	1.3	10.0/−4.2
Oregon	−0.1	1.4	5.0/−5.0
Washington	−0.5	2.2	5.0/−3.9

[1]Positive values indicate accretion; negative values indicate erosion. (After Dolan and others, 1985.)

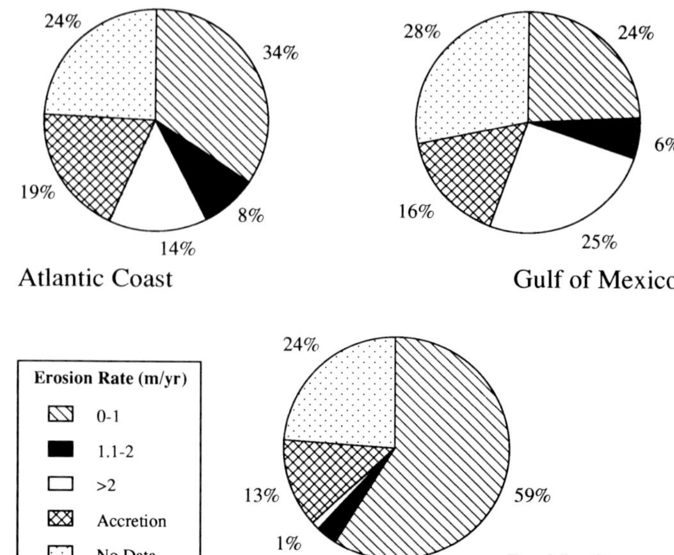

FIG. 1.—Pie charts of erosion rates on the U.S. open-ocean coastline. The large amount of shoreline classified as "no data" likely results in an artificially low figure for the total percentage of eroding shoreline (see text for discussion). (After Dolan and others, 1990a).

TABLE 2.—SHORELINE RATES OF CHANGE FOR COASTAL LANDFORM TYPES

Type/Region	Mean (m/yr)[1]	Standard Deviation	Range[1]
Barrier islands			
Atlantic Coast	−0.8	3.4	25.5/−24.6
Maine-New York	−0.3	2.6	4.5/−1.5
New York-North Carolina	−1.5	4.5	25.5/−24.6
North Carolina-Florida	−0.4	2.6	9.4/−17.7
Gulf of Mexico	−0.6	1.5	8.8/−4.3
Florida-Louisana	−0.5	1.7	8.8/−4.3
Louisiana-Texas	−0.8	1.2	0.8/−3.5
Sand beaches			
Atlantic Coast	−1.0	1.0	2.0/−4.5
Maine-Massachusetts	−0.7	0.5	−0.5/−2.5
Massachusetts-New Jersey	−1.3	1.3	2.0/−4.5
Gulf of Mexico	−0.4	1.6	8.8/−4.5
Pacific Coast	−0.3	1.0	0.7/−4.2
Sand beaches with rock headland	−0.3	1.9	10.0/−5.0
Pocket beaches			
Atlantic Coast	−0.5	—	−0.5/−0.5
Pacific Coast	−0.2	1.1	5.0/−1.1
Deltas	−2.5	3.5	8.8/−15.3
Mud flats			
Gulf of Mexico	−1.9	2.2	3.4/−8.1
Florida	−0.3	0.9	1.5/−1.5
Louisiana-Texas	−2.1	2.2	3.4/−8.1
Rock shorelines			
Atlantic Coast	1.0	1.2	1.9/−4.5
Pacific Coast	−0.5	—	−0.5/−0.5

[1]Positive values indicate accretion; negative values indicate erosion. (After Dolan and others, 1985.)

LOCAL FACTORS AFFECTING EROSION

On all types of sandy or unconsolidated coasts, beach erosion can be considered to be controlled by a dynamic equilibrium involving three major components: sediment supply, wave and tidal energy, and the position of, or changes in, sea level.

On barrier-island coasts, erosion rates must be examined in the context of the total barrier-island system. The concurrent development of transgressive and regressive barriers along the Texas coast, for example, is related to the longshore distribution of sediment sources and sinks (Morton, 1979; Nummedal, 1983).

Shoreline erosion on the Texas coast predominates at deltaic headlands, such as the Rio Grande and Brazos deltas. In the central portion of each embayment between the deltaic headlands, accretion or at least relatively slow erosion occurs (Morton, 1979). The eroding deltaic headlands are the source of sand for the accreting shoreline between river mouths.

Evolution of the Louisiana coast is similarly affected by regional-scale processes. Penland and others (1985) demonstrated a cyclic relation between Mississippi Delta lobe-switching events and barrier development and destruction. Extremely rapid shoreline-erosion rates prevail no matter what the stage of barrier development in this cycle. The high overall erosion rates are fundamentally due to subsidence-induced rapid rates of sea-level rise (0.50–1.00 cm/yr; Penland and others, 1985). However, the rates of shoreline recession, which vary from 5 to 15 m/yr (Penland and others, 1985), depend partly on the age of the individual barrier-island chains. Beaches on young barrier chains on the Mississippi Delta (e.g., Isles Dernieres) have relatively lower erosion rates because the sand supply (the eroding deltaic lobe) is close to sea level. As the lobe subsides, the sand source is effectively removed and shoreline-erosion rates increase.

Antecedent geology has also been shown to affect barrier development and hence shoreline-erosion rates. Barrier islands grounded on antecedent topographic features (e.g., Evans and others, 1985) may, at least temporarily, experience reduced erosion rates. In addition, the Holocene/Pleistocene stratigraphy underlying the barrier islands may play a major role in determining shoreline-erosion rates. Thick sequences of lagoonal or salt-marsh mud are responsible for the rapid shoreline-erosion rates of some islands along the Delmarva Peninsula (Demarest and Leatherman, 1985). The high beach-erosion rates are due to a combination of (1) ease of erosion of a muddy shoreface, and (2) a lack of sand contribution from the eroding shoreface (Demarest and Leatherman, 1985).

Waves associated with extratropical and tropical storms cause the most visible and obvious shoreline erosion, but often storm-caused erosion is substantially ''repaired'' by post-storm onshore and longshore sediment transport. Shoreline retreat may occur under both fairweather and storm conditions. The relative proportion of fairweather and storm retreat is not well documented, but it probably varies widely from beach to beach. In addition, different shorelines are adjusted to different wave climates. New England and Mid-Atlantic shorelines, for example, frequently retreat on an annual basis in response to ''northeaster'' storms. Erosion of Florida's Gulf of Mexico beaches, on the other hand, is most affected by hurricanes, often spaced decades apart.

The role of humans in causing coastal retreat ranges in scope from global to local. Global production of excess CO_2 is widely assumed to be leading to ''greenhouse effect''-related sea-level rise and accelerated erosion. Locally, the damming of rivers is cutting off a major source of sand for many beaches, which accelerates erosion. This problem is acute for California beaches (Griggs, 1987b). Flood prevention on rivers by the armoring of channel banks and construction of artificial levees accelerates erosion by inhibiting vertial accretion of river deltas and further reducing beach sediment supply.

The myriad sea walls, breakwaters, groins, and jetties that line developed shorelines divert offshore, slow down, trap, and otherwise reduce the regional beach sediment supplied by longshore currents, and thereby increase erosion rates. Seawall construction along formerly glaciated coasts has cut off the supply of sand normally contributed to the beaches by eroding bluffs. This phenomenon has led the state governments of Massachusetts and Maine to prohibit the construction of sea walls in front of bluffs.

A spectacular example of the role of jetties in affecting beach sand supply, and also a spectacular example of beach erosion and barrier-island migration, are afforded by the northern end of Assateague Island, Maryland (Leatherman, 1984). After the Ocean City Inlet opened during a hurricane in 1933, jetties were constructed to stabilize the inlet. Since that time, sand that would have been transported across the inlet to Assateague has been accumulating updrift of the jetties in front of Ocean City, or offshore at the end of the jetties. This diversion of sediment has caused the Assateague shoreline to retreat and the island itself to migrate landward. Today, the open-ocean surf zone of Assateague Island is landward of the island's 1933 lagoon shoreline.

The causes and mechanics of erosion are more complex on rocky coasts than on unconsolidated coasts. Processes involved in the erosion of rocky coasts are summarized in Trenhaile (1987). Factors controlling erosion on the California coast are addressed by Griggs and Savoy (1985).

The degree of cliff retreat in California depends on (1) orientation or exposure, (2) local wave climate, and (3) the physical properties of the cliff. Important rock properties include hardness, consolidation, and the presence or absence of internal weaknesses such as joints, fractures, and faults. Crystalline rocks such as granite tend to erode irregularly and very slowly, producing a highly irregular coastline. Sedimentary rock, broadly speaking, tends to erode somewhat more regularly and rapidly, producing relatively straight coasts.

Both marine and nonmarine processes are involved in cliff retreat (Norris, 1990; Emery and Kuhn, 1982). Marine erosion includes direct wave attack, a variety of chemical-solution processes, and flaking by salt-crystal expansion. Nonmarine cliff erosion is accomplished by chemical and mechanical processes usually involving rain water. Groundwater seepage along contacts and joints is often a critical part of cliff retreat. Along the California coast, the effects of seepage have been increased by lawn watering and septic-tank emplacement (Norris, 1990).

Mass movement in the form of slumps and various types of landslides is frequently the major process of retreat and is responsible for the very uneven spatial and temporal retreat rates of cliffs. Emery and Kuhn (1982) present a classification of coastal cliffs based in part on the relative importance of marine and nonmarine processes.

According to Griggs and Savoy (1985), erosion rates along the rocky coast of California are highly variable, ranging from negligible to 3 to 4 m/yr. Typically, shoreline-retreat rates are on the order of 10 to 20 cm/yr. Griggs and Savoy also note that the highly episodic nature of erosion rates here is strongly dependent on storminess. Frequent storms translate to high-erosion rates, and the degree of storminess itself is episodic.

PREDICTING FUTURE SHORELINE CHANGES

One of the most critical research problems in coastal geology today is predicting the effect of sea-level rise on shoreline-retreat rates. Coastal communities need accurate predictions of shoreline position on a time frame of decades, an impossibility at present.

Were sea level to rise catastrophically 10 m in the next year, there is no question that the shoreline would be inland at what is now the 10-m contour. Predicting shoreline behavior at a slower rate of sea-level rise, however, is much more difficult. The problem is complex, and standard engineering models such as the Bruun Rule (Bruun, 1962) are too rigid and are based on too many narrow assumptions to have wide application (Kraft and others, 1987; Orford, 1987; SCOR Working Group 89, 1991).

Methods of predicting future shoreline changes include extrapolation of present erosion rates, application of predictive models, and use of numerical models. Present erosion rates are often used by states to determine building setbacks (e.g., in North Carolina and South Carolina). The basic problem with this method, however, is that recent trends and their underlying causes may not in any way reflect the future. Dolan and others (1991) provide a review of the methods used to calculate rates of shoreline change using historical data.

A number of fairly simple predictive models are used to project future shoreline positions, including the Bruun Rule (Bruun, 1962; 1983), the Generalized Bruun Rule (Dean and Maurmeyer, 1983), the Edelman method (Edelman, 1970), and various others (e.g., Everts, 1987). The basis for these models is the assumption that the shoreface will retreat landward and upward due to a rise in sea level. Most of the models assume: (1) a constant wave climate, (2) a purely sandy shoreface, (3) no longshore loss of sand, (4) sand transport by incident waves only, and (5) no sand transport seaward beyond the shoreface by storms. These assumptions are never completely valid. Clearly, the slope of the surface over which a shoreline will retreat must also be a factor; this consideration, however, is not a part of any current models.

Pilkey and Davis (1987) applied several of these models to the barrier-island coast of North Carolina. South of Cape Lookout, the Bruun Rule, Generalized Bruun Rule and a model incorporating the slope of the migration surface all predicted similar recession for a given rise in sea level. North of Cape Lookout, consideration of the slope of the migration surface predicted a much greater recession than did Bruun-related models, perhaps because the islands are in an "out-of-equilibrium position" with respect to present sea level. That is, the recent (since about 4.5 ka) near-stillstand of relative sea level combined with a large sediment supply has permitted these islands to remain in place, while the back-barrier lagoons (Albemarle and Pamlico Sounds) have widened.

Numerical models of the coastal environment have become important tools for studying shoreline changes. Several approaches to modeling the coastal environment are given by Fox (1985) and Lakhan and Trenhaile (1989). It is important to note, however, that the coastal system consists of many poorly understood components (e.g., shoreface, tidal-inlet, and so forth) that cannot yet be modeled realistically. Wright and others (1991), for example, demonstrate that much more is involved in shoreface sediment transport than just the incident wave field considered in most models. It is doubtful that any existing models can predict shoreline-erosion rates with an accuracy useful to coastal communities.

REFERENCES

Bruun, P., 1962, Sea-level rise as a cause of shore erosion: Proceedings, American Society of Civil Engineers, Journal of the Waterways and Harbors Division, v. 88, p. 117–130.

Bruun, P., 1983, Review of conditions for uses of the Bruun Rule of erosion: Coastal Engineering, v. 7, p. 77–89.

Dean, R. G., and Maurmeyer, E. M., 1983, Models for beach profile responses, *in* Komar, P. D., ed., Handbook of Coastal Processes and Erosion: Boca Raton, Florida, CRC Press, p. 151–166.

Demarest, J. M., and Leatherman, S. P., 1985, Mainland influence on coastal transgression: Delmarva Peninsula, *in* Oertel, G. F., and Leatherman, S. P., eds., Barrier Islands: Marine Geology, v. 63, p. 19–33.

Dolan, R., Anders, F., and Kimball, S., 1985, Coastal Erosion and Accretion: National Atlas of the United States of America: Reston, Virginia, Department of the Interior, U.S. Geological Survey, 1 sheet.

Dolan, R., Fenster, M., and Holme, S., 1990a, Erosion of U.S. shorelines: Geotimes, v. 35, p. 22–24.

Dolan, R., Fenster, M., and Holme, S., 1991, Temporal analysis of shoreline recession and accretion: Journal of Coastal Research, v. 7, p. 723–744.

Dolan, R., Trossbach, S., and Buckley, M., 1990b, New shoreline erosion data for the mid-Atlantic coast: Journal of Coastal Research, v. 6, p. 471–477.

Edelman, T., 1970, Dune erosion during storm conditions: Proceedings, 12th International Conference on Coastal Engineering: American Society of Civil Engineers, New York, p. 1305–1307.

Emery, K. O., and Kuhn, G. G., 1982, Sea cliffs: their processes, profiles, and classification: Geological Society of America Bulletin, v. 93, p. 644–654.

Evans, M. W., Hine, A. C., Belknap, D. F., and Davis, R. A., 1985, Bedrock controls on barrier island development: west-central Florida coast, *in* Oertel, G. F., and Leatherman, S. P., eds., Barrier Islands: Marine Geology, v. 63, p. 263–283.

Everts, C. H., 1987, Continental shelf evolution in response to a rise in sea level, *in* Nummedal, D., Pilkey, O. H., and Howard, J. D., eds., Sea-Level Fluctuation and Coastal Evolution: Society of Economic Paleontologists and Mineralogists Special Publication 41, p. 49–57.

Fox, W. T., 1985, Modeling coastal environments, *in* Davis, R. A., ed., Coastal Sedimentary Environments: Springer-Verlag, New York, p. 665–705.

Griggs, G. B., 1987a, California's retreating shoreline: the state of the problem, *in* Magoon, O. T., ed., Coastal Zone '87, American Society of Civil Engineers, New York, p. 1370–1383.

Griggs, G. B., 1987b, The production, transport and delivery of coarse-grained sediment by California's coastal streams, *in* Kraus, N. C., ed., Coastal Sediments '87, American Society of Civil Engineers, New York, p. 1825–1838.

Griggs, G. B., and Savoy, L. E., 1985, Living With the California Coast: Duke University Press, Durham, North Carolina, 393 p.

Kraft, J. C., Chrzastowski, M. J., Belknap, D. F., Toscano, M. A., and Fletcher, C. H., 1987, The transgressive barrier-lagoon coast of Delaware: morphostratigraphy, sedimentary sequences and responses to relative rise in sea level, *in* Nummedal, D., Pilkey, O. H., and Howard, J. D., eds., Sea-Level Fluctuation and Coastal Evolution: Society of Economic Paleontologists and Mineralogists Special Publication 41, p. 127–143.

Lakhan, V. C., and Trenhaile, A. S., eds., 1989, Applications in Coastal Modeling: Elsevier Science Publishers, Amsterdam, 387 p.

Leatherman, S. P., 1984, Shoreline evolution of Assateague Island: Shore and Beach, v. 52, p. 3–10.

May, S. K., Dolan, R., and Hayden, B. P., 1983, Erosion of U.S. shorelines: EOS, Transactions, American Geophysical Union v. 64, p. 521–523.

May, S. K., Kimball, W. H., Grady, N., and Dolan, R., 1982, CEIS: The Coastal Erosion Information System: Shore and Beach, v. 50, p. 19–26.

Morton, R. A., 1979, Temporal and spatial variations in shoreline changes and their implications, examples from the Texas Gulf Coast: Journal of Sedimentary Petrology, v. 49, p. 1101–1111.

National Research Council, 1990, Managing Coastal Erosion: National Academy Press, Washington, D.C., 182 p.

Norris, R. M., 1990, Sea cliff erosion: A major dilemma: California Geology, August, p. 171–177.

Nummedal, D., 1983, Barrier islands, *in* Komar, P. D., ed., Handbook of Coastal Processes and Erosion: CRC Press, Boca Raton, Florida, p. 77–121.

Oertel, G. F., Ludwick, J. C., and Oertel, D. L. S., 1989, Sand accounting methodology for barrier island sediment budget analysis, *in* Stauble, D. K., ed., Barrier Islands: Process and Management: American Society of Civil Engineers, New York, p. 43–61.

Orford, J., 1987, Coastal processes: the coastal response to sea-level variation, *in* Devoy, R. J. N., ed., Sea Surface Studies: A Global View: Croom Helm, London, p. 415–463.

Penland, S., Suter, J. R., and Boyd, R., 1985, Barrier island arcs along abandoned Mississippi River deltas, *in* Oertel, G. F., and Leatherman, S. P., eds., Barrier Islands: Marine Geology, v. 63, p. 197–233.

Pilkey, O. H., and Davis, T. W., 1987, An analysis of coastal recession models: North Carolina coast, *in* Nummedal, D., Pilkey, O. H., and Howard, J. D., eds., Sea-Level Fluctuation and Coastal Evolution: Society of Economic Paleontologists and Mineralogists Special Publication 41, p. 59–68.

SCOR WORKING GROUP 89, 1991, The response of beaches to sea-level changes: A review of predictive models: Journal of Coastal Research, v. 7, p. 895–921.

TRENHAILE, A. S., 1987, The Geomorphology of Rock Coasts: Oxford University Press, New York, 384 p.

WOOD, W. L., HOOVER, J. A., STOCKBERGER, M. T., AND ZHANG, Y., 1988, Coastal Situation Report for the State of Indiana: Great Lakes Coastal Research Laboratory, Purdue University, West Lafayette, Indiana, 190 p.

WRIGHT, L. D., BOON, J. D., KIM, S. C., AND LIST, J. H., 1991, Modes of cross-shore sediment transport on the shoreface of the Middle Atlantic Bight: Marine Geology, v. 96, p. 19–51.

HOLOCENE MARINE COASTAL EVOLUTION OF THE UNITED STATES

RHODES W. FAIRBRIDGE
Columbia University and NASA-GISS, 2880 Broadway, New York 10025

ABSTRACT: The Holocene began about 10.7 ka following the glacial readvance of the Younger Dryas (Valders) interval that terminated the Pleistocene. World sea level had fallen to about −54 m. Nearly half the present United States continental shelf was then a coastal plain with vegetation ranging in nature from subarctic tundra to coniferous woodland. A fluctuating transgression (''Flandrian'' stage) followed, accompanied by rising temperatures. Reaching the present coastline about 6 ka, the transgressing seas drowned river valleys, creating estuaries and dendritic embayments. As barrier spits and islands developed, the estuaries and embayments became lagoons.

During the later Holocene, world sea level was modulated by numerous negative fluctuations, signaling cool intervals that are indicated by pollen analyses and neoglacial advances. Extra-warm cycles were characterized by higher storm frequency; in areas favorable for preservation, these conditions are recorded by distinctive sets of beach ridges.

Local paleogeography was dominated in most places by a tectonic ''groundswell,'' such as is presently found throughout the collapsing forebulge regions along subducting plate margins of southern Alaska and on the taphrogenic coast of California. More or less stable platforms are found only in northwest Alaska and Florida.

Present coasts are widely affected by the post-Little Ice Age warming, which has led to steric and glacio-eustatic sea-level rise. The warming has also slowed the velocity of the Gulf Stream and other geostrophic currents, reducing the Coriolis-controlled dynamic tilt and causing further sea-level rise along the mainland coasts.

INTRODUCTION

Paleogeography is an exercise in imaginative insight. One needs first to observe and understand the dynamic physical processes and landforms of the present and then transfer and re-apply these concepts to past landscapes. The purpose of this paper is to introduce the reader to the present coasts of the United States by means of a brief history of the last 10.7 millennia (the Holocene Epoch). This interval saw the arrival and settlement of the early humans, whose old midden sites (usually shellfish remains) yield informative traces of the former economy. Apart from modern anthropogenic effects, the Holocene appears to be typical of an interglacial cycle. In geology, physical evolution of the Earth surface does not destroy all traces of past events but builds on them (Fairbridge, 1989). Thus, each cycle is dependent on its predecessors and is never an exact replication of any earlier cycle, i.e., many of the present coasts are built on, or recycle materials of, earlier interglacial shores.

The Milankovitch orbital mechanisms provided by the Earth and Moon dictate that an interglacial-type climatic evolution is to be expected for the Holocene Epoch. Three principal parameters are involved—eccentricity, tilt, and precession—but each is constrained by different periodicities (at 93, 41, and 20 ka, respectively). The interaction of these parameters creates a continuously changing variable in effective solar radiation. As a result, no two interglacial stages are quite the same (Berger, 1989), and although one can look to former interglacials for analogous relations, the Holocene evidence must be treated quite distinctly and pragmatically. In this case, climatic modeling by Kutzbach (1984) is instructive.

The most striking point of the climatic modeling is that, within the latitudinal span of the conterminous United States and Alaska, the early Holocene was characterized by much warmer summers and somewhat colder winters than at present. The seasonal contrast was much greater than at present, especially in high latitudes, because of the precessional cycle. An important rule of climatology specifies that small temperature changes in low latitudes are greatly amplified at high latitudes; thus a 1°C change in summer mean temperature for Florida could be matched by a 10°C swing in Alaska. From time to time, at about 500- to 1,500-yr intervals, the atmosphere has cooled, as shown by paleobotany (Davis, 1984; Baker, 1984), glacial readvances in the mountains (Davis, 1988), and sea-level oscillations (Colquhoun and Brooks, 1986; and others).

HOLOCENE EUSTATIC RISE

During the early Holocene, a rapid increase in the level of Northern Hemisphere summer insolation led to a vast meltdown of the great mid-latitude ice sheets. Although the Laurentide Ice Sheet possessed by far the largest volume, punctuation of the melting history and consequent sea-level rise must have occurred in discrete steps as smaller glacial units melted. As a result, the fluvial input of meltwaters to the world ocean occurred not in the bell-curve form of a neatly predictable discharge, but rather in the form of steps. Superimposed were the effects of solar-induced climate oscillations. Rates of sea-level rise varied considerably, but at times reached 20 to 30 mm/yr, an order of magnitude higher than most 20th-century rates (Fairbanks, 1989).

In addition to the exogenetic forcing variables, one should also consider the role of the rising eustatic sea level itself, which probably caused large segments of ice shelves to become buoyant, break off, and float away into warmer latitudes. Evidence of such processes is also found in records from the last interglacial (Hollin, 1980). During the last glacial maximum about 18 ka, ice shelves extended to the edge of the continental shelf, especially around Antarctica (Greischar and Bentley, 1980), but by 6 ka or so they had been progressively trimmed back. Ice tongues must have broken off in large segments due to the buoyancy provided by step-like eustatic rises. An acceleration of the present rate of sea-level rise would justify concern over comparable surges (Anderson and Thomas, 1991) and NASA presently plans specific satellite surveillance.

Studies of Holocene climate change in general (e.g., Lamb, 1977; Rampino and others, 1987; Oliver and Fairbridge, 1987) disclose a complex history of fluctuations in

Quaternary Coasts of the United States: Marine and Lacustrine Systems, SEPM Special Publication No. 48
 ISBN 0-918985-98-6

temperature, precipitation and other variables. These fluctuations have led to "neoglacials" and discrete events such as the "Viking" warm phase (or "Little Climatic Optimum") and the Little Ice Age (Lamb, 1977).

These events were definitely non-local but world-wide phenomena, expressed variously in different regions. In closed basins, the rise and fall of lake levels is particularly striking (Dean and others, 1984; Harrison, 1989). In the subtropics, fluctuations of precipitation are more important than those of temperature, and the neoglacials are often documented (by tree rings or pollen records) as times of catastrophic drought (Stockton and others, 1985; Petersen, 1990). On open coasts in the lower latitudes, however, the minor fluctuations in precipitation or temperature might well pass unnoticed.

Early models of the Flandrian transgression (Paepe and Baeteman, 1979), based only on a few dated samples, depicted the transgression as either a smooth rise (Shepard, 1963) or as a generalized fluctuating sequence (Fairbridge, 1961a). The two are compared by Bird (1984) and by Kidson (*in* van de Plassche, 1986). Neither model is satisfactory at the present time.

The post-glacial sea-level rise did not follow a simple smooth curve or a broadly fluctuating one. Both models are incorrect, being based upon inadequate data, imprecise chronologies and, in part, conceptual limitations. In subsequent years, a succession of curves has been generated by various authors, reflecting reliance upon different points of view, data from different regions, and from different data sources.

The fluctuating eustatic concept of Fairbridge (1961a) was generated simply as a global working model, based upon geomorphologic and stratigraphic evidence of abrupt transgressions and regressions, and climatic indicators (primarily from Scandinavia). Sea-level data were integrated into this framework, and radiocarbon dates relating directly to sea level were subsequently added to the original curve to provide some chronologic support. As pointed out by Bird (1984), these well-known eustatic curves in fact agree quite well (Fig. 1) with the differences reflecting different degrees of analysis and regional aspects. Any scientist, however, who still uses either of these outdated conceptual models as a tool for explaining modern coastal morphology and late Holocene stratigraphy can be led to commit serious errors. The patterns of post-glacial sea level are very complicated and regionally varied (Tooley and Shennan, 1987).

Records of the post-glacial eustatic rise prior to 6 ka reveal four major fluctuations, each separated by extremely steep sections on the sea-level curve. During these times of rapid sea-level rise, mean annual rates at times must have exceeded 20 to 30 mm/yr over several decades (Fairbridge, 1961a; Mörner, 1969; Fairbanks, 1989). Abundant evidence also points to intervals of global cooling, usually for periods of less than 300 to 500 yr, implying a glacial readvance phase, at least in the mountains of the "sensitive (mid-) latitudes" (Fairbridge, 1961b) The amplitude of some of these eustatic fluctuations was quite large, on the order of several meters, especially prior to 6 ka. Given that 1 m of eustatic change represents 0.4×10^6 km^3 of ice, the observed fluctuations must have involved a volume of continental ice comparable to that of contemporary Greenland (2.6×10^6 km^3).

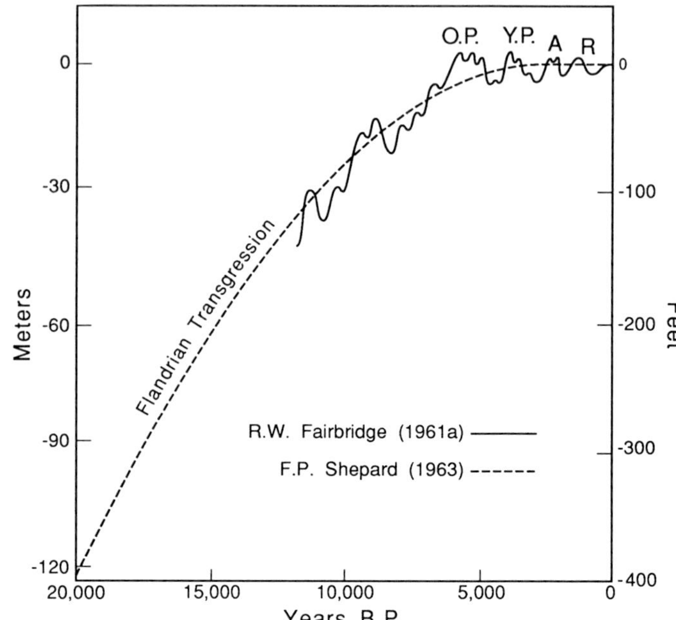

FIG. 1.—Two versions of the late and post-glacial eustatic curve (Flandrian transgression): the Fairbridge Curve (unbroken line), based on climatic, geomorphic and stratigraphic criteria; and the Shepard Curve (dashed line), based on an average of radiocarbon-dated materials (from Bird, 1984, fig. 15). In principle, the curves are the same, except that the data base for the former included detailed paleoclimatic information, supported by global field work in rather stable tectonic regions, and a philosophic conviction that sea level was highly susceptible to climatic forcing. The Fairbridge high sea-level stands are identified as follows: O.P. = Older Peron; Y.P. = Younger Peron; A = Abrolhos; R = Rottnest.

This conclusion presents serious problems, as yet unresolved. A meteorological scenario suggests that, during each warming pulse, the large volumes of meltwater reaching the Gulf of Mexico, the North Atlantic, and other ocean basins would favor enhanced evaporation, thus leading to renewed snow accumulation in mountainous areas. This periodic negative feedback would favor an autocyclic sequence of neoglacial readvances alternating with rapid ice retreats. During the Younger Dryas interval, isostatic tilting diverted the principal meltwater discharge from the Mississippi River to the St. Lawrence River, thus significantly cooling the North Atlantic. This mechanism, however, would not explain the Holocene fluctuations.

Four such cooling events marked the first half of the Holocene. The episodes have been chronologically identified in two ways: by ^{14}C-flux maxima (Stuiver and Braziunas, 1989), which reflect low solar activity and are identified in sidereal (or calendar) years B.C., and by approximate radiocarbon dating of coastal-peat deposits, expressed as isotopic age in uncalibrated radiocarbon years before A.D. 1950 (i.e., "present") and identified as "bp."

The first three cooling events occurred from 7,620 to 7,440 B.C. (9,600–9,300 bp), 7,080 to 6,940 B.C. (9,100–8,800 bp), and 6,480 to 5,900 B.C. (7,350–6,750 bp). A fourth but rather minor cooling occurred from 5,240 to 5,120 B.C. (6,190–6,140 bp). Around 4,550 B.C. (5,700 bp), eustatic sea level reached approximately its present elevation. However, in most areas the local sea level continued to rise due to subsidence, after a distinctive "kink" or flattening in the submergence curve. In places, this "kink" occurs as late as 4,000 bp, apparently due to a diachronistic shift in the forebulge adjustment zone. Some indication of the amplitude of these fluctuations is provided by seismic surveys of ravinement phenomena in the high-sedimentation area of the Mississippi Delta (Penland and others, 1988).

The early 19th-century observations of Holocene raised beaches and coastal terraces around North America (and elsewhere) were synthesized in a classic volume by Chambers (1848). The subject was neglected, however, for nearly a century until Daly (1934) studied the emerged reefs and wave-cut terraces of American Samoa. From their freshness and the lack of Pleistocene sea-level lowstand erosion, Daly deduced that the reefs are mid-Holocene terraces. This deduction has proven to be essentially correct.

A tempting hypothesis was that these emergent shores, widespread in the Pacific and Indian Oceans and commonly with several intermediate steps, were eustatic in origin and that the highest step corresponded to the oldest and warmest paleoclimatic oscillations, such as in the "Climatic Optimum." In northern latitudes—indeed, everywhere in formerly glaciated regions that were subsequently upwarped due to glacio-isostasy—warm cycles produced corresponding raised beaches (Fairbridge and Hillaire-Marcel, 1977). In these areas, each raised beach is a discrete geomorphic unit separated from the adjacent ones by an appropriate gap, i.e., a horizontal distance that varies with the local rate of uplift. Swamps are often found in these gaps, and many of the swamps have been bored, revealing pollen profiles that reflect dramatic fluctuations in climate (Birks and Birks, 1980). This botanical evidence in mid- to high latitudes indicates local paleotemperature departures of 2 to 5°C above and below the present summer mean. In northwest France, the paleobotanical work has been systematically integrated with studies of sea-level fluctuations; even at that latitude, the amplitude of the climatic fluctuations is quite large (Ters, 1987). In conclusion, a fluctuating Holocene eustatic curve simply cannot be avoided if one accepts the evidence of the rise and fall of high to mid-latitude temperatures and the advance and retreat of glaciers.

REGIONAL ASPECTS OF UNITED STATES HOLOCENE PALEOGEOGRAPHY

The introductary section of this paper deals largely with global aspects of Holocene paleogeographic constraints that apply particularly to the United States. In the field, however, categories of a regional nature exist that, in some cases, may overwhelm or mask the global trends (Putman, 1960; Inman and Nordstrom, 1971; Shepard and Wanless, 1971; Bird, 1984). The two categories of environmental setting that are of overriding importance are, primarily, *geotectonic province* and, secondarily, *climatic province*.

Geotectonic provinces may be broadly identified in the United States setting as follows (Table 1 and Fig. 2): (a) *active (convergent) plate margins and back-arc basins,* "Subduction Coasts;" (b) *taphrogenic (mainly strike-slip) plate margins,* "Taphrogenic Coasts;" (c) *passive plate margins (beyond glacial forebulge)* associated with thick geosynclinal sedimentary sequences, "Sedimentary Coasts;" (d) *passive plate margins (within former glacial forebulge),* also with thick sedimentary sequences; (e) *active plate margins and back-arc basins (within former glacial loading and forebulge areas);* (f) *semi-stable platforms,* dominated today by wind and meteorologic tides and by geostrophic current dynamics; (g) *oceanic fracture-zone volcanic complexes and seamount-related atolls.*

Superimposed on these seven structurally dependent foundations are the latitudinally related climatic provinces that often appear to play a dominant role on a short-term basis. Wave, climate, fetch, tidal range and currents are also important on this scale (see Davies, 1972); these processes are all susceptible to change through the course of time—seasonally, decadally and on a millennial scale.

The *climatic provinces,* with some generalization, may be designated as follows: (a) *high arctic;* (b) *sub-arctic;* (c) *west-coast westerly;* (d) *east-coast westerly;* (e) *subtropical and trade wind.* These provinces will be noted, under each regional heading, in the discussions that follow.

PRINCIPAL UNITED STATES COASTAL REGIONS

As indicated earlier, the description of any given region must consider primarily its tectonic framework and climatic zone, as well as sediment supply, exposure (wave energy, fetch), tidal characteristics, and current systems.

Alaska, North Coast

The north coast of Alaska is dominated by its *high arctic*-climate (above the Arctic Circle, 67°N) characteristics. The region encompasses all of the Alaskan north coast facing the Beaufort Sea, a sea that today is covered by ice for more than nine months of the year. The shoreline is subject to considerable ice push, especially during times of extraordinarily stormy summer-season onshore winds (Hume, 1965). In the early to mid-Holocene, the summer climate was considerably milder than at present, and the tree line was up to 500 km north of its present position. The July temperature was at least 10°C higher than at present, although winter, with its daily 24 hours of darkness, must have been extremely cold. Several generations of beach ridges suggest eustatic fluctuations; a sea-level fall to −2 m relative to present level, permitted an Eskimo village to be built at Birnik (Point Barrow) before 500 A.D., only to be submerged by Viking time, 1,000 A.D. (Hume, 1965).

Northwest Alaska (Facing the Chukchi Sea) and Western Alaska (Facing the Bering Sea)

These sub-arctic coasts are also frozen for considerable lengths of time and, under storm conditions, are exposed to a fetch of 500 km or more. In the late Pleistocene and earliest Holocene, eustatic lowering of sea level exposed the "Beringia" land bridge, which favored paleo-Indian migrations from eastern Asia (Hopkins, 1973), although protracted winter-spring coverage by sea ice makes a dry-

TABLE 1.—CATEGORIES OF COASTAL GEOTECTONIC PROVINCES OF THE UNITED STATES AND ITS TERRITORIES, WITH EXAMPLES

Categories	Examples
(a) ACTIVE (CONVERGENT) PLATE MARGINS and BACK-ARC BASINS, "Subduction Coasts"	Alaskan Peninsula, Aleutian Islands, southeastern Bering Sea coast
(b) TAPHROGENIC (MAINLY STRIKE-SLIP) PLATE MARGINS, "Taphrogenic Coasts"	California (in three distinctive sectors), Puerto Rico, U.S. Virgin Islands, most of the Alaskan Bering Sea coast
(c) PASSIVE PLATE MARGINS (BEYOND GLACIAL FOREBULGE) associated with thick geosynclinal sedimentary sequences, "Sedimentary Coasts"	Gulf Coast geosyncline, including the Mississippi Delta and smaller deltas such as the Rio Grande, and so forth; Alaskan North Slope geosyncline
(d) PASSIVE PLATE MARGINS (WITHIN FORMER GLACIAL FOREBULGE), also with thick sedimentary sequences	Cape Cod to East Georgia Embayment, with delta complexes such as the Delaware, Chesapeake Bay, and Savannah River; New England has undergone a transition from post-glacial isostatic uplift to forebulge adjustment
(e) ACTIVE PLATE MARGINS AND BACK-ARC BASINS (WITHIN FORMER GLACIAL LOADING AND FOREBULGE AREAS)	Northwest Washington, Puget Sound, and Oregon
(f) SEMI-STABLE PLATFORMS, DOMINATED TODAY BY WIND AND METEOROLOGIC TIDES AND BY GEOSTROPHIC-CURRENT DYNAMICS	Florida-Alabama and northwest Alaska (Chukchi Sea Coast)
(g) OCEANIC FRACTURE ZONE, VOLCANIC COMPLEXES AND SEAMOUNT-RELATED ATOLLS	Hawaiian Island chain, American Samoa, and Johnston Island, and Pacific dependencies

shod migration feasible at any time. In light of fishing traditions, a case can also be made for water-borne migrations. Wide strand plains with multiple beach ridges characterize sectors with an abundant supply of sand and gravel. Since some time before 4 ka, the summer-climate regimes gradually became colder (the precessional effect) and dominated by southerly or southwesterly winds with relatively mild conditions, or by cold northerly or northeasterly winds, corresponding to times when the polar front shifted southward. The beach-ridge sets display distinctive fanning patterns that correspond to these alternations, at intervals of 500 to 1,000 years (Mason, 1988).

Southern Alaska and Aleutian Chain

Dominated by its active plate-margin tectonics, this sector has a typically *west-coast westerly* climate. The quasi-permanent "Aleutian low" appears to have shifted systematically in concert with the polar frontal movements outlined earlier. Because of the steep offshore gradient typical of the western coast of the United States, the early Holocene eustatic rise contributed to the creation of the numerous narrow estuaries that made the west coast an attractive migration route for early humans. Large numbers of midden sites attest to the predominance of shellfish-gathering cultures, especially after the estuaries stabilized at about 6 ka. The active subduction sector provides unique opportunities to study early Holocene shorelines that have been tectonically uplifted (Black, 1974; Winslow and Johnson, 1989).

Pacific Northwest Coast (Northwest Washington, Oregon, Northern California and Puget Sound)

Located entirely within the climatic zone of the *prevailing westerlies,* this sector is dominated by processes associated with its position landward of the subducting Juan de Fuca Plate's active Cascadia convergence zone. The coasts of northwest Washington, Oregon, and northernmost California are steep and spectacularly cliffed, with pocket beaches and barred estuaries. By analogy with the subducting island arcs, the northwest Washington-Oregon belt corresponds to the non-volcanic, strongly overthrust "outer arc," whereas the volcanic "inner arc" equivalent is seen in the Cascade volcanic belt (e.g., Mounts Rainier, St. Helens, Hood) that extends from southern British Columbia to northern California.

The general geologic trend in the Pacific Northwest is Neogene subsidence near the coast and uplift inland, but vertical movements since the middle Pleistocene appear to have been generally minor, although variable in a north-south sense. The Olympic Mountains are faced by late Pleistocene uplift terraces at elevations of approximately 120 m and 60 to 30 m (West and McCrumb, 1988), but intertidal muds at Willapa Bay record a 190-ka sea level at a 2-m elevation, overlain by a probable Eemian (stage 5-e) formation. As noted later, similar terraces mark much of the Oregon coast.

The Holocene of the Pacific Northwest is poorly known. A wide, wave-cut platform fronting the cliffed sectors indicates multiple stages of erosion with a mean sea level within about 3 m of present level. Three areas of late Holocene beach ridges at 5 to 15 m elevation are found in western Washington, on the Long Beach peninsula, in the Cape Shoalwater-Westport area, and in the Ocean Shores peninsula of Point Grenville. Most of the ridges lie parallel to the present shore, but the Long Beach peninsula has two generations with the older ridges truncated (Cooper, 1958). The older ridges terminate landward along fossil cliffs that presumably date from about 6 ka.

Studies of peat-marsh stratigraphy in western Washington have disclosed at least 5 m of eight alternating layers of peat and intertidal mud (Atwater, 1987), which have been attributed to seismo-tectonic events. Burial events are cited at about 300, 1,600, 1,700, 2,700 and 3,100 bp (^{14}C), but alternatively they may correspond to eustatic and climatic fluctuations (Fairbridge, 1961a). Reference can also be made to archeological evidence for climatic variables (Strong, 1973), the pollen record (Baker, 1984), and tree rings (Schulman, 1956; Hughes and others, 1982; Yamaguchi, 1989). Similar peat stratigraphy and ages have been reported from the coastal marshes of Oregon (Peterson and Darienzo, 1991; Darienzo and Peterson, 1990). Atwater

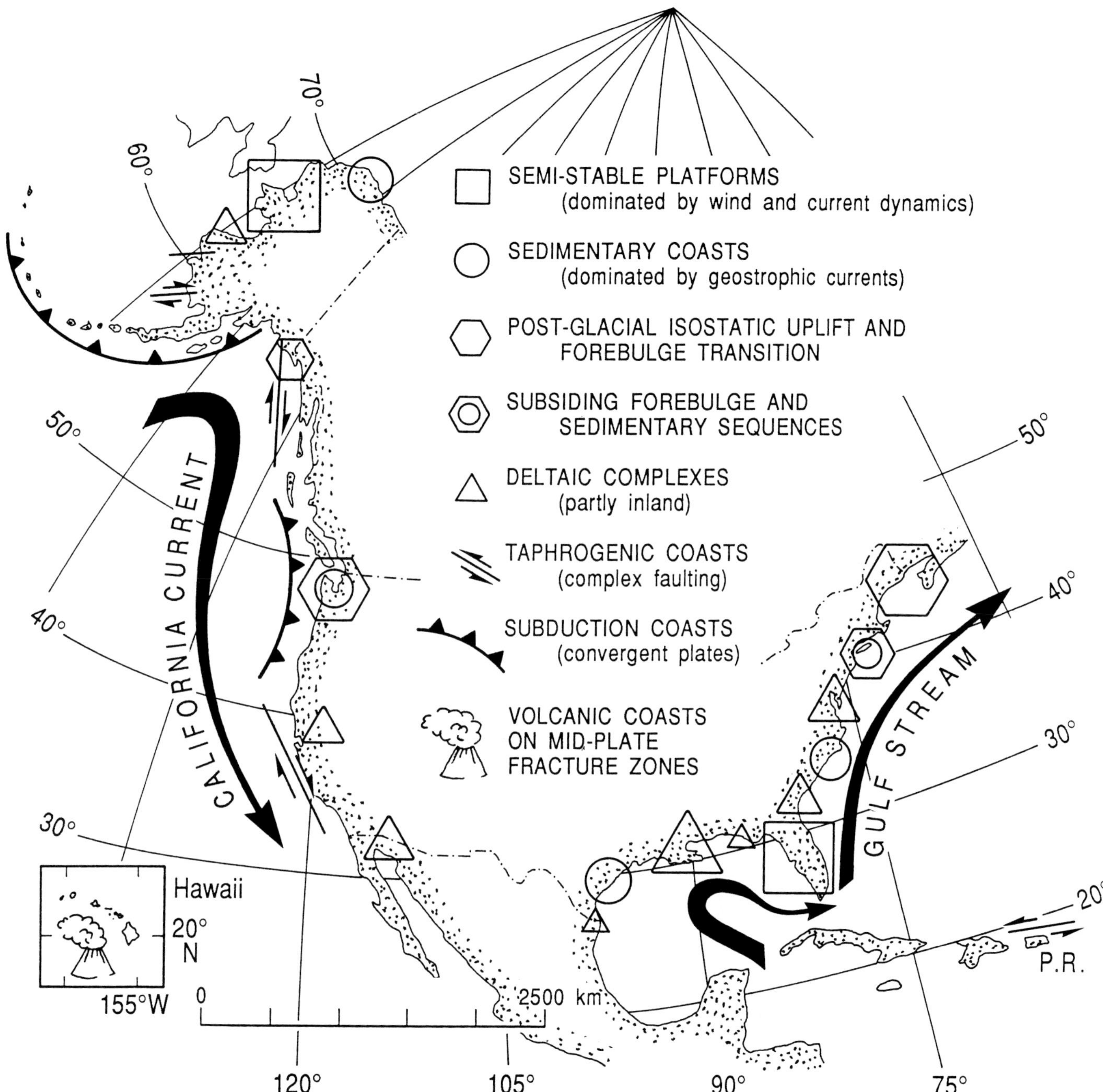

FIG. 2.—Principal regional characteristics controlling Holocene coastal history in the United States. Over the long term, tectonic setting has been the most important factor affecting local sea-level changes. Seven general categories are identified (with deltas added as a special case), although many sites are transitional. The status of some areas has changed during the course of the Holocene, e.g., Maine and southern New England, western Washington, and southeastern Alaska have passed gradually from post-glacial isostatic uplift into exponentially decaying forebulge subsidence. Most coasts are subject to 19th- and 20th-century sea-level rise because of post-Little Ice Age climatic warming, which led to steric expansion (Wigley and Raper, 1987), and more sluggish geostrophic currents, resulting in the reduction of dynamic ocean-surface tilt (decreased response to Coriolis effect). Base map: Lambert Equal-Area Zenithal Projection.

(1987) suggests that sand sheets overlapping the peat members might be related to seismicity and tsunamis may also be involved.

A succession of headlands and pocket beaches is characteristic of much of the Oregon coast as well, except that beaches here are more frequent. In the region between Coos Bay and Heceta Head is an uninterrupted 80-km-long beach backed by extensive dune fields (Cooper, 1958; Lund, 1973). The older dunes are well stabilized. The sand on many of the beaches and dunes appears to have been reworked from earlier Plio-Pleistocene terrace sands. The beaches and dunes, as well as fossil cliffs, truncate the uplifted Pleistocene marine terraces (Bird and Schwartz, 1985).

The shores of the partly dendritic inland waters of Puget Sound contrast sharply with those of the western open-ocean coast. The coastlines of Puget Sound are the product of an interesting mixture of post-glacial isostatic adjustments superimposed on the subsiding tectonic tendencies of a backarc basin. The city of Seattle sits neatly on top of one of the most extreme negative-gravity anomalies in North America. The shores of Puget Sound are marked by salt marshes that line the estuaries or by bluffs cut into terraced, late Pleistocene glacial drift. On the east side of the Olympic Peninsula lies the Hood Canal, an inundated glacial trough 104 km in length (Rau, 1980; Bird and Schwartz, 1985). By 13.5 ka, the ice had retreated from the Seattle area; there, isostatic rebound has been about 60 m, with some 100 m of rebound around Bellingham. At Cherry Point, a 30-m terrace is believed to be of Alleröd age and an 8-m terrace is possibly of Younger Dryas age (Rau, 1980). Archeological sites are rare, which seems to support geophysical evidence of subsidence around Seattle, but some apparently tectonically uplifted terraces are found there. One detailed Holocene sequence was obtained by Eronen and others (1987) from the north end of Hood Canal in a 7-m core that shows a 7,000-year succession of alternating intertidal or nearshore facies.

California Coast

Except for the small Klamath Mountains sector in northernmost California (which is tectonically part of the Cascadia sector), all of California belongs to a taphrogenic province, a complex of strike-slip and block-faulted terrains. The province can be broadly divided into three subprovinces: north, central, and south. The northern region parallels the San Andreas Fault and associated strike-slip sectors as far north as Cape Mendocino. The coast, steeply cliffed for the most part, includes only a few small pocket beaches, but terminates in the south at San Francisco Bay. The bay is a Plio-Pleistocene subsidence area, and is the outlet for the drainage systems of the Great Valley and Sacramento Valley. Here, the rising Flandrian transgression penetrated the area of the Golden Gate about 10 ka and initiated the sedimentation of a thick succession of fluviatile and salt-marsh deposits. According to Atwater and others (1977), average sea-level rise was about 20 mm/yr in the early Holocene and slowed to 1 to 2 mm/yr after about 6 ka. Superimposed on the eustatic effects has been tectonic and isostatic subsidence of about 5 m.

The central sector parallels the Southern Coast Ranges as far south as Point Arguello and is marked by spectacular cliffs and strongly upwarped Pleistocene marine terraces (e.g., Big Sur; Bradley and Griggs, 1976). Farther south are extensive beaches with dunes. At Morro Bay, a long barrier spit protects extensive salt marshes.

The southern sector includes two tectonic provinces: the seaward end of the Transverse Ranges near Santa Barbara, and the Peninsular Ranges, which extend southward into Baja California. The ranges are seismically very active, and intensive block faulting and folding have created abruptly bounded ranges separated by depressions, such as the Ventura Basin, the Los Angeles Basin, and others (Kern, 1977; Hanson and others, this volume; Muhs and others, this volume; Kern and Rockwell, this volume; Wehmiller, this volume). Whereas the headlands tend to be marked by upwarped Pliocene and Pleistocene terraces, active subsidence in the basins creates sandy coasts with barriers that cut off drowned estuaries (''sloughs''). Human activity in the present century has largely disrupted the previous seasonal opening and closing of the lagoons. The largest of these lagoons, south of Point Loma, has been developed as the harbor of San Diego. Interesting paleo-Indian artifacts have been collected offshore in this area (Masters and Flemming, 1983).

Climatically, the California coastline occupies a transition zone between the typical North Pacific *westerly* zone to the north and the Mediterranean to semi-arid *subtropical* zone to the south. The boundary between the two has repeatedly shifted abruptly north and south throughout the Holocene. This information comes not from Holocene coastal studies (which have been greatly neglected), but from offshore paleontological studies, most notably from the closed basin corresponding to the Santa Barbara Channel (Pisias, 1978). These results, combined with the palynological work of Heusser (1978), show that throughout the Holocene, the energy of the cold, south-setting California Current constantly fluctuated—a reflection of changing flow within the North Pacific gyre (Kuroshio and Aleutian Currents). When the gyre weakened, a northerly tongue of the warm, west-setting North Equatorial Current occasionally reached southern California, accompanied by distinctive warm-water faunas. Pollen carried seaward from the mainland also showed drastic fluctuations, suggesting that during the cool times, southern California enjoyed Mediterranean climate with winter rains supporting live oaks and other distinctive flora, comparable to those of today. During warmer periods, this flora would have been replaced by a cactus desert similar to that of present central Baja California, 300 to 500 km farther south.

The Santa Barbara Basin sediments show phases of terrigenous sedimentary inflow alternating with phases of stagnation; this pattern produced varved sediments undisturbed by bioturbation. Using faunal analysis, Pisias (1978) showed a difference of more than 15°C between the highest and lowest Holocene mean January sea-surface temperatures. Actualistic sedimentation and climatic patterns for the last two centuries were established by Soutar and Crill (1977). Isotopic studies of foraminifera suggest that the cool events correlate with enhanced upwelling and strengthened tradewind intensity. Intervals marked by frequent, warm, El Niño conditions ("Davidson Current") correlate with diminished upwelling and tradewind intensity. These conditions normally appear cyclically, but died out altogether during much of the Little Ice Age.

The Southern California paleoclimate history should be read in conjunction with that of the Gulf of California and coastal Mexico as far south as Mazatlán. From studies of the Nayarit beach ridges, Curray and others (1969) found that there had been cyclical reversals in the prevailing-wind direction, as shown by the fanning direction of the ridge sets. Similar structures are found in northwest Alaska (and Florida), but are evidently related to different air masses. The beach ridges have a periodicity compatible with the 11-year sunspot cycle, although the chronology is not yet adequate. However, sediment cores collected offshore in the Gulf of California also display varved sections, so that a precise year-by-year chronology should be possible.

Diatom productivity fluctuates markedly with upwelling intensity (Schrader and Baumgartner, 1983). High productivity corresponds to cool climates and sustained north-northwesterly winds (i.e., strong trade winds), reflecting times of low solar activity and high ^{14}C flux (in the southern California tree-ring series). Silicoflagellates characteristic of the cold Aleutian and California Currents appear. Decreased productivity is associated with warm, southerly winds (typical of El Niño years today; Philander, 1989) and weakened tradewind circulation. In the past, these times have matched low-^{14}C flux stages and thus high solar activity and global warming. The temperature curves are asymmetric and "saw-toothed," with cold events (sometimes called "La Niña") immediately following extra-warm ones (just as displayed by the ^{14}C-flux record).

These studies have now been extended back some 3,000 years in the Gulf of California (Juillet-Leclerc and Schrader, 1987). The records are very important for understanding both local and global paleoclimatic history, especially when taken in conjunction with other Holocene coastal records, such as those from Mexico, Alaska (Mason, 1988), Hudson Bay (Fairbridge and Hillaire-Marcel, 1977), and Florida (Stapor and others, 1991). The Greenland and Antarctic ice cores (Johnsen and others, 1970; Hammer and others, 1986; Lorius and others, 1985; Beer and others, 1988) also provide important corroborating evidence.

Northern and Central Atlantic Coast

This sector corresponds to the climate belt of the *east-coast westerlies,* which today extends from Maine to the Carolinas. With a wide continental shelf and great interplay between post-glacial isostatic recovery and marginal forebulge collapse, the east coast displays paleogeographic features very different from those of the west coast. Early Holocene deglaciation in New England continued the late Pleistocene retreat, probably following a Younger Dryas fluctuation. Evidence for these events, however, is sparse since local sea level in the Gulf of Maine was between approximately −40 to −60 m about 10.5 ka (Belknap and others, 1989; Kelley and others, this volume). A 10-m core obtained in Alpine, New Jersey, showed the Younger Dryas was marked by a temporary return to a boreal forest (*Picea, Abies, Larix, Betula, Alnus*) following an Alleröd-type warm cycle with deciduous hardwoods (Peteet and others, 1990).

In the early Holocene, the shape of the New England coastline changed rapidly and dramatically as crustal motions interacted with eustatic rise. The result was the introduction of maritime climate conditions to the formerly continental setting. Rapid sea-level rise created a fjärd-type coast (low-relief fjords), characteristic of southern Sweden, that persists to the present. On the continental shelf, fishermen have dredged teeth of mastodon, which became extinct about 6 ka. Freshwater peat found near the shelf margin (Emery and others, 1967) probably documents sea-level fluctuations. Although the early Holocene environment was warmer than that of today and would have been attractive to early humans, offshore middens are not known until around 6 ka, when local sea level was about −10 m and close to the present shore. Middens partially buried by peat swamps and dating from around 4 ka at 4 m below sea level seem to indicate persistent forebulge subsidence, and tide gauges along the New England coast do show continued sea-level rise. Evidence for cooling (but fluctuating) climates in the late Holocene has been found in the pollen data (Gajewski, 1988), and the cooling sea water has restricted the geographic extent of oyster-based human economies. In the Hudson estuary upstream from New York, near the zero isobase, giant oyster middens above present sea level have been dated back to about 7 ka (Brennan, 1974), clearly indicating much warmer conditions than today.

Along the coast from Cape Cod to the Carolinas, the Holocene is represented mainly by peat stratigraphy (Bloom, 1984; van de Plassche, 1990). Submergence seems to be controlled initially by the subsiding forebulge; later, the mean sea-level rise appears to be influenced more by a secular weakening of the Gulf Stream. Best known is the region of the lower Delaware estuary (Kraft, 1971; Fletcher and others, 1990, 1992), where extensive coring has revealed a nearly continuous Holocene profile of nearshore to littoral facies. Chesapeake Bay appears to be comparable and indeed has long been cited as the "classic" example of a youthfully drowned coast, the tectonic submergence of which continues today.

In the southern part of the region, notably in South Carolina (Colquhoun and Brooks, 1986) and Georgia (Howard and dePratter, 1977), the combination of a long-term coring campaign and archeological studies of midden sites has yielded the clearest record of fluctuating sea levels in the United States since the mid-Holocene. The results show several major fluctuations of as much as 2 to 3 m in amplitude. Many of the middens, perched on chenier-type beach ridges, are still preserved above mean sea level today. Wave-cut terraces and scarps are preserved at Old Island (Beaufort County), South Carolina (Stapor and Mathews, 1983).

In this southern sector of the Atlantic coast, subsidence rates have decreased with time. In the late Holocene (since 6 ka), the exponentially decreasing importance of the marginal-bulge collapse has been replaced by the post-Climatic Optimum secular cooling of the world oceans and other factors. It should be emphasized, however, that the cause of the continued submergence in the Carolinas is controversial. The hypothesis of forebulge collapse (Daly, 1934; Walcott, 1972) is well reasoned and is appropriate for the early to mid-Holocene. After that time, the rate of forebulge collapse should exponentially decay (Newman and others, 1971). However, tide-gauge records show that submergence continues today at rates of 2 to 4 mm/yr, a rate higher than could be expected due simply to steric expansion associated with post-Little Ice Age warming. Another geophysical hypothesis, also well reasoned, argues that the added load of glacial meltwater will depress the ocean crust by hydro-isostasy, leading to a secular emergence of mid-Holocene shorelines in the low latitudes (Clark and others, 1978; Chappell, 1984; Lambeck, 1990). This mechanism does not, however, seem to be appropriate for the localized submergence in the Carolinas, which appears to call for basement tectonic revival (an important seismic trend transects the embayment).

An additional and significant aspect of Holocene history is its fluctuating climate, often ignored in the geophysical reasoning. The systematic global temperature change from a glacial to an interglacial state involves a decrease of 25 to 30°C in the mean equator to pole temperature gradient (Fairbridge, 1989). It should be appreciated that this thermal-gradient curve displays very little change at its equatorial limit (2–3°C), but a huge variance in the polar regions of the Northern Hemisphere. Because of the continental nature of Antarctica, the variance is much less. A smaller decrease in the slope of the curve would be associated with secondary warming cycles. Any change in this gradient involves variations in the velocities and trajectories of the great geostrophic currents, and the dynamic slope across each of these currents (under the Coriolis effect) reflects the current velocity, e.g., for the Gulf Stream (Stommel, 1965). The summer-to-winter variance alone involves a fluctuation of more than 25 cm between Charleston and Bermuda. A similar seasonal shift between Miami and Bimini shows a linear correlation between volume transport and surface speed (Maul, 1986). The steeper the pole-to-equator thermal gradient, the higher the velocity of the current, and the steeper the transverse (Coriolis) tilt. Global warming leads to a corresponding weakening of geostrophic circulation, which in turn leads to a secular rise of sea level along the principal coastlines. In contrast, a global cooling accelerates the currents and steepens the transverse gradient, so that sea level falls along the coast. The same principle applies to all oceans in both hemispheres, except near the Equator where the Coriolis parameter is zero. This effect may account for the mid-Pacific atoll tide-guage records that displays no significant sea-level rise at the present time (Lyles and others, 1988).

Gulf Coast, Including the Mississippi Delta

The Gulf Coast province is a classic passive (divergent) plate margin, characterized by the thick sedimentary wedge of the Gulf geosyncline. The Mississippi Delta corresponds to an ancient, subsiding, northeast-southwest lineament system that intersects the east-west arc of the geosyncline. The early Holocene is marked by a rapid shift of the shoreline landward across the continental shelf; by 7 ka (^{14}C), the main deltaic depocenter (Maringouin Delta complex) had shifted to an area southwest of the present delta.

From 6.5 ka (^{14}C) onward, the Gulf coast became fairly well established at the line of a long series of barrier islands and spits that persist to this day (Wilkinson and Basse, 1978; Walker and Coleman, 1987). Numerous exploration bores along this line have disclosed the superposition of one beach system on another, up to 20 m thick. Radiocarbon dating of the shells, however, has proved a disappointment. The sediments had been so reworked that the normal stratigraphic ordering was, in places, reversed. In addition to upward aggradation, lateral migration (longshore drift) had also occurred on a large scale (Otvos, 1970, 1973, this volume). From offshore surveys along the Gulf coast, Shepard and Curray (1967) developed a general picture of early Holocene sea-level rise within a subsiding framework. Mid-Holocene dating of sites indicating higher than present sea level have been established only in one area just south of the Rio Grande River near Matamoras (Behrens, 1969).

For the Mississippi Delta, rapid crustal subsidence and sediment compaction, plus a high rate of fluvial discharge, have created one of the richest Holocene successions in the world. Initially, this record was interpreted in terms of Shepard's smoothly rising eustatic curve. A random series of delta lobes, breaking out successively in different directions (Kolb and Van Lopik, 1966), was superimposed on the sea-level-rise record. In all, Frazier (1967) identified four delta complexes and 16 lobes.

Radiocarbon dating of the individual lobes showed, however, that it was possible to re-interpret this excellent model in terms of sea-level fluctuations. At least seven principal-emergence phases are identified by Fairbridge (1983). On theoretical grounds it has been reasoned that when sea level rises rapidly across a delta plain, the point of debouchment of a river will be abruptly shifted upstream. In a low-gradient situation, e.g., 1:10,000 ratio, a rapid 1-m sea-level rise would thalassostatically shift the fluvial-crest surface some 10 km upstream, inviting a new breakout. The river has only a "short memory," and the old thalweg would rapidly infill with silt. A succession of such events would lead to repeated delta-lobe switching (Lowrie and Fairbridge, 1991).

The Vail terminology of sequence stratigraphy, if applied to the Holocene record of the Mississippi Delta, would suggest fifth-order eustatic fluctuations. The saw-toothed pattern of this scheme indicates gradual rises alternating with abrupt regressions; this characteristic applies even to many sixth-order cycles, including isotopic-flux rates (Beer and others, 1988).

The sequence approach would produce a succession of transgressive-system tracts in the delta (Haq and others, 1987). Recent studies of the delta succession appear to be greatly assisted by the new approach (Suter and others, 1987; Penland and others, 1988), and meticulous attention to facies has brought new clarity to the sequences. The lower boundary of each transgressive-sequence tract is ideally a thin sand sheet that spreads as much as 50 km landward with each major eustatic rise. From the historical record of the Yellow River or Huangho Delta in China, Grabau (1936) called this phenomenon "Marining;" however, the term was not widely adopted, having apparently appeared before its time, as did many of Grabau's ideas.

Rapid regression is represented by emergence and subaerial weathering or shallow freshwater-marsh facies. The accumulation of "organic blankets" (peat, and so forth) is favored by the delta-lobe switching (Kosters, 1989), so that the marshes (Coleman and Roberts, 1989) can persist for long periods in the same spot during a falling eustatic sea-level phase, because of the $\Delta_E - \Delta_I = 0$ rule (Fairbridge, 1983, 1987). The rule implies that if Δ_E (mean sea-level change) = -3 mm/yr and Δ_I (mean subsidence rate) = -3 mm/yr, then the "relative" or local sea level remains constant, its rate of change being zero. This fact helps to explain why the fluctuating eustatic curve of the Holocene has been so difficult to recognize in the Mississippi Delta (and equally so in other deltas; see, e.g., van de Plassche, 1982).

Florida and Alabama

An almost unique semi-stable platform area within the conterminous United States, the Florida-Alabama province is more *tropical* to *subtropical* than any other part of the continental United States; its southern salt marshes support the country's only widespread development of mangroves. The two coastlines of Florida are notably distinct. The east coast, with a nearly continuous belt of barrier beaches, changes south of Miami to a line of emergent late Pleistocene coral reefs, fronted today in most places by actively growing reefs. A set of beach ridges at Cape Canaveral is Holocene, but contains reworked material, making chronology difficult. The barriers and extensive shore-parallel lagoons (much modified by construction and maintenance of the Intracoastal Waterway) do not seem to have been studied in much detail.

A unique site at Fisher Island south of Miami Beach has furnished coral and mangrove-root samples at about +2 m dated at 6,020 to 5,690 bp (Hoffmeister and Multer, 1965); these roots document the highest Holocene submergence on the Florida east coast. In nearby Coral Gables, a beautifully developed 3-m-high intertidal notch marks the late Pleistocene Silver Bluff shoreline. Whereas the notch is certainly old (probably late Eemian), it probably was reworked by the mid-Holocene highstand. The notch is fronted by Holocene lagoonal deposits up to 2 m above sea level, but the deposits have never been studied.

Florida's west coast is much more extensively studied. There, a wide continental shelf subject to low-wave energy is furnished with an abundance of sand, in part derived from glacial-stage dunes that developed along much of that coast during the semi-arid cold maxima. These conditions created an ideal setting for the development of numerous "drumstick"-shaped barrier islands with small beach-ridge plains after sea level reached its present level about 6.5 ka (^{14}C). The sand is transported shoreward from the shelf (Davis and Kuhn, 1985). The sand-free salt-marsh coast has been investigated by Hine and others (1988).

The beach-ridge fanning direction is generally from northwest to southeast, implying an accumulation under cycles of enhanced winter storms from westerly frontal systems. These times of storms alternate with times of greater calm that result in beach ridges fanning from south to north, developed under the influence of the prevailing but weak trade winds that, here, swing around to blow from the southeast.

In the Apalachicola region, pre-Holocene (late Pleistocene) beach-ridge sequences were truncated by bluffs during the mid-Holocene sea-level rise of about +2 m (Stapor, 1975; Donoghue and Tanner, this volume). These "fossil cliffs" are fronted by multiple sets of younger beach ridges, each truncating the former. Growth of the beach ridges and their distinctive sets has been intermittent.

Farther west, the estuary of Mobile Bay has prograded about 35 km during the late Holocene, and its chronology has been partly confirmed by archeological work (Holmes and Trickey, 1974). At Tenshaw Lake, now totally isolated from the Gulf, midden sites contain *Rangea cuneata,* a typical brackish-water clam. These middens occur in four shell layers ranging in elevation from 0 to 1.5 m above sea level; they are separated by estuarine or fluvial silts, which represent stages when sea level was higher than present. The midden shells are radiocarbon dated at 4,100, 3,090, 2,040, and 1,080 bp. Each date corresponds to a time when sea level was somewhat lower than present, a necessary condition to provide for dry camp sites.

In southwest Florida, between St. Petersburg and Fort Myers, barrier islands have been comprehensively dated with more than 100 ^{14}C samples (Stapor and others, 1987, 1991). Inasmuch as most of the islands are oriented north-south, the fanning directions of the beach ridges display dramatic reversals in alternate stages, depending upon whether the climate of the time of formation was dominated by northwesterly winds or by southerly systems.

CONCLUSIONS

1. In most parts of the United States, the Holocene history and paleogeography of the coastal belt have been dominantly controlled by the late and post-glacial Flandrian transgression. This history has been greatly modified from region to region by tectonic and other factors.
2. Broadly speaking, this author interprets Holocene coastal history as being divided into two parts: the fluctuating global eustatic rise to about sidereal 6.5 ka (5,700 bp ^{14}C age), and, since then, the fluctuating eustatic fall of about 3 m that brought mean sea level to its present datum. Regional interpretation is greatly complicated by crustal motions, however, so that generalization is controversial.
3. Regional areas can be delineated, in a generalized way, based on tectonic and climatic criteria. Regions with important uplift histories are identified in three categories, as glacio-isostatic-rebound areas, as actively subducting plate-margin belts, and as positive segments of taphrogenic sectors.
4. Most coastal belts of the continental United States are subsiding areas and are thus known only from buried sedimentary sequences. Uplift areas, in contrast, are characterized mainly by erosional terraces. Only two areas are classified as semi-stable: northwest Alaska and the Florida-Alabama coasts. Dating of the late Holocene

beach-ridge sequences of the coasts is greatly aided by archeological data.

5. High-sedimentation, subsidence areas, such as the Mississippi Delta, disclose fifth-order sequence stratigraphy (Penland and others, 1988). Even finer "tuning" to a sixth order would make it reasonable to hope for eventual incorporation into a decade-scale planetary/^{14}C-flux/ climatic/eustatic framework.
6. The last millennium has been distinguished by three climatic phases with sea-level fluctuations of less than 1 m: the Viking warm phase about 1,000 A.D., the Little Ice Age, and the post-Little Ice Age warming since about 1,750 A.D. The last episode has led to a reduction of the global equator-to-pole thermal gradient, thus reducing tradewind energy and the velocity of geostrophic currents such as the Gulf Stream. A weakened Coriolis influence leads to reduction of the dynamic gradient; as a result, sea level is rising today along mainland shores almost worldwide, except at the Equator. The Coriolis factor is an additive to steric expansion and the widespread tectonic and compactional subsidence that forces sea-level rise in most places.

ACKNOWLEDGMENTS

This research was generated in connection with the writer's chairmanship of two working groups and membership on one committee: the International Union for Quaternary Science (INQUA): Holocene Commission, WG8-D (Cyclic Variations in Sedimentology and Shorelines); the International Geological Correlation Program (IGCP 274): Coastal Evolution (W.G. Astronomical Periodicities); and the INQUA executive committee of the International Geosphere-Biosphere Project. The report does not in any way reflect official opinions or policy. It may be quoted with appropriate acknowledgment.

Research was supported by the National Science Foundation (Critical Engineering Systems: Dr. E. Sabadell), Grant CES-8618525. Logistic support was supplied by the kindness of Dr. James Hansen, NASA-Goddard Institute for Space Studies. Early drafts of the manuscript were read, with beneficial comments and discussion, by Chip Fletcher, J. C. Kraft, Frank Stapor, and other friends.

REFERENCES

ANDERSON, J. B., AND THOMAS, M. A., 1991, Marine ice-sheet decoupling as a mechanism for rapid, episodic sea-level change: the record of such events and their influence on sedimentation: Sedimentary Geology, v. 70, p. 87–104.

ATWATER, B. F., 1987, Evidence for great Holocene earthquakes along the outer coast of Washington state: Science, v. 236, p. 942–944.

ATWATER, B. F., HEDEL, C. W., AND HELLEY, E. J., 1977, Late Quaternary depositional history, Holocene sea-level changes, and vertical crustal movement, southern San Francisco Bay, California: U.S. Geological Survey Professional Paper 1014, 15 p.

BAKER, R. G., 1984, Holocene vegetational history of the western United States, *in* Wright, H. E., Jr., ed., Late-Quaternary Environments of the United States, v. 2: University of Minnesota Press, Minneapolis, p. 109–127.

BEER, J., SIEGETHALER, U., BONANI, G., FINKEL, R. C., OESCHGER, H., SUTER, M., AND WÖLFLI, W., 1988, Information on past solar activity and geomagnetism from ^{10}Be in the Camp Century ice core: Nature, v. 331, p. 675–679.

BEHRENS, E. W., 1969, Recent emergent beach in eastern Mexico: Science, v. 152, p. 642–643.

BELKNAP, D. F., SHIPP, R. C., KELLEY, J. T., AND SCHNITKER, D., 1989, Holocene sea-level change in coastal Maine, *in* Anderson, W. A., and Borns, H. W., eds., Neotectonics of Maine: Maine Geological Survey Bulletin 40 Orono, p. 85–105.

BERGER, A., 1989, Pleistocene climate variability at astronomical frequencies: Quaternary International, v. 2, p. 1–14.

BIRD, E. C. F., 1984, Coasts: An Introduction to Coastal Geomorphology (3rd ed.), Basil Blackwood: Oxford, 320 p.

BIRD, E. C. F., AND SCHWARTZ, M. L., eds., 1985, The World's Coastline: Van Nostrand Reinhold, New York, 1071 p.

BIRKS, H. J. B., AND BIRKS, H. H., 1980, Quaternary Palaeoecology: University Park Press, London, Arnold, AND Baltimore, 289 p.

BLACK, R. F., 1974, Late Quaternary sea level changes, Umnak Island, Aleutians—their effects on ancient Aleuts and their causes: Quaternary Research, v. 4, p. 264–281.

BLOOM, A. L., 1984, Sea level and coastal changes, *in* Wright, H. E., Jr., ed., Late-Quaternary Environments of the United States, v. 2: University of Minneapolis Press, Minneapolis, p. 42–51.

BRADLEY, W. C., AND GRIGGS, G. B., 1976, Form, genesis, and deformation of central California wave-cut platforms: Geological Society of America Bulletin, v. 87, p. 433–449.

BRENNAN, L. A., 1974, The lower Hudson: a decade of shell middens: Archaeology of Eastern North America, v. 2, p. 81–93.

CHAMBERS, R., 1848, Ancient Sea-Margins, As Memorials of Changes in the Relative Level of Land and Sea: Chambers, Orr & Company, Edinburgh and London, 388 p.

CHAPPELL, J., 1984, Late Quaternary glacio- and hydro-isostasy on a layered Earth: Quaternary Research, v. 4, p. 429–440.

CLARK, J. A., FARRELL, W. E., AND PELTIER, W. R., 1978, Global changes in postglacial sea level: a numerical calculation: Quaternary Research, v. 9, p. 265–287.

COLEMAN, J. M., AND ROBERTS, H. H., 1989, Deltaic coastal wetlands: Geologie en Mijnbouw, v. 68, p. 1–24.

COLQUHOUN, D. J., AND BROOKS, M. J., 1986, New evidence from the southeastern U.S. for eustatic components in the late Holocene sea levels: Geoarchaeology, v. 1, p. 275–291.

COOPER, W. S., 1958, Coastal sand dunes of Oregon and Washington: Geological Society America Memoir 72, *169* p.

CURRAY, J. R., EMMEL, F. J., AND CRAMPTON, P. J. S., 1969, Holocene history of a strand plain, lagoon coast, Nayarit, Mexico, *in* Castañares, A. A., and Phleger, F. B., eds., Coastal Lagoons: A Symposium: Universidad Antónoma, Mexico City, p. 63–100 (reprint in Benchmark Papers in Geology, v. 42).

DALY, R. A., 1934, The Changing World of the Ice Age: Yale University Press, New Haven, 271 p.

DARIENZO, M. E., AND PETERSON, C. D., 1990, Episodic tectonic subsidence of late Holocene salt marshes, northern Oregon coast, central Cascadia Margin, USA: Tectonics, v. 9, p. 1–22.

DAVIES, J. L., 1972, Geographical Variation in Coastal Development, Geomorphology Text 4: Oliver and Boyd, Edinburgh, 204 p.

DAVIS, M. B., 1984, Holocene vegetational history of the eastern United States, *in* Wright, H. E., Jr., ed., Late-Quaternary Environments of the United States, v. 2: University of Minnesota Press, Minneapolis, p. 166–181.

DAVIS, P. T., 1988, Holocene glacier fluctuations in the American Cordillera: Quaternary Science Reviews, v. 7, p. 129–158.

DAVIS, R. A., Jr., AND KUHN, G. J., 1985, Origin and development of Anclote Key, west-peninsular Florida: Marine Science, v. 63, p. 153–171.

DEAN, W. E., BRADBURY, J. P., ANDERSON, R. Y., AND BARNOSKY, C. W., 1984, The variability of Holocene climate change: evidence from varved lake sediments: Science, v. 226, p. 1191–1194.

EMERY, K. O., WIGLEY, R. L., BARTLETT, A. S., RUBIN, M., AND BARGHOORN, E. S., 1967, Freshwater peat on the continental shelf: Science, v. 158, p. 1301–1307.

ERONEN, M., KANKAINEN, T., AND TSUKADA, M., 1987, Late Holocene sea-level record in a core from the Puget Lowland, Washington: Quaternary Research, v. 27, p. 147–159.

FAIRBANKS, R. G., 1989, A 17,000-year glacio-eustatic sea-level record: influence of glacial melting rates on the Younger Dryas event and deep-ocean circulation: Nature, v. 342, p. 637–642.

FAIRBRIDGE, R. W., 1961a, Eustatic changes in sea level, *in* Ahrens, L. H., Press, F., Rankama, K., and Runcorn, S. K., eds., Physics and Chemistry of the Earth: Pergamon Press, London, v. 4, p. 99–185.

FAIRBRIDGE, R. W., ed., 1961b, Solar Variations, Climatic Change, and Related Geophysical Problems: New York Academy of Science, Annals, v. 95, 740 p.

FAIRBRIDGE, R. W., 1983, Isostasy and eustasy, *in* Smith, D., and Dawson, A., eds., Shorelines and Isostasy: Academic Press, London, p. 3–25.

FAIRBRIDGE, R. W., 1987, The spectra of sea level in a Holocene time frame, *in* Rampino, M. R., Sanders, J. E., Newman, W. S., and Königsson, L. K., eds., Climate, History, Periodicity, and Predictability: Van Nostrand Reinhold, New York, p. 127–142.

FAIRBRIDGE, R. W., 1989, Quasi-equilibrium in Holocene climatic change: Quaternary International, v. 2, p. 83–89.

FAIRBRIDGE, R. W., AND HILLAIRE-MARCEL, C., 1977, An 8000 yr paleoclimatic record of the "Double-Hale" 45 yr solar cycle: Nature, v. 268, p. 413–416.

FLETCHER, C. H., KNEBEL, H. J., AND KRAFT, J. C., 1990, Holocene evolution of an estuarine coast and tidal wetlands: Geological Society of America Bulletin, v. 102, p. 283–297.

FLETCHER, C. H., KNEBEL, H. J., AND KRAFT, J. C., 1992, Holocene depocenter migration and sediment accumulation in Delaware Bay: a submerging marginal marine sedimentary basin: Marine Geology, v. 103, p. 165–183.

FRAZIER, D. E., 1967, Recent deltaic deposits of the Mississippi River: their development and chronology: Transactions, Gulf Coast Associations of Geological Societies, v. 17, p. 287–311.

GAJEWSKI, K., 1988, Late Holocene climate changes in eastern North America estimated from pollen data: Quaternary Research, v. 29, p. 255–262.

GRABAU, A. W., 1936, The great Huangho plain of China: Association of Chinese and American Engineers Journal, v. 17, p. 247–266.

GREISCHAR, L. L., AND BENTLEY, C. R., 1980, Isostatic equilibrium grounding line between the West Antarctic inland sheet and the Ross Ice Shelf: Nature, v. 283, p. 651–654.

HAMMER, C. U., CLAUSEN, H. B., AND TAUBER, H., 1986, Ice-core dating of the Pleistocene/Holocene boundary applied to a calibration of the ^{14}C time scale: Radiocarbon, v. 28, p. 284–291.

HAQ, B. U., HARDENBOL, J., AND VAIL, P. R., 1987, Chronology of fluctuating sea levels since the Triassic: Science, v. 235, p. 1156–1167.

HARRISON, S. P., 1989, Lake levels and climatic change in eastern North America: Climate Dynamics, v. 3, p. 157–167.

HEUSSER, L. E., 1978, Marine pollen in Santa Barbara Basin, California: Geological Society of America Bulletin, v. 89, p. 673–678.

HINE, A. C., BELKNAP, D. F., HUTTON, J. G., OSKING, E. B., AND EVANS, M. W., 1988, Recent geological history and modern sedimentary processes along an incipient, low-energy, epicontinental-sea coastline: northwest Florida: Journal of Sedimentary Petrology, v. 58, p. 567–579.

HOFFMEISTER, J. E., AND MULTER, H. G., 1965, Fossil mangrove reef of Key Biscayne, Florida: Geological Society of America Bulletin, v. 76, p. 845–852.

HOLLIN, J. T., 1980, Climate and sea level in isotope stage 5: an East Antarctic ice surge at ~95,000 BP?: Nature, v. 283, p. 629–633.

HOLMES, N. H., AND TRICKEY, E. B., 1974, Late Holocene sea-level oscillations in Mobile Bay: American Antiquity, v. 39, p. 122–124.

HOPKINS, D. M., 1973, Sea level history in Beringia during the past 250,000 years: Quaternary Research, v. 3, p. 520–540.

HOWARD, J. D., AND DE PRATTER, C. B., 1977, History of shoreline changes determined by archaeological dating: Georgia coast: Transactions, Gulf Coast Association of Geological Societies, v. 27, p. 251–258.

HUGHES, M. K., KELLY, P. M., PILCHER, J. R., AND LAMARCHE, V. C., eds., 1982, Climate from Tree Rings: Cambridge University Press, London, 223 p.

HUME, J. D., 1965, Sea-level changes during the last 2,000 years at Point Barrow, Alaska: Science, v. 150, p. 1165–1166.

INMAN, D. L., AND NORDSTROM, C. E., 1971, On the tectonic and morphologic classification of coasts: Journal of Geology, v. 79, p. 1–21.

JOHNSEN, S. J., DANSGAARD, W., CLAUSEN, H. B., AND LANGWAY, C. C., 1970, Climatic oscillations 1200–2000 A.D: Nature, v. 227, p. 482–483.

JUILLET-LECLERC, A., AND SCHRADER, H., 1987, Variations of upwelling intensity recorded in varved sediment from the Gulf of California during the past 3,000 years: Nature, v. 329, p. 146–149.

KERN, J. P., 1977, Origin and history of upper Pleistocene marine terraces, San Diego, California: Geological Society of America Bulletin, v. 88, p. 1553–1566.

KOLB, C. R., AND VAN LOPIK, J. R., 1966, Depositional environments of the Mississippi River deltaic plain—southeastern Louisiana, *in* Shirley, M. L., and Ragsdale, J. A., eds., Deltas: Houston Geological Society, Houston, p. 17–61.

KOSTERS, E. C., 1989, Organic-clastic facies relationships and chronostratigraphy of the Barataria interlobe basin, Mississippi Delta plain: Journal of Sedimentary Petrology, v. 59, p. 98–113.

KRAFT, J. C., 1971, Sedimentary facies patterns and geologic history of a marine transgression: Geological Society of America Bulletin, v. 82, p. 2131–2158.

KUTZBACH, J. E., 1984, Modeling of Holocene climates, *in* Wright, H. E. ed., Late-Quaternary Environments of the United States, v. 2: University of Minnesota Press, Minneapolis, p. 271–277.

LAMB, H. H., 1977, Climate History and the Future: Methuen, London, and Princeton University Press, 835 p.

LAMBECK, K., 1990, Glacial rebound, sea-level change and mantle viscosity: Quarterly Journal of Royal Astronomical Society, v. 31, p. 1–30.

LORIUS, C., JOUZEL, J., RITZ, C., MERLIVAT, L., BARKOV, N. I., KOROTKEVICH, Y. S., AND KOTLYAKOV, V. M., 1985, A 150,000-year climatic record from Antarctic ice: Nature, v. 316, p. 591–596.

LOWRIE, A., AND FAIRBRIDGE, R. W., 1991, Role of eustasy in Holocene Mississippi delta-lobe switching: Twelfth Annual Research Conference, Gulf Coast Section, p. 111–115.

LUND, E. H., 1973, Oregon coastal dunes between Coos Bay and Sea Lion Point: The Ore Bin, v. 35, p. 73–92.

LYLES, S. D., HICKMAN, L. E., JR., AND DEBAUGH, H. A., JR., 1988, Sea-level variations for the United States 1855–1986: National Oceanic and Atmospheric Administration, Rockville, Maryland, 182 p.

MASON, O. K., 1988, The sand ridge stratigraphy of northern Seward Peninsula, *in* Schaaf, J., ed., The Bering Land Bridge Natural Reserve: An Archaeological Survey: National Park Service Report AR-14, Anchorage, p. 364–409.

MASTERS, P. M., AND FLEMMING, N. C., eds., 1983, Quaternary Coastlines and Marine Archaeology: Academic Press, London, 660 p.

MAUL, G. A., 1986, Linear correlations between Florida Current volume transport and surface speed with Miami sea level and weather during 1964–70: Geophysical Journal of Royal Astronomical Society, v. 87, p. 55–66.

MÖRNER, N. A., 1969, The late Quaternary history of the Kattegatt Sea and the Swedish west coast; deglaciation, shore level displacement, isostasy and eustasy: Sverige Geologiska Undersökning, series C, no. 640, Arsbok, v. 63, p. 487.

NEWMAN, W. S., FAIRBRIDGE, R. W., AND MARCH, S., 1971, Marginal subsidence of glaciated areas: United States, Baltic and North Seas, *in* Ters, M., ed., Études sur le Quaternaire dans le Monde: Paris, VIII INQUA, 1969, p. 795–801.

OLIVER, J. E., AND FAIRBRIDGE, R. W., eds., 1987, The Encyclopedia of Climatology: Van Nostrand Reinhold, New York, 986 p.

OTVOS, E. G., JR., 1970, Development and migration of barrier islands, northern Gulf of Mexico: Geological Society of America Bulletin, v. 81, p. 241–246.

OTVOS, E. G., JR., 1973, Genetic and age problems of the Moreau Caminada Holocene coastal ridge complex, northeastern Louisiana: Southeastern Geology, v. 15, p. 37–43.

PAEPE, R., AND BAETEMAN, C., 1979, The Belgian coastal plain during the Quaternary, *in* Oele, E., Schüttenhelm, R. T. E., and Wiggers, R. J., eds., The Quaternary History of the North Sea: Uppsala University Symposium, v. 2, p. 143–146.

PENLAND, S., BOYD, R., AND SUTER, J. R., 1988, Transgressive depositional systems of the Mississippi Delta plain: a model for barrier shoreline and shelf sand development: Journal of Sedimentary Petrology, v. 58, p. 932–949.

PETEET, D. M., VOGEL, J. S., NELSON, D. E., SOUTHON, J. R., NICKMAN, R. J., AND HEUSSER, L. E., 1990, Younger Dryas climatic reversal in northeastern USA? AMS ages for an old problem: Quaternary Research, v. 33, p. 219–230.

PETERSON, C. D., AND DARIENZO, M. E., 1991, Discrimination of climatic, oceanic and tectonic forcing of marsh burial events from Alsea Bay, Oregon, *in* Roges, A. M., Kockelman, W. J., Priest, G., and Walsh, T. J., eds., Assessing and Reducing Earthquake Hazards in the Pacific Northwest: U.S. Geological Survey Open-File Report 91-441C, 53 p.

PETERSEN, K. L., 1990, Climate and the Dolores River Anasazi: University of Utah Anthropological Papers, v. 113, p. 152.

PHILANDER, S. G., 1989, El Niño, La Niña, and the Southern Oscillation: Academic Press, New York, 293 p.

PISIAS, N. G., 1978, Paleoceanography of the Santa Barbara Basin during the last 8,000 years: Quaternary Research, v. 10, p. 366–384.

PUTNAM, W. C., AXELROD, D. I., BAILEY, H. P., AND MCGILL, J. T., 1960, Natural Coastal Environments of the World: University of California, Los Angeles, Office of Naval Research Report Nonr-233(06) NR 388-013, 140 p.

RAMPINO, M. R., SANDERS, J. E., NEWMAN, W. S., AND KÖNIGSSON, L. K., eds., 1987, Climate: History, Periodicity, and Predictability: Van Nostrand Reinhold, New York, 588 p.

RAU, W. W., 1980, Washington Coastal Geology: Washington Department of Natural Resources Bulletin 72, Olympia, (unpaginated).

SCHRADER, H., AND BAUMGARTNER, T., 1983, Decadal variation of upwelling in the central Gulf of California, *in* Thiedel, J., and Suess, E., eds., Coastal Upwelling, Part B: Plenum, New York, p. 217–246.

SCHULMAN, E., 1956, Dendroclimatic Changes in Semi-arid America: University of Arizona Press, Tucson, 142 p.

SHEPARD, F. P., 1963, Thirty-five thousand years of sea level, *in* Clements, T., ed., Essays in Marine Geology: University of California Press, Los Angeles, p. 1–10.

SHEPARD, F. P., AND CURRAY, J. R., 1967, Carbon-14 determination of sea level changes in stable areas, *in* Sears, M., ed., Progress in Oceanography, Pergamon, Oxford, v. 4, p. 283–291.

SHEPARD, F. P., AND WANLESS, H. R., SR., 1971, Our Changing Coastlines: McGraw-Hill, New York, 579 p.

SOUTAR, A., AND CRILL, P. A., 1977, Sedimentation and climatic patterns in the Santa Barbara Basin during the 19th and 20th centuries: Geological Society of America Bulletin, v. 88, p. 1161–1172.

STAPOR, F. W., JR., 1975, Holocene beachridge plain development, northwest Florida: Zeitschrift für Geomorphologie, Supplement, v. 22, p. 116–144.

STAPOR, F. W., JR., AND MATHEWS, T. D., 1983, Higher-than-present Holocene sea-level events in wave-cut terraces and scarps: Old Island, Beaufort County, South Carolina: Marine Geology, v. 52, p. M53–M60.

STAPOR, F. W., JR., MATHEWS, T. D., AND LINDFORS-KEARNS, F. E., 1987, Episodic barrier island growth in southwest Florida: a response to fluctuating Holocene sea level?: Miami Geological Society, Memoir 3, p. 149–292.

STAPOR, F. W., JR., MATHEWS, T. D., AND LINDFORS-KEARNS, F. E., 1991, Barrier-island progradation and Holocene sea-level history in southwest Florida: Journal of Coastal Research, v. 7, p. 815–838.

STOCKTON, C. W., BOGGESS, W. R., AND MEKO, D. M., 1985, Climate and tree rings, *in* Hecht, A., ed., Paleoclimate Analysis and Modeling: J. Wiley, Interscience, New York, p. 71–161.

STOMMEL, H., 1965, The Gulf Stream (2nd ed): University of California Press, Berkeley, 202 p.

STRONG, E., 1973, Archaeological evidence of land subsidence on the northwest coast: The Ore Bin, v. 35, p. 109–114.

STUIVER, M., AND BRAZIUNAS, T. F., 1989, Atmospheric ^{14}C and century scale oscillations: Nature, v. 338, p. 405–408.

SUTER, J. R., BERRYHILL, H. L., JR., AND PENLAND, S., 1987, Late Quaternary sea-level fluctuations and depositional sequences, southwest Louisiana continental shelf, *in* Nummedal, D., Pilkey, O. H., and Howard, J. D., eds., Sea-Level Fluctuation and Coastal Evolution: SEPM Special Publication 41, p. 199–219.

TERS, M., 1987, Variations in Holocene sea level on the French Atlantic coast and their climatic significance, *in* Rampino, M. R., Newman, W. S., and Sanders, J. E., eds., Climate: History, Periodicity and Predictability: Van Nostrand Reinhold, New York, p. 204–237.

TOOLEY, M. J., AND SHENNAN, I., eds., 1987, Sea-Level Changes: Basil Blackwell, Oxford, 297 p.

VAN DE PLASSCHE, O., 1982, Sea-level change and water-level movements in the Netherlands during the Holocene: Mededelingen Rijks Geologische Dienst, v. 36, p. 1–93.

VAN DE PLASSCHE, O., ed., 1986, Sea-Level Research: A Manual for the Collection and Evaluation of Data: GeoBooks, Norwich, 618 p.

VAN DE PLASSCHE, O., 1990, Mid-Holocene sea-level change on the eastern shore of Virginia: Marine Geology, v. 91, p. 149–154.

WALCOTT, R. I., 1972, Late Quaternary vertical movements in eastern North America: quantitative evidence of glacio-isostatic rebound: Reviews of Geophysics and Space Physics, v. 10, p. 849–884.

WALKER, H. J., AND COLEMAN, J. M., 1987, Atlantic and Gulf Coastal Province, *in* Graf, W. L., ed., Geological Society of America, Decade of North American Geology Centennial, special vol. 2, p. 51–110.

WEST, D. O., AND MCCRUMB, D. R., 1988, Coastline uplift in Oregon and Washington and the nature of Cascadia subduction-zone tectonics: Geology, v. 16, p. 169–172.

WIGLEY, T. M. L., AND RAPER, S. C. B., 1987, Thermal expansion of sea water associated with global warming: Nature, v. 330, p. 127–131.

WILKINSON, B. H., AND BASSE, R. A., 1978, Late Holocene history of the central Texas coast from Galveston Island to Pass Cavallo: Geological Society of America Bulletin, v. 89, p. 1592–1600.

WINSLOW, M. A., AND JOHNSON, L. L., 1989, Prehistoric human settlement patterns in a tectonically unstable environment: Outer Shumagin Islands, southwestern Alaska: Geoarchaeology, v. 4, p. 297–318.

YAMAGUCHI, D. K., 1989, Tree-ring evidence for synchronous rapid submergence of the southwestern Washington coast (abst.): EOS, Transactions of the American Geophysical Union, v. 70, p. 1332.

PART II
ATLANTIC COASTAL SYSTEMS

SEA-LEVEL CHANGE AND LATE QUATERNARY SEDIMENT ACCUMULATION ON THE SOUTHERN MAINE INNER CONTINENTAL SHELF

JOSEPH T. KELLEY
Maine Geological Survey, State House Station #22, Augusta, Maine 04333
STEPHEN M. DICKSON
Maine Geological Survey, State House Station #22, Augusta, Maine 04333
DANIEL F. BELKNAP
University of Maine, Department of Geological Sciences, Orono, Maine 04469
AND
ROBERT STUCKENRATH, JR.
University of Pittsburgh, Applied Research Center, 365 William Pitt Way, Pittsburgh, Pennsylvania 15238

ABSTRACT: Sea-level changes have had an important influence on the distribution of late Quaternary inner continental-shelf sediment in the western Gulf of Maine. Previous stratigraphic models of sea-level change in the region were based on terrestrial observations and a large quantity of offshore high-resolution seismic-reflection data. These models, however, were not constrained by core data. Integration of new vibracore data with earlier observations indicates that nearshore regions were (1) probably deglaciated and subjected to glacio-marine conditions around 13.5 ka, (2) subaerially exposed by a fall in sea level sometime after 11 ka, and (3) flooded by a transgressing sea following an inferred lowstand of sea level between 11 and 9 ka. The greatest amount of sediment accumulated on the shelf during the initial transgression, under glacio-marine conditions. Sandy fluvial sediment accumulated in large quantities during the following regression and early Holocene transgression. Sediment influx from eroding bluffs of glacial origin was significant throughout the Holocene transgression, especially in regions lacking a fluvial source.

INTRODUCTION

The Gulf of Maine is distinguished from the rest of the United States east coast by its bedrock framework and glacial overprint. Within the past 14 ka, the inner continental shelf of the southwestern Gulf of Maine (Fig. 1) has experienced deglaciation accompanied by a marine transgression, emergence of at least part of the shelf, and resubmergence by another marine transgression. This brief but complex history has resulted in the deposition of a relatively thin, but complicated, Quaternary stratigraphic cover. Glacial deposits are often internally heterogeneous and unevenly distributed across the regionally variable, crystalline bedrock of the western Gulf of Maine. As a consequence of sea-level fluctuations, much of the glacial sediment was reworked twice by nearshore processes and was locally overlain by younger fluvial material. On the basis of data from bottom sampling, submersible dives, and seismic-reflection and sidescan-sonar profiling, models were developed for the timing of sea-level fluctuations and late Quaternary deposition. These models strongly rely on interpretations of seismic-reflection data, which can be ambiguous in a complex depositional setting. For example, the seismic data alone cannot improve our understanding of the chronology of late Quaternary events, particularly with respect to the time and depth of the early Holocene lowstand of sea level. Yet, resolution of sea-level fluctuations is necessary for understanding the geologic history of the region. The purpose of this paper is to test earlier models for the evolution of the inner shelf with new data from offshore vibracores, and to link the timing of sea-level fluctuations to sediment accumulation on the shelf.

PREVIOUS WORK

Bloom (1960, 1963) was the first to define the emergent glacio-marine sediment of southwestern Maine as the Presumpscot Formation, estimate its time of deposition in relation to the disappearance of glacial ice, and recognize that portions of the unit that are presently under water were once emergent. Shortly thereafter, Borns and Hagar (1965) established the Embden and North Anson Formations as (regressive) fluvial sand and gravel deposits that conformably overlie the Presumpscot Formation in terraces of the upper Kennebec River valley. They also noted that a large volume of Pleistocene valley fill was excavated by the river and carried downstream. Stuiver and Borns (1975) and later Smith (1985) bracketed the time of deposition of the Presumpscot Formation with radiocarbon dates to have been between 13.5 and 11.5 ka.

Slightly north of the study area, Ostericher (1965) obtained the first seismic-reflection records from Maine's inner continental shelf. He identified the Presumpscot Formation, as well as till and modern fluvial and marine deposits, from these records, and collected numerous cores to confirm his interpretations. He interpreted a coherent and widespread seismic reflector as the transgressive unconformity on the surface of the Presumpscot Formation, and in one core, collected a wood fragment from its surface at a core depth of 18 m. The wood sample yielded a radiocarbon date of 7,390±500 yrs and was subsequently cited by Knebel and Scanlon (1985) as conclusive evidence for a minimum depth and age of the lowstand of sea level for the western Gulf of Maine. On the basis of further mapping of the seismic reflector interpreted as the transgressive unconformity, they proposed that the lowstand of sea level along the central Maine shelf occurred at around −40 m relative to present sea level (Knebel and Scanlon, 1985; Knebel, 1986).

Schnitker (1974) had earlier inferred, on the basis of seismic-reflection profiles, that the "ultimate" lowstand off the mouth of the Kennebec River (Fig. 1) was at −65 m. He tentatively identified a "berm" at this depth on a seismic profile. Later workers (Belknap and others, 1989; Shipp, 1989, Shipp and others, 1989, 1991) recognized a similar

Quaternary Coasts of the United States: Marine and Lacustrine Systems, SEPM Special Publication No. 48

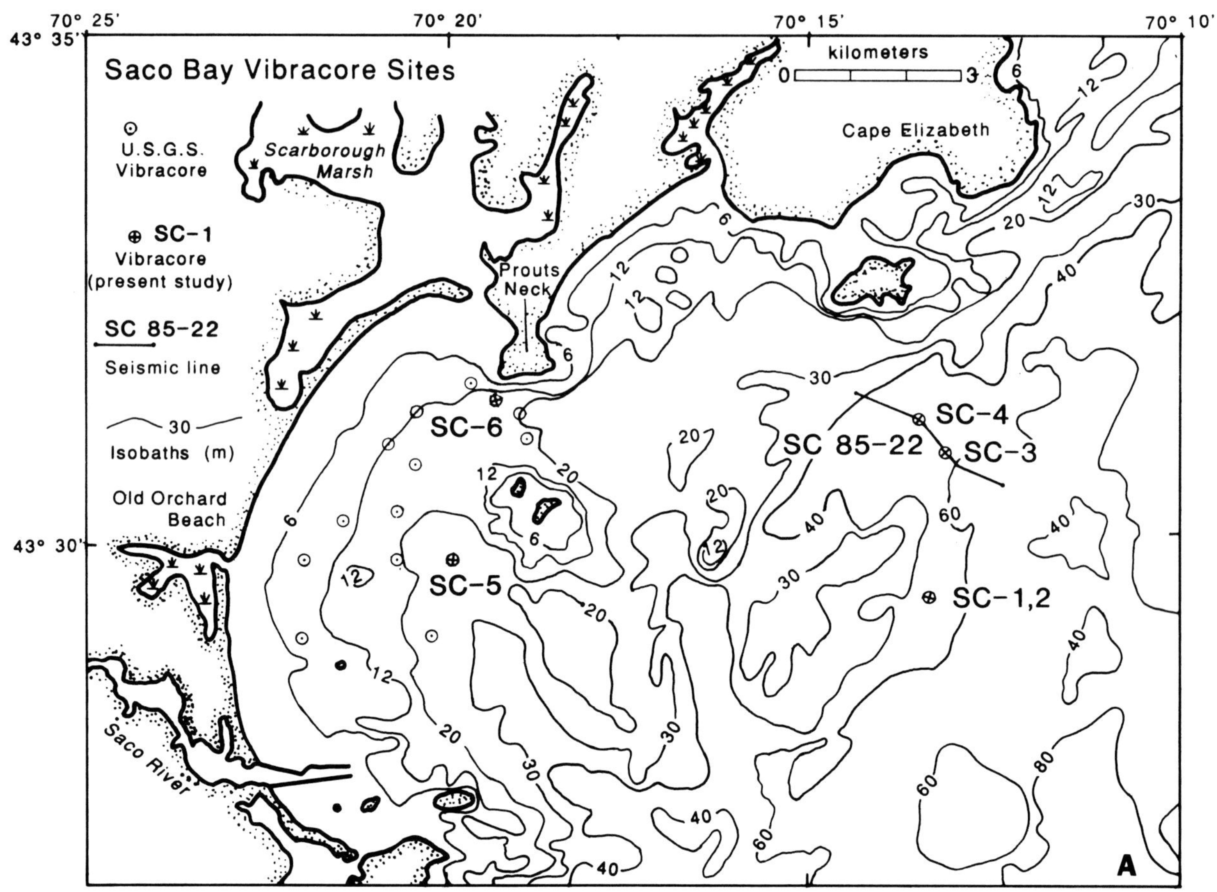

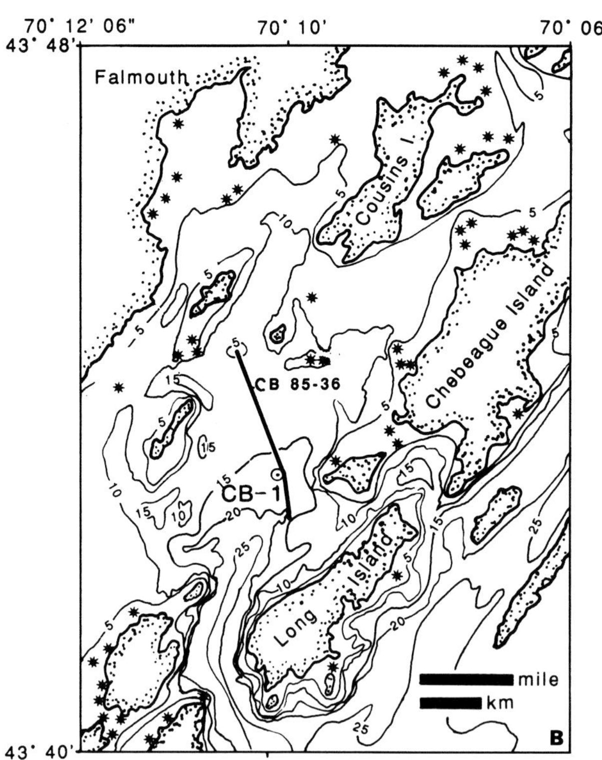

FIG. 2.—Continued.

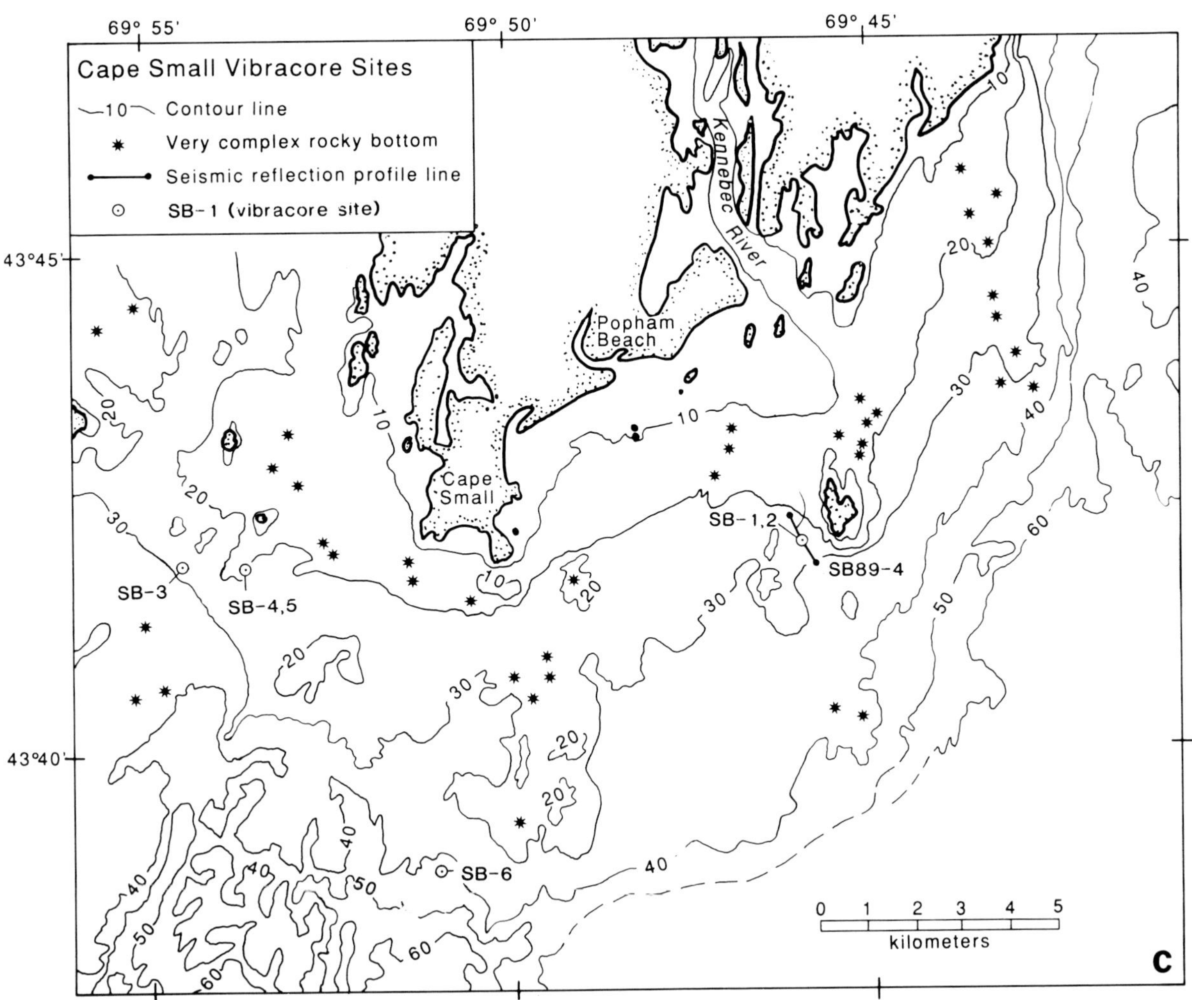

FIG. 2.—Bathymetric maps of the core locations. (A) Saco Bay. (B) Casco Bay; stars indicate areas of complex bathymetry. Note that the legend is the same as in (A). (C) Cape Small/Kennebec River mouth. Note that only one seismic line from each bay is presented in this report, although many other lines are presented by Kelley and others (1990).

low 5 m to a medium sand with several laminae of wood and charcoal (Fig. 3). The mud unit contained no datable macrofossils, but articulated *Mya arenaria* and *Macoma balthica* shells from 6.0 to 6.2 m were dated at 9,130±170 yrs, and a mixture of wood, charcoal, and shell fragments at 7.5 to 7.7 m yielded a date of 9,735±150 yrs (Fig. 3).

Cape Small

The Cape Small area is an exposed nearshore ramp seaward of the Kennebec River mouth (Figs. 1, 2C). Previous mapping (Belknap and others, 1989; Kelley and others, 1987b) has demonstrated that the sea floor in the Cape Small region is mantled with sand and gravel between occasional outcrops of bedrock. The ramp morphology probably represents a paleodelta of the Kennebec River (Belknap and others, 1986).

Cores SB-1 and SB-2 were collected from the same location west of Seguin Island in 19 m of water (Fig. 2c). Seismic profiles across the core site reveal a 30- to 40-m-thick acoustic unit with numerous coherent internal reflectors, interpreted as glacio-marine sediment, over bedrock (Fig. 4). Reflectors in the upper portion of the glacio-marine unit appear truncated by a seismic reflector that is strong and relatively flat. This reflector is overlain by a strongly reflecting unit, lacking internal reflectors, which pinches out in a seaward direction (Fig. 4).

Because both cores are from the same site and are very similar (Fig. 3), only the longer core, SB-2, is discussed. The upper 2 m of sediment from SB-2 are poorly sorted, muddy sands. A gradual coarsening of the sand occurs downcore between 2 and 5 m. Beds, defined by shells of articulated and fragmented *Mya arenaria* or *Mytilus edulis*, intertidal to shallow subtidal mollusks, or small laminae of fine or coarse micaceous sand, were common. Between 2.2 and 2.35 m, several *Mya arenaria* in life position were observed.

An abrupt change from gravelly sand above to sandy mud below occurs at 5 m in the core. Between 5 and 6.5 m, the sandy mud is very poorly sorted with many laminae of wood fragments. Below 6.5 m, the proportion of mud declines to less than 10 percent (except in rare muddy laminae), and

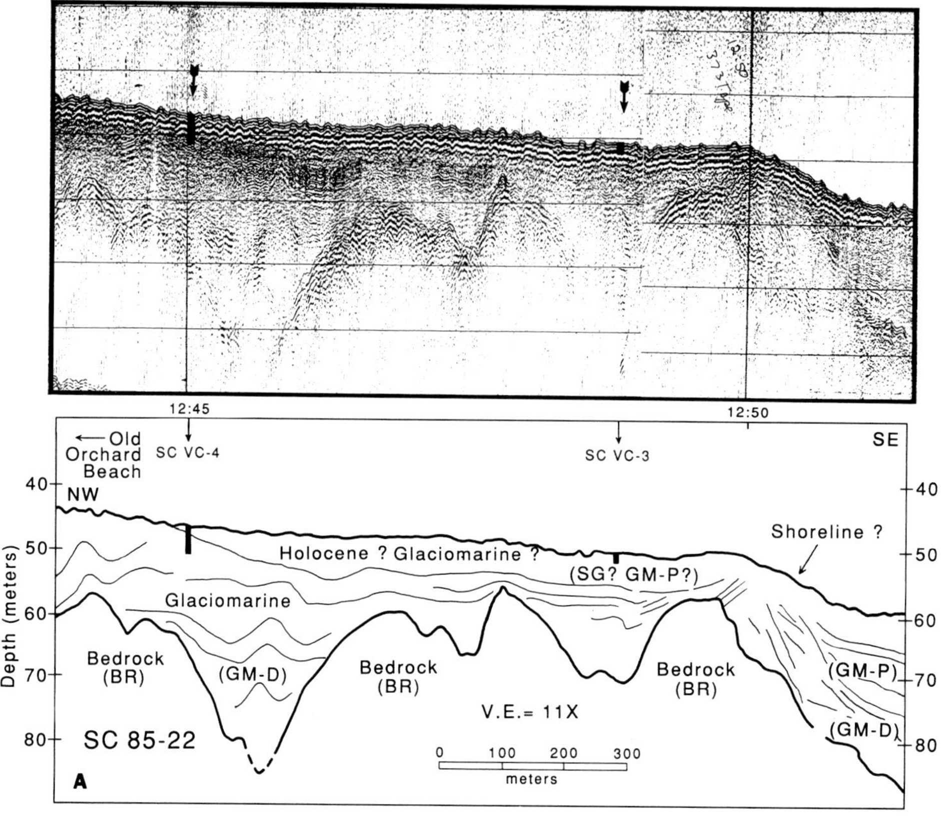

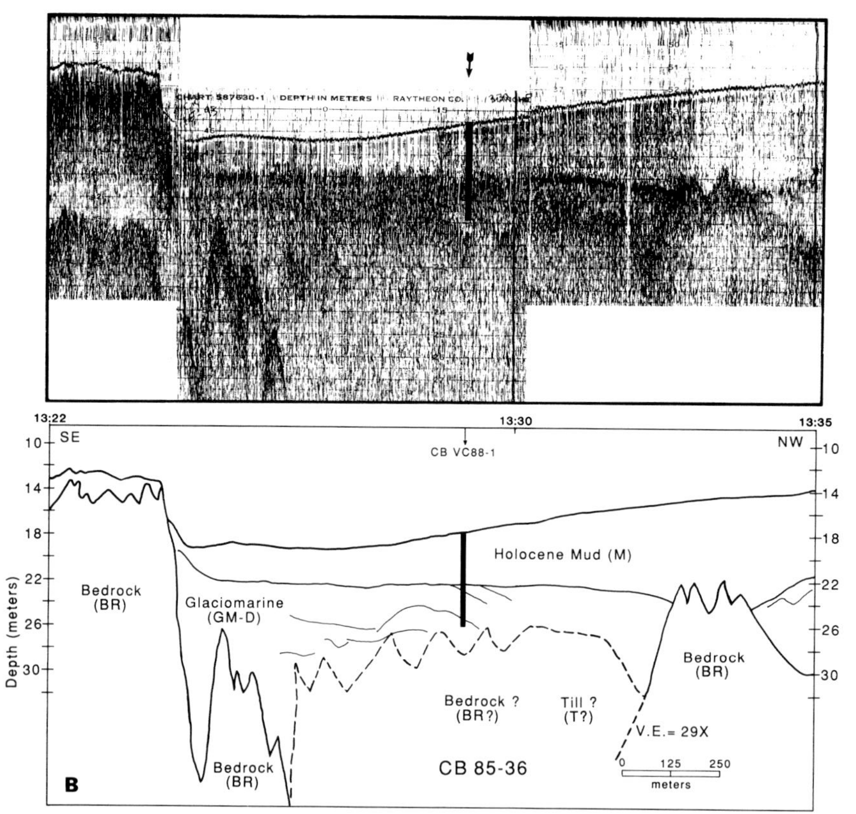

FIG. 4.—Continued.

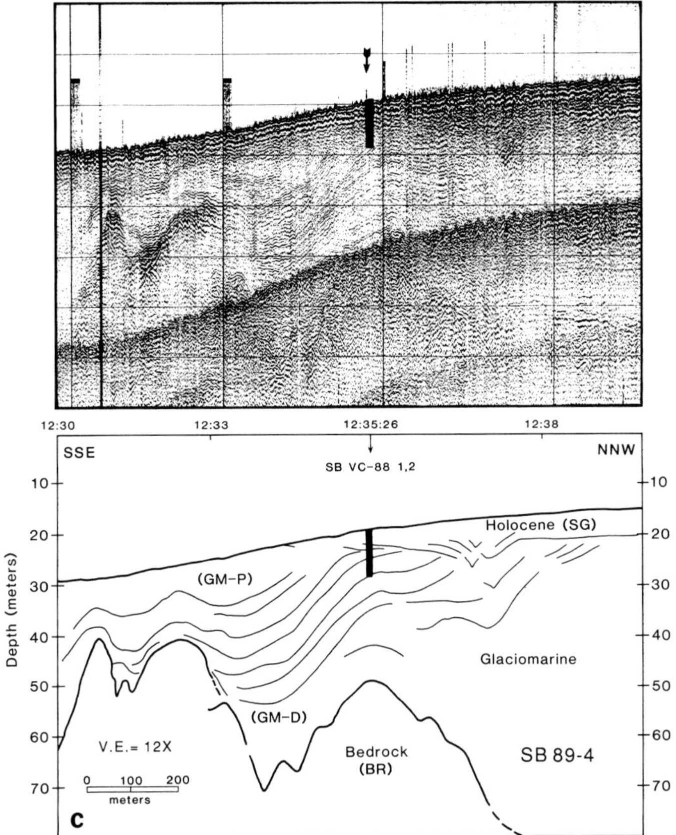

FIG. 4.—Seismic profiles and interpreted sections in relation to core positions. (A) Site of cores SC-3, 4. (B) Site of core CB-1. (C) Site of cores SB-1, 2 (modified from Kelley and others, 1990). Note that noise from seaweed clinging to transducer darkens the upper unit of CB-1, which is otherwise acoustically transparent.

TABLE 1.—RADIOCARBON DATES FROM THE SOUTHERN MAINE INNER CONTINENTAL SHELF

Lab Number	Core Number	Water Depth	Material	Radiocarbon Age (ka/bp)
PITT-0739	SC-2	51.3	*Artica isl.*	785±35
PITT-0740	SC-4	47.1	*Mya arenaria*	10,620±90
PITT-0741	SC-4	48.5	**Hiatella arc.*	5,915±155
PITT-0742	SC-6	22.8	*Mya truncata*	11.770±80
PITT-0743	SC-6	23.1	**Macoma balt.*	14,090±450
PITT-0744	SB-1	19.6	*Mytilus, Mya*	9,000±100
PITT-0745	SB-1	19.7	*Mya arenaria*	9,630±75
PITT-0746	SB-1	19.9	*Modiolus mod.*	9,700±65
PITT-0747	SB-1	21.7	*Mya arenaria*	9,260±100
PITT-0748	SB-2	19.85	**Mytilus ed.*	8,250±80
PITT-0749	SB-2	20.9	*Mya arenaria*	9,235±60
PITT-0585	SB-2	21.25	*Mya arenaria*	9,090±95
PITT-0586	SB-2	21.35	*Mya arenaria*	9,250±110
PITT-0587	SB-2	21.5	*Mya arenaria*	7,270±105
PITT-0751	SB-2	25.9	wood	11,550±160
PITT-0752	SB-2	26.8	*wood, shells	2,215±290
PITT-0753	SB-3	34.9	*Artica isl.*	1,300±35
PITT-0754	SB-3	36.0	*Mya arenaria*	2,570±50
PITT-0755	SB-3	39.1	*Mya arenaria*	2,950±210
PITT-0756	SB-6	43.1	*Artica isl.*	8,270±75
PITT-0737	CB-1	21.1	**Mya, Mytilus*	9,130±70
PITT-0738	CB-1	22.6	wood, shells	9,735±150

*Shell fragments from many organisms.

nebec River paleodelta (Belknap and others, 1989) were verified by cores SB-3, SB-4, and SB-5 (Fig. 3). Formerly interpreted sediment of glacial origin near Prouts Neck in Saco Bay (Kelley and others, 1987a, 1989c) is now recognized as a diamicton in SC-6 (Fig. 3).

The strong reflector interpreted as the transgressive unconformity in each of the bays (Belknap and others, 1989; Kelley and others, 1986; 1987a, 1987b) is confirmed. The reflector is a lithologic discontinuity of sand over mud in both Saco Bay (SC-4) and off the Kennebec River (SB-1, SB-2), and as mud over sand in Casco Bay (CB-1; Figs. 3, 4). The thin nature of the upper sand in Saco Bay (SC-5) was not recognized in earlier work (Kelley and others, 1986; 1987a) because the acoustical bubble pulse obscured any shallow reflectors. Earlier reports overestimated the volume of Holocene sand (Kelley and others, 1987a).

Lowstand Shoreline

Although the lowstand shoreline was not cored, SC-1, SC-2, SC-3, and SB-6 were all gathered from near the top of the inferred lowstand shoreline. Each of these cores contained well-sorted sand. Such textures are not being produced by modern shelf processes. It is not yet possible to determine if the sands are relict fluvial, or littoral, or palimpsest in origin, and if palimpsest, whether they represent reworked glacial sediment. The 785-yr date from an *Artica islandica* in SC-2 is almost certainly younger than the enclosing sand deposit, but it is not clear whether the 8,270-yr *Artica* date from SB-6 represents the time of sediment deposition. In either case, *Artica* is a relatively deep-water mollusk and only indicates that marine conditions existed at the time of accumulation (Fig. 5).

Sea-Level Indicators

Several of the fossils dated are intertidal or shallow subtidal mollusks whose ages are suitable for constructing a sea-level curve (Fig. 5). A large cluster of *Mya* and *Mytilus* shells occurs between 19.5 and 21.5 m in SB-1 and SB-2. Their proximity within the cores resembles extant communities of these mollusks landward of the core site in Maine's intertidal zone (Larson and Doggett, 1990). Several of the dates/depths appear incorrect as a consequence of small sample size (7,270 yrs), contamination during withdrawal of the core (9,260 yrs), and transport and mixing of fragments (8,250 yrs). Despite these anomalies, a significant difference exists between the cores, with SB-1, yielding older material than SB-2 at the same depth. Although the difference does not alter the sea-level curve very much, it remains a problem. Until resolved, we use the average of the 9,090-yr and 9,250-yr dates from *in situ* intertidal-community shells, 9.17 ka, as the best timeframe indicator of sea level at −21.3 m mean high water (MHW).

The radiocarbon dates from the Casco Bay core further support this sea-level position. The *Mya* and *Mytilus* shells are similar in their occurrence, as an intertidal community, to the fossils in the SB-1 and SB-2 cores; they were alive at about the same time (9.13 ka) and paleodepth (21.1 m; Fig. 5). The wood and shell fragments lower in the core are older, as expected from a partly terrestrial sample, and in reasonable agreement with the other dates. These sam-

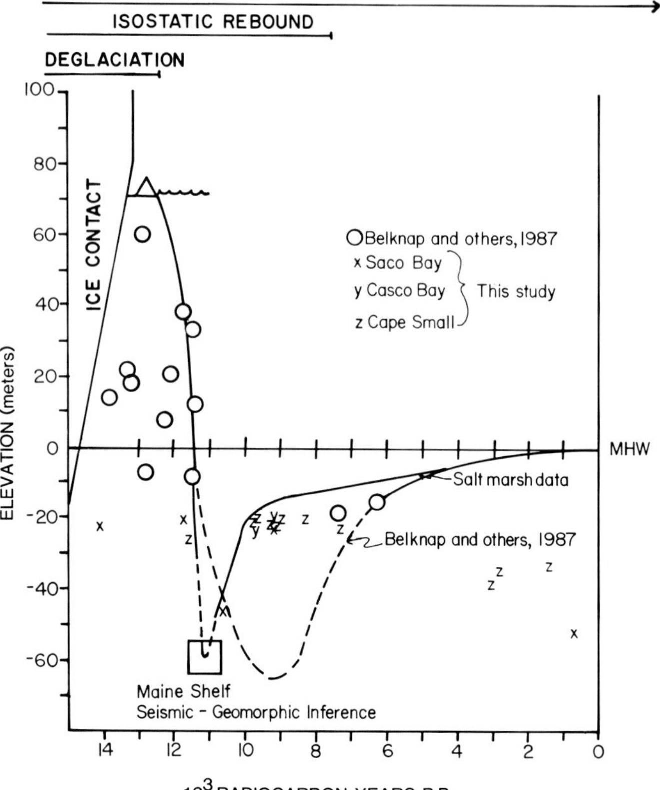

FIG. 5.—Sea-level curve for coastal Maine (modified from Belknap and others, 1987).

ples occur below or at the same depth as the prominent seismic reflector we have previously interpreted as the transgressive unconformity (Fig. 4; Kelley and others, 1986, 1987b, 1990). Taken together, the wood fragments and intertidal shells, in a sandy unit above glacio-marine sediment and associated with a transgressive setting, indicate deposition occurred at the base of an eroding bluff. Similar deposits are forming all along the margins of Casco Bay today (Friez and others, 1990; Hay, 1988).

Dates from the *Mya* samples of SB-3 (2,570 and 2,950 yrs) do not fit into the sea-level curve. The sea floor throughout the core area is intensely rippled and indicative of contemporary reworking (Belknap and others, 1988). It is plausible that the shells were transported into their present setting from shallower water.

The older dates are less certainly connected with sea level, but provide the link between dates from terrestrial outcrops and marine cores. The *Mya truncata* and *Macoma balthica* fossils from SC-6, and which are also common from emergent marine sediment in terrestrial outcrops (Belknap and others, 1987), provide excellent dates of glacio-marine conditions at 11.77 and 14.09 ka, respectively (Fig. 5). Similarly, the 11.55-ka date on the wood sample from SB-2 likely points to a time of deposition of regressive, marine estuarine sediments. Although our earlier seismic work interpreted all material below the unconformity as ice-proximal glacio-marine sediment (Belknap and others, 1986), clearly, because of the presence of wood fragments, only the lower portions of the unit could have glacial affinities.

The deepest *Mya* sample dated (10,620 yrs) is from the upper unit of SC-4 (Fig. 3). It is interpreted as part of a transgressive deposit, but the mollusk could have been living during the regression and its shell subsequently reworked during the transgression. Although we cannot rule out the latter possibility, we believe the deposit is transgressive because of the preservation of the shell. If true, the time of the lowstand shoreline would have to be moved to approximately 11 ka. This interpretation is supported by the stratigraphic setting above a seismic unconformity (Fig. 4). Sand within the upper unit also possesses a coarser primary-grain-size mode than sand from the lower portion of the core (Kelley and others, 1990; Fig. 12). In other cores from Saco Bay the grain-size mode is accompanied by a heavy mineralogy significantly different from that of the lower sand (Luepke and Grosz, 1986; Kelley and others, 1990). We interpret these textural and mineralogical differences to mean that the upper sand is a regressive/transgressive fluvial deposit originating from the Saco River, and that the lower muddy material is glacio-marine sediment. We believe the fossils inhabited the sediment during the transgression, although we cannot rule out the possibility that they are regressive.

Chronology of Sediment Accumulation on the Inner Shelf

The greatest amount of Holocene sediments accumulated on the inner shelf during deglaciation in the latest Pleistocene. Interbedded till and glacio-marine sediment accumulated along the present shoreline around 13.8 ka (Stuiver and Borns, 1975). Shortly prior to and following this time (i.e., from about 14.1 to 11.8 ka; SC-6, Table 1), glacio-marine conditions prevailed on the inner shelf. It is difficult to estimate the average thickness of glacial sediment due to the irregular bedrock topography, but till and glacio-marine material are commonly up to 40 m thick in low shelf valleys (Fig. 1; Kelley and others, 1987a; Shipp, 1989).

As sea level fell rapidly in response to isostatic uplift (Belknap and others, 1987; Fig. 5), rivers incised glacial deposits and transported sediment to the sea. During the regression, the Embden and North Anson Formations were deposited on terraces along the Kennebec River (Borns and Hager, 1965). Although these units have not been dated, and no known correlative deposits are found in the Saco River drainage, it seems likely that the regression occurred between 13 and 11 ka (Fig. 5). Hyland and others (1978) suggested that sea level had fallen to that of the present coastline by 12 ka, based on conventional radiocarbon dating of wood within the Presumpscot Formation. Recent AMS radiocarbon dates from the same location confirm a date of 11.7 to 11.5 ka (Anderson and others, 1990). Deltas greater than 50 m thick formed off large river mouths such as the Kennebec and Merrimack Rivers (Belknap and others, 1989; Oldale and others, 1983). No prominent deltas are known off smaller rivers such as the Saco, but there are thinner sand deposits (Kelley and others, 1987a).

These new data suggest that sea level rose more rapidly, following a period of isostatic adjustment, than current models predicted (Tushingham, 1989; Fig. 5). By 9.2 ka, sea level slowed its rate of rise, allowing communities of intertidal and shallow-water organisms to thrive, possibly in back-barrier (SB-1, SB-2) and estuarine bluff-toe locations (CB-1; Fig. 3). It is not clear whether fluvial sand was being deposited in significant quantities at that time, but between 9.2 ka and the present, fluvial-sand deposition slowed or stopped. The present sea floor off the Saco and Kennebec Rivers is covered with ripples and other indications of modern reworking (Belknap and others, 1988). The 9-ka shells are all from within 2 m of the modern sea floor, and glacial sediment crops out nearby (Kelley and others, 1987a, b).

The important difference between Casco Bay and Saco Bay is the lack of major fluvial-sediment input into Casco Bay (Kelley and others, 1986). Owing to drainage derangement by glaciation, no large rivers deliver sediment to Casco Bay. Bluff erosion has provided abundant muddy sediment to basins in the bay and on the adjacent shelf throughout the period of transgression. Sediment-budget estimates indicate that more bluff-supplied sediment has been produced than can be found in the bay today, implying significant export of material to the shelf (Hay, 1988).

CONCLUSIONS

Vibracores from the inner shelf off central and southern coastal Maine tested existing models of seismic stratigraphy and allowed refinement of models of late Quaternary coastal evolution in the area. In particular, identification of the ubiquitous latest Pleistocene or earliest Holocene unconformity was confirmed, and the Holocene section is actually thinner in Saco Bay than previously thought. Radiocarbon dates obtained from mollusk shells within the vibracores resulted in a revised model of late Quaternary local relative sea-level changes. Three dates indicate deposition of the glacio-marine Presumpscot Formation occurred between 14 and 11.5 ka. Thirteen other dates from shells located above a distinct unconformity require changes in the earlier regional sea-level curve (Belknap and others, 1987). These data suggest that local lowstand of the sea occurred between 11.5 and 10.5 ka at more than −50 m. No reliable dates were obtained for the postulated lowstand shoreline at −55 to −60 m (Shipp and others, 1991). Six late Holocene dates relate to modern seafloor processes, possibly including reworking from shallower environments.

Deposition of sandy nearshore-ramp and paleodelta deposits occurred between 11.5 and 8.5 ka. The thickness of the upper sandy units is directly related to availability of upland sources and size of the associated river system. Thus, the large Kennebec River system has a large paleodelta, the smaller Saco River system has a thinner sandy mantle offshore, and Casco Bay, with no significant streams draining sandy deposits, is dominated by mud. After 8.5 ka, the inner shelf was dominated by reworking of sand into the modern sandy shorelines of Saco Bay and the Kennebec River mouth, with some likely offshore movement into deeper water.

ACKNOWLEDGMENTS

The authors acknowledge support for this project from the Continental Margins Program of the U.S. Minerals Management Service and the Maine Geological Survey. Dr. R. Craig Shipp contributed significantly to the development of many of the ideas presented here as well as to the selection of the core sites. Ms. Julie Friez of the University of Maine performed the grain-size analyses. The manuscript was improved by reviews from Albert C. Hine, Robert McMaster, S. Jeffress Williams, and Harold W. Borns, Jr.

REFERENCES

Anderson, R. S., Miller, N. G., Davis, R. B., and Nelson, R. E., 1990, Terrestrial fossils in the marine Presumpscot Formation: implications for Late Wisconsinan paleoenvironments and isostatic rebound along the coast of Maine: Canadian Journal of Earth Science, v. 27, p. 1241–1246.

Belknap, D. F., Andersen, B. G., Anderson, R. S., Anderson, W. A., Borns, H. W., Jr., Jacobson, G., Jr., Kelley, J. T., Shipp, R. C., Smith, D. C., Stuckenrath, R. Jr., Thompson, W. B., and Tyler, D. A., 1987, Late Quaternary sea-level changes in Maine, *in* Nummedal, D., Pilkey, O. H., Jr., and Howard, J. D., eds., Sea-Level Fluctuation and Coastal Evolution. Society of Economic Paleontologists and Mineralogists Special Publication 41, p. 71–85.

Belknap, D. F., Kelley, J. T., and Robbins, D. H. W., 1988, Sediment dynamics of the nearshore Gulf of Maine: submersible experimentation and remote sensing, *in* Babb, I., and Deluca, M., eds., Benthic Productivity and Marine Resources of the Gulf of Maine: National Oceanic and Atmospheric Administration, National Undersea Research Program, Research Report 88-3, p. 143–176.

Belknap, D. F., and Shipp, R. C., 1991, Seismic stratigraphy of glacial-marine sediments, Maine inner shelf, *in* Anderson, J. B., and Ashley, G. M., eds. Glacial-Marine Sedimentation: Paleoclimatic Significance. Geological Society of America Special Paper 261, p. 137–157.

Belknap, D. F., Shipp, R. C., and Kelley, J. T., 1986, Depositional setting and Quaternary stratigraphy of the Sheepscot estuary, Maine: a preliminary report: Geographie Physique et Quaternaire, v. 40, p. 55–69.

Belknap, D. F., Shipp, R. C., Kelley, J. T., and Schnitker, D., 1989, Depositional sequence modeling of late Quaternary geologic history, west-central Maine coast, *in* Tucker, R. D., and Marvinney, R. G., eds., Studies in Maine Geology, v. 5, Quaternary Geology. Maine Geological Survey, Augusta, p. 29–46.

Birch, F. S., 1984, A geophysical study of sedimentary deposits on the inner continental shelf of New Hampshire: Northeastern Geology, v. 6, p. 207–221.

Birch, F. S., 1988, Sediments of the inner continental shelf: first and second year projects in New Hampshire, *in* Hunt, M. C., Ratcliff, D. C., Doenges, S., and Condon, C., eds. Proceedings, First Symposium on Studies Related to Continental Margins–A Summary of Year-One and Year-Two Activities. U.S. Department of Interior Minerals Management Service, p. 242–251.

Bloom, A. L., 1960, Late Pleistocene changes of sea level in southwestern Maine: Maine Geological Survey, Augusta, 143 p.

Bloom, A. L., 1963, Late-Pleistocene fluctuations of sea level and postglacial and crustal rebound in coastal Maine: American Journal of Science, v. 261, p. 862–879.

Borns, H. W., Jr., and Hager, D. J., 1965, Late-glacial stratigraphy of a northern part of the Kennebec River Valley, western, Maine: Geological Society of America Bulletin, v. 76, p. 1233–1250.

Friez, J. K., Schnitker, D., Belknap, D. F., and Kelley, J. T., 1990, Reconstruction of the late Quaternary paleoenvironmental evolution of the northwestern Gulf of Maine using benthic foraminifera: Geological Society of America Abstracts with Programs, v. 22, p. 18.

Gosner, K. L., 1971, Guide to Identification of Marine and Estuarine Invertebrates, Cape Hatteras to the Bay of Fundy. Wiley Interscience, New York.

Hay, B. W. B., 1988, The role of varying rates of local relative sea-level change in controlling the Holocene sedimentologic evolution of

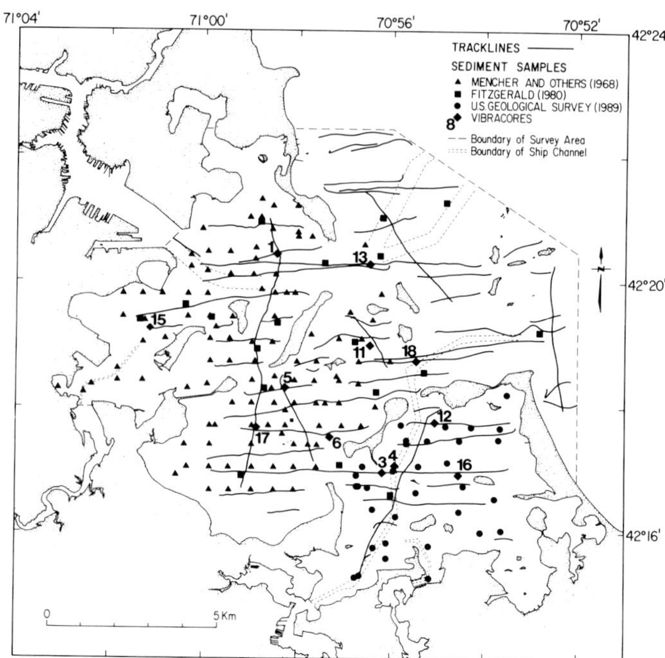

FIG. 1.—Map of Boston Harbor area showing locations of survey tracklines and sediment samples.

offshore from outcrops of bedrock and till on land (LaForge, 1932; Crosby, 1934; Phipps, 1964; Kaye, 1976, 1978, 1982) and laterally from sites in the harbor where these two units have been identified either by drilling (Metcalf and Eddy, Inc., 1989) or by seismic-refraction data (Weston Geophysical Engineers, Inc., 1969; Weston Geophysical Corporation, 1989).

The thickness of sediments above acoustic basement ranges from zero to more than 35 m. In general, the sediments are relatively thick (>20 m) within depressions in the acoustic basement (Figs. 2, 3), whereas they are thin (<5 m) or absent over basement highs and along mainland and insular shores.

Late Wisconsinan Ice-Proximal Drift

The lowermost unit observed on the subbottom profiles (Qdr in Fig. 2) unconformably overlies acoustic basement. This unit produces a strong surface return and typically has short, discontinuous, internal reflectors. It is found in isolated patches within basement lows and has a maximum thickness of about 18 m.

The stratigraphic position, seismic character, and distribution of the unit suggest that it is composed of upper Wisconsinan drift. Deposits of similar age and with similar characteristics have been identified at a number of locations on the inner shelf off New England (Oldale and Bick, 1987; Oldale and Wommack, 1987) and in the southwestern Gulf of Maine (Tucholke and Hollister, 1973). Correlative deposits also have been described on land in the Boston area (Kaye, 1982; Hanson, 1984) and in the area of late-glacial marine submergence of coastal Maine (Thompson, 1982; Smith, 1982). These studies indicate that the unit is a drift sequence composed of till, subaqueous outwash, and ice-contact sand and gravel that varies greatly in the degree of sorting and stratification. The sequence probably was deposited on the sea floor along the marine-based ice front (with ice and sea in contact) in the same manner as that described by Rust and Romanelli (1975) for subaqueous ice-contact and outwash deposits in the Ottawa Valley, Canada.

Late Wisconsinan Glacio-Marine Sediments

The succeeding acoustic unit (Qm in Figs. 2, 3) generally is characterized by continuous, closely spaced reflections that mimic the irregular underlying surface. In places, however, the internal reflections are weak or absent, and they sometimes become subhorizontal near the top of the unit. The unit is relatively thick (up to 25 m) within subbottom depressions, whereas it is thin or absent over subbottom highs.

The unit is composed primarily of stiff, bluish-gray to olive-gray clayey silts and silty clays that contain abundant thin (<10 cm) beds or lenses of sandy and gravelly sediments (Figs. 2, 3). The unit also includes scattered pebbles and shells, and the sediments are locally oxidized at the top (upper 1–2 m) of the section.

The acoustic, textural, and compositional characteristics of the unit are similar to those of fine-grained glacio-marine sediments of late Wisconsinan age that have been identified in nearby onshore and offshore areas. On land, late Wisconsinan glacio-marine muds with interbedded fine sands, scattered gravel, and some shells underlie much of the lowland surrounding the harbor (Kaye, 1982) and are present along the coasts of New Hampshire (Hanson, 1984) and southeastern Maine, where they comprise the Presumpscot Formation (Bloom, 1960; Thompson, 1982; Smith, 1982). Offshore, upper Wisconsinan glacio-marine muds are characterized in subbottom profiles by closely spaced reflections that are comformable with the underlying surface. Such deposits have been found in western Massachusetts Bay (Oldale and Bick, 1987), on the inner shelf off northern Massachusetts, New Hampshire, and Maine (Birch, 1984; Oldale and Wommack, 1987; Shipp, 1988), and in the southwestern Gulf of Maine (Tucholke and Hollister, 1973; Oldale, 1988).

Deposition of the glacio-marine muds accompanied the retreat of glacial ice from the area. The source of the fine-grained sediments probably was rock-flour-laden meltwater that was discharged either directly into the sea from the ice or through subaerial streams (Oldale, 1989; Oldale and others, 1990). The rhythmic-layered and draped character of the unit suggest that the muds were deposited rapidly and with little disturbance by physical and biologic processes. Coarse detritus within the deposit probably is due to iceberg rafting (Oldale and others, 1990). Deposition of the unit may have continued until as late as 11 ka (Stone and Borns, 1986; Oldale, 1989).

Holocene Fluvial and Estuarine Deposits

In some profiles, a discontinuous reflector (ru in Fig. 3) outlines small channels that were cut into the glacio-marine

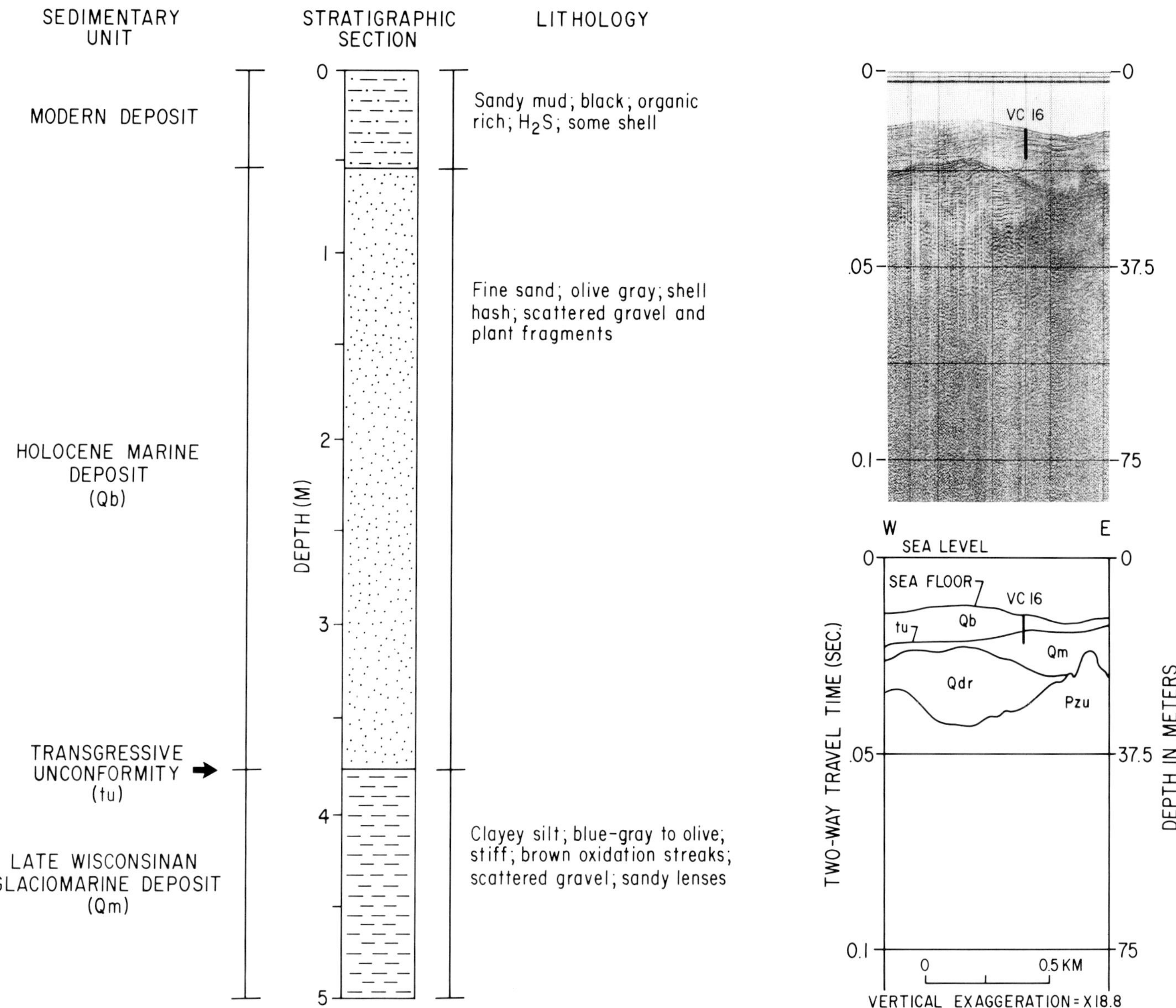

FIG. 2.—Stratigraphic section and subbottom profile at vibracore site 16 shown in Figure 1. Pzu = Paleozoic bedrock; Qdr = late Wisconsinan ice-proximal drift; Qm = late Wisconsinan glacio-marine sediments; tu = transgressive unconformity; Qb = Holocene marine deposits. Sound velocity in water and sediments is assumed to be 1,500 m/s.

sediments. These channels are located sporadically across the study area and were cut to maximum depths of about −20 m. Sediments within the channels (Qf in Fig. 3) either are acoustically transparent or contain short, weak reflectors. Maximum thickness of the unit is about 3 m.

The origin and texture of the channel-fill deposits can be inferred from correlative deposits located under the nearshore parts of Massachusetts Bay (Meisburger, 1976; Oldale and Bick, 1987). As local relative sea level fell during the immediate postglacial period (Kaye and Barghoorn, 1964), glacio-marine sediments across much of the region were subaerially exposed and fluvially eroded (Oldale and Bick, 1987; Oldale and Wommack, 1987). The resulting channels in the glacio-marine deposits then were filled with fluvial and estuarine sediments that range from gravel to clay as base level increased in response to the Holocene rise of sea level (Meisburger, 1976; Oldale and Bick, 1987).

Holocene Marine Deposits

The uppermost sedimentary unit outlined by the seismic records (Qb in Figs. 2, 3) discontinuously overlies a widespread erosional unconformity (tu in Figs. 2, 3). The unit ranges from moderately layered to acoustically transparent, is present both within depressions and on the flanks of subbottom highs, and has a maximum thickness of 5 m.

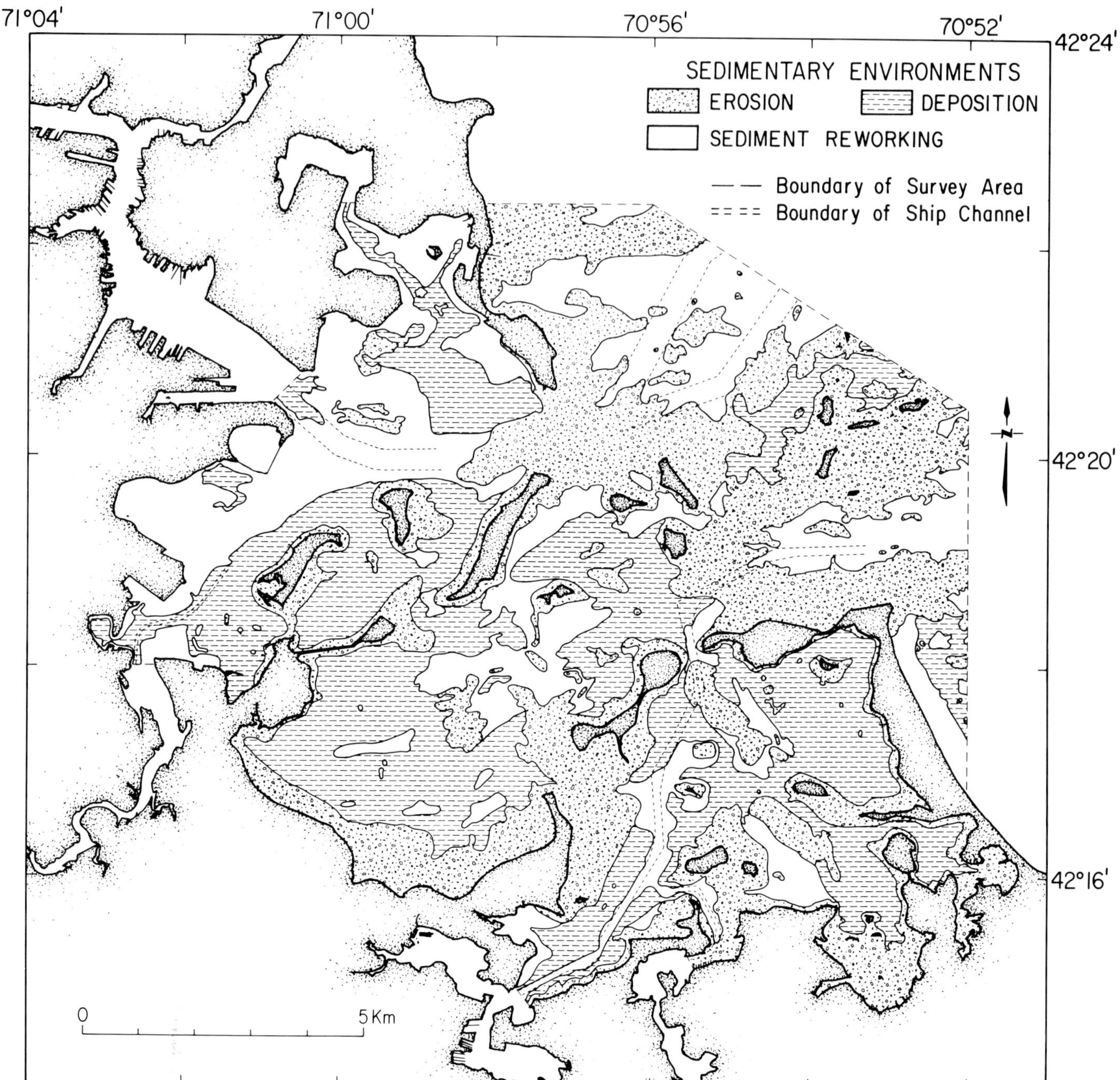

FIG. 5.—Distribution of sedimentary environments in the Boston Harbor area based on analyses of sidescan sonographs supplemented by available bathymetric, sedimentary, subbottom, and bottom-current data. From Knebel and others (1991).

especially effective erosional agents and cause frequent and widespread winnowing and transport of the bottom sediments (Bothner and Butman, 1988; Bothner and others, 1988; Butman and others, 1990). Inside the harbor, waves have eroded shallow areas near the shoreline in the southern part (Knebel and others, 1991) and, in many locations, they have cut bluffs into the till that composes the harbor islands (Phipps, 1964; Kaye, 1982). Across the deeper parts of the harbor floor, however, wave erosion probably is limited because of low sea states throughout the year (Bumpus and others, 1951; Fitzgerald, 1980). Here, exposures of bedrock and till atop scattered ridges and knolls probably are the result of scour by tidal currents that are locally enhanced by the topography.

Sonograph patterns of strong reflectivity, on the other hand, are the result of coarse-grained lag deposits (Knebel and others, 1989, 1991). The gravelly sands that produce these patterns are located primarily in large channels and

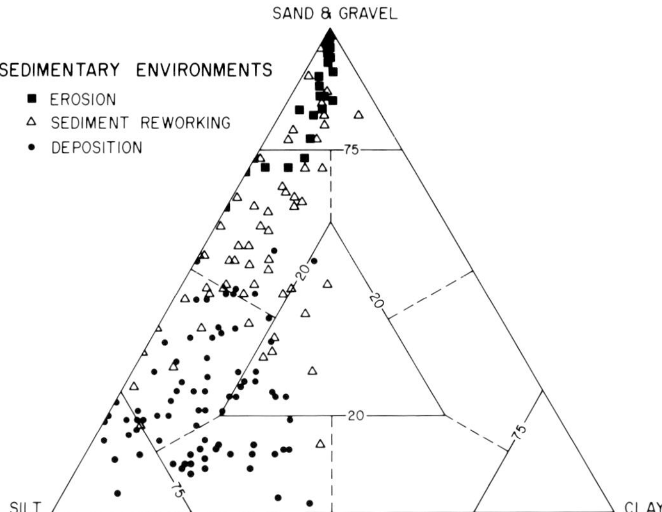

FIG. 6.—Texture of surface sediments within the three sedimentary environments identified in the Boston Harbor area. Original grain size data from Mencher and others (1968), Fitzgerald (1980), and U.S. Geological Survey (1989). Distributions of samples and sedimentary environments are presented in Figures 1 and 5.

depressions where the tidal flow is enhanced between islands and submerged topographic highs. In these constricted channels, near-bottom tidal currents typically exceed 51 cm/s (U.S. Coast and Geodetic Survey, 1953; Mencher and others, 1968; National Ocean Survey, 1977; Fitzgerald, 1980) and, thus, are strong enough to scour and winnow grain sizes finer than very coarse sand (Sternberg, 1972; Miller and others, 1977). Subbottom profiles and cores indicate that the lag deposits exist over subcrops of either pre-Wisconsinan till or gravelly Holocene marine deposits.

Environments of Deposition

Areas of deposition are depicted on the sonographs as smooth featureless surfaces that produce weak acoustic returns (light shading) from the sea floor (Knebel and others, 1989, 1991; Fig. 4C). Featureless patterns were found primarily over the subtidal flats in the southern part of the harbor and within bathymetric lows situated among islands and away from the main tidal channels (Fig. 5).

The sediments in depositional areas are fine grained (Fig. 6). They consist of gray-to-black silts, clayey silts, and sandy silts that contain relatively high concentrations of organic matter (Mencher and others, 1968; Fitzgerald, 1980; U.S. Geological Survey, 1989). These fine-grained deposits were 50 to 180 cm thick in vibracores collected at locations away from dredged channels.

The silty sediments in depositional areas have accumulated under relatively weak tidal currents. Deposits on the subtidal flats inside the harbor are located where tidal currents typically do not exceed 15 cm/s (U.S. Coast and Geodetic Survey, 1953; National Ocean Survey, 1977). In most other places, the deposits are restricted to sheltered depressions where near-bottom tidal-current speeds do not exceed 26 cm/s (National Ocean Survey, 1977). Such low-flow velocities are not strong enough to move the silty sediments that blanket these areas (Miller and others, 1977).

Environments of Sediment Reworking

Areas of reworked sediments are characterized on the sonographs as broad mosaics of light and dark patches produced by variable reflectivity (Knebel and others, 1989, 1991; Fig. 4D). Patterns with tonal patches were found across a variety of bathymetric features, including subtidal flats in the northwestern part of the harbor, the flanks of many islands and shoals, and isolated swales and depressions at the harbor approaches (Fig. 5).

The tonal patches probably are the result of relative changes in the grain size of the bottom sediments produced by a combination of intermittent erosional and depositional processes (Knebel and others, 1989, 1991). The textures of reworked areas range from clayey silts to sandy gravel (Mencher and others, 1968; Fitzgerald, 1980; U.S. Geological Survey, 1989), and they are intermediate between and overlap those typical of erosional and depositional environments (Fig. 6). Such diverse textures have produced similar tonal mosaics in sonographs obtained from a number of reworked shallow marine areas (Knebel, 1989, and references therein). These previous studies reveal that the highly reflective (dark) patches (Fig. 4D) represent erosional features created either by exposing a relatively coarse substrate or by winnowing away the finer sediments. The less reflective (light) patches, however, are parts of thin, discontinuous layers of relatively fine-grained sediments that have accumulated over or around the coarser grained deposits.

Bottom currents in reworked areas must necessarily fluctuate considerably in strength if the sediments there are intermittently eroded and deposited. Where measured, maximum near-bottom currents in reworked areas exhibit a wide range of speeds (20–67 cm/s) and are spatially variable (National Ocean Survey, 1977). In addition, a recent flow model of the harbor indicates that the maximum bottom stress due to tidal currents in reworked areas generally is intermediate between those of the other two sedimentary environments (Signell, pers. commun., 1991). Consequently, if this intermediate state of bottom stress is perturbed by other forces (such as wind-driven currents), then either erosion or deposition could result.

SUMMARY OF SEDIMENTARY HISTORY

Data from this study, when combined with previously collected acoustic, sedimentary, and geologic data, outline the sedimentary history of the Boston Harbor area during the complex period of glaciation and sea-level change since late Wisconsinan time. The last ice sheet retreated from the harbor area between 16 and 14 ka and was accompanied by a marine transgression with the ice and sea in contact. As the glacier withdrew toward the north, coarse-grained,

EVOLUTION AND HOLOCENE STRATIGRAPHY OF PLYMOUTH, KINGSTON, AND DUXBURY BAYS, MASSACHUSETTS

MICHAEL C. HILL[1] AND DUNCAN M. FITZGERALD
Boston University, Geology Department, Boston, Massachusetts 02215

ABSTRACT: Plymouth, Kingston, and Duxbury Bays (herein referred to as Plymouth Bay) form a large reentrant (area = 46 km^2) along the south shore of Massachusetts approximately 57 km south of Boston. The location of the barriers, which front the bay, is controlled by bedrock, drumlins, and other glacial deposits that provide the sediment sources and pinning points for sand accumulation and the development of barriers and spits. A continual sediment supply derived from the reworking of glacial deposits, combined with a barrier alignment that funnels sediment into the bay, created an ideal sink during the past 6,000 years that resulted in accumulation of up to 35 m of sediment. The sediment distribution of the bay fill is controlled, in part, by the location of the inlet, major channels, and degree of sheltering within the bay.

Analyses of 42 km of high-resolution seismic and sidescan-sonar profiles, 18.5 km of ground-penetrating radar transects, 336 bottom samples, and 15 vibracores indicate that the present configuration of the embayment and position of the major channels are closely tied to the paleotopography of the region. The existence of a major drainage valley, formed during the late Tertiary and operative during deglaciation, is also recognized. The back barrier is comprised of extensive intertidal flats (62 percent of the back barrier is exposed at mean low water [MLW]), shallow bays and channels, and intertidal and supratidal marsh.

Modification of Plymouth Bay and its barrier system during the Holocene has been a product of cyclic barrier progradation followed by destruction and subsequent landward translation of the shoreline. The variety of the back-barrier stratigraphy reflects not only the cyclic barrier transgression, but also the distance from the main inlet channel. The thin nature of the barrier spits and the existence of numerous washovers and flood-tidal-delta deposits associated with the many historical inlets that occurred along the spits are evidence that the barriers are in a transgressive phase. Radiocarbon dates of basal peats in the northern part of the study area indicate that relative sea level has been rising at a rate of about 1.1 mm/yr over the past 3,700 years. A radiocarbon date obtained 200 m seaward of the foredune ridge at an elevation of the beach face indicates that the barriers have been migrating landward at an average rate of 0.27 ± 0.05 m/year.

INTRODUCTION

There are few studies concerning the glacial history, evolution, and stratigraphy of embayments in New England except for the estuarine systems along the peninsular coast of Maine studied by Belknap and others (1987), Kelley (1987), Kelley and others (1986), and Knebel (1986). In this paper Plymouth, Kingston, and Duxbury Bays will be used as an example to illustrate the controls that barrier development, paleotopography, bay size and configuration, sediment supply, glacial features, inlet size, and tidal currents have on the evolution and stratigraphy of the back-barrier area. In Massachusetts, the coastline is dominated by three large reentrants: Castle Neck and Plum Island back barrier, Boston Harbor, and Plymouth Bay. Paleodrainage systems exert structural control over all of these bays, which are in various stages of sediment infilling.

The back barrier region of Plum Island has an abundant supply of sediment derived from a paleodelta and reworked glacial deposits. The sediment has allowed the construction of a large regressive barrier and the substantial infilling of the backbarrier region, producing a system of channels bounded by high marsh. Plymouth Bay is very shallow and dominated by extensive intertidal flats, whereas Boston Harbor, which is much deeper and larger than Plymouth Bay, has remained mostly open water. Barrier evolution in Boston Harbor and Plymouth Bay developed from eroding drumlins that migrated landward in response to rising sea-level and waning sediment supply until new pinning or anchor points were encountered that would foster the growth of the barrier. The rise in relative sea level combined with a fluctuating sediment supply produces cycles of barrier progradation followed by destruction and subsequent landward translation of the shoreline. As a result of the transgressive barriers fronting Plymouth Bay, the location of the inlet, and size of the bay, the back-barrier stratigraphy is variable. A conceptual model illustrates the evolution of Plymouth's shoreline and shows the progradational and destructional stages in the coast over the past 8,000 years along with the predicted future morphology of the region.

PHYSICAL SETTING

Plymouth Bay forms a large reentrant (area = 46 km^2) along the south shore of Massachusetts approximately 57 km south of Boston (Fig. 1). The embayment is bounded to the south by a transgressive spit, called Plymouth Beach or Long Beach, and to the north by a drumlin-anchored spit system consisting of Duxbury Beach and Saquish Neck. These barriers partially protect the shallow Plymouth, Kingston, and Duxbury Bays from storm-wave attack. The tidal range fluctuates between 1.7 and 4.2 m and the average deep-water wave height is 1.2 m (Coastal Engineering Research Center [CERC], 1973), making this a mixed-energy coast (Hayes, 1979; Nummedal and Fischer, 1978). The dominant wave energy is from the east-northeast and is associated with extratropical cyclones (CERC, 1973). The prevailing winds in the spring and summer are from the southwest and in the autumn and winter from the northwest (CERC, 1973).

Sources of sediment for the beach and back barrier have been and continue to be primarily the nearby Pleistocene deposits. Plymouth Spit, Duxbury Beach, and Saquish Neck and Head represent the western edge of a transgressive unconformity. Location of the present barriers that front the bay is controlled by bedrock, drumlin, and glacial-outwash seacliff anchors that provide the loci for sand accumulation

[1]Present Address: United States Environmental Protection Agency, Region I (HEE-CAN6), Boston, Massachusetts 02203-2211.

Quaternary Coasts of the United States: Marine and Lacustrine Systems, SEPM Special Publication No. 48

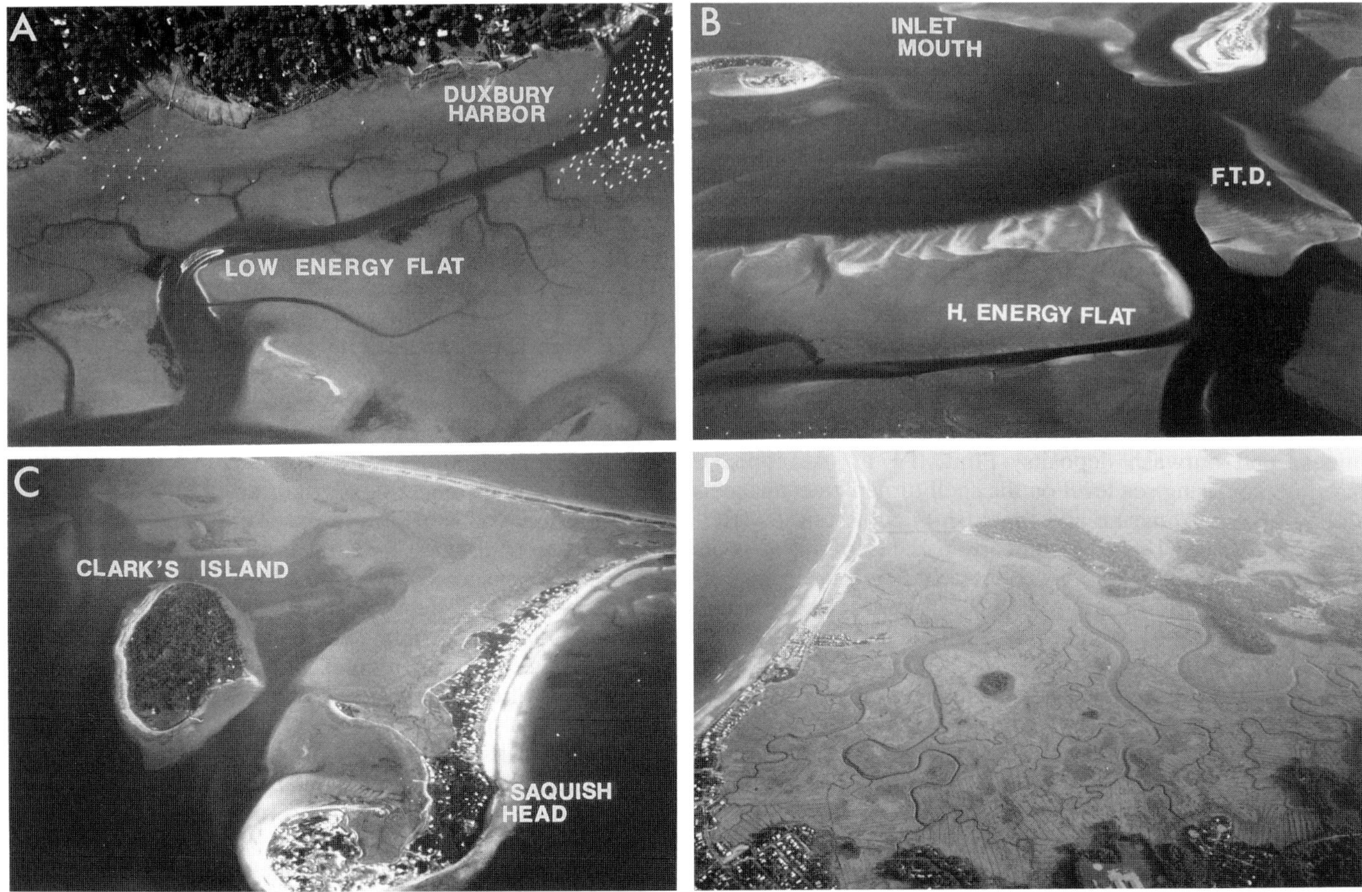

FIG. 3.—Four oblique aerial photographs depicting major environments and structures of the bays. (A) Low-energy intertidal flat adjacent to Duxbury Harbor. (B) High-energy intertidal flat (view to the southwest), tidal inlet, and the flood-tidal delta. (C) Saquish Head and Clark's Island, which are drumlins (view to the northeast). (D) Duxbury and Marshfield Marsh (view to south).

ing the Pleistocene. One paleochannel is responsible for the development and location of the present inlet and the other, between the Jones River outlet and Duxbury, exerts structural control on the dimensions of Plymouth Bay.

FACIES

As a consequence of the paleotopography, transgressive barriers, relative rise in sea level, and the physical setting, a sediment sink was created that allowed 13 surficial facies and subfacies to be deposited in Plymouth Bay (Fig. 4). The oldest Holocene unit in the study area is basal peat (maximum thickness 10 cm in the cores). The presence of basal peat is confirmed by vibracores in the northern part of the study area. This unit unconformably rests on till and consists of compacted organic remains of freshwater and high-salt-marsh plants.

The lagoonal facies (maximum thickness 4.0 m) is composed primarily of clay, silt, and very fine sand and is commonly light to olive gray. The lower contact is usually gradational, whereas the upper contact is unconformable with the overlying facies. Organic matter ranges from 1.1 to 3.1 percent. Shells of the soft-shelled clam, *Mya arenaria,* are also present and often etched. Primary sedimentary structures, when present, consist of flaser, wavy, linsen, and lenticular bedding. Bioturbation is extensive in sections.

The low-energy intertidal flat (maximum thickness 4.1 m) consists of mottled, light gray to dark olive gray, moderately sorted, fine, muddy sand to clayey silt. Upper and lower contacts are usually unconformable except where this facies grades into the high-energy intertidal flat. Bioturbation is pervasive, giving the facies a mottled appearance. Organic laminae and small pieces of wood are found in isolated areas. Some linsen and flaser bedding are also present. Shell-hash beds ranging in thickness from 20 to 97 cm typically cap the low-energy flat.

The intertidal mud flat consists of homogeneous clayey silt that is usually olive gray (0.2 to 3.7 m thick). The contacts are usually gradational with the underlying low-energy intertidal flat and overlying low-marsh facies. Sedimentary structures are poorly defined, but some fine parallel to wavy laminations are present (<1–3 mm thick). In places, sandy and sandy-mud lenses associated with wood fragments and shell fragments of *Mytilus edulis* and *Mya arenaria* are found at the base of this unit.

The high-energy intertidal flat (0.3 to 4.5 m thick) consists of light gray, medium to fine sand that is moderately

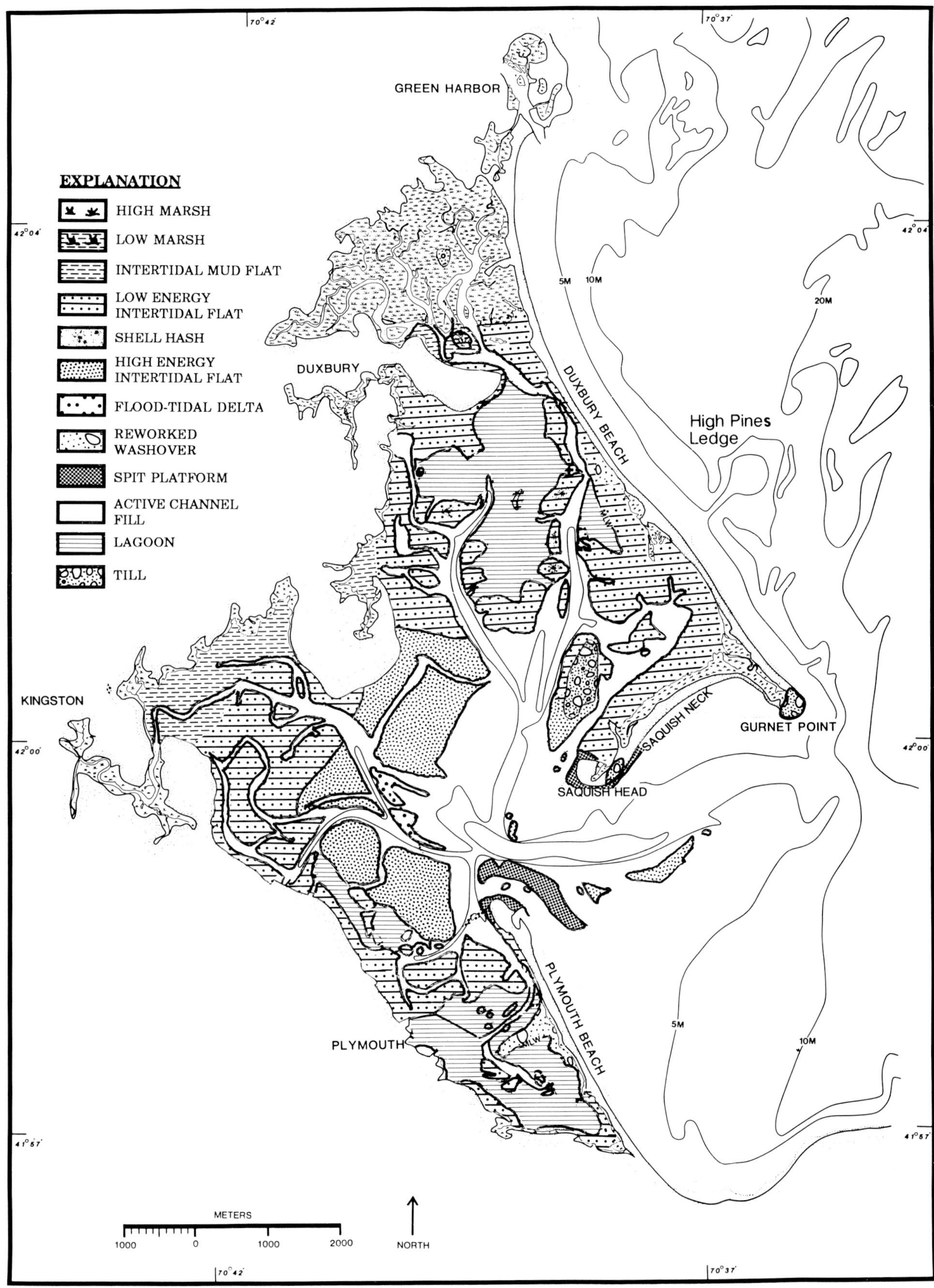

FIG. 4.—Map of Plymouth Bay depicting the varied lithotopes.

well sorted and well rounded with some isolated pebbles and scattered shells. Bioturbation is relatively high and destroys most of the sedimentary structures, but in some instances linsen and flaser bedding are preserved. The upper and lower contacts are usually unconformable, except where the unit grades into the low-energy intertidal flat. A subfacies of this unit is the flood-tidal delta (minimum thickness of 2.8 m in the cores and inferred thickness from seismic data of approximately 15 m), which consists of light gray, medium sand that is moderately well sorted. Sedimentary structures present are low-angle cross-stratification and muddy-sand lenses.

The low marsh (0.5 to 5.6 m thick) is characterized by the scattered, thick, soft, light brown hollow stems and rhizomes of *Spartina alterniflora*. The percentage of this halophyte to mud increases toward the top of this zone. The lithology of the unit consists of silty clay that ranges in color from olive gray to brown. The upper and lower contacts are gradational, respectively, with the high marsh and intertidal mud flat. Sedimentary structures are wavy and parallel laminae that consist of muddy sand, fine dark brown organic material, wood fragments, and occasional shell fragments.

The high marsh (0.2 to 1.7 m thick) is composed primarily of dark brown to tan to yellowish-orange, fibrous, tightly meshed roots and stalks of *Spartina patens* and has a fine texture. Other contributors of organic matter are *Juncus gerardi* and *Distichlis spicata*. The presence of ecophenotypes of *Spartina alterniflora* decreases toward the top of this unit as *Spartina patens* becomes more prevalent. The lower contact is gradational with the underlying low-marsh facies. The percentage of clastic sediments decreases upcore from the lower gradational contact. Scattered and sparse muddy-sand lenses (< 1 cm) occur throughout the section.

The channel-fill facies (0.3 to 1.7 m thick) ranges in color from a dark/brownish gray to a light gray. The texture ranges from a sandy mud to fine sand. Shell and wood fragments are often found at the base of the sequence. The lower contact is usually unconformable. Lenses (2 mm) of clay and fine-sand lenses (1 to 13 mm) are scattered throughout the section. Sedimentary structures consist of small-scale cross-laminae, finely laminated organic-rich layers, and linsen and flaser bedding. The abandoned-channel-fill facies (18 to 42 cm thick) always conformably overlies the channel-fill facies. The abandoned-channel-fill unit consists of homogeneous clay to very fine silt and is olive gray with scattered shell fragments.

The reworked-washover facies (0.1 to 1.7 m thick) ranges in texture from fine sand to cobbles. The sand is usually moderately sorted, whereas the gravel fraction is poorly sorted and light to a dark gray in color. The bottom contact is always unconformable and is characterized by a sharp erosional surface. Shell and wood fragments and pieces of various dune and marsh grasses are found scattered throughout the section.

The spit-platform facies is characterized by medium to coarse sand with occasional pebbles scattered throughout the section and is light gray in color. The lower contact is unconformable with the underlying sediments. The only primary sedimentary structure discernible is high-angle cross-stratification.

STRATIGRAPHY

The facies in the study area consist of Holocene sediments (unit H) that unconformably, as revealed from the seismic record, overlie till and/or bedrock and range in thickness from 0 to 35 m (Fig. 5). Antecedent topography strongly influences not only the bay configuration, but also the position of the barriers, which strongly influence the stratigraphy of the back barrier similar to many bay systems in Massachusetts (McCormick, 1969; Rendigs and Oldale, 1990). The basal unit in Plymouth Bay consists of diamict, drumlins, ground moraine and possibly bedrock. The glacial sediments were deposited during the Wisconsinan glaciation approximately 15 ka. As sea level rose, the first Holocene deposits (unconformably overlying the till) consisted of time-transgressive basal peats that formed in valleys and along the fringes of the embayment. Depending on the location and energy regimes, different sedimentary facies formed unconformably over the basal peat. Near stronger currents by tidal inlets and/or near the shoreline and shoaling areas where wave action can be relatively strong (capable of moving sand by bedload processes), high-energy intertidal flats developed. For example, inside the mouth of the present inlet is a flood-tidal delta that has probably been active since the formation of the bay, due to the inlet being anchored in a paleochannel and unable to migrate. Smaller ephemeral inlets that existed in the past deposited correspondingly smaller high-energy intertidal-flat and flood-tidal-delta sediments. In more protected regions low-energy intertidal and lagoonal facies developed due to bedload-deposition processes becoming less dominant and suspension processes becoming more active.

The coarsening-upward sequence found in most of the vibracores is a result of coarse-grained material introduced into the back barrier through overwashing and tidal inlets as the barriers transgress the bay and the ability of the large tidal channels to transport coarse-grained material into the back-barrier region (Fig. 6). In addition, it is likely that as the barriers transgress the back barrier (erosional transgression), a coarsening-upward sequence will result (Demarest and Kraft, 1987). The bay fill also coarsens toward the inlet due to increasing tidal and wave energy. The stratigraphy in the most landward (western) section of transect A-A′ consists of low-energy intertidal flats with two channel-fill facies approximately 1 m thick that represent shallow meandering channels (Fig. 6, core 1). Lagoonal facies underlie the present low-energy intertidal-flat facies and thicken in a western direction away from the inlet. Stronger currents and wave action at the inlet mouth east of Kingston Bay produce higher energy flats with their moderately well-sorted and fine to medium sand. Directly west of the inlet mouth, a large flood-tidal delta has developed that is the largest accumulation of medium to coarse sand in the Plymouth system (core 3). Close to Saquish Head (core 6), the high-energy intertidal flat interfingers with the spit-plat-

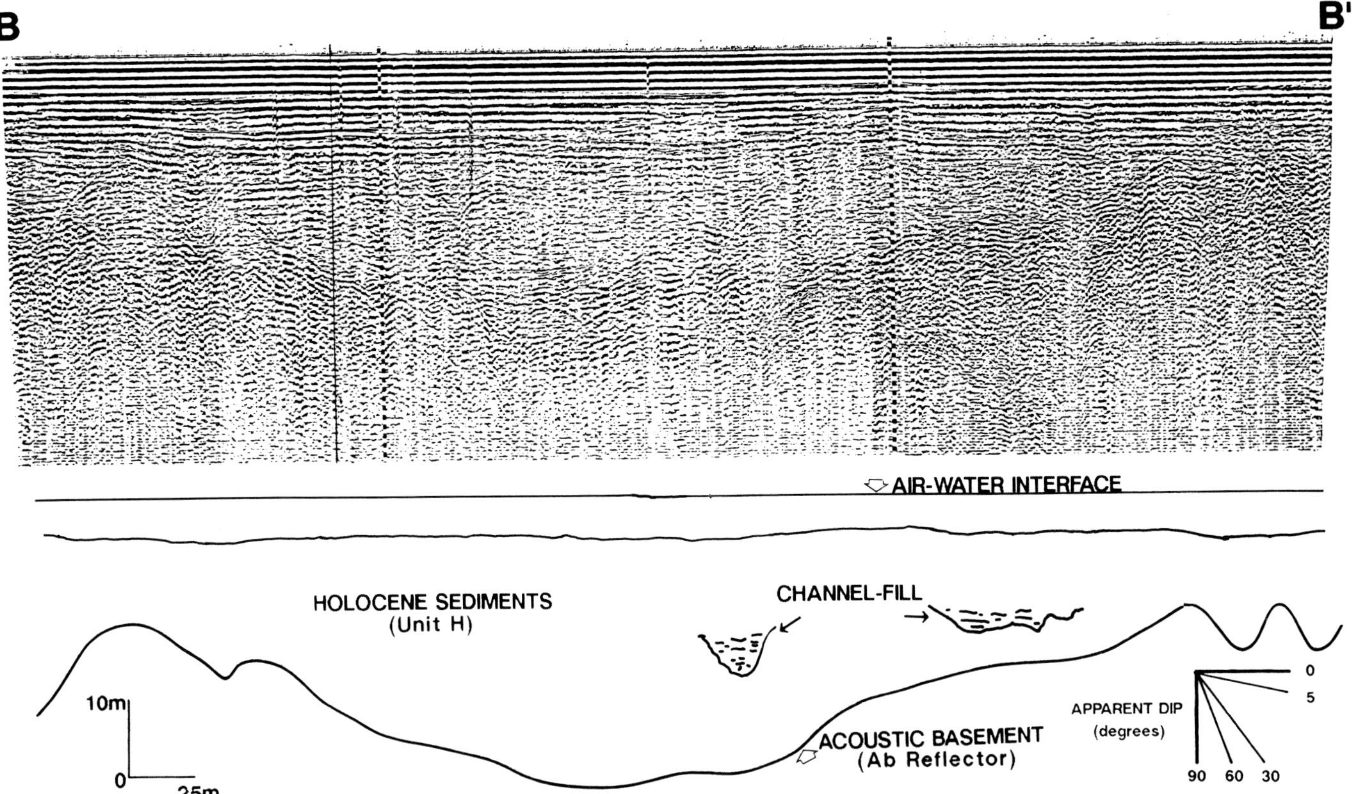

FIG. 5.—Seismic profile B-B′ showing thickness of Holocene sediments along the major channel to Duxbury Bay (see Fig. 1 for location).

form, reworked-washover, low-marsh and lagoonal facies. The presence of the reworked-washover facies associated with the buried low-marsh facies is indicative of the transgressive nature of Plymouth Bay. Saquish Head and Neck are actively being transgressed, as shown by the presence of *Spartina* sp. marsh grasses and peat now exposed seaward of these areas.

The lagoon facies reflects deposition in tranquil shallow water, where eel grass and seaweed proliferate. The lagoon environment is often sheltered by adjacent sand and mud shoals and/or promontories. The lagoon environment is also predominantly situated in and/or near the axis of the paleovalleys. Since the antecedent valleys are deeper than the surrounding areas and the sedimentation rates are not high, particularly when compared to the adjacent shallow areas, the flooded valleys remain as open water. The adjacent flats build vertically and remain as intertidal-sedimentation areas. Shallow channels rework much of the bay sediments as they migrate freely across the bay.

Saquish Head and the northern end of Plymouth Beach are recurved spits that have prograded in a landward direction as a result of wave refraction and flood-tidal currents. Boulder-retreat lags serve to indicate past positions of the barriers. Overwash processes during storms transport large quantities of poorly sorted sediment into the back barrier. Later, these deposits are reworked by tidal currents and wave action and in many instances these areas are vegetated by *Spartina* sp. marsh grasses.

Areas that are sheltered and farther away from high-energy environments are likely to develop intertidal mud flats that accrete vertically from suspension processes. Typically, but not always, low marsh develops on top of the mud flats once the elevation of the surface is within 2 m of mean high water. As long as wave energy is not sufficiently strong to disturb the plants and the substrate that supports *Spartina alterniflora* is secure, the low marsh will survive and grade into the high-marsh facies at an elevation of approximately mean high water. In the shadow zones behind the barriers, marsh systems can develop, such as those found landward of Gurnet Point, Saquish Neck and High Pines. However, depending on the energy regime of the area, a variety of facies can be encountered.

Figure 7 shows a schematic illustration of the various facies in Plymouth Bay. In general, sediments near the inlet mouth and other high-energy environments are coarse grained and composed of sand and fine gravel. Farther from the inlet, energy is much weaker and thus, basal peat and finer grained facies are preserved. Close to the inlet, the lower energy facies have been either eroded or not deposited. The mid-bay cross section is much thicker due to the existence of the paleovalley. The northern cross section reflects the development of a marsh in the upper stratigraphic units, whereas the lower stratigraphic units are indicative of a higher energy environment. As a consequence of the transgressive nature of the barriers and influx of coarse material into the back barrier (e.g., washover deposits and the existence of

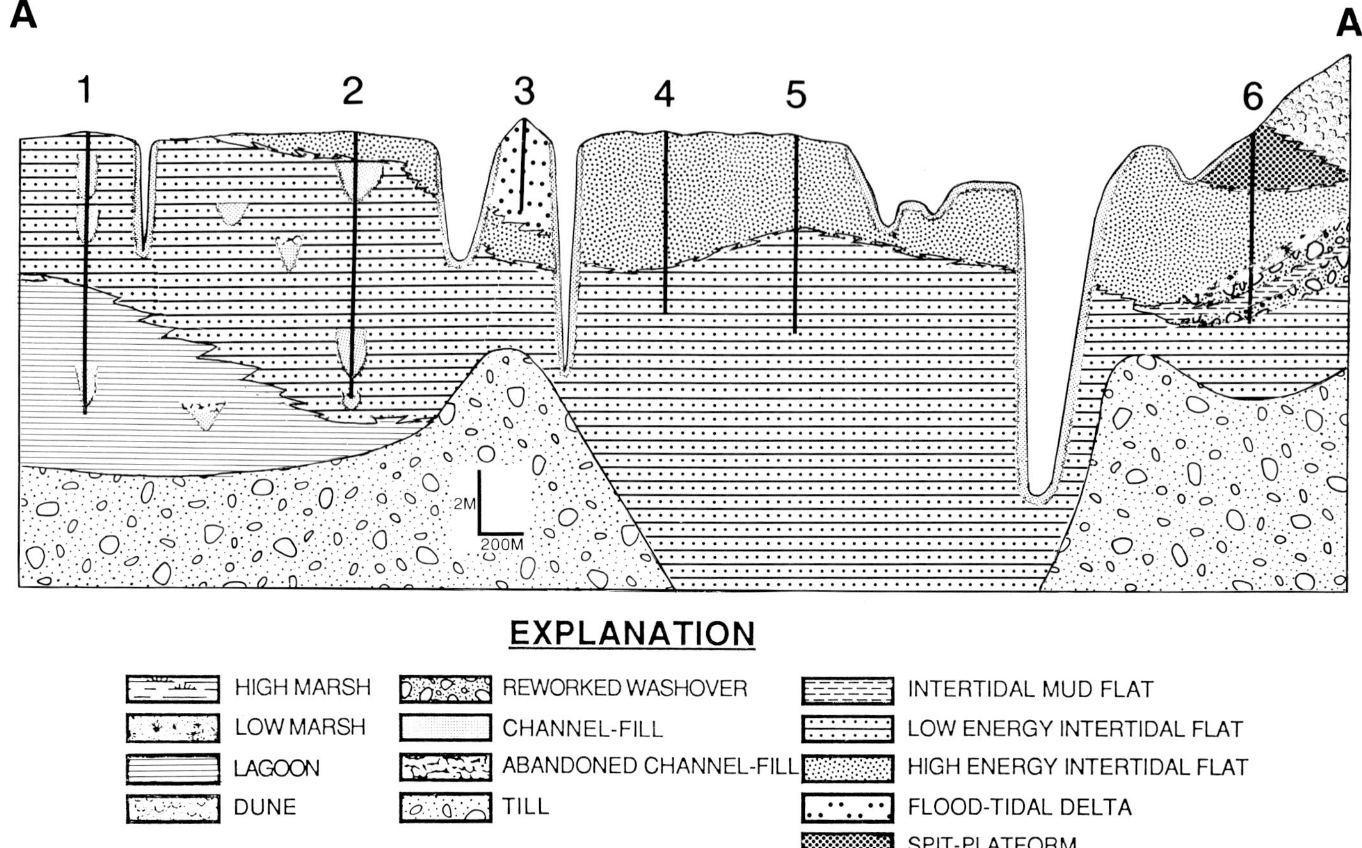

FIG. 6.—Six vibracores and detailed stratigraphic cross section across transect A-A′ adjacent to the tidal inlet.

numerous inlets present in the past), there is a considerable amount of variability in the stratigraphy.

BARRIER EVOLUTION AND EFFECTS ON BAY SEDIMENTATION

Barrier Development

Due to the diminishing sand supply along the outer coast, both the Duxbury and Plymouth barriers are eroding and migrating landward (Fig. 8). Although this transgression likely has been a continual process during the past 3,000 to 4,000 years, it is probable that the general trend has been interrupted by periods of barrier regression brought on by sea-level stillstands and/or increased supplies of sediment. Also, it is likely that most of this retreat occurs sporadically during major storms such as the February Blizzard of 1978 and the 1938 Hurricane. It is during these extreme events that extensive overwashing of the barriers takes place, ephemeral inlets are formed, and large quantities of beach sediment are deposited onto the marsh surface and carried into the bay. Evidence of the transgressive nature of the area are the cuspate-shaped deposits that occur along the backside of Plymouth Spit and Duxbury Beach. Other documentation includes historical photographs and maps, and ground-penetrating radar records that demonstrate the former presence of numerous tidal inlets. A radiocarbon date of a marsh peat exposed along Duxbury Spit indicates that the barrier is retreating at an average of 0.27 ± 0.05 m/yr (Fig. 1).

The entire area north of the study area is devoid of sediment and the coast lined by sea walls with little or no beach in front of the structures. The amount of sediment supplied to Plymouth Beach has steadily declined, due to the depletion of glacial sources and the accumulation of boulder-retreat lags in front of glacial deposits. The lags are so extensive along Rocky and Gurnet Points and Warren Cove that boulder beaches have developed. The boulder fortifications lessen the erosional strength of storm waves, thereby reducing the sediment supply. The boulder-retreat lags also serve to indicate the former extent and positions of drumlins. High Pines Ledge, a subtidal boulder-retreat shoal, is believed to be the site of a former drumlin and pinning point off mid-Duxbury Beach (Fig. 8). All of the boulder-retreat lags are evidence of landward migration and reworking of the glacial sediments.

Plymouth Beach is believed to have begun forming when rising sea level and the resulting wave attack eroded the 15-m-high cliffs along Warren Cove (Fig. 8). The eroded sediment was moved northwestward along the coast by longshore currents, causing spit building. A slight promontory along Warren Cove, which is still evident today, was likely the initial site of the formation of Plymouth Spit. In time, the spit built northwestward across Plymouth Bay and

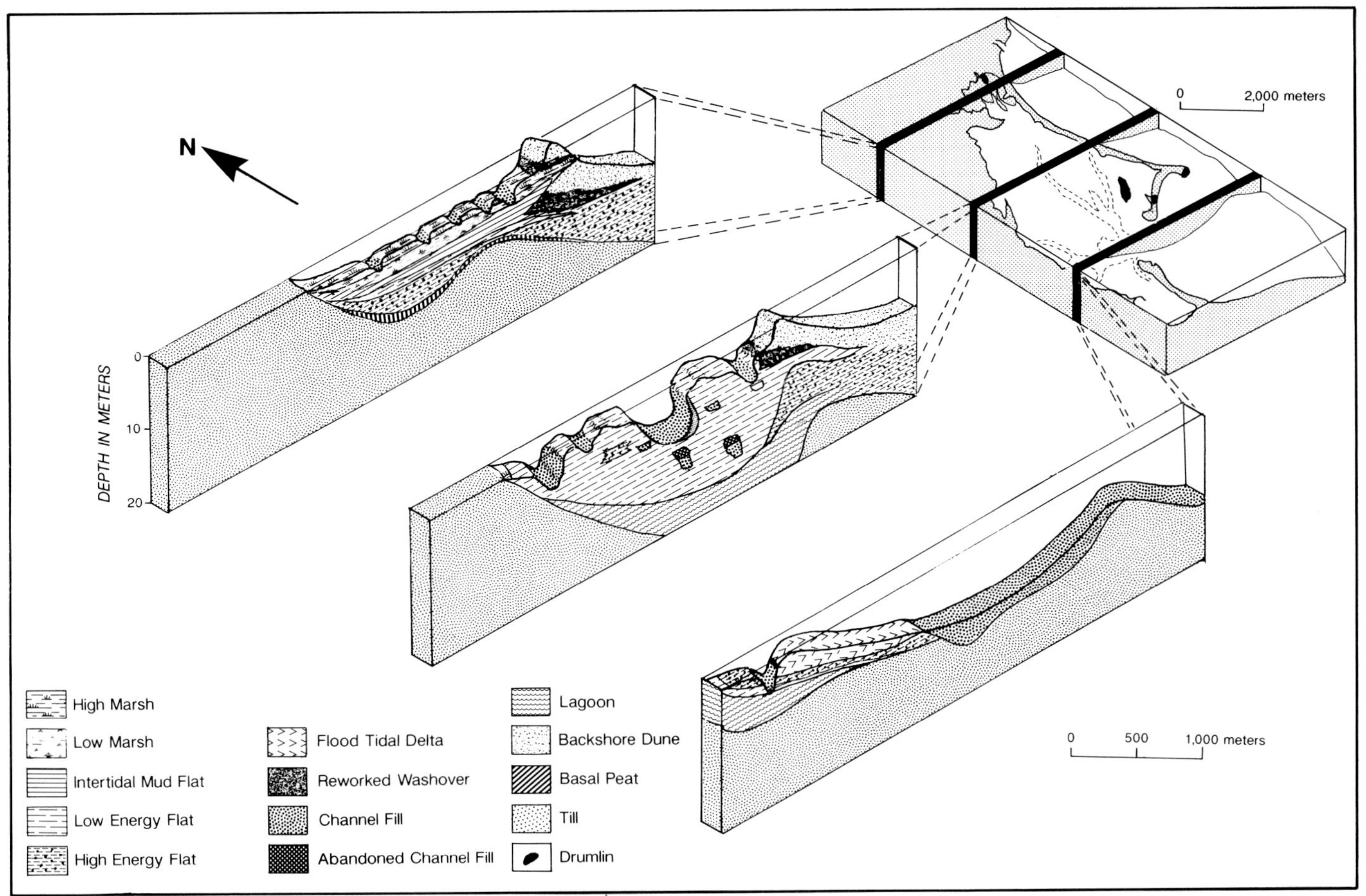

FIG. 7.—Schematic illustration of the stratigraphy in Plymouth, Kingston, and Duxbury Bays showing the different sequences of facies and finer grained sediments farther from the main inlet channel. The variability of the stratigraphy is caused, in part, from the cyclical nature of the barriers and the presence of drumlins.

as the supply of sediment from cliff erosion waned, the spit migrated landward.

Present longshore-transport patterns along the Duxbury barrier system, inferred to have been active in the past, explain the development of the spit systems. Sediment was transported from both Brant Rock in a southerly direction and Gurnet Point (a drumlin) in a northerly direction to High Pines. During the "Steamer Portland Storm of 1898," Gurnet Point was connected to Saquish Head by spit accretion. As Gurnet Point continued to erode and decrease in size, the supply of sand to Saquish Neck has also been reduced. Presently, the spit is thin and in danger of being breached.

Several points off Duxbury Beach consist of submerged mounds of till and bedrock that, at a lower sea level, would have been the pinning points of proto-Duxbury Beach. Farnham Rock and Howland Ledge, High Pines Ledge, and Gurnet Point also served as sediment sources and pinning points, and intervening bay beaches would have been between these headlands. With continued sea-level rise, the headlands were eroded and transgressed.

Based on long-term trends of this shoreline and the bay system, it is possible to make predictions about the future of the area and its configuration (Fig. 8). Sediment eroded from Warren Cove and areas to the south and east will continue to supply Plymouth Beach with a limited supply of sand. With the passage of time, the amount of available sediment will diminish, causing the southern end of Plymouth Beach to thin. If nourishment programs are not undertaken, it is likely that Plymouth Beach will be breached along its southern end. Although sea walls and revetments will forestall the transgression and landward migration of the spit, the diminished sediment supply will also eventually result in erosion of the thicker northern end of Plymouth Beach, which would create an island. Ultimately, Plymouth Beach will become a transgressive sand sheet similar to Slocum River Spit, Massachusetts (FitzGerald and others, 1986). However, when the spit reaches a new sediment supply or pinning point, Plymouth Spit will stabilize and reform.

The decrease in the amount of sediment available to the northern end of Duxbury Beach will result in the thinning and eventual breaching of that portion of the beach. As seen from historic maps, areas that are susceptible to washovers often develop into tidal inlets. Without the use of sand fencing and revegetation to close these washover channels artificially, inlets may develop along Duxbury and Plymouth

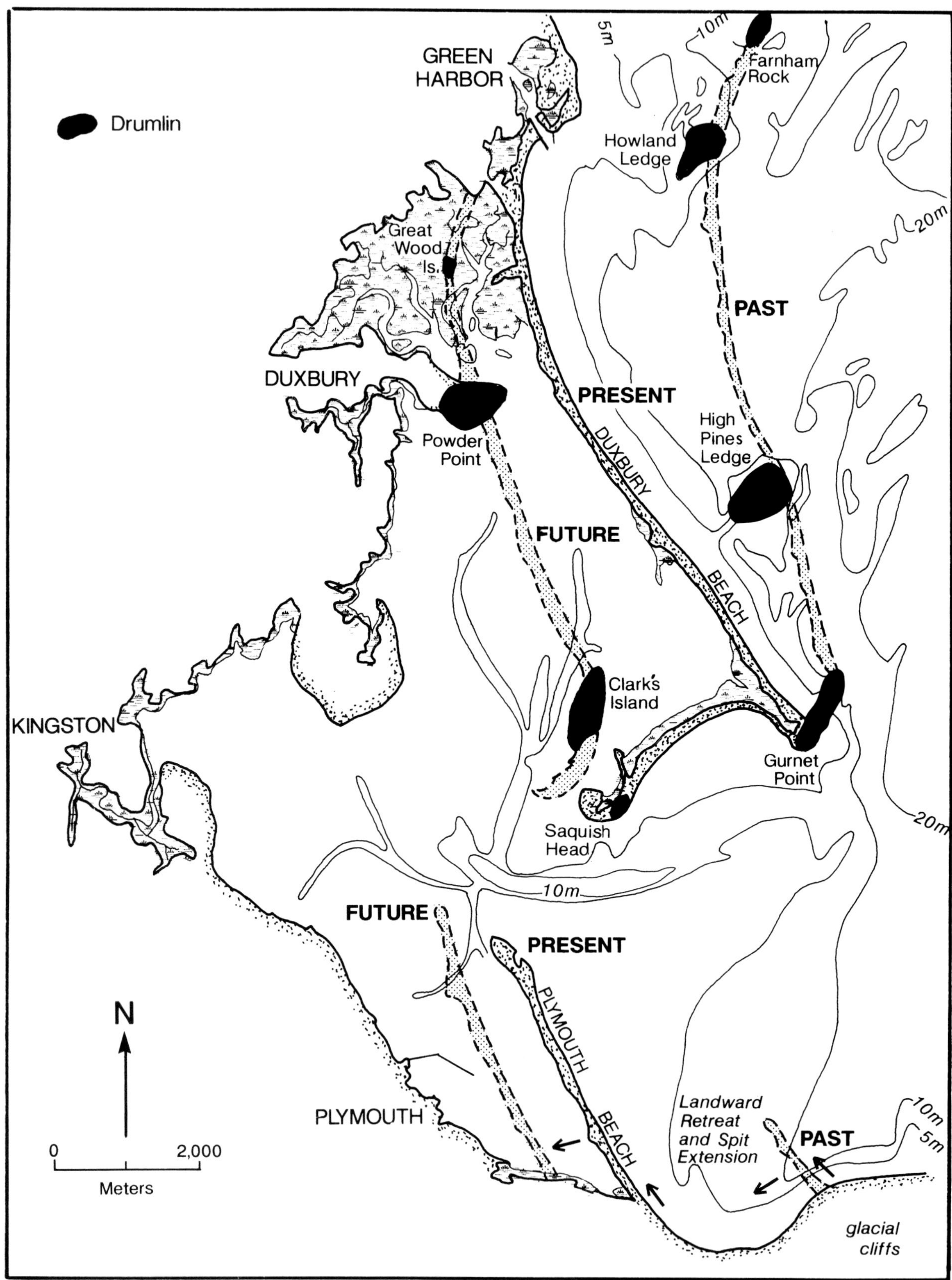

FIG. 8.—Evolutionary model of the area showing cyclic barrier progradation and the anchor points for the barriers and spits.

Beaches where marsh and shallow flats are not present in the back barrier. A similar case history has occurred on Conrads Beach along the eastern shore of Nova Scotia (cf. Boyd and others, 1987).

Eventually, continuing sediment depletion, rise in relative sea level, increase in washover intensity, and formation of tidal inlets will result in a chain of disconnected sand shoals, similar to Saquish Neck before it connected to Gurnet Point. When the transgressive shoreline reaches Great Wood Island, Powder Point, and Clark's Island, the barriers will temporally prograde and stabilize as the areas act as pinning points and new sources of sediment (Fig. 8).

During the same period, the reduction in the supply of sediment from Gurnet Point will hasten the retreat of Saquish Neck. At present, Saquish Neck is rotating in a counterclockwise direction and it is expected that it will become separated from Gurnet Point some time in the next 100 years. Saquish Head will once again be an island with a tidal inlet separating Saquish Head from Duxbury Beach. Eventually, Saquish Head and Neck will connect to Duxbury Beach, making Gurnet Point an island. With continued rise in sea level, Saquish Head and Gurnet Point will be eroded and transgressed to form a boulder-retreat shoal similar to the modern High Pines Ledge. Overall, due to the decreased sediment supply and rise in sea level, Plymouth Beach and Duxbury Beach (without human intervention) will continue to migrate landward.

This scenario of cyclic barrier progradation is similar to the models of Johnson (1925) and Boyd and others (1987), which summarize the evolution of a glacial-drumlin coast. In essence, the formation of new barriers requires that new sediment sources and a new set of headland-anchor points are encountered as the land is transgressed. In a regime of rising sea level, drumlins not only serve as a source of sediments, but also as anchor points to the barriers. Additional sediment is derived from the transgressive sand sheet of the former barriers, previous back-barrier deposits such as flood-tidal deltas, and sandy glacial deposits exposed on the shelf. As the transgression proceeds, large intertidal flats develop landward of the barriers and help stabilize the barriers. With continued rise in sea level, drumlins and other glacial sources and offshore sediment sources will be depleted. Once depleted of sediment, the barriers will be destroyed and become transgressive sand sheets. The sand sheets will migrate to new anchor and sediment sources where healthy barrier construction can commence, once again rejuvenating the system.

Effects on Bay Sedimentation

The cyclic nature of the barriers and presence of drumlins, antecedent topography, and reworked glacial deposits have resulted in a bay that has variable stratigraphy. Overwash processes introduce coarse-grained material into the back barrier, disrupting the general fining of sediments landward and farther from the inlet. The formation of inlets and associated flood-tidal deltas and growth and stabilization of the spits and barriers through the formation of spit platforms also deposit coarse-grained material into the back barrier. In addition, tidal channels in the bay are also capable of transporting large quantities of coarse-grained sediments into the back barrier. Finer grained tidal flats and marshes develop in more sheltered areas such as the area west of Gurnet Point and the northern part of the bay. In paleovalleys within the bay, which tend to remain as open water, lagoonal sediments accumulate.

CONCLUSIONS

1. Paleotopography has determined the trend of the overall configuration of the bay, the thickness (35 m) of Holocene deposits, and location of the main inlet channel.
2. The rising relative sea level coupled with new supplies of sediment and/or pinning points has allowed the barriers to transgress landward in a cyclical manner. The barriers and spits will migrate as transgressive sand sheets and will stabilize when new anchor points are encountered.
3. The barriers represent the edge of an erosional transgression.
4. The cyclic barrier transgression and size of Plymouth Bay, the influence of drumlins, and tidal currents capable of transporting coarse-grained material into the back barrier have produced a variable back-barrier stratigraphy. The presence of paleovalleys has provided the potential for preservation of the stratigraphy, creating the possibility that several different sequences will be found in the same bay environment.
5. Based on historical trends, predictions can be made concerning the future location of the barriers and the bay configuration.

ACKNOWLEDGMENTS

Funding for this research was provided by Minerals Management Service, U.S. Department of the Interior, in a subagreement with the Massachusetts Coastal Zone Management Office; Duxbury Beach Reservation, Inc.; and the Sigma Xi Chapter of Boston University. Kruegar Enterprises is acknowledged for radiocarbon dating analyses.

Kenneth Finkelstein and Kenneth Fink are thanked for their constructive review and comments.

REFERENCES

Belknap, D. F., Kelley, J. T., and Shipp, R. C., 1987, Quaternary stratigraphy of representative Maine estuaries: initial examination by high-resolution seismic reflection profiling, *in* FitzGerald, D. M., and Rosen, P. S., eds., Glaciated Coasts: Academic Press, San Diego, p. 177–207.

Belknap, D. F., and Kraft, J. C., 1985, Influence of antecedent geology on stratigraphic preservation potential and evolution of Delaware's barrier systems, *in* Oertel, G. F., and Leatherman, S. P., eds., Barrier Islands: Marine Geology, v. 63, p. 235–262.

Boyd, R., Bowen, A. J., and Hall R. K., 1987, An evolutionary model for transgressive sedimentation on the eastern shore of Nova Scotia, *in* FitzGerald, D. M., and Rosen, P. S., eds., Glaciated Coasts: Academic Press, San Diego, p. 87–114.

Coastal Engineering Research Center, 1973, Shore Protection Manual, U.S. Army Corps of Engineers, Washington, D.C., 3 vols (unpaginated).

Demarest, J. M., and Kraft, J. C., 1987, Stratigraphic record of Quaternary sea levels: implications for more ancient strata, *in* Nummedal, D., Pilkey, O. H., and Howard, J. D., eds., Sea-Level Fluctuation and Coastal Evolution: Society of Economic Paleontologists and Mineralogists Special Publication 41, p. 223–239.

FITZGERALD, D. M., SAIZ, A., BALDWIN, C. T., AND SANDS, D. M., 1986, Spit breaching at Slocum River Inlet: Buzzards Bay, Massachusetts: Shore and Beach, p. 11–17.

FLINT, R. F., 1971, Glacial and Quaternary geology: John Wiley and Sons, New York, 892 p.

HAYES, M. O., 1979, Barrier island morphology as a function of tidal and wave regime, *in* Leatherman, S. P., ed., Barrier Islands from the Gulf of St. Lawrence to the Gulf of Mexico: Academic Press, New York, p. 1–28.

JOHNSON, D., 1925, The New England–Acadian Shoreline: John Wiley and Sons, New York, 608 p.

KELLEY, J. T., 1987, An inventory of environments and classification of Maine's estuarine coastline, *in* FitzGerald, D. M., and Rosen, P. S., eds., Glaciated Coasts: Academic Press, Orlando, Florida, p. 151–176.

KELLEY, J. T., KELLEY, A. R., BELKNAP, D. F., AND SHIPP, R. C., 1986, Variability in the evolution of two adjacent bedrock-framed estuaries, *in* Wolf, D. A., ed., Estuarine Variability: Academic Press, Orlando, p. 21–42.

KNEBEL, H. J., 1986, Holocene depositional history of a large, glacially influenced estuary, Penobscot Bay, Maine: Marine Geology, v. 73, p. 215–236.

LARSON, G. J., 1982, Nonsynchronous retreat of ice lobes from southeastern Massachusetts, *in* Larson, G. J., and Stone, B. D., eds., Late Wisconsin Glaciation of New England: Kendall/Hunt Publishing Company, Dubuque, Iowa, p. 101–114.

MCCORMICK, C. L., 1969, Holocene stratigraphy of the marshes at Plum Island, MA, *in* Larson, F. D., and Hartshorn, J. H., eds., Coastal Environments of Northeastern Massachusetts and New Hampshire: Coastal Research Group Field Trip Guidebook, University of Massachusetts, Amherst, p. 368–390.

NUMMEDAL, D., AND FISCHER, I., 1978, Process-response models for depositional shorelines, the German and Georgia Bights: Proceedings, 16th Coastal Engineering Conference, American Society of Civil Engineers, p. 128–1231.

OLDALE, R. N., AND O'HARA, C. J., 1977, Geology of Cape Cod Bay, Massachusetts: U.S. Geological Survey Open-File Report 75–112 (series of maps).

OLDALE, R. N., AND O'HARA, C. J., 1980, New radiocarbon dates from the inner continental shelf of southeastern Massachusetts and a local sea-level curve for the past 12,000 years: Geology, v. 8, p. 102–106.

REDFIELD, A. C., AND RUBIN, M., 1962, The age of the salt marsh peat and its relation to sea level at Barnstable, Massachusetts: Proceedings, Natural Academy of Science, v. 48, p. 1728–1735.

RENDIGS, R. R., AND OLDALE, R. N., 1990, Maps showing the subbottom acoustic survey of Boston Harbor, Massachusetts: Miscellaneous Field Studies Map, U.S. Geological Survey Map MF-2124.

DEVELOPMENT OF PARABOLIC DUNES AND INTERDUNAL WETLANDS IN THE PROVINCELANDS, CAPE COD NATIONAL SEASHORE

MARJORIE G. WINKLER

Center for Climatic Research, University of Wisconsin, Madison, Wisconsin 53706

ABSTRACT: The Provincelands in the Cape Cod National Seashore developed about 5 ka from wind- and waterborne Outer Cape Cod outwash sands as sea level rose to submerge offshore banks. A late Holocene chronology of sand dunes in the Provincelands is established from radiocarbon-age determinations on the basal organics from interdunal bogs and ponds. The Provincelands ponds, thought to be lakes trapped behind hooked spits, are shown here to have developed from bogs similar to those in the extant parabolic-dune field. An older, stabilized dune field in the location of the ponds is hypothesized. Episodes of alternating dune movement and interdunal-wetland formation in the Provincelands correlate with climatic changes in the North Atlantic and elsewhere during the last 1.2 ka, as interpreted from ice core, glacier movement, sea-surface temperature, tree-ring, pollen, and historical data. The dune field data suggest a periodicity of change of 0.2 ka. Before about 1.2 ka, dunes in the Provincelands were active. Between about 1.15 and 0.9 ka, bogs formed which, at least in the west-central area of the Provincelands, developed into ponds having continuous organic sedimentation until today. This evidence indicates that the west-central dune field stabilized after 1.1 ka. Between 0.9 and 0.7 ka, dunes were active in the north and east-central Provincelands, but from about 0.7 to 0.5 ka, bogs formed within that dune field, and at least one bog in the west-central area became a pond, both events indicating a warm and wet climate. During the Little Ice Age (0.5 to ~0.2–0.1 ka), the Provincelands dunes were active, suggesting a cold, windy, and dry climate. Because of a warmer and wetter climate during the last 0.1 ka, bogs have formed again within the dunes.

INTRODUCTION

The Provincelands dunes in the Cape Cod National Seashore comprise a coupling of dry and wet environments formed by wind and water-level changes since the middle Holocene. Bogs have evolved within parabolic dunes at High Head near North Truro (Fig. 1A,B), and ponds are evident among other sand ridges in the west-central section of the Provincelands Hook (Fig. 1A). Examining the development of these wetlands enables us to trace the coastal geomorphology of outermost Cape Cod since the middle Holocene.

The study of dunes, stabilized in the past, offers important paleoclimatic information about the direction and intensity of ancient winds (Wells, 1983; Thorson and Bender, 1985; Kutzbach and Wright, 1985), aridity, and air-mass frequency. Many dune systems, both inland and coastal, were derived from outwash sands deposited during deglaciation of the Northern Hemisphere (Flint, 1971). Active sand dunes in North America today, including the Provinceland dunes, generally are being formed in the same direction as the resultant of modern winds of at least 6 m/s (13 mi/hr) in each region (Wells, 1983). The dune structure, therefore, provides wind direction and a measure of minimum wind intensity (Thorson and Bender, 1985), with the parabolic-dunes concave sides facing into the direction of the winds that formed them (Flint, 1971).

Ancient dune systems in North America date regionally from different time periods. Several dune complexes in the Sandhills of Nebraska were built during the late glacial and some during the early Holocene (Wright and others, 1985); parabolic dunes formed in the Anoka sand plain in east-central Minnesota during the middle Holocene (Keen and Shane, 1990), and in the central St. Lawrence lowland in Quebec, Canada, during the early Holocene (between 10 and 7.5 ka; Filion, 1987); and dunes in northwestern Quebec have formed episodically since 5 ka (Filion, 1984). In some areas dunes developed on well-vegetated land (Filion, 1987), but in others dune characteristics suggest that the land was open and unvegetated as they developed (Kutzbach and Wright, 1985; Filion, 1987).

The Provincelands originated from eolian deposits during the middle Holocene as sea level rose to cover the Georges Banks, winds intensified as they reached shore over open ocean, and the littoral drift on the north shore of the Outer Cape reversed, bringing eroded sand from the Atlantic escarpment (Zeigler and others 1965; Benedict and Leatherman, 1978). Sands from the outwash plains of the Outer Cape, which were deposited during deglaciation about 14 ka, were then redeposited by the wind, forming the proto-Provincelands by about 5 ka. The coastlines further accreted and were eroded by water and wind as sand ridges, dunes, hooks, and spits developed to form the present outline of the Provincelands (Fig. 1A).

The parabolic dunes in the High Head lowlands (Fig. 1A,B) are formed by deflation by the prevailing northwest winds (Leatherman and Godfrey, 1979) and move to the southeast. The construction and movement of these dunes are also affected by the more intense winds accompanying northeasterly storms (Leatherman and Godfrey, 1979). Surface roughness interrupts aerodynamic flow of sand. As a result, the wind-dune system of parabolic dunes is eventually disrupted by intradunal-bog development in the deflation hollow until a period of maximum winds, a fire, or anthropogenic disturbance creates a blowout and the sand moves once again.

Because the dunes are dynamic landforms, the intradunal bogs are ephemeral wetlands. They are affected by dune movement and the composition of the dune sands, the regional hydrology, and wind direction and intensity–all factors relating to climate as well as to physical disturbances, including those caused by humans.

The primary goal of this study is to provide a chronology for initiation of interdunal wetlands. These dates would constrain elements of the regional coastal geomorphology and would also delineate dune formation. Evidence of alternations of dunes and wetlands would facilitate reconstruction of the palaeoclimate of the northeastern coastal United States. Analyses of pollen, diatoms and other algae, and sediment in the cores from the interdunal wetlands doc-

Quaternary Coasts of the United States: Marine and Lacustrine Systems, SEPM Special Publication No. 48

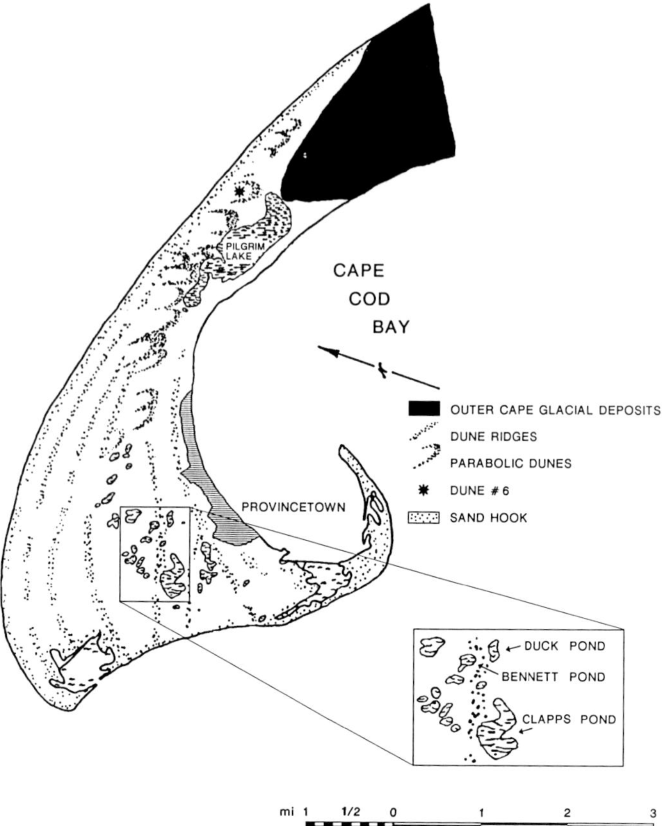

FIG. 1.—(A) Map of the Provincelands showing site of bogs within Dune #6 of the active parabolic dunes, and locations of Duck, Bennett, and Clapps Ponds (modified from Leatherman, 1979). (B) Photograph looking northeast from crest of Dune #6 showing parabolic-dune field with bogs in dune hollows along north shore of High Head, North Truro. Photo taken by S. Winkler.

ument the changing vegetation and hydrology of the bogs and ponds, but will only be discussed briefly here.

METHODS

A sediment core was taken from each of six bogs within the parabola of Dune #6 at High Head (Winkler, 1990a; Fig. 1). Sediment cores were also obtained from Duck and Bennett Ponds (Winkler, 1990b; Fig. 1A). The six intradunal bogs represented the changing vegetation within one dune valley (an area of about 0.2 km^2) and were chosen on the basis of vegetation diversity, vegetation height, and increasing organic-soil development, factors that were thought to reflect age of the bog. The youngest bogs were to the southeast nearest the crest of the dune, and the oldest bogs with more diverse vegetation were the farthest northwest. The bog cores were obtained with a modified PVC tube, 10 cm in diameter and 75 cm in length; the shallow-pond sediments were taken with a modified Livingstone piston corer, 5 cm in diameter. The core from Duck Pond was 55 cm long; the core from Bennett Pond was 45 cm long. All of the cores penetrated eolian sands.

Sediments were analyzed for percent organic matter and percent carbonate following procedures presented in Berglund and Ralska-Jasiewiczowa (1986) and Dean (1974). Counts of framboidal and crystalline pyrite and charcoal pieces in the sediment were made concurrently with pollen and algal counts. Pollen analysis emphasized the local wetland pollen and spore changes, although upland pollen was also counted. Processing and counting followed standard techniques (Gray, 1965; Faegri and Iversen, 1975). Charcoal, zooplankton, and diatoms in the sediments were also analyzed following procedures described in Winkler (1985a, 1988a, and 1990a,b).

The basal organic sediments from intradunal bogs 4 and 5 of Dune #6 were modern in age. The radiocarbon procedure cannot distinguish differences within the last 200 years because ^{14}C measurements from the recent past are not significantly different from that of the modern standard. The basal organic sediments from each pond were also radiocarbon dated and δ^{13}C was analyzed by mass spectrometer to identify marine-carbon contribution to the pond sediments.

The modern vegetation of the bogs suggests that they developed in a sequence from sparse lichen, rush, moss, and sedge-covered hollows, to cranberry bogs, to wetlands with woody shrubs, and, finally, to woody oases within the dunes that include pitch pine as well as *Ilex* and other lowland plants (LeBlond, 1986; Winkler, 1990a). Some of the younger bogs have developed in abandoned vehicle-track depressions (LeBlond, pers. commun., 1991).

RESULTS AND DISCUSSION

Results of Pollen, Diatom, and Other Stratigraphic Microanalyses

Dune #6, Intradunal Bog Core #5.—

Pollen was found only in the top 12 cm. Sediment below 12 cm consisted of unconsolidated eolian sands. The 12- to 6-cm interval represented moist conditions leading to consolidation of the basal sands and to bog initiation, according to the ^{14}C results, within the last 200 years.

Intradunal bog core #5 is from a small peat-filled catchment. Most of the pollen in the peat comes from local sources, which explains why total arboreal pollen is least abundant and pollen from shrubs and herbs dominate the pollen profile. Nonetheless, as in all other pollen diagrams from Cape Cod for the interval since European settlement in the early 1600s (Winkler, 1985b, 1989), oak and pine (almost entirely diploxylon *Pinus,* pitch pine) pollen are abundant and indeed represent the dominant trees in the Provincelands today (Fig. 2). The changes in pine-pollen percentages upcore suggest reforestation attempts (McCaffrey, 1973) in the face of dune expansion, with recent planting being most successful. The planting of beach grass, willow, and bayberry for dune stabilization (McCaffrey, 1973) is also indicated in the increased percentages of pollen from these taxa about a century or more ago. Pollen most representative of dune vegetation, i.e., the beach grass, composites (such as goldenrod and aster), Rosaceae (rose, strawberry, and beach plum), all increase in percentages in the top 2 cm of the core, suggesting the proximity of vegetated dunes. It is possible to provide time control for the vegetation changes from the pollen profile. *Castanea* (chestnut) pollen makes up 2.3 percent of the upland-pollen sum in the bottom sample, but is not found in overlying sediments (Fig. 2). The chestnut blight decimated trees in southern New England by 1913 (Brugam, 1978), and thus we know that the 12-cm level of the core was deposited before that time. The most abundant plants in older bogs today are cranberry and sheep laurel (both ericads), and bayberry (*Myrica*). *Myrica* is the dominant pollen in the bog core but is most abundant in the bottom samples, decreasing in percentages toward the top of the core as grass and *Ilex* (holly) pollen increase. Ericad pollen is most frequent in the middle samples, as is *Salix* (willow; Fig. 2).

Abundant *Typha* (cattail) and sedge pollen indicate a higher water table more than a century ago when bog growth started, as do increased regional pollen from trees other than pine and oak (maple, hemlock, birch, elm, hickory, beech, chestnut, and ironwood), and increased spores from ferns. Soot particles throughout the core attest to the post-Industrial age of the bog, as do the high percentages of *Ambrosia* (ragweed) pollen in the bottom sediment. *Ambrosia* pollen, however, in these coastal regions, may be from dune species (*A. chamissonis*) as well as from species that grow on disturbed soils. There was also increased *Artemisia* (sage) pollen, a dune taxon.

Along with other wetness indicators at initiation of the bog, a large number of pyrite (iron disulphide) spherules

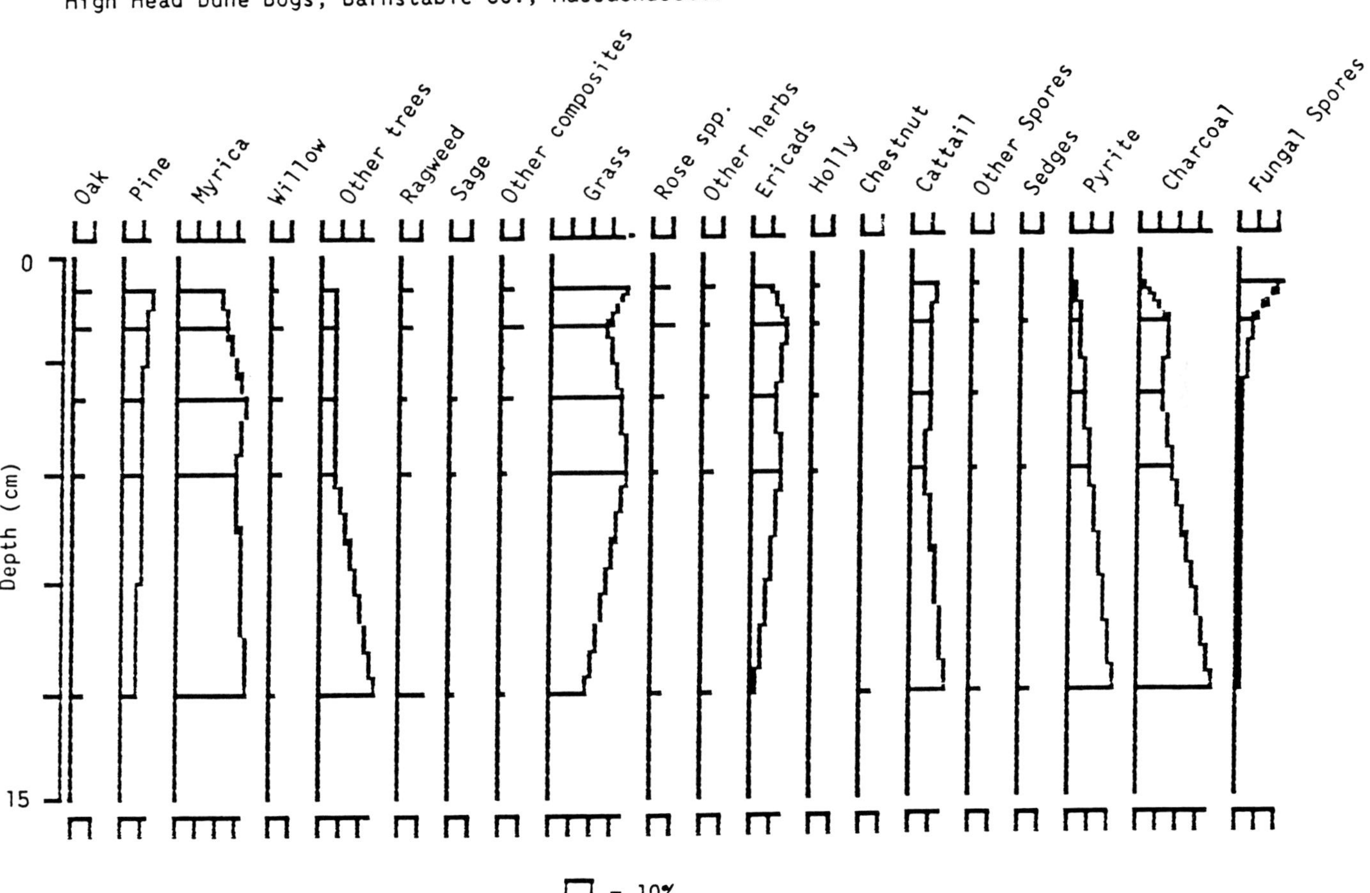

FIG. 2.—Percentage diagram for intradunal bog core #5 showing changes in major taxa and in six other variables: cattails, other spores, sedges, pyrite, charcoal, and fungal spores (plotted at right) From 200 to 660 pollen grains from trees, shrubs, and herbs were counted for each level; the pollen sum was used to calculate the percentages of each pollen taxon. In addition, between 150 and 350 grains and particles of the other variables were counted for each level and were included with the sum of shrub plus herb pollen to calculate percentages for these variables.

in the sediment at 12-cm depth indicates a saturated, anaerobic environment (Fig. 2). Abundant charcoal pieces in these basal organic sediments also indicate local fires (Fig. 2). Fungal hyphae are frequent in the top 5 cm of the sediments, as are several kinds of fungal spores (Fig. 2). Most of the identifiable fungal spores (conidia) are from *Helicoon pluriseptatum* and are found only in the top 2 cm. This species is also frequent in sediments at the bog/pond transition in the older interdunal ponds, Duck and Bennett, (discussed later). *Helicoon pluriseptatum* is a species of the air-water interface and is found today on the surface of peat bogs and on the moist surface film of leaves and wood in bogs (van Geel, 1976). It is also present as a subfossil in ancient highly organic peats, but not in minerotrophic peat (van Geel, 1976) as is the case in intradunal bog #5. Although some fungi and bacteria are known to break down mineral substrates (microbial weathering), here they are found primarily in the organic sediments, and are probably important in decomposition of organic material and in nutrient cycling in acid bogs.

Duck Pond, Provincetown.—

Pre-European settlement.—The ^{14}C date for the basal organic sediments of Duck Pond shows that wetland conditions originated at 960 ± 60 yrs. The $\delta^{13}C$ value is −28.4, well within the stable carbon isotope range of freshwater and upland-plant carbon. There is no indication of a marine-carbon contribution at initiation of this bog. The wetland pollen and spore stratigraphy (Fig. 3A) indicate that this site was a *Sphagnum* bog dominated by a groundcover of *Sphagnum* mosses, *Osmunda* ferns, and *Lycopodium* club mosses, surrounded by *Ilex* (holly) for the first 600 years of its development. This plant assemblage occurred at Duck Pond before European settlement of the Provincelands and is similar to the vegetation found in the interdunal bogs at High Head today. The regional forest was dominated by pitch pine (*Pinus rigida*) and oak (*Quercus* spp.). Charcoal abundances were high during this time, suggesting that the peatland burned periodically. Sarcodines (testate amoebae) were abundant below 15 cm in the core. These organisms are found frequently in acid, *Sphagnum* bogs where their resistant tests withstand periodic water-level changes.

Post-European Settlement (post-17th Century).—Ambrosia and other herb and grass pollen increase (Winkler, 1990b), indicating European settlement. There is a decrease in pollen from all trees except oak and an increase in birch (*Betula*) and bracken fern (*Pteridium*), possibly in response to post-settlement logging fires and forest clearance. Charcoal abundances, which had decreased in the sediment at the bog/pond transition (at about 15 cm), increase after that time.

The wetland changes are concurrent with the rise in *Ambrosia* pollen and show that there was an increase in water levels in the pond after European settlement. *Sphagnum* spores decrease and pollen from aquatic macrophytes (waterlilies, *Potamogeton,* and sedges) increase (Fig. 3A). Sarcodines become infrequent and cysts of chrysophyte algae increase in number. The wetland transition from *Sphagnum* bog to deep-water marsh is concurrent with settlement changes and may be anthropogenic at Duck Pond, probably due to disruption of water flow by road construction and land clearance.

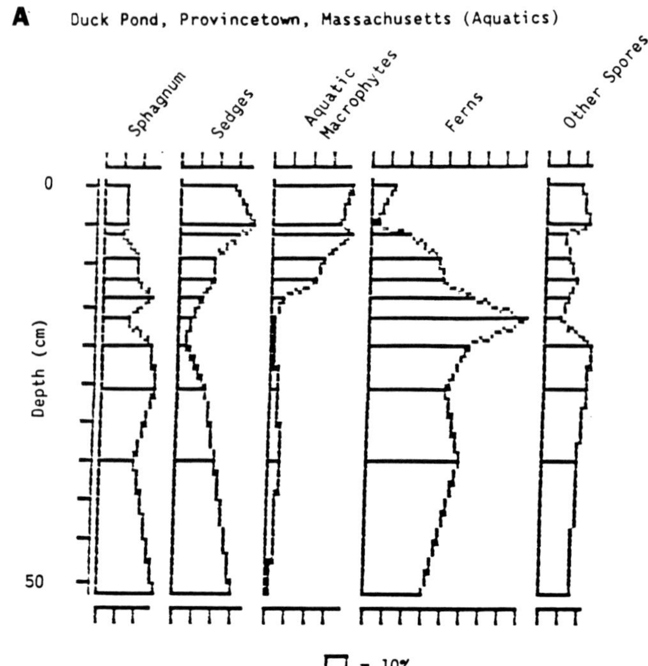

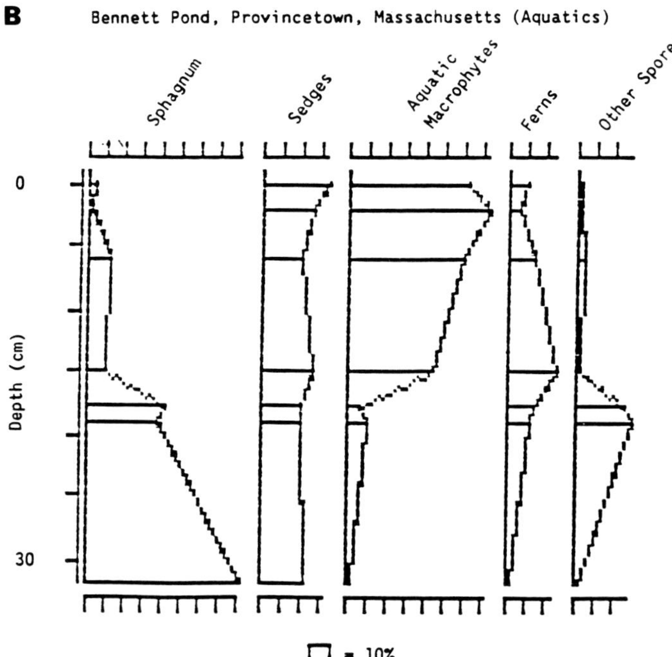

FIG. 3.—(A,B) Percentage diagrams of aquatic pollen and spores from Duck Pond and Bennett Pond, respectively, in the Provincelands. Between 20 and 203 grains of aquatics were counted for each level (basal sediments had sums <50). Aquatic sum was used for percentage calculation.

Percent organic-matter changes (a decrease from over 70 percent, when Duck Pond was a bog, to less than 50 percent at settlement) also document increasing water levels

(Winkler, 1990b). A further decrease in organic matter to 30 percent after settlement, as well as an increase in the percent of carbonates, indicates continued impacts from land-use changes.

Diatom changes upcore (Winkler, 1990b) are consistent with wetland vegetation changes. In the basal sediments, high percentages of *Pinnularia* diatoms commonly found in *Sphagnum* bogs, aerophilic diatoms, and other species that live on mosses occur. During the *Sphagnum*-bog stage, percentages of *Eunotia* diatoms (acid periphytic species) increase and, as the water level rose, planktonic diatoms such as *Tabellaria* and *Fragilaria* species become abundant. In the top 6 cm, diatoms that respond to more eutrophic conditions–*Melosira ambigua, Nitzschia, Gomphonema,* and *Synedra*–increase. Ratios of littoral to planktonic cladocera track the water-level changes indicated by the diatoms. Similarly, cladoceran biomass increases at the same time that diatom species changes indicate more eutrophic conditions (P. Sanford, pers. commun., 1991).

Bennett Pond.—

The vegetation changes shown by the pollen, and the evolution of the wetland shown by the aquatic taxa and diatom stratigraphy at Bennett Pond are similar to those at Duck Pond. At both places, *Sphagnum* bog changed to pond (Fig. 3), although the time of transition is different at each site. The change at Bennett Pond is evident from a decrease in organic matter (Winkler, 1990b), as well as a dramatic decrease in *Sphagnum* spores, a concurrent increase in pollen from sedges and aquatic macrophytes (Fig. 3B), a decrease in the number of bog-dwelling sarcodines (testate amoebae) in the sediments, and an increase in the number of algal chrysophyte cysts (abundant at times in deeper water). Bog initiation was radiocarbon dated at 910 ± 70 yrs and the transition to pond probably occurred at about 0.6 ka, well before European settlement (calculated from pollen data). The $\delta^{13}C$ value of the basal organics at Bennett Pond was −27.7, indicating no contribution of marine carbon at initiation of the interdunal bog, similar to the findings from Duck Pond, Provincetown.

The settlement horizon at about 12 cm is indicated by an increase in pollen from *Ambrosia* (ragweed), *Rumex,* other herbs, and grass, and a decrease in pine pollen. Charcoal is abundant in the bog peats and also at the time of European settlement. A change in organic matter at about 12 cm in the core is associated with erosion from land-use changes. High carbonates in the topmost 3 cm of Bennett Pond probably originated in windblown dust and silt from an adjacent landfill.

Diatom changes as well as wetland vegetation changes document water-level changes at Bennett Pond (Winkler, 1990b). As at Duck Pond, *Pinnularia* diatoms were dominant at bog formation followed by a transition to acid periphytic *Eunotia* and *Frustulia* species. As water level rose further, the diatoms that dominated the pond were *Fragilaria virescens* var *exigua,* an acid, planktonic species. In the top 3 cm, acid *Eunotia* diatoms dominate and *Fragilaria* (*F. lapponica*) increases, as do diatoms that grow well under eutrophic conditions, such as *Nitzschia, Gomphonema,* and *Synedra*.

The Chronology of the Interdunal Wetlands, the Parabolic Dunes, and the Provincelands

In order to determine the development of the Provincelands dunes, several hypotheses were explored.

1) Because of the coastal setting of the Provincelands, the parabolic dunes date from the middle Holocene, when sea-level rose to cover the Georges Banks, and have continued to be active since that time. Coastal winds caused by diurnal-temperature differences between the sea and land are constant (Houghton, 1985). This type of local-circulation system was active in the past as it is today and, today, prevailing winds from the northwest are causing advancement of the High Head parabolic-dune field to the southeast. There is evidence that the proximal part of the Provincelands developed as a barrier beach in the middle Holocene (Zeigler and others 1965; Fig. 4), so that both wind and sand, the ingredients for parabolic-dune growth, have been present from that time.

2) The High Head dune field (Figs. 1, 4) is a more recent landform, and dune movement was initiated as the vegetative cover of the Provincelands was disturbed after European settlement more than 300 years ago. Several podzols (conifer-forest soils) and trees buried by eolian sands within the dunes were radiocarbon dated as "modern" (within the last 200 years, McCaffrey and Leatherman,

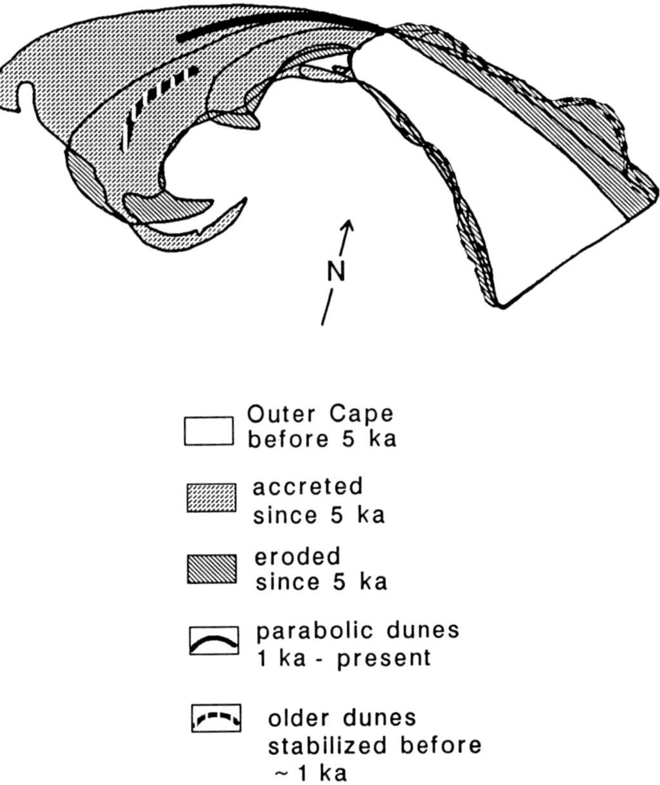

FIG. 4.—The changing coastline of Outer Cape Cod and the Provincelands since 5 ka. An older dune field is inferred within a hook that developed and stabilized before 1.2 ka, allowing intradunal bogs to evolve into ponds as water levels changed. General geomorphic changes modified from Zeigler and others. 1965.

1979), an indication of vegetated areas from early-settlement days subsequently inundated by sand. Documents from some of the first European settlers also told of forested dune lands (McCaffrey, 1973). However, buried-peat deposits within the area of the active dunes were found and described by Benedict and Leatherman (1978) and Leatherman (1979). The subsurface peats were also deposited on eolian sands (Leatherman, 1979), indicating that a moving dune field was present even before development of the buried peat.

Of the 40 or more vibracores taken in the interdunal swales throughout the Provincelands, most had very shallow surface peat overlying eolian sands (Leatherman, 1979), similar to the bog- and pond-core sediments obtained for this study (Winkler, 1990a). Benedict and Leatherman (1978) uncovered peat deposits buried by eolian sands about 1 m or less below the surface in four of the cores. The peats dated between 540 and 885 yrs. One of the cores had two peat deposits below the surface peat, each separated by 1 m of eolian sand. The middle peat layer was formed at 675±65 yrs (Leatherman, 1979). The core taken nearest the High Head dune site described in this study had a subsurface-peat deposit dated at 540±120 yrs (Leatherman, 1979). The dynamic nature of the present parabolic dune field and the ephemeral nature of the interdunal bogs is underscored by these dates, and by the buried "modern" forest soils described by McCaffrey (1973).

3) At least two groups of dunes comprise the Provincelands. Some were active during formation of the Provincelands Hook (Fig. 4), and some are relatively new forms (developed within the last millennium). Several ponds and wetlands have evolved within dune systems in the central and southwestern arm of the Provincelands (Fig. 1A). The basal organic sediment from some of the ponds has been radiocarbon dated. The oldest peat formed at least by 1155±200 yrs (the basal date for organic sedimentation in Clapps Pond; Zeigler and others, 1965). There is no information about wetland evolution at Clapps Pond, but, as presented earlier, other ponds, Duck and Bennett, within the outer hook of the Provincelands were initially wetlands, then bogs, and more recently, shallow ponds (Winkler, 1990b)–opposite to the classic theory of hydroseral development (see discussion in Winkler, 1988b). These wetlands formed on eolian sands at 960±60 and 910±70 yrs, respectively, and organics were deposited uninterruptedly for more than 900 years. Evidently, the dune field in this area of the Provincelands has not been active for the last 1 ka.

4) Dune building, dune movement, and wetland development in the Provincelands are episodic and are a function of large-scale climatic change as well as geomorphic changes and local events (Figs. 5, 6). Based on the evidence to date, wetlands were initiated at about 1 ka within an earlier and older dune field in the central and southwestern arm of the Provincelands hook, when the dune field in that region was disrupted. The disruption may have been due to several changing factors, all additive–a slowing of sea-level rise at about 2 ka (Redfield and Rubin, 1962), a slight change in the direction of the prevailing wind, an increase in precipitation resulting in dune stabilization by vegetation, and/or the slowing of sand-laden winds travelling across an ever-

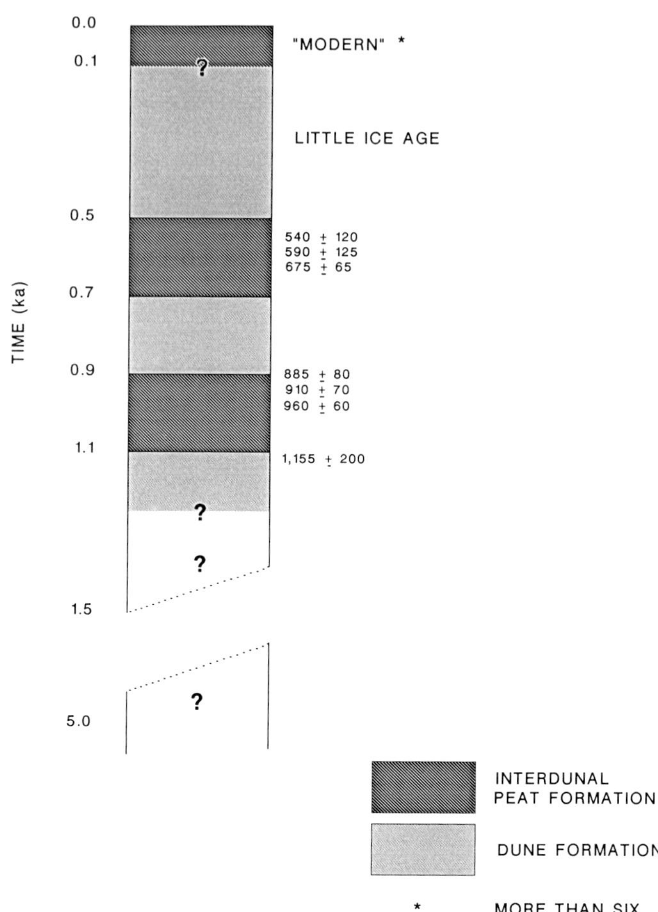

FIG. 5.—Clusters of radiocarbon dates for peatland development within the Provincelands wetlands provide evidence for alternating wet–warm and dry–cold climates within the last 1.2 ka. These changes correlate well with known episodes of climate change in the North Atlantic. The "modern" peat formation is based on radiocarbon dates from more than six samples of bog sediment, buried podzols, or buried wood.

broadening expanse (Fig. 4). Some of these factors could also have led to development of the present active dune field, which extends beyond the foredune along the northeast coast of the Provincelands today (Fig. 1A,B). According to radiocarbon dates on buried peats in those dunes, as a result of decreased wind intensity and increased precipitation, peatlands developed within the cusp of the dunes in the active dune field between about 0.7 and 0.5 ka (Fig. 5). During the Little Ice Age (between 0.5 and 0.1 ka, see Bryson and Murray, 1977; Lamb, 1969), however, dune advancement and burial of wetlands occurred in the High Head dune field, probably due to more intense, colder, and drier winds (expanded westerlies; Bryson and Murray, 1977). The Little Ice Age had an effect on climate in the North Atlantic (Fig. 6). Harbors in Iceland clogged with ice, vineyards failed in Europe, sailing lanes between Iceland and Greenland were closed by ice, glaciers expanded in the Alps, and famine was frequent in Europe as spring frosts lasted longer and fall frosts came earlier. Also during the Little Ice Age, European settlers arrived in the Provincelands and

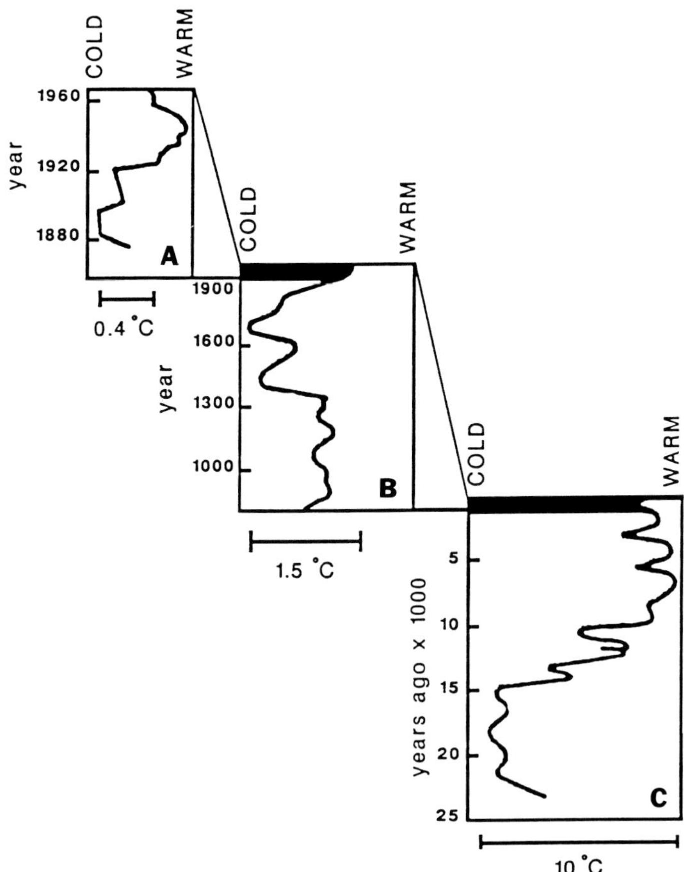

FIG. 6.—General trends in global climate (changing air temperatures) on different time scales. (A) Observed temperatures; (B) temperatures derived from historical records; (C) temperatures reconstructed from pollen and glaciers. The Provincelands dune movement and peat-formation stratigraphy correlate well with the climate changes depicted in (A) and (B) (see Fig. 5). Note change in temperature scales as well as time scales. After National Academy of Sciences Report (1975).

denuded the dunes. Dune movement increased until about a century ago when revegetation attempts (several earlier attempts at dune stabilization were unsuccessful; McCaffrey, 1973) coupled with a change to a warmer, less windy, and more moist climate, favored peatland development once more in the High Head dune fields.

Today in the Provincelands, after intense post-European settlement of the region and many land-use changes, a complex topography has resulted, with broken sand ridges and parabolic-dune fields (Fig. 1). Interdunal wetlands are evident and consist of bogs in various stages of development and shallow marshes and ponds (Fig. 1).

Similarities between the interdunal bogs of the Provincelands and those on the Nags Head, North Carolina, barrier beach suggest that these bogs evolved concurrently on the Atlantic Coastal Plain within the last hundred or so years, when a wetter and warmer climate favored wetland development in interdunal swales. Interdunal ponds on the Nags Head beach were similarly found to be modern in age, to have abundant charcoal in the basal sediments, to have water chemistry similar to that of ground water, and to be unaffected by marine influence (Burney and Burney, 1987; Kling, 1986).

Episodes of eolian activity in the Provincelands before 1 ka, between 0.9 and 0.7 ka, and between 0.5 and 0.2–0.1 ka in the active dune field (Fig. 5) also correlate with periods of dune expansion on the eastern coast of Hudson Bay in northwestern Quebec (Filion, 1984). Peak eolian activity in Hudson Bay similarly occurred before 1 ka and again at about 0.5 ka until about 0.1 ka. The synchrony of dune movement and interdunal-wetland development in the northeast region of the continent during the Little Ice Age and at other times is exciting evidence of the far-reaching extent of event-scale climatic changes.

CONCLUSIONS

1. Coastal geomorphology, dune formation, and bog and pond evolution in the Provincelands are dependent on climate. From the radiocarbon evidence to date, three episodes of eolian activity sufficiently strong to build and move dunes have occurred and three episodes of a warmer and wetter climate conducive to interdunal-wetland formation have occurred in the past 1.2 ka. Dune movement has been dated at before 1.1 ka, between 0.9 and 0.7 ka, and again between 0.5 and 0.1 ka during the Little Ice Age, when the North Atlantic climate was colder, drier, and windier. Interdunal wetlands developed between 1.1 and 0.9 ka, 0.7 and 0.5 ka (the medieval warm period), and within the past century or two.
2. The ponds in the central and southwestern region of the Provincelands originated as bogs similar in vegetation to those in the active dune field today. An older dune field is hypothesized to have existed in that region before 1.1 ka and was stabilized by at least 1 ka, and the wetlands (now ponds) at the Clapps, Duck, and Bennett sites that formed at that time have persisted until the present. A slowing of sea-level rise at about 2 ka may have mediated dune stabilization in this part of the Provincelands by favoring lateral expansion of the northwestern part of the coast.
3. A modern parabolic-dune-field is active today largely due to human disturbance. Because of a warm and more moist climate in the past century or so, soil is developing within the dunes and vegetation has become more diverse in the older bogs. Dune-stabilization efforts appear to have been successful, but the further development of the interdunal bogs will be affected by dune stabilization as well as by future wind, rain, and storm-frequency changes.

ACKNOWLEDGMENTS

I thank J. Portnoy, J. Overpeck, K. Gajewski, D. S., M. F., and S. S. Winkler, M. J. Linders, and K. Gregory for field assistance; P. Sanford for laboratory help and for cladoceran analysis; P. Behling for drafting figures; K. Hobler, M. Woodworth, and M. Kennedy at the Center for Climatic Research and J. Portnoy and D. Manski at Cape Cod National Seashore for logistical help; M. A. Soukup at the North Atlantic Regional Office of the National Park Service for scientific and financial support. Patricia R. Sanford and

L. J. Maher, Jr., kindly gave critical reviews of the manuscript. Radiocarbon dating was done by the Radiocarbon Laboratory of the Center for Climatic Research at the University of Wisconsin-Madison. The research was funded by Office of Scientific Studies, North Atlantic Regional Office of the National Park Service, Order No. CA-1600-6-0004 and by National Science Foundation grant ATM87-13227 to J. E. Kutzbach, Director, Center for Climatic Research, University of Wisconsin at Madison.

REFERENCES

Benedict, M. A., and Leatherman, S. P., 1978, Preliminary investigation of the geobotanical evolution of Provincetown Peninsula: University of Massachusetts-National Park Service Cooperative Research Unit Report No. 38, 56 p.

Berglund, B. E., and Ralska-Jasiewiczowa, M., eds., 1986, Handbook of Holocene Paleoecology and Paleohydrology: John Wiley, New York, 869 p.

Brugam, R. B., 1978, Pollen indicators of land-use change in southern Connecticut: Quaternary Research, v. 9, p. 349–362.

Bryson, R. A., and Murray, T. J., 1977, Climates of Hunger: The University of Wisconsin Press, Madison, Wisconsin, 171 p.

Burney, D. A., and Burney, L. P., 1987, Recent paleoecology of Nags Head Woods on the North Carolina Outer Banks: Bulletin of the Torrey Botanical Club, v. 114, p. 156–168.

Dean, W. E., Jr., 1974, Determination of carbonate and organic matter in calcareous sediments and sedimentary rocks by loss on ignition: comparison with other methods: Journal of Sedimentary Petrology, v. 44, p. 242–248.

Faegri, K., and J. Iversen, 1975, Textbook of Pollen Analysis: Hafner Press, New York, 295 p.

Filion, L., 1984, A relationship between dunes, fire and climate recorded in the Holocene deposits of Quebec: Nature, v. 309, p. 543–546.

Filion, L., 1987, Holocene development of parabolic dunes in the central St. Lawrence lowland, Quebec: Quaternary Research, v. 28, p. 196–209.

Flint, R. F., 1971, Glacial and Quaternary Geology: John Wiley, New York, 892 p.

Gray J., 1965, Extraction techniques, *in* Kummel, B., and Raup, D., eds., Handbook of Paleontological Techniques: Freeman and Company, San Francisco, p. 530–587.

Houghton, D. D., 1985, Handbook of Applied Meteorology: John Wiley, New York, 1461 p.

Keen, K. L., and Shane, L. C. K., 1990, Continuous record of Holocene eolian activity and vegetation change in the Lake Ann area, east-central Minnesota: Geological Society of America Bulletin, v. 102, p. 1646–1657.

Kling, G. W., 1986, The physicochemistry of some dune ponds on the Outer Banks, North Carolina: Hydrobiologia v. 134, p. 3–10.

Kutzbach, J. E., and Wright, H. E., Jr., 1985, Simulation of the climate of 18,000 years BP: results for the North American/North Atlantic/European sector and comparison with geologic record of North America: Quaternary Science Reviews, v. 4, p. 147–187.

Lamb, H. H., 1969, Climatic fluctuations, *in* Flohn, H., ed., World Survey of Climatology: Elsevier, New York, p. 173–249.

Leatherman, S. P., 1979, Evolution of the Province Lands, *in* Leatherman, S. P., ed., Environmental Geologic Guide to Cape Cod National Seashore, Field Trip Guide Book for the Eastern Section, Society of Economic Paleontologists and Mineralogists, National Park Service-Cooperative Research Unit, University of Massachusetts, Amherst, p. 193–206.

Leatherman, S. P., and Godfrey, P. J., 1979, Environmental effects of dune mining, Mount Ararat, the Province Lands, *in* Leatherman, S. P., ed., Environmental Geologic Guide to Cape Cod National Seashore, Field Trip Guide Book for the Eastern Section, Society of Economic Paleontologists and Mineralogists, National Park Service-Cooperative Research Unit, University of Massachusetts, Amherst, p. 223–232.

Leblond, R., 1986, Study of vegetation succession in dune wetlands of the Provincetown Peninsula: National Park Service Investigator's Annual Report, 2 p.

McCaffrey, C. A., 1973, An ecological history of the Province Lands, Cape Cod National Seashore: University of Massachusetts, Amherst, Report No. 1, National Park Service-Cooperative Research Unit, 71 p.

McCaffrey, C. A., and Leatherman, S. P., 1979, Historical land use practices and dune instability in the Province Lands, *in* Leatherman, S. P., ed., Environmental Geologic Guide to Cape Cod National Seashore, Field Trip Guide Book for the Eastern Section, Society of Economic Paleontologists and Mineralogists: University of Massachusetts, Amherst, National Park Service-Cooperative Research Unit, p. 207–222.

NAS, 1975, Understanding Climatic Change, a Program for Action: National Academy of Sciences Report, p. 130.

Redfield, A. C., and Rubin, M., 1962, The age of salt marsh peat and its relation to recent changes in sea level at Barnstable, Mass.: Proceedings, National Academy of Sciences, v. 48, p. 1728–1735.

Thorson, R. M., and Bender, G., 1985, Eolian deflation by ancient katabatic winds: a late Quaternary example from the North Alaska Range: Geological Society of America Bulletin, v. 96, p. 702–709.

Van Geel, B., 1976, A paleoecological study of Holocene peat bog sections, based on the analysis of pollen, spores and macro- and microscopic remains of fungi, algae, cormophytes and animals: Unpublished Ph.D. Dissertation, Hugo de Vries Laboratory, University of Amsterdam, Netherlands, 93 p.

Wells, G. L., 1983, Late-glacial circulation over North America revealed by aeolian patterns, *in* Street-Perrott, A., Beran, M., and Ratcliffe, R., eds., Variations in the Global Water Budget: Reidel, Boston, p. 317–330.

Winkler, M. G., 1985a, Charcoal analysis for paleoenvironmental interpretation: a chemical assay: Quaternary Research, v. 23, p. 313–326.

Winkler, M. G., 1985b, A 12,000-year history of vegetation and climate for Cape Cod, Massachusetts: Quaternary Research, v. 23, p. 301–312.

Winkler, M. G., 1988a, Paleolimnology of a Cape Cod kettle pond: diatoms and reconstructed pH: Ecological Monographs, v. 58, p. 197–214.

Winkler, M. G., 1988b, Effect of climate on development of two *Sphagnum* bogs in south-central Wisconsin: Ecology, v. 69, p. 1032–1043.

Winkler, M. G., 1989, Geologic, chronologic, biologic, and chemical evolution of the Acid Kettle Ponds within the Cape Cod National Seashore: Report to the National Park Service, North Atlantic Region, 145 p.

Winkler, M. G., 1990a, The evolution of modern and ancient interdunal bogs in the Provincelands of the Cape Cod National Seashore: Report to the National Park Service, North Atlantic Region, 39 p.

Winkler, M. G., 1990b, I. Recent changes in trophic status of Duck and Bennett Ponds in the Provincelands relative to possible landfill enrichment. II. Evolution of interdunal ponds in the Provincelands, Cape Cod National Seashore: Provincetown Ponds Final Report, National Park Service, North Atlantic Region, 38 p.

Wright, H. E., Jr., Almendinger, J. C., and Gruger, J., 1985, Pollen diagram from the Nebraska Sandhills and the age of the dunes: Quaternary Research, v. 23, p. 115–120.

Zeigler, J. M., Tuttle, S. D., Tasha, H. J., and Giese, G. S., 1965, The age and development of the Provincelands Hook, Outer Cape Cod, Massachusetts: Limnology and Oceanography, v. 10 (Supplement), p. R298-R311.

SEDIMENTOLOGIC AND MORPHOLOGIC EVOLUTION OF A BEACH-RIDGE BARRIER ALONG AN INDENTED COAST: BUZZARDS BAY, MASSACHUSETTS

DUNCAN M. FITZGERALD AND CHRISTOPHER T. BALDWIN
Geology Department, Boston University, Boston, Massachusetts 02215
NOOR A. IBRAHIM
Petroleum National of Malaysia, Kuala Lumpur, Malaysia
AND
STANLEY M. HUMPHRIES
IEP Consultants, Inc., Northboro, Massachusetts 01532

ABSTRACT: The sedimentologic development of Horseneck Beach on the northwest coast of Buzzards Bay, Massachusetts, is interpreted from 40 vibracores, 20 boreholes, 9 km of ground-penetrating radar and eight topographic profiles. The beach-ridge barrier is 4 to 10 m thick and consists primarily of fine sands with some coarse sand and gravel layers underlain by a coarsening-upward estuarine sequence (3–8 m thick) composed of silts and clays grading to fine sand. Underlying the estuarine deposits are glacio-fluvial sands and gravel up to 6 m thick, blanketing a Paleozoic bedrock surface.

Ground-penetrating-radar records and backhoe trenching indicate that progradation of the barrier occurs sporadically. An inferred-process model is proposed in which the accretionary phase represents a period of abundant sand supply when the beach widens, builds vertically, and is punctuated by a period of low-sediment supply and erosion. At the eastern end of the barrier, the erosional phase coincides with an influx of gravel, steepening of the beach profile, and ridge construction, which is probably controlled by the 100-year, or longer, storm frequency. During these infrequent events, sand and gravel are released from an offshore drumloid. Segregation of the sand and gravel results from differences in longshore sediment-transport rates.

INTRODUCTION

Beach-ridge barrier islands are regressive barriers consisting of a series of parallel to semiparallel, vegetated, dune ridges that are often separated from one another by low marshy swales. Individual ridges represent successive shoreline positions in the history of the seaward-barrier progradation. Beach-ridge barriers occur on depositional coasts throughout the world where sand supply was sufficiently abundant during the late Holocene to overcome the slowed rate of eustatic sea-level rise. The factors controlling barrier construction and their regressive *versus* transgressive nature include sea-level trends, sediment abundance, tectonic setting, and tidal range (Curray, 1964; Dickinson and others, 1972; Hayes and Kana, 1976; Elliot, 1986). Morton and Donaldson (1973), Wilkinson (1975), Halsey (1979), Belknap and Kraft (1985), and Boyd and others (1987) have also demonstrated how antecedent topography affects the development of barrier shorelines by controlling regional slope, the arrangement of bays and headlands, and location of tidal inlets.

Some of the better studied beach-ridge barriers of the world are those of the Netherland coast (Van Straaten, 1965; Jelgersma and others, 1979), the Costa de Nayarit, Mexico (Curray, 1969), and several locations in the United States, including Galveston Island, Texas (Bernard and others, 1962), the coast of South Carolina (Moslow and Colquhoun, 1981), and Bogue Banks, North Carolina (Steele, 1980). The above barriers are all located in depositional coastal-plain settings where there is an absence of bedrock, and the barrier chain extends many kilometers along the shoreline. In contrast, the beach-ridge barriers of northern Buzzards Bay, Massachusetts, occur along a highly indented shoreline in which bedrock ridges form the structural grain of the coast and the sediment supply has been meager. Within this stretch of shoreline, two areas of beach-ridge development exist: Horseneck Beach and Slocum River Beach (Fig. 1).

Whereas there are many published studies dealing with beach-ridge construction along depositional coasts, beach-ridge development along indented coasts has received little attention. Notable exceptions are works of Roy and others (1980) and Melville (1984) along the New South Wales coast of Australia. The intent of this paper is to describe the evolution and general stratigraphy of the Horseneck barrier and to propose a model for the formation and stacking of beach ridges along its eastern end.

PHYSICAL SETTING

The northern shoreline of Buzzards Bay, from Narragansett Bay, Rhode Island, to the Cape Cod Canal, Massachusetts, consists of a series of northwest-southeast-oriented bedrock ridges (average spacing = 1 km) with intervening valleys. This landscape was formed by fluvial erosion during the late Tertiary (McMaster and Asraf, 1973) and was further sculpted throughout the Pleistocene by four major periods of glaciation (Kaye, 1964). The sediment-starved eastern half of the shoreline is characterized by long peninsulas and deep embayments with few depositional features. Sediment sources are slightly more abundant along the west bay shoreline than that to the east and many of the embayments are fronted by barrier beaches and spit systems (Fig. 1). Although most of the present barriers are thin and transgressive, beach-ridge barriers are well developed at the mouths of Westport and Slocum River estuaries. A discussion of the morphologic development of the northwest Buzzards Bay shoreline can be found in the FitzGerald and others (1987).

The northwest Buzzards Bay coast is micotidal with semidiurnal tides having a mean range of 1.1 m and spring tidal range of 1.3 m (National Ocean Survey, 1989). A wind rose for nearby Block Island indicates that this region is affected by onshore winds during the spring and summer, when moderate to strong prevailing winds blow from the

Quaternary Coasts of the United States: Marine and Lacustrine Systems, SEPM Special Publication No. 48

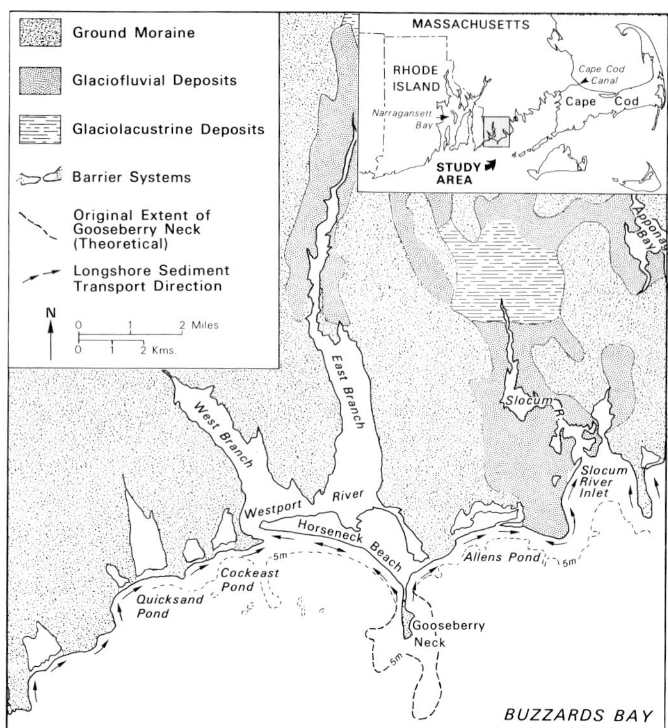

FIG. 1.—Location, surficial deposits, and longshore sediment-transport directions in the vicinity of the study area. Data sources from Stone and Piper (1982), Magee and FitzGerald (1980), and FitzGerald and others (1987).

southwest, and during periods of infrequent storms associated with extra-tropical cyclones.

Wave energy along the southwestern Buzzards Bay shoreline is generally low and highly variable. Wave data collected over a three-year period from a gauge located 9 km offshore of Gooseberry Neck indicate that the deep-water mean significant wave height for this region is 0.9 m and the mean wave period 7.5 s (Thompson, 1977). A more recent wave-hindcast study for the area utilizing 20 years of information by Jensen (1983) shows that the average nearshore (10 m water depth) wave height for this region is 0.75 m with a period of 6.0 s.

Because of the sheltering afforded by the Elizabeth Islands southeast of the study area (Fig. 1), the western Buzzards Bay shoreline is influenced primarily by waves approaching from the southwest. These waves are produced by weak, prevailing winds during the spring and summer, and infrequent southwesters that track across New York or through the Connecticut Valley during late fall and winter. Longer period swell (8 to 10 s) affects the coast as well. It should be noted that due to the overall low-wave and tidal energy of this coast, infrequent, large-magnitude (category >3) hurricanes may impose a strong imprint on the net sediment-transport patterns.

SEDIMENT SUPPLY

The supply of sediment to this coastline has been primarily from the reworking of deposits of repeated episodes of Pleistocene glaciation. As seen in the surficial-sediment map in Figure 1, the most prevalent glacial deposit in this area is ground moraine that forms a relatively thin sediment cover (<6 m thick) that thickens locally in valleys. Sieve analyses of the till indicate that the matrix contains an average of 51 percent sand and fine gravel (granules). Richer sand deposits occur within the drainage systems of the Slocum and East Branch of the Westport Rivers (Stone and Piper, 1982). These glacio-fluvial sources, along with the Gooseberry Neck drumloid (Fig. 2), were probably the primary contributors of sand to Horseneck Beach and the beach to the east. Additional sediment was likely transported onshore during the Holocene transgression from reworked glacial deposits in Buzzards Bay. It is certainly no coincidence that the beach-ridge barrier of Horseneck Beach is proximate to the most plentiful sand resources.

The present extent of Gooseberry Neck would not appear to be able to account for all the sand that forms Horseneck Beach or the spit systems that front Allens Pond. However, it is believed that it was once much larger, as indicated by the boulder retreat lag and surrounding shallows (5-m isobath; Fig. 1). In fact, the numerous subtidal-boulder regions offshore would suggest that Gooseberry Neck may have been only one of several glacial deposits that were gradually mined of sand during the transgression.

GLACIATION AND HOLOCENE SEA-LEVEL CHANGES

Stratigraphic evidence from Martha's Vineyard and elsewhere demonstrate that during the Pleistocene, this region experienced as many as four major periods of glaciation ranging from the Nebraskanan to Wisconsinan (Kaye, 1964; Dillon and Oldale, 1978; Oldale, 1982). The maximum southern excursion of the Wisconsinian ice sheet occurred between 18 and 17 ka and is delineated by a discontinuous terminal moraine extending west from Nantucket and Martha's Vineyard Islands to southern Long Island. Ice retreated from the Buzzards Bay area approximately 15 to 14.25 ka (Larson, 1982), leaving behind a series of recessional moraines and till sheets including the Buzzards Bay moraine, which encompasses the modern Elizabeth Islands.

A relative sea-level curve for the past 12 ka has been constructed by Oldale and O'Hara (1980) for southeastern Massachusetts based on radiocarbon dates of estuarine shells and basal peats (Fig. 3). Their curve extends to 5 ka and then is extrapolated to meet the relative sea-level curve of Redfield and Rubin (1962), which covers the past 3.4 ka. The curve indicates that relative sea level rose at a rate of 3 m/ka between 8 and 2.8 ka and has slowed to 1 m/ka since then. Data on the curve indicate that Buzzards Bay began flooding at approximately 7 ka. Some of the larger river systems including the Westport River began depositing estuarine sediments at about 7 to 5 ka (Ibrahim, 1986).

DATA BASE

The stratigraphy of Horseneck Beach is known from a study by Ibrahim (1986), in which he collected a total of 40 vibracores (3 to 9 m in length) taken along seven transects across the islands. This information was supple-

FIG. 2.—View of glacial scarp along exposed coast of Gooseberry Neck. Note shovel for scale.

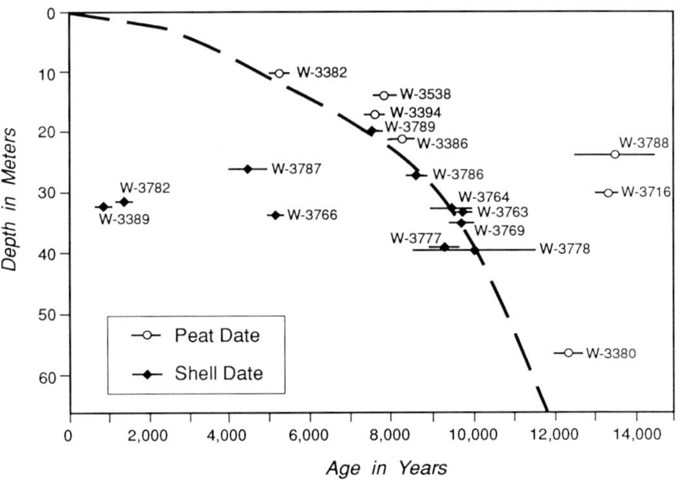

FIG. 3.—Holocene sea-level curve for southeastern New England (after Oldale and O'Hara, 1980).

mented with core logs of 19 deep boreholes (10 to 50 m in depth; Willey and others, 1983).

Details of beach-ridge development at Horseneck Beach have been interpreted from topographic profiles and from 5 km of ground-penetrating-radar transect data. The radar profiles were taken at the eastern half of the island along dip and strike transects (Fig. 4). The impulse-radar system utilizes a multifrequency radio wave in the range of 120 to 500 mHz and produces a record similar to that of single-channel, shallow, seismic-reflection profiles. The resolution of reflectors within the record is approximately 0.10 m. Vertical penetration in sediment above the saltwater wedge exceeds 25 m, but all signals are lost by the presence of saline ground water, as is the case with conventional shallow seismics. The application of this equipment in the study of sedimentary environments has been reported previously by FitzGerald and Baldwin (1986) and Leatherman (1987). A series of four trenches was excavated by a front-end loader along one of the strike transects to aid in the identification of individual reflectors. The trenches varied in length from 6 to 17 m and ranged in depth from 2 to 3 m. In addition, numerous holes were dug by hand along several of the transects to ascertain the depth to sand/gravel, sand/peat and other boundaries.

HORSENECK BEACH-RIDGE BARRIER

Morphology

Beach ridges are believed to make up most of the Horseneck Beach barrier complex (Figs. 4, 5), although much of the ridge morphology has been obliterated by buildings, roadways, parking lots, and secondary sand-dune formation. The present barrier system built across the mouth of the East and West Branches of the Westport River, the largest estuary along this stretch of coast. The progradation of

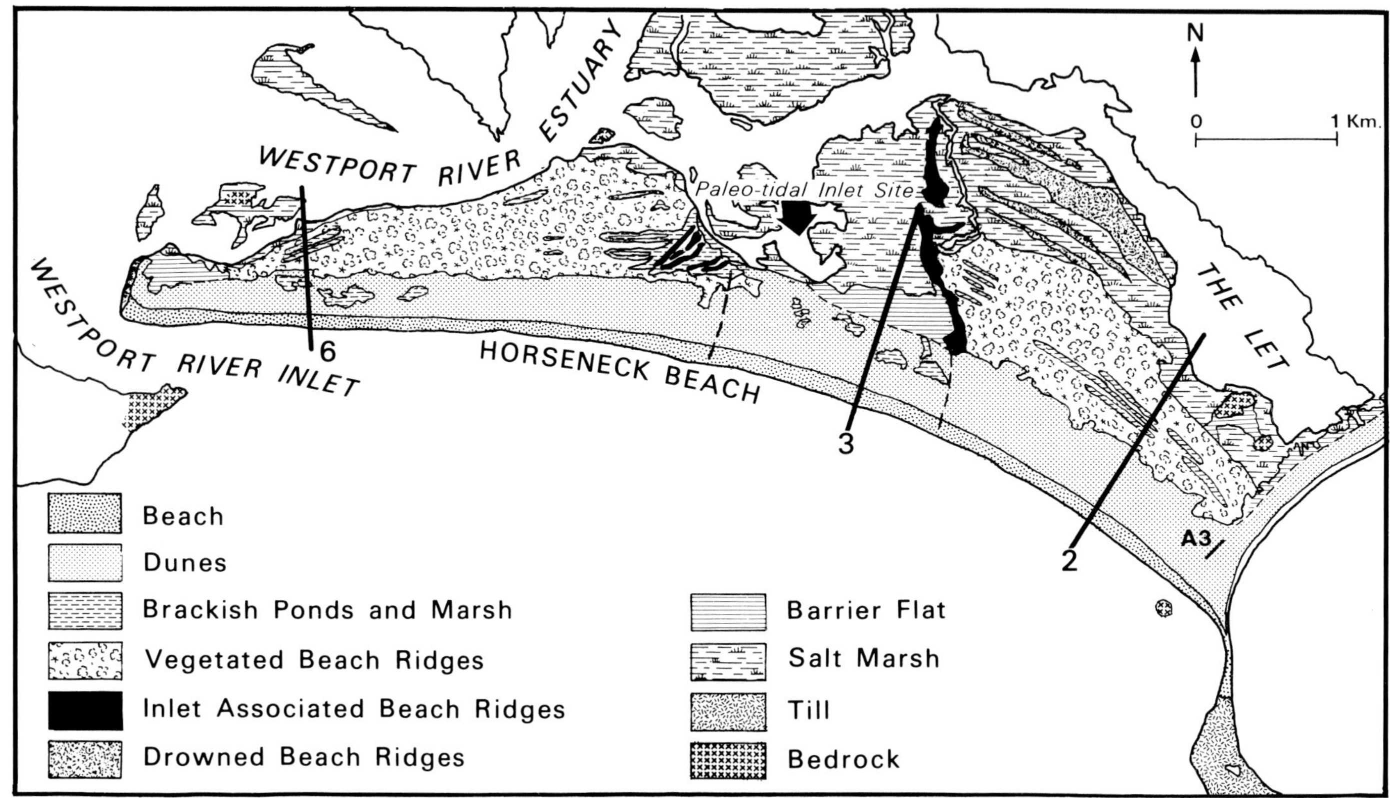

FIG. 4.—Geomorphic provinces of Horseneck Beach and location of the ground-penetrating radar transects. Lines 2, 3, and 6 indicate the locations of stratigraphic sections shown in Figure 6. Line A3 is site of ground-penetrating radar depicted in Figure 8.

FIG. 5.—Oblique aerial photograph of Horseneck Beach looking northwest. Note the beach ridges that begin as isolated and partially submerged ridges in the marsh and become more closely spaced toward the seaward side of the barrier. The impulse-radar record was taken in the parking lot in foreground.

Horseneck Beach involved the closure of at least two inlets (Magee and FitzGerald, 1980; Ibrahim, 1986). Marsh-covered flood-tidal delta deposits in the back barrier, and recurved ridges on the barrier itself mark the site of a paleo-inlet behind the middle of the barrier (Fig. 4). Ibrahim (1986) cored this section of the barrier and determined that the paleo-inlet was 2.5 m deep and closed approximately 485 yrs ago, as indicated by radiocarbon dates.

Historical records reveal that until recent time Horseneck Beach was an island, separated from the mainland on its eastern side by a tidal channel leading into the East Branch of the Westport River Estuary. Sometime between 150 and 200 years ago, a spit built across the channel and an open-water area called The Let was formed (Fig. 4) (Magee and FitzGerald, 1980). Today, the sand and gravel barrier that fronts The Let is thin and highly transgressive. While closure of the inlets along the Horseneck barrier was partly related to the abundant sediment supply, as evidenced by the progradation of beach ridges, it was also a product of sedimentation occurring in the back barrier. The formation of tidal deltas, intertidal-mud flats and marsh areas served to lessen the amount of open-water area and therefore the tidal prism. As the tidal prism decreased, so to did the equilibrium inlet cross section necessary to exchange waters between the ocean and estuary. The diminution of energy in the back barrier promoted sedimentation in that area.

Beach ridges tend to be small on the eastern end of Horseneck Beach, with an average height of 1 to 2 m, and gradually increase in elevation to the west, where heights reach 3 to 4 m. In the western end of the island, the destruction of the foredune ridge coupled with devegetation of dune grasses caused by storms and poor management of human traffic has led to sand mobilization. Presently, a 7-

to 8-m-high dune system is migrating landward over a forest.

The western end of the island abuts Westport River Inlet and appears to have formed through spit accretion, as indicated by marsh-covered flood-tidal-delta deposits behind the barrier. Ibrahim (1986) identification of spit-platform facies beneath the barrier sands in vibracores provides further evidence of spit building. The original recurved ridges associated with spit construction are not visible, presumably due to secondary-dune growth.

Constructional History

The depositional history of Horseneck Beach, which is summarized later, has been interpreted from seven stratigraphic cross sections of the barrier and several radiocarbon dates (Ibrahim, 1986). Three representative transects are shown in Figure 6, including transect #3 that shows a near-complete Holocene and late Pleistocene sequence overlying bedrock. The barrier lithosome, composed of beach-ridge, backshore, inlet-fill, foreshore, and shoreface facies, is a coarsening-upward sequence consisting of fine to medium sand with some layers of coarse sand and cobble-size gravel. The deposit thickens both in a seaward direction and toward the axis of the estuary, where it reaches a maximum thickness of about 10 m.

The sedimentation history of Horseneck Beach has been divided into five stages (Ibrahim, 1986) beginning with the deglaciation of the region, which took place at approximately 15.3 ka (Larson, 1982). As the ice retreated from the area, it left behind several recessional moraines and a pervasive lodgement till (Fig. 2). Many of the larger and deeper valleys, like the Westport River system, became conduits for meltwater streams and sites of glacio-fluvial sedimentation.

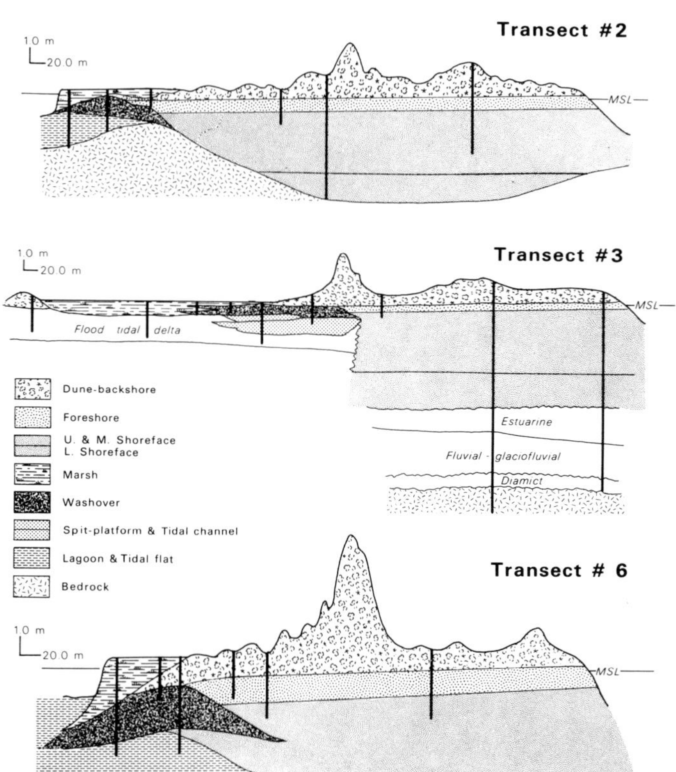

FIG. 6.—Stratigraphic cross sections of Horseneck Beach. Sites shown in Figure 4. Vertical black lines indicate core data used to construct the sections (from Ibrahim, 1986).

The second stage encompassed the period (14 to 7 ka) when local relative sea level was low and the inner continental shelf was exposed to surficial processes. In the Westport region, this interval saw the reworking of the glacial basal unit by fluvial, aeolian and soil-forming processes. The third stage (7 to 5 ka) marked the marine inundation of the Westport Embayment and its change from fluvial to estuarine conditions. This stage was reached earlier (at 7.5 ka) in deeper portions of nearby Narragansett Bay (McMaster, 1984). The estuarine facies deposited during this period are found under the middle of Horseneck Beach, are 1 to 3 m thick, and contain marine and brackish-water shells (Ibrahim, 1986).

During the fourth step (5 to 2.8 ka), the offshore ledges and glacial headlands, which had previously protected the embayment from open-ocean conditions, were gradually eroded and overtopped by rising sea level. The sand released from the glacial deposits contributed to the formation of a transgressive-barrier system. As the barrier migrated onshore, it partially reworked the underlying estuarine and glacial sediments, forming an erosional unconformity. Evidence of this stage is the washover deposits found along the entire back side of Horseneck Beach (Fig. 6; Ibrahim, 1986).

The last stage in the evolution of Horseneck Beach (2.8 ka to the present) marked a change in barrier dynamics, whereby increased sediment supply and a decrease in the rate of relative sea-level rise combined to produce a regressive sequence capped by prograding beach ridges. This final phase of barrier development resulted in the closure of two inlets and a widening of the barrier by 0.6 to 1.2 km. The depositional history of Horseneck Beach is similar to that of other regressive barriers in that barrier evolution begins with a transgressive-barrier stage. The juxtaposition of landward transgressive facies and seaward regressive facies also has been reported at Kiawah Island, South Carolina (Moslow and Colquhoun, 1981), along Bougue Banks, North Carolina (Steele, 1980) and the Netherland mainland barriers (Van Straaten, 1965).

BEACH-RIDGE DEVELOPMENT

Introduction

It is assumed that the addition of beach ridges to a regressive-barrier island is somehow coupled to the seaward progradation of the foreshore and shoreface (Swift, 1976). In fact, in most stratigraphic cross sections of regressive barriers, the time lines are drawn to indicate that progradation of the different parts of the barrier occurs contemporaneously (Fig. 7). While there are no data to suggest otherwise, it should be noted that these time surfaces are inferred and based solely on individual radiocarbon dates, which are used to date and define the typically sigmoidally bounded accretionary prisms that manifest themselves at the

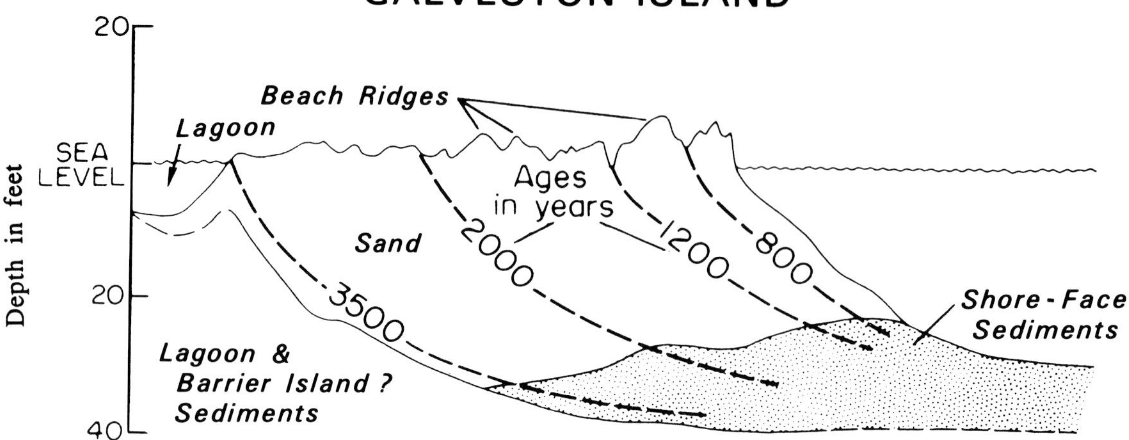

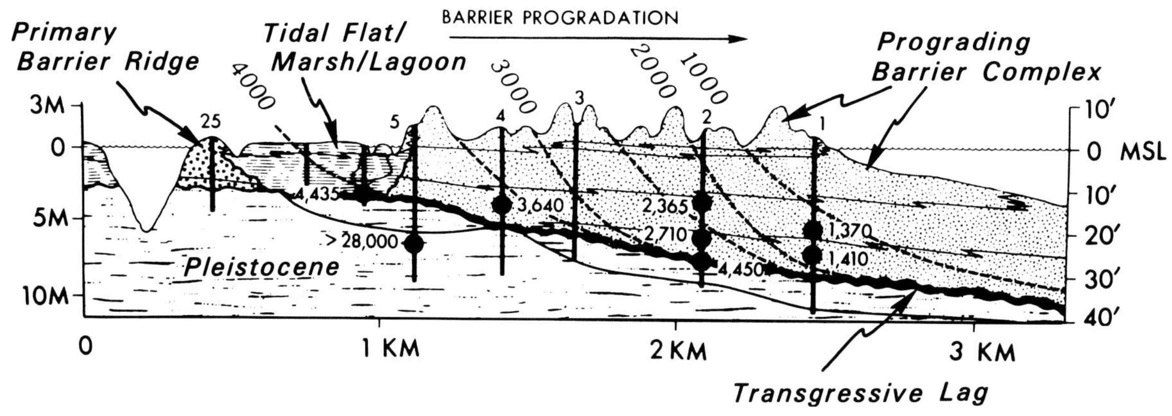

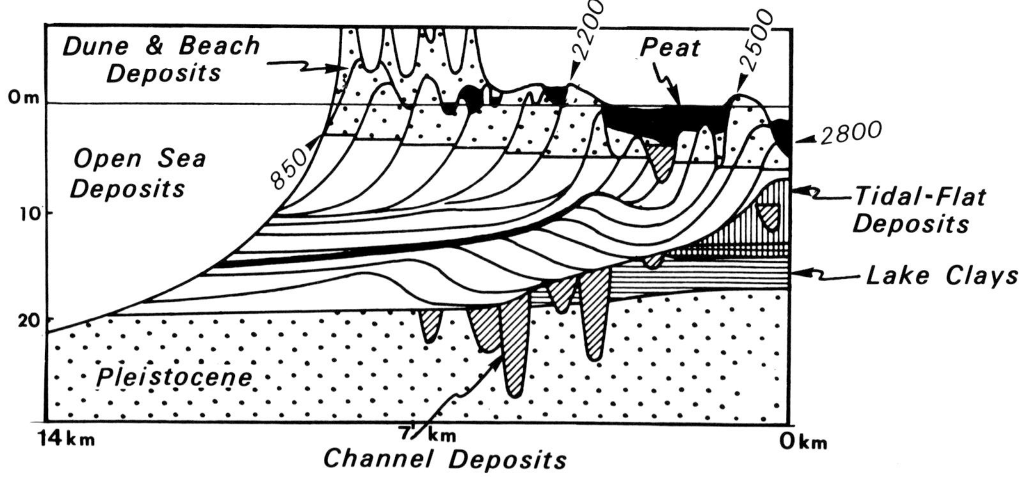

FIG. 7.—Regressive barrier-island cross sections of Galveston Island, Texas (after Bernard and others, 1962), Kiawah Island, South Carolina (after Moslow and Colquhoun, 1981) and the Netherlands coast (after Van Straaten, 1965).

landward end as beach ridges. Because the beach-ridge, foreshore and shoreface environments normally experience very different sedimentation processes, it would seem unlikely that a time surface would form a continuous bed that could be identified in cores or excavations, or that the prisms would remain so simple and continuous. At Horseneck Beach, an unusual bimodal sediment distribution has produced accretionary surfaces that have been recorded in the field using a ground-penetrating-radar system.

Geometric Properties

Progradation of the barrier is depicted in a ground-penetrating-radar profile (Fig. 8) along a dip section at the eastern end of the barrier (Fig. 4). The transect is dominated by strong, seaward-dipping (14° to 17°), sigmoidal reflectors that become almost horizontal at a depth of about 5 m. Each of the sharply defined radar facies is separated by an almost transparent unit that exhibits some horizontal to slightly seaward-dipping reflectors. Figure 8 shows four to five cycles of the paired facies described along a 100-m-long transect. Trenches excavated along the transect, also shown in Figure 8, reveal that the sharply defined reflectors are gravelly layers ranging in thickness from 5 to 190 cm (Fig. 9). The layers consist of well-rounded, cobble-size material (2 to 15 cm, long axis) in a matrix of sand and granules. The gravel layers exhibit a clast-supported framework, and thus, a portion of the sand matrix appears to have been deposited after the cobbles and subsequently sieved its way down through the gravel.

The trenches indicate that gravel forms ridges that interfinger with sand facies along their landward boundary (Fig. 9). The downward transition from cobble units to the underlying sand is sharp but not clearly erosive, that sieved through sand (see earlier discussion) being indistinguishable from that below. The underlying unit is fine to medium grained, moderately to well sorted, and buff colored with isolated occurrences of cobbles and granule layers. Horizontal to low-angle, seaward-dipping laminae, often accentuated by heavy minerals, are common sedimentary structures. These laminae give some indication of a partially erosive relation with the coarse-grained facies. Within the lower part of the exposed sigmoidal units (excavations were limited to around 3-m depth due to fresh ground water), the sandy laminae appear to be truncated by gravel facies, or at least they pass seaward with an abrupt change of angle (see T_1, Fig. 10).

Higher within each sigmoidal unit the gravel "feathers out," becoming more diluted with sand, and the seaward dip of the gravel/cobble layers merges with the sand laminae. In two excavations (trenches 3 and 4, Fig. 9) the gravel units reach their maximum elevation and eventually dip landward, with the landward-dipping portion being yet more diluted with sand until it gives way to a poorly defined lag-like surface with isolated clasts that appear to float in a clean matrix. In trench 3 the gravel unit feathers out into a peaty, medium sand, which thickens landward. In other trenches, similar peaty sand layers and lenses are intercalated with the overlying and apparently draping ferruginous micaeous sands. This last unit is truncated everywhere by a man-made surface so that no real morphologic surface feature can be correlated with the underlying facies. However, it should noted that trench 1 (Figs. 8 and 9) was dug at the extreme edge of the graded surface of the parking lot. North of this margin, the first clearly defined scrub-oak-covered beach ridge is present as a low (<3 m relief), continuous ridge that passes some 40 to 50 m to the west and terminates in a small dune blowout. A suite of similar ridges is present all the way to the back-barrier lagoon (Fig. 4). The clean sand that underlies the gravel is interpreted as berm, foreshore and beachface accretion, and the peaty sand and upper micaeous-sand unit is likely a paleo-soil that was formed in swale environments.

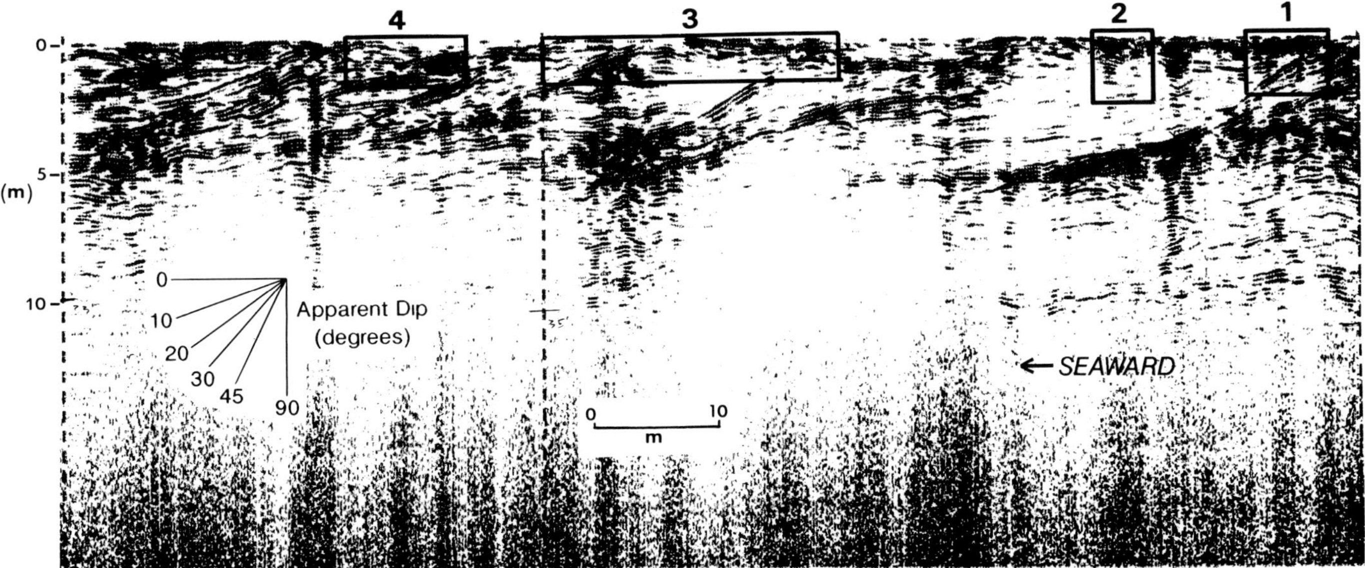

FIG. 8.—Ground-penetrating radar section A3 along east end of the barrier (see Fig. 4 for location). Dark sharp reflectors are gravel layers; the more transparent units are clean, well-sorted sands. Location of trenches shown in boxes 1-4.

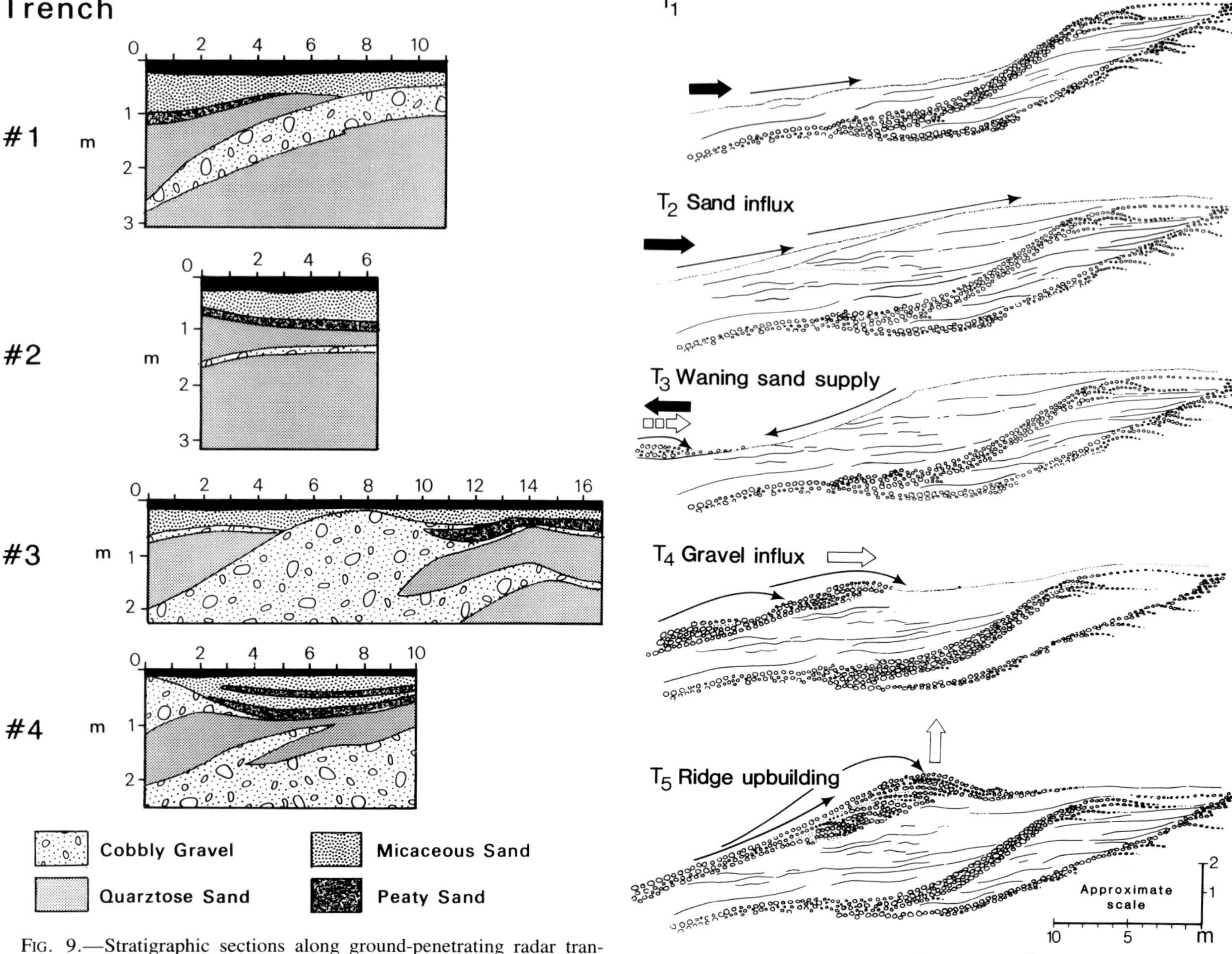

FIG. 9.—Stratigraphic sections along ground-penetrating radar transect A3 (from Ibrahim, 1986).

FIG. 10.—Model of sand deposition and gravel-ridge construction along eastern Horseneck Beach.

Model of Beach-Ridge Formation

The radar transects and trench data indicate that beach-ridge and barrier progradation in the eastern end of the barrier is dominated by discrete periods of mostly gravel accretion followed by sand deposition. A hypothetical sedimentation model has been constructed to explain the observed geometry and facies relations of the beach-ridge complex at Horseneck Beach in Figure 10. Stage T_1 depicts a period when a new supply of sand is being delivered to a mostly gravel beach. During stage T_2, the supply of sand is abundant and the beach experiences extensive progradation, resulting in horizontal to slightly seaward-dipping sand layers added to the front of the barrier. During this same period, overwash activity or dune formation in the rear of the beach covers the landward most part of the previously developed gravel ridge with sand.

At the beginning of stage T_3, the sand supply wanes and the beach erodes. At the end of this stage, the sand source has been exhausted, but the initiation of gravel transport to the beach prevents extensive erosion. During stage T_4, a sufficient volume of gravel is deposited along the beach to form a series of successively higher berms whose elevations are tied to various tidal heights and storm-surge levels. As the gravel ridge builds vertically (stage T_5), occasional wave overtopping produces a gravel pavement in the landward swale, which may be subsequently buried by aeolian sand deposition. The gravel layers that are deposited on the beach at this time are much steeper than the landward sand units formed during stages T_1 and T_2 (Fig. 8). This difference in stratification is due to differences in percolation properties of sand versus gravel. Gravel beaches maintain steeper equilibrium profiles than sand beaches, because a greater percentage of the uprushing wave swash percolates into the gravel beachface, thereby reducing the volume of the backwash.

In this model the transformation of the sand beach to a gravel beach is not the result of reworking of the sand facies and concentration of a gravel-lag deposit. If this were the case, gravel should be found throughout the sand facies, which it is not. Therefore, the gravel must be transported to the beach in a near absence of sand during discrete periods. The segregation of the coarse and fine components is the key to understanding the evolution of this beach system. A similar offlapping sequence of gravel and sand facies was reported by Bluck (1967) in a beach trench in South Wales.

Temporal Changes in Sediment Supply

To explain the progradation of the eastern end of Horseneck Beach requires an abundant long-term onshore and/ or longshore source of sediment. The stratigraphy of the site further requires that the type of sediment nourishing the barrier change periodically from sand to mostly gravel. Offshore of eastern Horseneck Beach is a drumloid called Gooseberry Neck, a feature believed to have been much larger prior to the Holocene transgression (Fig. 1). Gooseberry Neck is composed of boulders and smaller size gravel in a fine-grained matrix consisting of clay, silt and 50 percent sand. It is our hypothesis that this nearby source was the chief supply of sediment to the eastern end of Horseneck Beach (Fig. 11).

Whereas some sediment is eroded from Gooseberry Neck on a yearly basis by extra-tropical storms, the volume is likely quite small due to the presence of an extensive boulder retreat lag and a gravel beach, which serve to protect the bluff from direct wave attack (Fig. 2). Consequently, it is believed that the large-magnitude, infrequent storms (possibly 100-year or greater storm event) cause the most dramatic erosion of the neck. During these storms, large quantities of sand and gravel are released to the surf zone surrounding the peninsula (time 1, Fig. 11). After the passage of the hurricane, the storm deposits are reworked by lower energy waves and sediment is moved toward Horseneck Beach (time 2, Fig. 11).

Because sand is transported more easily than gravel under normal wave conditions (height <60 cm), it outdistances the gravel and is first to arrive at the barrier shoreline. This stage of the process produces a relatively smooth sedimentary surface (time 2, Fig. 11). In time, the storm deposit is depleted of sand and the supply of sand to the beach gradually diminishes. Eventually, the slower moving gravel is transported onshore, forming a cobble beach-and-ridge system (time 3, Fig. 11). This segregation of grain sizes, due to differences in transport behavior, explains the different sediment regimes and why barrier progradation at the eastern end of Horseneck Beach is dominated by sand deposition with distinct gravel accretionary events.

Observations of this sequence of events have not been made directly. However, a series of overflights and high-angle oblique photographs reveal gravel and cobble mantling the subtidal-sand surface at the extreme eastern end of Horseneck Beach and along the tombolo-like causeway to Gooseberry Neck island. The present beachface and incipient beach ridge along the front of the beach appear to show many of the features alluded to previously with ''dilute' gravel and cobble present on the upper beachface, along the berm crest, and in the indistinct (due to human tram-

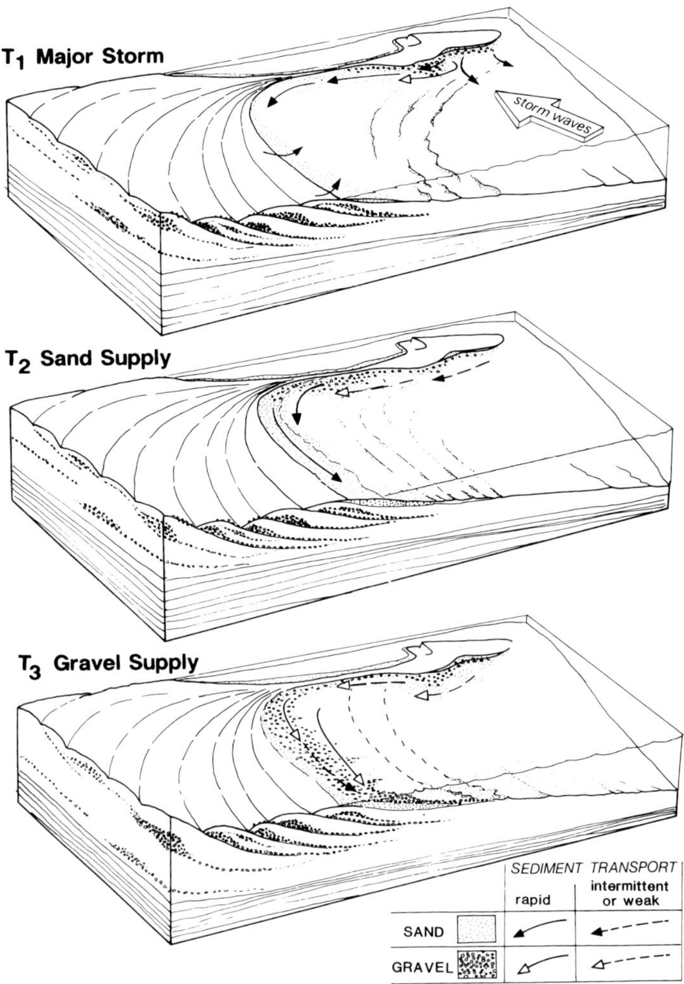

FIG. 11.—Sediment-dynamics model for sand versus gravel sedimentation and beach-ridge formation along eastern Horseneck Beach.

pling), landward-dipping surface of the back of the beach ridge. Whereas this particular ''beach ridge'' is neither dated nor documented, it is likely that it may related to the Blizzard of 1978, the storm of record for many parts of the Massachusetts coast (U.S. Army Corps of Engineers, 1979).

The exact geometry and facies relations of the accretionary wedges of Horseneck Beach are related to the frequency and magnitude of past storms. More extensive excavations of the eastern end of the barrier may reveal other types of deposits, including those that indicate the coincident arrival of both sand and gravel at the beach.

CONCLUSIONS

1. Beach-ridge barrier construction along the indented shoreline of northwestern Buzzards Bay is closely related to an abundant sediment supply consisting of glacio-fluvial deposits and sand-rich drumloidal deposits. Elsewhere along this section of coast, sand sources are minimal and barriers are thin and highly transgressive, or absent.

2. The Holocene evolution of Horseneck Beach began with estuarine filling (9 to 5 ka), followed by open-marine conditions, then a transgressive-barrier phase (5 to 2.8 ka), and ending with a regressive-beach-ridge-building period (2.8 ka to the present.
3. Ground-penetrating-radar records demonstrate that the eastern end of Horseneck Beach progrades through the addition of sigmoidal accretionary wedges. Stratigraphic data obtained from trenches further indicate that each accretionary unit represents a distinct cycle when the type of sediment nourishing the barrier changes from sand to gravel. During periods of abundant sand supply, the barrier progrades seaward and builds vertically by the addition of horizontal to slightly seaward-dipping sand layers. As the sand supply wanes, the foreshore and shoreface erode, and gravel dominates the supply of sediment to the beach. During this period, the beach steepens and gravel ridges are formed. Thus, each accretionary wedge consists of a laterally continuous, horizontal sand facies and a relatively thin, steeply seaward-dipping gravel facies. It should be emphasized that although the barrier has a bimodal grain size, it is dominated by sand.
4. The factors that control the type of sediment (sand, gravel, or sand and gravel) that nourishes the eastern end of Horseneck Beach, and the overall stratigraphy of the prograding beach-ridge complex are: a) the proximity of the drumloidal-sediment source and its wide range in grain size, b) frequency and magnitude of large storms that are capable of severely eroding Gooseberry Neck and releasing large volumes of sediment to the nearshore zone, and c) differences in the rate at which sand and gravel are transported in the surf zone, resulting in different arrival times to the eastern end of Horseneck Beach.

ACKNOWLEDGMENTS

This work was supported by grants and contracts from the Massachusetts Coastal Zone Management, the Towns of Westport and South Dartmouth, Massachusetts, and PETRONAS of Malaysia. The authors thank Rapi Muhammad Som and Azmi Mohd Yakzan for their assistance in collection of the trench data. Sytze van Heteren and Kenneth Finkelstein are acknowledged for their constructive, critical reviews of the manuscript. The figures in the paper were drafted by Eliza McClennen of Mapworks, Norton, Massachusetts.

REFERENCES

Belknap, D. F., and Kraft, J. C., 1985, Influence of antecedent geology on the stratigraphic preservation potential and evolution of Delaware's barrier systems: Marine Geology, v. 63, p. 235–262.

Bernard, H. A., LeBlanc, R. J., and Major, C. F., 1962, Recent and Pleistocene geology of southeast Texas, *in* Rainwater, E. H., and Zingula, R. P., eds., Geology of the Gulf Coast and Central Texas: Houston Geological Society, Field Trip Guidebook, p. 175–224.

Bluck, B. J., 1967, Sedimentation of beach gravels; examples from South Wales: Journal of Sedimentary Petrology, v. 37, 128–156.

Boyd, R., Bowen, A. J., and Hall, R. K., 1987, An evolutionary model for transgressive sedimentation on the eastern shore of Nova Scotia, *in* FitzGerald, D. M., and Rosen, P. S., eds., Glaciated Coasts: Academic Press, Inc., Boston, p. 88–113.

Curray, J. R., 1964, Transgressions and regressions, *in* Miller, R. L., ed., Papers in Marine Geology (Shepard Commemorative Volume): McMillan, New York, p. 175–203.

Dickinson, K. A., Beryhill, H. L., and Holmes, C. W., 1972, Criteria for recognizing ancient barrier coastlines, *in* Rigby, J. K., and Hamblin, W. K., eds., Recognition of Ancient Sedimentary Environments: Society of Economic Paleontologists and Mineralogists Special Publication 16, p. 192–214.

Dillon, W. P., and Oldale, R. N., 1978, Late Quaterary sea-level curve: reinterpretation based on glaciotectonic influence: Geology, v. 6, p. 56–60.

Elliot, T., 1986, Clastic shorelines, *in* Reading, H. G., ed., Sedimentary Environments and Facies: Blackwell, Oxford, p. 155–188.

FitzGerald, D. M., and Baldwin, C. T., 1986, Impulse radar as a new technique in the study of barrier island geology: Society of Economic Paleontologists and Mineralogists Coastal Research Group Meeting, Raleigh, North Carolina, 2 p.

FitzGerald, D. M., Baldwin, C. T., Ibrahim, N. A., and Sands, D. R., 1987, Development of the northwestern Buzzards Bay shoreline, Massachusetts, *in* FitzGerald, D. M., and Rosen, P. S., eds., Glaciated Coasts: Academic Press, Inc., Boston, p. 328–325.

Halsey, S. D., 1979, Nexus: new model of barrier island development, *in* Leatherman, S. P., ed., Barrier Islands from the Gulf Coast of Mexico: Academic Press, New York, p. 185–210.

Hayes, M. O., and Kana, T. W., 1976, Terrigeneous clastic depositional environments: Technical Report 11-CRD, Department of Geology, University of South Carolina, Columbia, 215 p.

Ibrahim, N. A., 1986, Sedimentological and morphological evolution of a coarse-grained regressive barrier beach, Horseneck Beach, MA: Unpublished M.S. Thesis, Boston University, Boston, 196 p.

Jelgersma, S., Oele, E., and Wiggers, A. J., 1979, Depositional history and coastal development in the Netherlands and the adjacent North Sea since the Eemian, *in* The Quaternary History of the North Sea: University of Uppsala Symposium, Uppsala, Sweden, p. 115–147.

Jensen, X. E., 1983, Atlantic Coast hindcasting shallow water significant wave information: Waterways Experiment Station Report #8, Vicksburg, Massachusetts, 75 p.

Kaye, C. A., 1964, Outline of Pleistocene geology of Martha's Vineyard, MA: U.S. Geological Survey Professional Paper 501-C, p. 134–139.

Larson, G. J., 1982, Nonsynchronous retreat of ice lobes from southeastern Massachusetts, *in* Larson, G. J., and Stone, B. D., eds., Late Wisconsinian of New England: Kendall/Hunt, Dubuque, Iowa, p. 101–115.

Leatherman, S. P., 1987, Coastal geomorphological application of ground penetrating radar: Journal of Coastal Research, v. 3, p. 397–399.

Magee, A. D., and FitzGerald, D. M., 1980, Investigation of the shoaling problems at Westport River Inlet and sedimentation processes at Horseneck and East Horseneck Beaches: Technical Report No. 3, Coastal Environmental Research Group, Department of Geology, Boston University, Boston, 118 p.

McMaster, R. L., 1984, Holocene stratigraphy and depositional history of the Narragansett Bay system, RI, USA: Sedimentology, v. 31, p. 777–792.

McMaster, R. L., and Asraf, A., 1973, Subbottom basement drainage system of inner continental shelf of southern New England: Geological Society of America Bulletin, v. 84, p. 187–190.

Melville, G., 1984, Headlands and offshore islands as dominant controlling factors during Late Quaternary barrier formation in Forster-Tuncurry area, New South Wales, Australia: Sedimentary Geology, v. 59, p. 243–271.

Morton, R. A., and Donaldson, A. C., 1973, Sediment distribution and evolution of tidal deltas along a tide-dominated shoreline, Wachapreague, VA: Sedimentary Geology, v. 10, p. 285–299.

Moslow, T. F., and Colquhoun, D. J., 1981, Influence of sea-level change on barrier island evolution: Oceanis, v. 7, p. 439–454.

National Ocean Survey, 1989, Tide Tables, East Coast of North and South America: U.S. Department of Commerce, Washington, D.C., 288 p.

Oldale, R. N., 1982, Pleistocene stratigraphy of Nantucket, Martha's Vineyard, the Elizabeth Islands and Cape Cod, Massachusetts, *in* Larson, G. J., and Stone, B. D., eds., Late Wisconsinian of New England: Kendall/Hunt, Dubuque, Iowa, p. 1–34.

OLDALE, R. N., AND O'HARA, C. J., 1980, New radiocarbon dates from the inner continental shelf of southeastern Massachusetts and a local sea-level curve for the past 12,000 years: Geology, v. 8, p. 102–106.

REDFIELD, A. C., AND RUBIN, M., 1962, The ages of salt marsh peat and its relation to recent changes in sea-level at Barnstable, MA: Proceedings, National Academy of Science, v. 48, p. 1728–1735.

ROY, P. S., THOM, B. G., AND WRIGHT, L. D., 1980, Holocene sequences on an embayed high energy coast: an evolutionary model: Sedimentary Geology, v. 26, p. 1–19.

STEELE, G. A., 1980, Stratigraphy and depositional history of Bogue Banks, North Carolina: Unpublished M.S. Thesis, Duke University, Durham, 201 p.

SWIFT, D. J. P., 1976, Coastal sedimentation, *in* Stanley, D. J. S., and Swift, D. J. P., eds., Marine Sediment Transport and Environmental Management: John Wiley & Sons, New York, p. 225–310.

STONE, B. D., AND PEPER, J. D., 1982, Topographic control of deglaciation of eastern Massachusetts, ice lobation and the marine incursion, *in* Larson, G. J., and Stone, B. D., eds., Late Wisconsinian of New England: Kendall/Hunt, Dubuque, Iowa p. 145–166.

THOMPSON, E. F., 1977, Wave climate at selected locations along U.S. coasts: U.S. Army Coastal Engineering Research Center, Technical Report No. 77-1, 364 p.

U.S. Army Corps of Engineers, 1979, Blizzard of 1979, Coastal Storm and Damage Study: New England Division, Waltham, Massachusetts, 135 p.

VAN STRAATEN, L. M. J. U., 1965, Coastal barrier deposition in south- and north-Holland, in particular in the areas of Scheveningen and Ijlmuiden: Mededelingen Van de Geologische Stichting, v. 17, p. 41–75.

WILKINSON, B. H., 1975, Matagora Island, Texas: the evolution of a Gulf Coast barrier complex: Geological Society of America Bulletin, v. 86, p. 959–967.

WILLEY, R. E., WILLIAMS, J. R., AND TASKER, G. D., 1983, Hydraulic data of the coastal drainage basins of southeastern Massachusetts, Narragansett Bay, and Rhode Island Sound: Massachusetts Hydrologic Report No. 25, U.S. Geological Survey, Boston, 42 p.

A RATIONAL THEORY FOR BARRIER-LAGOON DEVELOPMENT

GEORGE F. OERTEL
Department of Oceanography, Old Dominion University, Norfolk, Virginia, 23529
J. CHRIS KRAFT
Department of Geology, The University of Delaware, Newark, Delaware, 19716
MICHAEL S. KEARNEY
Department of Geography, The University of Maryland, College Park, Maryland
AND
H. J. WOO
Department of Oceanography, Old Dominion University, Norfolk, Virginia, 23529

ABSTRACT: The development of coastal-barrier lagoons is strongly governed by the primordial character of the lagoon floor, sea-level fluctuation and sediment input. During transgression, rising sea level affects lagoon capacity in several ways. When boundary conditions are laterally stable, flooding of the lagoon increases the capacity of the basin for sediment storage. However, shorelines associated with the mainland and shoreface are quite mobile during transgression. Relative movements of the inner and outer shorelines of a lagoon also affect lagoon capacity.

At wave-dominated coasts, the floors of barrier lagoons are initially smooth after lagoon formation, and basin infilling is dependent on changes in capacity relative to sea-level rise and sediment input from siltation, runoff, inlets and cross-island transport. However, lagoons are displaced laterally long before basins are filled. As a result, basin filling by upbuilding is rarely complete and open-water lagoons tend to stay open.

At tide-dominated coasts, the lagoon floors are irregular and reflect the antecedent topography of pre-transgressed landscape. The interfluve areas of the topography produced very shallow areas in the lagoon that form tidal flats or are colonized by marshes. Along many sections of coast, sediment supply is insufficient to keep pace with the rate of increasing lagoon capacity. Consequently, many of the lagoons along the middle Atlantic, Gulf and west coast of North America may have initiated as marsh lagoons and have evolved into open-water lagoons.

Thus, marsh lagoons (particularly in low-sediment-input areas of the middle Atlantic States) are not the climax stages of a long upfilling sequence, but are the initial stages of inundation in which marsh colonization occurs over shallow topographic surfaces.

INTRODUCTION

A new theory of barrier-lagoon evolution is proposed for transgressive coastal margins. Previous models of barrier-lagoon evolution do not stand up under morphostratigraphic scrutiny. Viable models of barrier-lagoon evolution must consider the morphostratigraphic influences of all six elements (the barrier island, the barrier lagoon, the mainland, the inlet and deltas, the barrier platform, and the barrier shoreface) of the barrier-island system (Oertel, 1985), as well as the characteristics of each of the barrier-lagoon sub-environments.

Barrier lagoons are wet basins formed by barriers separating the sea from the land. Early work on barrier lagoons was done in carbonate environments of the tropical seas around volcanic islands, where lagoons form behind barrier reefs (Darwin, 1842; Dana, 1885). However, along clastic coasts, barrier lagoons form behind barrier islands, barrier spits and bay-mouth barriers. Whereas the environments of deposition associated with carbonate and siliciclastic lagoons are dissimilar, they are both "true" barrier lagoons in that they separate the inner and outer shorelines of a dual-shoreline coast. Along siliciclastic coasts, barrier lagoons occur in two major landscape groups that are related to the *origin* of barrier islands (Gilbert, 1885; McGee, 1890). The relation of wave and tidal regime to barrier-island development (Hayes, 1979) also has an important influence on barrier-lagoon evolution. At wave-dominated coasts, the barrier lagoons are *initially* relatively deep, open-water features. At tide-dominated coasts, barrier lagoons have complicated bathymetry produced by extensive channels, tidal flats and marshes.

Since barrier lagoons are sedimentary basins, paradigms of basin stratigraphy for epeiric seas are often invoked for describing lagoonal basin fill and evolution. This is not universally appropriate, because many stratigraphic basins described from geologic records are not embayed by barriers. Stratigraphic basins are generally larger and exhibit more stable boundary conditions than their smaller counterparts along barrier coasts. More importantly, the evolution of barrier lagoons as transitional zones between the sea and land is influenced by characteristics of both environments.

Hypotheses for lagoon evolution during transgression must incorporate inundation of topography and the related hypsometric conditions that control the distribution of lagoon environments and lagoon evolution. The following discussion of barrier-lagoon evolution is primarily for clastic coasts and considers the relative importance of primary landscape features and secondary wave processes.

BARRIER-LAGOON EVOLUTION

Lagoon evolution is strongly influenced by the formation of barrier islands and lagoons. The two relevant concepts of lagoon formation are by (1) shoreface embayment by spit migration (Gilbert, 1885; 1890) or possibly by bar emergence (deBeaumont, 1845), and (2) coastal inundation by sea-level rise (McGee, 1890; Hoyt, 1967). The two different modes of formation start the evolution process with distinctly different lagoon floors. Halsey (1978, 1979) clearly demonstrated the importance of inundation and landscape topography on barrier-island formation. However, inundation and landscape topography also have strong influences on barrier-lagoon evolution. The primordial floors of barrier lagoons formed by coastal inundation (McGee, 1890) are shaped by the antecedent topography of the terrestrial landscape prior to inundation (Fig. 1). Thus, fringe marshes,

Quaternary Coasts of the United States: Marine and Lacustrine Systems, SEPM Special Publication No. 48
 ISBN 0-918985-98-6

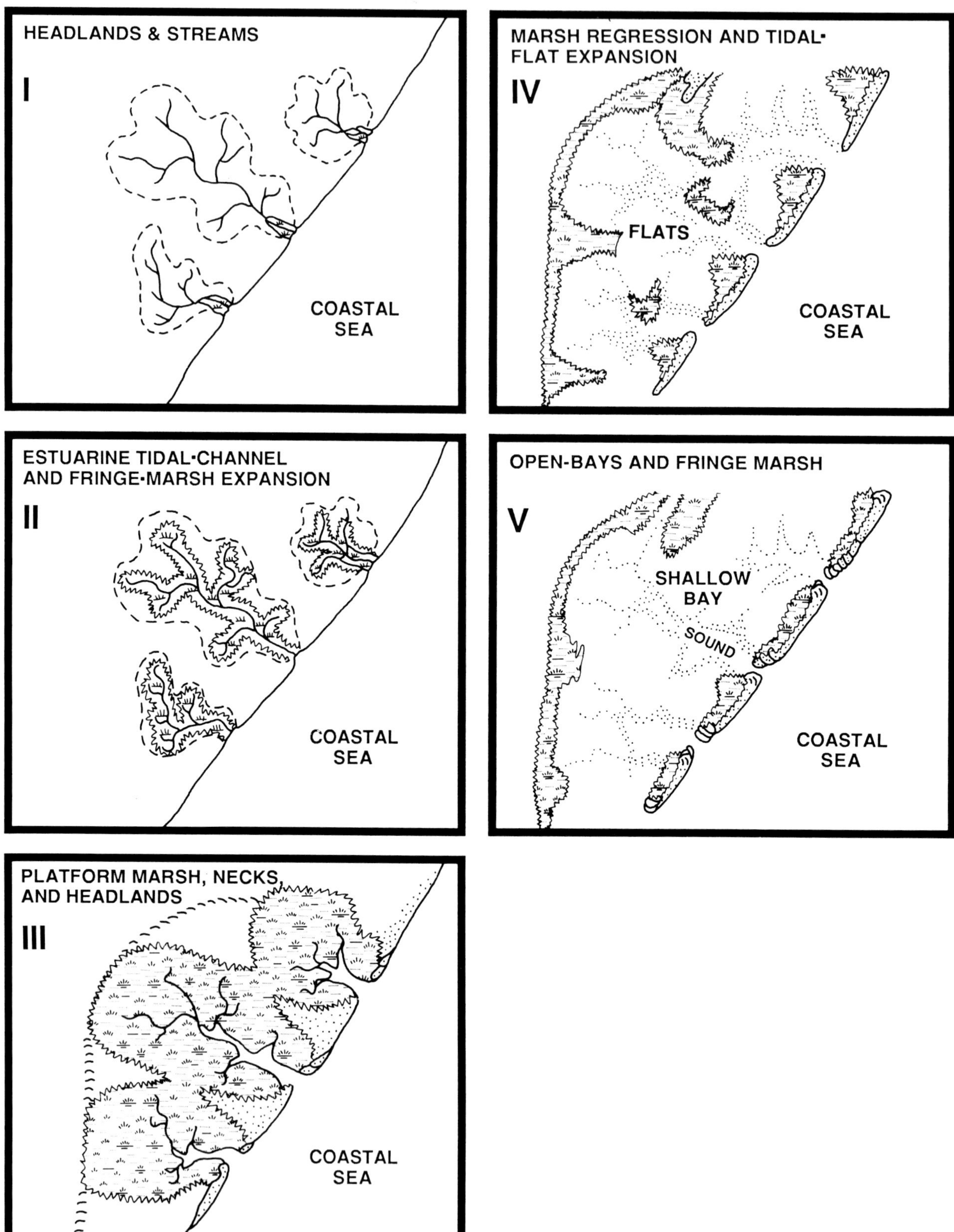

FIG. 1.—Landscape topographic model of barrier-lagoon development.

tidal flats and other environments controlled by elevation are generally not in shore-parallel zones, since they are related to drainage and landscape maturity. Fringe marshes occur at *all* suitable elevations between mean sea level and highest high water, including the margins of channels, interfluve areas and accretional banks. The extent of coverage is enhanced by elevated tidal ranges and low topographic slope. During transgression, the principal changes in the distribution of marshes and tidal flats are related to inundation of complex drainage surfaces rather than to the upbuilding of the lagoon floor. The drainage patterns of the antecedent landscape are enhanced and deepened by tidal currents. Deep tidal channels and inlets enhance repletion of the lagoon and further reinforce tidal dominance throughout the barrier-island system. Deep channels also restrict inlet jets from spreading into the lagoon. This prevents the formation of *net tidal-delta-flow* (NTD-flow) patterns and inhibits the formation of flood-tidal deltas.

Although lagoonal environments are determined by hypsometric conditions produced by both sedimentation and inundation of complicated topography, inundation appears to play a predominant role at many modern barrier lagoons. In the *landscape topographic model*, the initial inundation of the landscape occurs at stream valleys and, consequently, the deepest parts of the lagoons are in the thalwegs (not immediately behind barriers). During this stage, barrier lagoons are primarily confined to the margins of estuaries (Fig. 1, stage II). Marshes and tidal flats form along the relatively steep slopes of estuarine valleys. Clear examples of this stage are present in the cliff setting along the northern and central California coast, although similar examples are present in regions of more subtle topography (for example, the west side of the Delmarva Peninsula, Virginia). In this relatively early stage of development, coastal lagoons may not be continuous behind the barriers, but they may be separated from adjacent lagoons by terrestrial interfluves or "necks."

Continued sea-level rise causes fringe marshes and tidal flats to spread upward and laterally onto flatter and broader interfluve areas. In this stage, the tidal-channel network is inherited from the submerged terrestrial drainage and the distribution of platform marsh is adopted from the pattern of low-order interfluve areas within drainage basins (Fig. 1, stage III). At coastal reaches where sediment supply is limited, tidal channels broaden, forming sounds and subaqueous flats, and isolating marsh islands (Fig. 1, stage IV). Continued elevation of the water results in the total submergence of platform marsh, forming subtidal flats and shallow bays (Fig. 1, stage V). However, with sufficient sediment input, upward growth of the platform-marsh facies may keep pace with sea-level rise. The maintenance of platform marsh during sea-level rise is most common at deltas of major rivers such as the Mississippi or Savannah Rivers, although the effect of high tidal ranges along the southeast coast of the United States has enhanced the maintenance of extensive platform marshes in the tide-dominated barrier lagoons of South Carolina and Georgia. Whereas these marshes are rarely preserved, they may be transgressed by barriers and partially preserved as a continuous peat layer (not to be confused with a basal peat of an open-water basin-fill sequence). The tide-dominated coastal lagoons of the southern Delmarva Peninsula are more typical of coastal settings in a transgressive state with moderate- to low-sediment supply.

Previous models of barrier-lagoon evolution based on the work of Lucke (1934) are most appropriate for lagoon formation by shoreface embayment (Gilbert, 1885; 1890; de Beaumont, 1845). In the Lucke (1934) model, open-water lagoons evolve into marsh lagoons via basin fill. Basin fill is suggested to be associated with three main processes. Marshes prograde toward the center of the lagoon from the margins, marshes colonize expanding flood deltas, and the lagoon floor builds upward by siltation of suspended sediment (Fig. 2). The expansion of marsh islands results in the formation of tidal channels. Several tenuous assumptions are inherent in the Lucke (1934) model. First, rising sea level had to have had no impacts on the margins of the basin and basin capacity. Second, the initial basin had to

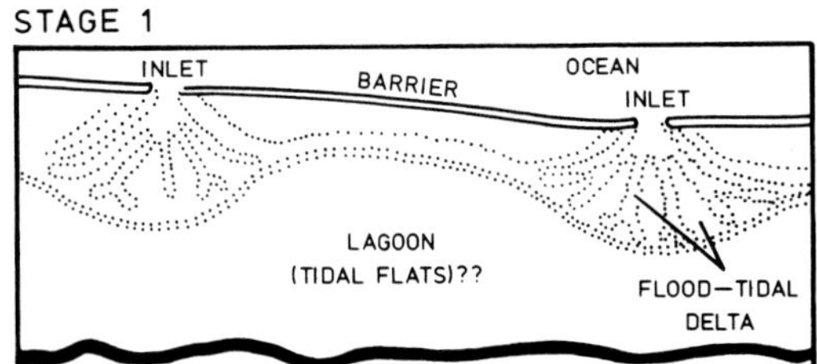

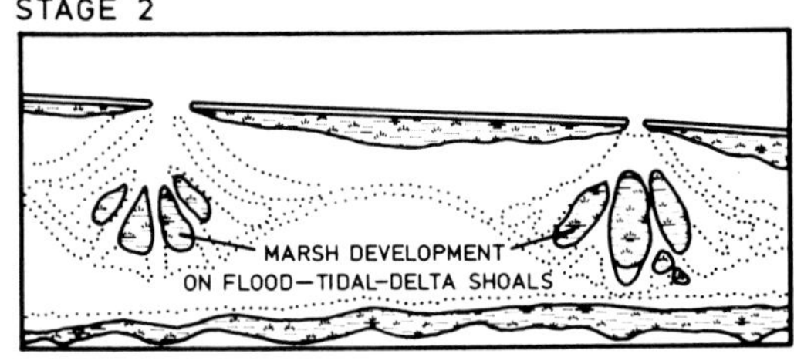

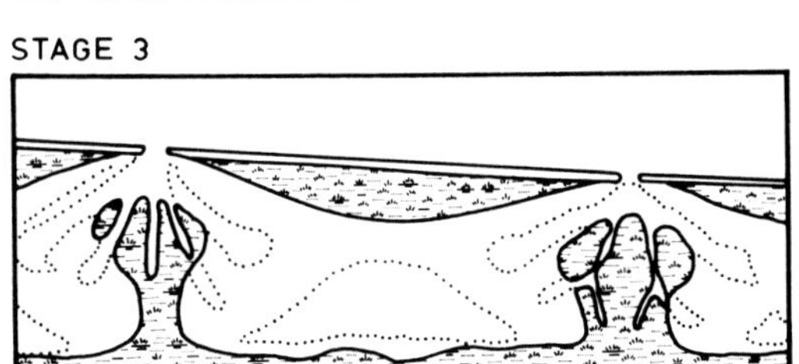

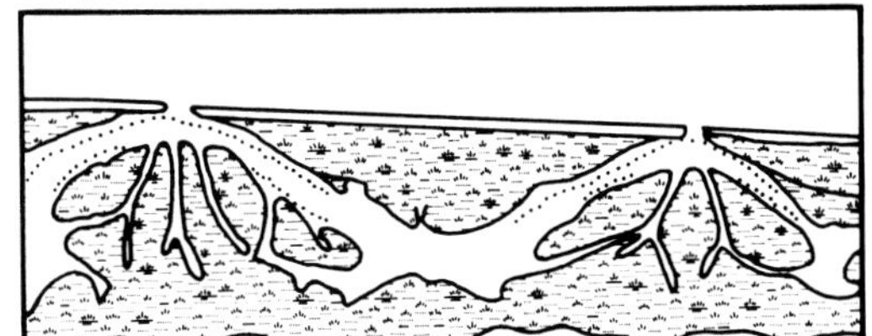

FIG. 2.—Schematic maps of a barrier lagoon showing stages of lagoon filling (modified after Lucke, 1934; and Oertel, 1987). This model is generally not applicable for most transgressive lagoons, because the irregular topography of the transgressed surface has a greater influence on hypsometric evolution than the flows of the inlet-flood field.

have had a deep, smooth floor where sediment was accumulating. Third, the initial basin had to have been open, causing a "flood field" to form a large flood delta. Since the lagoons were believed to be filling from an open-water state, the tidal flats and marsh islands were attributed to accretional patterns related to inlet flow.

Lagoons with smooth floors and large flood deltas are most likely to occur along wave-dominated coasts, where migrating spits and inlets (Hoyt and Henry, 1967; Kumar and Sanders, 1974; Fisher, 1967; 1968) have recently embayed a section of the shoreface. However, the rapid shoreline transgression during the Holocene may make these Lucke-based models valid only for small sections of coast, since sedimentation has not been able to keep pace with sea-level rise and rising-water levels have generally caused barrier lagoons to evolve from marsh lagoons with complicated drainage to open-water lagoons. In a rising-sea state, the initial Lucke-based models may become obscured with shoreface erosion and continued landward transgression of the barrier system.

Morphostratigraphic Evidence: Delmarva Peninsula

The regional dip of the southern Delmarva Peninsula is about 0.04°, the scarps forming the landward shoreline of the lagoon dip at about 1 to 1.5°, and the shoreface slope is about 0.1°. If the scarp were to continue under the entire 12.5 km of the lagoon floor (the width of the lagoon behind Hog and Cobb Islands, Virginia), then the primordial lagoon floor would have to be about 220 m deep adjacent to the barrier islands. This is clearly unrealistic and suggests that regional coastal-plain slopes (as illustrated by Leatherman, 1979a and Dolan and Lins, 1987) may be more appropriate for projecting the surface of the primordial lagoonal floors from the mainland fringe (Fig. 3). Leatherman (1979a) and Dolan and Lins (1987) depict a barrier lagoon as sliding up a smooth coastal-plain slope during transgression. However, if the primordial floor of the Hog/Cobb coastal lagoon had a slope of 0.04 to 0.05° (the average slope of the local Atlantic Coastal Plain), then the initial floor of the lagoon would have been about 10 m deep adjacent to the barrier islands (Fig. 4). Since the water depth of the lagoon is about 2 to 3 m, then this scenario suggests that about 7.5 m of sediment accumulated over the primordial lagoon floor. This is about 1.0×10^9 m^3 of sediment that would have to have been derived from the margins of the Cobb/Hog basin. Assuming that the drainage basins of the adjacent coastal plain were the primary sources of fill, then about 12.5 m of sediment would have to have been stripped off the surfaces of these basins. Since there are no major drainage basins draining into these lagoons, we feel that there is insufficient time (probably less than 3,000 years) to accomplish this magnitude of erosion from the upland.

Whereas stratigraphic studies in the lagoons of the southern Delmarva Peninsula (Newman and Munsart, 1968; Finkelstein and Ferland, 1987; Oertel and others, 1989; van de Plassche, 1990) have verified the presence of relatively thick (up to 12 m) fine-grained sequences, quantitative chronologic data for recognizing pre-transgressed boundaries in or below the muds have been limited. Oertel and others (1989) found that the thickest Holocene mud deposits occur in channel fills in the central parts of the Cobb Bay lagoon and not farthest from the mainland shore (Fig. 4). Van de Plassche (1990) found 4.4-ka peats 8 m deep on a steeply dipping (1.2°) surface relatively close to the mainland side of the lagoon. This was probably a fringe marsh on a pre-Holocene shoreface or on the margins of a buried channel. In other areas of the Cobb Bay lagoon where mud is thick, it was determined that only the upper 1 to 3 m of mud is Holocene. South of this region (between Wreck Island and Smith Island, Virginia), Pleistocene sediments occur within 2.5 m of the lagoon floor (Finkelstein and Kearney, 1988). Pollen assemblages dominated by boreal-forest species, such as spruce, dwarf birch and fir, corroborate a pre-Holocene age for mud below depths of 3 to 4 m. Stanley R. Riggs (pers. commun., 1991) noted similar findings for the Albermarle and Pamlico Sound lagoons of North Carolina. It is apparent that the primordial floors of these lagoons are not seaward-dipping planar surfaces that connect the coastal plain and the shoreface. In fact, this type of surface may only be present at lagoons formed when a section of the shoreface is engulfed by a spit (as described by Gilbert, 1885, p. 87–88). However, buried records of marsh from the highly dynamic barriers of the Atlantic, Gulf, and west coast of North America are probably quite patchy because of uneven antecedent topography.

In general, the concept that basal peats slide up a smooth coastal-plain ramp is not supported by records from the middle Atlantic region. Organic-rich sediment layers are not only formed by mainland fringe marshes. Peats or traces of buried marsh may be produced at back-barrier-fringe marshes, channel-fringe marshes, hammock marshes, marsh islands and platform marshes. The presence of different types of marshes buried in the sediment is suggested by several studies from the Delmarva Peninsula. Finkelstein and Ferland (1987) found progressively younger and seaward peats in the Metompkin Bay lagoon (dates progressed from 2.2 ka at 200 m from the mainland, to 1.66 ka about 1 km from the mainland, to 4.62 ka about 1.6 km from the mainland and 1.18 ka at 2.8 km). They attributed this difference to two different marsh types, a basal peat attached to the mainland (4.62 to 2.2 ka) and a modern platform marsh ($\approx$1.7 ka to present). Thus, the classical expectation of a continuous layer of basal peat in a transgressive sequence should be reconsidered (see discussion in Finkelstein and Kearney, 1989).

On the northern Delmarva Peninsula, Kraft and his colleages (Kraft, 1971; Kraft and others, 1979; Belknap and Kraft, 1985; Chrzastowski, 1986) also found that buried marsh deposits are discontinuous, with more than one marsh horizon likely at any site. Buried marsh muds below Rehoboth Bay were associated with the fringe of buried channels and back-barrier environments (Fig. 5). During transgression, these channel-fringe marshes migrated up the channel walls. At channels oriented parallel to the modern shoreline, a fringe marsh on the seaward side of the channel will migrate in a seaward direction. In these cases, younger peat dates were found seaward of older peat dates.

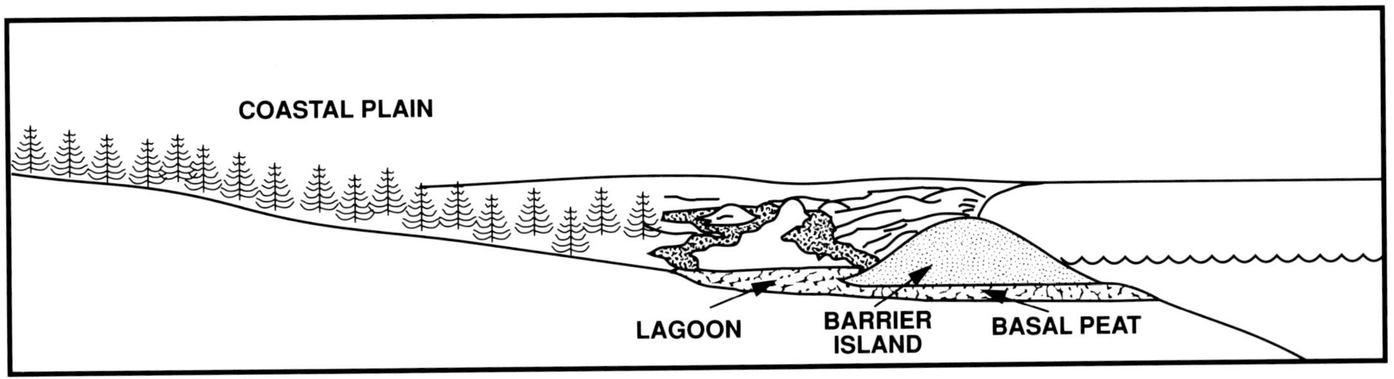

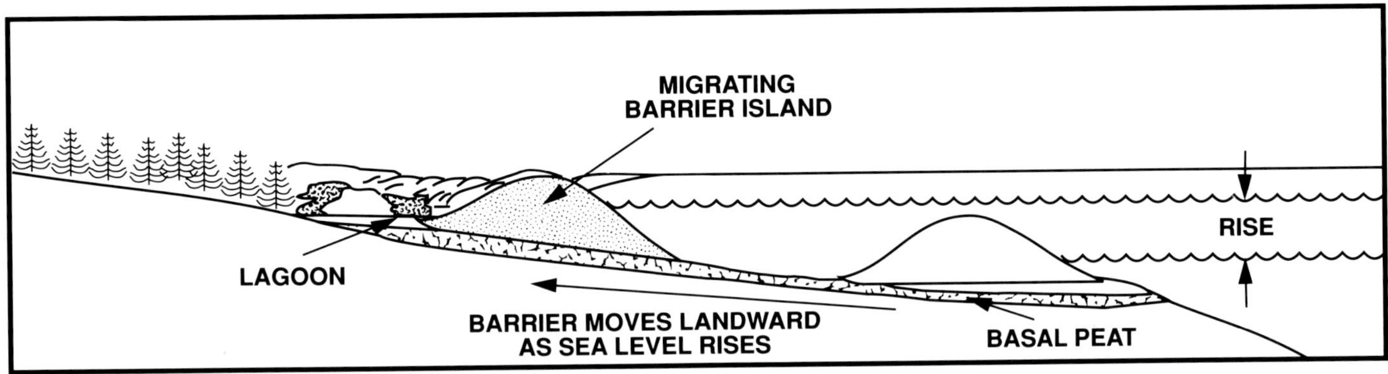

FIG. 3.—Diagrams of a model for barrier-lagoon transgression (modified after Leatherman, 1979a; and Dolan and Lins, 1987) showing the development and transgression of a basal-peat horizon (synthesized from sections from Psuty, 1980; Finkelstein, 1986; Finkelstein and Ferland, 1987; Byrnes, 1988). Stratigraphic evidence from the middle Atlantic coast does not verify a smooth basal surface and a laterally continuous peat horizon.

Evidence from the Landscape

Along tide-dominated coasts (e.g., Georgia, South Carolina, and the southern Delmarva Peninsula of Virginia), tidal currents have a strong influence on morphodynamics. Lagoons along these reaches have well-developed channel systems incised in the lagoon floors. The shallow regions between channels have tidal flats and marshes. The lagoons drain through large tidal inlets that are separated by short barrier islands. Draining tidal flow of lagoonal water to the inlet is strongly channelized. Ebbing-inlet jets overwhelm longshore currents and divert them many kilometers offshore (Oertel, 1988), forming large ebb-tidal deltas (Nummedal and others, 1977).

The large inlets and deep channels permit a very efficient exchange of the lagoonal-water mass with the sea, and tidal-phase lags are generally small. This phenomenon has two important impacts on lagoonal hydraulics. First, tidal repletion is high and tidal ranges at distal parts of the lagoon approach those of ocean tides. Since flow is primarily confined to channels, strong tidal currents inhibit sedimentation and often produce tidal scour in sections of the channel floor. Sediment scoured from the floors of inlet channels may also contribute to the size of the ebb deltas. Secondly, flood jets cannot spread into open embayments because flow is confined in tidal channels. As a result, NTD-flow patterns (Oertel, 1988) do not develop and flood-tidal deltas are often diminutive or missing. Regardless of whether tide-dominated barrier lagoons have large open-water areas or are marsh-choked, channelized tidal currents are prevalent throughout the system. These tide-dominated lagoonal processes are in contrast to the model suggested by Lucke (1934).

The distribution of channels, tidal flats and marsh islands in open-water lagoons also suggests a different scenario than simple infilling of an open-water lagoon. In the Hog Island lagoon (Virginia), marsh islands appear to be the last vestiges of channel-fringe marshes that are being submerged by rising sea level. The linear trend of arcuate marsh islands follows the meander pattern of a former coastal-plain stream on the floor of the lagoon. The channel and its fringe marsh are being submerged as the final stage of lagoon flooding is approached (Fig. 6). In the Rehoboth Bay lagoon, Delaware, Piney Island is a small marsh island in the middle of the bay. However, plats show an island with over 15 acres of upland existed less than 50 years ago. The distribution of many marsh environments in coastal lagoons appears to be related to the isolation of hammocks and channel-fringe marshes by rising water level. Thus, inundation, not shoaling, appears to be the main process producing the distribution pattern of tidal channels, marshes and tidal flats.

LAGOON CAPACITY AND FILL RATES

The capacity of barrier lagoons is primarily affected by sedimentation and boundary dynamics. Coarse-grained sediments are primarily transported from the lagoon bound-

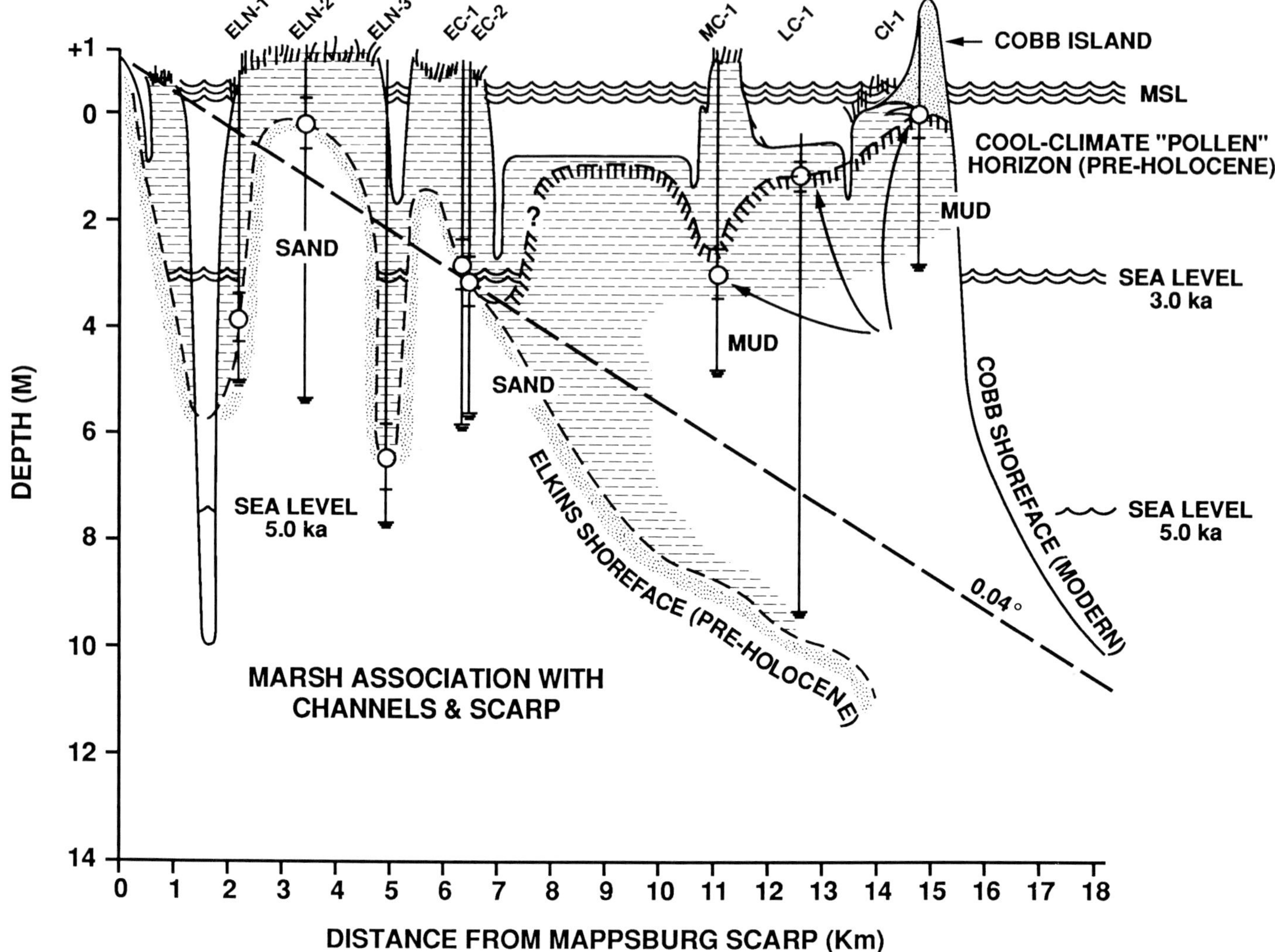

FIG. 4.—Section across Cobb Island barrier lagoon showing generally shallow nature of the pre-Holocene contact, except at the margins of deep channel fill. Sand facies surface may illustrate sections of both basal and ravinement unconformities. Dashed cool-climate pollen horizon is the approximate boundary between Holocene and pre-Holocene sediments.

aries, whereas fine-grained sediments settle from suspension and cause siltation and shoaling on the lagoon floor.

Along wave-dominated coasts, flood deltas provide a significant source of lagoon fill. Rapid inlet and island migration at open-water lagoons strands flood-tidal deltas along the back-barrier side of the migrating barrier islands. Marsh-covered remnants of flood deltas are present along many wave-dominated regions of the mid-Atlantic and Gulf coasts (Pierce, 1970; McGowen and Scott, 1975; Leatherman, 1979b). Marsh-covered remnants of flood deltas are relatively obvious subaerial features; however, major portions of inactive flood-tidal deltas are subaqueous (Kraft, 1971; Kraft and others, 1979; Kraft and others, 1987; Oertel and Kraft, in prep.). Sands from the relict flood deltas are often spread far onto the lagoon floor where they are subsequently mixed with muddy lagoon sediments (Fig. 7).

However, the main contribution to lagoonal filling may come from cross-island processes. *Subaqueous storm-surge platforms* (Boothroyd, and others, 1985) also form along the back-barrier margins of lagoons. During large storms, super elevation of water completely submerges portions of barrier islands, allowing wave bores to pump water and sand directly onto the lagoon floor (McGowen and Scott, 1975; Leatherman, 1979b). Cross-barrier flow persists only for a short period during storms, and often storm cuts and ephemeral channels become inactive when water levels return to normal. If the cut is sufficiently large, a significant amount of the lagoon water may be captured and drained into the sea. Subsequent tidal exchanges may produce a ''true'' tidal inlet with NTD flow (Oertel, 1988). However, most cuts heal rapidly following storms, and flood jets and ebb drains never develop. During relatively small storms, high-water level only partially submerges barrier islands. Nonetheless, swash may reach the island crest and flow down the back barrier into the lagoon. The resultant *washover fans* are primarily subaerial, although distal parts of

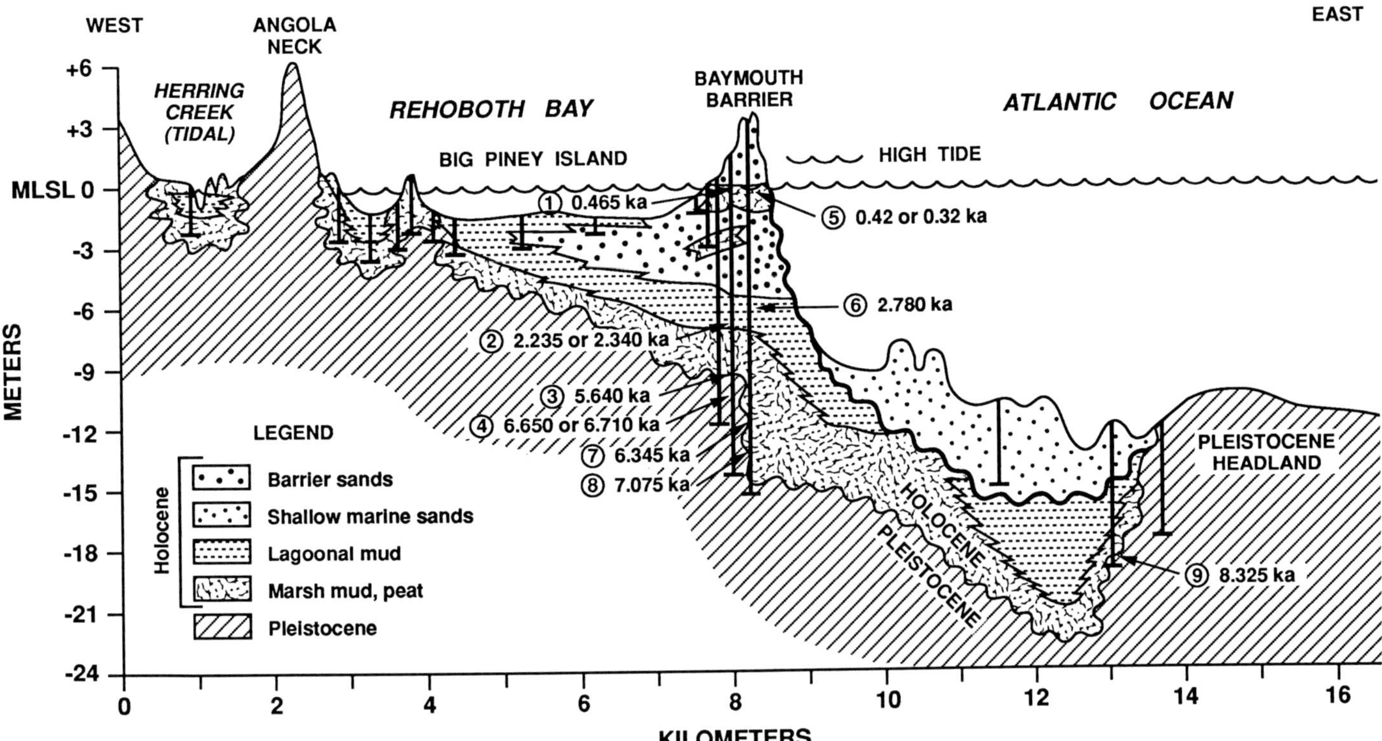

FIG. 5.—Cross section of Rehoboth Bay lagoon showing facies of mainland-fringe marsh, channel-fringe marsh and back-barrier-fringe marsh (modified after Kraft, and others, 1987).

these sand bodies may prograde into the lagoon forming scalloped-shaped back-barrier shorelines with subaqueous deltas.

Upbuilding of the basin floor by siltation is strongly influenced by fluvial sources, which contribute to the fine-grained sediment load and the eventual flux of particulates to the bed. Since basins with stable boundaries have finite capacities, they will eventually fill. In the Lucke model (Lucke, 1934), when the lagoon floor reaches mean sea level, fringe marshes spread laterally across the lagoon, transforming the basin from an open-water lagoon to a marsh-choked lagoon. The concept of open-water lagoons filling to marsh lagoons has been applied in several parts of the southern Delmarva Peninsula by Finkelstein and Ferland (1987) and Byrnes (1988). They suggested that muddy beds of infilling were continuous across the lagoon in nearly horizontal units (Fig. 8). Whereas the lithic units are often continuous across the lagoon, the Holocene portion of the muds are generally discontinuous because of irregular antecedent topography.

Nichols (1989) addressed the problem of stable boundary conditions in the Lucke model and illustrated that the capacity of a coastal lagoon is not constant but is affected by changes in relative sea level during transgressions and regressions. The term "deficit" was used to describe coastal lagoons where the capacity exceeded sediment supply. In his model, a basin could only be filled during "surplus" conditions, when the supply of sediment exceeded the capacity of the lagoon.

The open-water lagoons of the Atlantic and Gulf coasts of North America appear to be "deficit lagoons" and surfaces suitable for marsh initiation are only located along the lagoon margins. During Holocene sea-level rise, mainland fringe marshes expanded landward as rising water submerged the mainland margin. However, the distal edge of the fringe marsh sank into the lagoon as marsh accretion was unable to keep pace with rising sea level. This process slowly increased lagoon capacity as the central bays expanded at the expense of migrating fringe marsh (Fig. 9). Therefore, as suggested by Nichols (1989) and illustrated by Fletcher and others (1990), in estuarine lagoons, infilling is reversible depending on the balance between sediment supply and basin capacity.

In general, Holocene sediments in many lagoons along the middle Atlantic coast are thin, and thick accumulations are only associated with valley fills. The topography of the primordial floors of lagoons is rarely smooth with planar characteristics of an embayed shoreface. More commonly, the floors of barrier lagoons have the irregular topography of submerging coastal-plain drainage systems. The patchy distribution of buried marsh facies in the lagoon sediment column supports the suggestion that a large variety of marsh types was separated by irregular topography. Irregular topography and dynamic boundaries (causing state changes) directly conflict with the paradigms of classic basin-infilling models that would produce layer-cake stratigraphic sequences.

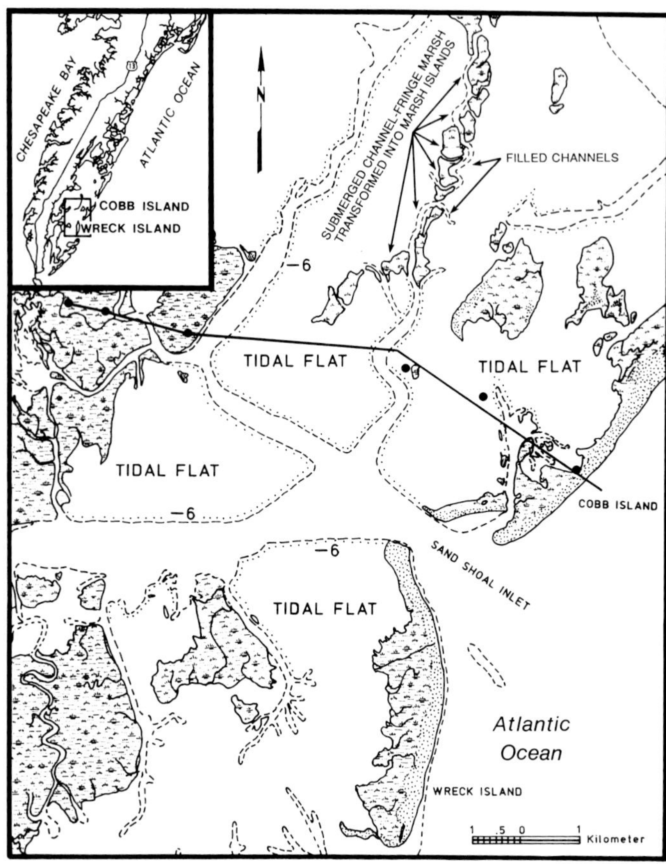

FIG. 6.—Map of a section of the Cobb Bay lagoon illustrating remnants of channel-fringe marsh that have been transformed into a chain of arcuate marsh islands in response to sea-level rise.

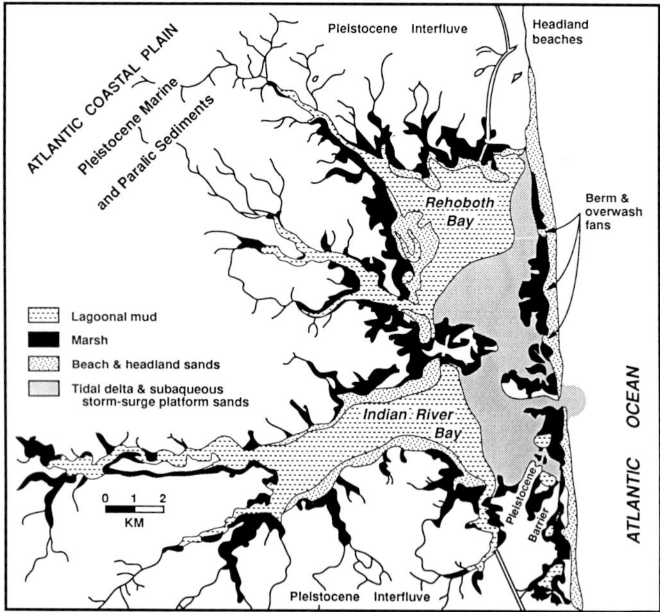

FIG. 7.—Map of Rehoboth/Indian River lagoon showing distribution of sandy sediments from cross-island-transport processes.

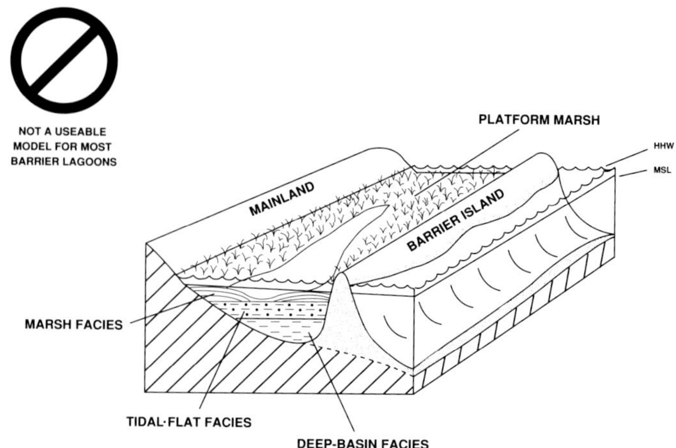

FIG. 8.—Block diagram illustrating a scenario for platform-marsh development by the infilling of an initially deep lagoonal basin to mean sea level. This model is generally not applicable for transgressive-barrier lagoons, because islands and lagoons migrate over the shallow irregular topography before deep basins can be filled by siltation processes.

A *fundamental element that has been missing from previous basin-infilling models is the effect that relative sea-level rise has on topography during rapid transgression. Inundation and related hypsometry are major factors controlling the distribution of lagoonal environments and lagoonal evolution.*

TIME AND LAGOON EVOLUTION

A major consideration for models of infilling barrier lagoons is the time scale with respect to sedimentation rate and sea-level rise. A useful paradigm developed by stratigraphers for sedimentation in large intercontinental basins or large gulfs (e.g., the Gulf of Mexico, the Mediterranean Sea, the Red Sea, and so forth) is that deep basins fill upward forming horizontal bedding. This layer-cake concept suggests that similar facies that are sampled from similar depths may be correlative. Basin infilling at these large scales is a relatively slow process involving 10^3 to 10^4 years. Regional subsidence at marginal or intercontinental basins may allow the filling process to continue for much greater than 10^4 years, producing sedimentary sections tens of meters in thickness. By comparison, Holocene coastal lagoons are generally quite small and dynamic. Boundary conditions at coastal lagoons are constantly changing in response to the Holocene transgression. Some sections of lagoon fill along the mid-Atlantic coast are believed to be on the order of 10 m thick (Shideler and others, 1984; Finklelstein and Ferland, 1987). However, at average lagoonal sedimentation rates of 0.1 to 2.0 mm/yr (Oertel and others, 1989), it would take 10^3 years to produce beds up to 2.0 m thick. Even during the relatively slow transgression rates of the late Holocene, many of the barrier lagoons of the middle Atlantic region have been displaced landward by the complete width of the lagoon in 10^3 years. The lateral shifts over terrestrial topography produce rapid changes in the shape

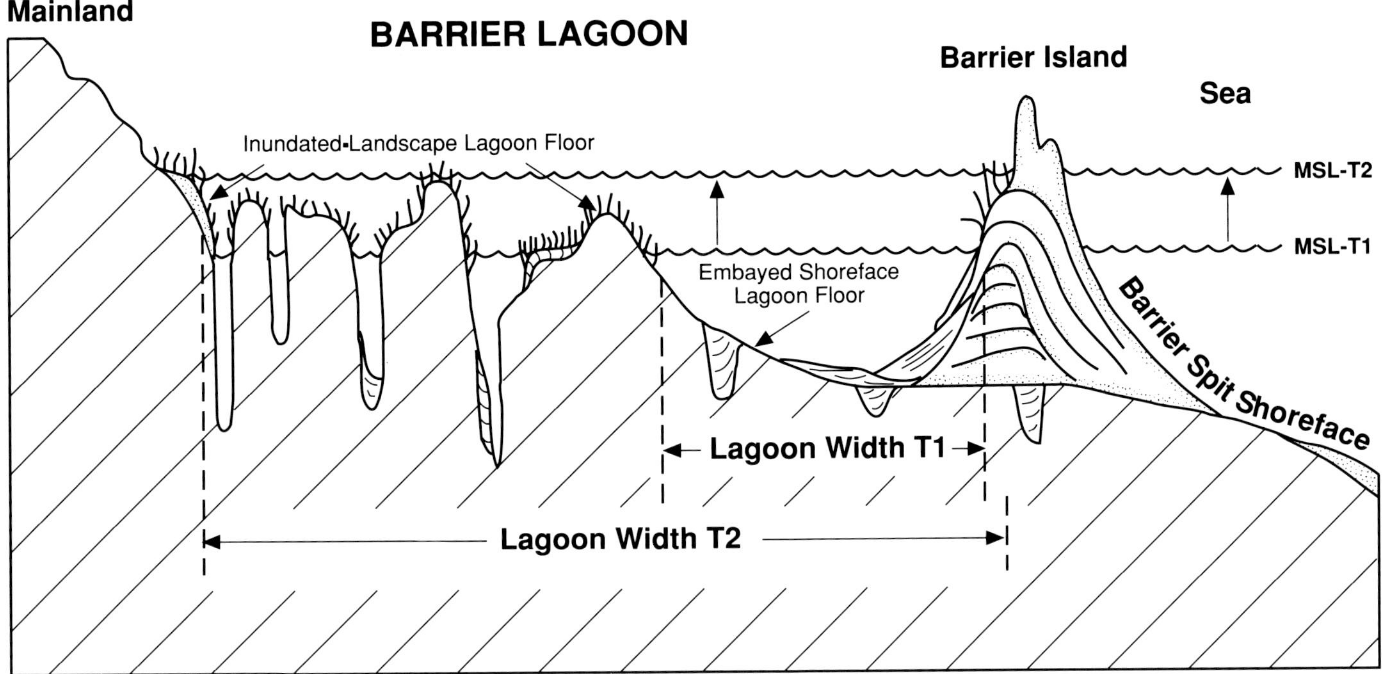

FIG. 9.—Sketch section of a barrier lagoon showing a scenario for lagoon expansion caused by sea-level rise and marsh shifting. Outer part of the lagoon floor shows smooth shoreface characteristics, inner section of floor shows irregular surface produced by inundated landscape topography.

of the basin floor, and probably have a greater impact on the morphodynamics of barrier lagoons than vertical sedimentation rates.

During transgression, the lagoon shifts over recently inundated landscape surfaces (Fig. 9). Long-term transgression may remove all remnants of the engulfed shoreface (and the lagoonal record of wave-dominated origin). When transgression proceeds beyond all traces of the initial shoreface-attached beach, then the mainland shoreline of the lagoon becomes irregular as it follows the valleys and interfluves of the subaerial-landscape surface. Many modern barrier lagoons along the wave-dominated section of the middle Atlantic coast probably formed farther seaward on the continental shelf at a time of lower sea level. The embayed shoreface that formed the initial floors of these lagoons has long since been buried or beveled off following transgression.

The landward migration of lagoons over the antecedent topography is not geometrically simple, since inner and outer shorelines need not move at the same rates. During the late Holocene (10-1 ka), coastal lagoons have been very mobile and sedimentary basins have expanded and contracted in response to shifts at the inner and outer shorelines. Whereas the inner shoreline primarily moves in response to inundation and sea-level rise, movement of the outer shoreline is also affected by erosion of the shoreface.

LAGOON HYPSOMETRY

The terms *marsh lagoon* and *open-water lagoon* describe end-member stages of lagoonal landscape and marsh distribution (Oertel and Dunstan, 1981). In open-water lagoons, the lagoonal tidal prism is directly related to open-water tidal range; however, in marsh lagoons with numerous tidal flats, marshes and intertidal channels, the hypsometric relation clearly changes during rising tides, when water spills out of the channels and over tidal-flat and marsh surfaces (Oertel and Dunstan, 1981; Oertel, 1987; Eiser and Kjerfve, 1986).

The many different environments in barrier lagoons are primarily linked to elevation and location within the lagoon. The relative coverage of each environment in a lagoon is dependent upon basin hypsometry, with deep open-water parts of the lagoons forming tidal channels, bays and sounds and shallow open-water areas with lower relief forming flats and shallow bays (Fig. 1).

Since muddy surfaces between mean sea level (MSL) and highest high water (HHW) are optimal surfaces for marsh colonization, the hypsometry of a lagoonal marsh is clearly dependent on tidal range and slope (Fig. 10). Microtidal marshes may be considerably smaller than mesotidal marshes because of limited surface area between MSL and HHW. Large marshes in microtidal areas clearly require very low slopes. However, meso- and macrotidal marshes are less slope dependent (Fig. 10). It is imperative that the evolution of these surfaces be known in interpreting the evolution of the lagoon.

Although Lucke-type models of barrier-lagoon evolution assume that deep basins fill upward until they reach depths suitable for marsh colonization, the inundation of the landscape surface during transgression is sufficient to produce conditions suitable for marsh colonization. Distinguishing between these two models of surface origin is particularly

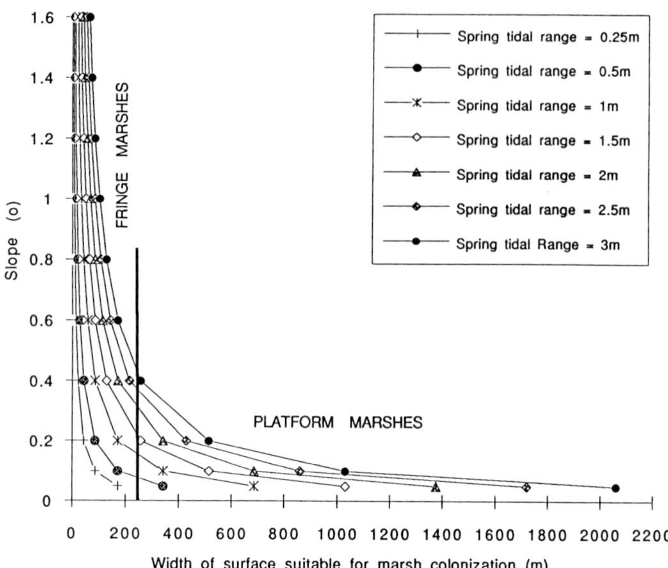

FIG. 10.—Graph showing the relation among tidal range, slope and width of surface suitable for marsh colonization. Slopes greater than 0.2° generally produce fringe marsh regardless of tidal range. In regions where spring tidal ranges are greater than 3 m or slopes are less than 0.05°, platform marshes are more common than fringe marshes.

important because marsh facies are commonly used for identifying former sea levels. Basal peats that are assumed to be produced by mainland fringe marshes have been used as a tracer of the Holocene transgression along the middle Atlantic coast of North America (Kraft, 1971; Finkelstein, 1986; Finkelstein and Ferland, 1987; Byrnes, 1988; Psuty, 1980). The facies record of the fringe marsh is believed to be the base (and beginning) of the sedimentary sequence of basin filling. Nevertheless, extensive marsh coverage (platform marshes) in barrier lagoons in the middle Atlantic area have been interpreted as the last or climax stage of basin infilling (Lucke, 1934; Finkelstein and Ferland, 1987; Ashely and Zeff, 1988). Detailed microfacies analysis is generally required to distinguish between facies of platform marsh and several different types of fringe marsh (Fig. 1).

CONCLUSIONS

Coastal-barrier lagoons have dynamic boundary conditions that do not permit the utilization of paradigms used in stratigraphic-basin analysis. Lagoons along wave-dominated coasts (that have formed by spits engulfing sections of shoreface) have simple topography. The lagoon floor inherits the beveled shoreface surface, which is shallowest along the mainland shore and deepest adjacent to the back barrier. With sufficient time and sediment supply, basins may fill upward until the lagoon floor becomes partially exposed at low water (forming tidal flats). Eventually, much of the lagoon floor may become exposed between high water and mean sea level, making conditions suitable for marsh colonization. Thus, a lagoon formed by a spit engulfing a section of shoreface is initially an open-water embayment and may fill to form a marsh-filled lagoon. *However, because lagoonal-boundary conditions have been so dynamic in response to the Holocene transgression, many North American barrier lagoons shifted laterally long before they were able to fill.* Wave-dominated conditions may continue to prevail along the ocean coast of the barrier shore, producing barrier spits and cross-island transfer deposits (washover fans and storm-surge platforms).

Lagoons that develop over complicated drainage of landscape topography have complicated hypsometry. Mainland shorelines are irregular and lagoons have deep-water pathways that are separated by pre-Holocene islands and flats. During initial inundation of the landscape, low-relief interfluve areas are transformed into extensive marsh platforms. Coastal inundation rather than sedimentation is the principal process producing new hypsometric conditions that produce tidal flats and surfaces suitable for marsh colonization. In major estuarine systems (for example, the Hudson, Delaware, and Chesapeake estuaries) much fine-grained sediment is retained in the estuary before it reaches the coast. The barrier-island systems along the main interfluve areas (between the major estuaries) have limited sediment supply during transgression and coastal inundation. Since siltation in barrier lagoons is unable to keep pace with sea-level rise, the distal edges of the fringe marsh become submerged and change to tidal flats and shallow bays. Thus, lagoons that form over landscape topography initially have large upland and marsh areas. New hypsometric conditions related to sea-level rise flood the hammocks, marshes and tidal flats, producing progressively increasing amounts of open-water area. Inlets are anchored in the antecedent-drainage pathways of the pre-transgressed landscape, and barrier islands are associated with the interfluvial areas between stream valleys.

In summary, landscape configurations, longshore and inlet dynamics, and sedimentary records all suggest that a viable theory for the evolution of barrier lagoons must differ from classic models used for stratigraphic-basin filling. Since the Holocene transgression has a major influence on much of the world's coasts, many of the world's coastal-barrier lagoons have transgressed landward a sufficient distance to be partially or totally affected by landscape topography. We consider the landscape topography model for lagoon evolution to be the principal theory for barrier lagoons of transgressive coasts.

ACKNOWLEDGMENTS

This manuscript is based on research that was primarily conducted along the Atlantic Coastal Plain of the United States over the past several decades. Many of our students and colleagues have directly and indirectly contributed to the evolution of the theories and alternative theories present above. Recent funding supporting morphostratigraphic efforts have come from the National Science Foundation Grant BSR-8702333-06. However, over the past score and several years, partial funding has come from various sources, including Old Dominion University, The University of Delaware, The University of Maryland, and the NOAA Sea Grant Programs of Delaware and Virginia. We also acknowledge the Virginia Coast Reserve of the Nature Conservancy for providing access to and encouraging research in Virginia's barrier-island systems.

REFERENCES

ASHLEY, G. M., AND ZEFF, M., 1988, Tidal channel classification for a low-mesotidal salt marsh: Marine Geology, v. 82, p. 17–32.

BELKNAP, D. F., AND KRAFT, J. C., 1985, Influence of antecedent geology on stratigraphic preservation potential and evolution of Delaware's barriers systems: Marine Geology, v. 63, p. 235–262.

BOOTHROYD, J. C., FRIEDRICH, N. E., AND MCGINN, S. R., 1985, Geology of microtidal coastal lagoons: Rhode Island: Marine Geology, v. 63, p. 35–76.

BYRNES, M. R., 1988, Holocene geology and migration of a low-profile barrier island system, Metompkin Island, Virginia: Unpublished Ph.D. Dissertation, Old Dominion University, Norfolk, Virginia, 302 p.

CHRZATOWSKI, M. J., 1986, Stratigraphy and geologic history of a Holocene lagoon: Rehoboth Bay and Indian River Bay, Delaware: Unpublished Ph.D. Dissertation, Department of Geology, University of Delaware, Newark, 337 p.

DANA, J. D., 1885, Origin of coral reefs and islands: American Journal of Science, Series 3, v. 30, p. 89–105, 169–191.

DARWIN, C., 1842, The structure and distributions of coral reefs: 214 pages reprinted in 1962, University of California Press, Berkeley-Los Angeles, California.

de BEAUMONT, E., 1845, Leçons de geologie practique, *in*, Bertrand, P., ed., Septieme Leçon: Paris, France, p. 223–252.

DOLAN, R., AND LINS, H., 1987, Beaches and barrier islands: Scientific American, v. 255, p. 67–77.

EISER, W. C., AND KJERFVE, B., 1986, Marsh topography and hypsometric characteristics of a South Carolina salt marsh basin: Estuarine, Coastal and Shelf Science, v. 23, p. 595–605.

FINKELSTEIN, K., 1986, Backbarrier contribution to a littoral sediment budget: Journal of Coastal Research, v. 2, p. 33–42.

FINKELSTEIN, K., AND FERLAND, M., 1987, Backbarrier response to sea-level rise, eastern shore of Virginia, *in* Nummedal, D., Pilkey, O. H., and Howard, J. D., eds., Sea-Level Fluctuation and Coastal Evolution: Society of Economic Paleontologists and Mineralogists Special Publication 41, p. 145–156.

FINKELSTEIN, K., AND KEARNEY, M. S., 1988, Late Pleistocene barrier-island sequence along the southern Delmarva Peninsula: implications for middle Wisconsinan sea levels: Geology, v. 16, p. 41–45.

FINKELSTEIN, K., AND KEARNEY, M. S., 1989, Late Pleistocene barrier island sequence along the southern Delmarva Peninsula: implications for middle Wisconsinan sea level. reply and discussion: Geology, v. 17, p. 86–87.

FISHER, J. J., 1967, Origin of barrier island chain shorelines: middle Atlantic states: Geological Society of America, Special Paper 115, p. 66–67.

FISHER, J. J., 1968, Barrier island formation: discussion: Geological Society of America Bulletin, v. 79, p. 1421–1426.

FLETCHER, C. H., KNEBEL, H. J., AND KRAFT, J. C., 1990, Holocene evolution of an estuarine coast and tidal wetlands: Geological Society of America Bulletin, v. 102, p. 283–297.

GILBERT, G. K., 1885, The topographic features of lake shore: Fifth Annual Report, U.S. Geological Survey, p. 69–123.

GILBERT, G. K., 1890, Lake Bonneville: U.S. Geological Survey, Monograph 1, 438 p.

HALSEY, S., 1978, Late Quaternary history and morphologic development of the barrier system along the Delmarva Peninsula of the Middle Atlantic Bight: Unpublished Ph.D. Dissertation, The University of Delaware, Newark, 592 p.

HALSEY, S., 1979, Nexus: a new model of barrier island development, *in* Leatherman, S.P., ed., Barrier Islands from the Gulf of St. Lawrence to the Gulf of Mexico: Academic Press, New York, p. 185–210.

HAYES, M. O., 1979, Barrier island morphology as a function of tidal and wave regimes, *in* Leatherman, S. P., ed., Barrier Islands from the Gulf of St. Lawrence to the Gulf of Mexico: Academic Press, New York, p. 1–27.

HOYT, J. H., 1967, Barrier island formation: Geological Society of America Bulletin, v. 78, p. 1125–1136.

HOYT, J. H., AND HENRY, V. J., 1967, Influence of island migration on barrier-island sedimentation: Geological Society of America Bulletin, v. 78, p. 77–86.

KRAFT, J. C., 1971, Sedimentary facies patterns and geologic history of a Holocene marine transgression: Geological Society of America Bulletin, v. 73, p. 2131–2158.

KRAFT, J. C., ALLEN, E. A., BELKNAP, D. F., JOHN, C. J., AND MAURMEYER, E. M., 1979, Processes and morphologic evolution of an estuarine and coastal barrier system, *in* Leatherman, S. P., ed., Barrier Islands from the Gulf of St. Lawrence to the Gulf of Mexico: Academic Press, New York, p. 149–184.

KRAFT, J. C., CHRZASTOWSKI, M. J. BELKNAP, D. F., TOSCANO, M. A., AND FLETCHER, C. H., 1987, Morphostratigraphy, sedimentary sequences and response to a relative rise in sea level along the Delaware coast, *in* Nummedal, D., Pilky, O. H., and Howard, J. D., eds., Sea-Level Fluctuation and Coastal Evolution: Society of Economic Paleontologists and Mineralogists Special Publication 41, p. 129–144.

KUMAR, N., AND SANDERS, J. E., 1974, Inlet sequence: a vertical sequence of sedimentary structures and textures created by the lateral migration of tidal inlets: Sedimentology, v. 21, p. 491–532.

LEATHERMAN, S. P., 1979a, Migration of Assateague Island, Maryland, by inlet and overwash processes: Geology, v. 7, p. 104–107.

LEATHERMAN, S. P., 1979b, The Barrier Island Handbook: National Park Service/University of Massachusetts, Amherst, 101 p.

LUCKE, J. B., 1934, A theory of evolution of lagoon deposits on shore lines of emergence: Journal of Geology, v. 42, p. 561–584.

MCGEE, W. J., 1890, Enchroachment of the sea: The Forum, v. 9, p. 437–449.

MCGOWEN, J. H., AND SCOTT, A. J., 1975, Hurricanes as geologic agents on the Texas coast, *in* Cronin, L. E., ed., Estuarine Research II: Academic Press, New York, p. 23–46.

NICHOLS, M. M., 1989, Sediment accumulation rates and relative sea-level rise in lagoons: Marine Geology, v. 88, p. 201–219.

NEWMAN, W. S., AND MUNSART, C. A., 1968, Holocene geology of the Wachapreague lagoon, eastern shore peninsula, Virginia: Marine Geology, v. 6, p. 81–105.

NUMMEDAL, D., OERTEL, G. F., HUBBARD, D. K., AND HINE, III, A. C., 1977, Tidal inlet variability–Cape Hatteras to Cape Canaveral, *in* Coastal Sediments '77: American Society of Civil Engineers, New York, p. 543–562.

OERTEL, G. F., 1985, The barrier island system: Marine Geology, v. 63, p. 1–18.

OERTEL, G. F., 1987, Backbarrier and inlet controls on inlet channel orientation, *in* Kraus, N. C., ed., Coastal Sediments '87: American Society of Civil Engineers, New York, p. 2022–2029.

OERTEL, G. F., 1988, Processes of sediment exchange between tidal inlets, ebb deltas and barrier islands, *in* Aubrey, D. G., and Weishar, L., eds., Hydrodynamics and Sediment Dynamics of Tidal Inlets: Springer-Verlag, New York, p. 297–318.

OERTEL, G. F., AND DUNSTAN, W. D., 1981, Suspended-sediment distribution and certain aspects of phytoplankton production off Georgia, U.S.A.: Marine Geology, v. 40, p. 171–197.

OERTEL, G. F., KEARNEY, M. S., LEATHERMAN, S. P., AND WOO, H. J., 1989, Anatomy of a barrier platform: outer barrier lagoon, southern Delmarva Peninsula, Virginia: Marine Geology, v. 88, p. 303–318.

PIERCE, J. W., 1970, Sediment budget along a barrier chain: Sedimentary Geology, v. 3, p. 5–16.

PSUTY, N. P., 1980, The forces that shape the islands, chapter 1, *in* Brown, P. M., and Renwick, H. L., eds., New Jersey's Barrier Islands: An Everchanging Public Resource: Barrier Islands Publication SNJ-DEP 80 5681, Center for Coastal and Environmental Studies, Rutgers State University of New Jersey, New Brunswick, p. 2–10.

SHIDELER, G., LUDWICK, J. C., OERTEL, G. F., AND FINKELSTEIN, K., 1984, Quaternary stratigraphic evolution of the southern Delmarva Peninsula coastal zone, Cape Charles, Virginia: Geological Society of America Bulletin, v. 95, p. 489–502.

VAN DE PLASSCHE, O., 1990, Mid-Holocene sea-level change on the eastern shore of Virginia: Marine Geology, v. 91, 149–154.

RECORD OF OXYGEN ISOTOPE STAGE 5 ON THE MARYLAND INNER SHELF AND ATLANTIC COASTAL PLAIN–A POST-TRANSGRESSIVE-HIGHSTAND REGIME

MARGUERITE A. TOSCANO

Department of Marine Science, University of South Florida, 140 Seventh Avenue South, St. Petersburg, Florida 33701

ABSTRACT: Stage 5 deposits (unit Q2) have been identified along the inner shelf of Maryland. Peak sea-level substage 5e deposits (lower Q2) consist of (buried) shoal-forming sands on the inner shelf, analogous to modern shelf sands, and relict barrier facies onshore. Substages 5d–5a, however, are represented by geographically widespread, thick, fossiliferous muds (upper Q2), whose age range was determined by amino-acid racemization. The time frame of mud deposition was further subdivided into stage 5 substages. Ostracode assemblage zones in unit Q2 record a consistent, repetitive sequence of four rapid and distinct climatic fluctuations that follow the isotopic excursions in stage 5. Sea levels remained high for the duration of stage 5, fluctuating within a maximum range of 30 m. The first physical evidence of substage 5d sea levels being higher than −23 m mean seal level (MSL) is contained within ostracode zone 2 in unit Q2. An anomalous lithofacies such as a thick mud deposited in a relatively high sea level requires alternative sediment sources and shelf sedimentation processes than those operating on the present and substage 5e transgressive sandy shelves.

Sea level during stage 5 along the Maryland coast consisted of a prolonged period (≈45,000 yrs) of water covering the shelf at less than maximum sea levels, following the substage 5e peak highstand (+6 m MSL) spanning 11,000 yrs. Stage 5 may be modeled as follows: Early stage 5 transgressive facies migrated across the shelf and were deposited at their peak position during substage 5e (at 125 ka), when sea level reached +6 m MSL. Late stage 5 sea levels maintained a range of −23 (extreme minimum) to +0.1 (present) m MSL and deposited thick, laminated, unbioturbated muds containing numerous radiations of *Mulinia lateralis* (Say). Onshore facies of late stage 5 include a barrier system (substage 5c at +0.1 m MSL) and tidal flats and/or marshes (of substages 5d, 5b, and 5a). It is inferred that at slightly depressed late stage 5 sea levels the Chesapeake and Delaware estuaries became fluvially dominated, as the estuarine portion of the drowned river valley was expelled, or shifted seaward out of the basin. Muds were thereby exported from these major fluvial systems onto the inner shelf and redistributed by shelf currents. Sources of excess fine sediment could include pre-stage 5 glacial sediment, as well as sediment eroded from exposed coastal areas at lowered sea levels. Deposition appears to have been exceedingly high, as evidenced by intact laminae of fine sand-silt in the muds and by the *in situ* fossil assemblages and their relation to sedimentary characteristics of the unit. The 45,000-yr-long period of relatively high sea level during stage 5 was unique in Pleistocene history, and was the controlling factor in establishing sea-level-related processes resulting in the muddy shelf environment, which characterizes only stage 5.

INTRODUCTION

Extensive deposits consisting of dewatered fossiliferous muds directly underlie modern shelf sands in the nearshore shelf of Maryland, attaining thicknesses of 6 m or more. The muds were determined, via amino-acid racemization (AAR) analyses, as having been emplaced during late oxygen isotope stage 5, subsequent to the peak transgressive phase of substage 5e (Toscano and others, 1989; Toscano and Kerhin, 1990; Toscano and York, 1992). Similar shelf deposits are not forming in any present transgressive sytem (the stage 5 muds are not analogous to present deltaic deposits) and therefore are not interpretable within the context of transgressive processes and stratigraphy. These muds represent a specific sea-level situation not occurring during the Holocene transgression, and are therefore without modern analogs. Geographically, late stage 5 (substages 5d–5a) shelf mud deposits extend from northern New Jersey to Cape Lookout, North Carolina, south of which they appear to pinch out in the shoreface. The thickness, depth range, and lateral extent of the muds indicate deposition during a prolonged period (≈45,000 yrs) of near-present sea levels, fluctuating between an extreme minimumn of −23 m MSL to +0.1 m MSL (Toscano and York, 1992). Equivalent, post-peak sea-level muds are not found in the Quaternary section of the Maryland shelf for time intervals older than stage 5. The pre-stage 5 Quaternary record (unit Q1) consists only of transgressive shelf sands, indicating that conditions of prolonged high sea level and sedimentation during stage 5 alone produced these anomalous nearshore deposits with no analogs.

PREVIOUS WORK

Quaternary Stratigraphy, Maryland Inner Shelf

Figure 1 shows the area of study and a shore-normal cross section of Tertiary and Quaternary units delimited by seismic-reflection profiling and vibracoring along the Maryland coast (Toscano and others, 1989). Tertiary shelf deposits are correlated to Tertiary onshore units via a major erosional surface, horizon M1, extended onshore using coastal-plain drillhole data. This erosional surface on top of Tertiary deposits is thought to represent subaerial exposure during a major regression due to the onset of late Pliocene glaciation at approximately 2.4 Ma, recorded in deep-sea cores (Toscano and York, 1992, Shackleton and Hall, 1984; Shackleton and others, 1984; Zimmerman and others, 1984). Five Quaternary units have been cored and mapped seismically above horizon M1. Unit Q1 (early?–middle Pleistocene) is a composite of sandy shelf sediments, analogous to those of the present shelf, of several pre-stage 5 interglacial high sea-level stands. Amino acid (Allo/Iso) ratios indicate that shell material in the upper 2 m of unit Q1 is from interglacial stages 9/11, 13, and one older interglacial (Toscano and York, 1992).

A shallowly incised, relatively flat erosional surface (horizon M2; Fig. 1) separates unit Q1 from unit Q2 above, and has been interpreted as the glacial stage 6 (190–130 ka; −130 m MSL) subaerial-shelf surface. Horizon M2 is correlative to the Eastville paleochannel of the Susquehanna River (ancestral Chesapeake Bay) of Colman and Mixon (1988), horizon R3 on the Virginia shelf (Shideler and others, 1972), and to the northern paleovalley of the

Quaternary Coasts of the United States: Marine and Lacustrine Systems, SEPM Special Publication No. 48
 ISBN 0-918985-98-6

(e.g., Salazar-Jiminez and others, 1982). Unit Q2 was therefore deposited in a dynamic open-shelf environment proximal to substantial sources of fine sediment. The preservation of laminae creates an ideal situation for geochronologic and paleoclimatic studies, but requires alternative explanations for sedimentation (sources and regimes) and estuarine/shelf circulation and dynamics than those normally inferred for either a typical mud facies or for a high sea-level shelf.

Amino-acid age determinations on *Mulinia lateralis* in unit Q2 muds (Fig. 3A) were highly consistent (averaging 8 percent variability) within the unit and among established stage 5 deposits on the adjacent coastal plain (Fig. 3B; Aminozone IIa of Wehmiller and others, 1988), some of which were re-analyzed using *Mulinia* (Toscano and others, 1989; Toscano and York, 1992). The variability of whole-valve *Mulinia* analyses prevents small time frames, such as the substages of stage 5, from being subdivided. However, extensive ostracode assemblage studies in unit Q2, and the application of the ostracode paleotemperature transfer function of Cronin and Dowsett (1990) to these assemblages, revealed several distinct climatic shifts in a repetitive stratigraphic sequence correlative among shelf vibracores (Fig. 4). This climatic sequence effectively follows climatic shifts inferred from the form of the oxygen isotope curve for stage 5 (e.g., Martinson and others, 1987).

Interglacial ostracode zone 1 characterizes ocean climate during deposition of substage 5e transgressive shelf sands in lower Q2, as well as for the composite interglacial shelf sands of unit Q1. The shelf of zone 1 (substage 5e) was not significantly warmer than the present shelf, although sea level was 6 m higher than present (Toscano and York, 1992). This is supported by CLIMAP reconstructions for sea-surface temperatures during stage 5 (Ruddiman and others, 1984). Zone 1 is sharply overlain (faunal unconformity) by cold zone 2, defined by the presence of subfrigid ostracode species occurring far south of their present latitudinal ranges. Zone 2 is thought to represent the initial post-5e drop in sea level due to reestablishment of interstadial conditions in substage 5d. The interstadial climatic trend of zone 2 was at least partially reversed, both as inferred from the negative shift of the isotopic curve and in the slightly warmer conditions indicated by zone 3 above. Zone 3, stratigraphically the thickest of the zones, may encompass only substage 5c (at a glacio-eustatic sea level of $\approx +0.1$ m MSL); however, no succeeding cold zone was found to account separately for interstadial substage 5b. Either substage 5b conditions were not significantly colder, or of lower sea level, than substage 5c, or sea level during substage 5b *was* sufficiently low (in this case lower than zone 2–5d sea level) as to leave no deposits in the study area. The lack of unconformities in cores and seismic data and the thickness of zone 3 argue for the inclusion of substage 5b within zone 3. Zone 4, the youngest and warmest in these stage 5 deposits, was tentatively assigned to substage 5a, with the implication that peak warmth of the entire stage 5 interval may have lagged well behind the peak of sea level that occurred initially, at 125 ka. Zone 4, of approximately 80-ka age (assuming 5a), would have to account for an $\approx$45,000-year time lag in peak ocean warmth. Zone 4 ideally requires further study to support such inferences, although a similar scenario can be inferred from the data of Ruddiman and others (1984).

The composite unit Q1 provides important perspectives for reinforcing the uniqueness of the late stage 5 muds of upper Q2. Amino acid racemization relative and calculated ages for both units Q1 and Q2 are shown in Figures 3A and 3B. If only the relative aspect of the ages is addressed, it is still clear from Figure 3B that there are at least two interglacial shelf deposits present in Q1, in aminozones IIc (stage 9/11) and IId (stage 13), as well as one older stage not shown. No major mud deposit analogous to upper Q2, neither in cores nor seismic data, separates these Q1 interglacial sands. In contrast, unit Q2 muds (aminozone IIa) grade upward from early Q2 peak transgressive shelf sands, separating substage 5e shelf sands from stage 1 (modern) shelf sands (see core data in Toscano and others, 1989). Unit Q2 muds were deposited at relatively high sea levels after the transgression in substage 5e peaked, indicating that conditions and processes resulting in mud deposition on shelves were directly tied to peculiarities of late stage 5 sea levels, and their effect on coastal and estuarine processes.

The depth range of each ostracode zone, or corresponding stage 5 substage, was utilized in evaluating the sea-level history for the region. Consistency in timing and height between sea-level standards and correlative strandline deposits on the Atlantic coastal plain provides tests of glacio-eustatic models and their general applicability. A well-constrained geochronology and sea-level history may then be the basis for evaluating the depositional paleoenvironments of anomalous deposits such as upper unit Q2.

Mid-Atlantic Stage 5 Sea-Level History

Cored, submerged, coastal-plain deposits such as those of the Maryland shelf provide access to lithostratigraphic and paleoclimatic evidence needed to address lowstand sea levels during late stage 5. Most glacio-eustatic records discuss sea-level maxima, leaving minima as question marks because of the lack, and inaccessibility, of lowstand deposits, even along tectonically rising coasts (e.g., New Guinea; Chappell, 1974; Bloom and others, 1974; Chappell and Veeh, 1978; Bloom and Yonekura, 1985). Minima in the mid-Atlantic area are defined by depth ranges of ostracode climatic zones (or stage 5 substages) and by the depth ranges of correlative onshore (including subsurface) deposits. The depth ranges of ostracode zones, placed in assigned time frames, are superimposed over the most recent glacio-eustatic curve of Bloom and Yonekura (1985; New Guinea; Fig. 5), and supported by recent data from the Bahamas addressing less-than-maximum late Pleistocene sea levels (Li and others, 1989; undersea cave deposits; Fig. 5). It is important to note that the revised New Guinea estimates and the most recent Bahamas data indicate higher sea levels during late stage 5 than do original estimates (e.g., Chappell, 1974; Bloom and others, 1974; Neumann and Moore, 1975; Matthews, 1973). Mid-Atlantic relict shorelines, which have been amino-acid dated, provide a realistic test for localized use of the revised New Guinea estimates, incorporating a reasonable error estimate of ± 1 m.

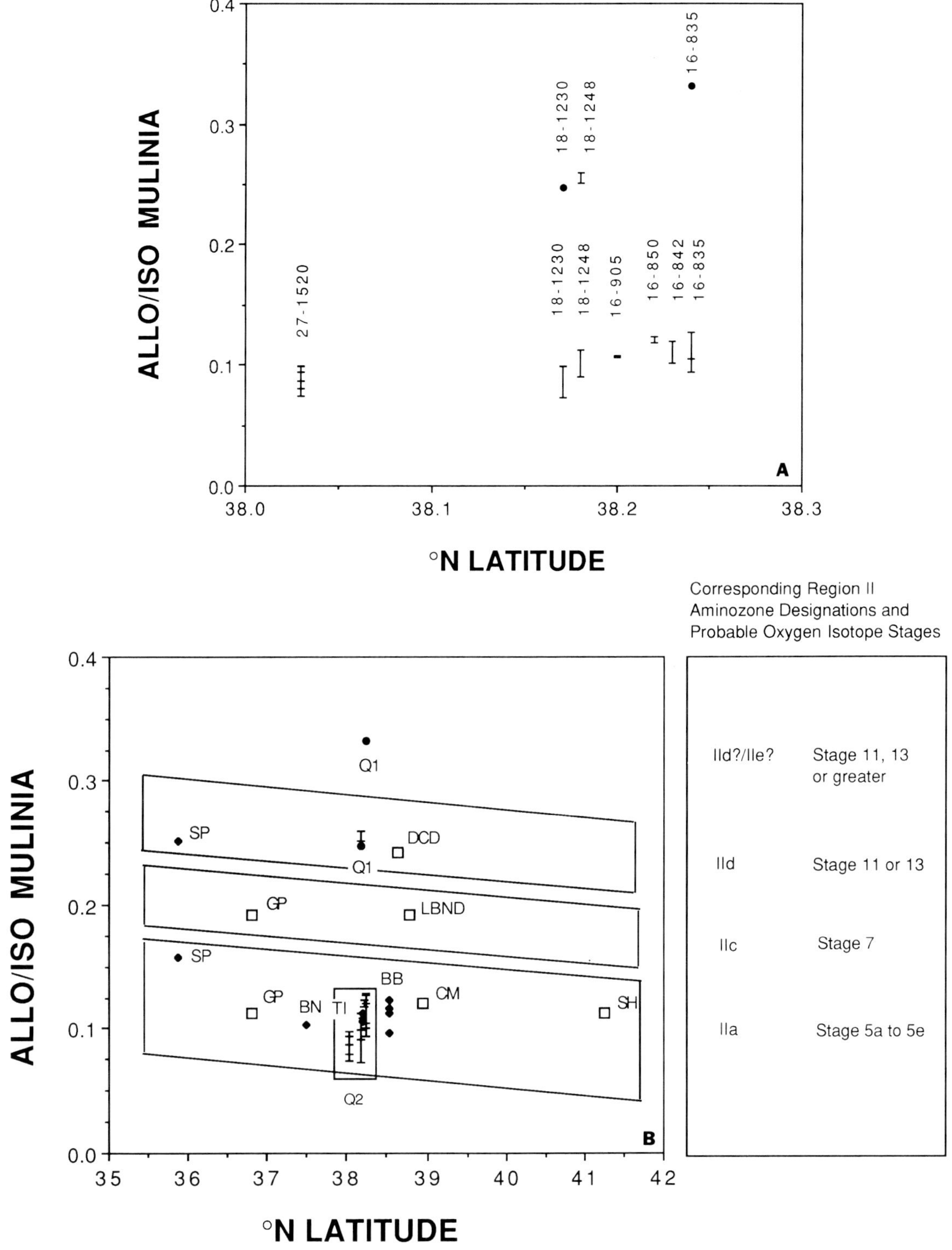

FIG. 3.—(A) *Mulinia lateralis* Allo/Iso values versus latitude for Maryland continental shelf core samples. Vertical range of data for each core is obtained by using the standard deviation ($\pm 1\sigma$) about the mean Allo/Iso value. Black circles represent Allo/Iso values for individual *M. lateralis* samples. From Toscano and York (1992). (B) Maryland shelf *Mulinia lateralis* Allo/Iso data shown in a regional aminozone framework. The aminozones outlined by the *M. lateralis* Allo/Iso data are correlated within the Region II framework of Wehmiller and others (1988). Region II includes southern New Jersey (CM-Cape May), Delaware (BB-Bethany Beach; DCD-Dirickson Creek Ditch; LBND-Lobiondo Pit), Maryland (TI-Tingles/Assateague Island), Virginia (BN-Bell Neck/Wachapreague Formation; GP-Gomez Pit/Tabb Formation), and is extended to northeastern North Carolina (SP-Stetson Pit). Maryland shelf data are plotted as in (A), with *M. lateralis* of unit Q2 enclosed within the rectangle. *M. lateralis* data from additional Atlantic Coastal Plain localities are indicated by the diamond-shaped symbols. Open-square symbols represent *Mercenaria* Allo/Iso values converted to equivalent *M. lateralis* Allo/Iso values using equations in York (1990). The size of the open-square symbols indicates a greater relative uncertainty in these converted values compared with the other Allo/Iso values shown. From Toscano and York (1992).

have been attempted in order to reconcile the large mud accumulations on shelves with stage 5 sea-level history and paleoclimate, sources of muds, and processes responsible for mud deposition on shelves. Information from the Maryland portion of the Delmarva shelf was used to reconstruct several intervals of the last interglacial, which could be extrapolated to adjacent shelf areas (Figs. 6–8).

Estuaries are dominant features of the mid-Atlantic bight, compartmentalizing the coastline from Long Island, New York, to the North Carolina coast. Ephemeral features, estuaries infill rapidly from their heads, over time scales of several thousand years (Schubel and Carter, 1984). Stable sea levels, which can be maintained over similar time scales, tend to promote estuarine infilling. As filling progresses, depth, volume, and surface area of the estuary decrease,

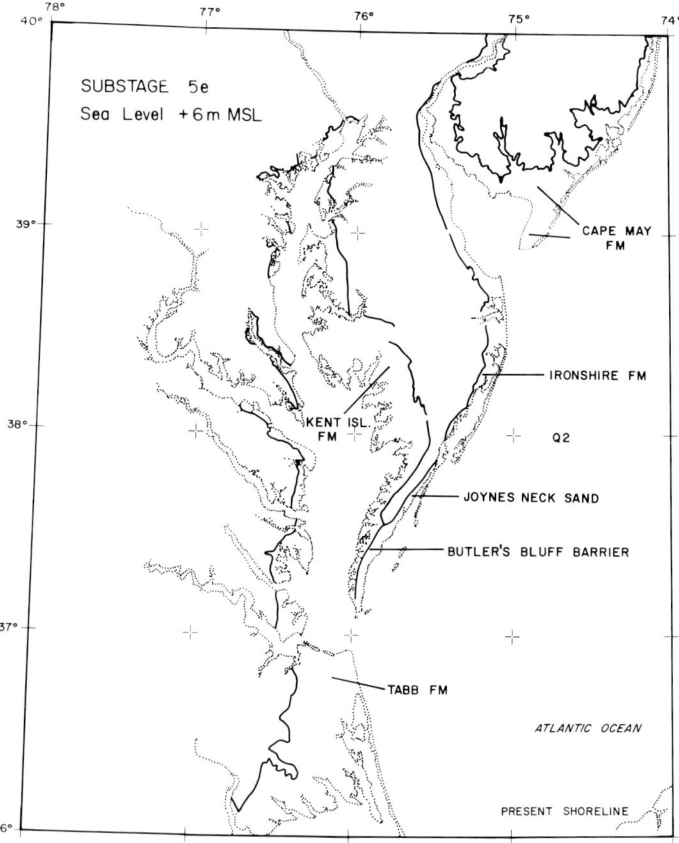

FIG. 6.—Paleogeography during substage 5e, the peak of the last interglacial. Sea level reached +6 m MSL, from glacio-eustacy and confirmed by relict shorelines surrounding the Delmarva Peninsula (see text). Shorelines are defined by the landward extent of the Kent Island Formation (Mixon, 1985) along Chesapeake Bay, and by the trend of relict barrier islands along the Atlantic coast (Nassawadox Formation-Butler's Bluff barrier, and Joynes Neck Sand, Mixon, 1985; Ironshire Formation, Owens and Denny, 1979; Bethany Paralic Unit of Omar Formation, McDonald, 1982). This shoreline trend can be extrapolated north to the Cape May Formation (New Jersey; locality CM, Fig 3b), and south to the Tabb Formation (Virginia, locality GP, Fig. 3b). Some substage 5e age interpretations of these units are from Toscano and York (1992). Estuarine circulation underwent a shift to a river-dominated system as described above due to stabilization of peak 5e sea level for several thousand years, and/or to the decrease in seal levels during the transition to substage 5d.

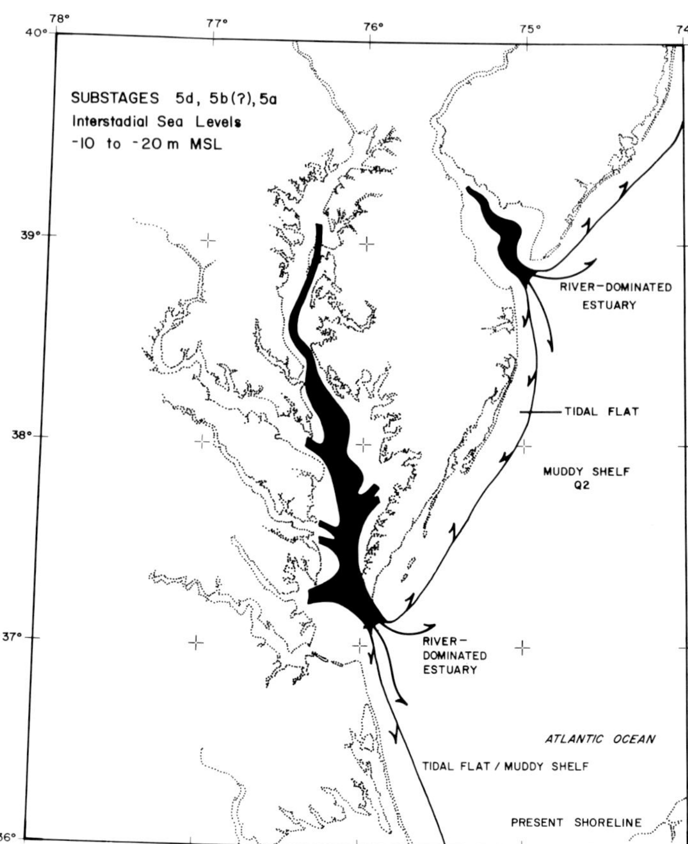

FIG. 7.—Paleogeography during isotope substage 5d (and also 5a), the first interstadial following the peak interglacial of substage 5e. Sea level dropped to no lower than −23 m MSL (and possibly only to −10 m) during this cold climatic interval. Estuary size decreased by at least 70 percent, becoming river dominated due both to basin shrinkage and increased glacially(?) controlled river discharge. High, episodic fluvial suspended-sediment transport passing the estuary mouth resulted in turbid shelf environments. The coastal zone was dominated by tidal flats. No barriers formed at this sea level. Headlands, which would have provided sediment sources for barrier formation, were eroded too far landward during substange 5e. Substages 5b and 5a are interpreted as accounting for similar paleoenvironments, although sea level was slightly higher. Marshes of substage 5a age under present barrier islands have often been cored.

effecting a decrease in intertidal estuarine volume. The resultant high river flow relative to tidal flow causes circulation patterns to shift to a more river-dominated system. The estuary shortens, possibly becoming short enough to be "expelled" from the basin, onto the shelf. Because most estuaries infill from their heads, the expulsion of the estuary onto the shelf reduces the filtering efficiency in the basin and sedimentation occurs directly on the shelf. High fluvial discharge (such as would likely occur during interglacials, or during summer glacial outwash periods) generally corresponds to large volumes of suspended sediment. Therefore, the fraction of fluvial sediment discharged directly onto the shelf would increase dramatically. Lowered sea level would result in decreased estuarine volume, increasing the likelihood of infilling and decreasing the time frame over which the expulsion process would happen.

Knebel and others (1988) produced a series of sea-level/ time slices for the Delaware estuary during the Holocene

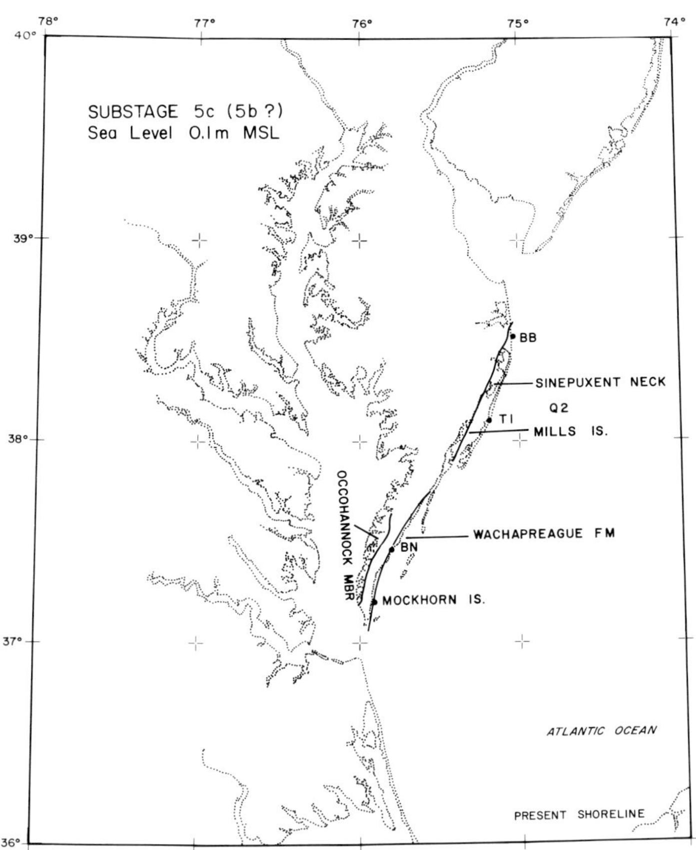

FIG. 8.—Paleogeography during substage 5c (continuing through 5b?). Sea level peaked at +0.1 m MSL, according to glacio-eustatic records. The Atlantic 5c coastline is traceable from the Wachapreague relict barriers of the southern Delmarva Peninsula (Mockhorn Island and locality BN, Mixon, 1985), several marsh islands in Chincoteague Bay behind Assateague Island (e.g., Mills Island, Halsey, 1978), through the ridges of Sinepuxent Neck (Owens and Denny, 1979), and veering offshore past Bethany Beach, Delaware (McDonald, 1982). Substage 5c interpretations are from Toscano and York (1992). Stratigraphic distinctions of substage 5c units along the Chesapeake Bay shorelines might include the Occohannock Member of the Nassawadox Formation and the Kent Island Formation (Mixon, 1985).

transgression. Portions of the Holocene history of Delaware Bay may serve as analogies for late stage 5, and may be extrapolated to the Hudson and Chesapeake estuaries. Reconstruction of the −20 m MSL sea-level (similar to low sea level in late stage 5) stage of transgression indicates a basin with approximately 30 percent of the present estuary area. The −10 m MSL reconstruction indicates that 60 percent of the present basin area was occupied by the estuary. The generally lowered (possibly stable) sea levels increased erodible exposures during late stage 5, and the possibility that smaller estuarine volume succumbed to river-dominated processes more quickly, likely combined to shift the locus of sediment filtering from the lower basin reaches out onto the adjacent shelves off the Hudson, Delaware, and Chesapeake estuary mouths.

Meade (1982) discussed the ultimate sources of sediment to coastal-plain rivers. An analogy may be made between Meade's example concerning the rapid, widespread coastal plain land clearing and farming, and resulting soil disturbance of the 1600s, to the similar disturbance of land affected by glaciation. The Hudson, Delaware, and Susquehanna Rivers (the headwaters of the Hudson, Delaware, and Chesapeake estuaries, respectively) originate in glaciated terrains in western New York State, and must have transported large quantities of glacially and proglacially derived sediment during stage 6 through 5.

An interesting concept of sediment introduction and movement along the mid-Atlantic shelf is offered by Darby (1990). The present sand cover of the shelves is apparently virtually all Hudson-derived and has been progressively moved southward to Cape Hatteras, North Carolina, over several interglacials. Darby's conclusion is based on heavy-mineral discriminant analyses and the dynamics and time constraints of sand transport over long distances on shelves. If the Hudson river is situated to intercept more of the glacially produced sediment yield and is the main conduit to the shelf, then it could also be responsible for the introduction of muds to the area. The difference would be that mud would more easily be transported over long distances, and the prolonged sea-level highstand of stage 5 would have provided sufficient time (during a single interglacial) to spread the mud to the North Carolina shelf. For purposes of this paper, however, major contributions of fine-grained sediment from both the Delaware and Susquehanna Rivers are implicitly assumed. Appropriate mineralogic analyses on the muds would have to be undertaken to assess the relative contributions of the three major estuaries, and may provide essential information on paraglacial processes in that region.

The paraglacial period of sediment transfer from the glacial-margin source areas can take thousands of years (Smith, 1985). For a lengthy period such as stage 5, containing at least one episode of colder climate (5d), the process might have been continuous, and aided by the depressed sea levels. The inferred history of stage 5 presented herein would be impossible to test by analogy to modern systems and processes, and must remain one of possibly several applicable scenarios.

Reconstructions

Substage 5e–(130–120 ka; Fig. 6).—

Rapid transgression to highest Pleistocene sea stand of +6 m MSL occurred, lasting approximately 11 ka. Estimated winter- and summer-ocean (shelf-bottom and SSTs) temperatures were similar to present. OST zone 1 of substage 5e, corresponds climatically to deep-sea core estimates of substage 5e ocean temperatures. Shelf sand ridges (lower Q2) trailed substage 5e barrier islands (Nassawadox Fm, Joynes Neck Sand, Ironshire Fm., Bethany Paralic Unit) emplaced at the +6-m peak of sea level, landward of present barriers. Headlands (sediment sources to barrier islands) eroded landward, keeping pace with barriers. Deposition of major estuarine facies (Kent Island Formation) began at this time. Stabilization of sea levels during substage 5e may have resulted in estuarine infilling, expulsion of the estuarine head onto the shelf, and initiation of mud deposition over shelf sands during late substage 5e. Mud deposition continued as sea level dropped during the transition to substage 5d.

Substage 5d–(120–110 ka; Fig 7).—

Lower unit Q2 (5e) shelf sands grade upward into unit Q2 muds deposited seaward of the estuary mouths. Estuarine shortening and expulsion from basins onto the shelf occurred due to sea-level lowering and increased discharge and sediment load from increased erosion and the possibility of resurgence of glaciation. Substage 5d may account for a sea-level minimum of −23 m MSL, but sea levels were probably no lower than −10 m, with river-dominated estuaries continuing to export muds onto the shelf. Significant climatic cooling is indicated by OST zone 2, by which the assignment of substage 5d is made. Marshes/tidal flats formed along the peninsula in coastal areas, with no barriers due to the probability that sources of coarse sediment for barrier-island formation were pushed (eroded) too far landward by the high sea levels of substage 5e.

Substage 5c–(110–103 ka; Fig. 8).—

Sea level rose to approximately present levels, according to glacio-eustasy and age (Fig. 3b) and morphology of local coastal deposits. Barriers formed just seaward of substage 5e barriers, and landward of, and oblique to, the present coastline. Mud continued to be deposited on the shelf. The peak of substage 5c could have been rapid and short lived so that river-dominated estuarine processes would have continued to maintain the locus of fine sedimentation on the shelves. However, the possibility that substage 5c represented another episode of stable sea level and estuarine infilling/sediment bypassing must also be considered until constraints can be placed on the time span of the 5c sea-level highstand. Climate is inferred as having been slightly cooler than present based on data of OST zone 3, which is the thickest of the four paleoclimatic subdivisions of unit Q2.

Substage 5b–(103–90 ka; Fig. 8).—

Substage 5b cannot be separately identified either by climate change or by a sea-level drop, as implied by oxygen isotope records (much like substage 5d). The lack of unconformities suggests continuation of 5c climate (OST zone 3), approximately stable (perhaps marginally lower) sea levels, and sedimentation patterns, which may have remained essentially similar since substage 5d time.

Substage 5a–(90–80 ka; Fig. 7).—

Eustatic sea level estimates of approximately −7 m MSL are consistent with the lack of an additional (substage 5a) relict barrier seaward of the substage 5c trend on the Atlantic coast (above or below sea level). Instead of barriers, marshes formed in tidal areas as during substage 5d. These marsh deposits have often been sampled from beneath Assateague Island, Maryland, and Mockhorn Island, Virginia. Mud continued to be discharged onto shelves due to lowered sea level. Climate appeared to be significantly warmer than present during late stage 5, not during the peak of sea level of substage 5e, if warm zone 4 accurately represents the time frame corresponding to substage 5a.

CONCLUSIONS

The slightly lowered sea levels of late stage 5 profoundly affected coastal, estuarine, and shelf processes and sedimentation. The prolonged relatively high sea level of late stage 5 was apparently unique, and resulted in a major depositional environment with neither modern nor known ancient analogs. Estuarine dynamics played an essential role in a lowered sea level to bring fine sediment onto shelves that are normally dominated by transgressive, high-energy sand transport. Stage 5 was the only time in Quaternary history, as known from the Maryland shelf, wherein large-scale deposits, other than peak transgressive facies, are preserved.

ACKNOWLEDGMENTS

The cooperation of the Maryland Geological Survey in making data from their Continental Margins study available for further investigations is greatly appreciated. Linda L. York and Thomas M. Cronin have been instrumental in providing amino-acid and ostracode analyses, respectively, and have contributed many insights to this study. Dr. John F. Wehmiller has been a source of support and interaction for five years, and has shared insights that greatly improved the scope and quality of the project. Albert C. Hine and Eugene A. Shinn provided critical reviews of the manuscript.

REFERENCES

BARD, E., HAMELIN, B., FAIRBANKS, R. G., AND ZINDLER, A., 1990, Calibration of the ^{14}C timescale over the past 30,000 years using mass spectrometric U-Th ages from Barbados corals: Nature, v. 345, p. 405–410.

BELKNAP, D. F., 1979, Applications of amino acid geochronology to stratigraphy of the late Cenozoic marine units of the Atlantic coastal plain: Unpublished Ph.D. Dissertation, Department of Geology, University of Delaware, Newark, Delaware, U.S.A., 550 p.

BELKNAP, D. F., AND KRAFT, J. C., 1981, Preservation potential of transgressive coastal lithosomes on the U.S. Atlantic shelf: Marine Geology, v. 42, p. 429–442.

BELKNAP, D. F., AND KRAFT, J. C., 1985, Influence of antecedent geology on stratigraphic preservation potential and evolution of Delaware's barrier systems. Marine Geology, v. 63, p. 235–262.

BLOOM, A. L., 1983, Sea level and coastal morphology of the United States through the late Wisconsin glacial maximum, *in* Porter, S. C., ed., Late Quaternary Environments of the United States, v. 1, The Late Pleistocene. University of Minnesota Press, Minneapolis, p. 215–229.

BLOOM, A. L., BROECKER, W. S., CHAPPELL, J., MATTHEWS, R. K., AND MESOLLELA, K. J., 1974, Quaternary sea level fluctuations on a tectonic coast: new ^{230}Th/^{234}U dates from the Huon Peninsula, New Guinea. Quaternary Research, v. 4, p. 185–205.

BLOOM, A. L., AND YONEKURA, N., 1985, Coastal terraces generated by sea level change and tectonic uplift, *in* Woldenberg, M. J., ed., Models in Geomorphology. Allen and Unwin, Inc., Winchester, Mass., p. 139–154.

CHAPPELL, J., 1974, Geology of Coral terraces, Huon Peninsula, New Guinea: A study of Quaternary tectonic movements and sea level changes. Geological Society of America Bulletin, v. 85, p. 553–570.

CHAPPELL, J., AND VEEH, H. H., 1978, ^{230}Th/^{234}U age support of an interstadial sea level of −40 m at 30,000 yr BP. Nature, v. 276, p. 602–604.

COLMAN, S. M., AND MIXON, R. B., 1988, The record of major Quaternary sea level changes in a large coastal plain estuary, Chesapeake

Bay, eastern United States. Palaeogeography, Palaeoclimatology, Palaeoecology, v. 68, p. 99–116.

CRONIN, T. M., AND DOWSETT, H. J., 1990, A quantitative micropaleontologic method for shallow marine paleoclimatology: applications to Pliocene deposits of the Western North Atlantic Ocean. Marine Micropaleontology, v. 16, p. 117–148.

DARBY, D. A., 1990, Evidence for the Hudson River as the dominant source of sand on the U.S. Atlantic shelf. Nature, v. 346, p. 828–831.

FAIRBANKS, R. G., 1989, A 17,000-year glacio-eustatic sea level record: influence of glacial melting rates on the Younger Dryas event and deep ocean circulation. Nature, v. 342, p. 637–647.

FIELD, M. E., 1980, Sand bodies on coastal plain shelves: Holocene record of the U.S. Atlantic inner shelf of Maryland. Journal of Sedimentary Petrology, v. 50, p. 505–528.

FIELD, M. E., AND DUANE, D. B., 1976, Post-Pleistocene history of the United States inner continental shelf: significance to the origin of barrier islands. Geological Society of America Bulletin, v. 87, p. 691–702.

FINKELSTEIN, K. L., AND KEARNY, M. S., 1988, Late Pleistocene barrier island sequence along the Delmarva Peninsula: implications for middle Wisconsin sea levels. Geology, v. 16, p. 41–45.

HALSEY, S. D., 1978, Late Quaternary geologic history and morphologic development of the barrier island system along the Delmarva Peninsula of the middle Atlantic bight. Unpublished Ph.D. Dissertation, Department of Geology, University of Delaware, Newark, Delaware, U.S.A., 592 p.

HARMON, R. S., SCHWARCZ, H. P., AND FORD, D. C., 1978, Late Pleistocene sea level History of Bermuda: Quaternary Research, v. 9, p. 205–218.

HINE, A. C., AND SNYDER, S. W., 1985, Coastal lithosome preservation: Evidence from the shoreface and inner continental shelf off Bogue Banks, North Carolina. Marine Geology, v. 63, p. 307–330.

KNEBEL, H. J., AND CIRCE, R. C., 1988, Late Pleistocene drainage systems beneath Delaware Bay. Marine Geology, v. 78, p. 285–302.

KNEBEL, H. J., FLETCHER, C. H., AND KRAFT, J. C., 1988, Late Wisconsinan–Holocene paleogeography of Delaware Bay; a large coastal plain estuary. Marine Geology, v. 83, p. 115–133.

LI, W.-X., LUNDBERG, J., DICKIN, A. P., FORD, D. C., SCHWARCZ, H. P., MCNUTT, R., AND WILLIAMS, D., 1989, High-precision mass spectrometric uranium-series dating of cave deposits and implications for paleoclimate studies. Nature, v. 339, p. 534–536.

MARTINSON, D. G., PISIAS, N. G., HAYS, J. D., IMBRIE, J., MOORE, T. C., AND SHACKLETON, N. J., 1987, Age dating and the orbital theory of the ice ages: Development of a high-resolution 0 to 300,00-year chronostratigraphy. Quaternary Research, v. 27, p. 1–29.

MATTHEWS, R. K., 1973, Relative elevation of late Pleistocene high sea level stands: Barbados uplift rates and their implications. Quaternary Research, v. 3, p. 147–153.

McDonald, K. A., 1982, Three-dimensional analysis of Pleistocene and Holocene Coastal sedimentary units at Bethany Beach, Delaware. Unpublished M.S. Thesis, Department of Geology, University of Delaware, Newark, Delaware, U.S.A., 205 p.

MEADE, R. H., 1982, Sources, sinks, and storage of river sediment in the Atlantic drainage of the United States. Journal of Geology, v. 90, p. 235–252.

MIXON, R. B., 1985, Stratigraphic and geomorphic framework of uppermost Cenozoic deposits in the southern Delmarva Peninsula, Virginia and Maryland. U.S. Geological Survey Professional Paper 1067-G, 53 p.

NEUMANN, A. C., AND MOORE, W. S., 1975, Sea level events and Pleistocene coral ages in the northern Bahamas. Quaternary Research, v. 5, p. 215–224.

OWENS, J. P., AND DENNY, C. S., 1979, Upper Cenozoic deposits of the central Delmarva Peninsula, Maryland and Delaware. U.S. Geological Survey Professional Paper 1067-A, 28 p.

RUDDIMAN, W. F., and 24 other CLIMAP Project Members 1984, The last interglacial ocean. Quaternary Research, v. 21, p. 123–224.

SALAZAR-JIMENEZ, A., FREY, R. W., AND HOWARD, J. D., 1982, Concavity orientations of bivalve shells in estuarine and nearshore shelf sediments, Georgia. Journal of Sedimentary Petrology, 52, 565–586.

SHACKLETON, N. J., BACKMAN, J., AND 15 OTHERS, 1984, Oxygen isotope calibration of the onset of ice-rafting and history of glaciation in the North Atlantic region. Nature, 307, 620–623.

SHERIDAN, R. E., DILL, C. E., AND KRAFT, J. C., 1974, Holocene sedimentary environments of the Atlantic inner shelf off Delaware. Geological Society of America Bulletin, 85, 1319–1328.

SHACKLETON, N. J., AND HALL, M. A., 1984, Oxygen and carbon isotope stratigraphy of deep sea drilling project hole 552A: Plio-Pleistocene glacial history. Initial Reports of the Deep Sea Drilling Project, 81, U.S. Government Printing Office, Washington, D.C., p. 599–609.

SHIDELER, G. L., LUDWICK, J. C., OERTEL, G. F., AND FINKELSTEIN K. L., 1984, Quaternary stratigraphic evolution of the southern Delmarva Peninsula coastal zone, Cape Charles, Virginia. Geological Society of America Bulletin, 95, 489–502.

SHIDELER, G. L., AND SWIFT, D. J. P., 1972, Seismic reconnaissance of post-Miocene deposits, middle Atlantic continental shelf—Cape Henry, Virginia to Cape Hatteras, North Carolina. Marine Geology, 12, 165–185.

SHIDELER, G. L., SWIFT, D. J. P., JOHNSON, G. H., AND HOLLIDAY, B. W., 1972, Late Quaternary stratigraphy of the inner Virginia continental shelf: a proposed standard section. Geological Society of America Bulletin, 83, 1787–1804.

SCHUBEL, J. R., AND CARTER, H. H. 1984, The estuary as a filter for fine-grained suspended sediment, *in* Kennedy, V. S. (ed.), The Estuary as a Filter: Academic Press, Inc., p. 81–105.

SMITH, N. D., 1985, Proglacial fluvial environment, *in* Ashley, G. M., Shaw, J., and Smith, N. D. (eds.) Glacial Sedimentary Environments: SEPM Short Course No. 16, 85–127.

STANLEY, S. M., 1970, Relationship of Shell Form to Life Habits of the Bivalvia (Mollusca). Geological Society of America Memoir 125, 296 p.

STUBBLEFIELD, W. L., MCGRAIL, D. W., AND KERSEY, D. G., 1984, Recognition of transgressive and post-transgressive sand ridges on the New Jersey continental shelf, *in* Tillman, R. W., and Siemers, C. T. (eds.) Siliclastic Shelf Sediments: Society of Economic Paleontologists and Mineralogists Special Publication 34, 1–24.

SWIFT, D. J. P., AND FIELD, M. E., 1981, Evolution of a classic sand ridge field, Maryland sector, North America inner shelf. Sedimentology, 28, 462–482.

SWIFT, D. J. P., KOFOED, J. W., SAULSBURY, F. P., AND SEARS, P., 1972, Holocene evolution of the shelf surface, south and central Atlantic shelf of North America, *in* Swift, D. J. P., Duane, D. B., and Pilkey, O. H. (eds.) Shelf Sediment Transport: Process and Pattern: Dowden, Hutchinson, and Ross, Stroudsberg, Pennsylvania, 499–574.

SWIFT, D. J. P., MCKINNEY, T. F., AND STAHL, L., 1984, Recognition of transgressive and post-transgressive sand ridges on the New Jersey continental shelf, Discussion, *in* Tillman, R. W., and Siemers, C. T. (eds.) Siliclastic Shelf Sediments: Society of Economic Paleontologists and Mineralogists Special Publication 34, 25–36.

THOM, B. G., 1973, The dilemma of high interstadial sea levels during the last glaciation. Progress in Geography, 5, 157–246.

TOSCANO, M. A., AND KERHIN, R. T., 1990, Subbottom structure and stratigraphy of the inner continental shelf of Maryland, *in* Hunt, M. C., Doenges, S., and Stubbs, G. S. (eds.), Studies related to continental margins—a summary of year three and year four activities: U.S. Minerals Management Service and Association of American State Geologists, 2nd Symposium, 124–142.

TOSCANO, M. A., KERHIN, R. T., YORK, L. L., CRONIN, T. M., AND WILLIAMS, S. J. 1989. Late Quaternary stratigraphy of the inner continental shelf of Maryland. Maryland Geological Survey Report of Investigations 50, 116 p.

TOSCANO, M. A., AND YORK, L. L., 1992, Quaternary stratigraphy and sea level history of the U.S. Middle Atlantic coastal plain. Quaternary Science Reviews, v. 11, p. 301–328.

WEHMILLER, J. F., BELKNAP, D. F., BOUTIN, B. S., MIRECKI, J. E., RAHAIM, S. D., AND YORK, L. L., 1988, A review of the aminostratigraphy of Quaternary mollusks from United States Atlantic coastal plain sites, *in* Easterbrook, D. L. (ed.) Dating Quaternary Sediments. Geological Society of America Special Paper 227, p. 69–110.

YORK, L. L., 1990, Aminostratigraphy of the U.S. mid-Atlantic Pleistocene coastal plain: Norfolk, Virginia to Charleston, South Carolina. Unpublished Ph.D. Dissertation, Department of Geology, University of Delaware, Newark, Delaware, U.S.A., 550 p.

ZIMMERMAN, H. B., SHACKLETON, N. J., BACKMAN, J., KENT, D. V., BALDAUF, J. G., KALTENBACK, A. J., AND MORTON, A. C., 1984, History of Plio-Pleistocene climate in the northeastern Atlantic, Deep Sea Drilling Project hole 552A. Initial Reports of the Deep Sea Drilling Projects, 81, U.S. Government Printing Office, Washington, D.C., p. 861–875.

PATTERNS AND RATES OF SEDIMENT ACCUMULATION IN THE CHESAPEAKE BAY DURING THE HOLOCENE RISE IN SEA LEVEL

STEVEN M. COLMAN
U.S. Geological Survey, Woods Hole, MA 02543
JEFFREY P. HALKA
Maryland Geological Survey, 2300 St. Paul St., Baltimore, MD 21218
AND
C. H. HOBBS, III
Virginia Institute of Marine Science, Gloucester Point, VA 23062

ABSTRACT: Holocene sediment thicknesses measured from seismic-reflection profiles, together with long-term rates of sediment accumulation calculated from these thicknesses and the history of relative sea level, indicate that the Chesapeake Bay has filled rapidly with sediment during Holocene submergence of the bay. Sediment-accumulation patterns indicate that both the Susquehanna River system and the continental shelf are important sources of sediment; averaged over Holocene time, sediment transported from the continental shelf through the mouth of the bay may be volumetrically more important than sediment derived from rivers.

Average Holocene rates of sediment accumulation show considerable spatial variability, presumably related to local variations in sediment sources, wave energy, and tidal currents. Nonetheless, these rates show several clear trends. Rates on the shallow marginal shelves of the bay tend to be low (0 to 2 mm/yr) and to increase only slightly toward the bay mouth. Rates in the deep channels are higher (1 to 5 mm/yr), have local maxima, and increase distinctly toward the bay mouth. At any given position in the bay, sediment-accumulation rates increase with depth to the base of the Holocene section.

Our estimates of average Holocene rates of sediment accumulation are clearly higher in many places than previous estimates, but they are somewhat less than short-term rates previously measured by a variety of methods. Short-term rates may be affected by anthropogenic changes in the basin and by recent acceleration of relative sea-level rise. In addition, most short-term rates are site specific, biased in their distribution, and fail to account for the large spatial variability observed in long-term rates. Maximum long-term rates of sediment accumulation are limited by the rate of submergence; many existing short-term measurements clearly cannot be extrapolated back in time.

Long-term rates of sediment accumulation confirm the ephemeral nature of estuaries and the close tie between sea level and estuarine history. These observations are important considerations for studies of the evolution of estuaries and the record of estuaries in the geologic record.

INTRODUCTION

Estuaries are recognized as important but short-lived features of the Earth's surface. They form extensively only in response to relatively rapid rises in sea level, and subsequently fill quickly with sediments (Schubel and Hirschberg, 1977; Nichols and Biggs, 1985). Because of their transient nature, they represent only a small proportion of geologic history, and yet evidence of their existence should be preserved in the geologic record because of the rapid accumulation of sediments in their basins.

Estimates of the present trapping efficiency of estuaries indicate that most of the sediments that enter the basins are retained there. For 18 United States estuaries evaluated by Biggs and Howell (1984), 11 had predicted trapping efficiencies for watershed-derived sediments in excess of 95 percent, and all but two exceeded 50 percent. Moreover, Emery and Uchupi (1972) estimated an average life span of only 9,500 years for the estuaries of the United States Atlantic and Gulf Coasts. Their calculation was based on estimates of fluvial sediment only and assumed that relative sea level will remain static at its present position. The value of such calculations is limited because the evolution of estuaries clearly depends on both the rate of submergence and the rate of accumulation of sediment. In addition, the rate of sediment accumulation itself is affected by the rate of relative sea-level change.

The life span of estuaries and lagoons and the preservation potential of their sediments both depend upon the comparison between relative sea-level rise and the rate of sediment accumulation (Nichols, 1989). The potential for preservation of estuarine deposits in the geologic record depends upon the same balance, along with other factors, such as subsequent erosion. During the Quaternary, multiple cycles of regression and transgression on the United States East Coast resulted in the preservation of regressive paleochannels filled with estuarine deposits in the subsurface beneath the present Chesapeake Bay (Colman and others, 1990) and beneath other estuaries (Knebel and Circe, 1988; Hine and Snyder, 1985). Each paleochannel and its fill represents a complete regressive-transgressive cycle. Thus, a single glacial-interglacial cycle appears sufficient to completely fill these estuaries with sediment.

Much of this paper deals with the relation between sea-level change and sediment accumulation. Rate of sediment accumulation is used in the sense of net vertical accumulation of sediment over an interval of time. This usage is different than instantaneous rates of sedimentation, as would be measured in a sediment trap. It includes the effects of actual deposition, resuspension, erosion, and compaction, integrated over the time interval of interest. Sea-level changes are always used in the sense of submergence (relative sea-level rise) or emergence (relative sea-level fall). As discussed in the section on sea-level history, the Chesapeake Bay has experienced submergence, a significant component of which has resulted from subsidence, throughout the Holocene.

Although the ability of estuaries to serve as effective sediment traps is widely recognized, deriving long-term (thousands of years) rates of sediment accumulation is difficult because of the spatial and temporal variability of sediment sources. Numerous techniques have been used to estimate

Quaternary Coasts of the United States: Marine and Lacustrine Systems, SEPM Special Publication No. 48

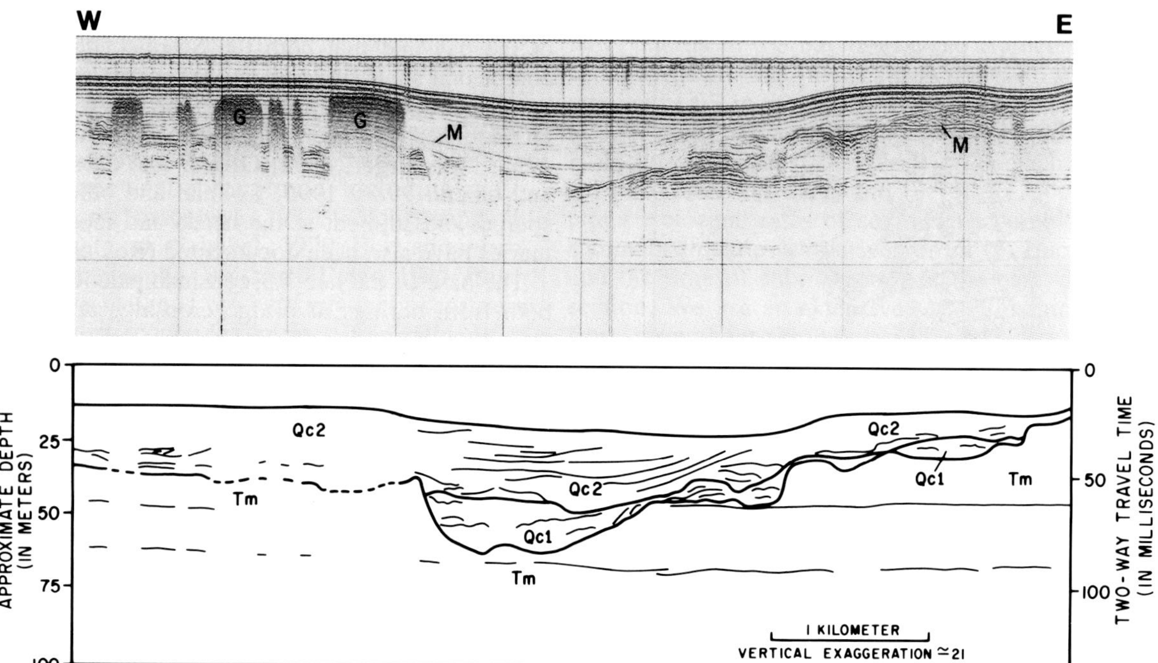

FIG. 2.—Example seismic-reflection profile and interpretive line drawing. Qcl, basal unit of post-glacial channel fill; Qc2, upper unit of post-glacial sediments; Qc, post-glacial deposits, undifferentiated; Tm, Tertiary marine strata; M, multiple reflection; G, reflections obscured by gas in sediments. Depth scale assumes a speed of sound in water and sediments of 1,500 m/s. Location shown in Figure 1.

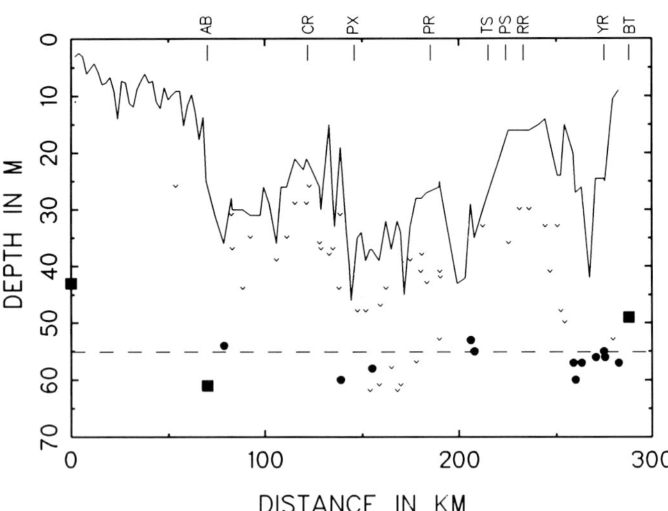

FIG. 3.—Profile of Cape Charles paleochannel and modern bathymetry above the paleochannel. Circles represent depth to base of paleochannel derived from seismic-reflection data; v = minimum values. Squares = depths to base of paleochannel in boreholes (Colman and others, 1990). Distances measured from mouth of the Susquehanna River along line of profile shown in Figure 1. Symbols along the top represent positions of the following: AB, Annapolis Bridge; CR, Choptank River; PX, Patuxent River; PR, Potomac River; TS, Tangier Sound; PS, Pokomoke Sound; RR, Rappahannock River; YR, York River; and BT, Chesapeake Bay Bridge-Tunnel.

The remainder of the sediments are estuarine and probably constitute 95 percent or more of the total volume of sediment above the unconformity. The volume of late Wisconsinan fluvial sediment that was included with the Holocene estuarine sediments is probably more than offset by the volume of estuarine sediment in shallow water that was excluded from the calculations.

A longitudinal profile of the volume of Holocene sediment (Fig. 5) shows major variations along the length of the bay. The profile is irregular and contains maxima and minima that appear to be cyclic. The maxima tend to occur just upstream from major tributary estuaries, but the reasons for this pattern are not clear. The net sediment flux through the mouths of the tributary estuaries, or even its sign, is poorly known (Hobbs and others, 1990). During normal hydrologic conditions, the tributary estuaries appear to be sediment traps (Schubel and Carter, 1976; Officer and Nichols, 1980; Hobbs and others, 1990), although the margins of some tributaries may be a source of sand to the bay (Hobbs and others, 1990). However, during floods, the tributary estuaries become sources of entirely freshwater riverine discharge to the bay (Nichols, 1977). Regardless of the role of the tributaries as sediment sources, their tidal circulation and fluvial discharge probably do affect the estuarine circulation of the main bay, and may thus indirectly influence sediment-accumulation patterns and rates in the main bay by altering tidal-flow patterns.

The volume profile shows a distinct overall trend of increasing Holocene sediment volume toward the bay mouth (Fig. 5). This relation clearly supports the contention that

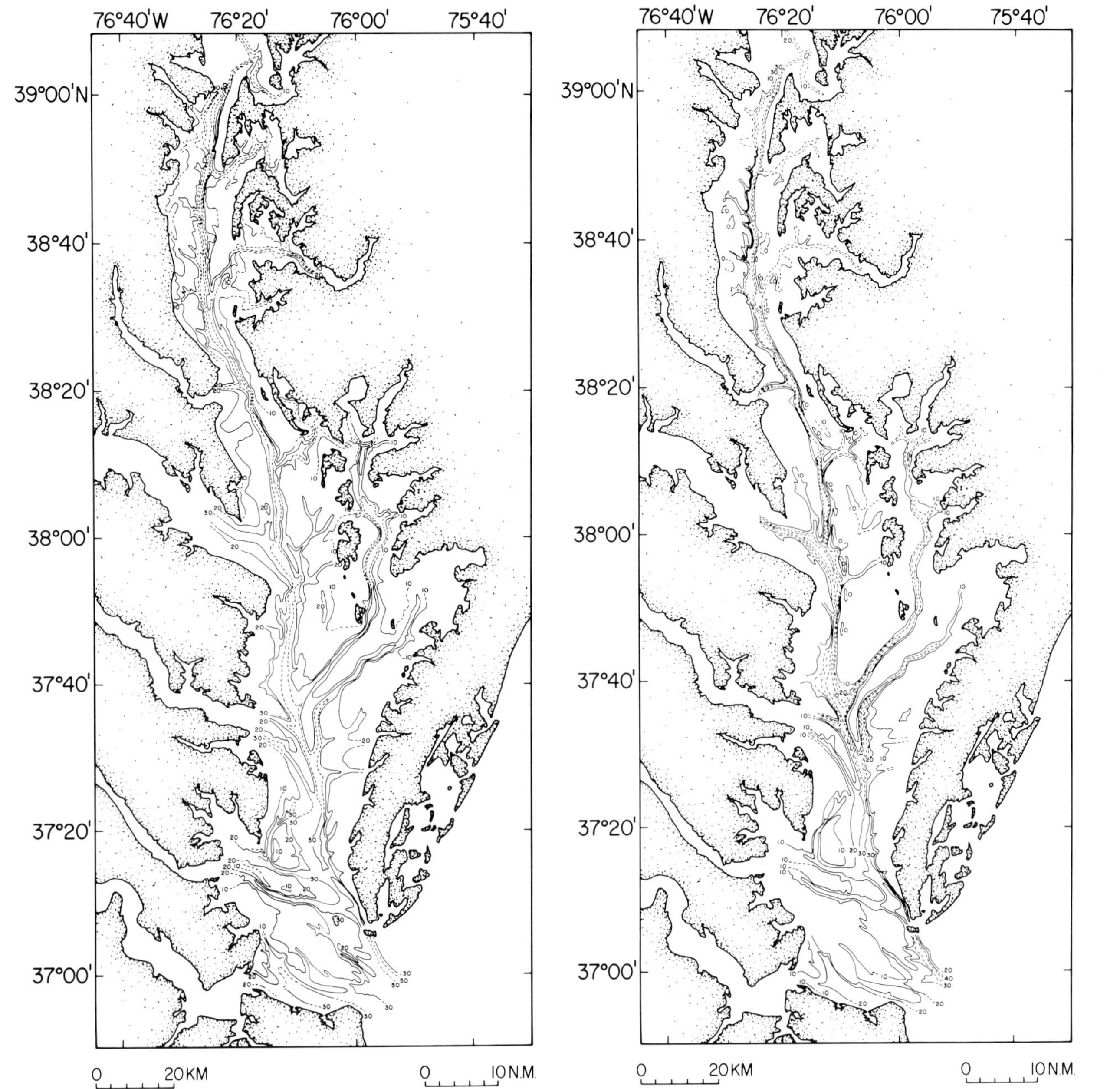

FIG. 4.—Contour maps of (A) depth to pre-Holocene erosion surface (contours at 10, 20, 30, and 50 m), and (B) thickness of Holocene sediments (contour interval 10 m). Generalized from Colman and Hobbs (1987, 1988) and Colman and Halka (1989a, 1989b).

much of the sediment in the lower bay has been transported into the bay through its mouth (Ryan, 1953; Meisburger, 1972; Officer and others, 1984; Hobbs and others, 1986, 1990; Colman and others, 1988). Officer and others (1984) suggested that the bay-mouth source was substantial, but was smaller (by about 60 percent) than the river source. In contrast, our data indicate that the greatest sediment volume is associated with the mouth of the bay, suggesting that the continental shelf may have been a more important source of sediment to the bay than the Susquehanna River, averaged over the course of the Holocene. In addition to our long-term data, a 100-year sediment budget based on bathymetric changes indicates that 2.7 to 7.6 times as much sand has been deposited in the bay than can be accounted for by sources other than the tributary estuaries and the continental shelf (Hobbs and others, 1990).

The greater volume of Holocene sediment at the bay mouth compared to that at the head contrasts with the short-term observations of (1) a maximum in the rate of sediment accumulation near the head of the bay (Fig. 3; Officer and others, 1984; Donoghue, 1990), and (2) the presence of a turbidity maximum in about the same location (Schubel and Carter, 1976, 1984). This contrast is probably the result of the fact that the configuration of the bay changes both spa-

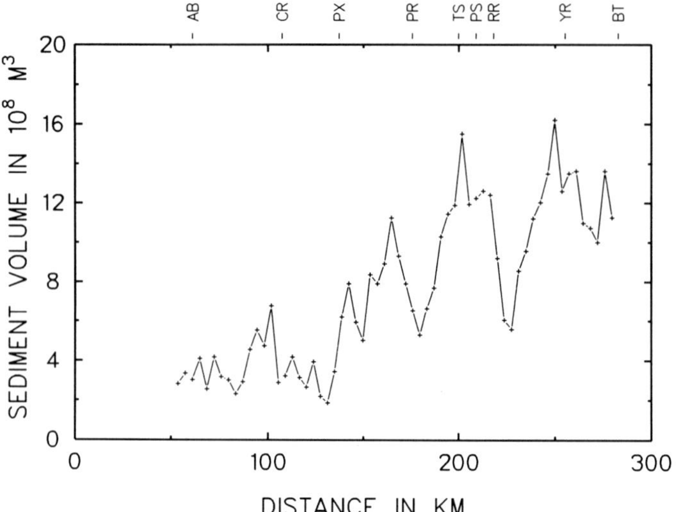

FIG. 5.—Longitudinal profile of Holocene sediment volume. Distances are measured on a north-south line from mouth of the Susquehanna River. Letter symbols as in Figure 3.

tially and with time. The head is much narrower than the rest of the bay (Fig. 1), so that deposition of a comparatively small volume of sediment in a restricted area may result in high local rates of sediment accumulation. As the bay has enlarged, deposition changed from higher rates focused in a narrow channel to slower rates accommodated in a larger cross-sectional area. In addition, the turbidity maximum and highest short-term rates of sediment accumulation have moved landward with time, and are presently located just north of the limit of the data in Figure 5.

Our calculations provide an estimate of 5.2×10^{10} m^3 for the volume of Holocene sediment deposited in Chesapeake Bay south of the Annapolis Bridge. This volume translates to an average accumulation rate of 5.2×10^6 m^3 yr^{-1} over the 10,000-year duration of the Holocene. The present mean-low-water volume of the bay below the Annapolis Bridge is 4.8×10^{10} m^3. At the rate calculated earlier, the present bay would fill with sediment in about 9,200 years in the absence of sea-level changes. This estimate is considered to be a minimum for several reasons. First, present rates of total sediment accumulation are probably higher than those earlier in the Holocene, because of the increasing importance of shoreline erosion as the estuary grew from a restricted river estuary to a large bay. Second, most of the sediment from the Susquehanna River is presently trapped north of the Annapolis Bridge (Biggs, 1970); as the northern end of the bay fills with sediment from the Susquehanna River, this source will become increasing important south of the Annapolis Bridge. For these reasons, the estimate of the time necessary to fill the rest of the bay under natural conditions and without additional sea-level change could be as little as 5,000 years. However, as noted for previous estimates of this type, such calculations have little relevance to realistic projections of the life of the estuary. They fail to account for the dynamic nature of the estuary, temporal changes in sediment accumulation, changes in relative sea level, and perhaps most importantly, feedbacks among these variables.

Several estimates of sediment budgets for all or part of the Chesapeake Bay have been made (Schubel and Carter, 1976; Officer and others, 1984; Hobbs and others, 1990) and these could be used to calculate rates of filling of the bay. However, the results of such calculations are difficult to interpret, for several reasons: (1) the flux of sediment through the mouth of the bay from the continental shelf, which may be the largest source, is poorly known; (2) estimates of the riverine source are based only on the short periods of modern measurements; and (3) all source estimates are affected by anthropogenic influences, such as land clearing, dam construction, and shoreline structures. In addition, all the problems associated with using long-term sediment accumulation rates to estimate rates of filling apply to these short-term calculations as well.

AVERAGE RATES OF SEDIMENT ACCUMULATION

Average Holocene rates of sediment accumulation can be calculated directly from point estimates of sediment thickness and estimates of the time interval represented by that thickness of sediment. Our seismic-reflection data provide abundant estimates of Holocene sediment thickness, and the duration of sedimentation can be estimated from the time of submergence of the base of the section, determined from sea-level curves.

We divided the 2,600 km of seismic-reflection profiles into 300-m segments and calculated the average thickness of Holocene sediment for each segment by subtracting the depth to the sea floor from the depth to the base of the Holocene section. When plotted as a function of latitudinal distance from the mouth of the Susquehanna River (Fig. 6), the data reveal a great deal of spatial variability. In spite of the scatter, the thicknesses have a pattern of maxima and minima similar to that for the sediment-volume data (Fig.

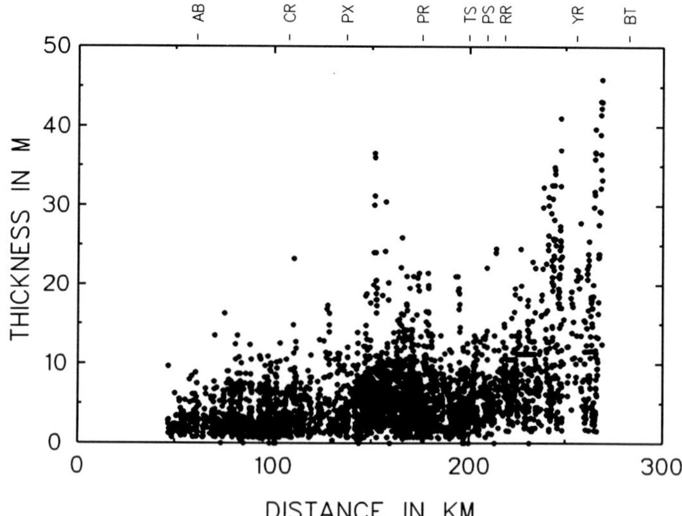

FIG. 6.—Longitudinal profile of Holocene sediment thickness. Distances are measured on a north-south line from mouth of the Susquehanna River. Letter symbols as in Figure 3.

5). The data also show a clear tendency for increased thickness toward the bay mouth.

The sediment thicknesses can be converted to long-term average rates of sediment accumulation if the time interval during which the sediments were deposited is known. Because the Holocene sediments beneath Chesapeake Bay are dominantly estuarine, they began to accumulate approximately when a given location was submerged by rising sea level. Except in local areas of tidal scour, they have accumulated continuously to the present. Consequently, the thickness of sediment divided by the time of submergence gives an approximate average rate of sediment accumulation. Clearly, the rates of accumulation have varied through time, and the rate of accumulation has lagged behind the rate of submergence, since the bay has not yet filled. However, the method produces a valid long-term average, subject to the following sources of error: (1) pre-existing fluvial sediments in the larger channels, which cause an overestimate of the sediment-accumulation rate; (2) possible delays between the time of submergence and the beginning of sedimentation, which could cause an underestimate of the sediment-accumulation rate; (3) local erosion, which removes sediment and causes an underestimate of the sediment-accumulation rate, and (4) compaction of the sediment with time, which produces an underestimate of the sediment-accumulation rate compared to short-term methods. These errors are partly self-canceling, and the first source of error is thought to be less than 10 percent (see previous discussion), so that the sum of the errors is inferred to be small. The calculated rates are clearly long-term averages that integrate temporal variations in actual rates of sediment accumulation.

The Chesapeake Bay area has experienced submergence throughout Holocene time. This submergence has resulted from a combination of the eustatic rise in sea level (Fairbanks, 1989) and local subsidence of the Earth's crust. Both historic records, based on leveling surveys (Holdahl and Morrison, 1974) and tide gauges (Hicks and Hickman, 1988), and geophysical modeling of glacial isostatic effects (Peltier, 1990) suggest that subsidence accounts for a substantial part of the submergence. Relative sea level for the Chesapeake Bay area is relatively well known for about the last 6,000 years from local studies that have dated basal salt-marsh peats submerged during the transgression (Newman and Rusnak, 1965; Ellison and Nichols, 1976). Other local studies have produced additional age estimates related to sea level (Hobbs, 1988; Donoghue, 1990). We used these data, combined with recent estimates of global eustatic sea level (Fairbanks, 1989), to produce a local composite sea-level curve for the Chesapeake Bay area for the last 10,000 years (Fig. 7). The curve shows that relative sea level at the beginning of the Holocene was at about 60 m below present sea level, just about the depth of the deepest parts of the late Wisconsinan channel of the Susquehanna River (Fig. 4A) beneath the present bay.

By comparison, Nichols and others (1991) recently compiled available data for the James River subestuary for the last 6,000 years and combined them with the regional data for the interval from 6,000 to 10,000 years. Because of regional variations in the rate of subsidence and errors related to sampling and dating, Nichols and others (1991) presented his sea-level-depth relation as an envelope rather than as a single curve. We attempted to minimize spatial variation in subsidence rates by only using sea-level data from the immediate Chesapeake Bay area, and used a single curve (Fig. 7) to simplify our calculation of sediment-accumulation rates. Our curve fits entirely within the envelope proposed by Nichols and others (1991).

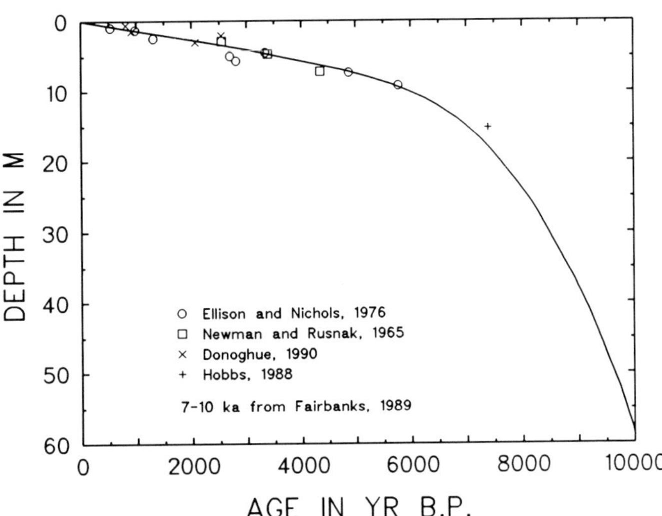

FIG. 7.—Holocene relative sea-level curve for the Chesapeake Bay area. See text for discussion.

Rates of sediment accumulation (Fig. 8A) derived from thicknesses (Fig. 6) and times of submergence (Fig. 7) show much the same pattern in longitudinal profile as do the sediment thickness data. Consequently, the rates show a high degree of scatter, but form distinct local maxima along the length of the bay. At any given point along the length of the bay, the data tend to cluster in the lower range of sediment-accumulation rates, and both the cluster and the maximum values increase toward the bay mouth.

Both Chesapeake Bay and the late Wisconsinan fluvial unconformity that underlies it have two main morphologic elements: broad, shallow, marginal shelves and deep channels (Figs. 1, 4A). We separated sediment-accumulation rates in these two morphologic elements by separating the data into areas where the fluvial unconformity at the base of the Holocene section was more than 20 m deep (channel areas) or less than 20 m deep (marginal shelves). Sediment-accumulation rates for the marginal shelves show considerably less scatter than that for the bay as a whole (Fig. 8B). In general, the rates on the shelves are in the lower part of the range for the bay as a whole, and they show little tendency to increase toward the bay mouth. In contrast, the rates for the channel areas (Fig. 8C) show a great deal of scatter, clear local maxima, and a tendency to increase toward the bay mouth. Only the clustering of rates low in the range for the bay as a whole is absent (Fig. 8C), and it is clear that this clustering (Fig. 8A) is a result of measurements from the marginal shelves (Fig. 8B).

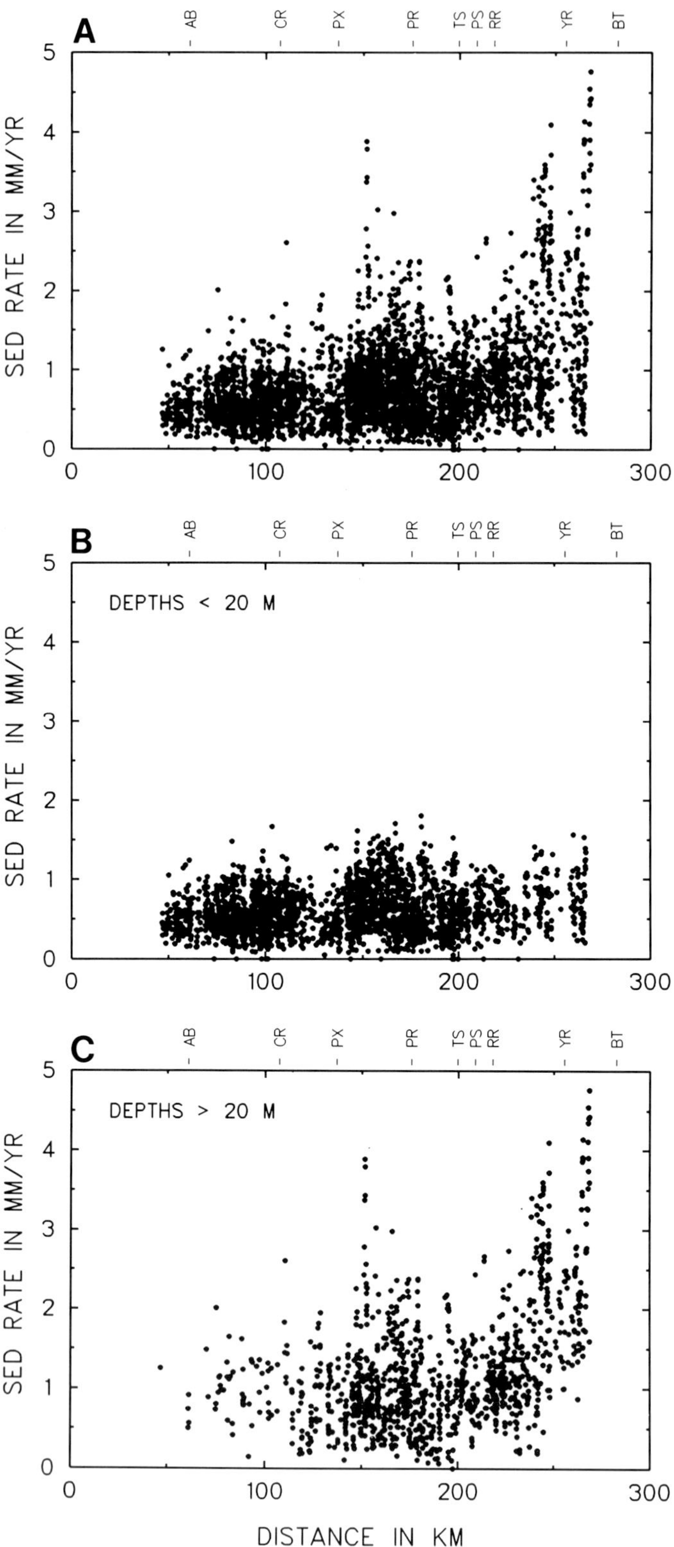

FIG. 8.—Longitudinal profile of average Holocene rates of sediment accumulation. (A) All data (B) Data for areas where depth to pre-Holocene erosion surface is <20 m (C) Data for areas where depth to pre-Holocene erosion surface is >20 m. Label "sed rate" on ordinate refers to sediment-accumulation rate. Distances are measured on a north-south line from mouth of the Susquehanna River. Letter symbols as in Figure 3.

The differences between the shallow shelves and the deep channels suggest a relation between sediment-accumulation rate and depth; a plot of the two variables confirms the relation (Fig. 9). As in other presentations of the data, the high degree of scatter suggests a large amount of spatial variation. The long-term average rates of sediment accumulation clearly tend to increase progressively with depth, so that they are higher, on average, in the deep channels than on the marginal shelves. However, at any given depth, a wide range of sediment-accumulation rates was observed. The theoretical maximum long-term rate of sediment accumulation at any given location is defined by the rate of submergence; it amounts to the depth to the pre-Holocene unconformity at that point divided by the time since the submergence of that depth. It is equivalent to the rate of sediment accumulation that would have kept up with relative sea-level rise to maintain a zero water depth. The value of this theoretical maximum rate of sediment accumulation as a function of depth is plotted as the dashed line in Figure 9. The maximum measured rates of sediment accumulation consistently approach this theoretical maximum, but they are between 0.5 and 0.9 mm/yr below it. These data appear to support Rusnak's (1967) contention that, once equilibrium is reached, the rate of estuarine filling equals the rate of relative sea-level rise. Donoghue (1990) concluded that, until the present millennium at least, rates of sediment accumulation were equal to the rate of relative sea-level rise in Chesapeake Bay; our data support this conclusion for at least some locations in the bay.

Our calculations indicate that the two most important variables affecting average Holocene rates of sediment accumulation in Chesapeake Bay are depth to the base of the Holocene and distance along the axis of the bay. Other local variables must be important to produce the observed scatter in the relations between sediment-accumulation rates and depth (Fig. 9) or distance (Fig. 8A). However, depth and distance clearly control the large-scale variation in sed-

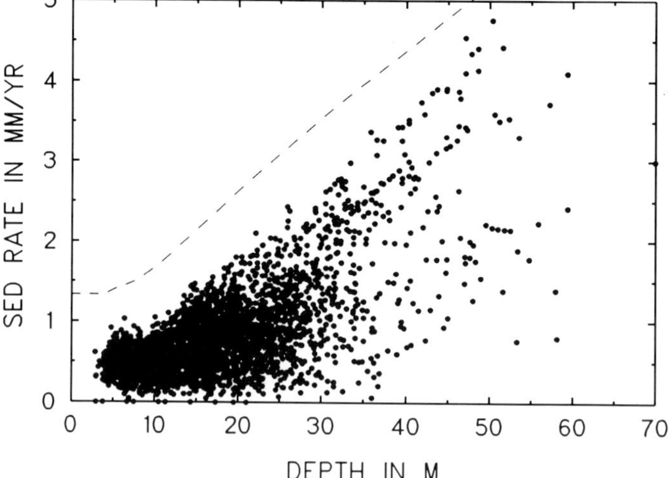

FIG. 9.—Average Holocene rates of sediment accumulation as a function of depth to base of the Holocene section. Label "sed rate" on ordinate refers to sediment-accumulation rate. Dashed line represents maximum possible average accumulation rate (see text).

iment-accumulation rates, so that sediment-accumulation rates can be viewed in terms of these two spatial variables (Fig. 10). The three-dimensional surface shown in Figure 10 obviously incorporates a great deal of averaging and smoothing, but it nevertheless shows the general relations well. Sediment-accumulation rates on shallow margins of the bay tend to be low and to increase only slightly toward the bay mouth. Sediment-accumulation rates in the deep channels are higher and increase distinctly toward the bay mouth. At any given position in the bay, sediment-accumulation rates increase with depth.

COMPARISON WITH OTHER MEASUREMENTS OF SEDIMENT-ACCUMULATION RATES

Average Holocene rates of sediment accumulation for Chesapeake Bay range from about 0 to 5 mm/yr (Fig. 10). On the marginal shelves, the rates are generally 0 to 2 mm/yr, whereas in the deeper channels, they range from about 1 to 5 mm/yr. These rates are rates of vertical change and do not take into account compaction of the sediments with time. Porosity and water contents of estuarine sediments in Chesapeake Bay vary with sediment type and with depth. With regard to sediment-accumulation rates, changes resulting from compaction with depth are the most important. For fine-grained sediments, the decrease in water content is typically 20 percent or less in the upper 1 to 2 m of sediment (Harrison and others, 1964; Biggs, 1970; Donoghue, 1990). Changes in water content or porosity in sand or below a depth of about 2 m in mud are generally small (Hobbs and others, 1990). Consequently, the long-term average rates calculated here may underestimate the initial rates of uncompacted sediment accumulation in fined-grained sediment by as much as 20 percent. This underestimation applies to the comparison between long-term and more recent rates of sediment accumulation and to the comparison between the recent sandy sediments near the bay mouth and the finer grained sediments farther up the estuary.

Donoghue (1990) has recently reviewed estimates of both short-term and long-term rates of sediment accumulation in Chesapeake Bay. The few previous estimates of long-term rates are mostly confined to tributary estuaries of the main bay. They range from 0.5 to 1.8 mm/yr, comparable to the rates we determined for the marginal shelves of the main bay. However, our calculations indicate that long-term rates of sediment accumulation in many of the deeper parts of the bay (Fig. 8C) are considerably higher than these previous estimates.

On the other hand, estimates of the rate of sediment accumulation in the Chesapeake Bay for the past few centuries are relatively abundant (Officer and others, 1984; Donoghue, 1990). These estimates have been related to distance along the axis of the bay with respect to sediment sources (Officer and others, 1984), but little attempt has been made to relate site-specific estimates of short-term rates to depth. Both our data (Fig. 9) and those from comparative bathymetry (Byrne and others, 1982; Kerhin and others, 1988; Hobbs and others, 1990) show that depth is an important variable controlling sediment-accumulation rates in the bay.

Short-term rates show extreme variation, but most range from about 1 to 10 mm/yr. The overall average of short-

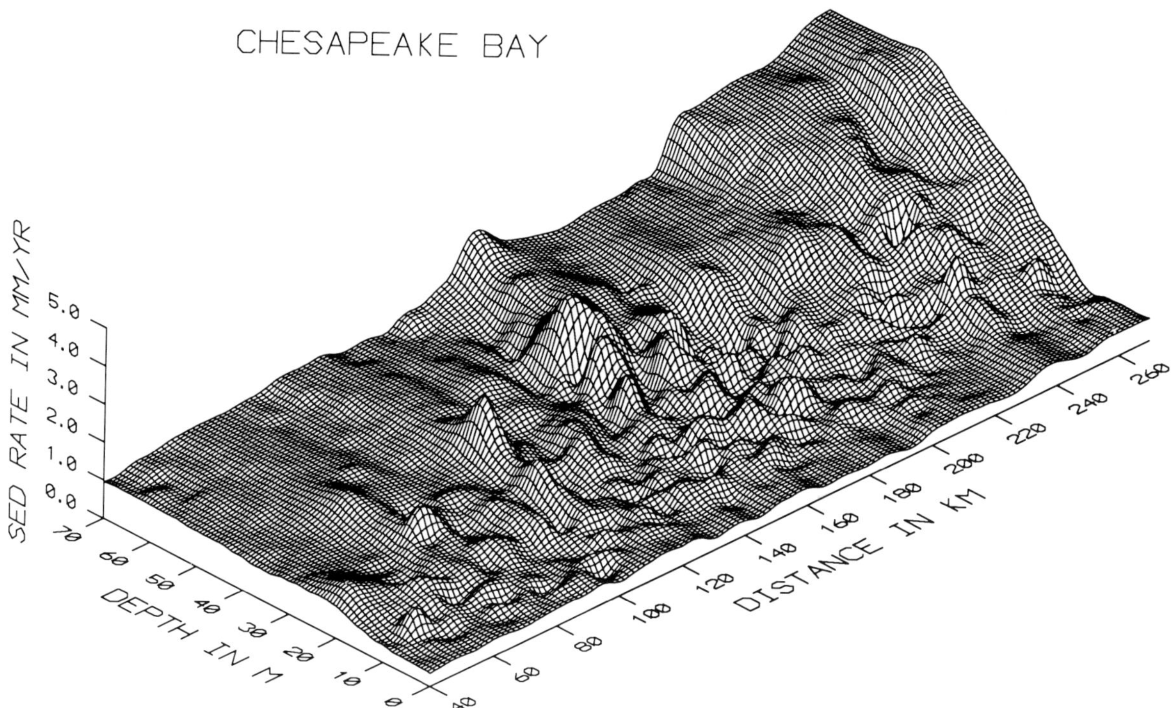

FIG. 10.—Average Holocene rates of sediment accumulation as a function of depth to base of Holocene section and distance along axis of bay. Label "sed rate" on ordinate refers to sediment-accumulation rate. Distances are measured on a north-south line from mouth of the Susquehanna River, as in Figures 5, 6, and 8.

PROVENANCE OF QUATERNARY BEACH DEPOSITS, VIRGINIA AND NORTH CAROLINA

DENNIS A. DARBY

Department of Geological Sciences, Old Dominion University, Norfolk, Virginia 23529

AND

ALLEN E. EVANS, JR.

Exxon Production Co., P.O. Box 61707, New Orleans, Louisiana 70161-1707

ABSTRACT: The elemental compositions of relatively unweathered Fe-Ti oxide grains, mostly ilmenite, separated from 83 samples collected from late Pleistocene to modern beach sands in Virginia and North Carolina were compared to those of 72 samples from five potential source rivers, the Roanoke, James, Potomac, Susquehanna, and Hudson Rivers. The composition of the Fe-Ti oxides from the toe of the Suffolk Scarp have a much different provenance than do younger beach deposits to the east. Based on discriminant analysis classification of the Fe-Ti oxide compositions with potential source rivers, the Suffolk Scarp beach is inferred to have been derived primarily from the James River; the younger beaches, including modern beach deposits of the Outer Banks, North Carolina, are inferred to have been primarily from the Susquehanna River with minor input by the Hudson River via longshore transport and reworking of shelf sands. The difference in provenance is due primarily to the origin of the Suffolk Scarp beach by erosion of older estuarine units in a protected-bay beach setting, whereas the younger beach deposits were derived from reworking of shelf sands, probably bay-mouth sand deposits (massifs), in an unprotected or barrier-beach setting. Subtle differences in the Fe-Ti oxide compositions among beach deposits are due to changes in the mix from the different river sources. Discrimination of the differences allows for a clearer understanding of the interrelation among those coastal-plain ridges and scarps that contain the beach sands.

INTRODUCTION

The United States Atlantic Coastal Plain contains several coast-parallel scarps and ridges representing Quaternary strandlines (Flint, 1940; Oaks and Coch, 1963; Cronin and others, 1981). At least one general consensus emerges from the numerous studies of the stratigraphy of these features: each successive ridge or scarp lower in elevation and closer to the present coast is younger in age. This succession is well represented in southeastern Virginia for a series of successively younger beach ridges east of and including the Suffolk Scarp (Fig. 1). Sedimentologic studies indicate that, regardless of whether the feature is named a scarp or a ridge, they all contain some well-sorted medium sands with features that might be indicative of beach or closely associated deposits, such as washover or shoreface deposits (Oaks and Coch, 1974; Darby, 1983). All of the beach sands from the scarps and ridges, except for those from Suffolk Scarp and Diamond Springs Scarp, have been interpreted as barrier-island beach deposits (Oaks and Coch, 1974). The latter two features contain beach deposits at the toe of an erosional scarp. The Diamond Springs Scarp also contains sands along its crest that have been interpreted as beach sands belonging to the same unit as beach sands in the Oceana Ridge (Oaks and Coch, 1974). However, the complex stratigraphy of the coastal plain makes the interrelations and origin of many of the beach deposits difficult to substantiate. In fact, the origin of the beach deposits, i.e., their depositional environment and provenance, have received far less attention than the stratigraphic scheme in which they occur (Oaks and Coch, 1974; Oaks and Dubar, 1974; Pebbles and others, 1984; Spencer and Campbell, 1987).

The use of detrital Fe-Ti oxide (primarily ilmenite) compositions for provenance has been demonstrated for some of the beach sands as well as shelf sands (Darby, 1984; Darby, 1990). This paper will use the same approach to provenance along with existing stratigraphic and sedimentological data in order to understand better the origin of the sands associated with late Pleistocene and modern beach ridges and deposits in southeastern Virginia and northeastern North Carolina.

RIDGE AND SCARP SANDS

Suffolk Scarp

The medium to coarse sands at the toe of the Suffolk Scarp (5 m to 6 m elevation) were assigned to the Norfolk Formation (Oaks and Coch, 1974). If correlative to fossiliferous sands beneath the Hickory Scarp, which were assigned to the same formation, now the Sedgefield Member of the Tabb Formation (Johnson, 1976; Cleaves and others, 1987; Mixon and others, 1989), then the sands are approximately 72 ka (Cronin and others, 1981). The Suffolk Scarp sands were originally described as the coarse sand facies of the Norfolk Formation and consist of 1- to 3-m-thick, medium to coarse sand and pebble gravel (Coch, 1968). The moderately sorted ($0.7\phi \pm 0.2\phi$; ± 1 standard deviation), medium-sand ($1.3\phi \pm 0.7\phi$) auger samples used in this study were overlain by a silty clay and silty sand deposit.

Hickory Scarp

The Hickory Scarp extends from the Diamond Springs Scarp near the Chesapeake Bay to Albemarle Sound (Darby, 1984, Fig. 1). Except for samples from three sand pits in the northern third of the area, all samples were obtained by hand auger, 1 to 1.5 m beneath the surface elevation (6.1 m). The sands range between 0.5ϕ and 3.5ϕ in mean size (Mz of Folk, 1974) with an average Mz of 2.0ϕ and are moderately sorted and near symmetrical in skewness. Low-angle ($<5°$) laminae with heavy-mineral concentrations are commonly found where exposures are good. Occasional fine pebbly lenses are also present (Darby, 1983). The sands typically grade into organic-rich muds and clays to the west and overlie 6 to 8 m of fossiliferous medium sand interpreted as bay deposits (Darby, 1983). The sands with beach characteristics are usually less than 2 m thick, are found at elevations between 3 and 5 m, and are overlain by pebbly mud 0.5 to 1.5 m thick.

Oceana Ridge

The Oceana Ridge was interpreted to be a spit-like barrier island that built southward from the Diamond Springs

Quaternary Coasts of the United States: Marine and Lacustrine Systems, SEPM Special Publication No. 48
 ISBN 0-918985-98-6

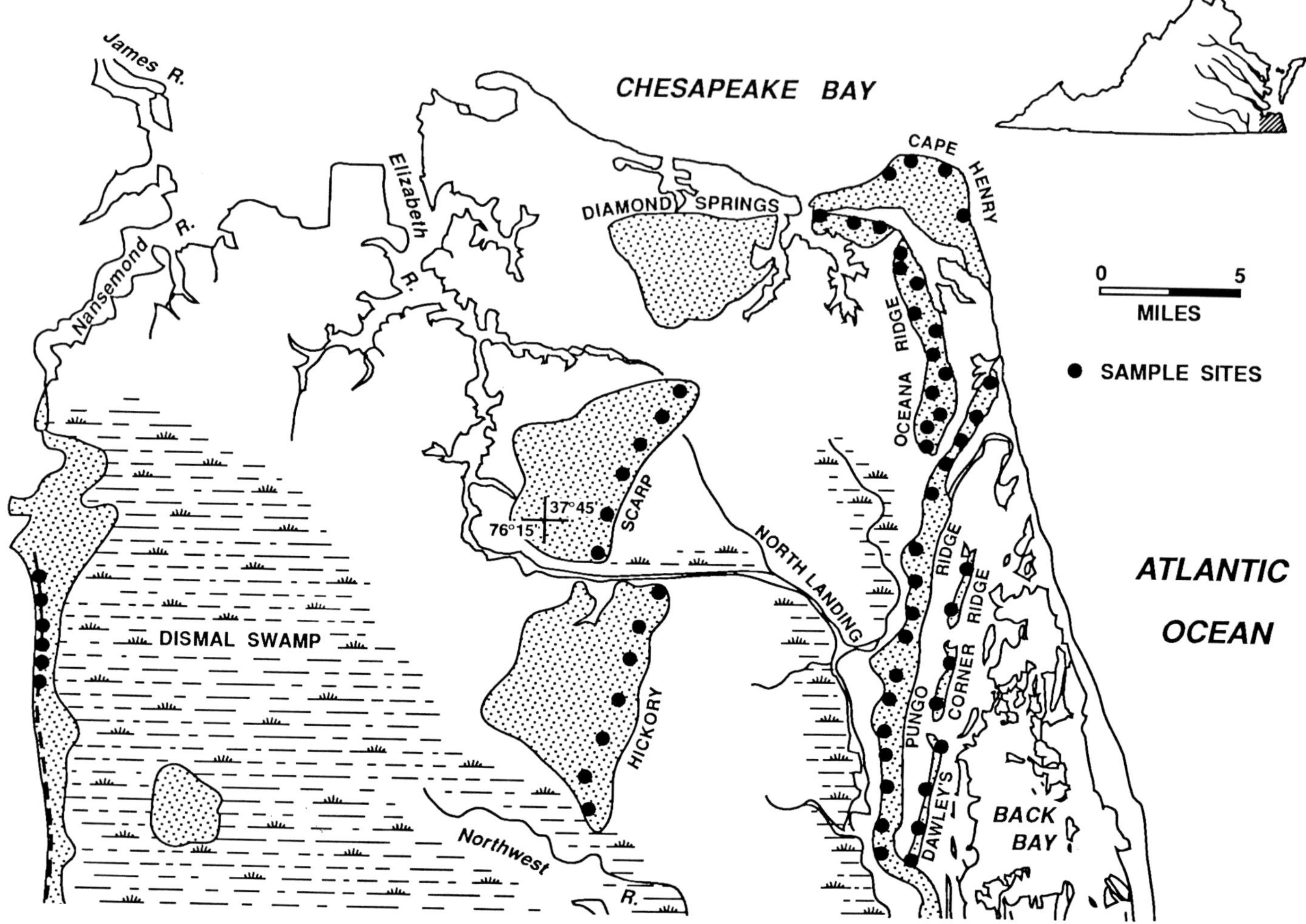

FIG. 1.—Map of scarps and ridges in southeastern Virginia with sample locations. Additional samples are shown in Figure 2. Morphology of sand deposits after Oaks and Coch (1974) and Mixon and others (1989).

Scarp area (Fig. 1; Oaks and Coch, 1974). The interpretation was based on ridge morphology, coarse texture, gently southeastward-dipping laminae (1–5°), and interfingering of ridge sands with silty muds to the west. Except for a sand pit sample at location 9 (Fig. 1), all samples were augered from depths of less than 1 m. They average 1.5ϕ (±0.3ϕ) mean size, 0.6ϕ (±0.2ϕ) sorting, and −0.1ϕ (±0.2ϕ) skewness. The sand unit is over 6 m thick with elevations up to 6 m (Oaks and Coch, 1974).

Pungo and Dawley Corners Ridges

The Pungo Ridge truncates the Oceana Ridge and closely parallels the Dawley Corners Ridge, which lies 2 km to the east (Fig. 1). Both ridges are between 1 and 3 m in elevation and contain moderately well-sorted (0.4–1ϕ), medium sand (1.8ϕ) with sparse pebbles. Here again, gently eastwardly inclined laminae with heavy-mineral concentrations are common. All samples from both ridges were obtained by hand auger at depths of less than 1.5 m.

Diamond Springs Scarp

Four auger samples were obtained from the crest of this largely erosional feature at elevations of about 5 m, or 1 m below the surface (Fig. 1). The texture of the sands is similar to that of modern beach sands and previously discussed ridge and scarp sands (mean size 1.3ϕ, sorting 0.7ϕ).

Cape Henry

The first dune/beach ridge landward of the modern beach and dune complex at Cape Henry was sampled at an elevation of about 3 m. The sand is similar to that of other ridge samples (mean size 2.0ϕ, sorting 0.6ϕ).

Outer Banks

Modern beach sands were collected at regular intervals along a 90-km segment of the Outer Banks north of Cape Hatteras (Darby, 1984).

POTENTIAL SOURCE RIVERS

Samples from five major potential source rivers were collected from unvegetated banks and mid-channel bars. A total of 72 samples from the Hudson River to the Roanoke River were used in order to ascertain the source of sand in the coastal-plain sand ridges (Fig. 2). In addition to samples from the lower reaches of the Susquehanna River, grab samples were used from the mid-bay channel of the Chesapeake Bay, where recent studies have shown deep erosion and exposure of older sediments (Colman and Halka, 1989). The Fe-Ti oxide composition of the grab samples is similar to those from the lower reaches of the Susquehanna River and differed from adjacent bay or estuary samples (Council, 1987; Darby, 1990).

Because modern beach sands along the Atlantic coast are partly derived from adjacent offshore sands (Swift, 1975), 22 surface grab samples from the Virginia Shelf were included in this study (Darby, 1990 Fig. 2).

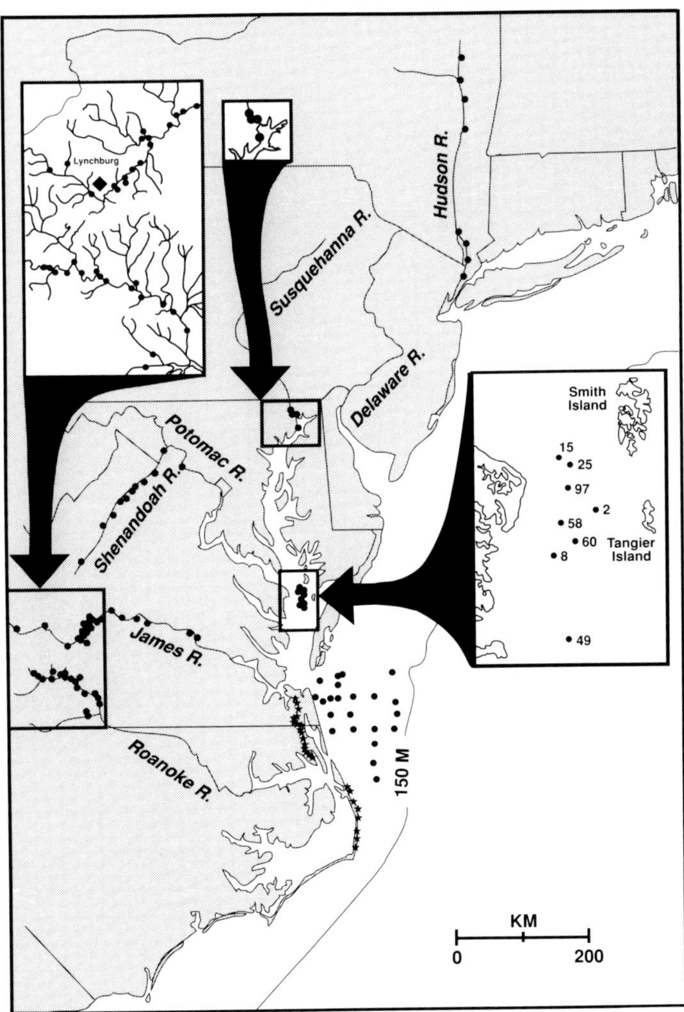

FIG. 2.—Potential source river sample locations including samples from the Virginia shelf (dots) and additional beach samples from the Hickory Scarp and modern Outer Banks beach (stars).

SAMPLE ANALYSIS

The Fe-Ti oxide grains were separated from the 2 to 4ϕ size fraction by a Frantz magnetic separator using 0.1 and 0.3 amp in order to obtain only the relatively unweathered grains (Darby, 1984; Darby, 1990). Non-opaque and obviously altered grains were removed with a fine brush under a dissecting microscope. The cleaned separates were fused with K-pyrosulphate, dissolved in nitric acid, and analyzed by flame atomic absorption (Darby and Tsang, 1987; Darby, 1990).

Examination of polished mounts of splits of two samples each from the Suffolk Scarp, Pungo Ridge, and Dawley Corners Ridge by ore microscope indicate that 97 to 99.6 percent of the grains are ilmenite with the remainder being titanomagnetite. Only one sample contained less than 1 percent hematite. Less than 3 percent of the ilmenite show hematite exsolution, but eroded linear voids thought to be previous hematite exsolution lamellae (Darby and Tsang, 1987) are found in all samples ranging from 15 to 34 percent of the ilmenite. There is no preference in abundance of voids between the younger or older beach deposits. Only 1 to 6 percent of the ilmenite grains contain leucoxene rims. Thus, microscopic examination indicate little, if any, alteration beyond stage one of Dimanche and Bartholome (1976). Because ilmenite dominates the Fe-Ti oxides, the magnetically separated fraction is referred to as ilmenite.

ILMENITE COMPOSITION

The average elemental composition of both beach and source river samples along with standard deviations show some variability even in the major elements, Fe and Ti (Table 1). The best procedure for examining the large data set is step-wise discriminant analysis (Klecka, 1975), an excellent statistical method for determining whether significant differences exist among groups of samples.

Beach Sands

Clearly, the Suffolk Scarp beach sand is distinctly different in its ilmenite composition from the other late Pleistocene beach sands, as demonstrated by a probability plot of the first two discriminant functions for the beach groups (Fig. 3). The robustness of the discriminant function analysis is demonstrated by the statistically significant low values of Wilk's lambda and high canonical correlations (Table 2). The most important variables in distinguishing the Suffolk Scarp sands are Mg, Cr, and Mn. The modern beach sand of the Outer Banks is also distinctly different, with Zn, V, Ti, and Cu most important. Because Zn was not analyzed in the Outer Banks samples (Darby, 1984), the discriminant analysis was run without this element and nearly the same amount of separation resulted.

The remaining beach sands clustered together and were analyzed separately in order to determine which were most similar or different. Surprisingly, there is no overlap between the Oceana Ridge sand and the Diamond Spring Scarp sand, and both were supposedly from the same unit, as suggested by Oaks and Coch (1974; Fig. 4). Instead, Diamond Springs Scarp beach sand is most similar to Cape Henry

TABLE 1.—MEAN Fe-Ti OXIDE COMPOSITIONS AND STANDARD DEVIATIONS (±1σ).

Sample GP.	No.	Al	Cr	Cu	Fe%	Mg	Mn	Ni	Ti%	V	Zn
Suffolk Scarp	6	NA	220	34	37.79	1007	11427	72	29.13	819	511
		—	79	4	1.31	486	921	11	1.11	77	25
Hickory Scarp	25	NA	456	37	36.66	2477	8854	63	27.04	839	348
		—	47	4	1.26	154	213	17	0.76	66	21
Oceana Ridge	13	NA	391	33	37.98	2602	9330	67	27.48	816	367
		—	46	4	1.77	135	233	11	1.74	57	28
Pungo Ridge	15	NA	383	34	36.44	2797	9209	73	28.19	821	317
		—	37	3	2.23	101	286	12	0.85	41	15
Dawley Corners Ridge	8	NA	339	36	35.64	2180	9544	67	26.73	810	351
		—	61	3	0.71	132	740	18	0.57	65	22
Diamond Springs Scarp	4	NA	345	29	37.70	2214	10741	29	25.97	799	376
		—	159	4	0.75	663	1029	6	0.34	2	57
Cape Henry Beach/Dune	4	NA	500	33	38.29	2237	9329	54	25.45	849	366
		—	125	2	0.69	250	578	15	0.29	60	8
Outer Banks Modern Beach	8	NA	207	55	37.75	2923	9293	50	24.73	152	NA
		—	49	13	0.72	291	450	11	0.14	113	—
Va. Shelf	22	1782	470	20	35.77	2042	9676	41	26.15	675	359
		956	244	10	0.26	482	1078	20	3.59	167	99
Hudson R.	8	1832	275	33	40.32	2229	6203	1	24.69	828	254
		755	102	35	0.94	510	1496	0.7	0.54	200	81
Susquehanna River*	11	1158	839	31	44.64	2468	8378	42	22.18	889	226
		169	632	32	0.43	256	1089	25	0.23	174	76
Potomac R.	10	1866	202	26	39.80	251	13032	47	20.66	1248	641
		466	74	17	0.36	107	2270	32	0.40	261	187
James River	25	1063	241	30	34.19	527	11269	50	20.55	878	574
		181	445	23	0.38	176	1235	19	0.34	262	134
Roanoke R.	18	NA	330	45	35.00	677	11383	51	17.41	918	555
		—	170	31	0.24	135	881	18	0.10	164	148

*Includes Samples from Chesapeake Bay mid-bay channel, interpreted to be exposed Susquehanna River sediments from the Holocene (Council, 1987; Colman and Halka, 1989; Darby, 1990). NA = not available. Values directly beneath each mean are one standard deviation. All values are in ppm (mg/kg) except those for Fe and Ti.

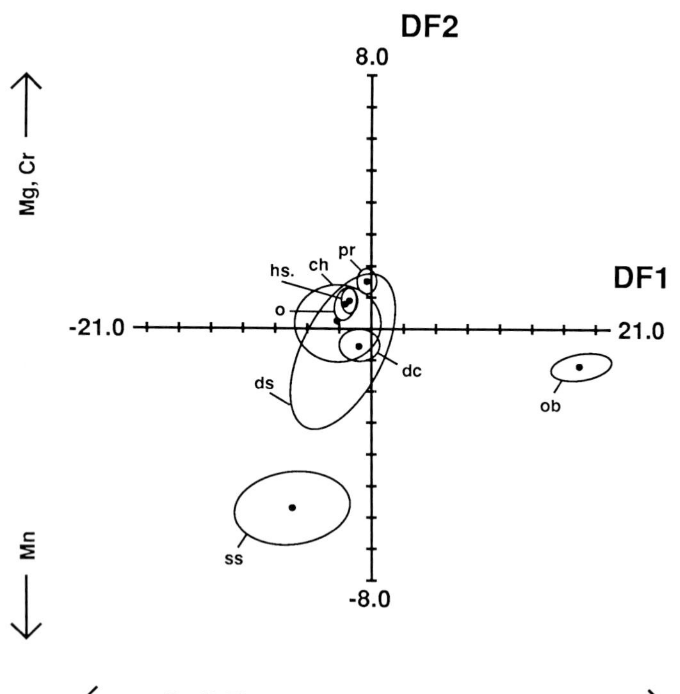

FIG. 3.—Probability ellipsoids (P = 0.95) of the coastal-plain beach deposits based on discriminant analysis of ilmenite compositions. Only the first two functions are plotted here, but they comprise 94 percent of the variation. The contribution of the most important variables is shown by arrows. ss = Suffolk Scarp, hs = Hickory Scarp, o = Oceana Ridge, pr = Pungo Ridge, dc = Dawley Corners Ridge, ds = Diamond Springs Ridge, ch = Cape Henry, and ob = Outer Banks, North Carolina.

TABLE 2.—STATISTICAL SIGNIFICANCE OF THE DISCRIMINANT FUNCTION ANALYSIS

Function	Canonical Correlation	Function Removed	Wilk's Lambda	Chi-Squared	Degree Freedom
		0	0.0038	303.82	40
1	0.9505	1	0.0393	176.43	27
2	0.9151	2	0.2414	77.45	16
3	0.7502	3	0.5523	32.35	7
4	0.6691				

sand, suggesting that the younger Cape Henry sand might have resulted in part from erosion of the Diamond Springs beach. On the other hand, Oceana sand is most similar in ilmenite composition to Pungo Ridge sands, as would be expected if the Pungo Ridge beach was partly derived by cannibalizing the Oceana Ridge, as postulated by Oaks and Coch (1974). Despite their juxtaposition, Pungo and Dawley Corners Ridges contain distinctly different ilmenite compositions (Fig. 4).

BEACH-SAND PROVENANCE

As previously demonstrated, ilmenite composition is quite useful in determining the source of sands (Darby, 1984; Darby, 1990). Tracing sands in this manner to source drainage basins requires that the ilmenite compositions in the potential source rivers be distinct. The largest rivers along the mid-Atlantic Coastal Plain have previously been shown to have distinct ilmenite compositions (Darby and Tsang, 1987; Darby, 1990), which is more obvious with a larger number of samples (Fig. 5).

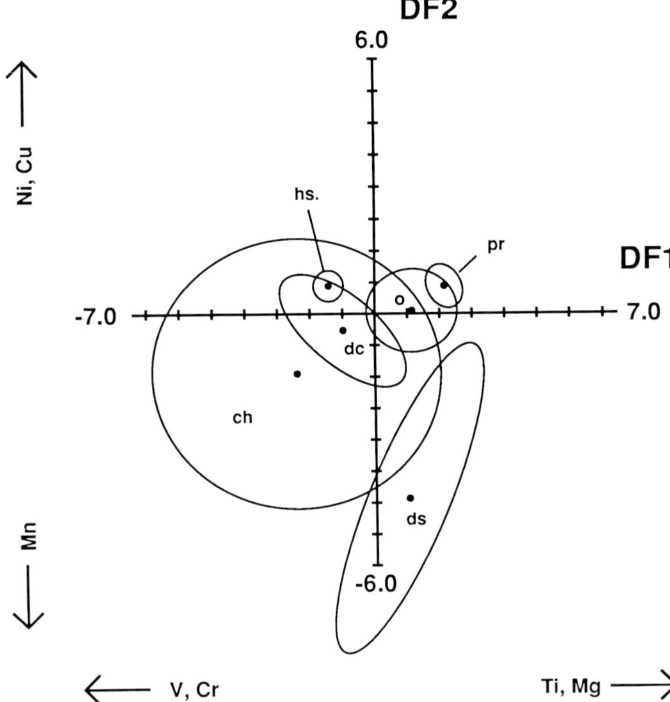

FIG. 4.—Probability ellipsoids (P = 0.95) for the beach deposits that overlapped one another in Figure 3 based on a separate discriminant analysis involving only these beach samples. Symbols are the same as in Figure 3.

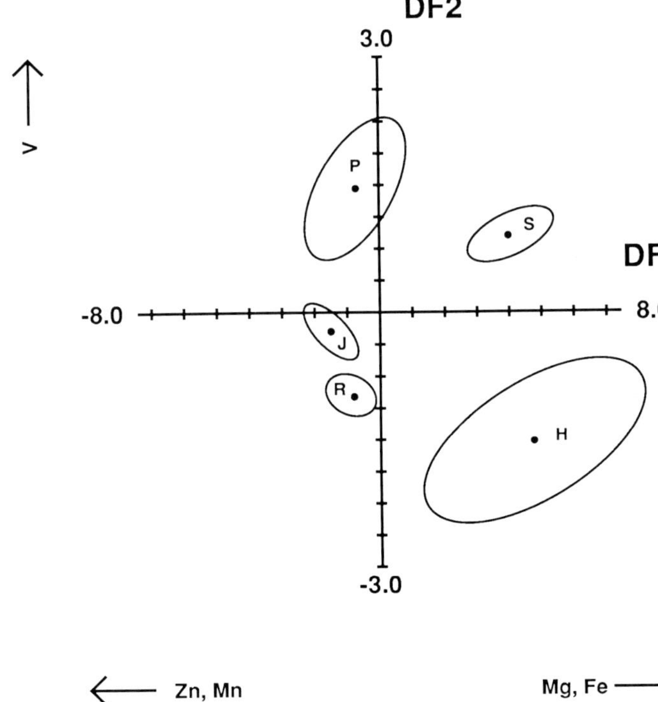

FIG. 5.—Probability ellipsoids (P = 0.95) for the potential source rivers. H = Hudson River, S = Susquehanna River, P = Potomac River, J = James River, and R = Roanoke River.

In order to determine the source for each beach sample, a step-wise discriminant function analysis was run leaving the beach samples to be classified with the most probable source river. The Suffolk Scarp beach sands classify primarily with those of the James River but at probabilities of less than 0.95 (Table 3), which indicates that some ilmenite and other clastic input was derived from the Potomac and possibly the Susquehanna Rivers (Table 3). All younger beach samples including those of the modern Outer Banks classify with the Susquehanna River at better than 0.95 average probability. This differs somewhat from an earlier interpretation of the Hickory Scarp and Outer Banks beach sands when Fe-Ti oxide compositions are used (Darby, 1984). The earlier study used mostly estaurine samples as potential source river groups. The discriminant function classification of the earlier samples indicated that Hickory Scarp and Outer Banks sand belonged to the James, Rappahannock, Potomac, and Roanoke River groups but at less than 0.95 probability (Darby, 1984). We have added many samples from the potential source rivers plus the Hudson River to define better the ultimate sources of coastal-plain beach sands. No estuarine samples were used as potential sources where the sampled sand could be partly eroded from Tertiary or Quaternary marine and fluvial units of the coastal plain. These units were ultimately derived from rivers prior to any coastal mixing and such mixing only obscures the ultimate provenance.

If the Virginia shelf sands are added as a source group, the discriminant analysis classifies all of the beach deposits except the Suffolk Scarp with the shelf ilmenite compositions (Table 3). Thus, the beach sands east of the Suffolk Scarp are more similar in ilmenite composition to modern shelf sands than to sands from individual source rivers, even the Susquehanna River. The Virginia shelf has been shown to be a mixture of sand from several rivers, primarily the Hudson River and other New England sources (glacial) as well as the Susquehanna River (Darby, 1990). The mixture suggests that the Hickory Scarp and younger beach sands were most likely derived from shelf sands similar in composition to those of the modern shelf but with a greater component from the Susquehanna River. Yet all of the beach sands can be distinguished by the ilmenite compositions. The differences among the beach deposits is due to the dif-

TABLE 3.—MEAN PROBABILITY OF BEACH SAMPLES BELONGING TO POTENTIAL SOURCE RIVER GROUPS BASED ON DISCRIMINANT ANALYSIS CLASSIFICATION

Beach Deposit	Hudson R.	Susquehanna R.	Potomac R.	James R.	Va. Shelf
Suffolk Scarp	0.0	0.16	0.21	0.62	0.16
Hickory Scarp	0.02	0.98	0.0	0.0	0.95
Oceana Ridge	0.0	0.99	0.0	0.0	0.95
Pungo Ridge	0.0	0.99	0.0	0.0	0.97
Dawley Corners	0.02	0.99	0.0	0.0	0.97
Diamond Springs	0.01	0.99	0.0	0.0	0.92
Cape Henry	0.02	0.98	0.0	0.0	0.92
Outer Banks	0.0	0.99	0.0	0.0	0.97
Virginia Shelf	0.36	0.40	0.13	0.08	0.0

The Roanoke River was not included because no samples classified with it. The mean probabilities listed for the Virginia shelf are from a separate discriminant analysis, which included all rivers and the shelf area.

ferent proportions in which various sources contribute to each beach.

Unlike the other beach deposits, the beach sands at the toe of the Suffolk Scarp group primarily with the James River, probably the result of local erosion of the scarp during beach-sand deposition. The eroded sands are from the mid-Pleistocene Windsor Formation, an estuarine to nearshore deposit (Oaks and Coch, 1974). The non-local sources of sand for the beach would be primarily via longshore drift. If the Suffolk Scarp beach sands correlate with fossiliferous sand underlying beach sands in the Hickory Scarp, as proposed by Oaks and Coch (1974), then the Suffolk Scarp beach would have been a protected beach, such as a bay beach. Both the elevation and interpreted bay environment of the Hickory Scarp sands suggest that a bay existed seaward of the Suffolk Scarp when the sands at the toe of this scarp were deposited. Most bay beaches have far less wave transport or longshore drift than barrier beaches. Therefore, local sources such as the James River drainage basin would dominate not only because the basin would have been a major contributor to the Windsor Formation, but also because it is closest to the Suffolk Scarp beach deposit.

The barrier-beach sands to the east received most of their sand from offshore or from erosion of former beaches that received sand from offshore (Swift, 1975; Swift and others, 1977; Shideler and others, 1972; Darby, 1990). The beach deposits received most of their sand from the Susquehanna River with some contribution from the Hudson River as evidenced by the discriminant classification of the sands with modern shelf sands off Virginia. The shelf sands have been shown to be derived primarily from both rivers (Darby, 1990). The fact that the beach samples did not classify with the Hudson River directly, as do many shelf samples, indicates that the Susquehanna is a more dominant source for the beach sands than for modern shelf sands. An explanation might be if the beach deposits were reworked from local bay-mouth massif sands and estuarine sands, which would be dominated by the Susquehanna River but still have significant input from the Hudson River and other New England sources, as postulated by Darby (1990).

CONCLUSIONS

The late Pleistocene beach sands at the toe of the Suffolk Scarp have a very different provenance than younger beach deposits eastward of this scarp. The most probable source for the Suffolk Scarp beach sands is the mid-Pleistocene Windsor Formation and possibly the nearby James River. The Windsor sediments were most likely derived largely from the James River with minor input from the Potomac and possibly the Susquehanna Rivers. The younger beach sands were derived primarily from the Susquehanna River with some input from the Hudson River and other New England sources via reworking of local shelf sands.

Whereas two different provenances and origins can be interpreted for the Suffolk Scarp and the younger beach deposits on the basis of ilmenite compositions, subtle differences among the younger beach deposits probably reflect minor differences in the mix of potential sources. Such differences question earlier stratigraphic connections between the beach sand in the crest of the Diamond Springs Scarp and that of the Oceana Ridge. These two beach deposits were more likely separate deposits, or at least were not part of the same longshore drift system. Our results support other previous interpretations for some of the beach deposits, such as the Pungo Ridge receiving sand from the truncated Oceana Ridge. We also suggest that the Holocene beach sampled at Cape Henry was largely derived from erosion of the Diamond Springs Scarp beach sand.

ACKNOWLEDGMENTS

We gratefully acknowledge the helpful suggestions of A. Basu, C. Nittrouer, and G. R. Whittecar, who reviewed earlier versions of this paper. We also thank the U.S. Geological Survey and the Virginia Institute of Marine Sciences for the 22 offshore samples.

REFERENCES

CLEAVES, E. T., GLASER, J. D., HOWARD, A. D., JOHNSON, G. H., WHEELER, W. H., SEVON, W. D., JUDSON, S., OWENS, J. P., AND PEEBLES, P. C., 1987, Quaternary geologic map of the Chesapeake Bay, 4° × 6° Quadrangle, United States: Richmond, G. M., Fullerton, D. S., and Weide, D. L., eds., U.S. Geological Survey Miscellaneous Investigations Series Map I-1420 (NJ-18).

COCH, N. K., 1968, Geology of the Benns Church, Smithfield, Windsor, and Chuckatuck quadrangles, Virginia: Virginia Divison of Mineral Resources Report of Investigations 28, 26 p.

COLMAN, S., AND HALKA, J., 1989, Map showing Quaternary geology of the southern Maryland part of the Chesapeake Bay: U.S. Geological Survey Map MF-1948C, 1:125,000.

COUNCIL, E. A., III, 1987, Provenance of sands within the lower Chesapeake Bay based on ilmenite composition: Unpublished M. S. Thesis, Old Dominion University, Norfolk, Virginia, 116 p.

CRONIN, T. M., SZABO, B. J., AGER, T. A., HAZEL, J. E., AND OWENS, J. P., 1981, Quaternary climates and sea levels of the U.S. Atlantic Coastal Plain: Science, v. 211, p. 233–240.

DARBY, D. A., 1983, Sedimentology, diagenesis, and stratigraphy of Pleistocene coastal deposits in southeastern Virginia: 15th Annual Virginia Geological Field Conference Guidebook, Old Dominion University, Norfolk, Virginia, 37 p.

DARBY, D. A., 1984, Trace elements in ilmenite: a way to discriminate provenance or age in coastal sands: Geological Society of America Bulletin, v. 95, p. 1208–1218.

DARBY, D. A., 1990, Evidence for the Hudson River as the dominant source of sand on the U.S. Atlantic shelf: Nature, v. 346, p. 828–831.

DARBY, D. A., AND TSANG, Y. W., 1987, The variation of ilmenite element composition within and among drainage basins: implications for provenance: Journal of Sedimentary Petrology, v. 57, p. 831–838.

DIMANCHE, F., AND BARTHOLOME, P., 1976, The alteration of ilmenite in sediments. Minerals Science Engineering: v. 8, p. 187–201.

FLINT, R. F., 1940, Pleistocene features of the Atlantic Coastal Plain: American Journal of Science, v. 238, p. 757–787.

FOLK, R. L., 1974, Petrology of Sedimentary Rocks: Hemphill Publishing Co., Austin, Texas, 182 p.

JOHNSON, G. H., 1976, Geology of the Mulberry Island, Newport News North, and Hampton Quadrangles, Virginia: Virginia Division of Mineral Resources Report of Investigations No. 41, 72 p.

KLECKA, W. R., 1975, Discriminant analysis, *in* Nie, N. H., Hull, C. H., Jenkins, J. G., Steinbrenner, K., and Bent, D. H., eds., Statistical Package for the Social Sciences, 2nd edition: McGraw-Hill Book Co., New York, p. 434–467.

MIXON, R. B., BERQUIST, C. R., JR, NEWELL, W. L., JOHNSON, G. H., POWARS, D. S., SCHINDLER, J. S., AND RADAR, E. K., 1989, Geologic map and generalized cross sections of the coastal plain and adjacent parts of the piedmont, Virginia: U.S. Geological Survey Miscellaneous Investigation Series Map I-2033.

Oaks, R. Q., Jr, and Coch, N. K., 1963, Pleistocene sea levels, southeastern Virginia: Science, v. 140, p. 979–983.

Oaks, R. Q., and Coch, N. K., 1974, Post-Miocene stratigraphy and morphology, southeastern Virginia: Virginia Division of Mineral Resources Bulletin, no. 82, 135 p.

Oaks, R. Q., and Dubar, J. R., 1974, Tentative correlation of Post-Miocene units, central and southern Atlantic Coastal Plain, *in* Oaks, R. Q., Jr, and Dubar, J. R., eds., Post-Miocene Stratigraphy, Central and Southern Atlantic Coastal Plain: Utah State University Press, p. 232–245.

Peebles, P. C., Johnson, G. H., and Berquist, C. R., Jr, 1984, The Middle and Late Pleistocene stratigraphy of the outer coastal plain, southeastern Virginia: Virginia Minerals, v. 30, p. 13–21.

Shideler, G. L., Swift, D. J. P., Johnson, G. H., and Holliday, B. W., 1972, Late Quaternary stratigraphy of the inner Virginia continental shelf: a proposed standard section: Geological Society of America Bulletin, v. 83, p. 1787–1804.

Spencer, R. S., and Campbell, L. D., 1987, The fauna and paleoecology of the Late Pleistocene marine sediments of southeastern Virginia: Bulletin of American Paleontology, v. 92, p. 1–124.

Swift, D. J. P., 1975, Barrier-island genesis: evidence from the central Atlantic Shelf, eastern U.S.A: Sedimentary Geology, v. 14, p. 1–43.

Swift, D. J. P., Nelson, T., McHone, J., Holliday, B., Palmer, H., and Shideler, G., 1977, Holocene evolution of the inner shelf off southern Virginia: Journal of Sedimentary Petrology, v. 47, p. 1454–1474.

COASTAL RESPONSE TO LATE PLIOCENE CLIMATE CHANGE: MIDDLE ATLANTIC COASTAL PLAIN, VIRGINIA AND DELAWARE

KELVIN W. RAMSEY
Delaware Geological Survey, University of Delaware, Newark 19716

ABSTRACT: The middle Atlantic Coastal Plain records the late Pliocene transition from typical marine deposition and coastlines of the Tertiary to that of fluvial-estuarine-marine of the Quaternary. The Early Pliocene Yorktown Formation (4.5–3.0 Ma) and Late Pliocene Chowan River Formation (approx. 2.8 Ma) record dominantly marine deposition in temperate to warm temperate climates with moderate- to low-sedimentation rates in a tectonically controlled basin. The Late Pliocene Bacons Castle Formation in Virginia (2.3–2.0 Ma) and the upper Beaverdam Formation of Delaware (2.3–2.0 Ma?) contain deposits with indicators of high rates of sedimentation, abundant sediment supply, an Appalachian source, and a cool terrestrial climate. Deposition was in response to a colder climate in the source area associated with Northern Hemisphere glaciation at approximately 2.4 Ma. Mechanical weathering and colluviation during a colder period moved sediment into the transport system that was later moved and deposited by high-discharge streams flowing into the coastal plain during subsequent warming. Deposition during the ensuing transgression kept pace with sea-level rise with some progradation of coastline. Later, Quaternary erosion and deposition in the region incised and partially reworked these late Pliocene deposits.

INTRODUCTION

The middle Atlantic Coast is an embayed coastline dominated by drowned river valleys (estuaries), such as the Chesapeake and Delaware Bays, with barrier-lagoon segments between the estuary mouths (Fig. 1). This coastline is the result of glacio-eustatic fluctuation of sea level, fluvial erosion during periods of low relative sea level (Colman and Mixon, 1988), and the influence of high-discharge glacial meltwater and headward erosion of tributary streams (White, 1979). A coastline similar to that of the present has existed in the region throughout the Quaternary with modifications mainly in the form of the southward progradation of the Delmarva Peninsula (Mixon, 1985), progradation of the shoreline south of the Chesapeake Bay and James River (Peebles, 1984), and fluvial and estuarine erosion and deposition along the rivers with rise of sea level (Peebles and others, 1984).

In contrast, the late Tertiary (Miocene-Early Pliocene) coastlines of the region were marginal to a large marine embayment (Salisbury Embayment) and consisted of scarped shorelines, in some places rocky, with small deltas where rivers flowed out of the Piedmont (Newell and Rader, 1982) and tidal flats in areas of inundated low relief. The transition between these two types of coastlines during the late Pliocene is the subject of this paper. The transition is attributed to increased sediment supply due to a change in climate in the source areas in the late Pliocene. Attention is focused on the Bacons Castle Formation of Virginia with some discussion of the Beaverdam Formation in Delaware.

STRATIGRAPHIC FRAMEWORK

Pliocene deposits are exposed over a wide area in the middle Atlantic Coastal Plain (Fig. 1). Of these deposits, the Yorktown Formation (early Pliocene) represents the last major phase of marine deposition and typical Testiary coastlines in the middle Atlantic Coastal Plain. Age estimates based on biostratigraphic and radiometric data indicate an age range bracketed by 4.8 to 2.4 Ma (Cronin and others, 1984; Fig. 2). Calibration with deep-sea isotopic records suggests that the age range is closer to 4.5 to 3.0 Ma (Krantz, 1991). At least three separate transgressions occurred during this interval (Ward and Blackwelder, 1980), lowstands being marked by unconformities and lithologic breaks within the Yorktown. Highest sea levels during deposition of the Yorktown were approximately 35 to 40 m above present (Krantz, 1991). The marine Yorktown extends from North Carolina throughout much of the coastal plain of Virginia and the southern Delmarva Peninsula of Virginia and Maryland (Ward and Blackwelder, 1980; Mixon and others, 1989).

The Chowan River Formation (late Pliocene) rests unconformably on the Yorktown Formation in southeastern Virginia and northeastern North Carolina. Biostratigraphic and radiometric age estimates for the unit range between 2.4 and 1.9 Ma (Cronin and others, 1984). Placement of the Chowan River in the late Pliocene is uncertain. Ramsey (1988) placed it at approximately 2.5 Ma on the basis of stratigraphic considerations. Groot (1991) placed the Chowan River at about 2.8 Ma based on palynology. Krantz (1991), however, proposed a possible younger date of 2.1 to 2.2 Ma on the basis of calibration with deep-sea isotopic records, although he does allow for the possibility of older ages between 2.9 and 2.6 Ma. The Chowan River Formation is a marine deposit capped by estuarine or lagoonal sediments in North Carolina (Blackwelder, 1981). A maximum sea-level high was approximately 15 m above present.

The Bacons Castle Formation rests unconformably on the Yorktown Formation in Virginia (Ramsey, 1988) and North Carolina and on the Chowan River Formation in North Carolina (Johnson and others, 1987). Unlike the Yorktown and Chowan River Formations, the Bacons Castle is generally nonfossiliferous. The Bacons Castle is considered to be late Pliocene on the basis of stratigraphic position (Ramsey, 1988) and palynological data (Groot, 1991). An age between 2.3 and 2.0 Ma is likely. The Bacons Castle is a transgressive unit consisting of fluvial to estuarine and tidal-flat deposits (Ramsey, 1988). Based on estimated late Cenozoic uplift rates for the region (0.025 m ka^{-1}, Pavich, 1984; 0.01 to 0.03 m ka^{-1}, Cronin, 1981; and 0.016 m ka^{-1}, Ramsey, 1988), and a 20-m difference in present maximum elevations between the Yorktown and Bacons Castle, maximum sea level for the Bacons Castle was the same as, to 10 m

Quaternary Coasts of the United States: Marine and Lacustrine Systems, SEPM Special Publication No. 48

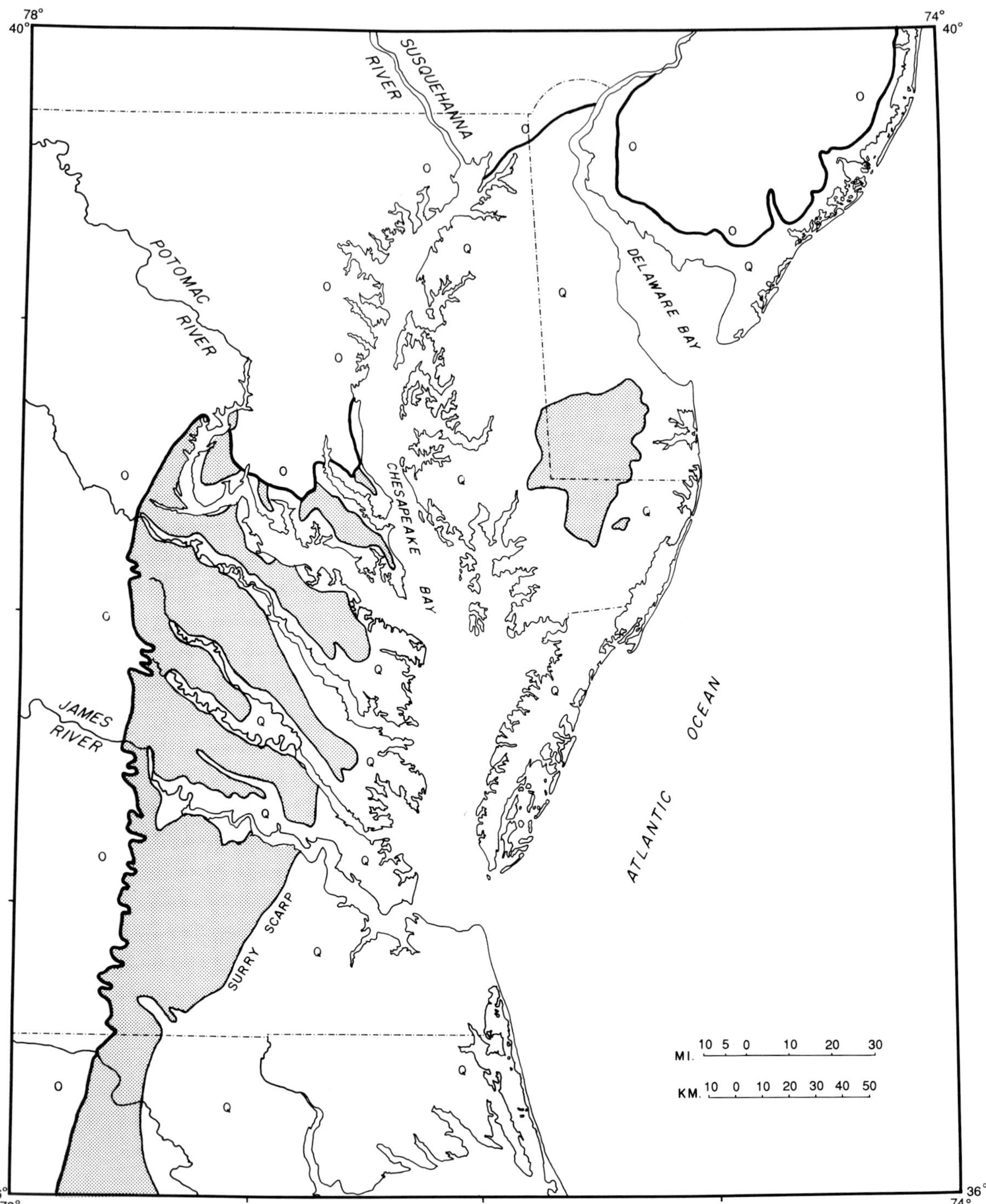

FIG. 1.—Location map. Shaded area represents area of deposits of Pliocene age exposed at the land surface. O—Rocks older than Pliocene. Q—Quaternary deposits. Note incision and deposition of Quaternary age deposits along major rivers and seaward of these deposits in Virginia. Data compiled from Mixon and others (1989); Owens and Denny (1978, 1979); Ramsey and Schenck (1990); McCartan (1989).

less than, maximum sea level for the Yorktown. A range of 30 to 40 m above present is likely. After Bacons Castle deposition, sea level dropped and a drainage system and coastline similar to those of today began to develop. The next major depositional phase produced the coast-parallel Surry Scarp (Fig. 1) and the deposits of the Windsor Formation in the early Pleistocene (1.5 Ma).

Direct correlation of Pliocene units across the Chesapeake Bay from Virginia to the Delmarva Peninsula in the subsurface has been difficult because of differences in

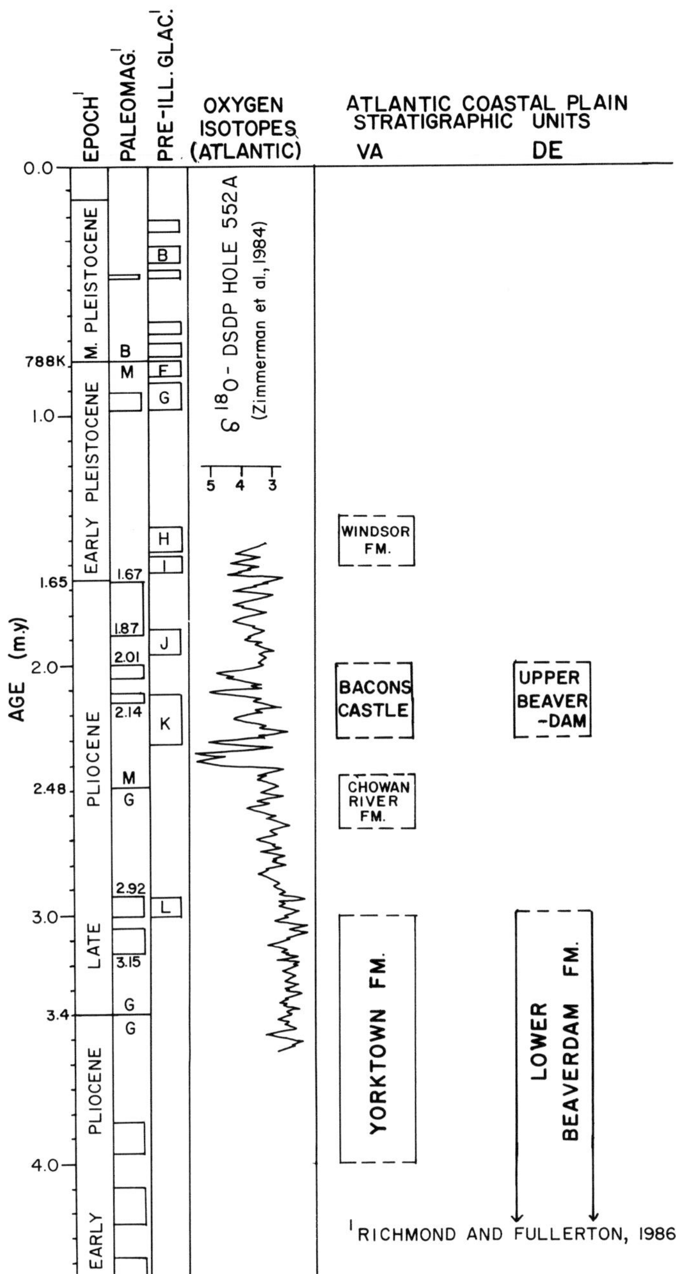

FIG. 2.—Stratigraphy of Pliocene deposits of the middle Atlantic Coastal Plain.

structure and tectonics (Brown and others, 1972). The marine Yorktown has been found on the Delmarva Peninsula only as far north as the Virginia-Maryland line (Mixon, 1985). North of this area lies a thick section of generally nonfossiliferous sand (Beaverdam Formation and Pocomoke aquifer) with scattered pollen-bearing clay beds (Hansen, 1981; Groot and others, 1990). No age equivalents of the Yorktown Formation, based on palynostratigraphic data, have been confirmed in Delaware. The lower Beaverdam in Delaware is probably in part correlative with the Yorktown to the south. The upper part of the Beaverdam Formation, however, is similar to that of the Bacons Castle Formation both in pollen assemblage content (Groot and others, 1990; Groot, 1991) and in depositional style. The Beaverdam was deposited in fluvial to estuarine environments (Groot and others, 1990). The upper Beaverdam is considered to be late Pliocene (between 2.4 and 2.0 Ma) on the basis of pollen assemblages (Groot, 1991; Fig. 2).

EARLY PLIOCENE DEPOSITION AND COASTLINES

A maximum-transgression coastline for the Yorktown Formation (Fig. 3) shows a relatively straight coastline south of the present James River, and to the north a tidal(?) delta filling the curve of the Salisbury Embayment. The configuration of the coastline is controlled in part by regional tectonics with uplift to the north and depression to the southeast (Newell and Rader, 1982, their fig. 7). Local deposition was influenced by position of structural troughs and highs that controlled distribution of lithofacies within the Yorktown (Newell and Rader, 1982; Johnson and others, 1987; Fig. 3). This type of structural control was common in the Salisbury Embayment throughout the Tertiary (Mixon and Newell, 1982; Newell and Rader, 1982). Sediment supply was moderate. Progradation of coastal over marine deposits occurred only to the north (Newell and Rader, 1982; Fig. 3) and may have been the result of tectonics rather than increased sediment supply. Sands in the marine Yorktown

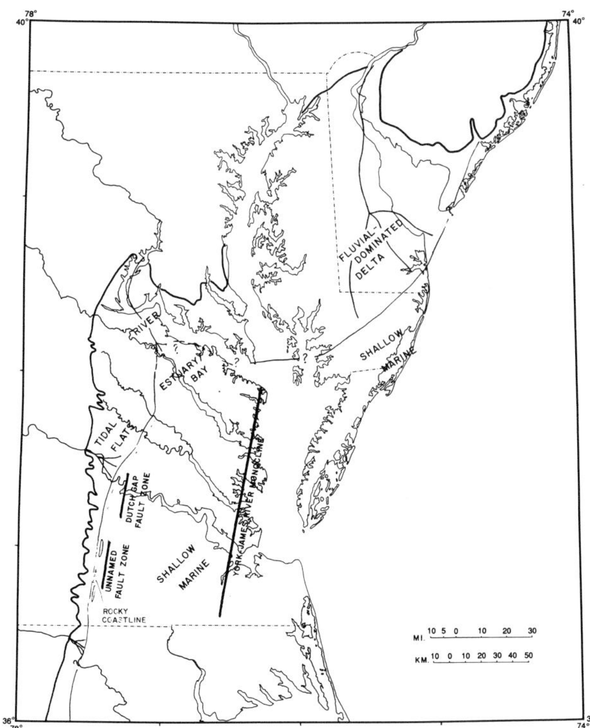

FIG. 3.—Coastal configuration of the middle Atlantic Coastal Plain during the early Pliocene at approximately maximum Yorktown transgression. Location of York-James River Monocline from Ward and Blackwelder (1980); Dutch Gap Fault zone from Dischinger (1987); and unnamed fault zone (Virginia Division of Mineral Resources, unpublished data).

are typically fine to medium grained (Ramsey, 1988). Sands in coastal environments of the Yorktown are more variable in texture, but are generally medium grained (Ramsey, 1988). Pebbles, where present, are siliceous, mostly quartz. Gravels are scattered and restricted to where rivers flowed out of the Piedmont and may be in part recycled from older fluvial, deltaic, and nearshore deposits (Ramsey, 1988). Sediment texture and mineralogy, then, suggest recycling of older deposits and limited contribution of "new" material to the basin. This along with the progradational (regressive) vertical sequence indicates tectonics to be a greater factor than sediment supply in influencing coastal configuration.

The straight, southern coastline (Fig. 3) was dominated by beaches along a scarped coastline of low cliffs. Vertical sequences suggest that barrier islands existed with tidal flats landward (Berquist and Goodwin, 1989). The tidal flats were eventually transgressed and overlain by beach and nearshore deposits (Berquist and Goodwin, 1989). In places, rocks of the Piedmont were exposed along the shore, the last time a rocky coastline existed that far to the south, along the Atlantic. To the north in Virginia (Fig. 3), the coastline was dominated by a prograded deltaic sequence with tidal flats of interbedded and interlaminated sand and mud, fluvial and estuarine gravelly and sandy deposits, and nearshore cross-bedded sands (Newell and Rader, 1982; Mixon and others, 1989). Barrier islands may also have been present, but have been removed by subsequent Bacons Castle erosion and deposition. On the Delmarva Peninsula probable contemporaneous deposits to the Yorktown are not fossiliferous. Sands and gravels of the lower Beaverdam (Groot and others, 1990) probably represent this part of the section, assuming continued deposition within this part of the Salisbury Embayment from the late Miocene to the early Pliocene as occurred throughout the rest of the basin. Otherwise, a mechanism for a shutdown of sediment supply for the Susquehanna and Delaware river systems for *only* the early Pliocene will have to be derived. Trends of deposition in the lower Beaverdam are similar to those in the late Miocene/early Pliocene Bethany Formation in a prograding deltaic system (Andres, 1986; Groot and others, 1990). Exact position of the shoreline is speculative. Coastal configuration and deposition during the early Pliocene in the region was largely controlled by tectonics overriding sediment supply as a deciding factor.

LATE PLIOCENE DEPOSITION AND COASTLINES

No coastal reconstruction for the Late Pliocene Chowan River Formation is given. The maximum extent of the Chowan River has been removed by subsequent erosion during the Late Pliocene Bacons Castle and younger Pleistocene transgressions. As in the case of the Yorktown Formation, deposition in the Chowan River was in part controlled by structure (Johnson and others, 1987). Estuarine deposits that form the upper part of the Chowan River (Colerain Beach Member) in the type area (Blackwelder, 1981) suggest that an embayment existed in North Carolina during the latter stages of Chowan River deposition. Only marine deposits are preserved in its northern extent in Virginia.

The Bacons Castle Formation rests unconformably on the Yorktown Formation. In much of the area of the contact, a paleosol formed on the marine Yorktown has been recognized (Ramsey, 1988; Macdonald and Ramsey, 1989), suggesting prolonged exposure of the Yorktown prior to Bacons Castle deposition. The contact is an eroded surface with paleovalleys and paleointerfluves (Ramsey, 1988). The paleovalleys roughly coincide to locations of present river systems flowing out of the Piedmont. Regional relief on this surface is about 28 m (12 to 40 m above present sea level); local relief is no more than 10 m. Although the relief on a regional scale may appear to be subtle, it played an important role in localizing and influencing deposition of lithofacies during the Bacons Castle transgression (Ramsey, 1988). This influence is in marked contrast to deposition controlled by regional tectonics or localized structural troughs (Newell and Rader, 1982) during Yorktown and Chowan River time. Contacts at the base of the Yorktown or Chowan River are marked by low regional relief, a lag of pebbles or cobbles, and no evidence of soil development (Ramsey, 1988; Johnson and others, 1987).

Sediment supply was higher for the Bacons Castle than for previous deposits. Gravels and gravelly sands are a significant component of the fluvial portion of the Bacons Castle. Extensive tidal-flat deposits within the unit are rarely bioturbated, suggesting high-sedimentation rates, and are sand dominated, indicating a constant supply of sand throughout deposition (Ramsey, 1988). The mineralogy of the Bacons Castle sands also suggests a different, or an additional, source of material than that of the Yorktown. The non-opaque heavy-mineral fraction of the Yorktown is amphibole rich (Lawrence, 1974; Ramsey, 1988), whereas that of the Bacons Castle is dominated by zircon. The feldspar component is also higher in the Bacons Castle than in the Yorktown (Coch, 1968; Ramsey, 1988). The pebble fraction of the Bacons Castle contains an abundance of chert (up to 15 percent) derived from an Appalachian source (Ramsey, 1988) that is present only rarely in the Yorktown.

Two reconstructions of coastlines are given for late Pliocene deposits of Virginia and Delaware. The first reconstruction (Fig. 4) is approximately midway through Bacons Castle deposition. South of the James River, fluvial sands and gravels were spread out in a braided river-dominated deltaic system that had filled the lows of the antecedent topography and had prograded to the southeast. The seaward extent of this system has been removed by subsequent erosion. Tidal deposits predominate from the present James River northward (Ramsey, 1988). These tidal deposits onlap onto antecedent highs of Yorktown and, to the north, even older deposits. The barrier islands shown are speculative; no clear evidence remains of their position. The extensive nature of the tidal deposits, however, suggests some seaward protection. Some fluvial deposition may have occurred along the ancestral Potomac River. No Chesapeake Bay existed at the time. The ancestral Susquehanna River, however, may have been near its present course. Coarse sands and gravels of the upper Beaverdam Formation (Groot and others, 1990) were deposited in a river-dominated deltaic system, with probable contribution from both the Susquehanna and Delaware systems. Lithologies of the Bea-

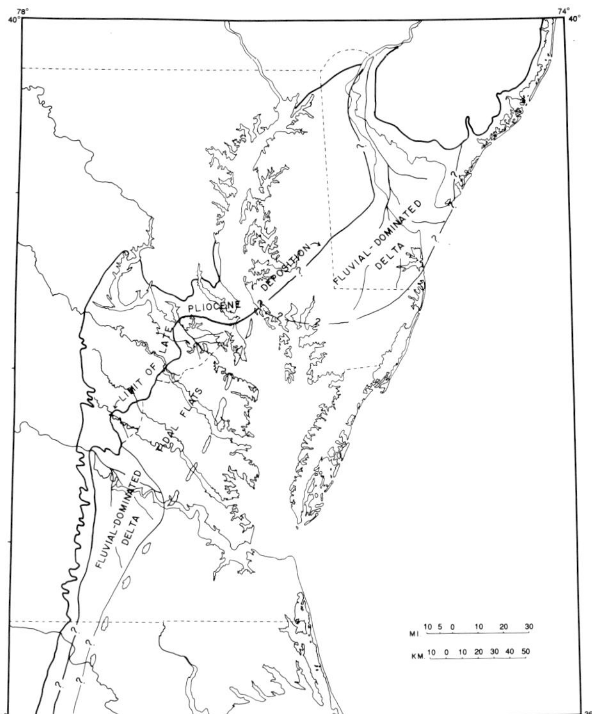

FIG. 4.—Coastal configuration of the middle Atlantic Coastal Plain during the late Pliocene at middle of Bacons Castle/upper Beaverdam transgression.

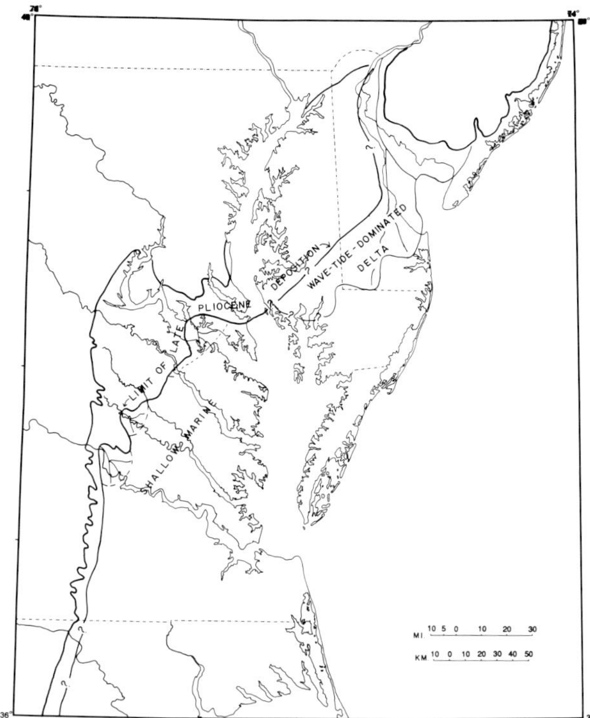

FIG. 5.—Coastal configuration of the middle Atlantic Coastal Plain during the late Pliocene at maximum Bacons Castle/upper Beaverdam transgression.

verdam and Bacons Castle are similar, dominated by medium to coarse sands with lesser amounts of mud. Gravels are common with pebble compositions, including abundant chert, indicating an Appalachian source.

The second reconstruction (Fig. 5) depicts the maximum transgression during Bacons Castle time. There are some similarities in coastlines between this and the maximum Yorktown transgression. A relatively straight coastline existed south of the James with a more crenulate coastline to the north. Unlike the Yorktown, the seaward component of the Bacons Castle consisted of a very shallow embayment with estuarine muddy sands deposited as a cap over older fluvial and tidal deposits. For the most part, sedimentation kept pace with transgression during Bacons Castle deposition (Ramsey, 1988) with an overall aspect of progradation. On the Delmarva Peninsula, fluvial conditions gave way to the sand-dominated estuarine system of the upper Beaverdam (Groot and others, 1990; Ramsey and Schenck, 1990). Only a generalized coastline for this area (Fig. 5) can be given because subsequent erosion and deposition have obscured facies relations within the Beaverdam. There are no indications of local structural control of deposition within either the Bacons Castle or upper Beaverdam Formations.

CLIMATE AND COASTAL CHANGE

Changes in coastal configuration between the early and late Pliocene were in response to increased sediment supply and deposition rates. The overall effect of late Pliocene deposition was progradation into the depositional basin of a mass of sediment later incised by erosion, and modified by deposition during the Quaternary. Two factors can be called upon to explain the influx of sediment during the late Pliocene, tectonics and climate.

The appearance of coarse clastics in late Tertiary coastal-plain deposits has been attributed to uplift in the Appalachians (Owens, 1970). Uplift in a source area affects the transport system by contributing to incision and downslope movement of sediment, but does not increase discharge unless relief and slope are increased. Models of uplift for the Appalachians (Hack, 1982; Pavich, 1984) argue for isostatic uplift in response to denudation. Mass removal and uplift could be accomplished with relatively little change in elevation, hence, little change in relief or slope. However, uplift rates of less than 40 m per million years have been suggested with a long-term average between 5 and 20 m per million years (Pavich, 1984, 1989). Uplift does not necessarily induce sediment production. In eastern Spain, uplift during the late Pliocene of 300 m per million years induced stream capture and incision. Sediment transport and deposition, however, only occurred during the glacial periods of the last 2 million years in response to climate change, not uplift (Harvey and Wells, 1987). A similar argument is here made for late Pliocene deposition in the middle Atlantic Coastal Plain.

Early Pliocene climate in the Atlantic Coastal Plain was relatively constant. Water temperatures in the marine environment increased somewhat during this period with the closure of the Isthmus of Panama (3.5–2.5 Ma) and reorientation of the Gulf Stream along eastern North America

(Hodell and others, 1985; Dowsett and Cronin, 1990). These temperatures are often cited for the climate of the region being warm temperate to subtropical during the early (Yorktown Fm) and early late (Chowan River Fm) Pliocene (Cronin and others, 1984). On the basis of palynology, however, terrestrial temperatures do not appear to have been much different than those of today, with warmer temperatures existing only during the later stages of the Chowan River (Groot and others, 1990; Groot, 1991). The Late Pliocene Bacons Castle and the upper part of the Beaverdam Formation reflect a significant cooling trend with temperatures cooler than those of today, indicated by a common component of *Picea* (spruce) not seen in older Pliocene deposits (Groot, 1991).

When these climatic indicators are taken into account with stratigraphic (Ramsey, 1988) and sea-level (Krantz, 1991) considerations, placement of the Bacons Castle and upper Beaverdam just after the cooling period associated with the onset of Northern Hemisphere glaciation (2.5–2.4 Ma) is reasonable. A sea-level lowstand after Yorktown deposition (about 3.0 Ma) and some fluctuation of sea level afterward (Chowan River, about 2.8 Ma) gave way to decreased sea levels with the glaciations of 2.6 and 2.4 Ma (Zimmerman and others, 1984; Krantz, 1991). This would allow approximately 500,000 years for soil formation on and incision of the Yorktown prior to Bacons Castle deposition. The production and transport of sediment for these deposits was forced by the climate change and its effect on the source area. Intensity of this late Pliocene glaciation on the North American continent is debatable. It has been suggested that intensity of glaciation in the Appalachians in the late Pliocene was like that of the late Wisconsinan glaciation (Braun, 1989).

Colder climate in the source area associated with the glaciation produced an increase in mechanical weathering (Conners, 1986), and creep and other mass-movement processes (Kite, 1987) contributed to lower angle slopes and collected sediment near the transport systems (Costa and Cleaves, 1984). Moderation of climate and increased rainfall, runoff, and discharge flushed the sediment down the transport system (Conners, 1986). The volume of sediment involved in the late Pliocene deposition was also enhanced by a long period of storage and production and saprolite and colluvium during the warmer Miocene and early Pliocene. A rearrangement of floral components (Groot, 1991) may also indicate a period of decreased vegetative cover during cooler conditions. These cycles of colluviation and alluviation continued throughout the Quaternary (Costa and Cleaves, 1984; Conners, 1986). The magnitude and volume of sediment distributed on the emerged coastal plain appears to have decreased from the Pliocene through the Quaternary. The decrease may be attributed to a decrease of total sediment available in the source area (Ramsey, 1988), or to increased bypassing and deposition of sediment on the outer shelf and slope during the Quaternary (Poag and Sevon, 1989). Quaternary coastal change on the middle Atlantic Coastal Plain consisted primarily of repeated shoreline progradation and incision of rivers and repeated formation of estuaries during each low- and highstand of sea level, respectively (Peebles and others, 1984; Mixon, 1985; Colman and Mixon, 1988; Mixon and others, 1989).

CONCLUSIONS

Tertiary deposition and coastlines in the middle Atlantic Coastal Plain were controlled by local and regional tectonic influence with a moderate supply of sediment distributed primarily in marine and nearshore environments. Coastal deposition and coastline configuration during the Quaternary were controlled by the position of through-flowing rivers incised into older deposits during sea-level lowstands and filled with sediment during sea-level rise. Progradation of coastlines occurred where sediment supply was sufficient and the transport system was capable of moving excess sediment. The transition between the two styles occurred during a relatively brief span of time (approx. 400,000–600,000 years) during the late Pliocene. The transport, volume, and texture of the transition deposits were influenced by climatic changes in the source area, and an available large volume of sediment in storage in the Appalachians. The time of this change occurred during the interval of 2.3 to 2.0 Ma in the late Pliocene. The coastline reflected progradation at a time of general sea-level rise. Further detailed investigation into these deposits may enhance our understanding of both earlier Tertiary and later Quaternary coastal deposition and processes within the region.

ACKNOWLEDGMENTS

Work on the Bacons Castle Formation was supported by the Department of Geology at the University of Delaware and by grants from the Geological Society of America and Sigma Xi. Tom Gardner, C. R. Berquist, and Richard Benson reviewed the manuscript and offered many helpful suggestions. Discussions with J. J. Groot, David Krantz, and Frank Pazzaglia helped iron out some of the difficulties in dating of the deposits. The years of discussion with Jerre Johnson on coastal-plain problems and stratigraphy are gratefully acknowledged.

REFERENCES

Andres, A. S., 1986, Stratigraphy and depositional history of the post-Choptank Chesapeake Group: Delaware Geological Survey Report of Investigations No. 42, 39 p.

Berquist, C. R., Jr., and Goodwin, B. K., 1989, Terrace gravels, heavy mineral deposits, and faulted basement along and near the Fall Zone in southeastern Virginia: Virginia Geological Field Conference, 21st Annual Meeting, Guidebook No. 5, Department of Geology, College of William and Mary, 33 p.

Blackwelder, B. W., 1981, Stratigraphy of upper Pliocene and lower Pleistocene marine and estuarine deposits of northeastern North Carolina and southeastern Virginia: U.S. Geological Survey Bulletin 1502-B, 19 p.

Braun, D. D., 1989, Glacial and periglacial erosion of the Appalachians: Geomorphology, v. 2, p. 233–256.

Brown, P. M., Miller, J. A., and Swain, R. M., 1972, Structural and stratigraphic framework, and spatial distribution of permeability of the Atlantic Coastal Plain, North Carolina to New York: U.S. Geological Survey Professional Paper 796, 79 p.

Coch, N. K., 1968, Geology of the Benns Church, Smithfield, Windsor, and Chuckatuck quadrangles, Virginia: Virginia Division of Mineral Resources Report of Investigations 17, 40 p.

Colman, S. M., and Mixon, R. B., 1988, The record of major Quaternary sea-level changes in large coastal plain estuary, Chesapeake Bay,

Eastern United States: Palaeogeography, Palaeoclimatology, Palaeoecology, v. 68, p. 99–116.

CONNERS, J. A., 1986, Quaternary geomorphic processes in Virginia, *in* McDonald, J. N., and Bird, S. O., eds., The Quaternary of Virginia–A Symposium Volume: Virginia Division of Mineral Resources Publication 75, p. 1–22.

COSTA, J. R., AND CLEAVES, E. T., 1984, Piedmont landscape of Maryland: a new look at an old problem: Earth Surface Processes and Landforms, v. 9, p. 59–74.

CRONIN, T. M., 1981, Rates and possible causes of neotectonic vertical crustal movements of the emerged southeastern U.S. Atlantic Coastal Plain: Geological Society of America Bulletin, v. 92, p. 812–833.

CRONIN, T. M., BYBELL, L. M., POORE, R. Z., BLACKWELDER, B. W., LIDDICOAT, J. C., AND HAZEL, J. E., 1984, Age and correlation of emerged Pliocene and Pleistocene deposits, U.S. Atlantic Coastal Plain: Palaeogeography, Palaeoclimatology, Palaeoecology, v. 47, p. 21–51.

DISCHINGER, J. B., Jr., 1987, Late Mesozoic and Cenozoic stratigraphic and structural framework near Hopewell, Virginia: U.S. Geological Survey Bulletin 1567, 48 p.

DOWSETT, J. H., AND CRONIN, T. M., 1990, High eustatic sea level during the middle Pliocene: evidence from the southeastern U.S. Atlantic Coastal Plain: Geology, v. 18, p. 435–438.

GROOT, J. J., 1991, Palynological evidence for late Miocene to early Pleistocene climate changes in the middle Atlantic Coastal Plain: Quaternary Science Reviews, v. 10, p. 147–162.

GROOT, J. J., RAMSEY, K. W., AND WEHMILLER, J. F., 1990, Ages of the Bethany, Beaverdam, and Omar Formations of Southern Delaware: Delaware Geological Survey Report of Investigations No. 47, 19 p.

HACK, J. T., 1982, Physiographic divisions and differential uplift in the Piedmont and Blue Ridge: U.S. Geological Survey Professional Paper 1265, 49 p.

HANSEN, H. J., 1981, Stratigraphic discussion in support of a major unconformity separating the Columbia Group from the underlying upper Miocene Aquifer Complex in eastern Maryland: Southeastern Geology, v. 22, p. 123–138.

HARVEY, A. M., AND WELLS, S. G., 1987, Response of Quaternary fluvial systems to differential epeirogenic uplift: Aguas and Feos river systems, southeast Spain: Geology, v. 15, p. 689–693.

HODELL, D. A., WILLIAMS, D. F., AND KENNETT, J. P., 1985, Late Pliocene reorganization of deep vertical water-mass structure in the western South Atlantic: faunal and isotopic evidence: Geological Society of America Bulletin, v. 96, p. 495–503.

JOHNSON, G. H., WARD, L. W., AND PEEBLES, P. C., 1987, Stratigraphy and paleontology of Pliocene and Pleistocene deposits of southeastern Virginia, *in* Whittecar, G. R., ed., Geologic excursions in Virginia and North Carolina: Guidebook to Southeast Section, Geological Society of America 1987 Field Trips, Norfolk, Virginia, p. 189–218.

KITE, J. S., 1987, Colluvial diamictons in the Valley and Ridge Province, West Virginia and Virginia, *in* Schulz, A. P., and Southworth, C. S., eds., Landslides of Eastern North America: U.S. Geological Survey Circular 1008, p. 21–23.

KRANTZ, D. E., 1991, A chronology of Pliocene sea-level fluctuations: the U.S. Middle Atlantic Coastal Plain record: Quaternary Science Reviews, v. 10, p. 163–174.

LAWRENCE, C. T. E., Jr., 1974, Optical and statistical analysis of heavy minerals from the Yorktown Formation: Virginia Journal of Science, v. 25, p. 93.

MCCARTAN, L., 1989, Geologic Map of St. Mary's County: Maryland Geological Survey County Geologic Map.

MACDONALD, R. H., AND RAMSEY, K. W., 1989, Development of a Pliocene paleosol in the coastal plain of Virginia: Geological Society of America, Abstracts with Programs, v. 21, p. 30.

MIXON, R. B., 1985, Stratigraphic and geomorphic framework of uppermost Cenozoic deposits in the southern Delmarva Peninsula, Virginia and Maryland: U.S. Geological Survey Professional Paper 1067-G, 53 p.

MIXON, R. B., AND NEWELL, W. L., 1982, Mesozoic and Cenozoic compressional faulting along the Atlantic Coastal Plain margin, Virginia, *in* Lyttle, P. E., ed., Central Appalachian Geology: NE–SE Geological Society of America Field Trip Guidebook, p. 29–54.

MIXON, R. B., BERQUIST, C. R., Jr., NEWELL, W. L., JOHNSON, G. H., POWARS, D. S., SCHINDLER, J. S., AND RADAR, E. K., 1989, Geologic map and generalized cross sections of the coastal plain and adjacent parts of the Piedmont, Virginia: U.S. Geological Survey Miscellaneous Investigations Series Map I-2033.

NEWELL, W. L., AND RADER, E. K., 1982, Tectonic control of cyclic sedimentation in the Chesapeake Group of Virginia and Maryland, *in* Lyttle, P. E., ed., Central Appalachian Geology: NE-SE Geological Society of America Field Trip Guidebook, p. 1–27.

OWENS, J. P., 1970, Post-Triassic tectonic movements in the Central and Southern Appalachians as recorded by sediments of the Atlantic Coastal Plain, *in* Fisher, G. W., Pettijohn, F. J., Reed, J. C., Jr., and Weaver, K. N., eds., Studies of Appalachian Geology, Central and Southern: Interscience Publishers, New York, p. 417–427.

OWENS, J. P., AND DENNY, C. S., 1978, Geologic Map of Worcester County: Maryland Geological Survey County Geologic Map.

OWENS, J. P., AND DENNY, C. S., 1979, Geologic Map of Wicomico County: Maryland Geological Survey County Geologic Map.

PAVICH, M. J., 1984, Appalachian Piedmont morphogenesis: weathering, erosion, and Cenozoic uplift, *in* Morisawa, M., and Hack, J. T., eds., Tectonic Geomorphology: Allen and Unwin, Boston, p. 299–319.

PAVICH, M. J., 1989, Regolith residence time and the concept of surface age of the Piedmont "Peneplain": Geomorphology, v. 2, p. 181–196.

PEEBLES, P. C., 1984, Late Cenozoic landforms, stratigraphy, and history of sea level oscillations of southeastern Virginia: unpublished Ph.D. Dissertation, Virginia Institute of Marine Science, College of William and Mary, 227 p.

PEEBLES, P. C., JOHNSON, G. H., AND BERQUIST, C. R., Jr., 1984, The middle and late Pleistocene stratigraphy of the outer coastal plain, southeastern Virginia: Virginia Minerals, v. 30, p. 13–22.

POAG, C. W., AND SEVON, W. D., 1989, A record of Appalachian denudation in post-rift Mesozoic and Cenozoic sedimentary deposits of the U.S. middle Atlantic Continental Margin: Geomorphology, v. 2, p. 119–157.

RAMSEY, K. W., 1988, Stratigraphy and sedimentology of a late Pliocene intertidal to fluvial transgressive deposit: Bacons Castle Formation, Upper York-James Peninsula, Virginia: Unpublished Ph.D. Dissertation, Department of Geology, University of Delaware, Newark, Delaware, 399 p.

RAMSEY, K. W., AND SCHENCK, W. S., 1990, Geologic map of southern Delaware: Delaware Geological Survey Open-File Report No. 32.

RICHMOND, G. M., AND FULLERTON, D. S., 1986, Summation of Quaternary glaciations in the United States of America, *in* Sibrava, V., Bowen, D. Q., and Richmond, G. M., eds., Quaternary Glaciations in the Northern Hemisphere: Quaternary Science Reviews, v. 5, p. 183–200.

WARD, L. W., AND BLACKWELDER, B. W., 1980, Stratigraphic revision of upper Miocene and lower Pliocene beds of the Chesapeake Group, middle Atlantic Coastal Plain: U.S. Geological Survey Bulletin 1482-D, 61 p.

WHITE, W. A., 1979, Influence of glacial meltwater in the Atlantic Coastal Plain: Southeastern Geology, v. 19, p. 139–156.

ZIMMERMAN, H. B., SHACKLETON, N. J., BACKMAN, J., BALDAUF, J. G., KALTENBACH, A. J., AND MORTON, A. C., 1984, History of Plio-Pleistocene climate in the northeastern Atlantic, Deep Sea Drilling Project Hole 552A: Initial Report of the Deep Sea Drilling Project, v. 81, p. 861–875.

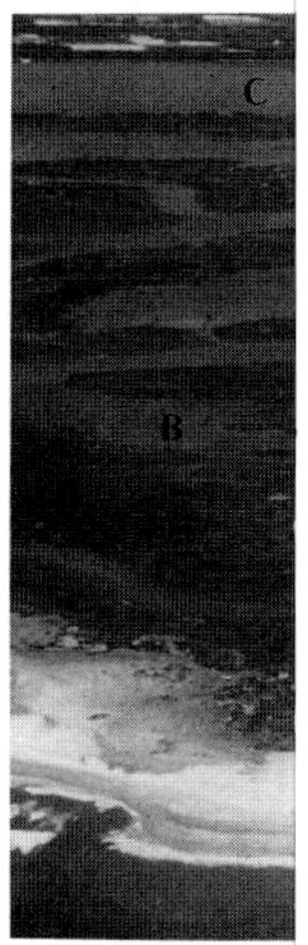

STRATIGRAPHY AND PRESERVATION POTENTIAL OF SEDIMENTS FROM ADJACENT HOLOCENE AND PLEISTOCENE BARRIER-ISLAND SYSTEMS, CAPE CHARLES, VIRGINIA

KENNETH FINKELSTEIN
Arthur D. Little, Inc., Acorn Park, Cambridge, Massachusetts 02140-2390

ABSTRACT: A total of 68 vibracores and 14 box cores in conjunction with high-resolution seismic records are used to describe the late Quaternary development of a twin-barrier-island complex. Based on the stratigraphy, lithology, radiocarbon dates, and microfossils, both a transgressive outer Holocene and inner Pleistocene barrier-island complex are recognized. The two subaerial subparallel barriers are a result of separate marine transgressions that occurred before and after late Wisconsin glaciation. The present landward migration of the Holocene barrier should put it atop the Pleistocene barrier in approximately 1,400 years.

Conformable and unconformable contacts separate the two barrier-island systems. The outer barrier island, Smith Island, is typical of the Holocene Virginia barrier-island chain; its dimensions are approximately 11 km in length, and width and height mostly in the range of 200 m and 1.5 m above mean sea level, respectively. The Pleistocene barrier, Mockhorn Island, lies approximately 7 km landward of Smith Island but within the back-barrier lagoon created by the Holocene marine transgression. Although Pleistocene shorelines are recognized on the mainland, these two chronologically distinct barrier systems are in contact, resulting in a geomorphology and upper Quaternary stratigraphy unique to the middle Atlantic coast.

Holocene sediments deposited in the back-barrier environment show a general shallowing and fining-upward sequence. A significant portion of these back-barrier deposits, at least, but probably greater than, 3 m thick, should be preserved below 75 to 100-cm-thick nearshore sands. However, inlet-fill deposits likely will not be preserved. With present sea-level rise and wave-base conditions continuing into the future, the lower Holocene and uppermost Pleistocene transgressive sequences have a strong preservation potential. The preserved succession of deposits consists of Pleistocene back-barrier mud and shoreface sand below Holocene back-barrier deposits. The stacking of transgressive barrier deposits, albeit those from different transgressions, might serve as a stratigraphic petroleum trap if preserved into the geologic record.

INTRODUCTION

Barrier-island stratigraphy has received considerable attention during the past four decades. Transgressive and regressive models, depicting the Holocene stratigraphy, are well documented. Some of the more seminal examples are Fisk (1959), Fischer (1961), and Kraft (1971). Most studies include a discussion describing the chronology of barrier development during a single transgression or regression. Moreover, only a speculation of the resulting stratigraphy after a subsequent cycle is usually available. In addition, when considering the landward retreating barriers of the United States Atlantic and Gulf Coasts, the potential for the preservation of transgressive barrier complex deposits on the inner shelf is usually not expected to be extensive (Belknap and Kraft, 1981), especially under wave-dominated conditions (Davis and Clifton, 1987). However, where multiple transgressive barrier shorelines parallel the coast, a portion of such sediments may overlap, resulting in the preservation of a segment of each specific transgressive unit.

This study examines the geomorphic, chronologic, and stratigraphic relations between adjacent Pleistocene and Holocene barrier-island complexes located on the eastern shore of Virginia (Fig. 1). A transgressive stratigraphic sequence underlies both Mockhorn Island (Finkelstein and Kearney, 1988), the inner Pleistocene island located in the midst of the modern back-barrier environment (Shideler, and others, 1984; Finkelstein and Kearney, 1988), and the more seaward Holocene barrier, Smith Island (Finkelstein and Ferland, 1987). The two transgressive barrier systems provide an opportunity to describe their present stratigraphic relation and to predict the resulting stratigraphy and preservation of sediments when the two barriers collide.

Although such stacking of transgressive deposits is not common and justifiably not well represented in the literature, it is noteworthy that several papers describe transgressive Pleistocene barriers found landward or underlying retreating Holocene barriers–in Georgia (Hoyt and Hails, 1967), Delaware (Demarest and others, 1981), and Australia (Thom and others, 1981). The general model of deposition that emerges from these studies is one of transgressive barrier complexes potentially preserved in a shore-normal direction; in the case of this study, a predicted stacking and resultant more permanent preservation of such barrier-island deposits is demonstrated.

PHYSICAL AND GEOLOGIC SETTING

The area of investigation comprises two spatially and chronologically distinct subparallel barrier-island complexes, Smith and Mockhorn islands, and their adjacent nearshore and back-barrier subenvironments (Fig. 2). Located on the southern Delmarva Peninsula, the study area is bounded by Myrtle Island on the north and the Chesapeake Bay entrance on the south; the east-west limits are the lower shoreface and the mainland back-barrier shoreline, respectively. This shoreline is defined by a sharp linear contact. Pleistocene fluvial and marine silts, sands, and gravels (Mixon, 1985) are as high as 3 m, 500 m from this shoreline. The two barrier islands are separated from each other and the mainland by broad expanses of marsh, mud flats, lagoon, and tidal creeks. The region is relatively unmodified by human influence.

Smith Island is approximately 11 km long, of which the northern 8 km of the island is 200 m in width and no higher than 1.5 m high above mean sea level (MSL). The southern portion is a broad beach-ridge area. The distance from Smith Island to the mainland ranges from 10.5 km at the north end to 5.5 km at the southern tip. On the average, Mockhorn Island is about 7.0 km landward of Smith Island. Smith island is rapidly retrograding at a rate historically ranging between 4 and 15 m/yr (Rice and others, 1976; Dolan and others, 1979; Rice and Leatherman, 1983) and averaging 5.6 m/yr since 1852 (Everts, 1987). Evidence of barrier

Quaternary Coasts of the United States: Marine and Lacustrine Systems, SEPM Special Publication No. 48

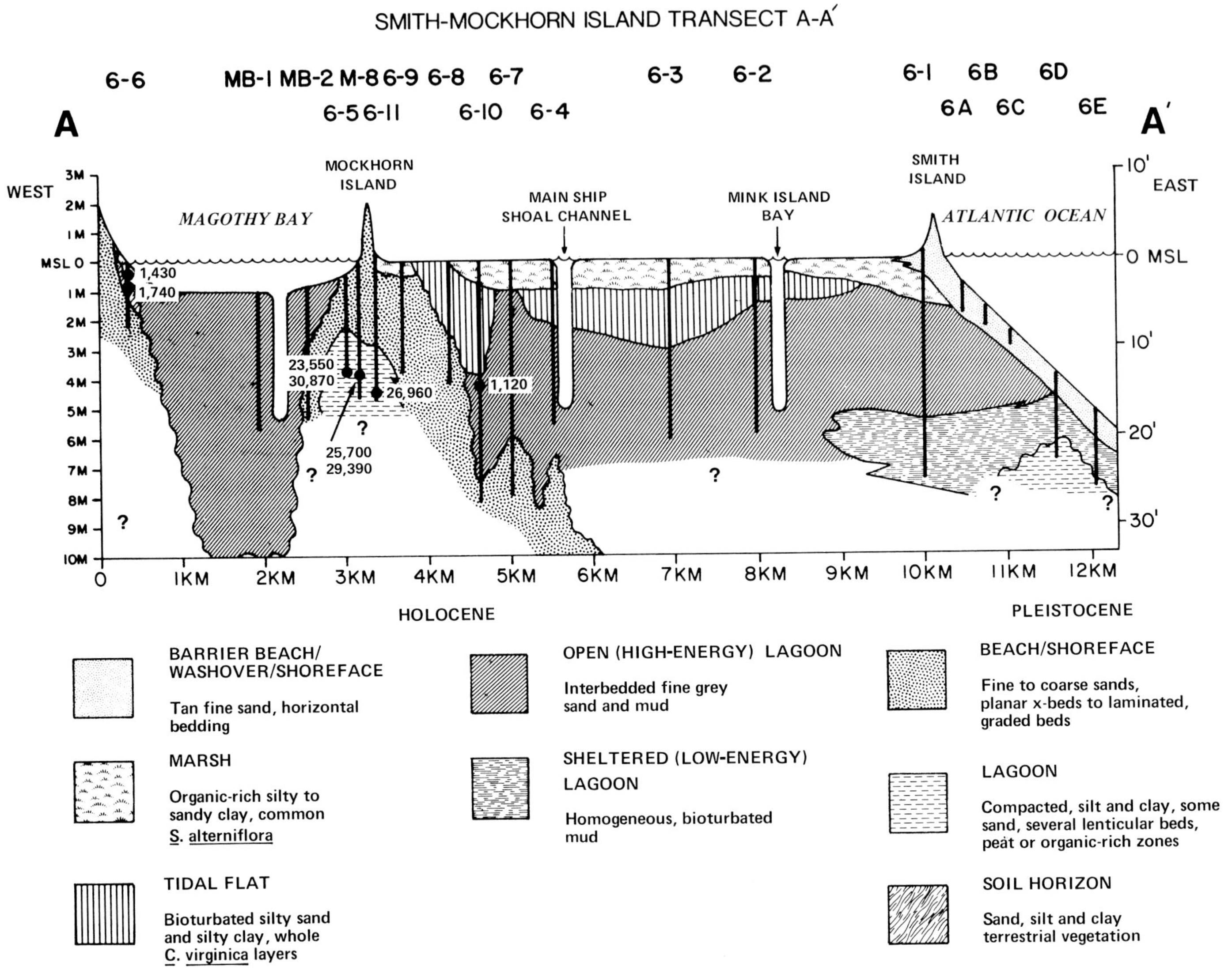

SMITH-MOCKHORN ISLAND TRANSECT A-A′

HOLOCENE

BARRIER BEACH/ WASHOVER/SHOREFACE
Tan fine sand, horizontal bedding

MARSH
Organic-rich silty to sandy clay, common S. alterniflora

TIDAL FLAT
Bioturbated silty sand and silty clay, whole C. virginica layers

OPEN (HIGH-ENERGY) LAGOON
Interbedded fine grey sand and mud

SHELTERED (LOW-ENERGY) LAGOON
Homogeneous, bioturbated mud

PLEISTOCENE

BEACH/SHOREFACE
Fine to coarse sands, planar x-beds to laminated, graded beds

LAGOON
Compacted, silt and clay, some sand, several lenticular beds, peat or organic-rich zones

SOIL HORIZON
Sand, silt and clay terrestrial vegetation

FIG. 3.—Stratigraphic cross section A-A′ extending from the lower shoreface/nearshore to the mainland shoreline. Note the thin Holocene shoreface/nearshore sands and the generally fining-upward Holocene back-barrier sequence. Given retreat of Smith Island, sea-level rise, and a shallow depth of shoreface erosion, the lower portion of both Holocene and Pleistocene transgressive stratigraphic sequences may be preserved. Radiocarbon dates shown are discussed in Finkelstein and Ferland (1987) and Finkelstein and Kearney (1988).

tics. Box cores provided representative characteristics for comparison purposes. A brief descriptive summary of each latest Pleistocene and Holocene deposit is presented in Table 1 and shown in a schematic drawing (Fig. 5). This Figure includes two older units, the Pliocene Yorktown Formation, previously reported by Shideler and others (1984) and Mixon (1985) as being in contact with overlying Pleistocene units such as the Wachapreague Formation. Mixon (1985) reports that the Wachapreague Formation is the youngest Pleistocene depositional surface on the Atlantic side of the Delmarva Peninsula, hence, it includes the Mockhorn Island barrier. However, Finkelstein and Kearney (1988) contend that Mockhorn Island was emplaced as a result of a later marine transgression.

Pleistocene Stratigraphic Units

Sand Facies (Beach, Shoreface, Flood-Tidal Delta).—

The most distinctive Pleistocene sand deposits are those found on Mockhorn Island. The interpretation as a barrier beach is based on the geomorphology, stratigraphic sequence, and sedimentary characteristics. The grain size of the sands is quite uniform and averages one-half phi size coarser than beach samples from the Holocene barrier, Smith Island. These sands vary between 2.0 to 3.0 m thick below MSL and, when combined with 1.5 to 2.0 m of upper intertidal and subaerial Mockhorn Island sands, amount to a deposit 3.5 to 5.0 m thick. The Mockhorn barrier-beach sands are ironstained, taking on an orange color compared

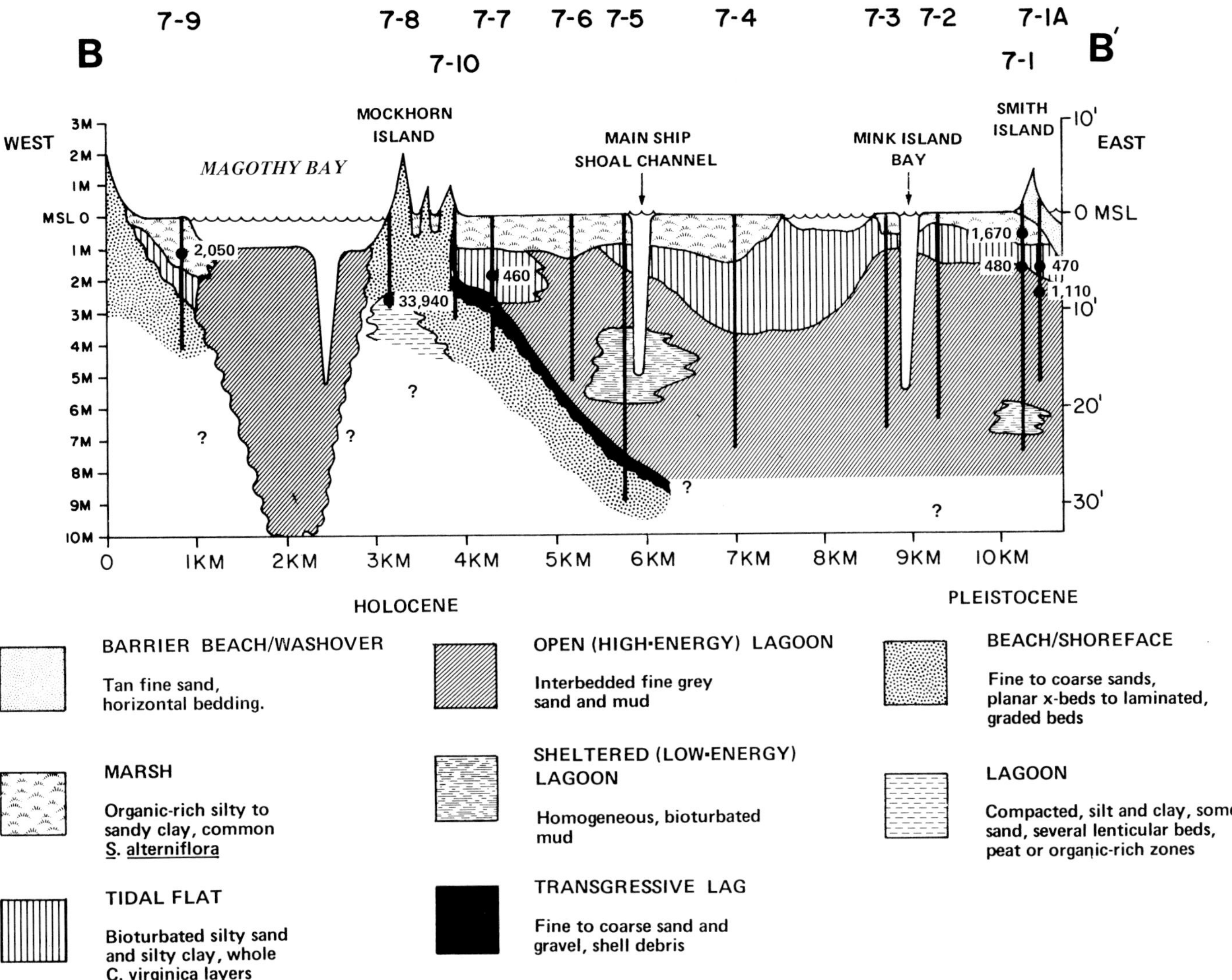

FIG. 4.—Stratigraphic cross section B-B′ extending from Smith Island to the mainland shoreline. Conclusions are similar to that given for Figure 3.

to the tan Smith Island barrier sands. The contact with the mud facies below is always sharp but conformable; it looks much like the modern barrier/lagoon and shoreface/lagoon contacts (Fig. 6). Heavy minerals are similar in concentration to those found on Smith Island and are well rounded, an indication of an oscillatory wave-dominated environment. Generally similar sands were found below Holocene muds in Magothy Bay from the erosion of Mockhorn Island sands during late Wisconsin sea-level lowstand.

Somewhat finer iron-stained subsurface sands dipping seaward of Mockhorn Island are interpreted as shoreface sediments. Physical sedimentary structures include graded and thin planar beds. These sediments lie below Holocene deposits. The Pleistocene-Holocene contact is usually gradational. However, a sharp upper contact with a Holocene transgressive lag is found locally.

Several sand bodies are found in the mud facies below the barrier sands. The stratigraphic position, gradational contact to the mud, and apparent rooting of the upper sand body indicate a back-barrier shoal origin, possibly a flood-tidal delta.

Mud Facies (Lagoon, Marsh).—

Fourteen vibracores (Finkelstein and Kearney, 1988) penetrated the Mockhorn Island barrier sands (Fig. 1) into a predominantly homogeneous mud-silt deposit. A marginal marine lagoonal environment is interpreted for this facies. This conclusion is based upon the stratigraphic re-

TABLE 1.—PRIMARY DEPOSITIONAL ENVIRONMENTS AND THEIR PHYSICAL AND BIOGENIC CHARACTERISTICS

Depositional Environment	Texture-Lithology	% Sand/Silt/Clay	Primary Sedimentary Structures	Thickness (m)	Contacts Top/Bottom	Biota
Pleistocene Barrier Sands	Med. well-sorted orange-tan sand. X = 1.81 ϕ		Heavy-mineral laminations. Planar, mostly horizontal beds. Round burrows.	3.5–5.0	None/Sharp but conformable	None
Shoreface	Fine, well-sorted orange-tan sand. X = 2.10 ϕ		Graded and thin planar beds.	1.0–2.0	Gradational or erosional/ Unknown	None
Lagoon	Dewatered mud, often mottled. Considerable mica.	20/44/36	Occasional lenticular sand beds. Interbedded peat and organic-rich zones.	>2.5	Sharp but conformable/ Unknown	A few calcareous and agglutinated foraminifera, a few dinoflagellates.
Holocene Barrier Beach and Washover	Fine, well-sorted tan sand. X = 2.35 ϕ		Low-angle and horizontal planar beds.	1.0–2.5	None/Sharp but conformable	Mollusk shells, foraminifera.
Shoreface/Nearshore	Fine grey sand. X = 2.90 ϕ	89/07/04	A few horizontal heavy-mineral beds. Several round burrows.	0.70–0.95	None/Sharp but conformable	Some shell fragments.
Salt Marsh	Dark grey organic-rich mud. Heavily vegetated.	27/55/18	None	0.5–1.0	Gradational/ Gradational	*S. alterniflora*, *S. virginica*, *D. Spicata*. Agglutinated forams and *L. irrorata*.
Tidal Flat	Dark grey, sandy mud. Oyster shells in layers up to 2 m thick.	40/40/20	Heavily bioturbated.	0.5–3.0	Gradational/ Gradational	Many *C. virginica* (oyster), some articulated. Many *I. obsoleta* and calcareous forams.
Open (High-Energy) Lagoon	Dark grey muddy sand. Analogous to mixed flat.	70/20/10	Flasar, wavy, lenticular and coarsely/thinly interlayered bedding. Increased bioturbation toward top.	1.0 6.0	Gradational/ Gradational	Some oyster shells, a few *M. lateralis*, ostracodes, and calcareous forams.
Sheltered (Low-Energy) Lagoon	Homogeneous mud with some plant debris.	20/48/32	Some sand laminae, mostly bioturbated.	0–3.0	Gradational/ Gradational	Calacreous forams, two ostracode spp., and marsh debris.
Transgressive Lag	Coarse sand, gravel, and shell debris.		None	0.2	Gradational/ Erosional	Some shell debris.
Back-barrier Beach	Fine to medium tan sand. X = 1.79 ϕ		Horizontal, planar bedding.	0.5–2.0	None/Gradational	Mollusk shells and salt-marsh vegetation.

lation with the sand facies, the textural characteristics of the mud, lenticular bedforms, and the few, but supportive, microfaunal assemblages, all of which compare favorably to back-barrier Holocene sediments of the area.

This mud is generally compacted and includes occasional lenticular beds and brown or black peat/organic-rich horizons composed of between approximately 10 to 40 percent total organic matter. It is at least 2.5 m thick; the lower contact was not penetrated. No macrofauna is seen in this facies and many of the microfossils (Table 1) may have been leached following deposition. The peat/organic-rich muds are believed to be the remnants of salt marshes. The peat/organic-rich sediments were radiocarbon dated. A mid-Wisconsin age was suggested by Finkelstein and Kearney (1988) but the absolute age was challenged by Colman and others (1989) and Toscano (1989). Despite this, the beds are clearly Pleistocene in age.

Pleistocene Stratigraphy

Previous late Pleistocene marine transgressions are recognized from barrier sequences from the Delaware and southeastern Virginia coastal plains (Demarest and others, 1981; Peebles, 1984). Likewise, Mockhorn Island exhibits a transgressive barrier stratigraphy. Such a stratigraphy of barrier beach over back-barrier lagoon reflects a landward retreating barrier island due to a rising sea level and/or an insufficient sediment supply.

The vibracore and seismic subbottom data show that Mockhorn Island is part of a continuous shore-perpendicular sand sheet, at least 3.5 m thick, extending eastward approximately 4 km. The barrier-island sands gradually dip seaward, transforming into a shoreface deposit beneath conformably and disconformably overlying modern back-barrier tidal-flat, lagoon, and marsh deposits. The Pleistocene surface is partially scoured, presumably by fluvial processes during a marine lowstand. Pleistocene lagoonal sediments lie below the Mockhorn Island sand but also disconformably below Holocene back-barrier sediments in lower shoreface/nearshore cores, 6D and 6E.

Mockhorn Island sands are presently being reworked by an encroaching sea due to the Holocene marine transgression. Erosion is apparent on the western shore of Mockhorn Island where Magothy bay provides a large fetch.

Mainland Pleistocene beach sands on the western side of Magothy Bay underlie Holocene salt-marsh or back-barrier beaches composed of Holocene and reworked Pleistocene sand. The Pleistocene-Holocene contact, approximately 2.0 m below MSL, is often recognized by Pleistocene pebbles over a 1.0- to 2.0-m-thick fine sand unit. The fine sands show many planar heavy-mineral beds and are interpreted as a former beach deposit.

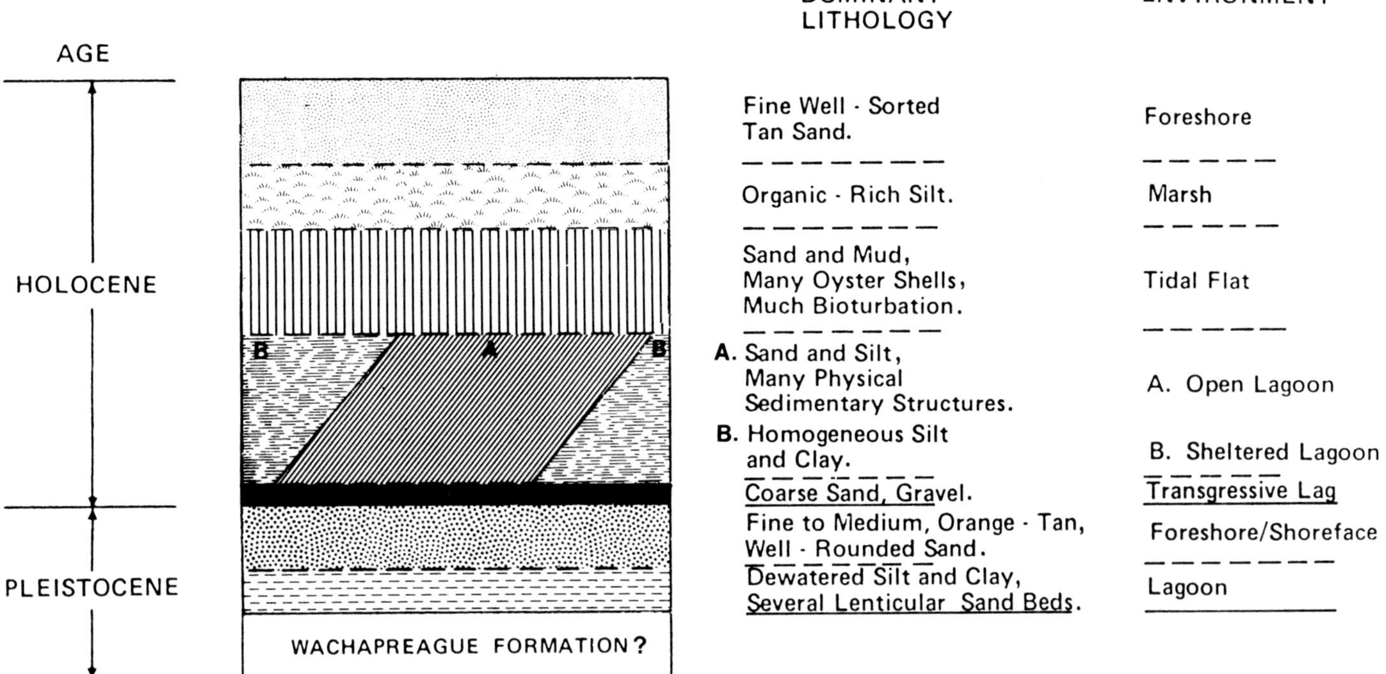

FIG. 5.—A schematic drawing depicting the stratigraphic position of individual sedimentary facies, depositional environments, and formations. This sequence was not seen in its entirety at any one location. It is a composite of the observed Holocene and Pleistocene stratigraphy.

Holocene Stratigraphic Units

Barrier Depositional Environments.—

1. The barrier-beach and washover: This environment is characterized by a relatively thin, texturally and mineralogically uniform, fine-grained, well-sorted, orthoquartzitic sand. A heavy-mineral assemblage composed of dominantly hornblende, garnet, and epidote is seen in the Smith Island barrier sands.

2. Shoreface: Smith Island shoreface sands are only approximately 1 m thick and are composed of approximately 90 percent sand. A trend of less mud and slightly coarser sand exists in the landward direction. Like the barrier-beach sands, the contact with the underlying back-barrier deposits is sharp but conformable.

Back-Barrier Depositional Environments.—

Back-barrier sediments identified in this study are at least 8 m thick. Descriptions of individual environments are similar to those previously reported in a regional study by Finkelstein and Ferland (1987) from southern Delmarva back-barrier cores. In summary, the back-barrier surficial environment is either salt marsh or tidal flat. Below the salt marsh, sediments coarsen downward, first into those deposits associated with tidal flats, and with further depth, primary sedimentary structures are encountered indicating a more active environment, suggesting an open, high-energy, environment. Exceptions are sheltered (low-energy) lagoonal deposits formed in inactive channels, sheltered back-barrier bays, or topographic lows of the Pleistocene surface. Although the contact with the underlying Pleistocene sediments may be identified by a coarse transgressive lag, in most places the contact is subtle. For example, mainland Holocene beach and salt marsh grade into Pleistocene sands. The barrier depositional environments, barrier beach, washover, or shoreface all are found atop back-barrier sediments, usually salt-marsh or tidal-flat environments.

Holocene Stratigraphy

Smith Island beach and washover sand exhibit abrupt, yet conformable, contacts with the back-barrier sediments found as much as 1.0 m below MSL. An erosional contact with back-barrier sediments is seen below shoreface/nearshore sands of approximately 1 m thickness. This contact is a transgressive discontinuity often called a ravinement surface (Belknap and Kraft, 1985). As Smith Island retrogrades and surface and subsurface deposits are reworked, nearshore sands are deposited on top of the ravinement surface. Some of this eroded material may be redeposited as washover or on the foreshore rather than as shoreface sands.

Most Holocene back-barrier environments change gradually both vertically and laterally. The most pertinent trends

FIG. 6.—Sharp but conformable contacts are seen between Pleistocene foreshore sands and back-barrier muds (left), Holocene foreshore sands and back-barrier muds (center), and Holocene lower shoreface/nearshore sands and back-barrier muds (right). These contacts are the ravinement surface associated with the late Pleistocene (left) and Holocene (center, right) marine transgressions. In the center photograph, note the muddy texture and shell debris (mud snails, oysters) of the tidal-flat environment.

are an overall fining-upward sequence of sediments from fine or muddy open-lagoon sand to a muddy tidal flat or salt marsh, and a dominance of biogenic sedimentary structures closer to the surface with a downward increase in primary sedimentary structures (cf., Finkelstein and Ferland, 1987; Fig. 5). This trend normally exists unless the locally encountered sheltered lagoon is penetrated.

The Holocene back-barrier sediments overlie Pleistocene sands associated with Mockhorn Island and the mainland. Scattered transgressive lag deposits are encountered between these two units. Elsewhere the contact is subtle. A basal Holocene peat, indicating a former sea level, might be expected above the Pleistocene sediments, yet this is rarely found. The lack of a distinct transgressive unconformity or a basal peat at the Pleistocene-Holocene contact diverges from the conceptual model of active transgression, i.e., the presence of an erosional back-barrier contact and/or transitional peat from the initial inundation of the sea. The back-barrier sediments surround Mockhorn Island with mostly salt marshes and tidal flats on the seaward side and open-lagoon sediments filling Magothy Bay (Figs. 1 and 2).

PRESERVATION POTENTIAL

A rapid sea-level rise may result in preservation of the entire transgressive sequence (Fischer, 1961, Belknap and Kraft, 1981, Davis and Clifton, 1987). Carter and others (1986) used seismic subbottom data to locate a preserved transgressive barrier sequence along the New Zealand inner continental shelf and based this preservation on a periodic sea-level rise rate as high as 1.2 cm/yr. Although Smith Island is migrating landward relatively rapidly, the relative sea-level rise rate is approximately four times less than that shown by Carter and others (1986) and therefore not nearly fast enough for such total preservation. In fact, shoreline erosion is presently destroying much of the back-barrier deposits as Smith island migrates landward. The Holocene barrier system demonstrates this as shown in Figures 3 and 4. After Smith Island retrogrades, barrier sands over back-barrier sediments are replaced by approximately 1.0 m of shoreface/nearshore sand above as little as 1.5 m of back-barrier sediments. The resulting nearshore stratigraphic section is characterized by the ravinement unconformity in close proximity to the basal unconformity (cf., Belknap and Kraft, 1985). Nevertheless, as the shoreline moves landward and upward, the potential for preservation of deeper and older deposits increases. In fact, high-resolution seismic subbottom data indicated 15 m of Holocene back-barrier sediments as close as 1 km seaward of Mockhorn Island (Shideler and others, 1984). Belknap and Kraft (1981) reported the greater probability for preservation of materials, such as lagoonal sediments, lower in the transgressive sequence.

Although sands filling inlets are often deposited below wave base, Finkelstein (1988) found that such preservation will likely not occur within the Virginia barrier-island chain. In this locality sedimentation associated with periodic transgressive events is needed to create a thicker sequence of preserved barrier complex deposits, although each transgression is, at least, partially self destructive. Much of the sediments associated with Mockhorn Island are buried beneath Holocene back-barrier deposits. The retreat of Smith Island is removing much of the Holocene back-barrier sequence but this island will eventually migrate atop the older barrier, Mockhorn Island. In order to stack and preserve transgressive sequences, periodic transgressive events are essential. Subsidence of the earlier barrier deposits and/or continuous sea-level rise during the subsequent transgression will provide a thicker sequence of sediments and a better potential for preservation.

The stacking of transgressive sequences was reported by Bridges (1976) but otherwise is not frequently recognized in the rock record. Ryer (1977) also noted several marine invasions leading to the stacking of distinct, but mostly regressive, depositional sequences. Other studies showing preserved transgressive barrier sequences are those of Hobday and Tankard (1978) and Cotter (1983).

The future migration of Smith Island is calculated in order to estimate the amount of Holocene and Pleistocene sediment preservation and the resulting facies geometry. The parameters relevant for this area are a migration rate of 5.0 m/yr (Everts, 1987), a historical sea-level rise/subsidence rate of 3.0 mm/yr (Hicks and others, 1983), a shoreline-erosion depth of 7.0 m below MSL (Finkelstein, 1986; Everts, 1987), and a distance of 7.0 km between the two barriers. At these rates it will take approximately 1,400 years for Smith Island to reach Mockhorn Island. Likely, Smith Island will join Mockhorn Island (cf., Demarest and Leatherman, 1985) rather than completely overtop it as sea level will rise 4.2 m based on historical trends. Nevertheless, by taking into account sea-level rise, Pleistocene sands and, likely, lagoonal muds, will be preserved below approximately 3.0 to 4.0 m or more (cf., Shideler and others, 1984, fig. 8) of Holocene back-barrier sediments and 1.0 m of nearshore sands (Figs. 3 and 4 and 7). The results will be the preservation of stratigraphically lower transgressive sequences of both Pleistocene and Holocene barriers. Specifically, with the continuation of present high sea-level rise rates, much of the Pleistocene mud facies and the shoreface portion of the sand facies, along with Holocene open-lagoon back-barrier sediments, should be preserved throughout the course of the Holocene marine transgression (Figs. 5 and 7).

DISCUSSION

Fischer (1961) and Curray (1964) observed that the ultimate preserved stratigraphic sequence and facies pattern are controlled by the rate of net deposition/erosion and its interaction with relative sea-level change through time. Furthermore, the preservation potential of transgressive deposits is a function of the depth of shoreface erosion. Ryer (1977) noted that the stratigraphic record of transgression will be a unit of offshore marine sediment disconformably overlying coastal-plain deposits. Such a stratigraphy is often expected as much of the transgressive sequence is destroyed with the retreat of the barrier. However, where cyclical marine transgression is common, a portion of each transgressive sequence may be preserved under specific conditions. Such a region exists on the Delmarva Peninsula, where the stratigraphy is clearly defined by transgressive deposits.

Four transgressive barrier shorelines in Delaware within the Omar Formation, as defined by Demarest and others (1981), are found landward of the Holocene barriers. Similarly, Mixon (1985) described Pleistocene transgressive barrier sequences on the southern Delmarva Peninsula–the Accomack Member of the Omar Formation and possibly part of the Nassawadox Formation (G. Johnson, College of William And Mary, pers commun., 1986). Like Mockhorn Island, each barrier originated during a period of rising sea level and was abandoned as sea level fell. The time for deposition of these barrier complexes is relatively short with only the deposits farthest inland, in each transgression, preserved. In some cases, regressive sand deposits are welded to the older Delaware barriers, an indication of a sea-level standstill or fall.

The Mockhorn and Omar barriers on the Delmarva Peninsula incorporate those deposits produced during several thousand years, culminating in a sea-level highstand. Each transgressive barrier complex is completely preserved, at least temporarily, in a shore-normal direction. The facies relation between Holocene back-barrier and barrier-island deposits and Pleistocene older barrier units is in response to continuous Holocene relative sea-level rise and landward migration of the Holocene barriers. Study of the geomorphology, stratigraphy, and migration rates of Smith and Mockhorn islands leads to a model that predicts subsequent barrier sediments to become stacked on top of previous transgressive deposits with the potential for partial preservation of the sequence. The principal factors that contribute to this model are (1) the large back-barrier area that provides for the deposition of sediment even as shoreface erosion migrates landward, (2) the modern back-barrier setting of the older barrier complex, and (3) sea-level rise (cf., Belknap and Kraft, 1981) that causes Pleistocene and lower Holocene back-barrier deposits to be more deeply buried. In addition, the low-gradient nearshore/offshore zones of the Atlantic provide a fairly shallow wave base, thus potentially removing only part of the Mockhorn and Smith Island transgressive sequence. This latter factor should result in more of the transgressive sequence preserved here than in nearby Delaware (Belknap and Kraft, 1985; Kraft and Belknap, 1986). It should be recognized that, although ravinement and basal unconformities are nearly in contact below the modern shoreface, they are expected to separate as the shoreface migrates landward.

This model is applicable to the abandoned mainland transgressive barriers mentioned above, e.g., the Omar Formation. They owe their origin to a sea level of 6 m above MSL plus or minus several meters; a similar future rise is suggested by Kraft and others (1987). This predicted sea-level rise, along with a gentle coastal-plain slope, should allow for an older barrier to be inundated by the lagoon associated with the coexisting retrograding barrier. Initially, the rise of Holocene sea level above the mainland shoreline west of the Mockhorn-Smith Island barriers will be delayed due to the relatively steep Pleistocene scarp (Demarest and Leatherman, 1985). Nevertheless, the future geomorphology of the southern Delmarva mainland result-

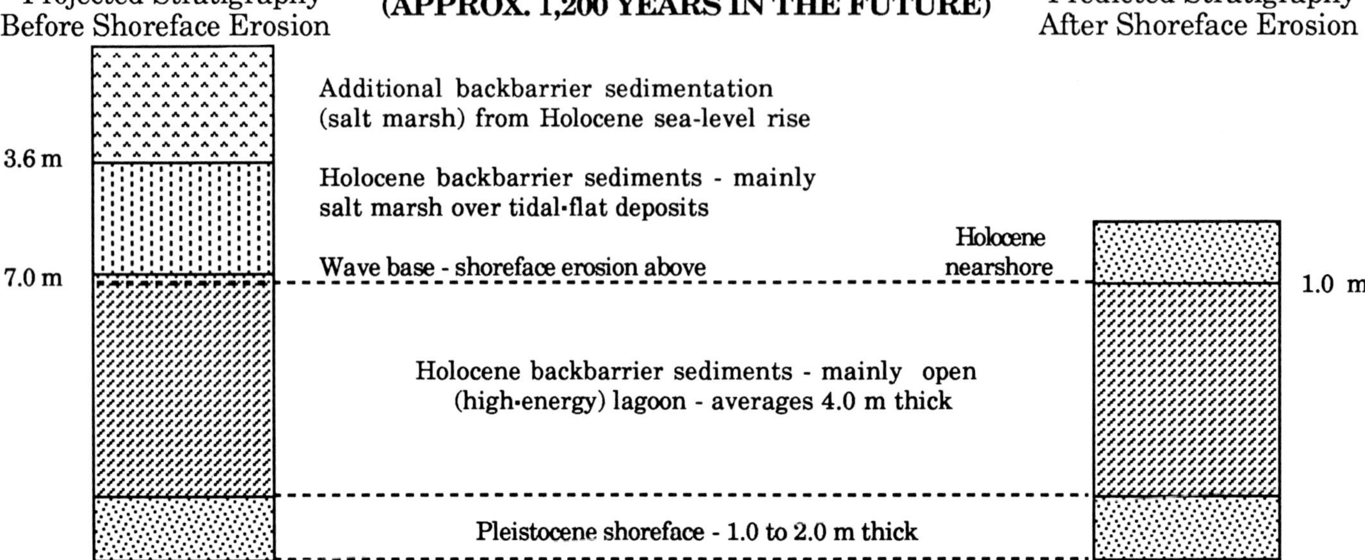

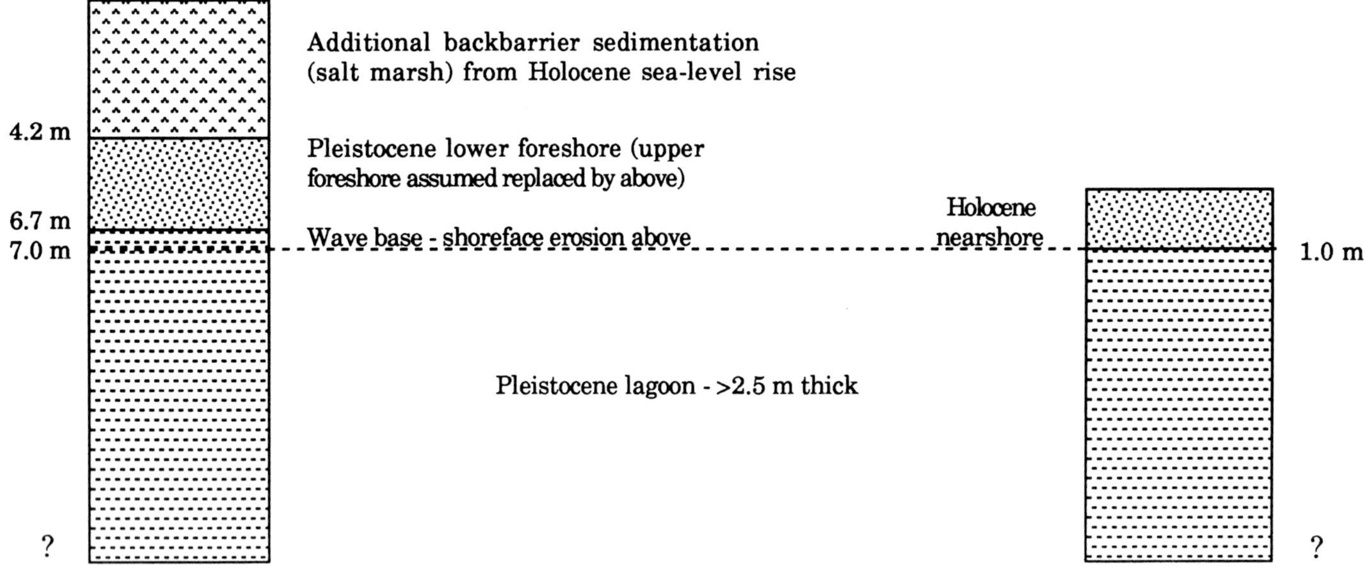

FIG. 7.—Preservation of transgressive sediments at hypothetical locations 1.0 km east of Mockhorn Island (top) and on the eastern flank of Mockhorn Island (bottom). Projected stratigraphy immediately before (left) and predicted stratigraphy after (right) the retreating shoreline passes the respective location. Preservation is based on a shoreline retreat rate of 5.0 m/yr, a sea-level rise rate of 3.0 mm/yr, a 7.0-km distance between the two barriers, and a wave base of 7.0 m below MSL. Sea level should rise 3.6 m in the 1,200 years of shoreline retreat at top and 4.2 m in 1,400 years at bottom. See Figures 3, 4, 5 and Table 1 for more information.

ing from marine transgression should look similar to that of Mockhorn and Smith islands with a similar opportunity for lower transgressive sequence preservation.

Finally, the projected stratigraphic sequence outlined in this study gives evidence for potential sources and traps for oil and gas. Holocene and Pleistocene back-barrier sediments, shown respectively above and below Pleistocene sands (Fig. 7), may form impermeable seals and serve as potential source beds. The Pleistocene sands may be thick and widespread enough to act as a stratigraphic trap. Onlapping offshore muds are the best source of hydrocarbons but they are infrequently found off the coast of Virginia. Everts (1985) did note that 64 percent of the Smith Island shoreface sediments is mud but it is likely that much of this material is reworked back-barrier deposits. Nevertheless, offshore muds that do accumulate will lie above thin lower shoreface/ nearshore sands. Therefore, subsequent barrier progradation is needed to make use of this source if it is, indeed, available. Such a model of preservation and potential oil accumulation may be applicable elsewhere. For example, in New South Wales, Australia, transgressive Holocene sequences are shown preserved below progradational barriers (Thom, 1984). In this case, sea level had reached its present position approximately 6.5 to 6 ka and, with sufficient sediment, transgressive deposits were overtopped by prograding bay barriers.

SUMMARY

Study of the stratigraphy and geomorphology of adjacent Pleistocene and Holocene barrier-island complexes, along with the offshore bathymetry and sea-level rise history of the area, suggests that a portion of each lower transgressive sequence may be preserved. Taking into account present sea-level rise, wave-base depth, and Holocene barrier retreat, at least 3.0 m of Holocene lagoonal deposits will be preserved below 1.0 m of nearshore sediments; below the Holocene basal unconformity, Pleistocene sand (shoreface) and lagoonal mud should be preserved. Conditions necessary for such preservation include a Pleistocene barrier located within the Holocene back-barrier environment, yet far enough away to allow sea-level rise and/or subsidence to bury further older barrier complex deposits before the two barrier islands join. Essentially, cyclical sea-level rise provides for a marine transgression with deposits partially preserved due to overtopping by the following transgression. Although the Holocene and latest Pleistocene depositional sequences were used to develop this model for preservation, older transgressive barrier complexes on the Delmarva mainland, and others elsewhere, may be analogous.

ACKNOWLEDGMENTS

This research was conducted as part of a Ph.D. study at the Virginia Institute of Marine Science, College of William and Mary. Many individuals provided insight, criticism, and field assistance during the course of the research, particularly L. D. Wright and G. H. Johnson. R. C. Mixon and J. P. Owens of the U.S. Geological Survey, Reston, provided initial comment on the stratigraphic and mineralogic interpretations. Rick Berquist of the Virginia Division of Mineral Resources helped run the seismic profiles. Financial support came from the U.S. Army Corps of Engineers' Coastal Engineering Research Center, a Grant-in-Aid of Research from Sigma XI, the Southeastern Section of the Geological Society of America, and the College of William and Mary minor research grant program. The National Oceanic and Atmospheric Administration provided resources for the completion of this paper.

D. F. Belknap, D. M. FitzGerald and L. A. Harris are thanked for very constructive reviews. E. R. Curry assisted with the final preparation of the figures.

REFERENCES

BELKNAP, D. F., AND KRAFT, J. C., 1981, Preservation potential of transgressive coastal lithosomes on the U.S. Atlantic Shelf: Marine Geology, v. 42, p. 429–442.

BELKNAP, D. F., AND KRAFT, J. C., 1985, Influence of antecedent geology on stratigraphic preservation potential and evolution of Delaware's barrier systems: Marine Geology, v. 63, p. 235–262.

BRIDGES, P. H., 1976, Lower Silurian transgressive barrier islands, southwest Wales: Sedimentology, v. 23, p. 347–362.

CARTER, R. M., CARTER, L., AND JOHNSON, D. P., 1986, Submergent shorelines in the SW Pacific: evidence for an episodic post-glacial transgression: Sedimentology, v. 33, p. 629–649.

COLMAN, S. M., MIXON, R. B., RUBIN, M., BLOOM, A. L., AND JOHNSON, G. H., 1989, Comment on "Late Pleistocene barrier island sequence along the southern Delmarva Peninsula: implications for middle Wisconsin sea levels": Geology, v. 17, p. 84–85.

COTTER, E., 1983, Shelf, paralic, and fluvial environments and eustatic sea-level fluctuations in the origin of the Tuscarora Formation (Lower Silurian) of central Pennsylvania: Journal of Sedimentary Petrology, v. 53, p. 25–49.

CURRAY, J. R., 1964, Transgressions and regressions, *in* Miller, R. L., ed., Papers in Marine Geology: MacMillan Press, New York, p. 175–203.

DAVIS, R. A., AND CLIFTON, H. E., 1987, Sea-level change and the preservation potential of wave-dominated and tide-dominated coastal sequences, *in* Nummedal, D., Pilkey, O. H., and Howard, J. D., eds., Sea-Level Fluctuations and Coastal Evolution: Society of Economic Paleontologists and Mineralogists Special Publication 41, p. 167–178.

DEMAREST, J. M., BIGGS, R. B., AND KRAFT, J. C., 1981, Time-stratigraphic aspects of a formation: interpretation of surficial Pleistocene deposits by analogy with Holocene paralic deposits, southeastern Delaware: Geology, v. 9, p. 360–365.

DEMAREST, J. M., AND LEATHERMAN, S. P., 1985, Mainland influence on coastal transgression: Delmarva Peninsula: Marine Geology, v. 63, p. 19–33.

DOLAN, R., HAYDEN, B., REA, C., AND HEYWOOD, J., 1979, Shoreline erosion rates along the middle Atlantic coast of the United States: Geology, v. 7, p. 602–606.

EVERTS, C. H., 1985, Sea level rise effects on shoreline position: Journal of Waterway, Port, Coastal and Ocean Engineering, v. 111, p. 985–999.

EVERTS, C. H., 1987, Continental shelf evolution in response to a rise in sea level, *in* Aubrey, D. G. and Weisher, L., eds., Hydrodynamics and Sediment Dynamics of Tidal Inlets, Lecture Notes on Coastal and Estuarine Studies, v. 29: Springer-Verlag, New York, p. 49–57.

FIELD, M. E., 1980, Sand bodies on coastal plain shelves: Holocene record of the U.S. Atlantic inner shelf off Maryland: Journal of Sedimentary Petrology, v. 50, p. 505–528.

FINKELSTEIN, K., 1986, Backbarrier contributions to a littoral sand budget: Journal of Coastal Geology, v. 2, p. 33–42.

FINKELSTEIN, K., 1988, An ephemeral inlet from the Virginia barrier island chain: stratigraphic sequence and preservational potential of infilled sediments, *in* Aubrey, D. G. and Weisher, L., eds., Hydrodynamics and Sediment Dynamics of Tidal Inlets, Lecture Notes on Coastal and Estuarine Studies, v. 29: Springer-Verlag, New York, p. 257–268.

FINKELSTEIN, K., AND FERLAND, M. A., 1987, Back-barrier response to sea-level rise, eastern shore of Virginia, *in* Nummedal, D., Pilkey, O. H., and Howard, J. D., eds., Sea-Level Fluctuations and Coastal Evolution: Society of Economic Paleontologists and Mineralogists Special Publication 41, p. 145–155.

FINKELSTEIN, K., AND KEARNEY, M. S., 1988, Late Pleistocene barrier island sequence along the southern Delmarva Peninsula: implications for middle Wisconsin sea levels: Geology, v. 16, p. 41–45.

FISCHER, A. G., 1961, Stratigraphic record of transgressing seas in light of sedimentation on Atlantic coast of New Jersey: American Association of Petroleum Geologists Bulletin, v. 45, p. 1656–1666.

FISK, B. N., 1959, Padre Island and the Laguna Madre flats of coastal South Texas, *in* Second Coastal Geographical Conference: National Academy of Science, National Research Council, p. 103–151.

HICKS, S. D., DEBAUGH, H. A., AND HICKMAN, L. E., 1983, Sea-level variations for the United States, 1855–1980: U.S. Department of Commerce, NOAA/NOS, 170 p.

HOBDAY, D. K., AND TANKARD, A. J., 1978, Transgressive-barrier and shallow-shelf interpretation of the lower Paleozoic Peninsula Formation, South Africa: Geological Society of America Bulletin, v. 89, p. 1733–1744.

HOYT, J. H., AND HAILS, J. R., 1967, Pleistocene shoreline sediments in coastal Georgia: deposition and modification: Science, v. 155, p. 1541–1543.

KRAFT, J. C., 1971, Sedimentary facies patterns and geologic history of a Holocene marine transgression: Geological Society of America Bulletin, v. 82, p. 2131–2158.

KRAFT, J. C., AND BELKNAP, D. F., 1986, Holocene epoch coastal geomorphologies based on local relative sea-level data and stratigraphic interpretations of paralic sediments: Journal of Coastal Research, v. 2, p. 53–59.

KRAFT, J. C., BELKNAP, D. F., AND DEMAREST, J. M., 1987, Prediction of effects of sea-level change from paralic and inner shelf stratigraphic sequences, *in* Rampino, M. R., Sanders, J. E., Newman, W. S., and Korigsson, L. K., eds. Climate: History, Periodicity, and Predictability: Van Nostrand Reinhold Co., New York, p. 166–192.

MIXON, R. B., 1985, Stratigraphic and geomorphic framework of uppermost Cenozoic deposits in the southern Delmarva Peninsula, Virginia and Maryland: U.S. Geological Survey Professional Paper 1067-G, 53 p.

PEEBLES, P. C., 1984, Late Cenozoic landforms, stratigraphy, and history of sea-level oscillations of southeastern Virginia and northeastern North Carolina: Unpublished Ph.D. Dissertation, Virginia Institute of Marine Science, Gloucester Point, Virginia, 142 p.

RICE, T. E., AND LEATHERMAN, S. L., 1983, Barrier island dynamics: the eastern shore of Virginia: Southeastern Geology, v. 24, p. 125–137.

RICE, T. E., NIEDORODA, A. W., AND PRATT, A. P., 1976, The coastal processes and geology, Virginia barrier islands, *in* The Virginia Coast Reserve Study: The Nature Conservancy, p. 108–382.

RYER, T. A., 1977, Patterns of Cretaceous shallow-marine sedimentation, Coalsville and Rockport areas, Utah: Geological Society of America Bulletin, v. 88, p. 177–188.

SHIDELER, G. H., LUDWICK, J. C., OERTEL, G. F., AND FINKELSTEIN, K., 1984, Quaternary stratigraphy evolution of the southern Delmarva Peninsula coastal zone, Cape Charles, Virginia: Geological Society of America Bulletin, v. 95, p. 489–502.

THOM, B. G., 1984, Transgressive and regressive stratigraphies of coastal sand barriers in southeast Australia: Marine Geology, v. 56, p. 137–158.

THOM, B. G., BOWMAN, G. M., AND ROY, P. S., 1981, Late Quaternary evolution of coastal sand barriers, Port Stephens-Myall Lakes areas, central New South Wales, Australia: Quaternary Research, v. 15, p. 345–364.

TOSCANO, M. A., 1989, Comment on "Late Pleistocene barrier island sequence along the southern Delmarva Peninsula: implications for middle Wisconsin sea levels": Geology, v. 17, p. 85–86.

DEPOSITIONAL PATTERNS RESULTING FROM HIGH-FREQUENCY QUATERNARY SEA-LEVEL FLUCTUATIONS IN NORTHEASTERN NORTH CAROLINA

STANLEY R. RIGGS

Department of Geology, East Carolina University, Greenville, North Carolina 27858

LINDA L. YORK AND JOHN F. WEHMILLER

Department of Geology, University of Delaware, Newark, Delaware 19716

AND

STEPHEN W. SNYDER

Department of Marine, Earth, and Atmospheric Science, North Carolina State University, Raleigh, North Carolina 27659

ABSTRACT: High-resolution seismic data suggest that portions of depositional sequences representing as many as 18 Quaternary sea-level highstands are preserved within 60 m of Quaternary deposits in northeastern North Carolina. Sediments deposited during at least seven of these Quaternary sea-level events have been defined within the upper 33 m in drill holes in Dare County. The complex stratigraphy was resolvable only after integrating detailed litho-, bio- and aminostratigraphic drillhole data with a high-resolution seismic framework.

High-frequency, sea-level cyclicity dominated the depositional patterns of the resulting Quaternary sediment sequences. As high-energy coastal systems moved repeatedly across the low-gradient continental shelf, sediment units that had previously been deposited in coastal and shelf environments were significantly modified. During each glacial episode, fluvial channels extensively dissected previously deposited coastal facies. Subsequent deglaciation and transgression flooded the channels, backfilling them with fluvial and estuarine sediments. The infilled channel facies were then partially truncated by shoreface erosion, which also eroded portions of previously deposited coastal sequences. During sea-level highstands, a new sequence of coastal facies was deposited over the ravinement surface cut into remnants of older and similar Quaternary sequences and the associated channel-fill systems. Thus, the resulting record consists of a series of imbricated coastal deposits of similar, but discontinuous, lithostratigraphic units with irregular geometries that only partially represent interglacial highstand deposition; the depositional sequences are highly punctuated and dominated by unconformity surfaces with extensive incised and backfilled channel deposits.

INTRODUCTION

Quaternary Sea-Level Fluctuations

The deep-sea sedimentary oxygen isotope record permits identification of at least 21 orbitally forced cycles of climatic fluctuation during the Quaternary (Shackleton and Opdyke, 1973; Hays and others, 1976; Imbrie and Imbrie, 1980; Zimmerman and others, 1984; Prell and others (1986), and Ruddiman and others, 1986). Isotopic curves and defined stages, interpreted as records of continental-ice volume, can be used to approximate sea-level changes with frequencies of 10,000 to 100,000 yrs (Shackleton, 1987) and amplitudes of approximately 50 to 130 m (Shackleton, 1987; Fairbanks, 1989) with associated flooding and draining of continental margins.

Such rapid and dramatic changes in paleoclimatic conditions would have had significant impacts upon continental-margin sedimentological systems. Changing climates would have caused modification and latitudinal migration of vegetative cover (Delcourt and Delcourt, 1981; Whitehead, 1981) with changes in associated weathering, sediment production, and fluvial processes (Meade, 1969; Copeland and others, 1984; Riggs and Belknap, 1988). Contemporaneous changes in physical and chemical conditions within the oceanographic realm would have shifted ecologic zones (Blackwelder, 1981; Cronin, 1980; Cronin and others, 1981) and changed patterns of authigenic/diagenetic sedimentation (Riggs, 1984; Schlee and others, 1988; Riggs and Mallette, 1990; Riggs and others, 1990). When combined with major fluctuations in sea level, the already complex depositional lithosomes are further complicated by shifting centers and patterns of deposition, as well as by submarine and subaerial erosional processes. Consequently, the coastal system repeatedly reworks and destroys much of what was previously deposited, leaving complex partial sections behind (Belknap and Kraft, 1981, 1985; Hine and Snyder, 1985; Riggs and Belknap, 1988). The results are depositional sequences unique to continental-margin environments.

Resolution and correlation of high-frequency Quaternary glacio-eustatic sea-level records within a continental-margin stratigraphic record are difficult for several reasons. First, predictive stratigraphic models that account for both the incomplete and homotaxial nature of these records are not well developed. Second, and most importantly, the chemical and physical tools necessary for the high-resolution studies are just now beginning to be applied simultaneously in regions where the complex stratigraphic records are preserved. Now with appropriate integration of high-resolution seismic stratigraphy, dating techniques, and litho- and biostratigraphic analyses, resolution to distinguish between major glacio-eustatic sea-level events should be within reach. In this paper we present the interpreted stratigraphy of northeastern North Carolina (Fig. 1), as deciphered using these diverse methods, and thereby demonstrate the internal consistency and consider the implications for the Quaternary depositional history of the region.

Modern Coastal System and Quaternary Section

The outer coastal plain of northeastern North Carolina contains a large portion of the modern coastal system (Fig. 1). The area includes the modern barrier islands, Roanoke Island, and a mainland area composed of extensive swamp forest and associated peat deposits. Behind the barrier islands, the land is surrounded by extremely variable estuaries that include the Albemarle and Pamlico Sounds, which are the drowned portions of the Roanoke and Tar Rivers, respectively (Riggs, 1985). Albemarle and Pamlico Sounds

Quaternary Coasts of the United States: Marine and Lacustrine Systems, SEPM Special Publication No. 48

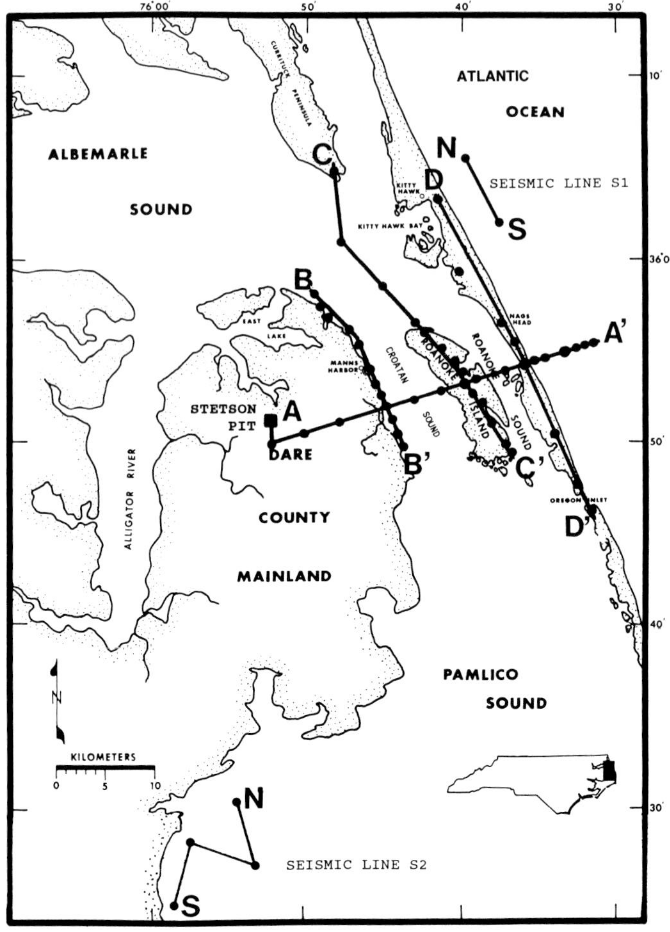

FIG. 1.—Map of the Dare County area, northeastern North Carolina showing the locations of Stetson Pit (square), core holes (circles), lithostratigraphic cross sections (A-A′, B-B′, C-C′, and D-D′), and high-resolution seismic sections (S1 and S2) utilized in this paper.

are connected by Croatan and Roanoke Sounds, two drowned lateral tributary streams that have flooded across the interstream divide, producing Roanoke Island. The Alligator River is a more landward lateral tributary that will breach across the interstream divide to the Pungo River with a small rise in sea level.

The subsurface sediments in northeastern North Carolina represent a gently eastward-dipping, seaward-thickening sequence of Mesozoic and Cenozoic sediments that form a sedimentary wedge from 1.5 to 2.0 km thick. Capping this wedge is a sequence of between 50 to 70 m of Quaternary sediments that rests unconformably on sediments of the Pliocene Yorktown Formation (Brown and others, 1972; Miller, J. A., 1982). The Quaternary sequence is considerably thicker than most coastal-plain sections because it was deposited within a regional depositional basin. Seismic data of Popenoe and Ward (1983) and Popenoe (1985) suggest that Quaternary sediments have filled the last remnants of the Aurora Embayment, a pre-Miocene depositional basin northwestward of the constructional Cape Lookout High. Ward and Strickland (1985) call the basinal feature the Albemarle Embayment, which is the terminology used in this paper.

The Quaternary section of northeastern North Carolina is ideal for recognition of high-frequency, sea-level cyclicity for three reasons: (1) the section that has infilled the Albemarle Embayment is relatively thick and well preserved (Eames, 1983; Popenoe, 1985; Riggs and Belknap, 1988; Wehmiller and others, 1988) compared to many other portions of the mid-Atlantic where the sediments are very thin or only locally preserved, such as in southern North Carolina (Hine and Snyder, 1985; Riggs and Belknap, 1988); (2) the data base on the Quaternary section for the emerged coastal plain is extensive with many drillhole records, significant biostratigraphic data and many numerical age estimates, including amino-acid racemization, U-series disequilibrium, and radiocarbon; (3) the extensive modern riverine/estuarine/barrier-island system offers an ideal opportunity to integrate the Holocene litho- and biostratigraphic facies with their partially preserved Quaternary counterparts through the combination of high-resolution seismic and chronostratigraphy.

RESOLUTION OF HIGH-FREQUENCY STRATA IN QUATERNARY SEQUENCES

Quaternary Seismic and Lithostratigraphy

An extensive lithostratigraphic data base has been developed for the coastal area of northeastern North Carolina (O'Connor and others, 1973; Riggs and O'Connor, 1974; Bellis and others, 1975; Hartness, 1977; Benton and others, 1978; O'Connor and others, 1978; Riggs and others, 1978; Pearson, 1979; Riggs, 1979; Benton, 1980; Hardaway, 1980; Copeland and others, 1983, 1984; Eames, 1983). This data base includes 30 4-in.-diameter continuous core holes that range from 20 to 30 m in depth, over 200 vibracores that range from 3 to 12 m in length, and 927 km (500 nmi) of high-resolution seismic data. These studies established that physical characteristics of the modern North Carolina coastal system are dictated by the prior Quaternary history, as well as by the morphology and lithology of subsurface Quaternary units. The older units define specific subenvironments within the estuaries and determine the geometry of the coastal system, type of shoreline, nature of the bottom sediments (Riggs, 1985; Wells, 1989; Wells and Kim, 1989), and indirectly influence tidal and current patterns within the water column (Pietrafesa and others, 1986).

Eames (1983) developed the basic lithofacies interpretations presented in Figures 2 through 5 and defined five lithologic depositional sequences (Table 1). The youngest depositional sequence consisted of Holocene sediments and includes the modern coastal system. All depositional sequences contain multiple lithofacies representing many different coastal environments. At any one time within a depositional sequence, all of these systems are present. In response to the ongoing interglacial sea-level rise, each system systematically migrates upward and landward, but not all facies are necessarily deposited (Riggs, 1985). If all facies are deposited, they may not be preserved as the transgression continues, or they may not be sampled with

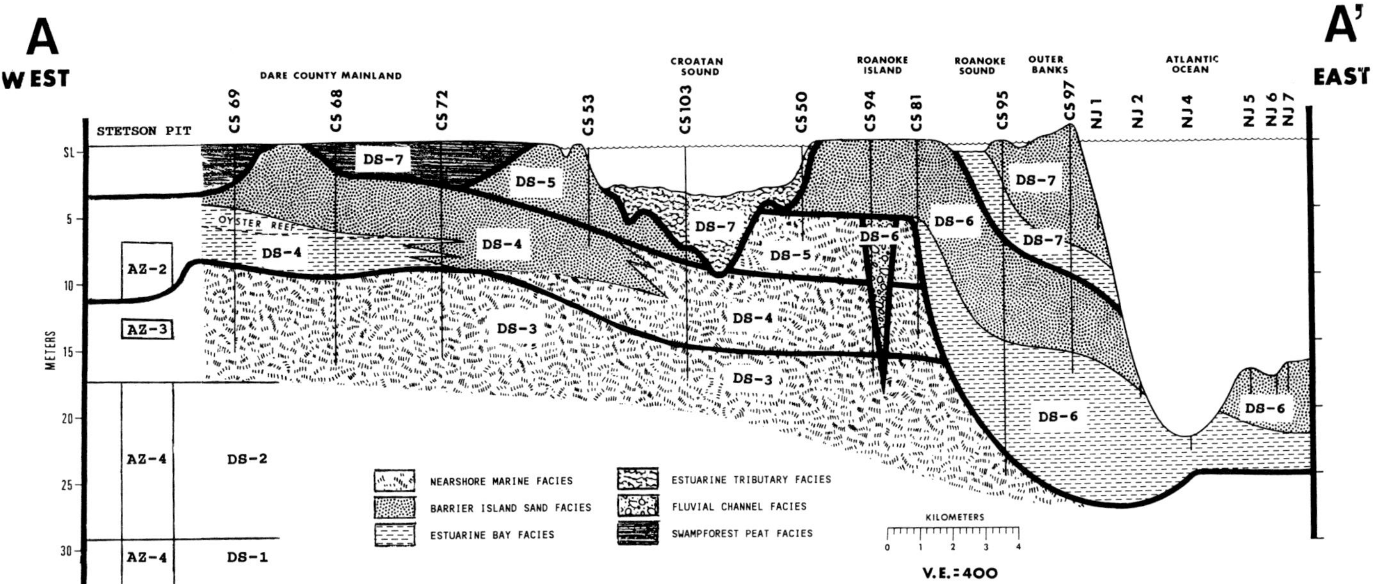

FIG. 2.—A west to east geologic cross section extending from Stetson Pit (A), east into the Atlantic Ocean (A') (see Fig. 1 for section location). Stetson Pit section is situated about 3 km north of the section line; irregularities in lithologic contacts reflect paleotopography. This section shows (1) seven stacked Quaternary depositional sequences (labeled DS-1 through DS-7), (2) major sediment facies of each depositional sequence, and (3) aminozone designations on associated mollusks (labeled AZ-2 through AZ-4). DS-7 is the Holocene depositional sequence. (Modified from O'Connor and others, 1973; Riggs and O'Connor, 1974; and Eames, 1983).

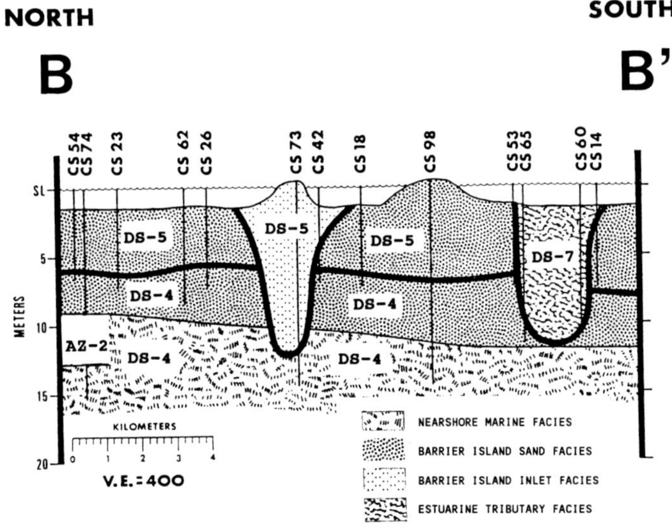

FIG. 3.—A north-to-south geologic cross section along western shoreline of Croatan Sound and eastern side of mainland Dare County. Section extends from Mashoes on northeastern tip of land (B) to south of Manns Harbor (B') (see Fig. 1 for section location). This section shows (1) portions of at least three stacked Quaternary depositional sequences (labeled DS-4, DS-5, and DS-7), (2) major sediment facies of each depositional sequence, and (3) aminozone designations on associated mollusks (labeled AZ-2). DS-7 is the Holocene depositional sequence. (Modified from O'Connor and others, 1973; Riggs and O'Connor, 1974; and Eames, 1983).

limited drill holes. Consequently, not all facies have been recognized within each depositional sequence.

The important role of the fluvial-channel facies within the stratigraphic record of the coastal system in northeastern North Carolina was recognized by O'Connor and others (1973) and Riggs and O'Connor (1974). However, the extent to which fluvial processes modified previous depositional sequences, or the role of other modifying processes leading to the partial preservation of coastal lithofacies, was not fully appreciated. Figure 6 is an interpreted, shore-parallel, high-resolution seismic line that is located seaward of the modern barrier island (Fig. 1). The profile in Figure 6 demonstrates the complexity of channel processes within a set of fluvial/estuarine channel facies of the paleo-Roanoke-Albemarle system.

Based upon drillhole samples and seismic depositional patterns, the paleo-Roanoke-Albemarle channel system is characterized by the following features: (1) multiple channels are superimposed upon each other with varying degrees of preservation; (2) the narrow and deep channels have been backfilled by fining-upward fluvial sediments with clinoform and chaotic depositional patterns; and (3) the fluvial channels grade upward into broad upper portions that have been filled by organic-rich muds with estuarine fauna and flora and horizontal depositional patterns. As Holocene sea level rose, the Roanoke River channel complex was backfilled, first by fluvial then by estuarine sediments. The upper portion of the estuarine fill is presently being truncated by the ravinement surface of the modern barrier system as it migrates landward. Where truncated, the channel-fill facies crop out on the sea floor. The sandy estuarine muds produce abundant, resistant, bathymetric ridge systems that rise slightly above the surrounding sediments.

The same channel complex shown in Figure 6 occurs below the modern barrier (Fig. 5) and constitutes the modern estuarine-sediment system within Albemarle Sound (Fig. 4). The simplified sections in Figures 4 and 5 are based

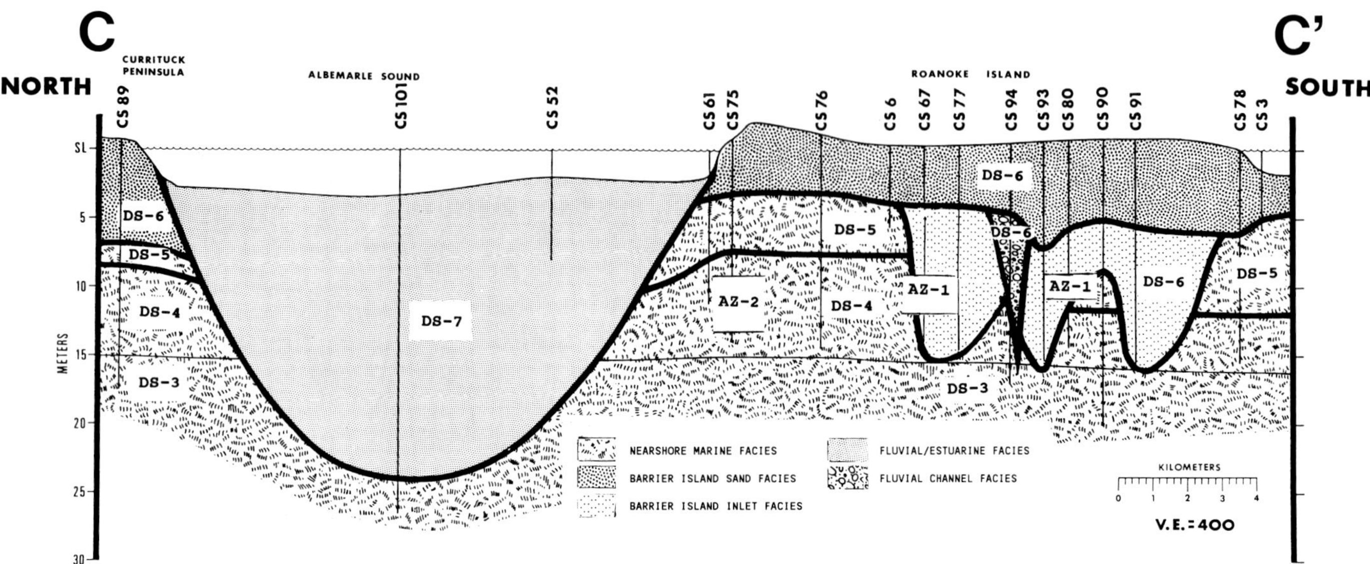

FIG. 4.—A north-to-south geologic cross section extending from end of Currituck Peninsula (C), south across Albemarle Sound and Roanoke Island (C′) (see Fig. 1 for section location). This section shows (1) portions of at least five stacked Quaternary depositional sequences (labeled DS-3 through DS-7), (2) major sediment facies of each depositional sequence; and (3) aminozone designations on associated mollusks (labeled AZ-1 and AZ-2). DS-7 is the Holocene depositional sequence. (Modified from O'Connor and others, 1973; Riggs and O'Connor, 1974; and Eames, 1983).

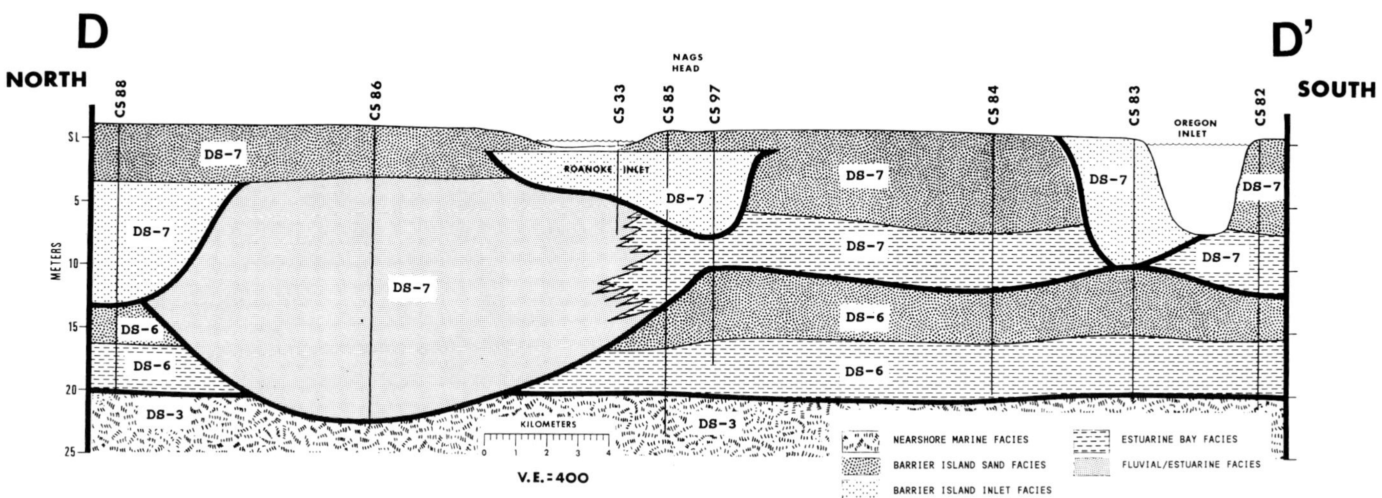

FIG. 5.—A north-to-south geologic cross section, down the center of the present barrier island, extending from north side of Kitty Hawk Bay on the north (D), southward through Nags Head, and to south side of Oregon Inlet (D′) (see Fig. 1 for section location). This section shows (1) portions of at least three stacked Quaternary depositional sequences (labeled DS-3, DS-6, and DS-7) and (2) major sediment facies of each depositional sequence. DS-7 is the Holocene depositional sequence. (Modified from O'Connor and others, 1973; Riggs and O'Connor, 1974; and Eames, 1983).

solely on a few drillhole records. The actual complexity of channel morphologies and sediment facies can only be inferred from the seismic profile in Figure 6; understanding the potentially complex history of the channel system needs to be tested with a series of closely spaced drill holes. Another important, yet unanswered, question is the time component represented by both the channel system and the individual channels. The degree of preservation of individual channels is obviously a function of the relative age; the most complete represents the most recent fluvial event. However, is the entire channel system a response to complex, small-scale fluctuations during the ongoing Holocene sea-level rise, or are some of the older channels from earlier Quaternary sea-level events?

All channels in Figure 6 are interpreted to be of fluvial rather than inlet origin. Inlet channels do occur as integral components within and slightly below barrier-island sand facies; however, they apparently are only temporarily preserved in the Quaternary record landward of the modern ravinement surface. Inlet channels do occur in seismic records and have been drilled within three of the Quaternary depositional sequences in Dare County: the modern barrier

TABLE 1.—NUMERICAL AGE ASSIGNMENTS FOR FOUR AMINOZONES BY YORK (1990), THEIR CORRELATION TO THE DEPOSITIONAL SEQUENCES OF EAMES (1983), AND THE MODIFIED DEPOSITIONAL SEQUENCES FOR DARE COUNTY AS USED IN THIS PAPER.

Aminozones York (1990)	Numerical Age Assignment		Depositional Sequences: Eames (1983)	Depositional Sequences: This Paper
	Stetson Pit	Dare Co. Cores		
	Holocene	Holocene	V	DS-7
AZ-1		78- 51 ka	IV	DS-6
	?	?	III	DS-5
AZ-2	120- 70 ka	120- 70 ka	II	DS-4
AZ-3	530-330 ka	530-330 ka	I	DS-3
AZ-4	1.8-1.1 MA			DS-2
AZ-4	1.8-1.1 MA			DS-1

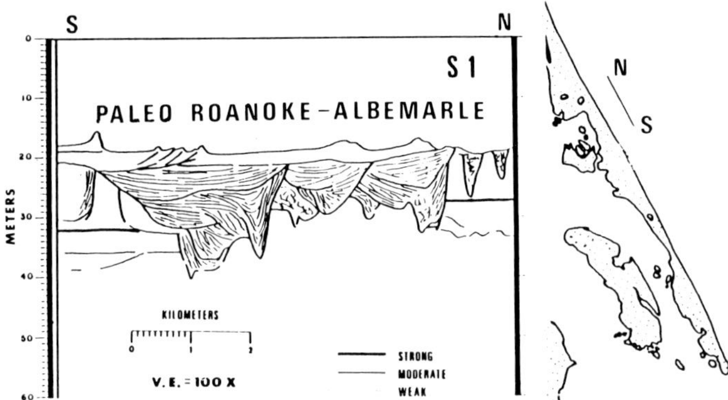

FIG. 6.—Interpreted stratigraphic line drawing of high-resolution seismic line S1 running shore parallel and offshore of Kitty Hawk Bay (see Fig. 1 and map on right side of lower panel for location of seismic line). A subbottom velocity of 1,700 m/sec (2-way) was used to generate the approximate vertical scale. This section shows a set of multiple, stacked, paleochannel systems in various stages of preservation, interpreted to be erosional and depositional infill products of the paleo-Roanoke River and Albemarle Sound fluvial/estuarine system. However, it is not yet known whether each cut and infill event represents a single sea-level event during the Quaternary, or if the complex system is a product of changing paleo-hydraulic responses during the last Holocene sea-level transgression. (Modified from Eames, 1983).

facies on the Outer Banks (Fig. 5) and the barrier-sand facies occurring on Roanoke Island (Figs. 2 and 4) and along Manns Harbor (Fig. 3). These inlet channels, as well as the associated barrier-sand facies, probably will not be preserved in the long-term Quaternary sediment record as the present transgression continues due to the deep level of shoreface erosion, as evidenced in Figure 2. Also, Hine and Snyder (1985) demonstrated that all preserved Quaternary channels occurring on the inner continental shelf of Onslow Bay are the more deeply incised fluvial rather than inlet channels.

Ongoing seismic work by two of the co-authors (SWS and SRR) within northeastern North Carolina has demonstrated that the stratigraphic units summarized in Table 1 represent only a small portion of the Quaternary record of the region. Figure 7 shows a small but typical piece of interpreted stratigraphy produced from high-resolution seismic data collected from the western Pamlico Sound shoreline (Fig. 1). This Quaternary section consists of multiple stacked sequences of coastal and estuarine lithosomes cut by extensive fluvial paleochannels on unconformity surfaces. Seismic data and infill histories for two channel complexes are presented in detail in Figures 8 and 9. Note that most sequence boundaries contain fluvial paleochannels that formed during subsequent sea-level lowstands. Until these complex paleo-fluvial/estuarine systems can be drilled and studied from a detailed and integrated seismic-, litho- and chronostratigraphic approach, we will not understand whether each cut-and-infill event represents a single sea-level event or if the complex system is a product of changing paleo-hydraulic responses during one sea-level event.

Figure 10 is a preliminary subdivision of Quaternary sequences (early = E, middle = M, and late = L) in Figure 7 based on identification of unconformity surfaces and seismic correlations to type sections, existing ages, and previous age assignments in the Cape Lookout and Pamlico Sound areas by Snyder and others (1982, 1984) and Belknap (1984). Based upon preliminary correlations, as many as 18 different depositional sequences can be recognized: four in the early (E1 to E4), nine in the middle (M1 to M9), and five in the late Quaternary sections (L1 to L5). The seismic lines are presently being tied into the lithostratigraphic sections to be discussed later.

Amino-Acid Racemization (AAR)

The racemization of amino acids in fossil mollusks has been used for stratigraphic and chronologic analyses of Quaternary sections in many regions of the world, as recently reviewed by Miller and Brigham-Grette (1989) and Wehmiller (1990). Szabo (1985), Wehmiller and others (1988) and Hollin and Hearty (1990) summarized the majority of the aminostratigraphic data for the entire Atlantic Coastal Plain and discussed some of the conflicting interpretations of those results. Because of the issues identified in those papers, detailed field studies of sections (like those in Dare County, North Carolina) are necessary to verify the stratigraphic consistency of racemization dating methods.

AAR is a chemical dating method (see Colman and others, 1987, for a summary of dating-method nomenclature) that can be used for relative age assignment (aminostratigraphy) or numerical age assignment (aminochronology) if appropriately calibrated samples are available. The method relies upon the conversison of L-amino acids present in living samples into an equilibrium mixture of D- and L-amino acids after death. Rates of racemization, and hence, times to reach this equilibrium value, are temperature dependent. In temperature ranges typical of the mid-Atlantic Coastal Plain (i.e., 14° to 18°C), racemic equilibrium (D/L values = 1.0 for most amino acids) is reached in about 1.5 my. AAR offers great potential because of the small-sample size required for analysis (<10 mg of carbonate) and the extensive age range (50 percent or more of the Quaternary) in which different apparent ages can be resolved.

Quaternary Amino- and Biostratigraphy

Molluscan aminostratigraphic data have been developed for seven localities in Dare County, North Carolina. First, aminostratigraphic, faunal (ostracode), floral (pollen), and

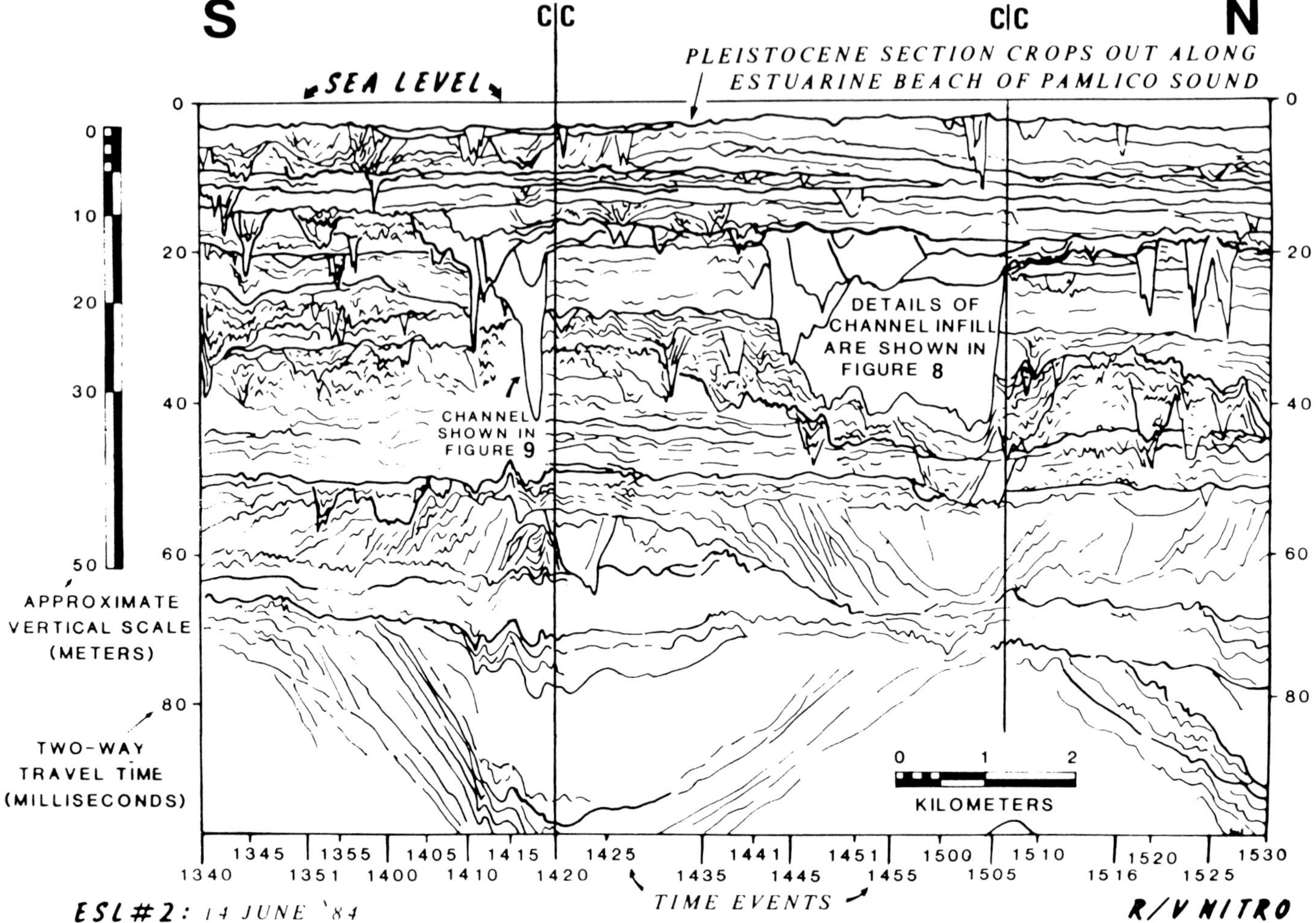

FIG. 7.—Interpreted stratigraphic line drawing of high-resolution seismic line S2 along the estuarine shoreline of western Pamlico Sound (see Fig. 1 for location of seismic line). A subbottom velocity of 1,700 m/sec (2-way) was used to generate the approximate vertical scale. The vertical lines labeled cc indicate coarse changes as shown on Figure 1. The Quaternary section consists of at least 18 stacked depositional sequences of coastal and estuarine lithosomes cut by extensive paleo-fluvial channels. Seismic data and infill histories for two channel systems are shown in Figures 8 and 9. Figure 10 presents the preliminary Quaternary chronostratigraphic assignments.

lithostratigraphic data will be considered for the Stetson Pit area in the north-central portion of mainland Dare County (Fig. 1). Stetson Pit, an aggregate quarry, and a series of associated drill holes (with depths to 33 m below mean sea level, or MSL) have been the focus of numerous biostratigraphic and chronostratigraphic studies during the past decade. Stetson Pit and associated drill cores were sampled for mollusks for aminostratigraphic analysis and ostracode and pollen samples for faunal and floral analyses by York (1984, 1990) and York and others (1989). Paleomagnetic analyses produced uncertain results and were not subsequently used. Some other pertinent studies on Stetson Pit, and upon which the present paper is based, include those of W. Miller (1982), Szabo (1985), Wehmiller (1982), Wehmiller and Belknap (1982, 1987), and Wehmiller and others (1988). Stetson Pit contains a fairly detailed depositional record through a series of Pleistocene interglacials. Stetson Pit data will then be compared to six regional lithostratigraphic sections upon which aminostratigraphic data have been obtained.

Ostracode data suggest the presence of five informal zones based upon the distribution and abundance of temperature-sensitive species and five peaks in shoreline species that are indicative of paleoshoreline positions (York and others, 1989). Pollen data suggest the presence of six pollen zones with five interpreted to represent interglacial conditions (York and others, 1989).

Aminostratigraphic analysis of Stetson Pit (Fig. 1) by York (1984, 1990) and York and others (1989) suggests that there are at least three distinct aminozones based upon the Allo/Iso ratios of both *Mulinia lateralis* and *Mercenaria* sp. Table 1 summarizes the numerical age assignments of these three aminozones, as well as the additional zones defined within the regional drill holes. Because mollusk samples are not uniformly distributed throughout sedimentary sequences, the physical boundary of any aminozone is con-

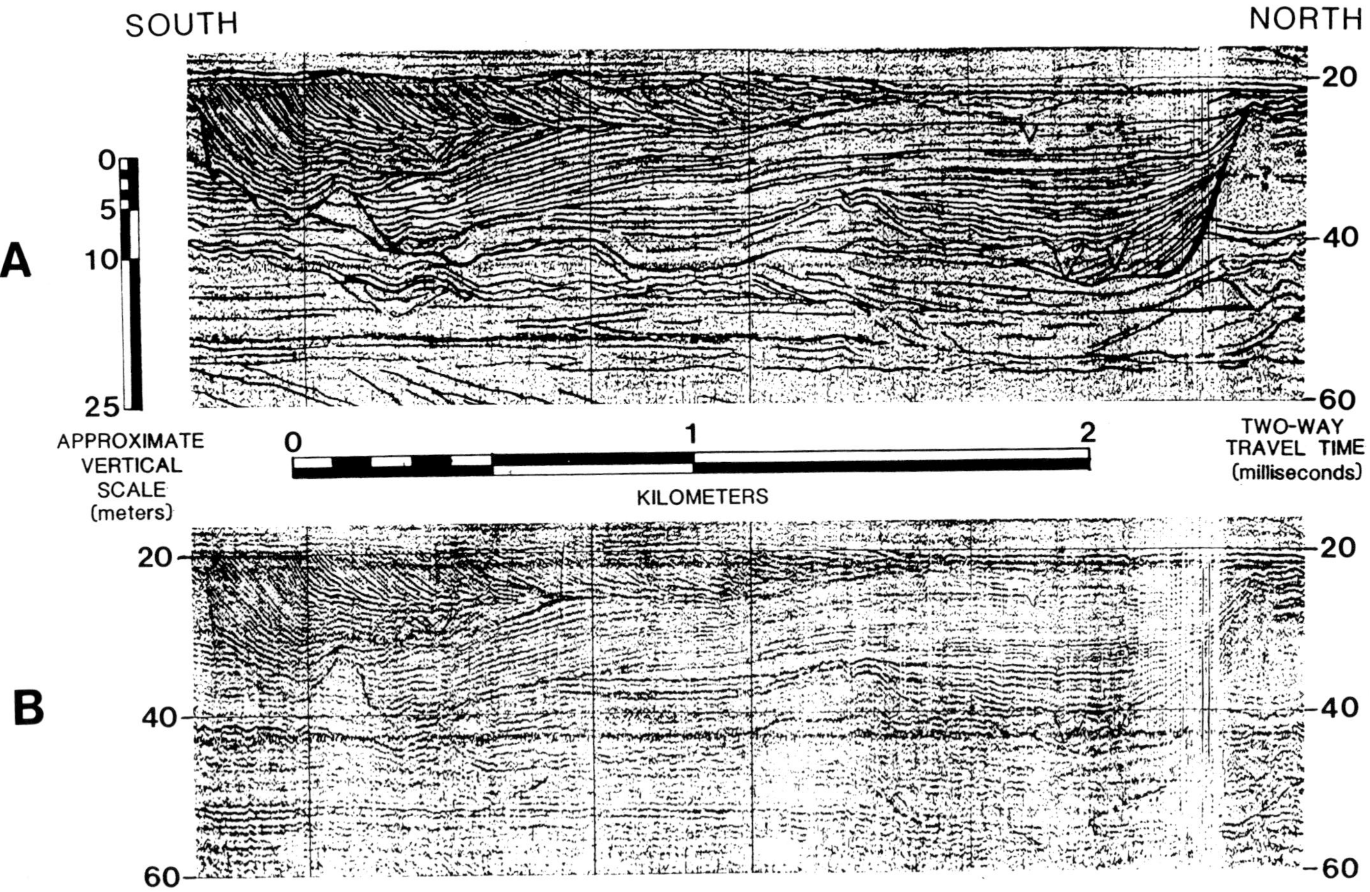

FIG. 8.—Example of high-resolution seismic data on the bottom panel with a preliminary interpreted section in the top panel. The middle Quaternary fluvial paleochannel is from western Pamlico Sound and is labeled M8A in Figure 10. Note that channel-fill lithosomes contain several cut and progradational fill events.

strained by sample availability. Solitary coral $^{230}Th/^{234}U$ dates have been used to constrain the regional aminostratigraphy, as summarized in Wehmiller and others (1988). Dates of 72±4 ka and 221±35 ka for samples from Stetson Pit and Ponzer (about 60 km southwest of Stetson Pit), respectively, are particularly relevant to this calibration. AAR age estimates for aminozones with no independent age control are based on extrapolation of an appropriate nonlinear kinetic model or racemization (Wehmiller and others, 1988).

The lowest aminozone AZ-4 (−17.4 to −33 m MSL) has an age estimate between 1.1 Ma and 1.8 Ma. The upper boundary of AZ-4 (−17.4 m) is a major lithologic contact with abundant abraded and leached shells occuring above the contact (Fig. 2). Aminozone AZ-3 (−13 to −14.2 m MSL) has an age estimate of 330 ka to 530 ka, represents an interglacial episode, and occurs within a sediment unit located directly above the lithologic contact at −17.4 m MSL (Fig. 2). York and others (1989) interpret this unconformity contact to represent as much as 800,000 years. The upper contact of AZ-3 is possibly coincident with a minor change in patterns of sedimentation occurring at −11.2 m (Fig. 2). The upper aminozone AZ-2 (−7.2 to −11.2 m MSL) has an age estimate between 70 ka and 120 ka and occurs within a sediment unit with minor lithologic boundaries at the top and bottom of this zone (−7.2 and −11.2 m MSL).

Stratigraphically above aminozone AZ-2 in the Stetson Pit cores (−3 to − 7.2 m in MSL), the sediments contain a peak in shoreline ostracode species and a pollen assemblage representing a transition from interglacial to glacial vegetation. This stratigraphic interval is interpreted to be the barrier-island facies for AZ-2 (Fig. 2). The uppermost portion of the section (+1 to −3 m MSL) contains a pollen assemblage representing Holocene swamp-forest vegetation, is barren of ostracodes, and most certainly represents Holocene nonmarine deposition.

York (1990) also recognized three distinct aminozones in six cores in eastern Dare County (Fig. 3 and 4). Only one aminozone was found in each core with each aminozone occurring in two different cores. Statistical analysis indicates that the *Mulinia* samples from each aminozone are from different populations and that the aminozones are statistically distinct from one another. Aminozones AZ-3 and AZ-2 can be correlated with the two upper aminozones (AZ-3 and AZ-2) identified in Stetson Pit, respectively. However, none of the cores was deep enough to reach aminozone AZ-4. In addition, aminozone AZ-1 was identified in these cores but was not recognized within the Stetson Pit

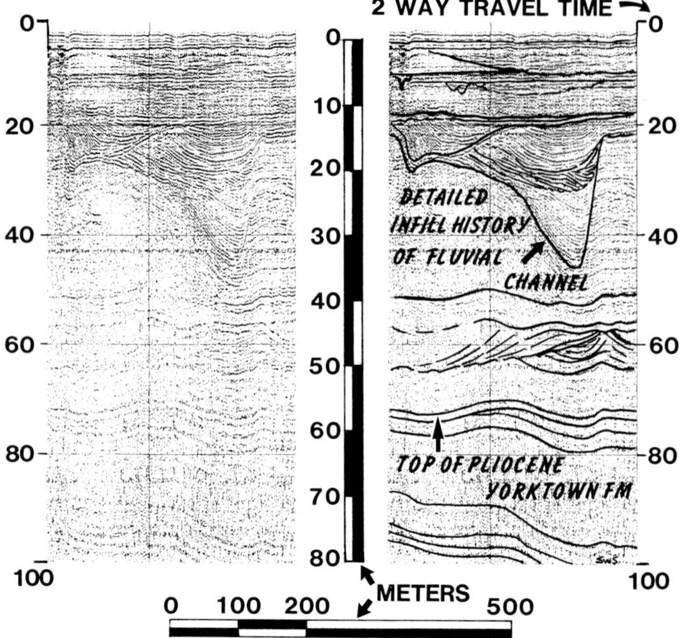

FIG. 9.—Example of high-resolution seismic data in left panel with a preliminary interpreted section in right panel. The middle Quaternary fluvial paleochannel system is from western Pamlico Sound and is labeled M8B in Figure 10. Data demonstrate multiple Quaternary sea-level events showing complex patterns of fluvial erosion and subsequent infill deposition.

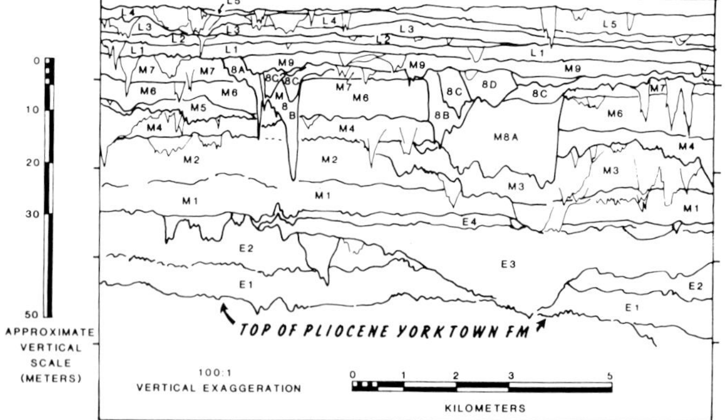

FIG. 10.—Preliminary subdivision of Quaternary depositional sequences in Figure 7 based on identification of unconformity surfaces and the law of superposition. Section is subdivided into early (E), middle (M), and late Quaternary (L) units based on preliminary seismic correlations to type sections, existing age dates, and previous AAR assignments in the Cape Lookout and Pamlico Sound areas by Snyder and others (1982, 1984) and Belknap (1984). Based upon the preliminary interpretation, up to 18 different depositional sequences can be recognized: four in the early (E1 to E4), nine in the middle (M1 to M9), and five in the late Quatenary (L1 to L5). Note that all sequence boundaries contain fluvial paleochannels.

cores. Aminozone AZ-1 has a numerical age assignment between 51 and 78 ka (Table 1).

QUATERNARY HISTORY OF DEPOSITIONAL SEQUENCES

The present seismic data base suggests that there are many more depositional sequences within the Quaternary section than we presently can differentiate using drillhole data. However, we presently have lithostratigraphic and aminochronologic information to differentiate seven specific sequences, which are summarized in this section and in Table 2.

Depositional Sequences DS-1 and DS-2 (Early Quaternary)

Depositional sequences DS-1 and DS-2 have only been recognized in the lower portion of drill cores in Stetson Pit (−17.4 to −33 m MSL) (Fig. 2). Both DS-1 and DS-2 are totally within aminozone AZ-4, have been assigned numerical age estimates between 1.1 and 1.8 Ma, and consequently are considered to be early Quaternary (Table 2), based upon the nomenclature of Richmond and Fullerton (1986). Lithologically, both units occur below a major hiatal contact at −17.4 m MSL and consist of coarsening-upward sediments that grade from silty clays into muddy sands and contain a molluscan assemblage dominated by whole *Mulinia* and angular *Mercenaria* fragments. The pattern of significant changes in concentration of temperature-sensitive and shoreline species of ostracodes and pollen species led York and others (1989) and York (1990) to conclude that aminozone AZ-4 actually consisted of two interglacial sea-level events. The faunal and floral changes coincide with a 3-m-thick zone (−26.4 to −29.2 m MSL) that contains abundant whole and fragmented shells of *Rangia cuneata*, a brackish-water mollusk.

The lithologic and faunal and floral data suggest that the contact between DS-1 and DS-2 occurs at the base of the *Rangia* zone (−29.2 m MSL). Only an inner-shelf marine facies has been recognized in DS-1 (−33 to −29.2 m MSL). DS-2 is represented lithologically by an estuarine facies (−29.2 to −26.4 m MSL) followed by the inner-shelf ma-

TABLE 2.—SUMMARY OF AMINOZONES AND DEPOSITIONAL SEQUENCES FOR THE DARE COUNTY AREA, THEIR CHRONOLOGIC ASSIGNMENT, AND PRELIMINARY CORRELATION TO THE DEEP-SEA OXYGEN ISOTOPE STAGES. AGES ARE BASED UPON DATA OF RICHMOND AND FULLERTON (1986). OXYGEN ISOTOPE STAGES ARE BASED UPON DATA OF ZIMMERMAN AND OTHERS (1984).

Geochronologic Units		Age years	Aminozones	Depositional Sequence	Correlation to Oxygen Isotope Stages
QUATERNARY	Holocene			DS-7	1
		10,000			
	Late		AZ-1	DS-6	very late 5 or 3
				DS-5	mid- to late 5
			AZ-2	DS-4	early 5
		132,000			
	Middle		AZ-3	DS-3	9, 11, or 13
		788,000			
	Early		AZ-4	DS-2	23 to 29
			AZ-4	DS-1	23 to 29
		1,650,000			
Pliocene					

rine facies (−26.4 to −17.4 m MSL). No other coastal facies have been preserved in this locality. Depositional sequences DS-1 and DS-2 (and aminozone AZ-4) should dip slightly and extend eastward, occurring just below the depth of core penetration in Figures 2 through 5.

Depositional Sequence DS-3 (Middle Quaternary)

Within Stetson Pit, DS-3 includes aminozone AZ-3 (−13 to −14.2 m MSL) and corresponds to a lithostratigraphic unit that extends from −11.2 m MSL to the major hiatal surface at −17.4 m MSL. The unit is a silty, fine- to medium-grained sand with interbeds of dark gray and abraded, fossil shell hash. The sequence includes a suite of primarily thermophilic (warm-water) ostracode species, a peak of shoreline ostracode species, and a pollen assemblage representing interglacial, temperate conditions (York and others, 1989). The estimated numerical age for aminozone AZ-3 is between 530 and 330 ka, or middle Quaternary (Table 2).

Depositional sequence DS-3 extends eastward and is correlated with a large portion of depositional sequence I of Eames (1983), the lower portion of most of the Eames core holes, based upon the lithologic similarity of inner-shelf sediments and the absence of obvious hiatal surfaces. The widespread sequence is generally a marine unit of muddy quartz sand with a rich molluscan assemblage (*Ensis, Mercenaria,* and *Mulinia*) and a major soil profile along the upper surface.

Oaks and others (1974) mapped the Hickory Scarp in southeastern Virginia and extended it to the Camden Peninsula on the north shore of Albemarle Sound. Oaks and DuBar (1974) correlated the Hickory Scarp to a relict shoreline in the center of Tyrrell and Hyde Counties to the west of the Alligator River. We hypothesize that the Hickory Scarp could represent the barrier-island system that was contemporaneous with the nearshore marine sediments of depositional sequence DS-3. The sand ridge is a very subtle surficial expression of a more extensive subsurface barrier-island facies that has been almost completely buried by the upward growth of the Holocene swamp forest. On the seaward side, depositional sequence DS-3 has been truncated by the shoreface erosion of sequence DS-6 (Fig. 2).

Depositional Sequence DS-4 (Late Quaternary)

Within Stetson Pit, depositional sequence DS-4 includes aminozone AZ-2 (−7.2 to −11.2 m MSL) with a numerical age estimate of 70 ka to 120 ka (Table 2). DS-4 consists of gray-green silty sand to sandy silt with abundant *Crassostrea* and *Mercenaria* fragments. The estuarine muddy sand is equivalent to the lower facies of depositional sequence II of Eames (1983; Table 1) and grades laterally eastward into sandy estuarine muds with extensive oyster reefs. Throughout eastern mainland Dare County, the lower estuarine sediments of DS-4 coarsen upward and grade from dark, muddy fine sands with abundant *in situ* reefs of *Crassostrea* into clean sands with abundant abraded shell hash. The DS-4 barrier-island sand facies is shallowest down the central portion of the Dare County mainland with only local and very minor subaerial expression; it has been almost totally buried by modern swamp-forest peats of DS-7 (Fig. 2).

The barrier-island sand facies migrated landward over the nearshore marine facies that is still preserved under Roanoke Island, Croatan Sound, and the easternmost portion of the Dare County mainland (Figs. 2, 3, and 4). The nearshore marine facies consists entirely of very fossiliferous, medium-grained sands with a diverse suite of marine mollusks that grade upward into an open estuarine facies as the barrier-island facies developed and migrated westward into the Dare County area (Fig. 2).

Depositional Sequence DS-5 (Late Quaternary)

Depositional sequence DS-5 does not contain any recognized aminozones. On mainland Dare County, DS-5 consists of only scattered preserved remnants of estuarine origin behind a major barrier-island sand facies that occurs along the western side of the Croatan Sound shoreline (Figs. 2 and 3). The highest portion of the barrier-island sand ridge rises just above sea level and slightly above the Holocene swamp-forest peats to form a sand substrate for the towns of Manns Harbor and Mashoes. The barrier-sand facies is probably correlative with the major shoreline feature that extends down the Currituck Peninsula and forms the Powells Point Ridge of Oaks and Dubar (1974). Eastward, below Croatan Sound and Roanoke Island, DS-5 consists entirely of nearshore marine sediments deposited in an inner-shelf environment. On the seaward side, depositional sequence DS-5 was truncated along with DS-4 by the ravinement surface associated with the formation of sequence DS-6 (Figs. 2 and 5).

No AAR data are available for DS-5, but its age is stratigraphically constrained between AZ-2 and AZ-1 with a maximum range between 120 ka and 51 ka. Consequently, DS-5 most likely represents a small transgression within the middle to latest portion of isotope stage 5.

The dissected paleotopography of the barrier-island sand facies of sequences DS-4 and DS-5 forms the irregular sand surface upon which extensive Holocene peat facies have been deposited throughout the mainland portion of Dare County. The modern swamp-forest peats have almost totally buried the older sequences (Fig. 2). The paleotopography has determined the distribution and thickness of the Dare County Holocene peat deposits (O'Connor and others, 1973; Riggs and O'Connor, 1974; Eames, 1983).

Depositional Sequence DS-6 (Late Quaternary)

Depositional sequence DS-6 consists of a fairly well-preserved sequence of estuarine and barrier-island facies. DS-6 dominates the subaerial and surficial sediments on Roanoke Island and may be correlative with the fossil dune features on Collington Island to the north (Fig. 1). Migration of the sequence of estuarine and barrier-island facies formed the major ravinement surface below Roanoke Island that truncated sequences DS-3, DS-4, and DS-5 (Fig. 2). Shells from the various barrier-inlet channels within the sequence (Fig. 4) occur within aminozone AZ-1 with a numerical age estimate of 51 ka to 78 ka (Table 1). Assuming that the AAR age estimates are correct, aminochronology would place the formation of this major coastal sequence at least as young as the very late portion of isotope stage 5 or possibly of stage 3 age (Table 2). Several other methods are presently being utilized to refine the age assignment. Aminzone AZ-1 has not been recognized in the Stetson Pit section.

Depositional sequence DS-6 drops fairly steeply into the subsurface below Roanoke Sound and the modern barrier island. Facies of DS-6 crop out or occur below very thin layers of modern sands on the lower portion of the lower forebeach (Fig. 2), where they are being truncated by shoreface erosion (Pearson, 1979; Riggs, 1979). They also crop out extensively across the modern sea floor and constitute core sediments within major bathymetrically high features known as the Albemarle and Platt Shoals (Fig. 2; Pearson, 1979; Riggs, 1979).

Four major facies have been recognized within sequence DS-6. The basal unit is a fossiliferous estuarine mud dominated by *Crassostrea* and other estuarine fauna. The estuarine sediments grade upward into a complex sequence of clean barrier-island sand facies that have migrated up and over the estuarine muds (Fig. 2). Associated with the barrier-island sand facies is an extensive sequence of inlet and possibly fluvial-channel deposits (Fig. 4). The paleochannels occurring below subaerial portions of Roanoke Island, where seismic traces are not available, have poorly known extents and distribution patterns that are based only upon drillhole data.

Depositional Sequence DS-7 (Late Quaternary-Holocene)

Depositional sequence DS-7 includes the sedimentary facies produced by the Holocene rise in sea level, or isotope stage 1 (Table 2). The modern barrier-island system, sediments filling the associated estuarine systems, peats associated with modern marshes and swamp forests, and sediments within inland fluvial systems are integral parts of DS-7 (Figs. 2–5).

The Holocene is the latest interglacial sea-level event controlling sedimentation on the North Carolina continental margin. Changing climatic conditions resulting from deglaciation are readily apparent in the preserved sediment sequence in northeastern North Carolina (Copeland and others, 1983, 1984; Riggs and others, 1989, 1991). Beginning about 18 ka, cool, semi-arid climates resulted in sediment-choked, braided streams discharging coarse terrigenous sediments across the subaerially exposed continental shelf. As climates warmed, glaciers receded, sea level rose, and extensive wetland pocosins developed, first with boreal and then temperate vegetation (Whitehead, 1981). Increased vegetative cover decreased the volume and size of fluvial-sediment loads to predominantly suspended silt and clay. The leading edge of the transgression flooded topographic lows, forming the present embayed estuarine system. The old fluvial channels, deeply incised within their floodplain systems, filled first with coarse fluvial sand and gravel followed by thick accumulations of organic-rich, estuarine mud.

The modern coastal system is an instantaneous time slice within the evolutionary continuum of Quaternary deposition and erosion. Modern rivers are delivering predominantly suspended sediments that are being partially trapped in the estuaries and partially discharged through inlets to the continental shelf (Meade, 1972; Benton and others, 1978; Riggs and others, 1989, 1991). Fluvial sands have not been transported into the modern estuaries since flooding began (Hartness, 1977; Riggs and others, 1989, 1991); all modern sands are derived either internally from estuarine shoreline erosion or from adjacent barrier islands (O'Connor and others, 1973; Riggs and O'Connor, 1974; Bellis and others, 1975; Hardaway, 1980). In northern North Carolina, sea level is continuing to rise at the rate of 10 to 25 cm/century (Benton, 1980; Riggs and others, 1989, 1991). This low rate of rise causes a general flooding of the land, first up the topographically low river valleys and then laterally across the uplands. Because of the low regional slope, the relatively slow rate of flooding produces rapid rates of lateral erosion that average 1 m/yr and locally may be 2 to 6 m/yr (Bellis and others, 1975; Riggs and others, 1978). The entire coastal system maintains its integrity through time as it migrates upward and landward by a systematic evolutionary succession: the incised drainages are drowned as the estuaries and barrier system displace the fluvial system landward (Riggs, 1985).

SUMMARY

A complex portrait is now emerging for the Quaternary section based upon detailed stratal pieces from throughout northeastern North Carolina. The combined litho-, seismic, bio-, and aminostratigraphic subsurface data suggest that there have been multiple and significant changes in shoreline position, depositional environments, water-mass characteristics, and terrestrial climate. Seismic data suggest that at least 18 major sea-level events have impacted the outer coastal zone of North Carolina. Utilizing shallow subsurface core data from the upper 33 m, we have been able to recognize and decipher portions of at least seven of these depositional events, as summarized in Table 2. Each of these seven sequences is interpreted to represent a major interglacial sea-level transgression of the coastal system that truncated and modified previously deposited sequences, producing sets of coastal facies with irregular erosional geometries that are, in part, dependent upon the paleotopography of the transgression surface. Correlation of the kinetic-model age estimates proposed for the aminozones with global chronology provides a basis for a preliminary correlation of the seven depositional sequences to the deep-sea oxygen isotope stages, as summarized in Table 2.

Thus, major Quaternary sea-level fluctuations have produced an extremely complex sediment record reflecting intricate patterns of migrating depositional regimes and associated erosional events. Glacial periods brought lowered sea level, subaerial processes and abundant fluvial sediment across the continental shelf as evidenced by extensive fluvial channeling. The periods of lowered sea level are characterized by braided streams, as suggested by the underfit, character of modern coastal-plain streams (Flint, 1971), extensive gravel deposits that occur throughout the upper coastal plain (Daniels and Gamble, 1974), and fluvial sand and gravel sediments in basal sections of cored channel deposits. The paleochannels have been sequentially backfilled with fluvial and estuarine sediments.

After each glacial episode, transgressive seas produced ravinement surfaces that migrated landward by shoreface erosion, eliminating large portions of the previously deposited sediments. Quaternary sediments occurring within the abundant, deeply scoured fluvial channels underlying the ravinement surface are preserved on the continental shelf. Important factors in preservation potential include balances between wave energy, sediment budget, antecedent topography, and relative rates of sea-level rise and margin subsidence. During high sea-level stands, a new sequence of temperate, coastal and shelf sediments is deposited over an unconformity cut into previously deposited and highly eroded Quaternary sequences composed of the same lithofacies.

The resulting sediment sequence consists of multiple sets of stacked deposits with similar lithologies and "depositional geometries" that have been severely modified by subsequent erosional processes. Thus, erosional processes associated with sea-level cyclicity on stable or slowly subsiding continental margins lead to only partial preservation of repetitive lithologies. Both attributes make it difficult to resolve high-frequency stratal events utilizing scattered drill holes and coastal-plain outcrops or employing conventional bio- and lithostratigraphic analysis alone. However, a detailed multi-disciplinary approach has begun to successfully resolve these high-frequency stratal events within the Quaternary record within northeastern North Carolina.

ACKNOWLEDGMENTS

Thanks are due to Albert Hine, Kelvin Ramsey, and Marguerite Toscano for reviewing the manuscript and supplying many constructive criticisms and helpful suggestions. This research was in part supported by the following grants: National Science Foundation grants EAR8407024 and EAR 8915747 to JFW; National Science Foundation grant OCE-8342777 to SRR; and National Oceanographic and Atmospheric Administration/University of North Carolina Sea Grant College Program grants to SRR.

REFERENCES

Belknap, D. F., 1984, Amino acid racemization in "mid-Wisconsin" C-14 dated formations on the Atlantic Coastal Plain: Geological Society of America, Abstracts with Programs, v. 16, p. 2–3.

Belknap, D. F., and Kraft, J. C., 1981, Preservation potential of transgressive coastal lithosomes on the U.S. Atlantic shelf: Marine Geology, v. 42, p. 429–442.

Belknap, D. F., and Kraft, J. C., 1985, Influence of antecedent geology on evolution of Delaware's barrier system: Marine Geology, v. 63, p. 235–262.

Bellis, V., O'Connor, M. P., and Riggs, S. R., 1975, Estuarine shoreline erosion in the Albemarle-Pamlico region of North Carolina: University of North Carolina Sea Grant Publication No. UNC-SG-75-29, 67 p.

Benton, S. B., 1980, Holocene evolution of a nanotidal brackish marsh-protected bay system, Roanoke Island, North Carolina: Unpublished M.S. Thesis, Department of Geology, East Carolina University, Greenville, North Carolina, 179 p.

Benton, S. B., Riggs, S. R., and O'Connor, M. P., 1978, Suspended sediment concentration of North Carolina drowned river valley estuaries: Elisha Mitchell Jour., v. 94, no. 2, p. 5.

Blackwelder, B. W., 1981, Late Cenozoic marine deposition in the U.S. Atlantic Coastal Plain related to tectonism and global climate: Palaeogeography, Palaeoclimatology, Palaeoecology, v. 34, p. 87–113.

Brown, P. M., Miller, J. A., and Swain, F. M., 1972, Structural and stratigraphic framework, and spatial distribution of permeability of Atlantic coastal plain, North Carolina to New York: U.S. Geological Survey Professional Paper 796, 79 p.

Colman, S., Pierce, K. L., and Birkeland, P. W., 1987, Suggested terminology for Quaternary dating methods: Quaternary Research, v. 28, p. 214–319.

Copeland, B. J., Hodson, R. G., and Riggs, S. R., 1984, Ecology of the Pamlico River, North Carolina: An Estuarine Profile: U.S. Department of Interior, Fish and Wildlife Service, Washington D.C., FWS/OBS-82-06, 83 p.

Copeland, B. J., Hodson, R. G., Riggs, S. R., and Easley, J. E., 1983, Ecology of Albemarle Sound, North Carolina: An Estuarine Profile: U.S. Department of Interior, Fish and Wildlife Service, Washington D.C., FWS/OBS-83-1, 68 p.

Cronin, T. M., 1980, Biostratigraphic correlation of Pleistocene marine deposits and sea levels, Atlantic Coastal Plain of the southeastern U.S.: Quaternary Research, v. 13, p. 213–229.

Cronin, T. M., Szabo, B. J., Ager, T. A., Hazel, J. E., and Owens, J. P., 1981, Quaternary climates and sea levels of the U.S. Atlantic Coastal Plain: Science, v. 211, p. 233–240.

Daniels, R. B., and Gamble, E. E., 1974, Surficial deposits of the Neuse-Cape Fear divide above the Surry Scarp, North Carolina, *in* Oaks, R. Q., and DuBar, J. R., eds., Post-Miocene Stratigraphy Central and Southern Atlantic Coastal Plain: Utah State University Press, Logan, p. 88–101.

Delcourt, P. A., and Delcourt, H. R., 1981, Vegetation maps for North America: 40,000 yr B.P. to the present, *in* Romans, R. C., ed., Geobotany II: Plenum Publishing Co., New York, p. 123–165.

Eames, G. B., 1983, The late Quaternary seismic stratigraphy, lithostratigraphy, and geologic history of a shelf-barrier-estuarine system, Dare Co., North Carolina: Unpublished M.S. Thesis, Department of Geology, East Carolina University, Greenville, North Carolina, 196 p.

Fairbanks, R. G., 1989, A 17,000-year glacio-eustatic sea level record: influence of glacial melting rates on the Younger Dryas event and deep-ocean circulation: Nature, v. 342, p. 637–642.

Flint, R. F., 1971, Glacial and Quaternary Geology: John Wiley and Sons, New York, 892 p.

Hardaway, C. S., 1980, Shoreline erosion and its relationship to the geology of the Pamlico River Estuary: Unpublished M.S. Thesis, Department of Geology, East Carolina Univ., Greenville, North Carolina, 116 p.

Hartness, T. S., 1977, Distribution and clay mineralogy of organic-rich mud sediments in the Pamlico River Estuary, North Carolina: Unpublished M.S. Thesis, Department of Geology, East Carolina University, Greenville, North Carolina, 45 p.

Hays, J. D., Imbrie, J., and Shackleton, N. J., 1976, Variations in the Earth's orbit: pacemaker of the ice ages: Science, v. 194, p. 1121–1132.

Hine, A. C., and Snyder, Stephen W., 1985, Coastal lithosome preservation: evidence from the shoreface and inner continental shelf off Bogue Banks, North Carolina: Marine Geology, v. 63, p. 307–330.

Hollin, J. T., and Hearty, P. J., 1990, South Carolina interglacial sites and stage 5 sea levels: Quaternary Research, v. 33, p. 1–17.

Imbrie, J., and Imbrie, J. Z., 1980, Modeling the climatic response to orbital variations: Science, v. 207, p. 943–953.

Meade, R. H., 1969, Landward transport of bottom sediments in estuaries of the Atlantic Coastal Plain: Journal of Sedimentary Petrology, v. 39, p. 222–234.

Meade, R. H., 1972, Transport and deposition of sediments in estuaries: Geological Society of America Memoir 133, p. 91–120.

Miller, G. H., and Brigham-Grette, J., 1989, Amino acid geochronology: resolution and precision in carbonate fossils: Quaternary International, v. 1, p. 111–128.

Miller, J. A., 1982, Stratigraphy, structure, and phosphate deposits of the Pungo River Formation of North Carolina: North Carolina Division of Land Resources, Geological Survey Section, Bulletin 87, 32 p.

Miller, W., III, 1982, The paleoecologic history of Late Pleistocene estuarine and marine fossil deposits in Dare County, North Carolina: Southeastern Geology, v. 23, p. 1–13.

Oaks, R. Q., Coch, N. K., Sanders, J. E., and Flint, R. F., 1974, Post-Miocene shorelines and sea levels, southeastern Virginia, *in* Oaks, R. Q., and DuBar, J. R., eds., Post-Miocene Stratigraphy, Central and

Southern Atlantic Coastal Plain: Utah State University Press, Logan, p. 53–87.

OAKS, R. Q., AND DUBAR, J. R., 1974, Tentative correlation of post-Miocene units, central and southern Atlantic Coastal Plain, *in* Oaks, R. Q., and DuBar, J. R., eds., Post-Miocene Stratigraphy, Central and Southern Atlantic Coastal Plain: Utah State University Press, Logan, p. 232–246.

O'CONNOR, M. P., RIGGS, S. R., AND BELLIS, V., 1978, North Carolina estuarine shoreline relative erosion potential: University of North Carolina Sea Grant College Publication, Raleigh, North Carolina, 2 p.

O'CONNOR, M. P., RIGGS, S. R., AND WINSTON, D., 1973, Recent estuarine sediment history of the Roanoke Island area, North Carolina, *in* Nelson, B. W., ed., Environmental Framework of Coastal Plain Estuaries: Geological Society of American Memoir 133, p. 453–464.

PEARSON, D. K., 1979, Surface and shallow subsurface sediment regime of the nearshore inner continental shelf, Nags Head and Wilmington areas, North Carolina: Unpublished M.S. Thesis, Department of Geology, East Carolina University, Greenville, North Carolina, 120 p.

PIETRAFESA, L. J., JANOWITZ, G. S., CHAO, T. Y., WIESBERG, R. H., ASKARI, F., AND NOBLE, E., 1986, The physical oceanography of Pamlico Sound: University of North Carolina Sea Grant College Program, Raleigh, North Carolina, Working Paper 86-5, 125 p.

POPENOE, P., 1985, Cenozoic depositional and structural history of the North Carolina margin from seismic-stratigraphic analyses, *in* Poag, C. W., ed., Geologic Evolution of the United States Atlantic Margin: Van Nostrand Reinhold Co., New York, p. 125–188.

POPENOE, P., AND WARD, L. W., 1983, Description of high-resolution seismic reflection data collected in Albemarle and Croatan sounds, North Carolina: U.S. Geological Survey Open-File Report 83-513, 3 p.

PRELL, W. L., IMBRIE, J., MARTINSON, D. G., MORLEY, J. J., PISIAS, N. G., SHACKLETON, N. J., AND STREETER, H. E., 1986, Graphic correlation of oxygen isotope stratigraphy application to the late Quaternary: Paleoceanography, v. 1, p. 137–162.

RICHMOND, G. M., AND FULLERTON, D. S., 1986, Introduction to Quaternary glaciations in the United States of America: Quaternary Science Reviews, v. 5, p. 3–10.

RIGGS, S. R., 1979, A geologic profile of the North Carolina coastal—inner continental shelf system, *in* Langfelder, J., ed., Ocean Outfall Wastewater Disposal Feasibility and Planning: North Carolina State University Press, p. 90–113.

RIGGS, S. R., 1984, Paleoceanographic model of Neogene phosphorite deposition, U.S. Atlantic continental margin: Science, v. 223, p. 123–131.

RIGGS, S. R., 1985, Evolution of coupled fluvial/estuarine/barrier coastal systems through the late Quaternary of northeastern North Carolina: Geological Society of America, Abstracts with Program, v. 17, p. 700.

RIGGS, S. R., AND BELKNAP, D. F., 1988, Upper Cenozoic processes and environments of continental margin sedimentation: eastern United States, *in* Sheridan, R. E., and Grow, J. A., eds., The Atlantic Continental Margin, U.S.: Geological Society of America, The Geology of North America, v. I–2, p. 131–176.

RIGGS, S. R., BRAY, J. T., POWERS, E. R., HAMILTON, J. C., AMES, D. V., YEATES, D. D., OWENS, K. L., LUCAS, S. L., WATSON, J. R., AND WILLIAMSON, H. M., 1991, Heavy metal pollutants in organic-rich muds of the Neuse River estuary: Albemarle-Pamlico Estuarine Study, U.S. Environmental Protection Agency and North Carolina Department of Natural Resources and Community Development, Report No. 90-07, 174 p.

RIGGS, S. R., AND MALLETTE, P. M., 1990, Patterns of phosphate deposition and lithofacies relationships within the Miocene Pungo River Formation, North Carolina continental margin, *in* Burnett, W. C., and Riggs, S. R., eds., Neogene to Modern Phosphorites: Cambridge University Press, Cambridge, England, Phosphate Deposits of the World, v. 3, p. 424–443.

RIGGS, S. R., AND O'CONNOR, M. P., 1974, Relict sediment deposits in a major transgressive coastal system: University of North Carolina Sea Grant College Publication No. UNC-SG-74-04, 37 p.

RIGGS, S. R., O'CONNOR, M. P., AND BELLIS, V., 1978, Estuarine shoreline erosion in North Carolina: University of North Carolina Sea Grant College Publication, Poster Series, 5 p.

RIGGS, S. R., POWERS, E. R., BRAY, J. T., STOUT, P., HAMILTON, C., AMES, D., MOORE, R., WATSON, J., LUCAS, S., AND WILLIAMSON, M., 1989, Heavy metal pollutants in organic-rich muds of the Pamlico River estuarine system: Albemarle-Pamlico Estuarine Study, U.S. Environmental Protection Agency and North Carolina Department of Natural Resources and Community Development, Report No. 89–06, 108 p.

RIGGS, S. R., SNYDER, STEPHEN W., SNYDER, SCOTT W., AND HINE, A. C., 1990, Stratigraphic framework for cyclical deposition of Miocene phosphorites in the Carolina Phosphogenic Province, *in* Burnett, W. C., and Riggs, S. R., eds., Neogene to Modern Phosphorites: Cambridge University Press, Cambridge, England, Phosphate Deposits of the World, v. 3, p. 381–395.

RUDDIMAN, W. F., RAYMO, M., AND MCINTYRE, A., 1986, Matuyama 41,000-year cycles: North Atlantic Ocean and Northern Hemisphere ice sheets: Earth and Planetary Science Letters, v. 80, p. 117–129.

SCHLEE, J. S., MANSPEISER, W., AND RIGGS, S. R., 1988, Paleoenvironments offshore Atlantic U.S. margin, *in* Sheridan, R. E., and Grow, J. A., eds., The Atlantic Continental Margin, U.S.: Geological Society of America, The Geology of North America, v. I-2, p. 365–385.

SHACKLETON, N. J., 1987, Oxygen isotopes, ice volume and sea level: Quaternary Science Reviews, v. 6, p. 183–190.

SHACKLETON, N. J., AND OPDYKE, N. C., 1973, Oxygen isotope and paleomagnetic stratigraphy of equatorial Pacific core V28-V238: Oxygen isotope temperatures and ice volumes on a 105 and 106 year scale: Quaternary Research, v. 3, p. 39–55.

SNYDER, STEPHEN W., BELKNAP, D. F., HINE, A. C., AND STEELE, G. A., 1982, Seismic stratigraphy, lithostratigraphy, and amino acid racemization of the Diamond City Formation: reinterpretation of a reported ''mid-Wisconsin high'' sea-level indicator from the North Carolina Coastal Plain: Geological Society of America, Abstracts with Programs, v. 14, p. 84.

SNYDER, STEPHEN, W., HINE, A. C., AND BELKNAP, D. F., 1984, Stratigraphic consequences of erosional transgression: a Holocene model applied to the Quaternary record: Society of Economic Paleontologists and Mineralogists, Annual Midyear Meeting, Abstracts, v. 1, p. 76.

SZABO, B. J., 1985, Uranium-series dating of fossil corals from marine sediments of the United States Atlantic Coastal Plain: Geological Society of America Bulletin, v. 96, p. 398–406.

WARD, L. W., AND STRICKLAND, G. L., 1985, Outline of Tertiary stratigraphy and depositional history of the U.S. Atlantic Coastal Plain, *in* Poag, C. W., ed., Geologic Evolution of the United States Atlantic Margin: Van Nostrand Reinhold Co., New York, p. 125–188.

WEHMILLER, J. F., 1982, A review of amino acid racemization studies in Quaternary mollusks: stratigraphic and chronologic applications in coastal and interglacial sites, Pacific and Atlantic coasts, United States, United Kingdom, Baffin Island, and tropical islands: Quaternary Science Reivews, v. 1, p. 83–120.

WEHMILLER, J. F., 1990, Amino acid racemization: applications in chemical taxonomy and chronostratigraphy of Quaternary fossils, *in* Carter, J. G., ed., Skeletal Biomineralization: Patterns, Processes, and Evolutionary Trends: Van Nostrand Reinhold, New York, v. 1, p. 583–608.

WEHMILLER, J. F., AND BELKNAP, D. F., 1982, Amino acid age estimates, Quaternary Atlantic Coastal Plain: comparison with U-series dates, biostratigraphy, and paleoclimatic control: Quaternary Research, v. 18, p. 311–336.

WEHMILLER, J. F., AND BELKNAP, D. F., 1987, Aminostratigraphy of coastal U.S. Quaternary marine deposits—deciphering the timing of interglacial high sea levels and the thermal histories of coastal regions, *in* Rampino, M. R., Sanders, J. E., Newman, W. S., and Konigsson, L. K., Climate, History, Periodicity, and Predictability: Van Nostrand Reinhold, New York, p. 157–165.

WEHMILLER, J. F., BELKNAP, D. F., BOUTIN, B. S., MIRECKI, J. E., RAHAIM, S. D., AND YORK, L. L., 1988, A review of the aminostratigraphy of Quaternary mollusks from United States Atlantic Coastal Plain sites, *in* Easterbrook, D. L., ed., Dating Quaternary Sediments: Geological Society of America Special Paper 227, p. 69–110.

WELLS, J. T., 1989, A scoping study of the distribution, composition, and dynamics of water-column and bottom sediments: Albemarle-Pamlico estuarine system: Albemarle-Pamlico Estuarine Study, U.S. Environmental Protection Agency and North Carolina Department of Natural Resources and Community Development, Report No. 89-05, 39 p.

WELLS, J. T., AND KIM, S. Y., 1989, Sedimentation in the Albemarle-Pamlico lagoonal system: synthesis and hypothesis: Marine Geology, v. 88, p. 263–284.

WHITEHEAD, D. R., 1981, Late-Pleistocene vegetational changes in northeastern North Carolina: Ecological Monographs, v. 51, p. 451–471.

YORK, L. L., 1984, Aminostratigraphy of Stetson Pit and Ponzer areas of North Carolina by Pleistocene mollusk analysis: Unpublished M.S. Thesis, Department of Geology, University of Delaware, Newark, 188 p.

YORK, L. L., 1990, Aminostratigraphy of U.S. Atlantic coast Pleistocene deposits: Maryland continental shelf and North and South Carolina coastal plain: Unpublished Ph.D. Dissertation, University of Delaware, Newark, 580 p.

YORK, L. L., WEHMILLER, J. F., CRONIN, T. M., AND AGER, T. A., 1989, Stetson Pit, Dare County, North Carolina: an integrated chronologic, faunal, and floral record of subsurface coastal sediments: Palaeogeography, Palaeoclimatology, Palaeoecology, v. 72, p. 115–132.

ZIMMERMAN, H. B., SHACKLETON, N. J., BACKMAN, J., KENT, D. V., BALDAUF, J. G., KALTENBACK, A. J., AND MORTON, A. C., 1984, History of Plio-Pleistocene climate in the northeastern Atlantic, Deep Sea Drilling Project hole 552A, *in* Initial Reports of Deep Sea Drilling Project 81, U.S. Government Printing Office, Washington, D.C., p. 861–875.

A LATE HOLOCENE SEA-LEVEL FLUCTUATION IN SOUTH CAROLINA

PAUL T. GAYES
Center for Marine and Wetland Studies, University of South Carolina-Coastal Carolina College, Myrtle Beach, South Carolina 29526
DAVID B. SCOTT AND ERIC S. COLLINS
Centre for Marine Geology, Dalhousie University, Halifax, Nova Scotia B3H 3J5 Canada
AND
DOUGLAS D. NELSON
Center for Marine and Wetland Studies, University of South Carolina-Coastal Carolina College, Myrtle Beach, South Carolina 29526

ABSTRACT: A highstand of relative sea level occurred at 4.2 ka in Murrells Inlet on the northern coast of South Carolina. The highstand was followed by a sea-level fall of 2 m until 3.6 ka and then a slow, steady sea-level rise of 10 cm/century to the present. Although a mid-Holocene highstand has been suggested by others, it has not been well constrained.

Strong differential submergence between Murrells Inlet and Santee Delta, South Carolina, has occurred over the last 4 ka, probably as a result of sediment loading by, and subsidence of, the Santee Delta system. The occurrence of the 4.2-ka highstand corresponds to the range (7–4 ka) of the mid-Holocene Hypsithermal. The rate and magnitude of the relative sea-level fluctuation are similar to those projected for future flooding and suggest that evaluation of the Hypsithermal highstand may provide an insight to projected greenhouse-effect-related change for the future.

INTRODUCTION

Both Colquhoun and Brooks (1986) and DePratter and Howard (1981) have identified Holocene sea-level fluctuations along the southeast coast of the United States that suggest a highstand in relative sea level occurred at about 4 ka. The rates and magnitudes of these fluctuations are larger than those proposed by many greenhouse-scenario predictions for future sea-level rise (Houghton and others, 1990; Daniels, 1992). Some of these predictions suggest that a 12 cm rise in sea level will accompany a 4°C rise in average temperature (Kuhn, 1989). However, the resolution of these studies is limited by poor preservation of deposits that record the fluctuation and by methodological constraints. In the present study, high-resolution foraminiferal zonation and closely spaced vibracores were used to delineate the relative sea-level fluctuation preserved within a small tidal-marsh inlet on the north coast of South Carolina.

Mid-Holocene highstand paleoshorelines have been identified in several areas in the South Atlantic (e.g., Brazil: Dominguez and others, 1987; West Africa: Giresse, 1989) and have been suggested to be directly tied to the Hypsithermal climatic episode (approximately 7 to 4 ka). However, unlike the fluctuation documented in a limited area of the southeastern United States, these mid-Holocene shorelines are 5 to 10 m above present sea level and are readily accessible. The only report of a mid-Holocene sea-level fluctuation north of the present study area, within the present St. Lawrence River valley, is by Dionne (1988). Dionne's study shows that a high sea level occurred between 5.8 and 4.4 ka on a rapidly emerging coastline.

Short-lived fluctuations, of limited amplitude, are less likely to survive erosion by meandering tidal creeks, migrating inlets, and shoreface retreat, and thus are less likely to become preserved along transgressive coastlines than long-lived fluctuations. Researchers working to the north (e.g., North Carolina: Moslow and Heron, 1981; Virginia: van de Plassche, 1990; Connecticut: van de Plassche and others, 1989) or south (southwest Florida: Scholl and others, 1969) have not reported short-lived fluctuations. The non-recognition of a short-lived fluctuation in the North Atlantic may indicate that such an event may have been a highly localized rise in relative sea level, or that the evidence of the fluctuation was destroyed by coastal processes. The record of the mid-Holocene relative sea-level oscillation in South Carolina was limited to a small, local, tidal-marsh system, where the deposits had not yet been reworked by creeks. Reworking of sequences containing relative sea-level records has hampered efforts to produce detailed mid- to late Holocene relative sea-level histories at other sites. The Holocene record on the adjacent shelf of North Carolina and South Carolina is poorly preserved (Hine and Snyder, 1985). Preservation of deposits recording detailed relative sea-level changes is also limited within relatively protected coastal systems (Gayes and Bokuniewicz, 1991).

In this paper we present a relative sea-level curve from tidal-wetland deposits of Murrells Inlet, South Carolina, that documents a mid-Holocene relative sea-level highstand. A second data set from the Santee River Delta shows differential Holocene submergence along the coast that may have resulted from sediment loading by the Santee Delta.

STUDY AREA

Murrells Inlet is a small bar-built estuary located along the northern coast of South Carolina (Fig. 1). Approximately 11.6 km^2 of marsh exist between the barrier beaches and the Pleistocene upland. Tides are semi-diurnal with a mean range of 1.4 m. Murrells Inlet presently receives very limited freshwater input as no streams currently drain into it. The inlet presently exists within an embayment produced by the paleo-Pee Dee River valley.

PREVIOUS WORK

Among regional works, Colquhoun and Brooks (1986) have reported on sea-level fluctuations from basal peat, intercalated peat, and archeological evidence closest to the Murrells Inlet site. Their data, largely confined to between Winyah Bay and the Savannah River, all south of Murrells Inlet, suggest a rapidly fluctuating eustatic sea level between 6 ka and the present. Their strongest evidence for a highstand is dated at about 4 ka, which agrees with that of

Quaternary Coasts of the United States: Marine and Lacustrine Systems, SEPM Special Publication No. 48

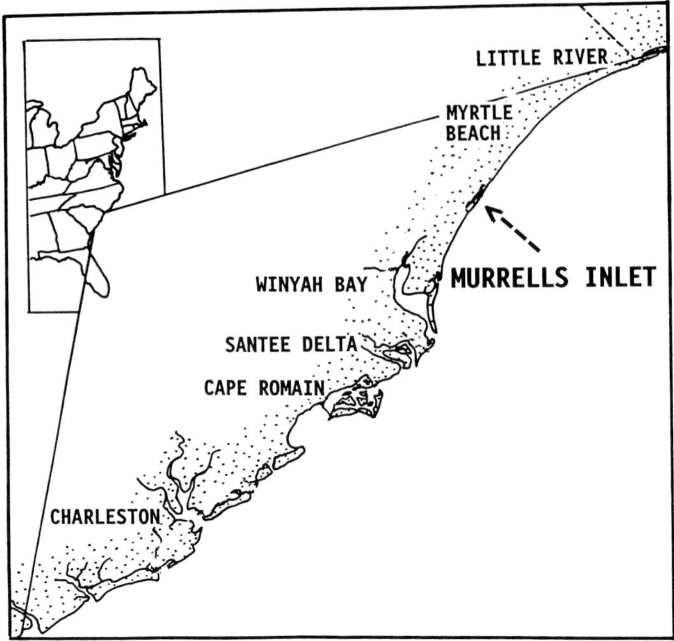

FIG. 1.—General location map of study area of Murrells Inlet, South Carolina. Core coverage extends from Santee Delta to Little River.

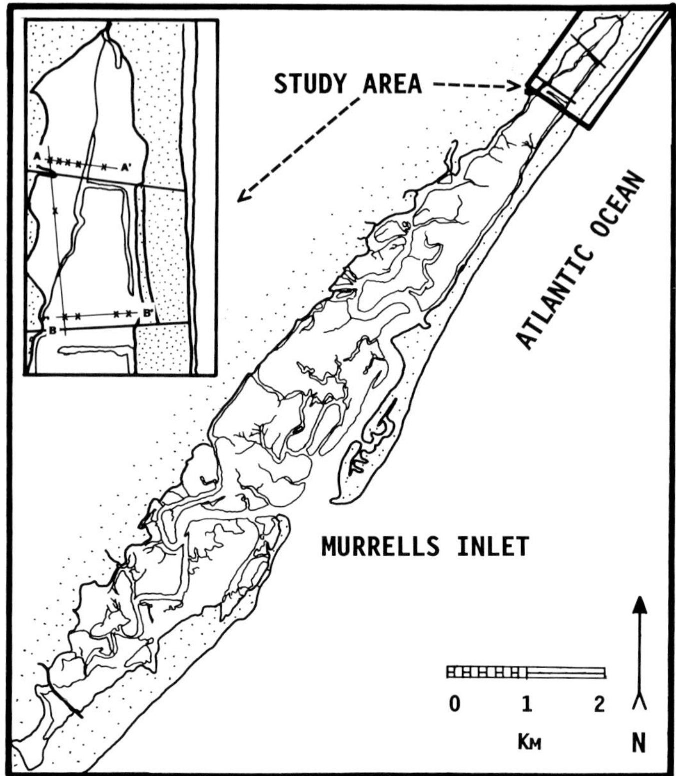

FIG. 2.—Detailed map of Murrells Inlet. Inset shows area of detailed preservation of Holocene submergence record and specific core locations (x). Cross sections for transects A-A′, B-B′, and A-B are shown in Figures 3–5.

the present study. Depratter and Howard (1981) identify a single highstand from 4.5 to 3.0 ka along the Georgia coast defined by the presence of tree stumps rooted in salt-marsh deposits that were in turn overlain by younger salt-marsh deposits.

The 4.5- to 3.0-ka highstand has not been recorded elsewhere along the North Atlantic seaboard except by Dionne (1988) in the St. Lawrence Valley, and is notably absent in studies conducted in adjacent coastal areas (e.g., Florida: Scholl and others, 1969; North Carolina: Moslow and Heron, 1981; Virginia: van de Plassche, 1990; Delaware: Belknap and Kraft, 1977; Connecticut: van de Plassche, and others, 1989; van de Plassche, 1991). In addition, evidence for the highstand has not been found on the Caribbean island of Barbados, where a record of eustatic sea-level rise might be expected to exist (Fairbanks, 1989).

METHODS

Field Methods

We collected 107 vibracores from the northern South Carolina coast with 50 cores concentrated in Murrells Inlet, South Carolina (Figs. 1, 2). Most of these provide reconnaissance information from Little River to Santee Delta and a general stratigraphy of Murrells Inlet. However, only 10 of the cores in Murrells Inlet and five cores in Santee provide any detailed sea-level data. The limited preservation of evidence of local sea-level change diminishes the resolution of sea-level curves drawn in most areas. Vibracores were collected using a cement vibrator connected to 9-m-long, 7.62-cm-diameter aluminum irrigation pipe. Core locations and elevations were surveyed to nearby state benchmarks. Sediment compaction occurring during coring was measured for each vibracore taken. ''Rodding'' or advancement of the core without sampling was not experienced during the coring process. In the upper 1 m of the critical sequence in core 103, we were able to determine elevations of unit boundaries using non-compacted hand-augered samples, but in deeper sequences only the units resting on non-compactible units yielded precise elevation measurements. Compaction values during coring ranged from 10 cm to 140 cm. Compaction during deposition was more difficult to measure and remains a serious problem that limits the accuracy of our paleo-sea-level curve. Marsh peats overlying hard substrates minimize compaction, but such a sequence was not common in the study area.

We assumed that the older, eroded and desiccated freshwater peats were pre-compacted and thus could be considered ''non-compactible'' substrates. Evidence for pre-compaction is the uniform depth of the 9-ka peat (cores 73, 106, 8, Figs. 3–5) across the Murrells Inlet marsh at approximately −3 m. Two of the most critical sea-level data points in the curve are derived from sediments that overlie older freshwater peats. A third data point recording the highstand was obtained from a sandy organic mud that overlies a beach ridge and was relatively unaffected by compaction.

Cores were cut into 1.5-m sections, transported to the laboratory and split. Split cores were described, and foraminiferal and radiocarbon samples were taken within two

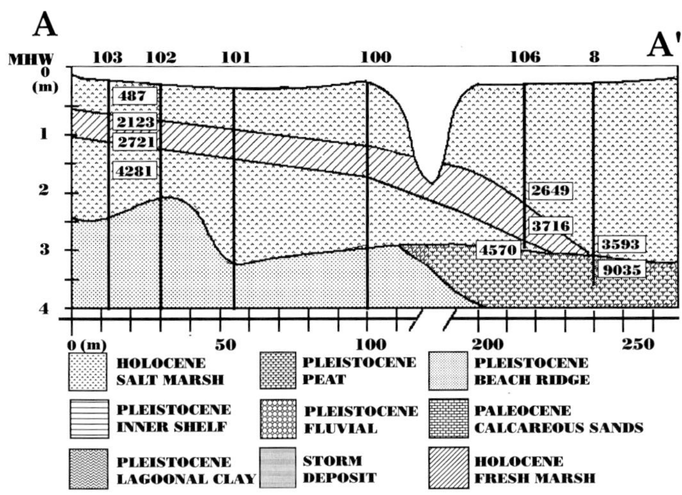

FIG. 3.—Cross section A-A′, with radiocarbon dates, showing Holocene fresh- and saltwater-marsh sequences overlying Pleistocene beach-ridge deposits. Specific depths of radiocarbon dates are provided in Table 1. Legend for this figure also applies to Figures 4 and 5. Elevations are relative to mean high water (MHW).

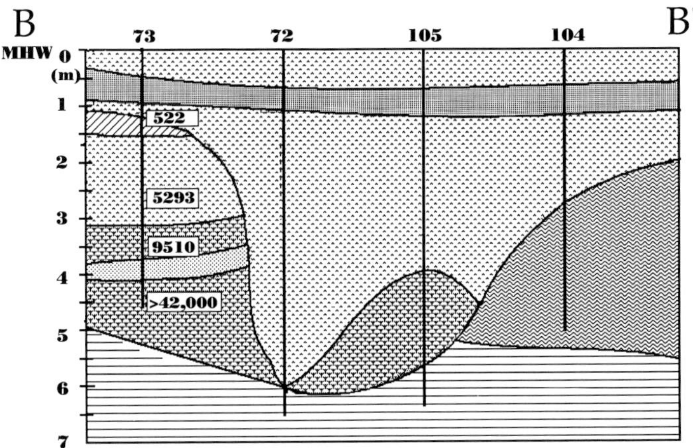

FIG. 4.—Cross section B-B′, with radiocarbon dates, showing extensive erosion of Holocene relative sea-level record. Submergence history was preserved in core 73 but modern tidal-creek migration had reworked the Holocene and underlying Pleistocene deposits at sites of cores 72, 105, and 104. A large storm layer was found across the entire transect area.

weeks of collection. Following sampling, the cores were archived at 4°C for future studies.

Laboratory Methods

Foraminiferal assemblages were examined to characterize environments and sea-level positions at selected intervals. The relation of marsh foraminifera to sea level is well documented for other areas (e.g., Scott and Medioli, 1978, 1980, 1986). Although no published surficial foraminiferal data exist for Murrells Inlet, a study in progress specifically addresses this area. Similar studies have been conducted in Georgia (Goldstein and Frey, 1986) and the Mississippi Delta (Scott and others, 1991) that permit discrimination of marsh

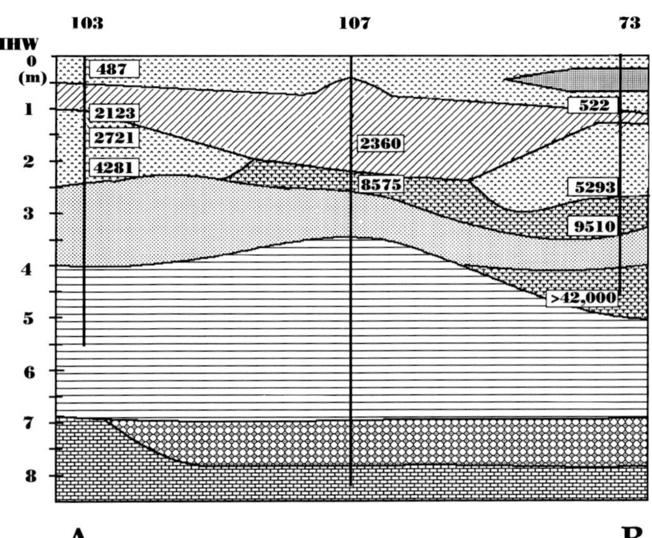

FIG. 5.—Cross section A-B, with radiocarbon dates, showing Holocene fresh- and saltwater-marsh sequences overlying Pleistocene beach-ridge deposits in the northern, more protected area of the marsh. The sea-level record is progressively eroded by tidal creeks toward the open ocean to the south.

levels in the study area with some degree of vertical accuracy. Processing methods for foraminifera are the same as those described in Scott and Medioli (1980, 1986).

Radiocarbon samples were selected based on lithologic data and foraminiferal content of selected horizons. The radiocarbon dates are listed in Table 1 with depth below mean high water (MHW) corrected for compaction. ^{14}C values were corrected for sideral year (Stuiver and Reimer, 1987).

RESULTS

Of the 50 cores taken in Murrells Inlet, only 10 had significant preservation of a detailed Holocene submergence record (Figs. 3–5). Figure 3 shows a nearly complete record of the mid-Holocene sea-level reversal with salt-marsh peat overlain by freshwater peat, which is in turn overlain by salt-marsh peat (e.g., Fig. 3, core 103). The strata are largely reworked by the migration of tidal creeks in a transect approximately 200 m to the south (e.g., Fig. 4, cores 104 and 105). Within most of the Murrells Inlet estuary (40 cores over 80 percent of the surface area), the strata are completely reworked, and thus the detailed record of relative sea-level change in the late Holocene has been erased.

Core Data

Core 73.—

Core 73 contained sediments with a wide range of late Holocene to late Pleistocene ^{14}C dates. At the base of the core, an old freshwater peat (>42 ka) was overlain by a second freshwater peat (9.51±0.285 ka). These late Pleistocene to early Holocene peats are overlain by a Holocene sequence. This sequence then rests on older undated Pleistocene lagoonal and shelf deposits. The Holocene sequence

TABLE 1.—RADIOCARBON DATES, LABORATORY NUMBERS, ERROR BARS, CORRECTED DEPTH IN CORE (I.E.., CORING-COMPACTION CORRECTION), MATERIAL DATED, AND C^{14} VALUES.

Core No.*	Core Depth Below MHW (cm)	Lab #	Material Dated	Radiocarbon Age	Calibrated Age**	Del ^{14}C (‰)
1	420–422	GX-15529	Salt-Marsh Peat	2,040±345	2,009 (2,369–1,590)	−24.9
2	564–566	GX-15530	Salt-Marsh Peat	2,510±250	2,721 (2,869–2,329)	−22.2
3	472–474	GX-15531	Salt-Marsh Peat	2,890±230	3,011 (3,369–2,779)	−23.8
6	525–527	GX-15532	Salt-Marsh Peat	2,285±210	2,335 (2,709–2,059)	−23.0
7	371–373	GX-15533	Salt-Marsh Peat	2,680±195	2,782 (3,339–2,339)	−26.4
8	310–315	GX-15987	Salt-Marsh Peat	3,340±240	3,593 (3,889–3,359)	−22.6
8	327–329	GX-15988	Peat	9,035±245	NA	−27.8
103	42–47	GX-16475	Salt-Marsh Peat	405±145	487 (550–310)	−26.4
103	85–90	GX-16476	Peat	2,140±230	2,123 (2,369–1,860)	−28.4
103	95–100	GX-16477	Wood	2,510±140	2,721 (2,769–2,369)	−28.2
103	115–120	GX-16478	Salt-Marsh Peat	3,850±145	4,281 (4,449–4,089)	−28.9
106	261–266	GX-16567	Salt-Marsh Peat	2,475±135	2,649 (2,759–2,359)	−21.4
106	288–291	GX-16568	Peat	3,460±155	3,716 (3,929–3,559)	−27.5
106	308–311	GX-16569	Salt-Marsh Peat	4,090±235	4,570 (4,879–4,289)	−24.4
107	87–91	GX-16571	Peat	2,355±140	2,360 (2,709–2,299)	−27.8
107	216–224	GX-16572	Peat	8,575±270	NA	−27.3
73	93–97	GX-16479	Salt-Marsh Peat	475±180	522 (660–320)	−18.0
73	298–305	GX-15989	Salt-Marsh Peat	4,560±270	5,293 (5,599–4,859)	−21.7
73	332–335	GX-16480	Peat	9,510±285	NA	−29.0
73	434–439	GX-16481	Peat	>42,000	NA	−27.4

*Cores 1–7 from Santee Delta; all others from Murrells Inlet.
**Ages determined using CALIB Rev. 2.1 radiocarbon-age calibration program (Stuiver and Reimer, 1987).

begins with salt-marsh peat dated at 5.293 (5.599–4.859) ka, which is the deepest and oldest sea-level data point observed in Murrells Inlet. The peat sequence continues to the surface of the core but is interrupted by a sand layer at 90 to 40 cm (Fig. 4) that was observed in other cores along the same transect. Preliminary foraminiferal results suggest that there may be a freshwater sequence at 120 to 150 cm, but the interval has not been dated due to severe compaction in this part of the core. This freshwater interval might correlate with a similar interval in core 103. A sample just below the sand unit (93–97 cm, Table 1) is dated at 0.522 (0.66–0.32) ka. This core segment, together with the segment at 298 to 305 cm dated at 5.293 (5.599–4.859) ka, has relatively abundant high-marsh foraminifera (*Trochammina* spp.) that provide evidence for accurate former sea levels.

Core 103.—

In core 103 the Holocene sequence overlies a Pleistocene beach-ridge deposit. Within the upper 2 m of the core, salt-marsh deposits reach a level of 110 cm below the present marsh surface and are dated at 4.281 (4.449–4.089) ka. This level has relatively common middle- and high-marsh foraminifera (*Ammostauta inepta, Haplophragmoides* spp.). The salt-marsh deposit was extremely sandy and not highly compactible. At 100 cm below the present marsh surface, freshwater peat (no foraminifera) with large tree roots dated at 2.721 (2.769–2.369) ka marks the re-establishment of freshwater conditions as sea level fell. The final transgressive phase, recorded at −40 cm and dated at 0.487 (0.550–0.310) ka, contained abundant high-marsh foraminifera (*Haplophragmoides* spp., *Trochammina* spp., *Arenoparella, Tiphotrocha*).

Core 8.—

In core 8 freshwater peat (9.035±245 ka) is overlain by salt-marsh peat dated at 3.593 (3.889–3.359) ka at 310 to 315 cm below the present surface. Some foraminifera (*Trochammina* spp.) were observed in the salt-marsh peat. Foraminifera occur in most of the samples above this level to the top of the core. The highstand record appears to have been eroded away, but the lower limit of the subsequent sea-level fall between 4.2 and 3.6 ka may be preserved. The 9.035-ka peat contains a surface that can be correlated to a peat of similar age in core 73.

Core 106.—

Although the sites of cores 106 and 8 were approximately 10 m apart, comparison of the two cores reveals the difficulty of locating and accurately delineating the mid-Holocene highstand. In core 106 the contact between the 9-ka peat and the Holocene salt-marsh peat occurs at −3 m. The peat, dated at 4.570 (4.879–4.289) ka, was deposited before the highstand and fills burrows of marine organisms in the underlying freshwater peat. No foraminifera have been examined from this core. The peat is probably not high-marsh peat, and thus, does not provide an accurate former sea-level record, but it does support a sea level higher than −3 m at 4.6 ka. Overlying the salt-marsh peat is a freshwater peat at −2.8 m, dated at 3.716 (3.929–3.559) ka based on the dark brown color and presence of burrowing from above. This peat records the fall of sea level to a lowstand at −3 m in core 8 at 3.6 ka. Just above −3 m, salt-marsh peat fills burrows in the underlying freshwater peat. The transgression after 3.6 ka is recorded at −2.56 m and dated at 2.649 (2.759–2.359) ka. All the levels above the lowest marine level have vertical errors associated with unknown post-depositional compaction. However, the levels could not have been lower relative to sea level, only higher, if compaction was a large factor.

Sea-Level Curve

Cores 8, 73, 103, and 106 provide the basis for the sea-level curve drawn in Figure 6. Marine-peat dates from Santee Delta, 70 km to the south of Murrells Inlet, are systematically deeper than peats of similar age found in Murrells Inlet and are also plotted on Figure 6.

The Santee dates show some scatter, but they appear to be lower by about 2 m than corresponding Murrells Inlet data points. Although sea level appears to have fluctuated less at Santee than in Murrells Inlet, this could be the result of a lack of reliable sea-level data older than 3.1 ka from Santee. It is possible that a fluctuation exists between 3.0 and 2.7 ka in the Santee curve, but sufficient control is lacking to document it. The marsh foraminiferal faunas encountered in Santee are atypical and will require additional refinement to permit distinguishing high- from low-marsh subenvironments.

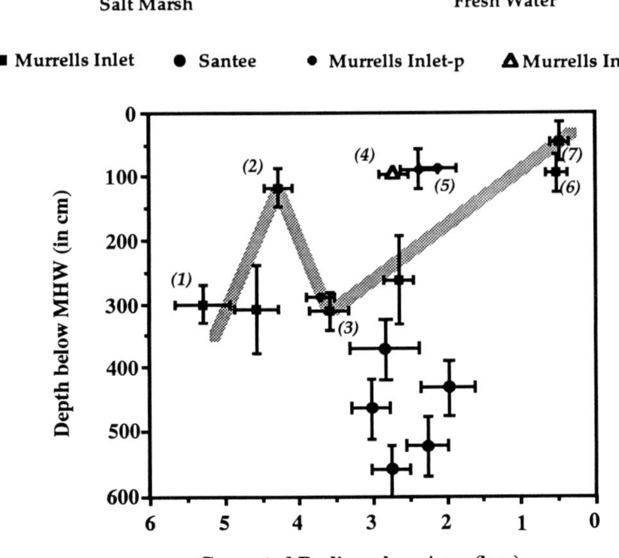

FIG. 6.—Sea-level curve interpreted from radiocarbon dates from cores 8, 73, 103, and 106; p = peat; w = wood. Numbered dates are key intervals containing foraminifera. (1) 298–305 cm, core 73; (2) 115–120 cm, core 103; (3) 310–315 cm, core 8; (4) 95–100 cm, core 103; (5) 85–90 cm, core 103; (6) 93–97 cm, core 3; (7) 42–47 cm, core 103.

General Comments on Foraminiferal Results

No detailed foraminiferal data are presented here, and the error bars on Figure 6 reflect present uncertainties until study of surficial transects is completed.

No foraminifera or freshwater thecamoebians were observed in any of the freshwater peats. These organisms do exist in these environments (Scott and others, 1991), but it appears that subaerial exposure and transgressive erosion/oxidation of the peats may have destroyed the largely organic-walled rhizopods. Such alteration contributes to the complex late Holocene history of this area.

DISCUSSION AND SUMMARY

Although DePratter and Howard (1981) and Colquhoun and Brooks (1986) had reported a sea-level fluctuation similar to that reported here, they were unable to constrain the event unequivocally. DePratter and Howard (1981) used undifferentiated marsh peat as an indicator of a large time period (4.5–3.0 ka) during which the fluctuation was supposed to have occurred. Colquhoun and Brooks (1986) proposed several periods of fluctuation both before and after 4.0 ka; however, their sea-level indicators were suspect for some intervals. Using a series of closely spaced cores, marsh foraminiferal zonations, and many ^{14}C dates, we believe we have constrained a late Holocene sea-level highstand to between 5.3 and 3.6 ka. The highstand consists of a transgressive phase with a 2-m rise between 5.3 and 4.3 ka and a regressive phase with a 2-m fall from 4.3 to 3.6 ka. The rates of both rise and fall of sea level during this period are 50 cm/100 yr, high mid-Holocene rates. Similar rates are predicted for a projected sea-level rise influenced by a greenhouse effect (Houghton and others, 1990). The rate of rise since 3.6 ka is much slower (10 cm/100 yr), and no good evidence exists in our cores to show that this rate has increased in the last 100 years in response to global warming. However, it is also unlikely that such a small change could be detected in our core material.

Eustatic sea-level rise must exceed the background relative sea-level rise resulting from isostatic adjustments for the mid-Holocene highstand to be detected. It appears that Murrells Inlet may be near the northern limit of the mid-Holocene highstand, because this highstand has not been detected in Virginia (van de Plassche, 1990). However, the highstand is also not observed by Fairbanks (1989) in Barbados, where it should be apparent. One possible explanation is that the rather small fluctuation might not have been detected in the Barbados data, especially in the last 6.0 ka where the data appear to show a lot of scatter.

Data points from the Santee Delta plot well below those of the Murrells Inlet curve. The delta is 70 km south of Murrells Inlet (Fig. 1), yet a 2-m difference exists between sea-level points from the two sites dating at 3.0 ka. Such a differential is to be expected for a prograding delta where sediment loading can be significant, as occurs on a much larger scale in the Mississippi Delta.

In summary, the rapid fluctuation of sea level that occurred in South Carolina and Georgia between 5.3 and 3.0 ka was similar to, if not greater than, the projected future change in sea level proposed as a result of possible greenhouse effects. Thus, it appears appropriate that more effort should be devoted to defining and delineating this mid-Holocene climatic event and potentially associated relative sea-level change since it can serve as a fossil analog to what might happen at a specific location in the future.

ACKNOWLEDGMENTS

Several undergraduate students from Coastal Carolina College assisted with the field work described here. These students were: Todd Ward, Donna Radcliffe, Jim Matthews, Colten Bowles, Mike Pearson, Neal Gielstra, Kirk Koneski, and Sally Peace. Donna Radcliffe assisted in drafting the illustrations. Ms. Amy Nelson of USC-Coastal Carolina College also provided assistance in the field and laboratory. Harold Krueger of Geochron kindly provided us with C^{14} dates in a ''very timely fashion.'' Discussions of southeast U.S. archeological data with Jim Michie, USC-Coastal Carolina College, were particularly helpful. Radiocarbon analyses were performed at Geochron Laboratories in Cambridge, Massachusetts.

Funding for this project was provided by grants from the South Carolina Higher Education Commission-Cutting Edge program (Gayes and Nelson) and the Minerals Management Service Grant #14-12-0001-30432 (Gayes and Nelson). This paper was greatly improved by reviews from Chip Fletcher, Chris Kraft, and Orrin Pilkey and the editorial efforts of Barbara Lidz. Special thanks to Dave Zilkoski (National Geodetic Survey) who supplied us with unpublished information on recent crustal movements along the South Carolina coast.

REFERENCES

BELKNAP, D. F., AND KRAFT, J. C., 1977, Holocene relative sea level changes and coastal stratigraphic units on the northwest flank of the Baltimore Canyon Trough geosyncline: Journal of Sedimentary Geology, v. 47, p. 610–619.

COLQUHOUN, D. J., AND BROOKS, M. J., 1986, New evidence from the southeastern United States for eustatic components in the late Holocene sea levels: Geoarcheology, v. 1, p. 275–291.

DANIELS, R. C., 1992, Sea-level rise on the South Carolina coast: two case studies for 2100: Journal of Coastal Research, v. 8, p. 56–71.

DEPRATTER, C. B., AND HOWARD, J. D., 1981, Evidence for a sea-level lowstand between 4500 and 2400 years b.p. on the southeast coast of the United States: Journal of Sedimentary Petrology, v. 51, p. 1287–1296.

DIONNE, J. C., 1988, Holocene relative sea-level fluctuations in the St. Lawrence Estuary, Quebec, Canada: Quaternary Research, v. 29, p. 233–244.

DOMINQUEZ, J. M., MARTIN, L., AND BITTENCOURT, A. C., 1987, Sea-level history and Quaternary evolution of river mouth-associated beach-ridge plains along the east-southeast Brazilian coast—a Summary, *in* Nummendal, D., Pilkey, O. H., and Howard, J. D., eds., Sea-Level Fluctuations and Coastal Evolution: Society of Economic Paleontologists and Mineralogists Special Publication 41, p. 115–128.

FAIRBANKS, R. G., 1989, A 17,000-year glacio-eustatic sea level record: influence of glacial melting rates in the Younger Dryas event and deep ocean circulation: Nature, v. 342, p. 637–642.

GAYES, P. T., AND BOKUNIEWICZ, H. J., 1991, Estuarine paleoshorelines in Long Island Sound, New York, *in* Gayes, P. T., Lewis, R., and Bokuniewicz, H. J., eds., Quaternary Geology of Long Island Sound and Adjacent Coastal Areas: Journal of Coastal Research, Special Edition 11, p. 39–55.

GIRESSE, P., 1989, Geodynamique des lignes de revage Quaternaire du continent Africain applications, *in* Scott, D. B., Pirazzoli, P. A., and Honig, C. A., eds., Late Quaternary Sea-Level Correlations and Applications: Kluwer Publication, NATO ASI Series C, v. 256, p. 121–152.

GOLDSTEIN, S. J., AND FREY, R. W., 1986, Salt marsh foraminifera, Sapelo Island, Georgia: Senckenbergiana Maritima, v. 18, p. 97–121.

HINE, A. C., AND SNYDER, S. W., 1985, Coastal lithosome preservation evidence from the shoreface and inner continental shelf off Bougue Bank, North Carolina: Marine Geology, v. 63, p. 307–330.

HOUGHTON, J. T., JENKINS, G. J., AND EPHRAUMS, J. J., eds., 1990, Climate Change: Report Prepared for the Intergovernmental Panel on Climate Change: Working Group 1, University of Cambridge Press, Cambridge, 364 p.

KUHN, M. A., 1989, The role of land ice and snow in climate, *in* Berger, A., Dickinson, R.E., and Kidson, J. W., eds., Understanding Climate Change: Geophysical Monograph 52, IUGG, v. 7, American Geophysical Union, Washington, D.C., p. 17–30.

MOSLOW, T. F., AND HERON, S. D., 1981, Holocene depositional history of a microtidal cuspate foreland cape: Cape Lookout, North Carolina: Marine Geology, v. 41, p. 251–270.

SCHOLL, D. W., CRAIGHEAD, SR., F. C., AND STUIVER, M., 1969, Florida submergence curve revisited: its relation to coastal sedimentation rates: Science, v. 163, p. 562–564.

SCOTT, D. B., AND MEDIOLI, F. S., 1978, Vertical zonations of marsh foraminifera as accurate indicators of former sea levels: Nature, v. 272, p. 528–531.

SCOTT, D. B., AND MEDIOLI, F. S., 1980, Quantitative studies of marsh foraminiferal distributions in Nova Scotia: implications for sea-level studies: Cushman Foundation for Foraminiferal Research, Special Publication 17, 58 p.

SCOTT, D. B., AND MEDIOLI, F. S., 1986, Foraminifera as sea-level indicators, *in* van de Plassche, O. ed., Sea-Level Research; A Manual for the Collection and Evaluation of data, Geo Books, Norwich, United Kingdom, p. 435–456.

SCOTT, D. B., SUTER, J. R., AND KOSTERS, E. C., 1991, Marsh foraminiferal and arcellaceans of the lower Mississippi Delta: controls on spatial distributions: Micropaleontology, v. 37, p. 373–392.

STUIVER, M., AND REIMER, P. J., 1987, User's Guide to the programs CALIB and DISPLAY 2.1: Quaternary Isotope Center, University of Washington, Seattle, 13 p.

VAN DE PLASSCHE, O., 1990, Mid-Holocene sea-level change on the eastern shore of Virginia: Marine Geology, v. 91, p. 149–154.

VAN DE PLASSCHE, O., 1991, Late Holocene sea-level fluctuations in Long Island Sound, *in* Gayes, P. T., Lewis, R., and Bokuniewicz, H. J., eds., Quarternary Geology of Long Island Sound and Adjacent Coastal Areas: Journal of Coastal Research, Special Edition 11, p. 159–181.

VAN DE PLASSCHE, O., MOOK, W. G., AND BLOOM, A. L., 1989, Submergence of Coastal Connecticut 6,000-3000 (C-14) Years: Marine Geology, v. 86, p. 349–354.

EVOLUTION OF QUATERNARY SHOAL COMPLEXES OFF THE CENTRAL SOUTH CAROLINA COAST

WALTER J. SEXTON
Athena Technologies, Inc., 3504 Devine St., Columbia, South Carolina, 29205
MILES O. HAYES
Research Planning, Inc., Columbia
AND
DONALD J. COLQUHOUN
Department of Geological Sciences, University of South Carolina

ABSTRACT: Subtidal shoals composed of Pleistocene and early to middle Holocene deposits, on the north-central South Central inner shelf, have been modified by wave and tidal processes since being flooded during the Holocene sea-level rise. The shoals, which extend seaward up to 20 km, are composed of a wide variety of sediment types, including silt- to gravel-size terrigenous clastic sediments, abundant shell material, and peat. Typically, the shoals terminate on their seaward ends at abrupt dropoffs, with relief commonly exceeding 14 m over a distance of 0.5 km. Three skoals, located off the Santee/Pee Dee Delta, Cape Romain, and the entrance to Bulls Bay, were studied by means of vibracores, scuba observations, box cores, bottom-sediment samples, and extensive bottom fathometer profiles.

The shoal complex seaward of the modern Santee/Pee Dee Delta is the largest of the three studied, most probably because (at a lower sea-level stand) the Santee and Pee Dee Rivers were joined, forming a large deltaic deposit. Shore-parallel scarps on the shoals correlate well with updip stratigraphic units and most probably give evidence of pauses during the most recent sea-level rise.

The shoal seaward of the modern Cape Romain cuspate foreland contains stacked packages of progradational mid- to lower shoreface (barrier-island) deposits. This progradational, downdrift-trending shoreline sequence is backed by a 6- to 7-km-wide back-barrier region and was formed during a pause or a drop of the most recent sea-level rise. All evidence indicates that the present transgressive Cape Romain barrier islands have migrated landward *en mass* away from the submerged shoal.

The Bulls Bay shoal, which probably originated as an abandoned delta lobe of the ancestral Santee River, shows the highest degree of modification and is characterized by abundant ridge topography and high shell content. The reworked ridges have lengths of 1 to 2 km and elevations of 2 to 4 m.

Shoal surfaces show evidence of significant wave and current modifications, such as abundant bedforms and erosional scarps. Southwesterly-facing scarps were cut presumably by storms (hurricanes). However, bathymetry, sediment-distribution patterns, bedforms and gross morphology of the shoals indicate that northeasterly waves dominate normal shelf processes.

INTRODUCTION

Bathymetry and sediment composition indicate that shoals occurring on the north-central South Carolina coast have remarkably different origins and evolutionary histories (Fig. 1). This paper presents their detailed bathymetry, sediment composition and stratigraphy. The data yield several meaningful clues regarding the impact of modern marine processes and changes in sea level on the evolution of the shoals. The study area extends from Bull Island to the Santee/Pee Dee Delta (Fig. 1) and encompasses over 1,700 km^2 of nearshore inner shelf. The inner-shelf area had not been previously investigated, although the stratigraphy and sedimentology of the depositional environments of the adjacent shoreline have been studied in detail. Studies by Aburawi (1972), Stephen and others (1975), Hodge (1981), Ruby (1981), Eckard (1986), Hayes and others (1986), and Hayes and Sexton (1989) provide an extensive geologic data base for the adjacent supratidal areas.

Much of the past and present research on the continental shelf of the the United States has focused on its dynamic nature, with particular reference to ridge-and-swale topography. These studies, for the most part, have been carried out on the mid-Atlantic continental shelf (e.g., Duane and others, 1972; Stubblefield and Swift, 1976; Swift and others, 1978; Huthnance, 1982; Stubblefield and others, 1983; Swift and others, 1984). Three studies to date, those of Hoyt and Henry (1971), Swift and others (1972), and Rine and others (1986) made reference to the origin of the shoals discussed in this paper.

Hoyt and Henry (1971) concluded that the shoal off the Santee/Pee Dee Delta is of deltaic origin, being essentially a relict deposit. They stated that during the most recent fall in sea level (late Pleistocene, approximately 50 to 20 ka), river gradients were sufficiently steep to increase river-sediment loads. The increase in sediment supply resulted in significant deltaic accumulations at several locations across the shelf. During the most recent rise in sea level, the deltaic deposits were flooded, and the present shoal complexes were established. Those authors noted that since submergence, physical processes have eroded the outer margin of the shoals (''headland erosion''), leaving a strong imprint of marine processes on the deposits (Hoyt and Henry, 1971).

Swift and others (1972) believe that it is possible that the shoals seaward of Cape Romain and the Santee/Pee Dee Delta are deltaic in origin, but that inner-shelf physical processes are presently reworking and depositing marine sediments at these sites. They classify the shoals off the Santee/Pee Dee Delta and Cape Romain as cape-associated retreat massifs. Generally speaking, they believe that the shoals are more marine in origin than was suggested by the drowned-delta theory of Hoyt and Henry (1971). Both of these papers cite the lack of evidence to verify their theories and state the need for quantitative field data.

PHYSICAL AND GEOLOGICAL SETTING

Maximum water depths in the study area are 20 m, with the shoal tops seaward of Bull Island and Cape Romain averaging 5 m. The shoal seaward of the Santee/Pee Dee Delta averaged 7 m in depth (Fig. 1).

Quaternary Coasts of the United States: Marine and Lacustrine Systems, SEPM Special Publication No. 48

The study area is influenced by prevailing winds from the south-southeast and storm winds, which generally blow from the northeast (Brown, 1977). Even though northeasterly winds are less frequent than the prevailing south-southwesterly winds, their greater strength results in significant waves, wave-energy flux, and longshore transport directed to the south/southwest in the study area (Finley, 1975).

Wave information for the study area was obtained from two sources on Atlantic coast wave hindcasting: Corson and others (1982) and Jensen (1983). Deep-water waves (100 m) for the study area have an average wave height of 1.3 m, with the largest wave hindcasted at 6.4 m using a 20-year data base (1956–1975; Corson and others, 1982). Shallow-water waves (10 m), which were generated through deep-water wave transformation, have a predicted maximum wave height of 4.5 m and a mean wave height of 0.7 m (Jensen, 1983). Average wave heights on the beaches along the South Carolina coast are 50 to 60 cm (Finley, 1976; Kana, 1977). Nearshore coastal currents have not been studied in detail, although work by Schwing and others (1983) did record summer longshore current velocities of 25 cm/sec.

Winds of hurricane strength (exceeding 75 mph) have a frequency of occurrence of one event every 7 years (Myers, 1975). These winds are relatively rare in the study area but could cause geomorphic as well as sediment textural changes to the sea floor on the inner shelf (Hayes, 1965). It is felt by the authors that Hurricane *Hugo* (1989) most probably had a pronounced impact on the study area, although no field measurements have been made.

The dominant geologic feature within the study area is the Santee/Pee Dee Delta. Throughout the late Pleistocene, the delta had a significant impact on the nearby coastline, including Cape Romain and Bulls Bay. The clastic sediments in the study area are both mineralogically and texturally immature in comparison with sediments from other portions of the South Carolina coastline (Griffen, 1984). The present delta is underlain by buried alluvial valleys that were incised to depths of 20 m during late Pleistocene low-

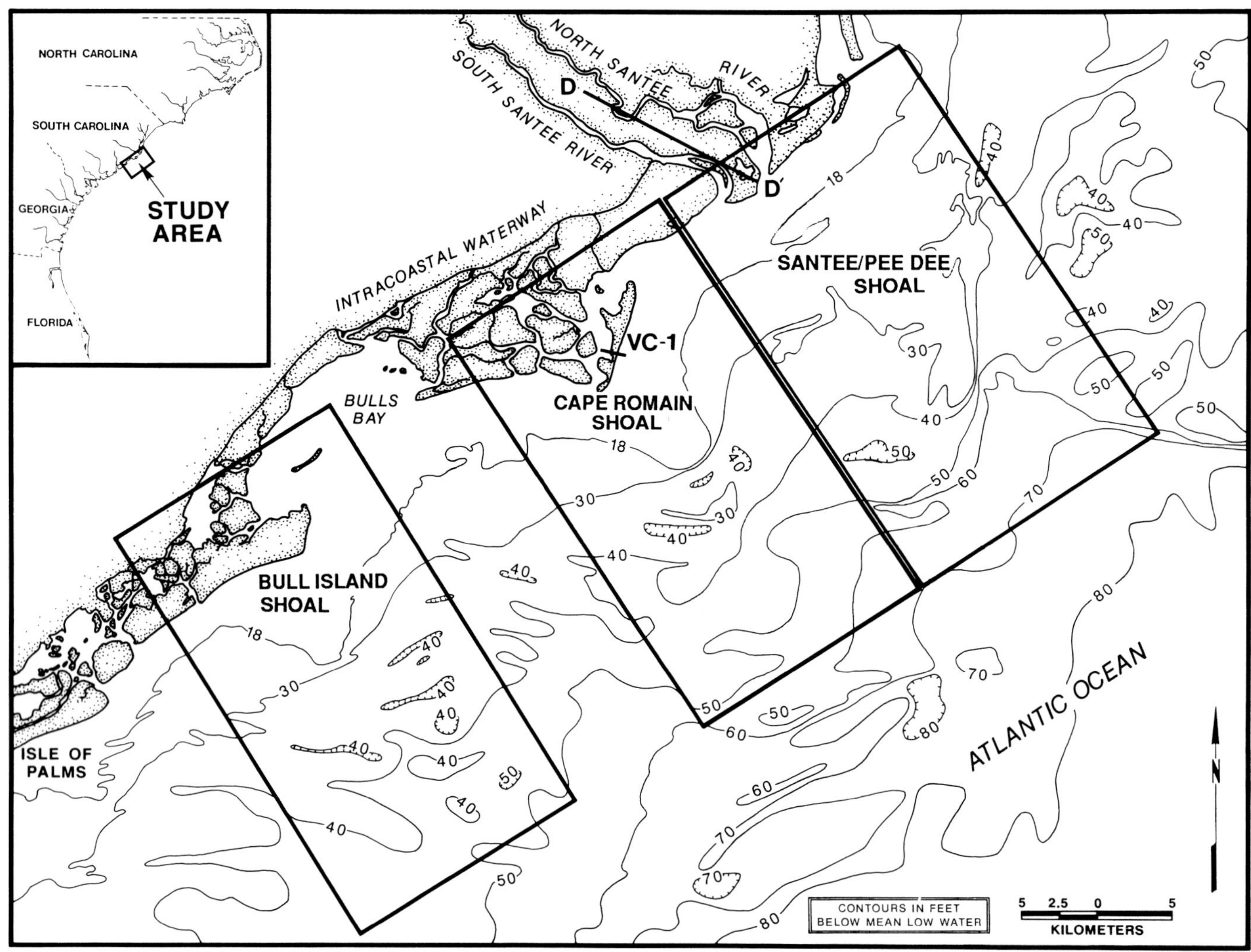

FIG. 1.—Location of the study area showing the nearshore inner continental shelf between Santee/Pee Dee Delta and Isle of Palms, South Carolina. Note the shoals (boxed areas) located seaward of Bull Island, Cape Romain and the Santee/Pee Dee Delta.

stands in sea level (Eckard, 1986; Hayes and Sexton, 1989). Predominantly river sands and gravel and fine-grained floodplain sediments have been deposited within the buried valley under the modern Santee River (Fig. 2). The near-surface sediments are organic-rich, freshwater-swamp deposits and peat on the upper portions of the delta, and marginal-marine to open-marine deposits at the delta front (Fig. 2). Morphologically, the present delta front is composed of several progradational barrier islands and three large tidal inlets (Fig. 1). A laterally continuous peat layer that overlies an extensive floodplain deposit occurs immediately beneath the marine deposits (Fig. 2).

Cape Romain is located 15 km to the southwest of the Santee River (Fig. 1). All the beaches associated with the cape are transgressive (erosional), except for those on the larger recurved spits (Hubbard and others, 1977). The Holocene stratigraphy of this rapidly retreating cuspate foreland was studied in detail by Ruby (1981), who proposed a scenario for its evolution from 30 ka to the present. He showed that during the fall and subsequent rise in sea level between 30 ka and the present, the general area where Cape Romain is today was always located to the southwest of the mouth of the Santee River. Therefore, the cape is not necessarily related in origin to an abandoned delta lobe, as was suggested by Hoyt and Henry (1971). A stratigraphic cross section across Cape Island, which is located on the northeastern arm of Cape Romain, is presented in Figure 3. The relict marsh deposit at the base of the section probably postdated the maximum lowstand, when sea level dropped to near the edge of what is now the continental shelf (Ruby, 1981). A transgressive shell lag dated between 3.7 and 4.42 ka overlies the relict marsh deposit. The sediments above this shell lag have been deposited since the formation of the Cape Romain cuspate foreland. Initially, a large lagoon existed behind the islands that formed the Cape, and open-lagoon deposition predominated. As the area aggraded and infilled, tidal-flat deposition was most common and was followed by an encroaching salt marsh. During the infilling of the lagoon, the islands that had formed the front of the cape became transgressive and migrated landward over the lagoon fill (Ruby, 1981).

Bulls Bay and Bull Island are located to the southwest of Cape Romain (Fig. 1). These areas have not been studied in detail, but the authors believe Bulls Bay overlies an ancestral Santee River valley (Colquhoun, 1972).

METHODS

Sediment sampling and bottom profiling were conducted off a research vessel equipped with a Loran-C navigation system for positioning. Eighty-seven grab samples were collected with a modified clam shell or Peterson sampler (Fig. 4). Samples were analyzed for grain size, percent mud, and percent calcium carbonate. The grab samples collected off the Santee/Pee Dee Delta and Cape Romain also were analyzed for the content of heavy minerals and feldspars, as well as for grain shape using fourier analysis (reported by Keane, 1986). Sediment analysis was performed using the methods and statistics of Folk (1974). A total of 150 km of fathometer traces was run (Fig. 4). In addition to the fathometer profiling, several scuba dive stations were established to describe the sea floor.

Sediment cores were collected by vibracoring or box coring. Vibracoring was accomplished using a system similar to that described by Lanesky and others (1979). Cores were logged for physical and biogenic structures, lithology, texture and faunal assemblages. Photographs and X-ray radiographs were made of selected sections of each core.

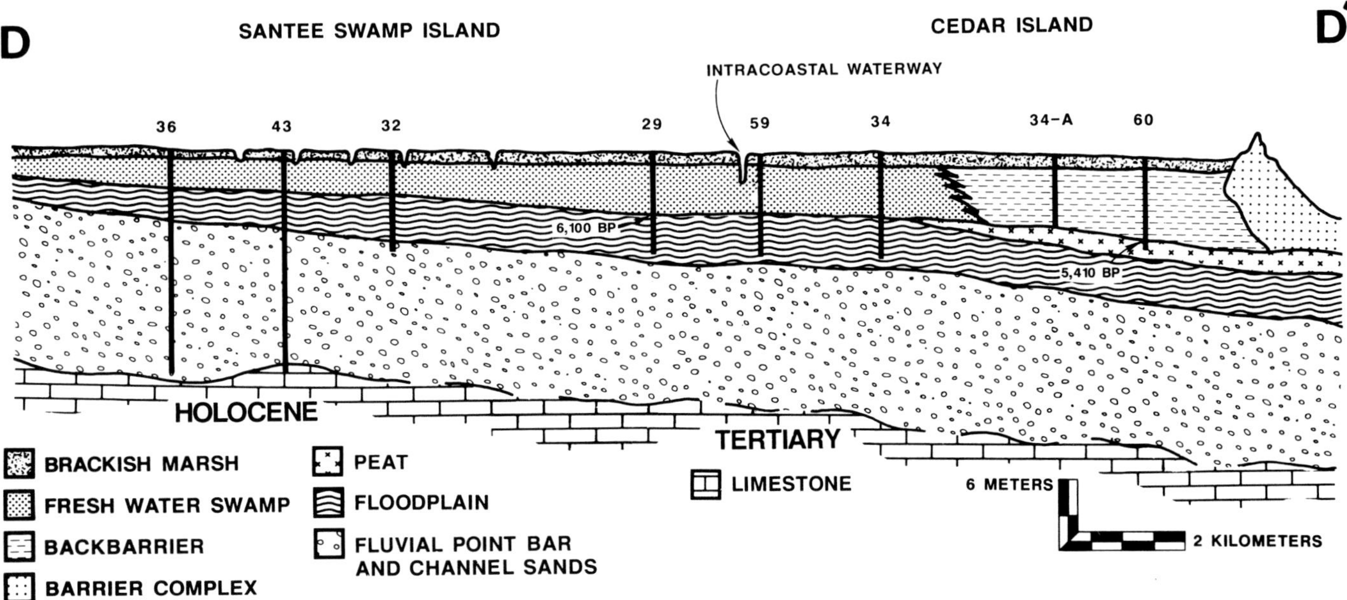

FIG. 2.—Shore-perpendicular cross section illustrating the Holocene stratigraphy in the vicinity of the North Santee River (transect D-D′; Fig. 1). Floodplain deposits formed at a lower stand of sea are overlain by freshwater peat and marine back-barrier and barrier-island sediments along the delta front, which illustrates an aggrading alluvial stage followed by a passive transgression. From Eckard (1986).

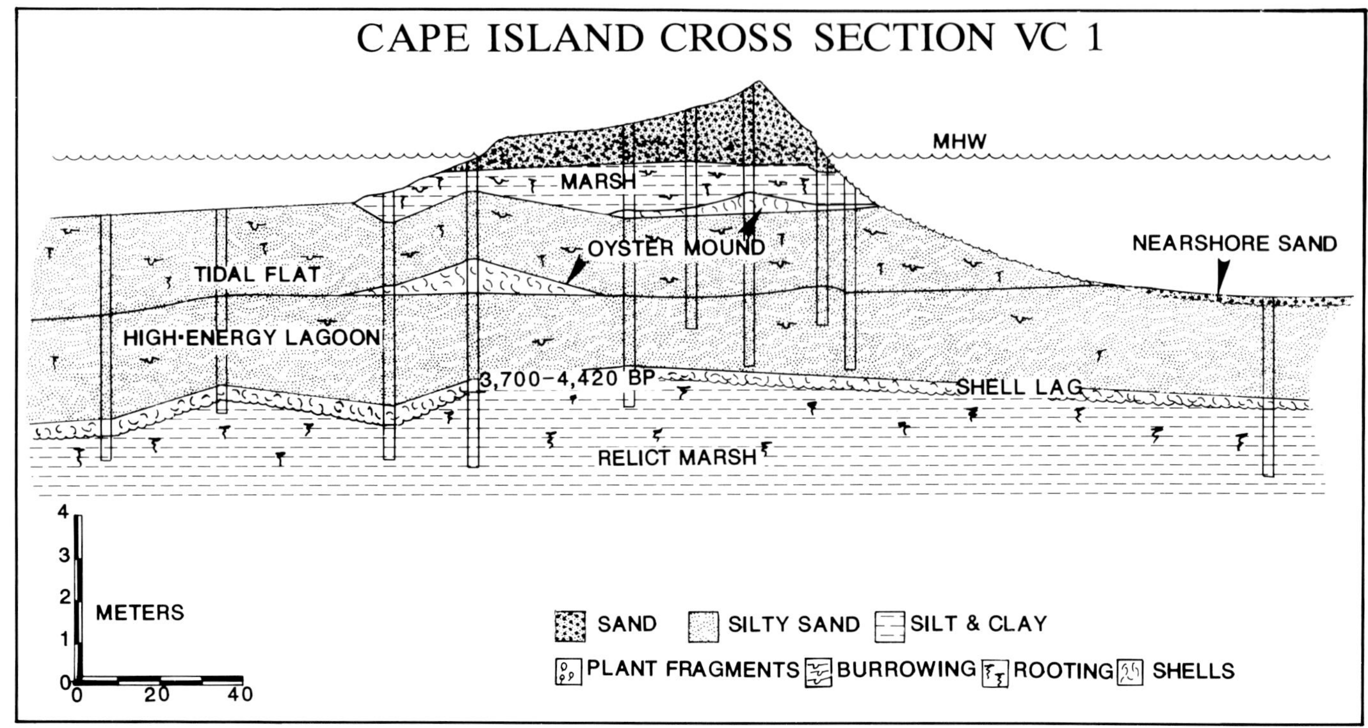

FIG. 3.—A cross section of the mid-portion of Cape Island on Cape Romain (transect VC-1; Fig. 1) showing the transgressive barrier-island sedimentary sequence. The sandy transgressive barrier is perched on top of back-barrier marsh and tidal-flat deposits. After Ruby (1981).

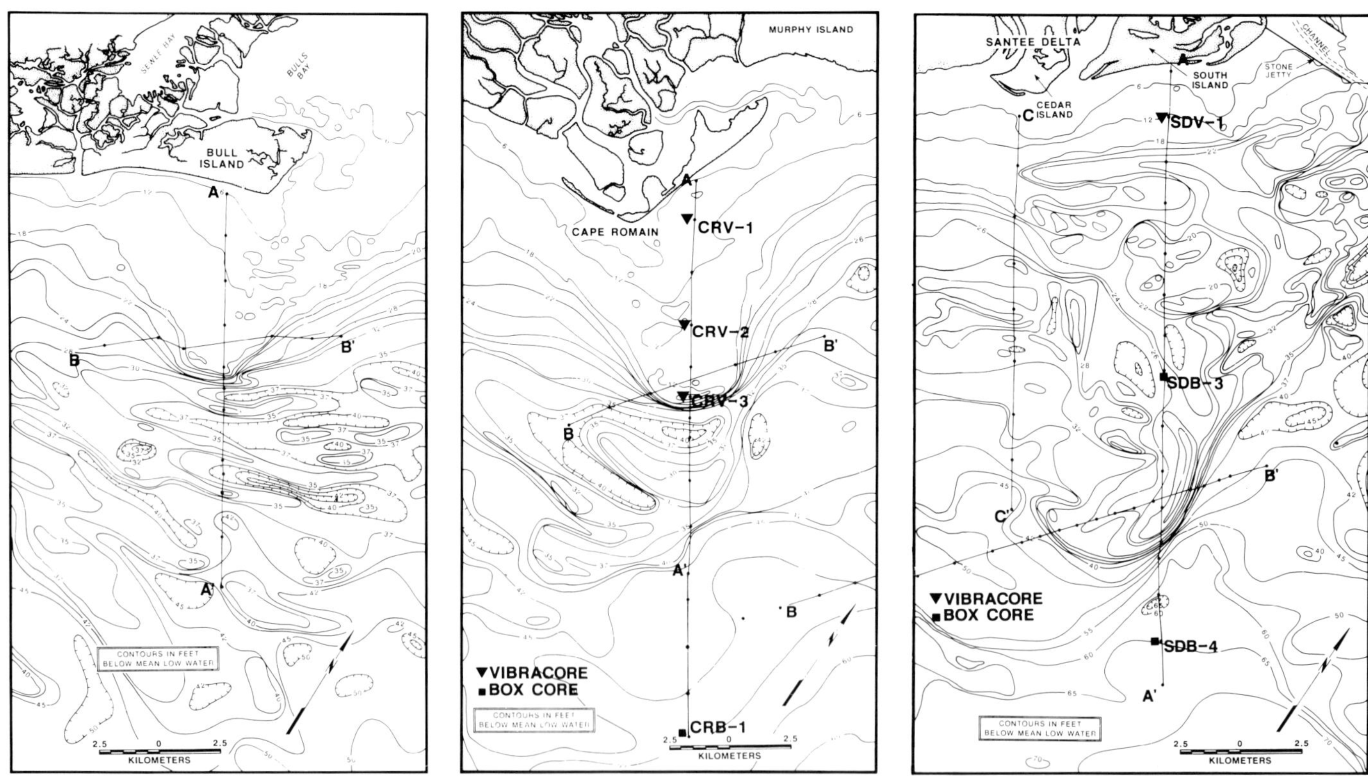

FIG. 4.—Detailed maps of the nearshore inner shelf of the three shoals studied: Bull Island, Cape Romain, and Santee/Pee Dee. Maps located on Figure 1.

SEAFLOOR MORPHOLOGY

The inner-shelf morphology of the three shoals is summarized separately.

Bull Island Shoal

Two transects were established on the Bull Island shoal (Fig. 4), off the northern end of Bull Island. Water depths over the shoals range from 5.5 m to 8 m mean low water (MLW). The detailed bathymetric map shown in Figure 4 was constructed from data points from NOAA shipboard smooth sheets presented on the 1985 nautical chart. The relative flatness of the Bull Island shoal is indicated by the wide spacing of contours down the center of the shoal. Water depths increase rapidly down to 11.3 m just seaward of the apex of the shoal. From that point seaward, out to the 15.2-m contour, the bottom topography is composed of a series of ridges and troughs. Bathymetric profile A-A′ (Fig. 5) starts at the 2-m MLW contour seaward of the northern end of Bull Island. The profile is very flat with a gentle seaward slope out to the 5-km point. Seaward of the 5-km mark, bottom topography is much more irregular. Seaward of the apex of the shoal, water depths range from 9.1 to 13.7 m MLW, and the bottom morphology is composed of ridge-and-swale topography. The ridges, which do not show a regular spacing, have heights ranging from 2 to 3 m, and widths of approximately 0.5 km. The precise shapes of these ridges were not determined, but they appear to be essentially parallel to shore. The ridge topography seaward of the shoal is clearly shown on trace A (sec. 1), and it appears that some of the ridges may be asymmetric in a seaward direction.

Fathometer trace B-B′ (Fig. 4) crosses the seaward end of the Bull Island shoal in a northeast-southwest direction and exhibits a much different profile than trace A-A′. Both the southwest and northeast sides of the trace are smooth, gradually rising up onto the shoal. The absence of ridge topography on trace B indicates that the ridges on the shoal are oriented essentially parallel to shore.

Cape Romain Shoal

Two fathometer transects were run across the Cape Romain shoal (Fig. 4), which extends 11 km seaward of the present Cape Romain beaches. The shoal lies in average water depths of 3.9 m MLW and the shoal crest is nearly intertidal (Fig. 5). The shoal has a width of 4 to 5 km, measured from the 6.1-m MLW contour.

Transect A-A′ (Fig. 5) is 18.4 km long, starts several hundred meters seaward of Cape Island beach, and runs down the long axis of the shoal (Fig. 4). The top of the shoal is flat except for two sets of two nearshore bars (Fig. 5). On the outer margin of the shoal, water depths are so shallow that breaking waves are commonly present on the front of the shoal. This active shoal face is shown in detail on Figure 5. The water depth off the shoal front increases by 10 m over a distance of less than 1 km. Seaward of the shoal front is a very large, low-relief sediment mound (Figs. 4 and 5). The mound is 4.5 km wide and 3.1 m high. The entire surface of the mound is smooth, with the seaward side being only slightly steeper than the landward side. From the base of the mound seaward, only slight changes in the seafloor morphology were recorded (Fig. 4).

Profile B-B′ is a shore-parallel transect that cuts across the seaward end of the Cape Romain shoal (Fig. 4). Located on the southwest side of the shoal (Fig. 4) is a pronounced near-vertical scarp with 6 m of relief over a horizontal distance of several meters. The profile of the top of the shoal is relatively flat until a subtle bar is encountered on the northeast side. The profile outlines a smooth and gradual descent off the shoal into deeper water in a northeast direction.

Santee/Pee Dee Delta Shoal

One of the major differences in this shoal in comparison with the two previously described is that it occurs in deeper water, 7.6 m as opposed to 4.5 m, and extends much farther seaward, 20 km vs. 10 km (Fig. 4).

Trace A-A′ (Fig. 4) was run perpendicular to the shoreline and down the long axis of the Santee/Pee Dee Delta shoal. The first 2 km of the profile show a smooth bottom. At the 2-km mark on the profile (Fig. 6), the slope of the sea floor increases, and over the next 3 km, water depths double to 6.1 m. Ridges are common between the 5-km mark and the 16-km mark on the profile. A scarp with a near-vertical face and a relief of 3.6 m occurs at the 11-km mark (Fig. 6). Ridges 1.6 m high and 0.7 km wide are present seaward of the scarp out to the 16-km mark. The ridges are slightly asymmetric in a seaward direction, similar to those observed seaward of the Bull Island shoal. Scuba observations in the bottom area of abundant ridge topography revealed a moderate bottom current on a calm day. Also observed at this station were abundant (*Diapatra cuprea*) burrows, and landward-oriented ripple bedforms with a spacing of 35 cm and a height of 8 cm were common. From the 16-km mark seaward to the end of the shoal, which occurs at the 21-km mark, water depths gradually become shallower, from 8.5 m to 6.7 m, and the sea floor is smooth. At the shoal front, water depths increase from 6.7 m to 16.8 m over a horizontal distance of less than 1 km. For 6.5 km seaward of the active shoal margin, several ridges are present. From this point seaward, ridge features do not occur in deeper water.

Trace B (Figs. 4 and 6) runs parallel to shore and across the top of the Santee/Pee Dee Delta shoal and continues nearly to trace A seaward of Cape Romain (Fig. 4). On the southwest side of trace B, the sea floor is smooth and water depths are 15.2 m (Fig. 4). To the northeast (Fig. 6) is a low area that is 1.5 m below the surrounding sea floor. The 2-km-wide depression has a small scarp with a total relief of 3 m on its northeastern margin. From the small scarp at the 4.5-km mark to the 9.5-km mark, several ridges are present with widths of 0.4 km and heights of 1.8 m. At the 9.5-km mark on trace B is another scarp with 3 m of relief. The nearly flat terrace runs from the 9.5-km mark on the profile to the 12-km mark. At the 12-km mark on trace B is a third scarp that has 4.6 m of relief and extends onto the top of the Santee/Pee Dee Delta shoal. Over an 8-km distance (4.5-km mark to the 12-km mark on trace B; Fig.

BULL ISLAND - TRACE A

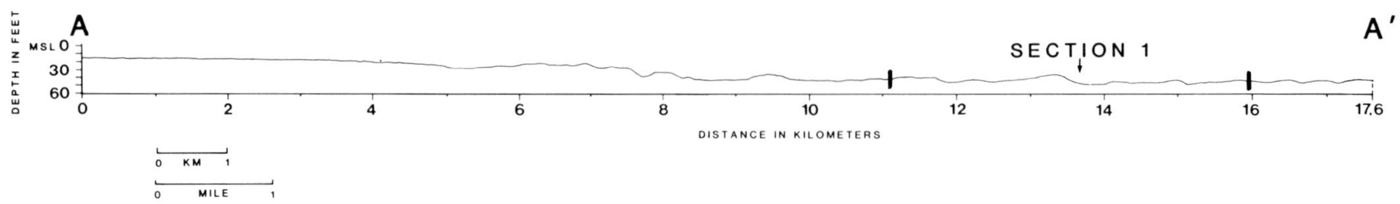

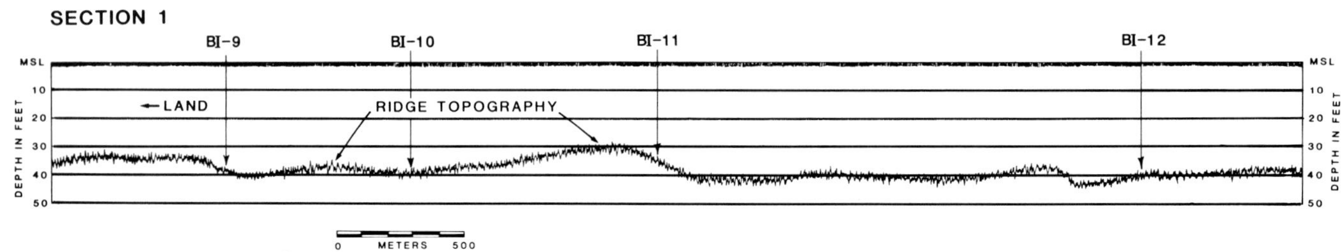

CAPE ROMAIN - TRACE A

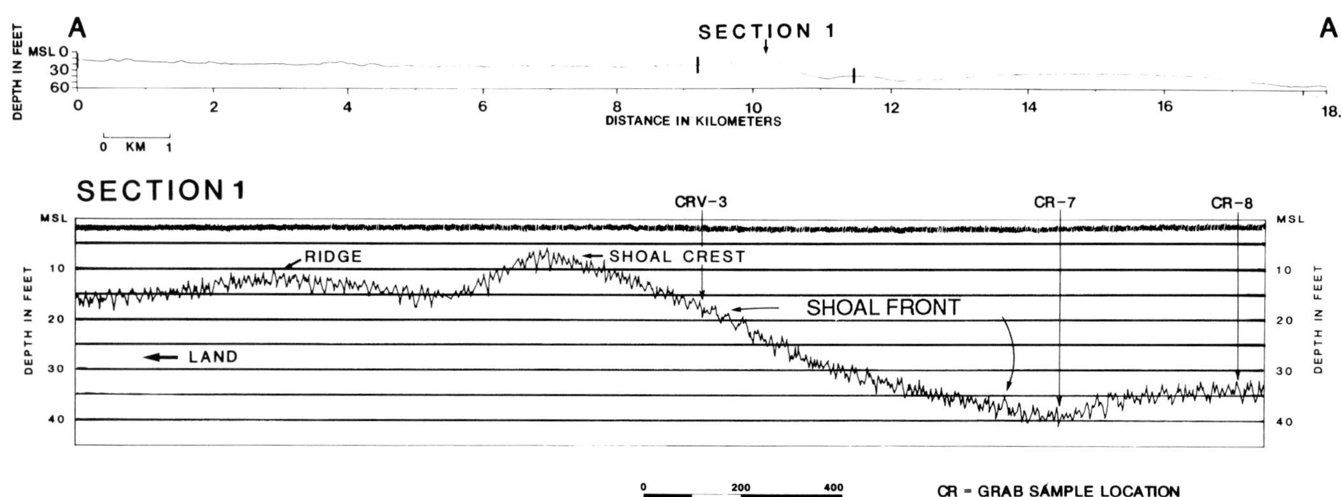

FIG. 5.—Fathometer traces of the Bull Island and Cape Romain shoals. Traces located on Figure 4.

6), two terraces occur over which water depths decrease from 18.3 m to 9.1 m. The top of the Santee/Pee Dee Delta shoal is essentially flat and is 6 km wide. Water depths also slightly decrease from 9.1 m to 6.7 m when transversing the shoal top from a southwest to northeast direction. The northeast side of the shoal is essentially flat and does not have multiple scarps/terraces.

Santee Delta profile trace C is located seaward of Cedar Island and runs perpendicular to the coastline (Fig. 4). The bathymetry at the beginning of trace C (closest to shore) reveals a smooth, flat bottom with a gentle seaward slope, which represents the shoreface associated with Cedar Island. Seaward of the shoreface, flat areas are followed by areas with ridge topography. The ridge features, like most of the other ridges observed in the study area, do not have consistent spacing.

SEAFLOOR SEDIMENTS

Bull Island Shoal

From just seaward of the active surf zone out to approximately the 4.6-m contour below MLW along transect A-A′ (Figs. 4 and 5), the seafloor sediments are very well-sorted, fine-grained sand with near-symmetrical size-distribution curves and low calcium-carbonate percentages (<5 percent). Seaward of the 15-ft MLW contour (Fig. 4), the

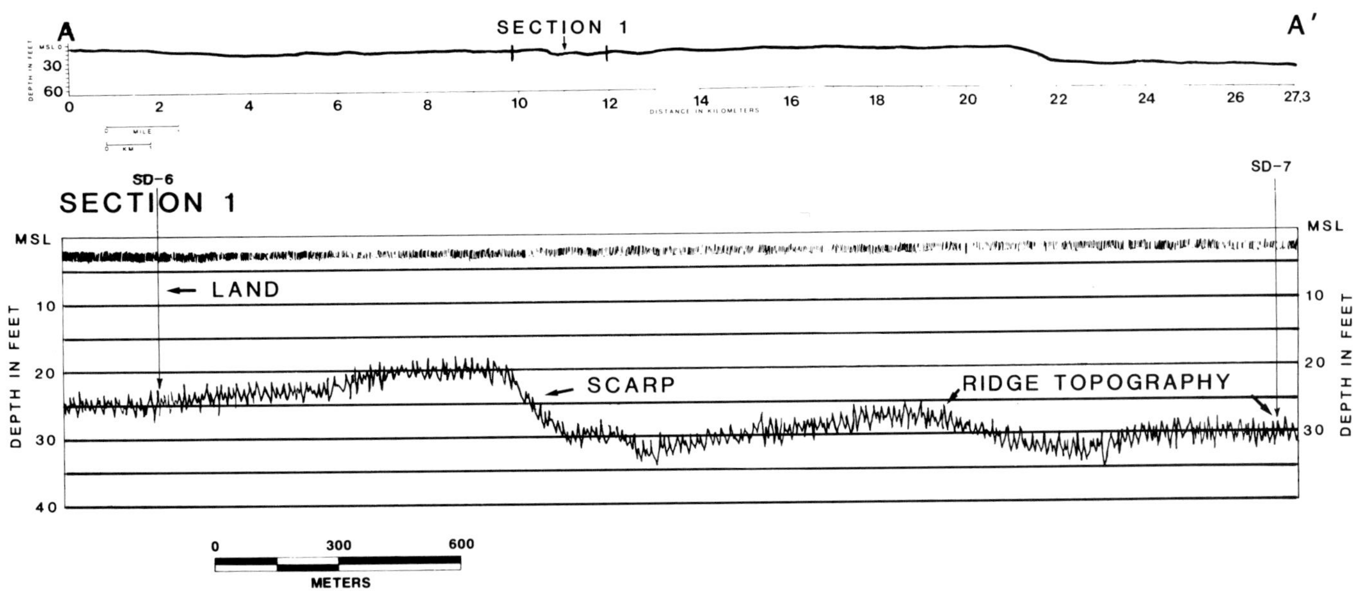

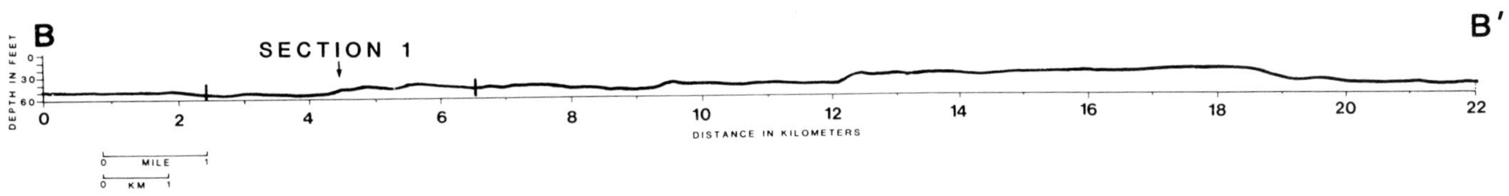

FIG. 6.—Fathometer traces for the Santee/Pee Dee Delta shoal. Traces located on Figure 4.

sediments on top of the shoal and the shoal's seaward margin are well-sorted, medium-grained sand with much higher (14 to 23 percent) concentrations of calcium-carbonate (shell material; Fig. 7). From the base of the seaward margin of the shoal out to the end of transect A-A′, sediments alternate from well-sorted fine-grained sand to well-sorted medium-grained sand. In this area, the sea floor has multiple ridges and swales. The tops of the ridges are covered with medium-grained sand and the troughs contain fine-grained sand (Fig. 7). All samples seaward of the shoal's margin contain greater than 10 percent calcium carbonate (shell debris). Sediments generally coarsen in a seaward direction.

Cape Romain Shoal

The beaches on Cape Island are comprised of medium- to coarse-grained, well-sorted, quartz-rich (90 to 95 percent) sand (Ruby, 1981). The seafloor sediments seaward of the beach change in an offshore direction from fine-grained to medium-grained to coarse-grained sand, i.e., sediments coarsen in a seaward direction. The sediments are generally moderately well sorted, with calcium-carbonate (shell) concentrations in excess of 10 percent of the total sample weight. The seaward front of the shoal is covered with medium-grained sand containing less than 6 percent shell debris. All but one of the samples collected seaward of the shoal's seaward margin are moderately sorted coarse-grained sand with calcium-carbonate (shell) concentrations varying from a high of 25 percent to a low of 9.1 percent.

The highest (3.1 percent) mud content is in the vicinity of the Cape Romain shoals.

Santee/Pee Dee Delta Shoal

The first two grab samples collected along transect A-A′ (out to 2.5 km seaward of South Island) reveal well-sorted, fine-grained sand (Figs. 4, 6, and 7) with low shell concentrations and the highest mud concentrations for this transect and shoal area. Similar to that of the Bull Island shoal, the grain size of the seafloor sediments changes dramatically from fine-grained sand to coarse-grained sand in the vicinity of the 4.6-m contour below MLW (Fig. 7). The abrupt grain-size change probably marks the toe of the active progradational barrier-island shoreface. In effect, it de-

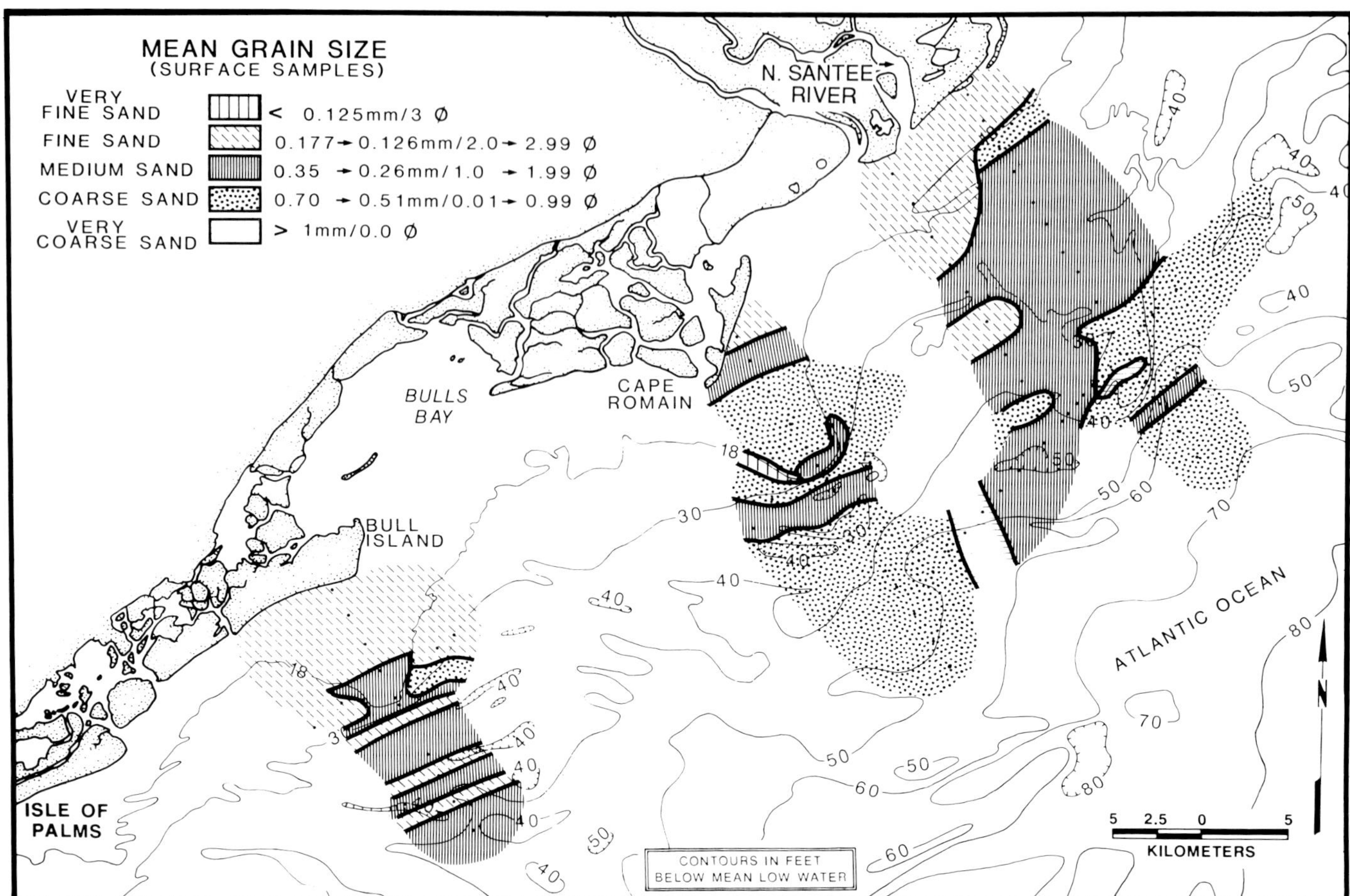

FIG. 7.—Grain-size map for the terrigenous-clastic fraction of the grab samples collected in the study area. Note the band of very coarse sand between the Cape Romain and Santee/Pee Dee Delta shoals.

fines the zone of onshore/offshore sediment transport. The rest of the shoal out to the last 2 or 3 km, is characterized by moderately sorted medium-grained sand containing low percentages of calcium carbonate (shell material, <5.5 percent). On the seaward end of the shoal, sediments coarsen to coarse-grained to very coarse-grained sand, shell content increases and sorting decreases. Seaward of the shoal, sediments remain coarse grained and are moderately sorted, but calcium-carbonate (shell) concentrations increase to 20 percent of total sample weight. Mud concentrations remain below 1.3 percent.

Transect B-B′ runs parallel to the shoreline across the seaward end of the shoal (Figs. 4, and 6). The low-lying area 2.5 to 4 km along the transect (Fig. 6) contains the coarsest grained samples collected during the study. The very coarse-grained sands are poorly sorted with relatively low concentrations of calcium-carbonate (shell) material (7.6 and 4.4 percent; Fig. 7). The southwest side of the shoal is terraced, but no change in the seafloor sediment character, medium-grained, moderately sorted sand was observed. Transect C-C′, located seaward of the end of the ebb-tidal delta of the North Santee Inlet Fig. 4), contains the finest grained sediments in the vicinity of either the Santee/Pee Dee Delta or Cape Romain shoals. The fine-grained sand zone probably reflects distal delta-front sedimentation. The zone extends to the 9.1-m contour below MLW and contains less than 4 percent calcium carbonate (shell material). Sediments on the beach landward of the transect on Cedar Island are also fine-grained sand, but comprise a zone that extends farther offshore and into twice the water depth than was observed at either South Island or Bull Island. Farther seaward along this transect, sediments coarsen from fine-grained sand to medium-grained sand. The transect ends in moderately sorted, coarse-grained sand.

In summary, there is a wide diversity of sediment types within the study area (Fig. 7), but medium and coarse-grained sand are most common. The sediments in the vicinity of the Bull Island shoal are finer grained (and have higher shell concentrations) than those of the Cape Romain or the Santee/Pee Dee Delta shoals.

SHOAL STRATIGRAPHY

Four vibracores and three box cores were taken (Fig. 4 and 5) on the Cape Romain and Santee/Pee Dee Delta shoals. The offshore cores were compared with the onshore cores from the earlier studies cited previously. Ruby (1981) investigated the late Pleistocene/Holocene depositional history of the northeast side of Cape Romain, and Eckard (1986) and Hayes and Sexton (1989) summarized several years of coring studies of the lower delta plain and delta front of the Santee River portion of the modern Santee/Pee Dee Delta. These studies provided essential late Pleistocene/Holocene subsurface data.

Three vibracores were collected from the Cape Romain shoal (Fig. 4). Core CRV-1 was collected 2 km seaward of Cape Island and closely resembles the most seaward core collected during Ruby's (1981) study (Fig. 3). At the top of the core is a thin marine sand (18 cm), rich in shallow-marine shell species. It is underlain by a horizontal to wavy, bedded, silty sand with no shell material and little to no bioturbation (Fig. 8). A shell lag, underlain by mixed mud and shell, characterizes the base of the core. A second core, CRV-2, was taken 6.25 km seaward of Cape Island (Fig. 4), slightly seaward of the center of the shoal. This core differs significantly from CRV-1 (Fig. 8) in that it has a 90-cm-thick marine-sand bed at the top underlain by 2.5 m of moderately burrowed fine-grained sand, mud and shell material (Fig. 8). Burrows are commonly filled with sand. The core bottomed in a shell-rich zone (lag) comprised predominantly of oyster shells (*Crassostrea virginica*). The top of the core is in the same stratigraphic position as the high-energy lagoonal sequence in the onshore core (Fig. 3). The sand was deposited in an open-marine setting, possibly a marine-influenced bay or shoreface associated with a progradational barrier island. Below the 45-cm-thick shell lag, mud and organic content increase and shell content decreases (Fig. 8). This section of the core is somewhat burrowed, and is interpreted to represent a brackish-water depositional setting. The third core, which was 80 cm long, was collected from the hard-packed mixed sand and shell on the terminus of the shoal front (Figs. 4, and 5). Shell material consists of open-marine species. The top 45 cm of the core are clean and oxidized, indicating recent reworking, whereas the lower 40 cm are reduced and bioturbated.

One vibracore and one box core were collected from the Santee/Pee Dee Delta shoal (Figs. 4, and 6). The vibracore (SDV-1) was taken 2.5 km seaward of South Island immediately northeast of the mouth of the Santee River (Fig. 4). The top of the core consists of 1.2 m of clean moderately burrowed marine sand and shell, which is underlain by an extensively burrowed mixture of fine-grained sand, mud, and shell. A 30-cm-thick shell lag of mostly open-marine species underlies the mixed-sediment horizon (Fig. 8). Freshwater deposits ranging from rooted mixed mud and organic material to a woody peat are present from the base of the shell lag to the bottom of the core. A box core (SDB-3) was taken in 7 m of water 13 km seaward of South Island on transect A (Fig. 4) The core reveals clean medium-grained sand with abundant low-angle bedding planes. The low-angle beds (5 to 7°) are thought to represent moderate-size bedforms (megaripples) or hummocky cross-stratification (Walker, 1982; Harms and others, 1975).

DISCUSSION

The three shoals studied are located in the nearshore inner-shelf environment in water depths ranging from the surf zone to less than 20 m. Swift and Niedoroda (1985) defined the shelf as predominantly storm dominated. The nearshore inner shelf is essentially the interface between the shoreface and shelf. Sediments reflect the dynamic character of the study area ranging from mud to very coarse-grained sand. The largest active river delta on the east coast of North America, the Santee/Pee Dee and its ancestral equivalents are thought to have been responsible, at least in part, for deposition of the shoals presently seaward of the Santee/Pee Dee Delta, Cape Romain, and Bull Island.

The Bull Island shoal has predominantly fine- and medium-grained sand on its surface (Fig. 7), with the percentage of shell material mixed with the clastic sediment increasing in a seaward direction. The topography of this shoal differs from that on the other two shoals. The Bull Island shoal does not have a pronounced front or scarp at its seaward end or elsewhere, and abundant ridges occur seaward of the shoal (Figs. 4 and 5). All these factors indicate that the Bull Island shoal is either different in origin, or showing evidence of more marine reworking than the other two shoals.

Ridges seaward of the Bull Island shoal appear to be similar to the ridges on the southern Virginia and North Carolina inner shelf described by Swift and others (1978) as a "shoal retreat massif." The retreat massifs are commonly found in front of bays or estuaries and represent estuarine mouths or ebb-tidal deltas reworked during a rise in sea level. One characteristic common to the Bull Island ridges that is different from other east coast ridges is that the sediments on the Bull Island ridge crests are commonly coarser grained than those in the associated troughs. Also, very little mud is found in the troughs. The grain-size pattern and lack of silt and clay indicate that this setting is too dynamic for the deposition of mud, and that these features are molded by active marine processes. The sediment mass of the shoal, however, is most probably a delta lobe of the ancestral Santee River formed during a Pleistocene lowstand. The sediments are thought to be the primary source of sand for the progradational barrier-island chain located to the southwest (Fig. 1).

The shoal seaward of Cape Romain has the greatest relief of the three shoals, with near-vertical scarps occurring on two of the three sides (Figs. 4 and 5). The surface sediments on this shoal are coarser grained as a whole (mixed coarse- and medium-grained sand) than those on the other two shoals. Both this shoal and the Bull Island shoal extend into 10 m of water and have the same general dimensions (Fig 1), whereas the Santee/Pee Dee Delta shoal is twice as long (20 km) and also extends to the 15-m bathymetric contour. Sediments coarsen in a seaward direction across the top of the shoal from fine- to coarse-grained sand. Shell content ranges from 4 to 30 percent, with the amount of shell material generally increasing seaward. The bathymetry seaward of the Cape Romain shoal is characterized by a few moderate-size ridges and one very large ridge 4.5 km long, 2.0 km wide and with a relief of 3 m. The large ridge could be a drowned barrier island or possibly a reworked remnant of a shoal that was overstepped during sea-level rise (Figs. 4 and 5). The large ridge has a similar size and

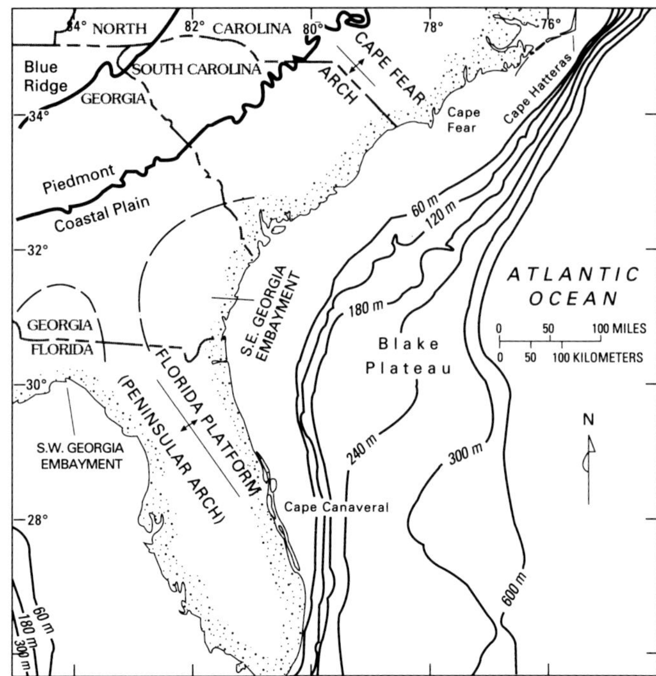

FIG. 1.—Map showing state boundaries, major physiographic provinces, and structural features of the southeastern United States (modified from Paul and Dillon, 1979).

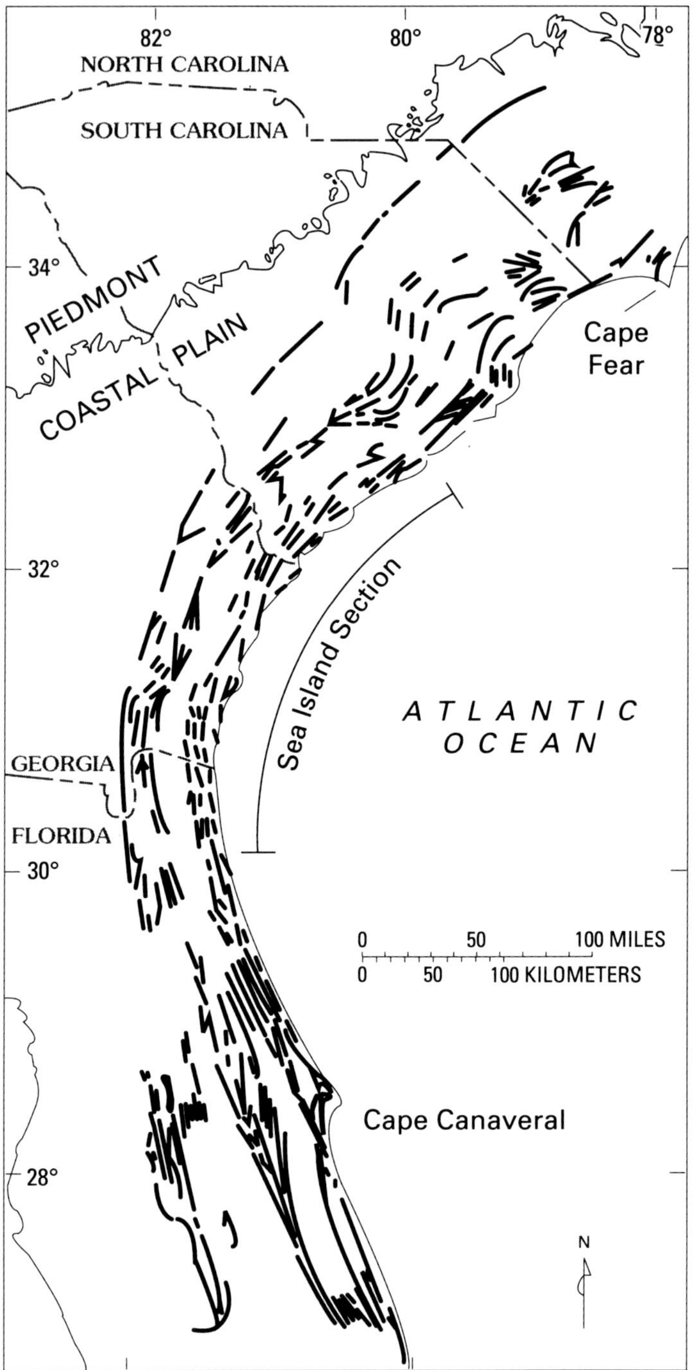

FIG. 2.—Map showing the trend of shoreline features (scarps, beach ridges, and barriers) in the southeastern United States that parallel or subparallel the present coast and are cut by southeastward-trending river valleys. FLA, Florida; GA, Georgia; SC, South Carolina; NC, North Carolina (modified from Winker and Howard, 1977).

shaped or long and linear. Back-barrier sediments may or may not be present. Three sets of welded, Pleistocene barriers define the south half of a large Savannah River paleodelta. The age-equivalent paleodelta of the Altamaha River is not well defined. In this area southward to the St. Marys River, Pleistocene barriers are expressed as low, north-northeast-trending, closely spaced, linear ridges surrounded by back-barrier sediments.

Emergent Pliocene and Pleistocene sediments in southeast Georgia are largely nonfossiliferous and locally dominated by fluvial components. Most deposits are undated because of the low-fossil content. The few fossil-bearing outcrops are small, isolated lenses of marl, shell hash, or shells in a matrix of organic-rich sand or clay. Microfossil data from cuttings and cores, which are quite extensive near the Georgia coast (Herrick and Wait, 1955; Herrick, 1961; Herrick and Vorhis, 1963; unpublished data from Georgia Geologic Survey), indicate the presence of late early, early late, and latest late Pliocene strata. Most fossils are from sediments considered to be biostratigraphically equivalent to the Duplin Formation (3.5–2.8 Ma) in the Cape Fear area of the Carolinas. Fossils from pre-Wisconsinan Pleistocene sediments are not common. Where present, they generally represent a back-barrier assemblage with minimal species diversity.

Purpose and Scope

The purposes of this paper are to (1) present the available paleontologic data for the emergent Pliocene and Pleistocene strata of southeastern Georgia and adjacent parts of southeastern South Carolina (fossil localities are shown on Fig. 4; data are given in Fig. 5); (2) discuss the stratigraphic relations of these units as seen in the field and in cores; and (3) discuss possible mechanisms that would account for the lack of fossils in Pliocene and Pleistocene sediments in this area of the southeastern Atlantic Coastal Plain. It is our intent to draw attention to a large area of the southeastern

Atlantic Coastal Plain for which few age data are available. It is an area that is critical to unravelling the late Cenozoic history of the southeastern United States, particularly the relations between the Gulf of Mexico and the Atlantic Ocean.

PREVIOUS WORK AND AVAILABLE DATA

Since the early 1900s, numerous geologists have studied the late Cenozoic geology of the Atlantic Coastal Plain in the southeastern United States. Some of early regional publications include those by Veatch and Stephenson (1911), Clark and others (1912), Cooke (1936, 1943, 1945), Cooke and Mossom (1929), and a map of Tertiary and Quaternary formations in Georgia (MacNeil, 1947). Although some early publications referred to fossils in Miocene and younger sediments (Hodgson, 1846; Dall, 1896, 1898; Aldrich, 1911; Richards, 1943; Edwards, 1944; and others), most correlations of units and/or shorelines were based solely upon their topographic and geomorphic positions within the landscape (Cooke, 1936, 1943).

The hiatus in publications in the middle and late 1940s was largely the result of World War II. Interest in the

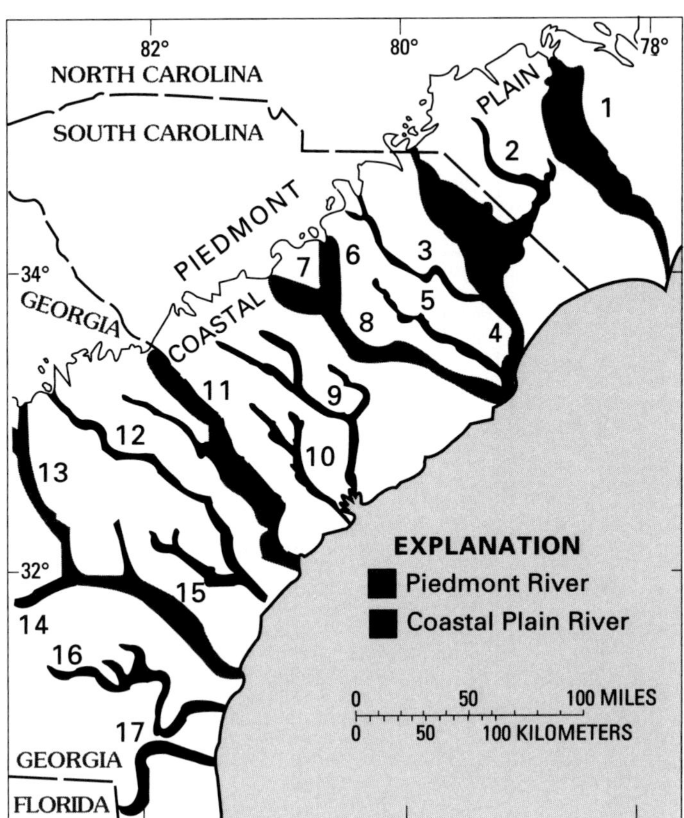

FIG. 3.—Map showing the major eastward- and southeastward-flowing streams in the southeastern United States: (1) Cape Fear; (2) Little Pee Dee and (3) Lynches that form (4) Great Pee Dee; (5) Black; (6) Wateree and (7) Congaree that form (8) Santee; (9) Edisto; (10) Combahee; (11) Savannah; (12) Ogeechee; (13) Oconee and (14) Ocmulgee that form (15) Altamaha; (16) Satilla; (17) St. Marys. Stipled pattern shows river valley that drains part of the Blue Ridge and/or Piedmont physiographic provinces as well as the Coastal Plain. Black shows river valley that drains only the Coastal Plain (modified from Hayes, 1989).

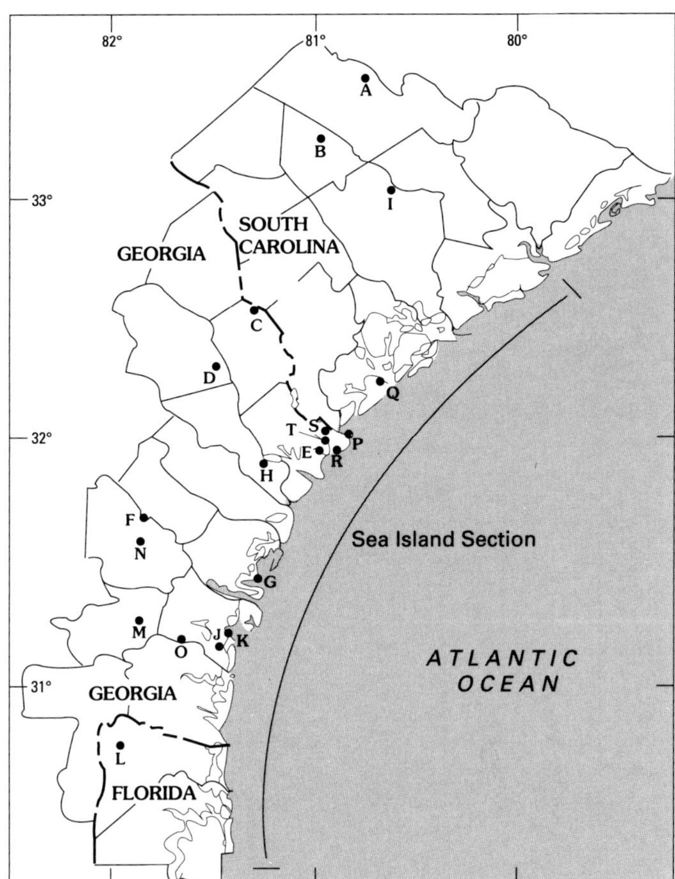

FIG. 4.—Location of fossil localities referred to in the text and in other figures. Counties outlined in black. (A) Outcrop, Orangeburg County, South Carolina, N 33°32′12″, W 80°46′24″ (Pooser, 1965; Colquhoun, 1965). (B) Outcrop, Bamberg County, South Carolina, N 33°17′54″, W 81°01′06″ (new data). (C), Outcrop, Porters Landing, Effingham County, Georgia, N 32°34′24″, W 81°21′39″ (Veatch and Stephenson, 1911; new data). (D) Outcrop, Bulloch County, Georgia, N 32°20′31″, W 80°30′24″ (new data). (E) Core, CH1, Chatham County, Georgia, N 31°59′47″, W 81°02′51″ (Huddlestun, 1988; new data). (F) Outcrop, Doctortown, Wayne County, Georgia, N 31°39′19″, W 81°49′49″ (Veatch and Stephenson, 1911; Herrick, 1976; new data). (G) Core, Sapelo Island, McIntosh County, Georgia, N 31°23′46″, W 81°16′32″ (Woolsey, 1976). (H) Core, BR1, Bryan County, Georgia, N 31°51′20″, W 81°12′45″ (new data) (I) Outcrop, Colleton County, South Carolina, N 33°06′01″, W 80 39′57″ (Cooke, 1936; Blackwelder and Ward, 1979). (J) Dredgings, Turtle River, Glynn County, Georgia, N 31°11′31″, W 81°32′24″ (Veatch and Stephenson, 1911; new data). (K) Dredgings, Brunswick Canal, Glynn County, Georgia, N 31°13′16″, W 81°30′14″ (Veatch and Stephenson, 1911). (L) Core, Cassidyl, Nassau County, Florida, N 30°38′10″, W 81°56′05″ (Huddlestun, 1988). (M) Outcrop, Brantley County, Georgia, N 31°03′31″, W 81°51′30″ (Veatch and Stephenson, 1911). (P) Five wells, Chatham County, Georgia: 1) core, PC1, Petit Chou Island, N 31°56′39″, W 80°55′39″ (Huddlestun, 1988; new data); 2) cuttings, GGS-772, Fort Screven, N 32°01′22″, W 80°51′01″ (Huddlestun, 1988); 3) cuttings, GGS-381, Fort Pulaski, N 32°01′51″, W 80°54′04″ (Huddlestun, 1988); 4) core, House Creek, N 31°57′40″, W 80°54′51″, (Huddlestun, 1988); 5) core, CH10, Tybee Island, N 31°59′16″, W 80°51′05″ (Huddlestun, 1988; new data). (Q) Cuttings, BFT-315, Hilton Head Island, Beaufort County, South Carolina, N 32°15′58″, W 80°43′13″ (Herrick, 1976; Huddlestun, 1988). (R) Core CH13, Chatham County, Georgia, N 30°58′26″, W 80°59′54″ (Huddlestun, 1988; new data). (S) Core CH14, Chatham County, Georgia, N 32°04′29″, W 80°09′17″ Huddlestun, 1988). (T) Three outcrops, Chatham County, Georgia, 1) southwest side of Skidaway Island on Burnside River, N 31°55′12″, W 81°04′25″; 2) east side of Isle of Hope, Skidaway River, N 31°58′53″, W 81°03′18″; 3) intersection, White Bluff Road and White Bluff Creek, N 31°59′03″, W 81°07′48″ (Veatch and Stephenson, 1911).

Locations:
Pleistocene (N23)
H. Core, Bryan Co., GA
J. Turtle River, Glynn Co., GA
K. Brunswick Canel, Glynn Co. GA
O. Outcrop, Glynn Co., GA
P.1. Well No. PC 1, Chatham Co., GA
R. Core, CH 13, Chatham Co., GA
T.1. Outcrop, Chatham Co., GA
2. Outcrop, Chatham Co., GA
3. Outcrop, Chatham Co., GA
Late Pliocene (PL5, N20-22)
L. Core, Cassidy #1, Nassau Co., FL
M. Outcrop, Brantly Co., GA
N. Core, Wayne #1, Wayne Co., GA
S. Core, CH 14, Chatham Co., GA
Late Pliocene (PL3, N19-21)
A. Outcrop, Orangeburg Co., SC
B. Outcrop, Bamberg Co., SC
C. Outcrop, Effingham Co., GA
D. Outcrop, Bulloch Co., GA
E. Test Hole, CH 1, Chatham Co., GA
F. Outcrop, Wayne Co., GA
G. Test Hole, McIntosh Co., GA
I. Outcrop, Colleton Co., SC
J. Turtle River, Glynn Co. GA
Early Pliocene (PL1, N18-19)
P.2. Well No. GGS-772, Chatham Co., GA
3. Well No. GGS-381, Chatham Co., GA
4. Well, House Ck., Chatham Co., GA
5. Well No. Ch 10, Chatham Co., GA
Q. Well No. BFT-315, Beaufort Co., SC

Q	5	4	3	P.2	J	I	G	F	E	D	C	B	A	S	N	M	L	3	2	T.1	R	P.1	O	K	J	H	FAUNA
																											Foraminifer
-	-	-	-	-	-	-	-	X	-	-	-	-	-	-	-	-	-	-	-	-	-	-	-	-	-	-	*Textularia articulata*
-	-	-	-	-	-	-	-	X	-	-	-	-	-	-	-	-	-	-	-	-	-	-	-	-	-	-	*Textularia candeiana*
X	-	-	-	-	-	-	-	X	-	-	-	-	X	-	-	-	-	-	-	-	-	-	-	-	-	-	*Textularia gramen*
-	-	-	-	-	-	-	-	X	-	-	-	-	X	-	-	-	-	-	-	-	-	-	-	-	-	-	*Textularia mayori*
-	-	-	X	-	-	-	-	X	-	-	-	-	X	-	-	-	-	-	-	-	-	-	-	-	-	-	*Nodosaria catesbyi*
X	-	-	-	-	-	-	-	-	-	-	-	-	X	-	-	-	-	-	-	-	-	-	-	-	-	-	*Lagena clavata*
-	-	-	-	-	-	-	-	X	-	-	-	-	-	-	-	-	-	-	-	-	-	-	-	-	-	-	*Lagena costata amphora*
X	-	-	-	X	-	-	-	-	-	-	-	-	-	-	-	-	-	-	-	-	-	-	-	-	-	-	*Lagena laevis*
-	-	-	-	-	-	-	-	-	-	-	-	-	X	-	-	-	-	-	-	-	-	-	-	-	-	-	*Lagena perlucida*
-	-	-	-	X	-	-	-	X	-	-	-	-	X	-	-	-	-	-	-	-	-	-	-	-	-	-	*Lagena semistriata*
X	-	-	-	-	-	-	-	-	-	-	-	-	-	-	-	-	-	-	-	-	-	-	-	-	-	-	*Lagena substriata*
X	-	-	-	X	-	-	-	-	-	-	-	-	-	-	-	-	-	-	-	-	-	-	-	-	-	-	*Lagena sulcata*
-	-	-	-	X	-	-	-	-	-	-	-	-	-	-	-	-	-	-	-	-	-	-	-	-	-	-	*Lagena tenuis*
-	-	-	-	-	-	-	-	-	-	-	-	-	-	-	-	-	-	-	-	-	X	-	-	-	-	-	*Lagena* sp.
X	-	-	X	X	-	-	-	X	-	-	-	-	-	-	-	-	-	-	-	-	-	-	-	-	-	-	*Lenticulina americana*
-	-	-	-	-	-	-	-	X	-	-	-	-	-	-	-	-	-	-	-	-	-	-	-	-	-	-	*Lenticulina mayi*
-	-	-	-	-	-	-	-	-	-	-	-	-	X	-	-	-	-	-	-	-	-	-	-	-	-	-	*Robulus* (=Lenticulina) cf. *nikobarensis*
-	-	-	-	-	-	-	-	X	-	-	-	-	-	-	-	-	-	-	-	-	-	-	-	-	-	-	*Plectofrondicularia* cf. *longistriata*
-	-	-	-	-	-	-	-	X	-	-	-	-	-	-	-	-	-	-	-	-	-	-	-	-	-	-	*Globulina caribaea*
-	-	-	-	-	-	-	-	X	-	-	-	-	X	-	-	-	-	-	-	-	-	-	-	-	-	-	*Globulina gibba*
X	-	-	-	X	-	-	-	X	-	-	-	-	-	-	-	-	-	-	-	-	-	-	-	-	-	-	*Globulina inaequalis*
X	-	-	-	X	-	-	-	X	-	-	-	-	-	-	-	-	-	-	-	-	-	-	-	-	-	-	*Guttulina austriaca*
-	-	-	-	-	-	-	-	X	-	-	-	-	-	-	-	-	-	-	-	-	-	-	-	-	-	-	*Guttulina caudata*
-	-	-	-	-	-	-	-	X	-	-	-	-	X	-	-	-	-	-	-	-	-	-	-	-	-	-	*Guttulina pseudocostatula*
-	-	-	-	-	-	-	-	-	-	-	-	-	-	-	-	-	-	-	-	-	X	-	-	-	-	-	*Guttulina* sp.
-	-	-	-	-	-	-	-	X	-	-	-	-	X	-	-	-	-	-	-	-	-	-	-	-	-	-	*Pseudopolymorphina rutila*
-	-	-	X	-	-	-	-	-	-	-	-	-	-	-	-	-	-	-	-	-	-	-	-	-	-	-	*Pseudopolymorphina* sp.
-	-	-	-	-	-	-	-	X	-	-	-	-	-	-	-	-	-	-	-	-	-	-	-	-	-	-	*Sigmomorphina pearceyi*
X	-	-	X	X	-	-	-	X	-	-	-	-	X	-	-	-	-	-	-	-	-	-	-	-	-	-	*Sigmomorphina terquemiana*
-	-	-	-	-	-	-	-	X	-	-	-	-	-	-	-	-	-	-	-	-	-	-	-	-	-	-	*Sigmomorphina undulosa*
-	-	-	-	-	-	-	-	-	-	-	-	-	X	-	-	-	-	-	-	-	-	-	-	-	-	-	*Sigmomorphina williamsona*
-	-	-	-	-	-	-	-	X	-	-	-	-	-	-	-	-	-	-	-	-	-	-	-	-	-	-	*Laryngosigma williamsoni*
X	-	-	X	X	-	-	-	X	-	-	-	-	-	-	-	-	-	-	-	-	-	-	-	-	-	-	*Oolina hexagona scalariformis*
-	-	-	-	-	-	-	-	-	-	-	-	-	X	-	-	-	-	-	-	-	-	-	-	-	-	-	*Oolina melo*
-	-	-	-	-	-	-	-	-	-	-	-	-	X	-	-	-	-	-	-	-	-	-	-	-	-	-	*Oolina quadrata*
-	-	-	-	-	-	-	-	-	-	-	-	-	X	-	-	-	-	-	-	-	-	-	-	-	-	-	*Oolina scalariforma*
-	-	-	-	-	-	-	-	-	-	-	-	-	X	-	-	-	-	-	-	-	-	-	-	-	-	-	*Fissurina lacunata*
-	-	-	-	-	-	-	-	X	-	-	-	-	-	-	-	-	-	-	-	-	-	-	-	-	-	-	*Fissurina lucida*
-	-	-	-	-	-	-	-	X	-	-	-	-	X	-	-	-	-	-	-	-	-	-	-	-	-	-	*Fissurina marginatoperforata*
-	-	-	-	-	-	-	-	X	-	-	-	-	-	-	-	-	-	-	-	-	-	-	-	-	-	-	*Fissurina orbignyana lacunata*
-	-	-	-	-	-	-	-	X	-	-	-	-	-	-	-	-	-	-	-	-	-	-	-	-	-	-	*Parafissurina marginata*
-	-	-	-	-	-	-	-	-	-	-	-	-	X	-	-	-	-	-	-	-	-	-	-	-	-	-	*Buliminella* cf. *bassendorfensa*
X	-	-	X	X	-	-	-	X	-	-	-	-	-	X	-	-	-	-	-	-	-	-	-	-	-	-	*Buliminella curta*
X	-	-	-	X	-	-	-	X	-	-	-	-	X	X	-	-	-	-	-	-	X	-	-	-	-	-	*Buliminella elegantissima*
X	-	-	-	X	-	-	-	X	-	-	-	-	X	-	-	-	-	-	-	-	-	-	-	-	-	-	*Bolivina advena*
-	-	-	-	-	-	-	-	X	-	-	-	-	X	-	-	-	-	-	-	-	-	-	-	-	-	-	*Bolivina marginata* (marginata)
-	-	-	-	-	-	-	-	X	-	-	-	-	X	-	-	-	-	-	-	-	-	-	-	-	-	-	*Bolivina marginata multicostata* B. (marginata) *multicostata*
-	-	-	X	X	-	-	-	X	-	-	-	-	X	-	-	-	-	-	-	-	-	-	-	-	-	-	*Bolivina paula*
X	-	-	-	X	-	-	-	X	-	-	-	-	X	-	-	-	-	-	-	-	-	-	-	-	-	-	*Bolivina plicatella*
-	-	-	-	-	-	-	-	X	-	-	-	-	-	-	-	-	-	-	-	-	-	-	-	-	-	-	*Bolivina* cf. *suteri*
-	-	-	X	X	-	-	-	-	-	-	-	-	-	-	-	-	-	-	-	-	-	-	-	-	-	-	*Bolivina* sp.
-	-	-	-	X	-	-	-	-	-	-	-	-	X	-	-	-	-	-	-	-	-	-	-	-	-	-	*Cassidulinoides bradyi*
-	-	-	-	-	-	-	-	X	-	-	-	-	-	-	-	-	-	-	-	-	-	-	-	-	-	-	*Nodogenerina* (=Stilostomella) *advena*
X	-	-	X	X	-	-	-	X	-	-	-	-	-	-	-	-	-	-	-	-	-	-	-	-	-	-	*Bulimina elongata*

FIG. 5.—Pliocene and Pleistocene fossil data from localities in southeastern South Carolina, southeastern Georgia, and extreme northeastern Florida. New data on ostracodes, macrofossils, diatoms, and foraminifera. Other fossil data from references, given by locality in Figure 4.

Species	Q.	5.	4.	3.	P.2.	J.	I.	G.	F.	E.	D.	C.	B.	A.	S.	N.	M.	L.	3.	2.	T.1.	R.	P.1.	O.	K.	J.	H.
Bulimina marginata	-	-	-	-	X	-	-	-	-	-	-	-	-	-	-	-	-	-	-	-	-	-	-	-	-	-	-
Reusella spinulosa	-	-	-	-	X	-	-	-	X	-	-	-	-	-	-	-	-	-	-	-	-	-	-	-	-	-	-
Uvigerina auberiana	X	-	-	X	X	-	-	-	X	-	-	-	-	-	-	-	-	-	-	-	-	-	-	-	-	-	-
Uvigerina canariensis	-	-	-	-	-	-	-	-	X	-	-	-	-	-	-	-	-	-	-	-	-	-	-	-	-	-	-
Uvigerina parvula	-	-	-	-	-	-	-	-	-	-	-	-	-	X	-	-	-	-	-	-	-	-	-	-	-	-	-
Uvigerina suberiana	-	-	-	-	-	-	-	-	-	-	-	-	-	X	-	-	-	-	-	-	-	-	-	-	-	-	-
Uvigerina subperegrina	X	-	-	X	X	-	-	-	X	-	-	-	-	-	-	-	-	-	-	-	-	-	-	-	-	-	-
Angulogerina (=Trifarina) *occidentalis*	X	-	-	X	X	-	-	-	X	-	-	-	-	X	-	-	-	-	-	-	-	-	-	-	-	-	-
A. (=T.) *occidentala*																											
Angulogerina (=Trifarina) sp.	-	-	-	-	-	-	-	-	-	-	-	-	-	-	-	-	-	-	-	-	-	X	-	-	-	-	-
Discorbis duplinensis	-	-	-	-	-	-	-	-	-	-	-	-	-	X	-	-	-	-	-	-	-	-	-	-	-	-	-
Discorbis terquemis	-	-	-	-	-	-	-	-	-	-	-	-	-	X	-	-	-	-	-	-	-	-	-	-	-	-	-
Discorbis turritis	-	-	-	-	-	-	-	-	-	-	-	-	-	X	-	-	-	-	-	-	-	-	-	-	-	-	-
Discorbis valvulatus	-	-	-	-	X	-	-	-	-	-	-	-	-	-	-	-	-	-	-	-	-	-	-	-	-	-	-
Discorbis vilardeboanus	-	-	-	-	X	-	-	-	-	-	-	-	-	-	-	-	-	-	-	-	-	-	-	-	-	-	-
Buccella depressa	-	-	-	-	-	-	-	-	-	-	-	-	-	X	-	-	-	-	-	-	-	-	-	-	-	-	-
Buccella mansfieldi	X	-	-	X	X	-	-	-	X	-	-	-	-	-	X	-	-	-	-	-	-	-	-	-	-	-	-
Bucella sp.	-	-	-	-	-	-	-	-	-	-	-	-	-	-	-	-	-	-	-	-	-	X	-	-	-	-	-
Conorbina orbicularis	-	-	-	-	-	-	-	-	X	-	-	-	-	-	-	-	-	-	-	-	-	-	-	-	-	-	-
Rosalina floridana	-	-	-	-	-	-	-	-	X	-	-	-	-	-	-	-	-	-	-	-	-	-	-	-	-	-	-
Rosalina subaraucana	X	-	-	X	-	-	-	-	X	-	-	-	-	-	-	-	-	-	-	-	-	-	-	-	-	-	-
Rosalina turrita	X	-	-	-	X	-	-	-	X	-	-	-	-	-	-	-	-	-	-	-	-	-	-	-	-	-	-
Cancris (sagra) *sagra*	X	-	-	-	X	-	-	-	X	-	-	-	-	-	-	-	-	-	-	-	-	-	-	-	-	-	-
Cancris sagra communis	-	-	-	-	-	-	-	-	X	-	-	-	-	X	-	-	-	-	-	-	-	-	-	-	-	-	-
C. (sagra) *communis*																											
Valvulineria sp.	X	-	-	X	X	-	-	-	X	-	-	-	-	-	-	-	-	-	-	-	-	-	-	-	-	-	-
Ammonia beccarii	X	-	-	X	X	-	-	-	X	-	-	-	-	-	-	-	-	-	-	-	-	X	-	-	-	-	-
Elphidium advena	-	-	-	-	-	-	-	-	X	-	-	-	-	X	-	-	-	-	-	-	-	-	-	-	-	-	-
E. *advenum*																											
Elphidium clavatum	X	-	-	X	X	-	-	-	X	-	-	-	-	-	-	-	-	-	-	-	-	-	-	-	-	-	-
Elphidium gunteri	X	-	-	-	X	-	-	-	X	-	-	-	-	X	-	-	-	-	-	-	-	-	-	-	-	-	-
Elphidium incertum	-	-	-	-	X	-	-	-	X	-	-	-	-	X	-	-	-	-	-	-	-	-	-	-	-	-	-
Elphidium limatulum	-	-	-	-	-	-	-	-	-	-	-	-	-	X	-	-	-	-	-	-	-	-	-	-	-	-	-
Elphidium matagordanum	-	-	-	-	-	-	-	-	-	-	-	-	-	X	-	-	-	-	-	-	-	-	-	-	-	-	-
Elphidium poeyanum	X	-	-	-	X	-	-	-	X	-	-	-	-	X	-	-	-	-	-	-	-	-	-	-	-	-	-
Elphidium varium	-	-	-	-	-	-	-	-	X	-	-	-	-	-	-	-	-	-	-	-	-	-	-	-	-	-	-
Elphidium sp.	-	-	-	-	-	-	-	-	-	-	-	-	-	-	-	-	-	-	-	-	-	X	-	-	-	-	-
Chiloguembelina cubensis	-	-	-	-	-	-	-	-	-	-	-	-	-	X	-	-	-	-	-	-	-	-	-	-	-	-	-
Hastigerina aequilateralis aequilateralis	-	X	X	X	X	-	-	-	-	X	-	-	-	-	-	-	-	X	-	-	-	X	X	-	-	-	-
Globigerinella siphonifera (=H. aequilateralis ...)																											
Globigerinella (=Hastigerina) *aequilateralis praesiphonifera*	-	X	X	X	X	-	-	-	-	X	-	-	-	-	-	-	-	X	-	-	-	X	X	-	-	-	-
Hastigerina sp.	-	-	-	-	-	-	-	X	-	-	-	-	-	-	-	-	-	-	-	-	-	-	-	-	-	-	-
Globorotalia inflata	-	-	-	-	-	-	-	-	-	-	-	-	-	-	-	-	-	-	-	-	-	X	X	-	-	-	-
Globorotalia menardii menardii (dextral)	X	X	X	X	X	-	-	X	X	X	-	-	-	-	-	-	-	-	-	-	-	-	X	-	-	-	-
G. *menardii*																											
Globorotalia menardii menardii (sinistral)	-	-	-	-	-	-	-	X	-	-	-	-	-	-	-	-	-	-	-	-	-	X	X	-	-	-	-
Globorotalia menardii miocenica	-	-	-	-	-	-	-	-	-	-	-	-	-	-	-	-	-	X	-	-	-	-	-	-	-	-	-
Globorotalia margaritae margaritae	-	X	X	X	X	-	-	-	-	-	-	-	-	-	-	-	-	-	-	-	-	-	-	-	-	-	-
Globorotalia puncticulata	-	-	-	-	-	-	-	-	-	-	-	-	-	-	-	-	-	X	-	-	-	-	-	-	-	-	-
Globigerina apertura	-	X	X	X	X	-	-	-	-	X	-	-	-	-	-	X	-	X	-	-	-	-	X	-	-	-	-
Globigerina bulloides	-	X	X	X	X	-	-	-	-	-	-	-	-	X	-	X	-	X	-	-	-	X	-	-	-	-	-
Globigerina cf. *bulloides*	-	-	-	-	-	-	-	-	-	X	-	-	-	-	-	-	-	-	-	-	-	-	-	-	-	-	-
Globigerina calida	-	-	-	-	-	-	-	-	-	X	-	-	-	-	-	-	-	-	-	-	-	X	-	-	-	-	-
Globigerina decorapertura	-	-	-	-	-	-	-	-	-	-	-	-	-	-	-	-	-	X	-	-	-	-	-	-	-	-	-
Globigerina cf. *decorapertura*	-	-	-	-	-	-	-	-	-	X	-	-	-	-	-	X	-	-	-	-	-	-	-	-	-	-	-
Globigerina falconensis	-	-	-	-	-	-	-	-	-	-	-	-	-	-	-	-	-	-	-	-	-	X	-	-	-	-	-
Globigerina cf. *falconensis*	-	-	-	-	-	-	-	-	-	-	-	-	-	-	-	X	-	X	-	-	-	-	X	-	-	-	-
Globigerina nepenthes	-	X	X	X	X	-	-	-	-	-	-	-	-	-	-	-	-	-	-	-	-	-	-	-	-	-	-
Globigerina quinqueloba	-	-	-	-	-	-	-	-	-	-	-	-	-	-	-	-	-	-	-	-	-	X	-	-	-	-	-
Globigerina rubescens	-	-	-	-	-	-	-	-	-	-	-	-	-	-	-	-	-	-	-	-	-	X	-	-	-	-	-
Globigerina cf. *rubescens*	-	X	X	X	X	-	-	-	-	-	-	-	-	-	-	-	-	-	-	-	-	-	X	-	-	-	-
Globigerina triloculinoides	-	-	-	-	-	-	-	-	-	-	-	-	-	X	-	-	-	-	-	-	-	-	-	-	-	-	-
(=Globigerinoides quadrilobatus quadrilobatus)																											
Globigerinoides conglobatus	-	-	-	-	-	-	-	-	-	-	-	-	-	-	-	-	-	-	-	-	-	X	-	-	-	-	-
Globigerinoides cf. *conglobatus*	-	X	X	X	X	-	-	-	-	X	-	-	-	-	-	-	-	-	-	-	-	-	-	-	-	-	-
Globigerinoides obliquus obliquus	-	X	X	X	X	-	-	X	-	X	-	-	-	-	-	X	-	X	-	-	-	-	X	-	-	-	-
Globigerinoides obliquus extremus	-	X	X	X	X	-	-	-	-	X	-	-	-	-	-	-	-	-	-	-	-	-	-	-	-	-	-
Globigerinoides quadrilobatus quadrilobatus	-	X	X	X	X	-	-	-	-	X	-	-	-	-	-	-	-	X	-	-	-	X	X	-	-	-	-
Globigerinoides quadrilobatus sacculiferus	-	-	-	-	-	-	-	-	-	X	-	-	-	-	-	-	-	-	-	-	-	X	-	-	-	-	-
Globigerinoides ruber	X	-	-	-	X	-	-	X	X	X	-	-	-	-	-	X	-	X	-	-	-	X	X	-	-	-	-
G. *rubra*																											
Globoquadrina altispira	-	X	X	X	X	-	-	X	-	X	-	-	-	-	-	-	-	-	-	-	-	-	-	-	-	-	-

FIG. 5.—Continued.

Q.	5.	4.	3.	P.2.	J.	I.	G.	F.	E.	D.	C.	B.	A.	S.	N.	M.	L.	3.	2.	T.1.	R.	P.1.	O.	K.	J.	H.	
X	X	X	X	X	-	-	-	X	X	-	-	-	-	-	-	-	-	-	-	-	-	-	-	-	-	-	*Neogloboquadrina* (acostaensis)*acostaensis*
																											Globorotalia acostaensis
-	X	X	X	X	-	-	X	-	X	-	-	-	-	-	-	-	-	-	-	-	-	-	-	-	-	-	*Neogloboquadrina* (acostaensis) *humerosa*
																											Globoquadrina humerosa
-	-	-	-	-	-	-	X	-	-	-	-	-	-	-	-	-	-	-	-	-	-	-	-	-	-	-	*N. ... humerosa,* aberrant form
-	-	-	-	-	-	-	-	-	-	-	-	-	-	-	-	-	X	-	-	-	X	X	-	-	-	-	*Neogloboquadrina dutertrei*
-	-	-	-	-	-	-	-	-	-	-	-	-	-	-	X	-	-	-	-	-	-	-	-	-	-	-	*Neogloboquadrina* cf. *dutertrei*
																											Sphaeroidinella cf. *dehiscens*
-	X	X	X	X	-	-	-	-	-	-	-	-	-	-	-	-	-	-	-	-	-	-	-	-	-	-	*Sphaeroidinellopsis seminulina*
-	-	-	X	X	-	-	X	X	-	-	-	-	-	-	-	-	-	-	-	-	-	-	-	-	-	-	*Sphaeroidinellopsis subdehiscens* (=paenedehiscens)
-	X	X	X	X	-	-	-	-	-	-	-	-	-	-	-	-	-	-	-	-	X	-	-	-	-	-	*Orbulina universa*
-	X	X	X	X	-	-	-	-	-	-	-	-	-	-	-	-	-	-	-	-	X	X	-	-	-	-	*Globigerinita glutinata*
-	X	X	X	X	-	-	-	-	-	-	-	-	-	-	-	-	-	-	-	-	-	-	-	-	-	-	*Globigerinata uvula*
-	-	-	-	-	-	-	-	X	-	-	-	-	-	-	-	-	-	-	-	-	-	-	-	-	-	-	*Eponides antillarum*
-	-	-	-	X	-	-	-	-	-	-	-	-	-	-	-	-	-	-	-	-	-	-	-	-	-	-	*Eponides* cf. *regularis*
-	-	-	-	-	-	-	-	-	-	-	-	-	X	-	-	-	-	-	-	-	-	-	-	-	-	-	*Eponides repandus*
-	-	-	-	-	-	-	-	X	-	-	-	-	-	-	-	-	-	-	-	-	-	-	-	-	-	-	*Poroeponides lateralis*
-	-	-	-	-	-	-	-	X	-	-	-	-	-	-	-	-	-	-	-	-	-	-	-	-	-	-	*Amphistegina lessonii*
-	-	-	-	-	-	-	-	-	-	-	-	-	X	-	-	-	-	-	-	-	-	-	-	-	-	-	*Amphistegina* sp.
X	-	-	X	X	-	-	-	X	-	-	-	-	-	-	-	-	-	-	-	-	-	-	-	-	-	-	*Planulina depressa*
-	-	-	-	-	-	-	-	-	-	-	-	-	X	-	-	-	-	-	-	-	-	-	-	-	-	-	*Planulina* cf. *depressa*
X	-	-	-	-	-	-	-	X	-	-	-	-	-	-	-	-	-	-	-	-	-	-	-	-	-	-	*Cibicides americanus*
X	-	-	X	X	-	-	-	X	-	-	-	-	X	-	-	-	-	-	-	-	-	-	-	-	-	-	*Cibicides duplinensis*
X	-	-	X	X	-	-	-	X	-	-	-	-	X	-	-	-	-	-	-	-	-	-	-	-	-	-	*Cibicides lobatulus*
X	-	-	X	-	-	-	-	X	-	-	-	-	-	-	-	-	-	-	-	-	-	-	-	-	-	-	C. *lobatulus* var.
-	-	-	-	X	-	-	-	X	-	-	-	-	X	-	-	-	-	-	-	-	-	-	-	-	-	-	*Cibicides sapeloensis*
-	-	-	-	-	-	-	-	-	-	-	-	-	X	-	-	-	-	-	-	-	-	-	-	-	-	-	*Cibicidella variabilis*
-	-	-	-	X	-	-	-	-	-	-	-	-	-	-	-	-	-	-	-	-	-	-	-	-	-	-	*Planorbulina mediterranensis*
X	-	-	-	-	-	-	-	X	-	-	-	-	-	-	-	-	-	-	-	-	-	-	-	-	-	-	*Cymbaloporetta squamossa*
-	-	-	X	X	-	-	-	-	-	-	-	-	-	-	-	-	-	-	-	-	-	-	-	-	-	-	*Virgulina* (=Fursenkoina) *fusiformis*
-	-	-	-	-	-	-	-	-	-	-	-	-	X	-	-	-	-	-	-	-	-	-	-	-	-	-	*Virgulina* (=Fursenkoina) *gunteri*
-	-	-	-	-	-	-	-	-	-	-	-	-	X	-	-	-	-	-	-	-	-	-	-	-	-	-	*Virgulina* (=Fursenkoina) *pontoni*
X	-	-	-	X	-	-	-	X	-	-	-	-	X	-	-	-	-	-	-	-	-	-	-	-	-	-	*Virgulina* (=Fursenkoina) *punctata*
-	-	-	-	-	-	-	-	-	-	-	-	-	-	X	-	-	-	-	-	-	-	-	-	-	-	-	*Virgulinella gunteri*
X	-	-	X	X	-	-	-	X	-	-	-	-	X	-	-	-	-	-	-	-	-	-	-	-	-	-	*Cassidulina crassa*
X	-	-	X	X	-	-	-	-	-	-	-	-	-	-	-	-	-	-	-	-	-	-	-	-	-	-	*Cassidulina laevigata laevigata*
-	-	-	-	-	-	-	-	X	-	-	-	-	X	-	-	-	-	-	-	-	-	-	-	-	-	-	*Cassidulina laevigata carinata*
																											C. *carinata*
-	-	-	-	-	-	-	-	X	-	-	-	-	-	-	-	-	-	-	-	-	-	-	-	-	-	-	*Cassidulina subglobosa*
X	-	-	-	X	-	-	-	X	-	-	-	-	X	-	-	-	-	-	-	-	-	-	-	-	-	-	*Nonion grateloupi*
X	-	-	X	X	-	-	-	-	-	-	-	-	-	-	-	-	-	-	-	-	-	-	-	-	-	-	*Nonion pizarrense*
-	-	-	-	-	-	-	-	X	-	-	-	-	X	-	-	-	-	-	-	-	-	-	-	-	-	-	*Astrononion glabrellum*
-	-	-	X	X	-	-	-	X	-	-	-	-	X	X	-	-	-	-	-	-	-	-	-	-	-	-	*Florilus atlantica*
																											Nonionella atlantica
-	-	-	-	-	-	-	-	X	-	-	-	-	-	-	-	-	-	-	-	-	-	-	-	-	-	-	*Gyroidina orbicularis*
X	-	-	X	X	-	-	-	X	-	-	-	-	X	-	-	-	-	-	-	-	-	-	-	-	-	-	*Hanzawaia concentrica*
-	-	-	-	-	-	-	-	X	-	-	-	-	-	-	-	-	-	-	-	-	-	-	-	-	-	-	*Robertina* cf. *subteres*
																											Coelenterata
-	-	-	-	-	X	-	-	-	-	-	X	-	-	-	-	-	-	-	-	-	-	-	-	-	-	-	*Septastrea crassa*
-	-	-	-	-	-	-	-	-	-	-	X	-	-	-	-	-	-	-	-	-	-	-	-	-	-	-	*Astrangia* sp.
																											Bryozoa
-	-	-	-	-	-	-	-	-	-	-	X	-	-	-	-	-	-	-	-	-	-	-	-	-	-	-	*Lunulites* (=Trochopora) sp.
																											Pelecypoda
-	-	-	-	-	-	-	-	X	-	X	X	-	-	-	-	-	-	-	-	-	-	-	-	-	-	-	*Nucula proxima*
-	-	-	-	-	-	-	-	X	-	X	-	-	-	-	-	-	-	-	-	-	-	-	-	-	-	-	*Nuculana acuta*
-	-	-	-	-	-	-	-	-	-	-	X	-	-	-	-	-	-	-	-	-	-	-	-	-	-	-	*Leda acuta* (=*Nuculana acuta*)
-	-	-	-	-	-	-	-	-	-	-	X	-	-	-	-	-	-	-	-	-	-	-	-	-	-	-	*Arca improcera*
-	-	-	-	-	X	-	-	-	-	-	-	-	-	-	-	-	-	-	-	-	-	-	-	-	-	-	*Arca incongrua*
-	-	-	-	-	X	-	-	-	-	-	-	-	-	-	-	-	-	-	-	-	-	-	-	-	-	-	*Arca lienosa*
-	-	-	-	-	X	-	-	-	-	-	-	-	-	-	-	-	-	-	-	-	-	-	-	-	-	-	*Arca limula*
-	-	-	-	-	X	-	-	-	-	-	-	-	-	-	-	-	-	-	-	-	-	-	-	-	-	-	*Arca plicatura*
-	-	-	-	-	X	-	-	-	-	-	-	-	-	-	-	-	-	-	-	-	-	-	-	-	-	-	*Arca transversa*
-	-	-	-	-	-	-	-	-	-	-	X	-	-	-	-	-	-	-	-	-	-	-	-	-	-	-	*Arca* sp.
-	-	-	-	-	-	-	-	-	-	-	-	-	X	-	-	-	-	-	-	-	-	-	-	-	-	-	*Arca* (=Barbatia) cf. *marylandica*
-	-	-	-	-	-	-	-	-	-	-	-	-	-	-	-	-	-	-	-	-	-	-	-	X	-	-	*Barbatia adamsi*
-	-	-	-	-	-	-	-	X	-	X	-	-	-	-	-	-	-	-	-	-	-	-	-	-	-	-	*Anadara improcera*
-	-	-	-	-	-	-	-	-	-	-	-	-	-	-	-	-	-	-	-	-	-	-	-	-	X	-	*Glycymeris americana* G. *parilis*
-	-	-	-	-	X	-	-	-	-	-	-	-	X	-	-	-	-	-	-	-	-	-	-	-	X	-	*Glycymeris subovata subovata*
-	-	-	-	-	-	-	-	-	-	X	-	-	-	-	-	-	-	-	-	-	-	-	-	-	-	-	*Glycymeris subovata plagia*
-	-	-	-	-	-	-	-	-	-	-	-	-	-	-	-	-	-	-	-	-	-	-	-	X	-	-	*Glycymeris* sp.
-	-	-	-	-	-	-	-	X	-	-	-	-	-	-	-	-	-	-	-	-	-	-	-	-	-	-	*Mytilus* sp.

FIG. 5.—Continued.

Q.	5.	4.	3.	P.2.	J.	I.	G.	F.	E.	D.	C.	B.	A.	S.	N.	M.	L.	3.	2.	T.1.	R.	P.1.	O.	K.	J.	H.	
-	-	-	-	-	-	-	-	-	-	-	-	-	-	-	-	X	-	-	-	-	-	-	-	-	-	-	*Modiolaria* (=Musculus) sp.
-	-	-	-	-	-	-	-	X	-	-	-	-	-	-	-	-	-	-	-	-	-	-	-	-	-	-	*Atrina* sp.
-	-	-	-	-	-	-	-	-	-	-	X	-	-	-	-	-	-	-	-	-	-	-	-	-	-	-	*Pteria colymbus*
-	-	-	-	-	-	-	-	-	-	-	X	-	-	-	-	-	-	-	-	-	-	-	-	-	-	-	*Amusium mortoni*
-	-	-	-	-	-	-	-	-	-	-	-	-	X	-	-	-	-	-	-	-	-	-	-	-	-	-	*Amusium* cf. *mortoni*
-	-	-	-	-	-	-	-	-	-	-	X	-	-	-	-	-	-	-	-	-	-	-	-	-	-	-	*Pseudamusium* (=Palliolum) sp.
-	-	-	-	-	-	-	-	-	-	-	-	-	X	-	-	-	-	-	-	-	-	-	-	-	-	-	*Chlamys* sp.
-	-	-	-	-	-	-	-	X	-	-	-	-	-	-	-	-	-	-	-	-	-	-	-	-	-	-	*Argopecten vicenarius* (?)
-	-	-	-	-	-	-	-	-	-	X	-	-	-	-	-	-	-	-	-	-	-	-	-	-	-	-	*Leptopecten irremotis*
-	-	-	-	-	-	-	-	X	-	X	X	X	-	-	-	-	-	-	-	-	-	-	-	-	-	-	*Carolinapecten eboreous*
																											Pecten eboreus
-	-	-	-	-	X	-	-	-	-	-	X	-	-	-	-	-	-	-	-	-	-	-	-	-	-	-	*Chesapecten jeffersonius jeffersonius*
																											Pecten jeffersonius
-	-	-	-	-	X	-	-	-	-	X	X	-	X	-	-	-	-	-	-	-	-	-	-	-	-	-	*Chesapecten jeffersonius septenarius*
																											Pecten septenarius
																											Chlamys jeffersonia septenaria
-	-	-	-	-	X	-	-	-	-	-	-	-	-	-	-	-	-	-	-	-	-	-	-	-	-	-	*Chesapecten madisonius*
																											Pecten madisonius
-	-	-	-	-	X	-	-	-	-	-	X	-	-	-	-	-	-	-	-	-	-	-	-	-	-	-	*Pecten* n. sp.
-	-	-	-	-	-	-	-	X	-	-	X	-	-	-	-	-	-	-	-	-	-	-	-	-	-	-	*Plicatula marginata*
-	-	-	-	-	-	-	-	X	-	X	X	X	-	-	-	-	-	-	-	-	-	-	-	-	-	-	*Anomia simplex*
-	-	-	-	-	-	-	-	-	-	-	X	-	-	-	-	-	-	-	-	-	-	-	-	-	-	-	*Placuanomia* (=Placunanomia) sp.
-	-	-	-	-	-	-	-	-	-	-	X	-	-	-	-	-	-	-	-	-	-	-	-	-	-	-	*Placunanomia plicata*
-	-	-	-	-	X	-	-	-	-	-	-	-	-	-	-	-	-	-	-	-	-	-	-	-	-	-	*Phacoides* (=Lucina) *amiantus*
-	-	-	-	-	X	-	-	-	-	-	X	-	-	-	-	-	-	-	-	-	-	-	-	-	-	-	*Phacoides* (=Lucina) *anodonta*
-	-	-	-	-	-	-	-	-	-	-	X	-	-	-	-	-	-	-	-	-	-	-	-	-	-	-	*Phacoides* (=Lucina) *cribarius*
-	-	-	-	-	X	-	-	-	-	-	X	-	-	-	-	-	-	-	-	-	-	-	-	-	-	-	*Phacoides* (=Lucina) *multilineatus*
-	-	-	-	-	X	-	-	-	-	-	-	-	-	-	-	-	-	-	-	-	-	-	-	-	-	-	*Phacoides* (=Lucina) *radians*
-	-	-	-	-	-	-	-	X	-	-	-	-	-	-	-	-	-	-	-	-	-	-	-	-	-	-	*Lucinisca cribrarius*
-	-	-	-	-	-	-	-	X	-	-	-	-	-	-	-	-	-	-	-	-	-	-	-	-	-	-	*Bellucina tuomeyi*
-	-	-	-	-	-	-	-	-	-	X	-	-	-	-	-	-	-	-	-	-	-	-	-	-	-	-	*Stewartia* (=Megaxinus) *anodonta*
-	-	-	-	-	-	-	-	X	-	X	-	-	-	-	-	-	-	-	-	-	-	-	-	-	-	-	*Parvilucina crenulata*
-	-	-	-	-	-	-	-	X	-	-	-	-	-	-	-	-	-	-	-	-	-	-	-	-	-	-	*Parvilucina multilineata*
-	-	-	-	-	-	-	-	X	-	-	-	-	-	-	-	-	-	-	-	-	-	-	-	-	-	-	*Cavilinga trisulcata*
-	-	-	-	-	X	-	-	-	-	-	-	-	-	-	-	-	-	-	-	-	-	-	-	-	-	-	*Diplodonta acclinis*
-	-	-	-	-	-	-	-	-	-	X	-	-	-	-	-	-	-	-	-	-	-	-	-	-	-	-	*Chama congregata*
-	-	-	-	-	-	-	-	X	-	-	-	-	-	-	-	-	-	-	-	-	-	-	-	-	-	-	*Chama striata*
-	-	-	-	-	X	-	-	-	-	-	-	-	-	-	-	-	-	-	-	-	-	-	-	-	-	-	*Echinochama* (=Arcinella) *arcinella*
-	-	-	-	-	-	-	-	-	-	X	-	-	-	-	-	-	-	-	-	-	-	-	-	-	-	-	*Pseudochama corticosa*
-	-	-	-	-	X	-	-	-	-	-	-	-	-	-	-	-	-	-	-	-	-	X	-	-	-	-	*Crassostrea virginica*
-	-	-	-	-	X	-	-	-	-	-	X	-	-	-	-	-	-	-	-	-	-	-	-	-	-	-	*Ostrea disparilus*
-	-	-	-	-	-	-	-	-	-	-	X	-	-	-	-	-	-	-	-	-	-	-	-	-	-	-	*Ostrea raveneli*
-	-	-	-	-	-	-	-	X	-	X	-	-	-	-	-	-	-	-	-	-	-	-	-	-	-	-	*Ostrea raveneliana*
-	-	-	-	-	-	-	-	X	-	X	X	X	-	-	-	-	-	-	-	-	-	-	-	-	-	-	*Conradostrea sculpturata*
																											Ostrea sculpturata
-	-	-	-	-	X	-	-	-	-	-	-	-	-	-	-	-	-	-	-	-	-	-	-	-	-	-	Carditamera arata
-	-	-	-	-	-	-	-	-	-	-	X	-	-	-	-	-	-	-	-	-	-	-	-	-	-	-	*Carditamera* sp.
-	-	-	-	-	X	-	-	-	-	X	X	-	-	-	-	-	-	-	-	-	-	-	-	-	-	-	*Cyclocardia granulata*
																											Venericardia granulata
-	-	-	-	-	-	-	-	X	-	-	-	-	-	-	-	-	-	-	-	-	-	-	-	-	-	-	*Pleuromeris decemcostata*
-	-	-	-	-	-	-	-	-	-	X	-	-	-	-	-	-	-	-	-	-	-	-	-	-	-	-	*Pleuromeris decemcostata* ssp.?
-	-	-	-	-	-	-	-	-	-	-	X	-	-	-	-	-	-	-	-	-	-	-	-	-	-	-	*Venericardia perplano*
-	-	-	-	-	-	-	-	-	-	-	X	-	-	-	-	-	-	-	-	-	-	-	-	-	-	-	*Venericardia tridentata*
-	-	-	-	-	X	-	-	X	-	X	X	-	-	-	-	-	-	-	-	-	-	-	-	-	-	-	*Astarte concentrica*
-	-	-	-	-	X	-	-	-	-	-	-	-	-	-	-	-	-	-	-	-	-	-	-	-	-	-	*Astarte cuneformis*
-	-	-	-	-	X	-	-	-	-	-	X	-	-	-	-	-	-	-	-	-	-	-	-	-	-	-	*Astarte distans* var. *floridana*
-	-	-	-	-	X	-	-	-	-	-	-	-	-	-	-	-	-	-	-	-	-	-	-	-	-	-	*Astarte undulata* var.
-	-	-	-	-	X	-	-	-	-	-	-	-	-	-	-	-	-	-	-	-	-	-	-	-	-	-	*Astarte undulata* var. *vaginulata*
-	-	-	-	-	X	-	-	-	-	X	X	-	-	-	-	-	-	-	-	-	-	-	-	-	-	-	*Marvacrassatella undulata*
																											Crassatella undulata
																											Crassatellites undulata
-	-	-	-	-	-	-	-	-	-	-	X	-	-	-	-	-	-	-	-	-	-	-	-	-	-	-	*Marvacrassatella* cf. *undulata*
-	-	-	-	-	X	-	-	-	-	-	-	-	-	-	-	-	-	-	-	-	-	-	-	-	-	-	*Crassatellites* sp.
-	-	-	-	-	-	-	-	-	-	-	X	-	-	-	-	-	-	-	-	-	-	-	-	-	-	-	*Crassinella dupliniana*
-	-	-	-	-	-	-	-	X	-	X	-	-	-	-	-	-	-	-	-	-	-	-	-	-	-	-	*Crassinella duplinensis* (?)
-	-	-	-	-	-	-	-	X	-	X	-	-	-	-	-	-	-	-	-	-	-	-	-	-	-	-	*Crassinella lunulata*
-	-	-	-	-	X	-	-	-	-	-	-	-	-	-	-	-	-	-	-	-	-	-	-	-	-	-	*Cardium robustum*
-	-	-	-	-	-	-	-	-	-	-	X	-	-	-	-	-	-	-	-	-	-	-	-	-	-	-	*Cardium* sp.
-	-	-	-	-	X	-	-	X	-	-	X	-	X	-	-	-	-	-	-	-	-	-	-	-	-	-	*Mulinia congesta*
-	-	-	-	-	X	-	-	-	-	-	-	-	-	-	-	-	-	-	-	-	-	-	-	-	-	-	*Mulinia congesta* var. *contracta*
-	-	-	-	-	X	-	-	-	-	-	-	-	-	-	-	-	-	-	-	-	-	-	-	-	-	-	*Mulinia congesta* var. *elongata*
-	-	-	-	-	-	-	-	X	-	-	-	-	-	-	-	-	-	-	-	-	-	-	-	-	-	-	*Mulinia lateralis*
-	-	-	-	-	-	-	-	X	-	-	-	-	-	-	-	-	-	-	-	-	-	-	-	-	-	-	*Mulinia* sp.
-	-	-	-	-	X	-	-	-	-	-	-	-	-	-	-	-	-	-	-	-	-	-	-	-	-	-	*Rangia clathrodon*
-	-	-	-	-	-	-	-	-	-	-	-	-	-	-	-	X	-	-	-	-	-	-	-	-	-	-	*Rangia cuneata*

FIG. 5.—Continued.

Q.	5.	4.	3.	P.2.	J.	I.	G.	F.	E.	D.	C.	B.	A.	S.	N.	M.	L.	3.	2.	T.1.	R.	P.1.	O.	K.	J.	H.	
-	-	-	-	-	-	-	-	X	-	-	-	-	-	-	-	-	-	-	-	-	-	-	-	-	-	-	*Tellina declivis*
-	-	-	-	-	-	-	-	-	-	-	X	-	-	-	-	-	-	-	-	-	-	-	-	-	-	-	*Tellina* (Angulus) *umbra*
-	-	-	-	-	-	-	-	X	-	-	X	-	-	-	-	-	-	-	-	-	-	-	-	-	-	-	*Tellina* sp.
-	-	-	-	-	-	-	-	-	-	-	X	-	-	-	-	-	-	-	-	-	-	-	-	-	-	-	*Strigilla* sp.
-	-	-	-	-	X	-	-	-	-	-	-	-	-	-	-	-	-	-	-	-	-	-	-	-	-	-	*Donax* sp.
-	-	-	-	-	-	-	-	X	-	-	-	-	-	-	-	-	-	-	-	-	-	-	-	-	-	-	*Semelina* sp.
-	-	-	-	-	-	-	-	-	-	-	X	-	-	-	-	-	-	-	-	-	-	-	-	-	-	-	*Venus tridacnoides*
-	-	-	-	-	-	-	-	-	-	-	X	-	-	-	-	-	-	-	-	-	-	-	-	-	-	-	*Venus tridacnoides* var. *rileyi*
-	-	-	-	-	-	-	-	X	-	-	-	-	-	-	-	-	-	-	-	-	-	-	-	-	-	-	*Gouldia metastriata*
-	-	-	-	-	-	-	-	-	-	-	X	-	-	-	-	-	-	-	-	-	-	-	-	-	-	-	*Transennella caloosana*
-	-	-	-	-	-	-	-	X	-	-	-	-	-	-	-	-	-	-	-	-	-	-	-	-	-	-	*Dosinia* sp.
-	-	-	-	-	-	-	-	-	-	-	-	-	-	-	-	X	-	-	-	-	-	-	-	-	-	-	*Gemma purpurea*
-	-	-	-	-	X	-	-	-	-	-	-	-	-	-	-	-	-	-	-	-	-	-	-	-	-	-	*Chione alveata*
-	-	-	-	-	-	-	-	-	-	-	X	-	-	-	-	-	-	-	-	-	-	-	-	-	-	-	*Chione athleta*
-	-	-	-	-	X	-	-	-	-	-	-	-	-	-	-	-	-	-	-	-	-	-	-	-	-	-	*Chione cancellata*
-	-	-	-	-	X	-	-	-	-	-	-	-	-	-	-	-	-	-	-	-	-	-	-	-	-	-	*Chione* aff. *cortinaria*
-	-	-	-	-	-	-	-	-	-	-	-	-	X	-	-	-	-	-	-	-	-	-	-	-	-	-	*Chione latilirata*
-	-	-	-	-	-	-	-	X	-	-	-	-	-	-	-	-	-	-	-	-	-	-	-	-	-	-	*Lirophora* (=C. Lirophora) sp.
-	-	-	-	-	-	-	-	-	-	-	-	-	X	-	-	-	-	-	-	-	-	-	-	-	-	-	*Mercenaria mercenaria*
-	-	-	-	-	-	-	-	-	-	X	-	-	-	-	-	-	-	-	-	-	-	-	-	-	-	-	*Mercenaria rileyi*
-	-	-	-	-	-	-	-	-	-	X	-	-	-	-	-	-	-	-	-	-	-	-	-	-	-	-	*Mercenaria tridacnoides*
-	-	-	-	-	-	-	-	-	-	-	X	-	-	-	-	-	-	-	-	-	-	-	-	-	-	-	*Mercenaria* sp.
-	-	-	-	-	-	-	-	X	-	-	-	-	-	-	-	-	-	-	-	-	-	-	-	-	-	-	*Sphenia dubia*
-	-	-	-	-	X	-	-	-	-	-	X	-	-	-	-	-	-	-	-	-	-	-	-	-	-	-	*Corbula inaequalis*
-	-	-	-	-	-	-	-	X	-	-	-	-	-	-	-	-	-	-	-	-	-	-	-	-	-	-	*Caryocorbula conradi*
-	-	-	-	-	-	-	-	-	-	X	-	-	-	-	-	-	-	-	-	-	-	-	-	-	-	-	*Caryocorbula cuneata*
-	-	-	-	-	-	-	-	-	-	X	-	-	-	-	-	-	-	-	-	-	-	-	-	-	-	-	*Gastrochaena* sp.
-	-	-	-	-	-	-	-	X	-	-	-	-	-	-	-	-	-	-	-	-	-	-	-	-	-	-	*Pandora* sp.
																											Gastropoda
-	-	-	-	-	-	-	-	-	-	-	-	-	-	-	-	X	-	-	-	-	-	-	-	-	-	-	*Neritina* sp?
-	-	-	-	-	-	-	-	-	-	-	X	-	-	-	-	-	-	-	-	-	-	-	-	-	-	-	*Calliostoma armillatum*
-	-	-	-	-	-	-	-	-	-	-	X	-	-	-	-	-	-	-	-	-	-	-	-	-	-	-	*Calliostoma mitchelli*
-	-	-	-	-	-	-	-	X	-	-	-	-	-	-	-	-	-	-	-	-	-	-	-	-	-	-	*Calliostoma tuomeyi* (?)
-	-	-	-	-	-	-	-	-	-	X	-	-	-	-	-	-	-	-	-	-	-	-	-	-	-	-	*Calliostoma* sp.
-	-	-	-	-	X	-	-	-	-	-	-	-	-	-	-	-	-	-	-	-	-	-	-	-	-	-	*Fissuridea* (=Diodora) *carolinensis*
-	-	-	-	-	-	-	-	-	-	-	X	-	-	-	-	-	-	-	-	-	-	-	-	-	-	-	*Fissuridea* sp.
-	-	-	-	-	-	-	-	X	-	-	-	-	-	-	-	-	-	-	-	-	-	-	-	-	-	-	*Diodora nucula*
-	-	-	-	-	-	-	-	-	-	X	-	-	-	-	-	-	-	-	-	-	-	-	-	-	-	-	*Turbonilla* sp.
-	-	-	-	-	-	-	-	X	-	-	-	-	-	-	-	-	-	-	-	-	-	-	-	-	-	-	*Epitonium leai* (?)
-	-	-	-	-	-	-	-	-	-	-	X	-	-	-	-	-	-	-	-	-	-	-	-	-	-	-	*Scala* (=Epitonium) ssp.
-	-	-	-	-	X	-	-	-	-	-	-	-	-	-	-	-	-	-	-	-	-	-	-	-	-	-	*Neverita duplicata*
-	-	-	-	-	-	-	-	-	-	-	-	-	-	-	-	X	-	-	-	-	-	-	-	-	-	-	*Neverita* sp?
-	-	-	-	-	-	-	-	-	-	-	-	-	-	-	-	X	-	-	-	-	-	-	-	-	-	-	*Potamides cancelloides*
-	-	-	-	-	-	-	-	-	-	-	-	-	-	-	-	X	-	-	-	-	-	-	-	-	-	-	*Potamides saltillensis*
-	-	-	-	-	-	-	-	-	-	-	-	-	-	-	-	X	-	-	-	-	-	-	-	-	-	-	*Amnicola expansilabris*
-	-	-	-	-	-	-	-	-	-	-	-	-	-	-	-	X	-	-	-	-	-	-	-	-	-	-	*Amnicola georgiensis*
-	-	-	-	-	-	-	-	-	-	-	-	-	-	-	-	X	-	-	-	-	-	-	-	-	-	-	*Amnicola saltillensis*
-	-	-	-	-	-	-	-	X	-	-	-	-	-	-	-	-	-	-	-	-	-	-	-	-	-	-	*Polinices duplicatus*
-	-	-	-	-	-	-	-	X	-	-	-	-	-	-	-	-	-	-	-	-	-	-	-	-	-	-	*Calyptraea* sp.
-	-	-	-	-	-	-	-	-	-	X	-	-	-	-	-	-	-	-	-	-	-	-	-	-	-	-	*Crucibulum* sp.
-	-	-	-	-	X	-	-	-	-	-	-	-	-	-	-	-	-	-	-	-	-	-	-	-	-	-	*Crepidula fornicata*
-	-	-	-	-	X	-	-	-	-	-	-	-	-	-	-	-	-	-	-	-	-	-	-	-	-	-	*Solarium* (=Architectonica) *granulatum*
-	-	-	-	-	-	-	-	-	-	-	X	-	-	-	-	-	-	-	-	-	-	-	-	-	-	-	*Omphalius* (=Cassiope) *exoletus*
-	-	-	-	-	-	-	-	-	-	-	X	-	-	-	-	-	-	-	-	-	-	-	-	-	-	-	*Turritella duplinensis*
-	-	-	-	-	X	-	-	-	-	-	-	-	-	-	-	-	-	-	-	-	-	-	-	-	-	-	*Turritella plebeia*
-	-	-	-	-	X	-	-	-	-	-	X	-	-	-	-	-	-	-	-	-	-	-	-	-	-	-	*Turritella variablilis*
-	-	-	-	-	-	-	-	X	-	X	-	-	X	-	-	-	-	-	-	-	-	-	-	-	-	-	*Turritella* sp.
-	-	-	-	-	X	-	-	-	-	-	-	-	-	-	-	-	-	-	-	-	-	-	-	-	-	-	*Nassa* (=Buccitriton) *acuta*
-	-	-	-	-	X	-	-	-	-	-	-	-	-	-	-	-	-	-	-	-	-	-	-	-	-	-	*Nassa* (=Buccitriton) *vibex*
-	-	-	-	-	X	-	-	-	-	-	-	-	-	-	-	-	-	-	-	-	-	-	-	-	-	-	*Ilyanassa obsoleta*
-	-	-	-	-	X	-	-	-	-	-	-	-	-	-	-	-	-	-	-	-	-	-	-	-	-	-	*Fulgur carica* (=Busycon *caricum*)
-	-	-	-	-	X	-	-	-	-	-	-	-	-	-	-	-	-	-	-	-	-	-	-	-	-	-	*Columbella avara* var
-	-	-	-	-	X	-	-	-	-	-	-	-	-	-	-	-	-	-	-	-	-	-	-	-	-	-	*Astyris lunata*
-	-	-	-	-	X	-	-	-	-	X	X	-	-	-	-	-	-	-	-	-	-	-	-	-	-	-	*Ecphora quadricostata*
-	-	-	-	-	-	-	-	X	-	X	-	-	-	-	-	-	-	-	-	-	-	-	-	-	-	-	*Ecphora* sp.
-	-	-	-	-	-	-	-	-	-	X	-	-	-	-	-	-	-	-	-	-	-	-	-	-	-	-	*Ptychosalpinx* sp.
-	-	-	-	-	X	-	-	-	-	-	-	-	-	-	-	-	-	-	-	-	-	-	-	-	-	-	*Marginella contracta*
-	-	-	-	-	-	-	-	-	-	X	-	-	-	-	-	-	-	-	-	-	-	-	-	-	-	-	*Marginella* sp.
-	-	-	-	-	X	-	-	-	-	-	-	-	-	-	-	-	-	-	-	-	-	-	-	-	-	-	*Olivella mutica*
-	-	-	-	-	-	-	-	-	-	X	-	-	-	-	-	-	-	-	-	-	-	-	-	-	-	-	*Oliva canaliculata*
-	-	-	-	-	X	-	-	-	-	-	-	-	-	-	-	-	-	-	-	-	-	-	-	-	-	-	*Oliva literata*
-	-	-	-	-	X	-	-	-	-	-	-	-	-	-	-	-	-	-	-	-	-	-	-	-	-	-	*Terebra* (=Strioterebrum) *dislocata*
-	-	-	-	-	X	-	-	-	-	-	-	-	-	-	-	-	-	-	-	-	-	-	-	-	-	-	*Drillia abundans*
-	-	-	-	-	X	-	-	-	-	-	-	-	-	-	-	-	-	-	-	-	-	-	-	-	-	-	*Conus* sp.

FIG. 5.—Continued.

Q.	5.	4.	3.	P.2.	J.	I.	G.	F.	E.	D.	C.	B.	A.	S.	N.	M.	L.	3.	2.	T.1.	R.	P.1.	O.	K.	J.	H.	
-	-	-	-	-	-	-	-	X	-	-	-	-	-	-	-	-	-	-	-	-	-	-	-	-	-	-	*Acteocina canaliculata*
-	-	-	-	-	-	-	-	-	-	-	-	-	-	-	-	X	-	-	-	-	-	-	-	-	-	-	*Planorbis antiquatus*
-	-	-	-	-	-	-	-	-	-	-	-	-	-	-	-	X	-	-	-	-	-	-	-	-	-	-	*Paludestrina plana*
-	-	-	-	-	-	-	-	X	-	-	-	-	-	-	-	-	-	-	-	-	-	-	-	-	-	-	*Longchaeus suturalis*
																											Scaphopoda
-	-	-	-	-	-	-	-	X	-	X	X	-	-	-	-	-	-	-	-	-	-	-	-	-	-	-	*Cadulus thallus*
-	-	-	-	-	X	-	-	-	-	-	-	-	-	-	-	-	-	-	-	-	-	-	-	-	-	-	*Dentalium carolinense*
																											Ostracoda
-	-	-	-	-	-	-	-	-	-	-	-	-	X	-	-	-	-	-	-	-	-	-	-	-	-	-	*Haplocytheridea bassleri*
-	-	-	-	-	-	-	-	X	-	-	-	-	-	-	-	-	-	-	-	-	-	-	-	-	-	-	*Hulingsina rugipustulosa*
-	-	-	-	-	-	-	-	X	-	X	-	-	-	-	-	-	-	-	-	-	-	-	-	-	-	-	*Hulingsina* sp.
-	-	-	-	-	-	-	-	X	-	-	-	-	-	-	-	-	-	-	-	-	-	-	-	-	-	-	*Cytherura forulata*
-	-	-	-	-	-	-	-	X	-	-	-	-	-	-	-	-	-	-	-	-	-	-	-	-	-	-	*Cytherura* sp.
-	-	-	-	-	-	-	-	X	-	X	-	-	-	-	-	-	-	-	-	-	-	-	-	-	-	-	*Cytheropteron yorktownensis*
-	-	-	-	-	-	-	-	X	-	X	-	-	-	-	-	-	-	-	-	-	-	-	-	-	-	-	*Paracytheridea altila*
-	-	-	-	-	-	-	-	X	-	X	-	-	-	-	-	-	-	-	-	-	-	-	-	-	-	-	*Peratodytheridea* sp.
-	-	-	-	-	-	-	-	X	-	-	-	-	-	-	-	-	-	-	-	-	-	-	-	-	-	-	*Proteoconcha gigantica*
-	-	-	-	-	-	-	-	-	-	X	-	-	-	-	-	-	-	-	-	-	-	-	-	-	-	-	*Proteoconcha tuberculata*
-	-	-	-	-	-	-	-	-	-	-	-	-	X	-	-	-	-	-	-	-	-	-	-	-	-	-	*Aurila conradi conradi*
-	-	-	-	-	-	-	-	-	-	X	-	-	-	-	-	-	-	-	-	-	-	-	-	-	-	-	*Campylocythere laeva*
-	-	-	-	-	-	-	-	-	-	X	-	-	-	-	-	-	-	-	-	-	-	-	-	-	-	-	*Loxoconcha* cf. L. *edentonensis*
-	-	-	-	-	-	-	-	-	-	X	-	-	-	-	-	-	-	-	-	-	-	-	-	-	-	-	*Loxoconcha reticularis*
-	-	-	-	-	-	-	-	X	-	-	-	-	-	-	-	-	-	-	-	-	-	-	-	-	-	-	*Cytheromorpha newportensis*
-	-	-	-	-	-	-	-	-	-	X	-	-	X	-	-	-	-	-	-	-	-	-	-	-	-	-	*Cytheromorpha warneri*
-	-	-	-	-	-	-	-	X	-	X	-	-	-	-	-	-	-	-	-	-	-	-	-	-	-	-	*Malzella conradi*
-	-	-	-	-	-	-	-	-	-	-	X	-	-	-	-	-	-	-	-	-	-	-	-	-	-	-	*Malzella evexa*
-	-	-	-	-	-	-	-	-	-	X	-	-	-	-	-	-	-	-	-	-	-	-	-	-	-	-	*Actinocythereis captionis* (large form)
-	-	-	-	-	-	-	-	-	-	-	-	-	X	-	-	-	-	-	-	-	-	-	-	-	-	-	*Murrayina barclayi*
-	-	-	-	-	-	-	-	-	-	-	-	-	X	-	-	-	-	-	-	-	-	-	-	-	-	-	*Murrayina martini*
-	-	-	-	-	-	-	-	-	-	X	X	-	-	-	-	-	-	-	-	-	-	-	-	-	-	-	*Orionina vaughni*
-	-	-	-	-	-	-	-	-	-	X	-	-	-	-	-	-	-	-	-	-	-	-	-	-	-	-	*Puriana carolinensis*
-	-	-	-	-	-	-	-	-	-	X	X	-	-	-	-	-	-	-	-	-	-	-	-	-	-	-	*Puriana rugipunctata*
-	-	-	-	-	-	-	-	X	-	-	-	-	-	-	-	-	-	-	-	-	-	-	-	-	-	-	*Puriana* sp.
-	-	-	-	-	-	-	-	X	-	-	-	-	-	-	-	-	-	-	-	-	-	-	-	-	-	-	*Pumilocytheridea* sp.
-	-	-	-	-	-	-	-	-	-	X	-	-	-	-	-	-	-	-	-	-	-	-	-	-	-	-	*Pseudocytheretta* (=Pseudocythereis) sp.
-	-	-	-	-	-	-	-	-	-	X	X	-	-	-	-	-	-	-	-	-	-	-	-	-	-	-	*Tetracytherura* (=Eocytheropterinae) *choctawhatcheensis*
-	-	-	-	-	-	-	-	-	-	-	X	-	-	-	-	-	-	-	-	-	-	-	-	-	-	-	*Muellerina* cf. *ohmerti*
-	-	-	-	-	-	-	-	X	-	X	X	-	-	-	-	-	-	-	-	-	-	-	-	-	-	-	*Bensonocythere* ssp.
-	-	-	-	-	-	-	-	-	-	-	X	-	-	-	-	-	-	-	-	-	-	-	-	-	-	-	*Balanus* ssp.
																											Chondrichphyes
-	-	-	-	-	-	-	-	-	-	-	-	-	-	-	-	-	-	-	-	-	-	-	-	X	X	-	*Lamna* sp.
-	-	-	-	-	-	-	-	-	-	-	-	-	-	-	-	-	-	-	-	-	-	-	-	X	X	-	*Galeocerdo* sp.
-	-	-	-	-	-	-	-	-	-	-	-	-	-	-	-	-	-	-	-	-	-	-	-	X	X	-	*Carcharodon* sp.
-	-	-	-	-	-	-	-	-	-	-	-	-	-	-	-	-	-	-	-	-	-	-	-	X	-	-	*Dasyatis* sp.
-	-	-	-	-	-	-	-	-	-	-	-	-	-	-	-	-	-	-	-	-	-	-	-	-	X	-	*Pastinacea* sp.
-	-	-	-	-	X	-	-	-	-	-	X	-	-	-	-	-	-	-	X	-	-	-	X	X	-	-	(shark ssp)
																											Reptilia
-	-	-	-	-	-	-	-	-	-	-	-	-	-	-	-	-	-	-	-	-	-	-	-	X	-	-	*Chelonia couperi*
-	-	-	-	-	-	-	-	-	-	-	-	-	-	-	-	-	-	-	-	X	-	-	-	X	-	-	*Terrapene canaliculata*
-	-	-	-	-	-	-	-	-	-	-	-	-	-	-	-	-	-	-	-	-	-	-	-	X	X	-	*Crocodylus* sp.
																											Mammalia
-	-	-	-	-	-	-	-	-	-	-	-	-	-	-	-	-	-	-	-	-	-	-	-	X	-	-	*Physter ? vetus* or *Physterula ? neolassicus*
-	-	-	-	-	-	-	-	-	-	-	-	-	-	-	-	-	-	-	-	-	-	-	-	X	X	-	*Castoroides ohioensis*
-	-	-	-	-	-	-	-	-	-	-	-	-	-	-	-	-	-	-	X	X	-	-	-	X	-	-	*Elephas columbi*
-	-	-	-	-	-	-	-	-	-	-	-	-	-	-	-	-	-	X	-	X	-	-	-	X	X	-	*Mammut americanum*
-	-	-	-	-	-	-	-	-	-	-	-	-	-	-	-	-	-	-	-	-	-	-	-	-	X	-	*Mammut floridanum*
-	-	-	-	-	-	-	-	-	-	-	-	-	-	-	-	-	-	-	-	X	-	-	-	X	X	-	*Bison* cf. *bison*
-	-	-	-	-	-	-	-	-	-	-	-	-	-	-	-	-	-	-	-	-	-	-	-	X	X	-	*Cervus* sp.
-	-	-	-	-	-	-	-	-	-	-	-	-	-	-	-	-	-	-	-	-	-	-	-	X	X	-	*Tapirus haysii*
-	-	-	-	-	-	-	-	-	-	-	-	-	-	-	-	-	-	-	-	-	-	-	-	-	X	-	*Megatherium americanum*
-	-	-	-	-	-	-	-	-	-	-	-	-	-	-	-	-	-	-	-	X	-	-	-	X	X	-	*Megatherium mirabile*
-	-	-	-	-	-	-	-	-	-	-	-	-	-	-	-	-	-	X	-	X	-	-	-	X	-	-	*Mylodon harlani*
-	-	-	-	-	-	-	-	-	-	-	-	-	-	-	-	-	-	-	-	-	-	-	-	X	-	-	*Chelonia couperi*
-	-	-	-	-	-	-	-	-	-	-	-	-	-	-	-	-	-	-	-	-	-	-	-	X	-	-	*Equus complicatus*
-	-	-	-	-	-	-	-	-	-	-	-	-	-	-	-	-	-	-	-	X	-	-	-	X	-	-	*Equus leidyi* (=fraternus)
-	-	-	-	-	-	-	-	-	-	-	-	-	-	-	-	-	-	-	-	-	-	-	-	X	X	-	*Equus littoralis*
-	-	-	-	-	-	-	-	-	-	-	-	-	-	-	-	-	-	-	-	-	-	-	-	-	X	-	*Equus tau?*
-	-	-	-	-	-	-	-	-	-	-	-	-	-	-	-	-	-	-	X	-	-	-	-	-	-	-	*Equus* sp.

FIG. 5.—Continued.

Q.	5.	4.	3.	P.2.	J.	I.	G.	F.	E.	D.	C.	B.	A.	S.	N.	M.	L.	3.	2.	T.1.	R.	P.1.	O.	K.	J.	H.	
-	-	-	-	-	-	-	-	-	-	-	-	-	-	-	-	-	-	-	-	-	-	-	-	-	X	-	*Cetacea* sp.
-	-	-	-	-	-	-	-	-	-	-	-	-	-	-	-	-	-	-	-	-	-	-	X	-	-	-	(mammal remains)
																											FLORA
																											Diatomaceae
-	-	-	-	-	-	-	-	-	-	-	-	-	-	-	-	-	-	-	-	-	-	X	-	-	-	-	*Cussia* sp.
-	-	-	-	-	-	-	-	-	-	-	-	-	-	-	-	-	-	-	-	-	-	-	-	-	-	X	*Rhaphoneis amphiceros*
-	-	-	-	-	-	-	-	-	-	-	-	-	-	-	-	-	-	-	-	-	-	X	-	-	-	-	*Rhaphoneis angularis*
-	-	-	-	-	-	-	-	-	-	-	-	-	-	-	-	-	-	-	-	-	-	-	-	-	-	X	*Rhaphoneis* cf. *angularis*
-	-	-	-	-	-	-	-	-	-	-	-	-	-	-	-	-	-	-	-	-	-	-	-	-	-	X	*Rhaphoneis rhombica*
-	-	-	-	-	-	-	-	-	-	-	-	-	-	-	-	-	-	-	-	-	-	-	-	-	-	X	*Rhaphoneis surirella*
-	-	-	-	-	-	-	-	-	-	-	-	-	-	-	-	-	-	-	-	-	-	-	-	-	-	X	*Actinocyclus ochotensis*
-	-	-	-	-	-	-	-	-	-	-	-	-	-	-	-	-	-	-	-	-	-	-	-	-	-	X	*Actinocuclus octonarius*
-	-	-	-	-	-	-	-	-	-	-	-	-	-	-	-	-	-	-	-	-	-	-	-	-	-	X	*Actinocyclus tenellus*
-	-	-	-	-	-	-	-	-	-	-	-	-	-	-	-	-	-	-	-	-	-	X	-	-	-	-	*Actinoptychus minutus*
-	-	-	-	-	-	-	-	-	-	-	-	-	-	-	-	-	-	-	-	-	-	-	-	-	-	X	*Actinoptychus senarius*
-	-	-	-	-	-	-	-	-	-	-	-	-	-	-	-	-	-	-	-	-	-	-	-	-	-	X	*Actinoptychus splendens*
-	-	-	-	-	-	-	-	-	-	-	-	-	-	-	-	-	-	-	-	-	-	-	-	-	-	X	*Aulacodiscus argus*
-	-	-	-	-	-	-	-	-	-	-	-	-	-	-	-	-	-	-	-	-	-	-	-	-	-	X	*Cocconeis sublittoralis*
-	-	-	-	-	-	-	-	-	-	-	-	-	-	-	-	-	-	-	-	-	-	-	-	-	-	X	*Coscinodiscus eccentricus*
-	-	-	-	-	-	-	-	-	-	-	-	-	-	-	-	-	-	-	-	-	-	-	-	-	-	X	*Coscinodiscus marsinatus*
-	-	-	-	-	-	-	-	-	-	-	-	-	-	-	-	-	-	-	-	-	-	-	-	-	-	X	*Coscinodiscus nitidus*
-	-	-	-	-	-	-	-	-	-	-	-	-	-	-	-	-	-	-	-	-	-	-	-	-	-	X	*Coscinodiscus perforatus*
-	-	-	-	-	-	-	-	-	-	-	-	-	-	-	-	-	-	-	-	-	-	-	-	-	-	X	*Coscinodiscus radiatus*
-	-	-	-	-	-	-	-	-	-	-	-	-	-	-	-	-	-	-	-	-	-	-	-	-	-	X	*Coscinodiscus stellaris*
-	-	-	-	-	-	-	-	-	-	-	-	-	-	-	-	-	-	-	-	-	-	-	-	-	-	X	*Cyclotella striata*
-	-	-	-	-	-	-	-	-	-	-	-	-	-	-	-	-	-	-	-	-	-	-	-	-	-	X	*Cyclotella* sp.
-	-	-	-	-	-	-	-	-	-	-	-	-	-	-	-	-	-	-	-	-	-	-	-	-	-	X	*Cymatosira cf. immunis*
-	-	-	-	-	-	-	-	-	-	-	-	-	-	-	-	-	-	-	-	-	-	-	-	-	-	X	*Cymatosira* sp.
-	-	-	-	-	-	-	-	-	-	-	-	-	-	-	-	-	-	-	-	-	-	-	-	-	-	X	*Diploneis bombus*
-	-	-	-	-	-	-	-	-	-	-	-	-	-	-	-	-	-	-	-	-	-	-	-	-	-	X	*Eupodiscus radiatus*
-	-	-	-	-	-	-	-	-	-	-	-	-	-	-	-	-	-	-	-	-	-	-	-	-	-	X	*Melosira granulata*
-	-	-	-	-	-	-	-	-	-	-	-	-	-	-	-	-	-	-	-	-	-	-	-	-	-	X	*Navicula clavata*
-	-	-	-	-	-	-	-	-	-	-	-	-	-	-	-	-	-	-	-	-	-	-	-	-	-	X	*Navicula hennedyii*
-	-	-	-	-	-	-	-	-	-	-	-	-	-	-	-	-	-	-	-	-	-	-	-	-	-	X	*Nitzschia angularis*
-	-	-	-	-	-	-	-	-	-	-	-	-	-	-	-	-	-	-	-	-	-	-	-	-	-	X	*Nitzschia granulata*
-	-	-	-	-	-	-	-	-	-	-	-	-	-	-	-	-	-	-	-	-	-	-	-	-	-	X	*Nitzschia plana*
-	-	-	-	-	-	-	-	-	-	-	-	-	-	-	-	-	-	-	-	-	-	-	-	-	-	X	*Opephora* sp.
-	-	-	-	-	-	-	-	-	-	-	-	-	-	-	-	-	-	-	-	-	-	-	-	-	-	X	*Paralia sulcata*
-	-	-	-	-	-	-	-	-	-	-	-	-	-	-	-	-	-	-	-	-	-	-	-	-	-	X	*Podosira stelliger*
-	-	-	-	-	-	-	-	-	-	-	-	-	-	-	-	-	-	-	-	-	-	-	-	-	-	X	*Thalassiosira* spp.
-	-	-	-	-	-	-	-	-	-	-	-	-	-	-	-	-	-	-	-	-	-	-	-	-	-	X	*Triceratium farus*
-	-	-	-	-	-	-	-	-	-	-	-	-	-	-	-	-	-	-	-	-	-	-	-	-	-	X	*Triceratium* sp.
-	-	-	-	-	-	-	-	-	-	-	-	-	-	-	-	-	-	-	-	-	-	X	-	-	-	-	*Biddulphia seticulosa*

FIG. 5.—Continued.

southeastern Coastal Plain revived: Richards (1950); DuBar (1962, 1971, 1974); Pooser (1965); Hoyt and Hails (1967, 1974); Herrick (1961, 1964, 1965, 1976); Colquhoun (1965, 1974); Colquhoun and Pierce (1971); Colquhoun and Brooks (1986); and Colquhoun and others (1968, 1987). Each of these studies added data from, and interpretation of, Pliocene and Pleistocene sediments in the region.

In the late 1970s numerous researchers began using combinations of paleontologic, geomagnetic, chemical, and isotopic analyses to date and correlate Pliocene and Pleistocene deposits in the southeastern Atlantic Coastal Plain, resulting in time-stratigraphic data for much of the Cape Fear area of the Carolinas (Akers, 1972; Akers and Koeppel, 1973; Liddicoat and others, 1979, 1981; Cronin and Hazel, 1980; Liddicoat, 1982; McCartan and others, 1982, 1984; Wehmiller and Belknap, 1982; Cronin and others, 1984; Colquhoun and Brooks, 1986; Huddlestun, 1988; Ward and Huddlestun, 1988; Wehmiller and others, 1988; Owens, 1989; Dowsett and Poore, 1990; McCartan and others, 1990; Cronin, 1990; and others). The lack of fossil material in Pliocene and Pleistocene sediments in southeastern South Carolina and Georgia, however, has resulted in relatively few late Cenozoic time-stratigraphic data for the Southeast Georgia embayment.

STRATIGRAPHY

A compilation of identified Pliocene and early Pleistocene units in southeastern Virginia, North Carolina, South Carolina, and Georgia is given in Figure 6. Time-stratigraphic data for Pliocene and Pleistocene sedimentary sequences in southeastern Georgia are insufficient to make definitive correlations with named units in other Atlantic Coast states. Some isolated deposits of fossiliferous Pliocene sediments, however, have been biostratigraphically correlated to named units in the Carolinas and in southeastern Virginia.

In the following sections we briefly discuss the stratigraphy of Pliocene and Pleistocene sediments in the Atlantic Coastal Plain of southeastern South Carolina and Georgia. We give a general description of the sediments and discuss some of the similarities and differences between probable age-equivalent deposits in this area and in the Cape Fear region of the Carolinas. Due to the paucity of fossil data, the lack of isotopic, geomagnetic, and biogeochemical-age data, and the limited geologic mapping, we do not at this time suggest the use of specific formation names for Pliocene and Pleistocene sedimentary sequences in this part of the Southeast Georgia embayment. Current and future mapping and age determinations in the area should result in a stratigraphic framework that can be compared to the framework already established for the Cape Fear region of the Carolinas (McCartan and others, 1982, 1984, 1990; Owens, 1989).

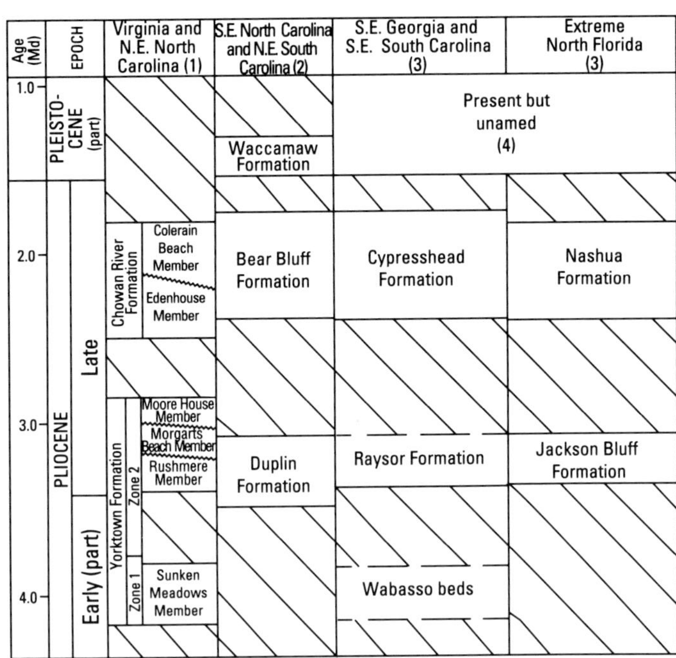

(1) T.M. Cronin (written communication, 1990), L.W. Ward (written communication, 1990)
(2) McCartan and others (1982, 1984), Owens (1989)
(3) Huddlestun (1988), Ward and Huddlestun (1988)
(4) Satilla Formation of Huddlestun (1988) and Veatch and Stephenson (1911)

FIG. 6.—Pliocene and Pleistocene stratigraphy from southeastern Virginia to extreme northern Florida.

Pliocene

Wabasso Beds.—

Three different ages of Pliocene marine sediments have been paleontologically identified in the Atlantic Coastal Plain of Georgia. The oldest sediments include variably phosphatic and calcareous sand with intermittent clay beds. Available samples are from cores (18–23 m [60–75 ft] depths) from several of the coastal islands from Beaufort, South Carolina to Little Tybee Island, Georgia (Fig. 4, loc. P and Q). Woolsey (1976) and Herrick (1976) considered these beds to be a facies of the early late Pliocene Duplin Formation. Huddlestun (1988) informally named the sediments the Wabasso beds and suggested that the foraminifera were indicative of a fully open-marine environment and a late early Pliocene age. The presence of *Globigerina nepenthes* in these sediments suggests an age no younger than 4.2–4.0 Ma. The presence of *Globigerina margaritae margaritae* suggests that the Wabasso beds are no older than about 5.7 Ma (J. E. Hazel, pers. commun., 1991). Huddlestun (1988, p. 100) concluded that "the co-occurrence of *Globigerina nepenthes* and *Globigerina margaritae margaritae* is indicative of Zone PL1 of Berggren (1973)."

The age range and environment of deposition of the Wabasso beds suggest that these sediments represent a major late early Pliocene transgression, and that they are probably time equivalent to the Sunken Meadows Member of the Yorktown Formation of southeastern Virginia and northeastern North Carolina (Fig. 6). Available data, however, do not preclude the possibility that the Wabasso beds are upper Miocene.

Duplin and Raysor Formations.—

The Duplin Formation was first described as the Duplin beds of Miocene age from an area near the Cape Fear River in Duplin County, North Carolina (Dall, 1898). Veatch and Stephenson (1911) first used the name in Georgia. Cooke (1936) proposed the name Raysor Marl for sediments in southeastern South Carolina thought to be older than Duplin; he later abandoned the name Raysor and included the sediments in the Duplin (Cook, 1945). Mansfield (1944) thought that Duplin sediments south of central North Carolina were equivalent to the youngest part of his uppermost unit (zone 2) of the Yorktown Formation in southern Virginia.

In the middle 1970s the Duplin was recognized to be definitely Pliocene (Akers, 1972; Woolsey, 1976). The name Raysor was reinstated by Blackwelder and Ward (1979), who at the same time abandoned the name Duplin and replaced it with the Yorktown Formation in eastern North Carolina and South Carolina, with the Raysor in south-central and southeastern Georgia, and with the Jackson Bluff Formation (Puri and Vernon, 1964) in Florida. Owens (1989), citing the need for more detailed mapping of the area between southern Virginia (type area of the Yorktown) and the Cape Fear region, reinstated the name Duplin for southeastern North Carolina and northeastern South Carolina.

Huddlestun (1988) and Ward and Huddlestun (1988) assigned an early late Pliocene age to the Raysor, which they considered consistent with Zone PL3 of Berggren (1973). Although the Duplin and Raysor Formations may, and probably do, include sediments deposited during several depositional cycles, these cycles cannot presently be differentiated either by paleontologic analysis or by stratigraphic position. Therefore, we consider sediments in southeastern Georgia that contain fossils consistent with Zone PL3 of Berggren (1973) to be equivalent wholly or in part to the Duplin and Raysor Formations and to the Rushmere Member of the upper part of the Yorktown Formation (Ward and Huddlestun, 1988; Cronin, and others, 1984; T. M. Cronin, pers. commun., 1990; L. W. Ward, pers. commun., 1990), and to the Jackson Bluff Formation of the eastern Gulf Coastal Plain of Florida (Huddlestun, 1988) (Fig. 6).

As applied in Georgia (Veatch and Stephenson, 1911), the Duplin was restricted to the unconformably bound, tan to white marl, shells, and clay exposed along the right bank of the Savannah River between Porters Landing and Cedar Bluff Landing about 80 km (50 mi) upstream from the river mouth (Fig. 4, loc. C). Veatch and Stephenson (1911) also referred to Duplin fossils from dredged spoil near the mouth of the Altamaha River (Fig. 4, loc. J and K) in southeastern Georgia (also discussed by Darby and Hoyt, 1964) and from an outcrop of organic-rich clay near Doctorstown, about 64 km (40 mi) upstream from the mouth of the Altamaha River (Fig. 4, loc. F). Recently, early late Pliocene fossils have been identified from organic-rich sand present as spoil adjacent to a stock pond in Bulloch County, Georgia, about 80 km (50 mi) upstream from the mouth of the Ogeechee River (Fig. 4, loc. D). The fossiliferous Raysor Formation has been reported from 15 to 16 m (49–52 ft) depth in the Chatham 1 core (Fig. 4 , loc. E) in eastern Chatham County, where it was described as a "richly foraminiferal, phosphatic, argillaceous, finely sandy, calcarenitic limestone" (Huddlestun, 1988, p. 114).

The majority of probable early late Pliocene sediments in Georgia are nonfossiliferous. The shell beds between Cedars Landing and Porters Landing on the Savannah River, in Bulloch County along the Ogeechee River, and at Doctorstown on the Altamaha River, crop out between 26 and 30 m (85 and 100 ft) in altitude and can be traced upstream for several kilometers where they grade laterally into fine- to medium-grained, well-sorted, well-rounded, quartz sand with thin lenses of carbonaceous, micaceous silt and clay. Also, a fine-grained, well-sorted, well-rounded, quartzose marine sand crops out between 46 and 64 m (150 and 210 ft) in altitude in the highly dissected terrain between the Ogeechee and Savannah Rivers, and at the surface and in the shallow subsurface in south-central Georgia, near the apex or western limit of the Southeast Georgia embayment. In the area between the Ogeechee and the Savannah Rivers, the marine sand is unconformably overlain by cross-bedded, medium- and coarse-grained, subangular, fluvial quartz sand and pebbles. In south-central Georgia, the marine sand is exposed at the surface, except near the Ocmulgee and Altamaha Rivers, where it is overlain by fluvial sands and gravels.

Lack of exposures, limited subsurface data, and truncation by younger geomorphic features (i.e., wave-cut scarps) prohibit definitive correlation, but we tentatively suggest that: (1) the fossil-rich early late Pliocene sediments that crop out along the Savannah and the Altamaha Rivers grade updip into a nonfossiliferous marine sand; and (2) the outcropping nonfossiliferous marine sand present throughout southeastern Georgia (exposed at altitudes between 46 and 64 m [150–210 ft]) is the updip equivalent of the isolated fossil-bearing early late Pliocene deposits identified on Figure 4.

The Cypresshead and Bear Bluff Formations.—

Latest late Pliocene sediments that are stratigraphically above the Duplin Formation have been referred to as the Bear Bluff Formation in the Carolinas (Dubar, 1971; Dubar and others, 1974; McCartan and others, 1982; Owens, 1989) and the Cypresshead Formation by Huddlestun (1988) in Georgia. The Bear Bluff Formation was defined by DuBar (1971) from a locality in northeastern South Carolina; its definition was later modified by Dubar and others (1974) and again by Owens (1989) to include a variety of marine facies. It is commonly fossiliferous, especially in channel deposits where fossil-rich carbonate sediments can be 20 to 45 m thick.

In Georgia, Huddlestun (1988, p. 119) gave the name Cypresshead Formation to fossil-poor sediments that are stratigraphically above his Raysor Formation (Fig. 6). He described the Cypresshead as "a prominently thin- to thick-bedded and massive, planar- to cross-bedded, variably burrowed and bioturbated, fine-grained to pebbly, coarse-grained sand formation in the terrace region of eastern Georgia." He called the Cypresshead a regionally extensive unit composed primarily of quartz sand with prominent clay beds that are locally dominant.

The mineralogy of latest late Pliocene sediments in southeastern South Carolina and Georgia is variable. Sediments are arkosic near the Savannah and Altamaha Rivers and quartzose away from the rivers. That part of the Cypresshead Formation between 30 and 49 m (100–160 ft) in altitude includes quartzose to arkosic fluvial marine, back-barrier, barrier, and shelf sediments and is probably lithostratigraphically and chronostratigraphically equivalent to the Bear Bluff Formation, as described by Owens (1989).

McCartan and others (1982) placed the age of the Bear Bluff Formation in east-central Carolina between 2.4 and 1.8 Ma. Huddlestun (1988) reported a planktonic foraminiferal assemblage from the Nashua Formation (probable Cypresshead equivalent) in northeastern Florida just south of the Georgia line (Fig. 4, loc. L). He considered the assemblage to be equivalent to Zone PL5 of Berggren (1973) and to the Bear Bluff Formation in northeastern South Carolina (Fig. 6).

No age data are available for Cypresshead sediments in Georgia outcrops. Subsurface microfossil data are from pods or discontinuous lenses of sediment overlain by Pleistocene barrier and/or back-barrier deposits. These data include: (1) a small assemblage of juvenile planktonic foraminifera from depths of 16.3 to 17.0 m (53.5–56.0 ft) in a core from Wayne County, Georgia (Fig. 4, loc. N), which suggests an age no younger than Pliocene based upon the presence of *Glogigerina apertura* and *Globigerinoides obliquus* (Huddlestun, 1988); (2) the presence of the benthic foraminifer *Virgulinella gunteri* between 12 and 14 m (39 and 45 ft) depths in the Chatham 14 core (Fig. 4, loc. S) in eastern Chatham County, also indicating a Pliocene age; and (3) a diatom assemblage at 14 to 18 m (45–59 ft) depths in the Bryan 1 core (Fig. 4, loc. H) in eastern Bryan County, suggesting a probable latest late Pliocene to early Pleistocene age. The assemblage consists of fairly typical Pleistocene forms with possible Pliocene forms in the genus *Rhaphoneis*.

Assuming that the samples from these cores represent the same stratigraphic unit, then the age of the Cypresshead in southeastern Georgia ranges from the latest late Pliocene into early Pleistocene. At present, however, we consider the Cypresshead to be latest late Pliocene and age equivalent (wholly or in part) to the Bear Bluff Formation in northeastern South Carolina (Fig. 6).

Pleistocene

Pleistocene sediments in southeastern Georgia comprise sequences of largely nonfossiliferous fluvial marine, back-barrier, barrier, and shallow-shelf sand with minor amounts of clay. Generally, the sediments are micaceous, burrowed, and highly weathered. Historically, Pleistocene units have been differentiated on the basis of concepts and/or models that relate mode of deposition to preserved landform (Cooke, 1930a, b, 1931, 1943; Hoyt, 1967; Hoyt and Hails, 1974). Veatch and Stephenson (1911) proposed that fluvial and marine Pleistocene sediments in southeastern Georgia be referred to as the Satilla Formation, a name of discontinued use in succeeding years. Herrick (1965) thought that Pleistocene sediments in southeastern Georgia belonged to one deltaic sequence and should include all near-surface and surface sediments below 82 m (270 ft). He did not name the unit. Huddlestun (1988) also concluded that there should be a "one formation" designation for all Pleistocene units. He reintroduced the name Satilla Formation, restricted it to marine and marginal-marine sediments, and expanded the age range to include both Pleistocene and Holocene deposits. Recent field work by the first author suggests an alternative to the "one-formation" concept that more closely agrees with the stratigraphy given by McCartan and others (1984, 1990) for the Charleston area of South Carolina.

Paleontologic, isotopic, and paleomagnetic data are available for Pleistocene sediments in the Cape Fear area of the Carolinas northeast of the Edisto River (Richards, 1936, 1943; DuBar and Chaplin, 1963; DuBar and Furbunch, 1965; Colquhoun and others, 1968; Liddicoat and others, 1979, 1981; Liddicoat, 1982; McCartan and others, 1982, 1984; Szabo, 1985; Owens, 1989). Few fossil data are available from probable age-equivalent sediments between the Edisto and St. Marys Rivers. Richards (1969, p. 9) commented on the lack of Pleistocene fossils from Georgia: "While only 15 species (of mollusks) have been found in the Pleistocene deposits of Georgia, mostly near Savannah, a much more extensive fauna is known to occur in both South Carolina and Florida, and presumably lived in the Pleistocene seas of Georgia." Despite numerous investigations since the 1960s, only a few additional Pleistocene fossil localities have been identified in this area.

Published data on Pleistocene invertebrate fossils in Georgia include well locations and locations of drainage ditches and road cuts from in and near Savannah. Richards (1969) commented that by that time no fossils could be seen in the ditches near Savannah, but fossils could be seen along the Skidaway River southeast of Savannah. Fossiliferous outcrops, which are currently being studied, may be age equivalent to those along the Skidaway River. Data on Pleistocene invertebrate fossils are given in Figure 5.

Pleistocene sediments near the coast in southeastern Georgia have yielded a variety of vertebrate fossils (a partial list is given in Fig. 5). Cooke (1943) suggested that the bone beds recognized in Georgia might be equivalent to what was then referred to as the Melborne bone bed in Florida. In the Savannah area, bones of a giant ground sloth (*Megatherium*) were found in 1823 and reported by Hodgson (1846) and Lyell (1855). Hay (1923) included a faunal list based upon fossil bones from the Brunswick Canal. Cooke (1943) reprinted Hay's species list, which included giant beaver, elephant, mastodon, buffalo, deer, tapir, horses, ground sloths, crocodiles, and several types of fish. Cooke (1943) noted that several of the species found at Brunswick, and some not reported from Brunswick, such as the box tortoise *Terrapene canaliculata,* had been reported from the Savannah area. Cooke (1943) also mentioned a fossil-bone locality at Hayners Bridge near Savannah, 4 km (2.5 mi) west of Isle of Hope, that contained *Mammut americanum* (mastodon) and *Mylodon harlani* (sloth). Hurst (1957) summarized known occurrences, to that date, of vertebrate fossils from coastal Georgia and included fossils from the Brunswick canal.

Differences in vertebrate and invertebrate faunal and/or floral assemblages have not been sufficient to differentiate Pleistocene units. Differences in species abundance and shell morphology are evident in foraminifera associated with sedimentary sequences above and below 9 m (30 ft) in altitude. Foraminifera from surface and near-surface sediments located below 9 m (30 ft) are more delicate, glassy, and less encrusted than the thicker, more calcitic foraminifera associated with older Pleistocene and late Pliocene sediments. The work of McCartan and others (1982, 1984, 1990) and Owens (1989) suggests that deposits associated with surfaces between 6 and 9 m (20 and 30 ft) in altitude in the South Carolina Coastal Plain are about 200 ka. Weathering and soil-profile data from similar deposits, in the same altitude range, in the Savannah area suggest an age around 500 ka. More data are needed to determine the age(s) of these sediments in Georgia. Until such data are available, changes in species abundance and shell morphology of foraminifera can be used to differentiate between Pleistocene sediments older and younger than 500 ka.

DISCUSSION

Herrick (1965, p. 6) observed: "Except for the extreme coastal area, the Pleistocene deposits of Georgia are uniformly nonfossiliferous." Despite numerous investigations since then, that statement still stands. It is also true of the Pliocene, except possibly for the early late Pliocene sediments. The question is, "Why?". In this section we suggest some factors that may control or contribute to the differences in depositional history and fossil content of age-equivalent Pliocene and Pleistocene sediments in the Atlantic Coastal Plain of the Carolinas and Georgia.

Local Drainages

Outcrop and subsurface data from the latest late Pliocene Cypresshead Formation and younger sediments in Georgia indicate a significant freshwater or fluvial influence on unit lithology and faunal composition. As discussed by Hayes (1989), the coastal area of southeastern South Carolina and Georgia form the head or apex of the Georgia Bight, the arcuate stretch of coastline that extends from Cape Hatteras, North Carolina, to Cape Canaveral, Florida. The head of the bight is locally referred to as the Sea Islands area and extends from the Edisto River, South Carolina, to the St. Marys River on the Florida/Georgia line (Fig. 3), which is roughly the width of the Southeast Georgia embayment (Fig. 1). Compared to the flanks of the Georgia Bight, the Sea Islands area has an order of magnitude higher freshwater discharge and suspended-sediment influx. This is manifest in the high turbidity of coastal waters and the "brown" beaches that are common to the Sea Islands area. The relatively low-wave height and high-tidal range of this area also contribute to the high turbidity of nearshore waters. We suggest that from the latest late Pliocene to the present, the combination of high turbidity and relatively low salinity (particularly at time of high discharge from the region's rivers) has restricted the number and variety of invertebrate marine fauna common to the area.

Regional and Local Structures

We also suggest that the large fluvial input to the area's ecosystems has, in large part, been controlled or affected by slow uplift associated with local and regional geologic structures (Winker and Howard, 1977; Cronin, 1981; Markewich, 1985; Markewich and others, 1986; Soller, 1988; Dowsett and Cronin, 1990). Although sediment load cannot be directly related to warping or regional uplift, styles and rates of deformation of both regional and local structures can affect stream orientation, drainage density, shoreline configuration, and direction of longshore currents.

In the southeastern Atlantic Coastal Plain, prominent regional structures that have affected Cenozoic sediment distribution include the Cape Fear arch, Peninsular arch, and the Southeast Georgia embayment (Fig. 1; Gohn, 1988). Smaller structures, not identified on Figure 1, have also been important. The Beaufort arch is a coastal "high" that has affected the distribution of Cenozoic sediments in the area between Beaufort, South Carolina, and Savannah, Georgia (Heron and Johnson, 1966; Colquhoun and others, 1969; Woolsey, 1976). Data suggest that these structures have also influenced local drainages throughout the Pliocene and Pleistocene. The southward migration of the Pee Dee and Savannah Rivers throughout the Pleistocene and Holocene (Markewich, 1985, and in prep.; and Soller, 1988) has been in response to the Cape Fear and the Beaufort arches, respectively. The Beaufort arch may also have deflected south-flowing longshore currents seaward. Both the southward migration of the rivers and the deflection of currents would have maintained, and/or increased through time, the length of barrier-protected shoreline that is dominated by fine-grained, micaceous, fluvial marine sediments.

Erosion

Another possible reason for the lack of fossiliferous material in Pliocene and Pleistocene sediments of southeastern South Carolina and Georgia is the pattern of marine erosion. Rarely are there stacked sequences of Pliocene and/or Pleistocene sediments in this part of the emergent southeastern Atlantic Coastal Plain. There are no known Pleistocene fossil localities seaward of Pleistocene barriers in this part of the Georgia Bight. Therefore, there are no data on the assemblages of marine invertebrates indigenous to the shallow shelf during the Pleistocene. This lack of Pleistocene open-marine, nearshore marine invertebrates suggests either that the invertebrates were never present, or that they were removed by a subsequent transgression(s).

Weathering

Weathering has greatly affected the preservation of shells in the near-surface sediments. Minimum depths of oxidation range from about 3 m (10 ft) in marine sand younger than 50 ka to about 12 m (40 ft) in uppermost Pliocene fluvial marine sands and clayey sands, to about 19 m (60 ft) in early late Pliocene marine sands. Most of the area's surface and near-surface waters have a low pH (3.5–5.5) and a high organic content (unpublished data). In this area, there is no near-surface limestone to buffer the low-pH shallow ground water. Modification of the pH of near-surface ground water and/or protection from the effects of ground water are needed for preservation of fossils. Most shell-bearing sediments were originally deposited near the surface, or they are remnant of older units that have been truncated and shallowly buried. The shallow depth of burial has resulted in dissolution of the shells or local dolomitization. Dissolution can occur quickly once the sediments are above the local water table. Individual thin halves of shells or "ghosts" of shells that can be seen in drainage ditches when first dug are often not evident six months later.

Shoreline Evolution

As seen in Figure 5, most fossiliferous material has been collected from early late Pliocene sediments. We do not believe that this is an artifact of sampling. Field evidence suggests that both the shoreline configuration and the spatial distribution of streams discharging into the Atlantic were different in early late Pliocene time. We suggest that: (1) the drainage density of the eastern part of the Atlantic Coastal Plain was less in the early late Pliocene than during the latest late Pliocene and Pleistocene; (2) the shoreline distance between the paleodeltas of the region's large rivers (such as the Pee Dee and the Savannah) was greater in the early late Pliocene than in the latest late Pliocene and Pleistocene; (3) in the early late Pliocene the Atlantic Ocean and Gulf of Mexico were connected by a shallow platform across south-central Georgia and north-central Florida; and (4) when the Atlantic and Gulf were connected, the Oconee and Ocmulgee rivers (Fig. 3) had not yet joined to form the Altamaha River. Some of these ideas date to the turn of the century. Some are new. They are presently being tested by ongoing field mapping and sample analyses.

Data are insufficient to comment on the paleogeography of the late early Pliocene. The Wabasso beds provide the only record of that period in this area of the Southeast Georgia embayment.

SUMMARY

Pliocene and Pleistocene sediments in southeastern Georgia are largely nonfossiliferous barrier and back-barrier fine- and very fine-grained quartz sand with minor amounts of silt and clay. Emerged, repetitious sedimentary sequences are embayed one into another and define a "stepped" low-altitude, low-relief terrain characterized by barrier ridge, back-barrier flats, and shallow-shelf plains. Some of the sedimentary sequences have been recognized as formations, but few data are available on which correlations can be made with probable age-equivalent sediments in the Cape Fear area of the Carolinas. The fossil-poor character of the sediments is largely the result of environmental influences that have been active in the area since the latest Pliocene. These influences include: (1) a large freshwater influence resulting from the numerous rivers that empty into this part of the Atlantic Coast; (2) the effects of local and regional geologic structures on the area's drainages and nearshore currents; (3) the erosive nature of Pliocene and Pleistocene transgressive events; and (4) the intensity and rapidity of weathering. Sediment type and fossil content have also been affected by changes in shoreline configuration, drainage pattern, and drainage density during the Pliocene and Pleistocene.

ACKNOWLEDGMENTS

The authors extend their thanks to the following individuals for their cooperation and help: T. M. Cronin, U.S. Geological Survey, Reston, Virginia, identified ostracodes from recently sampled localities in the area; L. W. Ward, Curator of Invertebrate Paleontology, Virginia Museum of Natural History, Martinsville, identified many of the macrofossils; W. H. Abbott, formerly with Mobile Oil Corporation, Houston, Texas, provided unpublished data on diatoms; J. E. Hazel, Louisiana State University, Baton Rouge, freely discussed the difficulties in identifying lower Pliocene sediments in the southeastern Atlantic Coastal Plain; J. P. Owens, U.S. Geological Survey, Reston, Virginia, and D. J. Colquhoun, University of South Carolina, Columbia, shared their knowledge of the Pliocene and Pleistocene of the area. Lyle Campbell of the University of South Carolina, Spartenburg, freely discussed the taxonomy and stratigraphy of marine invertebrates in the southeastern Atlantic Coastal Plain. We also thank T. M. Cronin, J. P. Owens, and E. D. Kuzmin for their constructive suggestions on earlier versions of the manuscript and M. J. Hogan for drafting the included figures.

REFERENCES

AKERS, W. H., 1972, Planktonic foraminifera and biostratigraphy of some Neogene formations, northern Florida and Atlantic Coastal Plain: Tulane Studies in Geology and Paleontology, v. 9, p. 1–139.

AKERS, W. H., AND KOEPPEL, P. E., 1973, Age of some Neogene formations, Atlantic Coastal Plain, United States and Mexico, *in* Smith, L. A., and Hardenbol, J., eds., Proceedings of Symposium on Calcareous Nannofossils, Gulf Coast Association of Geological Sciences and Gulf Coast Section, Society of Economic Paleontologists and Mineralogists, 23rd Annual Convention, Houston, Texas, 1973, p. 80–93.

ALDRICH, T. H., 1911, Notes on some Pliocene fossils from Georgia, with descriptions of new species: Nautilus, v. 24, p. 131–132 and 138–140.

BERGGREN, W. A., 1973, The Pliocene time scale: calibration of planktonic foraminiferal and calcareous nannoplankton zones: Nature, v. 243, p. 391–397.

BLACKWELDER, B. W., AND WARD, L. W., 1979, Stratigraphic revision of the Pliocene deposits of North and South Carolina: South Carolina State Development Board, Division of Geology, Geologic Notes, v. 23, p. 33–49.

CLARK, W. B., MILLER, B. L., STEPHENSON, L. W., JOHNSON, B. L., AND PARKER, H. N., 1912, The Coastal Plain of North Carolina: North Carolina Geologic and Economic Survey, v. 3, 552 p.

COLQUHOUN, D. J., 1965, Terrace sediment complexes in central South Carolina: Atlantic Coastal Plain Geological Association Field Conference Guidebook, University of South Carolina, Columbia, 62 p.

COLQUHOUN, D. J., 1974, Cyclic surficial stratigraphic units of the middle and lower Coastal Plains, central South Carolina, *in* Oaks, R. Q., Jr., and DuBar, J. R., eds., Post-Miocene Stratigraphy, Central and South Atlantic Coastal Plain: Utah State University Press, Logan, p. 179–190.

COLQUHOUN, D. J., AND BROOKS, M. J., 1986, New evidence from the southeastern U.S. for eustatic components in the late Holocene sea levels: Geoarchaeology: An International Journal, v. 1, p. 275–291.

COLQUHOUN, D. J., FRIDDELL, M. S., WHEELER, W. H., DANIELS, R. B., GREGORY, J. P., MILLER, R. A., AND VAN NOSTRAND, A. K., 1987,

Quaternary Geology, Savannah 4° × 6° Quadrangle, United States, *in* Richmond, G. M., Fullerton, D. S., and Weede, D. L., eds., Quaternary Geologic Atlas of the United States: U.S. Geological Survey Miscellaneous Investigations Series Map I-1420 (NI-17).

Colquhoun, D. J., Heron, S. D., Jr., Johnson, H. S., Jr., Pooser, W. K., and Siple, G. E., 1969, Up-dip Paleocene-Eocene stratigraphy of South Carolina reviewed: South Carolina State Development Board, Division of Geology, Geologic Notes, v. 13, p. 1–26.

Colquhoun, D. J., Herrick, S. M., and Richards, H. G., 1968, A fossil assemblage from the Wicomico Formation in Berkeley County, South Carolina: Geological Society of America Bulletin, v. 79, p. 1211–1220.

Colquhoun, D. J., and Pierce, W., 1971, Pleistocene transgressive-regressive sequences on the Atlantic Coastal Plain: Quaternaria, v. 15, p. 35–50.

Cooke, C. W., 1930a, Correlation of coastal terraces: Journal of Geology, v. 38, p. 577–589.

Cooke, C. W., 1930b, Pleistocene shorelines: Washington Academy of Sciences Journal, v. 20, p. 389–589.

Cooke, C. W., 1931, Seven coastal terraces in the southeastern states: Washington Academy of Sciences Journal, v. 21, p. 503–513.

Cooke, C. W., 1936, Geology of the Coastal Plain of South Carolina: U.S. Geological Survey Bulletin 867, 196 p.

Cooke, C. W., 1943, Geology of the Coastal Plain of Georgia: U.S. Geological Survey Bulletin 941, 121 p.

Cooke, C. W., 1945, Geology of Florida, Florida Geological Survey Bulletin 29, 339 p.

Cooke, C. W., and Mossom, S., 1929, Geology of Florida, Florida Geological Survey 20th Annual Report, 1927–1928, p. 29–227.

Cronin, T. M., 1981, Rates and possible causes of neotectonic vertical crustal movements of the emerged southeastern United States Atlantic Coastal Plain: Geological Society of America Bulletin, v. 92, p. 812–833.

Cronin, T. M., 1990, Evolution of Neogene and Quaternary marine Ostracoda, United States Atlantic Coastal Plain: evolution and speciation in Ostracoda, IV: U.S. Geological Survey Professional Paper 1367-C, 43 p.

Cronin, T. M., Bybell, L. M., Poore, R. Z., Blackwelder, B. W., and Liddicoat, J. C., 1984, Age and correlation of emerged Pliocene and Pleistocene deposits, U.S. Atlantic Coastal Plain: Palaeogeography, Palaeoclimatology, Palaeoecology, v. 47, p. 21–51.

Cronin, T. M., and Hazel, J. E., 1980, Ostracode biostratigraphy of Pliocene and Pleistocene deposits of the Cape Fear Arch region, North and South Carolina, *in* Shorter contributions to paleontology, 1979: U.S. Geological Survey Professional Paper 1125-B, 25 p.

Dall, W. H., 1896, Diagnosis of new Tertiary fossils from the southern United States: U.S. National Museum Proceedings, v. 18, p. 31–46.

Dall, W. H., 1898, A table of North American Tertiary horizons, correlated with one another and those of Europe with annotations: U.S. Geological Survey 18th Annual Report, part 2, p. 323–348.

Darby, D. G., and Hoyt, J. H., 1964, An upper Miocene fauna dredged from tidal channels of coastal Georgia: Journal of Paleontology, v. 38, p. 67–73.

Dowsett, H. J., and Cronin, T. M., 1990, High eustatic sea level during the middle Pliocene: evidence from the southeastern U.S. Atlantic Coastal Plain: Geology, v. 18, p. 435–438.

Dowsett, H. J., and Poore, R. Z., 1990, A new planktic foraminifera transfer function for estimating Pliocene–Recent paleoceanographic conditions in the North Atlantic: Marine Micropaleontology, v. 16, p. 1–23.

DuBar, J. R., 1962, New radiocarbon dates for the Pamlico Formation in South Carolina and their stratigraphic significance: South Carolina Development Board, Division of Geology, Geologic Notes, v. 6, p. 21–24.

DuBar, J. R., 1971, Neogene stratigraphy of the lower Coastal Plain of the Carolinas: Atlantic Coastal Plain Geological Association, 12th Annual Field Conference Guidebook, Myrtle Beach, South Carolina, 128 p.

DuBar, J. R., 1974, Summary of the Neogene stratigraphy of southern Florida, *in* Oaks, R. Q., and DuBar, J. R., eds., Post-Miocene Stratigraphy, Central and Southern Atlantic Coastal Plain: Utah State University Press, Logan, p. 206–231.

DuBar, J. R., and Chaplin, J. R., 1963, Paleoecology of the Pamlico Formation (late Pleistocene), Nixonville Quadrangle, Horry County, South Carolina: Southeastern Geology, v. 4, p. 127–165.

DuBar, J. R., and Furbunch, H. W. C., 1965, The Waccamaw Formation (Pliocene?) and its microfauna, Intracoastal Waterway, Horry County, South Carolina: South Carolina Development Board, Division of Geology, Geologic Notes, v. 9, p. 1–24.

DuBar, J. R., Johnson, H. S., Jr., Thom, B., and Hatchell, W. O., 1974, Neogene stratigraphy and morphology, south flank of the Cape Fear arch, North and South Carolina, *in* Oaks, R. Q., Jr., and DuBar, J. R., eds., Post-Miocene Stratigraphy, Central and Southern Atlantic Coastal Plain: Utah State University Press, Logan, p. 139–173.

Edwards, R. A., 1944, Ostracoda from the Duplin marl (upper Miocene): Journal of Paleontology, v. 18, p. 505–528.

Gohn, G. S., 1988, Late Mesozoic and early Cenozoic geology of the Atlantic Coastal Plain: North Carolina to Florida, *in* Sheridan, R. E., and Grow, J. A., eds., The Geology of North America, The Atlantic Continental Margin, U.S.: Geological Society of America, Decade of North American Geology, v. I-2, p. 107–130.

Hay, O. P., 1923, The Pleistocene of North America and its vertebrated animals from the sites east of the Mississippi River and from the Canadian Provinces east of longitude 95°: Carnegie Institute of Washington Publication 322, 307 p.

Hayes, M. O., 1989, Modern clastic depositional environments, South Carolina: 28th International Geological Congress Field Trip Guidebook T371, Charleston to Columbia, South Carolina: American Geophysical Union, Washington, D.C., 85 p.

Heron, S. D., and Johnson, H. S., Jr., 1966, Clay mineralogy, stratigraphy, and structural setting of the Hawthorn Coosawatchie District: Southeastern Geology, v. 7, p. 51–53.

Herrick, S. M., 1961, Well logs of the Coastal Plain of Georgia: Georgia Geological Survey Bulletin 70, 462 p.

Herrick, S. M., 1964, Foraminiferal fauna of late Miocene age from Georgia and South Carolina (abs.), *in* Abstracts for 1964: Geological Society of America Special Paper 82, p. 301–302.

Herrick, S. M., 1965, A subsurface study of Pleistocene deposits in coastal Georgia: Georgia Geological Survey Information Circular 31, 8 p.

Herrick, S. M., 1976, Two foraminiferal assemblages from the Duplin marl in Georgia and South Carolina: Bulletin of American Paleontology, v. 70, p. 125–163.

Herrick, S. M., and Vorhis, R. C., 1963, Subsurface geology of the Georgia Coastal Plain: Georgia Geological Survey Information Circular 25, 78 p.

Herrick, S. M., and Wait, R. L., 1955, Interim report on the results of test drilling in the Savannah area, Georgia and South Carolina: U.S. Geological Survey Open-File Report 55, 38 p.

Hodgson, W. B., 1846, Memoir on the *Megatherium* and other extinct gigantic quadrapeds of the coast of Georgia, with observations on its geologic features: Bartlett and Welford, New York, 47 p., 3 pl.

Hoyt, J. H., 1967, Barrier island formation: Geological Society of America Bulletin, v. 78, p. 1125–1136.

Hoyt, J. H., and Hails, J. R., 1967, Deposition and modification of Pleistocene shoreline sediments in coastal Georgia: Science, v. 155, p. 1541–1543.

Hoyt, J. H., and Hails, J. R., 1974, Pleistocene stratigraphy of southeastern Georgia, *in* Oaks, R. Q., Jr., and DuBar, J. R., eds., Post-Miocene Stratigraphy, Central and Southern Atlantic Coastal Plain: Utah State University Press, Logan, p. 191–205.

Huddlestun, P. F., 1988, A revision of the lithostratigraphic units of the Coastal Plain of Georgia: the Miocene through Holocene: Georgia Geologic Survey Bulletin 104, 162 p.

Hurst, V. J., 1957, Prehistoric vertebrates of the Georgia Coastal Plain: Georgia Mineral Newsletter, v. 10, p. 77–93.

Liddicoat, J. C., 1982, Paleomagnetic and amino acid dating of sediment in the Atlantic Coastal Plain: Geological Society of America Abstracts with Programs, v. 14, p. 546.

Liddicoat, J. C., Blackwelder, B. W., Cronin, T. C., and Ward, L. W., 1979, Magnetostratigraphy of upper Tertiary and Quaternary sediment in the central and southeastern Atlantic Coastal Plain: Geological Society of America Abstracts with Programs, v. 11, p. 187.

Liddicoat, J. C., McCartan, L., Weems, R. E., and Lemon, E. M., 1981, Paleomagnetic investigation of Pliocene and Pleistocene sediments in the Charleston, South Carolina area: Geological Society of America Abstracts with Programs, v. 13, p. 12–13.

Lyell, C., Jr., 1855, A second visit to the United States of North America, 3rd ed., v. 1: John Murray, London, p. 313–314.

MANSFIELD, W. C., 1944, Stratigraphy of the Miocene of Virginia and the Miocene and Pliocene of North Carolina, *in* Gardner, J., ed., Mollusca from the Miocene and Lower Pliocene of Virginia and North Carolina, Part 1: Pelecypoda: U.S. Geological Survey Professional Paper 199-A: p. 1–19.

MARKEWICH, H. W., 1985, Geomorphic evidence for Pliocene-Pleistocene uplift in the area of the Cape Fear Arch, North Carolina, *in* Morisawa, M., and Hack, J. T., eds., Tectonic Geomorphology: Allen and Unwin, Boston, p. 279–298.

MARKEWICH, H. W., PAVICH, M. J., MAUSBACH, M. J., STUCKEY, B. N., JOHNSON, R. G., AND GONZALEZ, V., 1986, Soil development and its relation to the ages of morphostratigraphic units in Horry County, South Carolina: U.S. Geological Survey Bulletin 1589-B, 61 p.

MACNEIL, F. S., 1947, Geologic map of the Tertiary and Quaternary formations of Georgia: U.S. Geological Survey Oil and Gas Investigations Preliminary Map 72.

MCCARTAN, L., LEMON, E. M., JR., AND WEEMS, R. E., 1984, Geologic map of the area between Charleston and Orangeburg, South Carolina: U.S. Geological Survey Miscellaneous Investigations Series Map I-1472.

MCCARTAN, L., OWENS, J. P., BLACKWELDER, B. W., SZABO, B. J., BELKNAP, D. F., KRIAUSAKUL, N., MITTERER, R. M., AND WEHMILLER, J. F., 1982, Comparison of amino acid racemization geochronometry with lithostratigraphy, biostratigraphy, uranium-series coral dating, and magnetostratigraphy in the Atlantic Coastal Plain of the southeastern United States: Quaternary Research, v. 18, p. 337–359.

MCCARTAN, L., WEEMS, R. E., AND LEMON, E. M., JR., 1990, Quaternary stratigraphy in the vicinity of Charleston, South Carolina, and its relationship to local seismicity and regional tectonism: U.S. Geological Survey Professional Paper 1367-A, 39 p.

OWENS, J. P., 1989, Geologic map of the Cape Fear region, Florence 1° × 2° Quadrangle and northern half of the Georgetown 1° × 2° Quadrangle, North Carolina and South Carolina: U.S. Geological Survey Miscellaneous Investigations Series Map I-1948-A.

PAUL, C. K., AND DILLON, W. P., 1979, The subsurface geology of the Florida-Hatteras shelf, slope and inner Blake Plateau: U.S. Geological Survey Open-File Report 79-448, 96 p.

POOSER, W. C., 1965, Biostratigraphy of Cenozoic Ostracoda from South Carolina: Arthropoda, Article 8: University of Kansas Publications, Lawrence, p. 1–80.

PURI, H. S., AND VERNON, R. O., 1964, Summary of the Geology of Florida and a guidebook to the classic exposures: Florida Geologic Survey Special Publication 5 (revised ed.), 312 p.

RICHARDS, H. G., 1936, Fauna of the Pleistocene Pamlico Formation of the southern Atlantic Coastal Plain: Geological Society of America Bulletin, v. 47, p. 1611–1656.

RICHARDS, H. G., 1943, Pliocene and Pleistocene fossils from the Santee-Cooper area, South Carolina: Academy of Natural Sciences, Philadelphia, Notulae Naturae, v. 118, 7 p.

RICHARDS, H. G., 1950, Geology of the Coastal Plain of North Carolina: Transactions of the American Philosophical Society, v. 4, part 1, p. 1–83.

RICHARDS, H. G., 1969, Illustrated fossils of the Georgia Coastal Plain: reprint of paper written in the 1950s for the Georgia Mineral Newsletter, Georgia Department of Mines, Mining and Geology, 47 p.

SOLLER, D. R., 1988, Geology and tectonic history of the lower Cape Fear River valley, southeastern North Carolina: U.S. Geological Survey Professional Paper 1466-A, 60 p.

SZABO, B. J., 1985, Uranium-series dating of fossil corals from marine sediments of southeastern United States Atlantic Coastal Plain: Geological Society of America Bulletin, v. 96, p. 398–406.

VEATCH, O., AND STEPHENSON, L. W., 1911, Preliminary report on the geology of the Coastal Plain of Georgia: Georgia Geological Survey Bulletin 26, 466 p.

WARD, L. W., AND HUDDLESTUN, P. F., 1988, Age and stratigraphic correlation of the Raysor Formation, late Pliocene, South Carolina: Tulane Studies in Geology and Paleontology, v. 21, p. 59–75.

WEHMILLER, J. F., AND BELKNAP, D. F., 1982, Amino acid age estimates, Quaternary Atlantic Coastal Plain: comparisons with U-series dates, biostratigraphy, and paleomagnetic control: Quaternary Research, v. 18, p. 311–336.

WEHMILLER, J. F., BELKNAP, D. F., BOUTIN, B. S., MIRECKI, J. E., RAHAIM, S. D., AND YORK, L. L., 1988, A review of the aminostratigraphy of Quaternary mollusks from United States Atlantic Coastal Plain sites, *in* Easterbrook, D. J., ed., Dating Quaternary Sediments: Geological Society of America Special Paper 227, p. 69–110.

WINKER, C. D., AND HOWARD, J. D., 1977, Correlation of tectonically deformed shorelines on the southern Atlantic Coastal Plain: Geology, v. 5, p. 123–127.

WOOLSEY, J. R., JR., 1976, Neogene stratigraphy of the Georgia coast and the inner continental shelf: Unpublished Ph.D. Dissertation, University of Georgia, Athens, 222 p.

PART III
GULF OF MEXICO COASTAL SYSTEMS

HOLOCENE COASTAL DEVELOPMENT ON THE FLORIDA PENINSULA

RICHARD A. DAVIS, JR.
Department of Geology, University of South Florida, Tampa, Florida 33620
ALBERT C. HINE
Department of Marine Science, University of South Florida, St. Petersburg, Florida 33701
AND
EUGENE A. SHINN
U. S. Geological Survey, Center for Coastal Studies, 600 4th St. South, St. Petersburg, Florida 33701

ABSTRACT: The Florida peninsula contains five distinct coastal sections, each resulting from its own spectrum of coastal processes and sediment availability during a slowly rising, late Holocene sea level. The east coast barrier system is wave-dominated and has a large cuspate foreland (Cape Canaveral) near its middle. The Florida Keys and reef tract represent the only coastal carbonate system in the continental United States. An open-marine mangrove coast characterizes the low-energy, tide-dominated southwest part of the State. The central Gulf barrier system displays a mixed-energy morphology in a microtidal, low-energy setting. The open-coast marsh system of the Big Bend area that is north of the barrier system is also tide dominated, and is developed on a sediment-starved carbonate platform.

The oldest preserved coastal Holocene section is the Florida Keys area where, at about 6 to 8 ka, sequences accumulated during the Holocene. Most of the remainder of the peninsular coast is characterized by terrigenous sequences less than 3 ka. The younger sequences accumulated almost exclusively from reworking of older strata without benefit of additional sediment supply from land.

INTRODUCTION

The Florida peninsula represents a wide variety of coastal morphologies that have developed in response to a spectrum of processes and geologic settings. The peninsula can be subdivided into five geomorphic elements: (1) the east coast barrier system, (2) a limestone island arc, the Florida Keys, (3) the mangrove coast of southwest Florida, (4) the central Gulf barrier-island system, and (5) the marsh coast of the Big Bend area (Fig. 1). Each is well defined and internally similar, except for the east coast barrier system where the portion north of Cape Canaveral forms the south part of the Georgia Bight and is adjacent to a broad shelf, in comparison to the southern portion.

The emphasis in this paper will be on the regional development of coastal geomorphology and stratigraphy. The aim is to provide a general discussion of how the various process-response systems produce the Holocene record along these coastal segments. Attention will be directed toward the rates of sea-level change, geologic setting and sediment availability, which directly or indirectly play important roles in producing the present coast.

GENERAL SETTING

The Florida peninsula is a tectonically stable carbonate platform upon which the surficial, dominantly terrigenous, Quaternary sediments have accumulated in an asymmetric fashion; relatively thick on the east coast and relatively thin on the west coast. The carbonate platform was active from initial inception in the Mesozoic until the late Paleogene when the Suwannee Straits/Gulf Trough seaways were finally terminated (Chen, 1965; McKinney, 1984; Pinet and Popenoe, 1985). The seaway system was a dynamic barrier that prevented siliciclastic sediments shed off the southern Appalachain Mountains from inundating and smothering the carbonate-producing environments to the south. Eventually, the seaways were infilled and peninsular Florida was largely covered by siliciclastics from the north via fluvial and longshore-transport processes. The combination of antecedent topography on the carbonate bank and multiple sea-level fluctuations heavily influenced distribution of the dominately quartz-sand cover.

The east coast is bounded by a narrow and steeply sloping shelf to the north that narrows greatly to the south. South of Cape Canaveral and continuing through the Florida Keys, the adjacent shelf is narrow and steep, permitting relatively high-wave energy to impact the coastal zone with mean annual wave heights of up to 70 cm (Nummedal and others, 1977). The west coast shows a marked contrast in that the shelf is about 200 km wide with shoreface gradients that range from 1:1,300 in the central area to less than 1:3,000 in both the north and south. Such conditions coupled with the limited fetch of the Gulf of Mexico cause wave energy reaching the coast to be very low; mean annual wave heights of 10 to 25 cm are typical (Tanner, 1960).

Tidal range is microtidal throughout the Florida peninsula except on the northeast section, where spring ranges exceed 2 m. There is, however, considerable variation in the effect of tidal processes on coastal development. It has been demonstrated that regardless of absolute tidal ranges, it is the relative influence of tides and waves that molds coastal morphology (Davis and Hayes, 1984); the Florida peninsula provides several examples of this important relation between wave and tidal energies. The tide-dominated sections are the mangrove coast of southwest Florida and the marshy coast of the Big Bend area (Fig. 1), where spring ranges are no more than 1.3 m. A broad, gently sloping shoreface produces very low-wave energies in these areas and tidal effects dominate. Wave-dominated conditions are present throughout most of the east coast and the Florida Keys. Mixed-energy coastal morphology prevails in the northeast part of the peninsula where tidal range is greatest and in the west-central barrier system where tides are just under 1 m.

Littoral drift is dominantly to the south on both the Atlantic and Gulf sides of the peninsula; however, there are

Quaternary Coasts of the United States: Marine and Lacustrine Systems, SEPM Special Publication No. 48

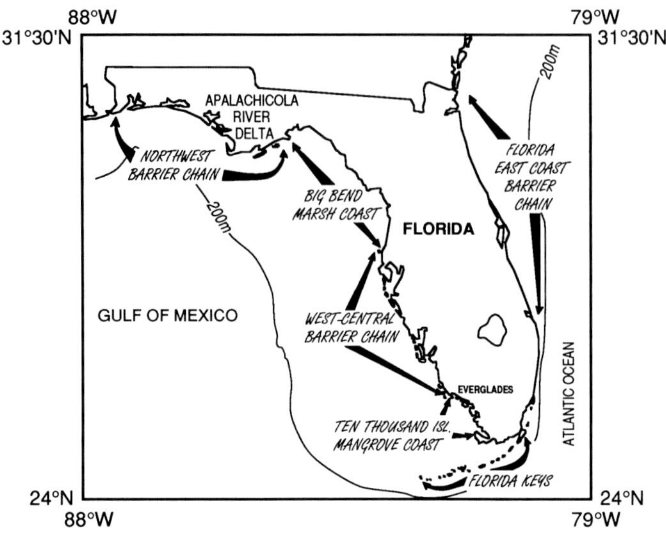

FIG. 1.—Outline map of Florida showing the six major coastal sections on the peninsula considered in this paper.

numerous local reversals due to shoreline orientations, inlets, and offshore bathymetry that influence approaching waves. The general pattern of littoral drift is a result of strong northerly winds associated with winter frontal systems that dominate longshore currents, even though the prevailing winds have a distinct southerly component.

HOLOCENE SEA LEVEL

Holocene sea-level data are scarce throughout much of the Florida peninsula with the best from the southwest coast and the Florida Keys. The period of interest begins about 8 ka (Fig. 2). The rate of rise has been modest over that time with a rather distinct slowing at 3.5 to 3.0 ka (Scholl and others, 1969; Hine, 1977; Enos and Perkins, 1979; Robbin, 1984). It is estimated that rates of sea-level rise (Fig. 2) were 25 cm/100 yrs before about 3.0 ka and only 4 cm/100 yrs since that time (Wanless, 1982). Both of these rates are slow relative to the conditions of the early Holocene, when sea level rise was about 1 m/100 yrs.

The combination of the slow rate of sea-level rise and the general dearth of terrigenous sediment supplied to the

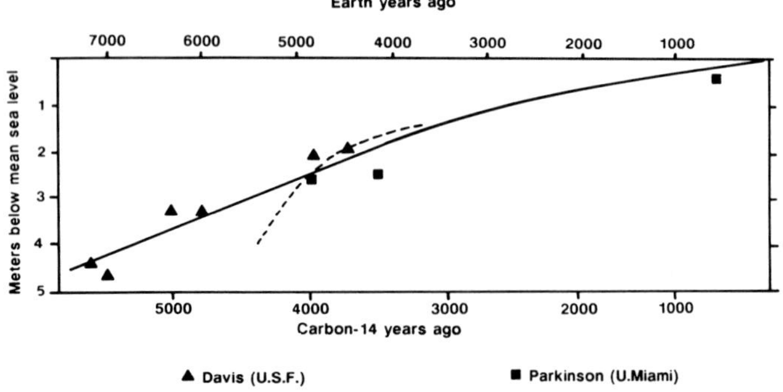

FIG. 2.—Generalized late Holocene sea-level curve after Scholl and Stuiver (1967) with additional dates from southwest Florida.

Florida coast has resulted in most of the terrigenous portion of the Holocene coastal sequence being developed from a reworking of older Quaternary sediments. As a result, the Holocene stratigraphy is rather thin overall and tends to be restricted to the present coast with the inner shelf lacking in Holocene terrigenous sediments. The carbonate system of the Florida Keys on the other hand, has benefitted from the slow sea-level rise with thriving reefs accumulating relatively thick sequences (Shinn and others, 1989).

EAST COAST BARRIER SYSTEM

Geomorphology

The Florida east coast (Figs. 1 and 3) contains the longest barrier-island/tidal-inlet system (550 km) of any state in the United States. Of 22 inlets, all but one (Matanzas) have been significantly influenced by engineering activities (Marino and Mehta, 1986). Located approximately in the center of this coast is Cape Canaveral, a very large cuspate foreland that projects 20 km seaward of the main coastal trend. Seaward of the cape is a well-developed cape-retreat massif (Swift and others, 1972; Field and Duane, 1974). Cape Canaveral controls shoreline orientation and sedimentation patterns for at least 125 km of the Florida east coast.

At the north end of this system, barrier morphology is influenced by a mixed regime of wave and tidal energies. The barriers form the south limb of the Georgia Bight, a regional embayment having spring-tidal ranges of up to 3 m. This range is higher than that along the North Carolina coast (1 m) and most of the Florida east coast (<1 m), which have wave-dominated barrier-island features (Nummedal and others, 1977; Hayes, 1979). At the south end of the barrier system is Key Biscayne, beyond which is the northern end of the Florida Keys. Although the Key Largo Limestone dominates the northernmost islands of the Keys, boreholes have shown that siliciclastic sediments have extended farther south on the Florida Platform in the past (Enos and Perkins, 1977).

The present east coast of Florida is dominated by Holocene quartz-sand barrier islands. There are restricted but significant areas, however, (e.g., Anastasia Island) where carbonate-rich Pleistocene barriers face the open Atlantic Ocean (Fig. 4). These coastal areas are dominated by the Anastasia Formation, a cemented, molluscan grainstone having been formed in the beach/shallow-nearshore environment (Stauble and McNeill, 1985). Exposures of this formation along with modern beach rock and worm reefs (Kirtley and Tanner, 1968) create the only rocky coasts in east Florida outside of the Florida Keys. Beyond these localized anomalies, the Florida east coast barriers are typically anchored by the underlying Pleistocene Anastasia Formation and develop high foredunes locally that prevent overwashing and landward migration. The dunes represent an important type of microtidal, wave-dominated barrier-island system that contrasts markedly with the wave and overwash-dominated, barrier-island model presented by Hayes (1979). The barriers along the east coast of Florida represent variations of Godfrey's (1976) vegetated-barrier models.

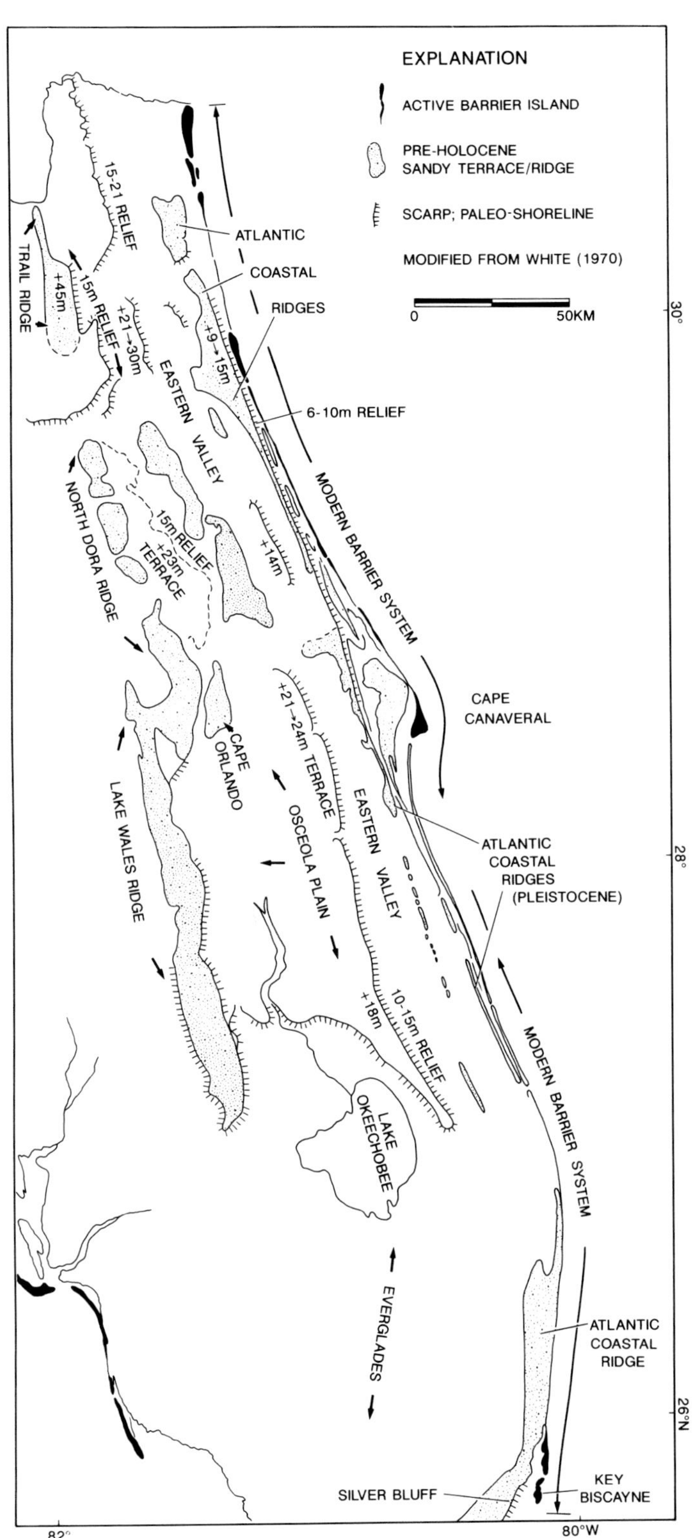

FIG. 3.—Physiographic map of east-central Florida coast and adjacent interior (after White, 1970).

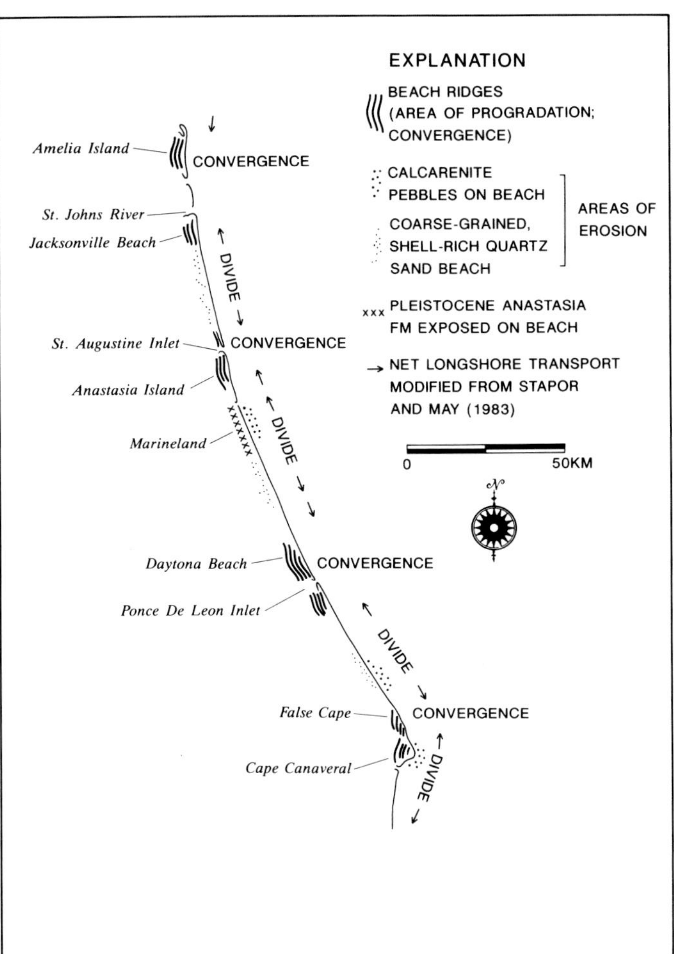

FIG. 4.—Variation in beach-sand texture and mineralogy and hypothesized littoral drift cells for the coast of Florida north of Cape Canaveral (from Stapor and May, 1983).

Sedimentary Processes

The Florida east coast faces the open Atlantic and receives long-period swell, but because of its lower latitude, is not influenced as regularly by intense extratropical storms as are the Outer Banks, Long Island, and Cape Cod to the north. The relatively broad shelf off the northeast Florida coast and the presence of the Bahama Banks off the central/southeast Florida coast also provide protection. The mean significant wave height during winter months is only approximately 1.2 m (Stauble and Da Costa, 1987). The persistent easterly, onshore winds, as a result of being influenced by the northern limit of the easterly Trade Winds, provide the mechanism for dune growth and development. The relatively low frequency of high-intensity storms (including hurricanes) coupled with a regional abundance of sand, both on the Florida coastal plain to the west and the submerged shelf to the east, have formed a relatively stable, barrier-island chain.

Most of the islands support a prominent, thickly vegetated foredune ridge, commonly backed by other vegetated dune ridges. The lack of overwash fans, as compared to other wave-dominated barrier coasts, is striking. In addition, narrow, low zones prone to overwash or new inlet formation are not widespread.

North of Cape Canaveral the regional longshore sand-transport system consists of a series of littoral drift cells defined by convergences and divergences (Fig. 4; Stapor and May, 1983; Stauble and DaCosta, 1987). The convergences are areas of abundant sand supply resulting in beach-ridge growth, whereas the divergences are areas of erosion indicated by the presence of calcarenite pebbles presumably from the Pleistocene Anastasia Formation. Where these divergences occur, serious local erosion problems may be present (Stauble and DaCosta, 1987). Wave-refraction diagrams suggest offshore wave energy focused at the beach by topographic variations on the shelf (Mehta and Brooks, 1973). This process could account for the littoral cells. Interestingly, longshore-transport diagrams based upon shoreline orientation and SSMO (Synoptic Shipboard Meteorological Observations) data indicate that although gross longshore transport is relatively high (8×10^5 yds/yr), there is negligible net transport, indicating that this coastal sector is in equilibrium with the seasonally bimodal wave climate (Walton, 1976; Jones and Mehta, 1978). The lack of inlets can be attributed to the low net transport and small tidal prisms.

Quaternary History

The physiography of eastern/central peninsular Florida is dominated by ancient shorelines indicated by ridge-and-swale topography (Vernon, 1951; Puri and Vernon, 1964; Alt and Brooks, 1965; White, 1970). The central part of the State is marked by the Tertiary Mount Dora–Lake Wales Ridge (Fig. 3). These highlands were islands and, at times, were submerged during the extreme sea-level highstands of the Neogene. At the coast, a Plio-Quaternary ridge complex exists that includes the Atlantic Coastal Ridge (White, 1958). Between the highlands and coastal ridge is a broad lowland (Eastern Valley) along which drainage is north-south. Only in a few areas do streams flow eastward by cutting through the coastal ridges, possibly passing through ancient inlet sites. Where the streams meet the ocean, they form the only shore-normal, tidal-inlet/estuarine systems on the east coast having been formed from drowned river valleys. The remaining estuarine systems are flooded, shore parallel, back-barrier Pleistocene and Holocene lagoons.

FLORIDA KEYS

Pleistocene Substrate

The Key Largo Limestone, a member of the Miami Limestone, is a fossilized (120–130 ka) Pleistocene coral reef (Sanford, 1909; Hoffmeister and Multer, 1968). The linear trend of the Key Largo Limestone forms the upper Keys and extends southward above sea level to the beginning of the lower Keys at the prominent break in the trend, as shown on Figure 5. The lower Keys, composed of the Pleistocene Miami Oolite (Sanford, 1909), each trend perpendicular to the upper Keys (Fig. 6) and have been interpreted as preserved tidal bars because of comparisons with active sand bars in the Bahamas (Hoffmeister and others, 1967). Although the oolite was previously interpreted to be younger, recent core drilling has shown the youngest Miami Oolite and Key Largo Limestone to be contemporaneous (Kindinger, 1986; Shinn and others, 1989). Older Key Largo Limestone underlies the Pleistocene oolite west of Key West. The oolite extends westward to form the platform under the Holocene carbonate muds and high-energy sands of the Marquesas/Quicksands area (Fig. 6A; Shinn and others, 1990). The Miami Limestone grades northward into terrigenous clastics and freshwater limestone both laterally and vertically (Perkins, 1977).

The upper surface of all Pleistocene limestones throughout the area, whether oolitic or coralline, is karstic and generally calcrete coated. The deposits were leached, cemented and coated with calcrete when sea level fell more than 100 m between 120 and 100 ka. At least five similar subaerial unconformities have been located within the Miami Limestone and used for stratigraphic correlation (Perkins, 1977). The youngest calcrete crust has been traced offshore beneath Holocene sediments and reefs (Robbin and Stipp, 1979; Robbin, 1981; Shinn and others, 1989; Lidz and Shinn, 1991).

Development of Holocene Reefs

The Florida reef tract forms a belt of Holocene sediments and reefs between the Pleistocene Keys and the edge of the platform (18-m contour) to the east (Fig. 5). The arcuate 240-km-long, 5- to 8-km-wide belt extends from near Miami to the area south of Halfmoon Shoal, west of Key West. Near Key West the reef tract trends to the west (Fig. 5). The senescent outermost reefs off Miami contain an admixture of terrigenous and carbonate sands and extend northward nearly 100 km. The reefs flourished between 7 and 3 ka (Lighty, 1977). The terrigenous sediments, which increase in abundance near shore, have been transported by southerly longshore drift as discussed previously.

Three reef types occur in the Florida reef tract: (1) linear platform-margin banks, (2) subcircular inshore patch reefs, and (3) newly discovered outlier reefs. The linear banks are similar to barrier reefs, but they are discontinuous and separated by topographic lows. They occur at or near the platform margin, and recent work (Lidz and Shinn, 1991) shows that their location is controlled by underlying Pleistocene bedrock. At approximately 8 ka, when sea level was about 10 to 12 m below present, a string of isolated platform-margin bedrock islands of Key Largo Limestone existed (Fig. 7; Shinn and others, 1989; Lidz and Shinn, 1991). The islands were probably outlier reefs. Holocene fringing reefs grew around the bedrock islands and as sea level rose, coral grew up and over them, forming discontinuous bank reefs. The dominant framework coral was *Acropora palmata*.

Thousands of subcircular patch reefs are scattered in the 5- to 8-km-wide belt between the outer-bank reefs and the subaerially exposed Keys. Surface observations and core drilling show that these patch reefs are composed of as much as 8 m of massive head corals (Shinn and others, 1989).

The outlier reefs (Fig. 8) in the middle of the lower Keys are typically pre-Holocene and their composition resembles that of the Key Largo Limestone (Lidz and others, 1991).

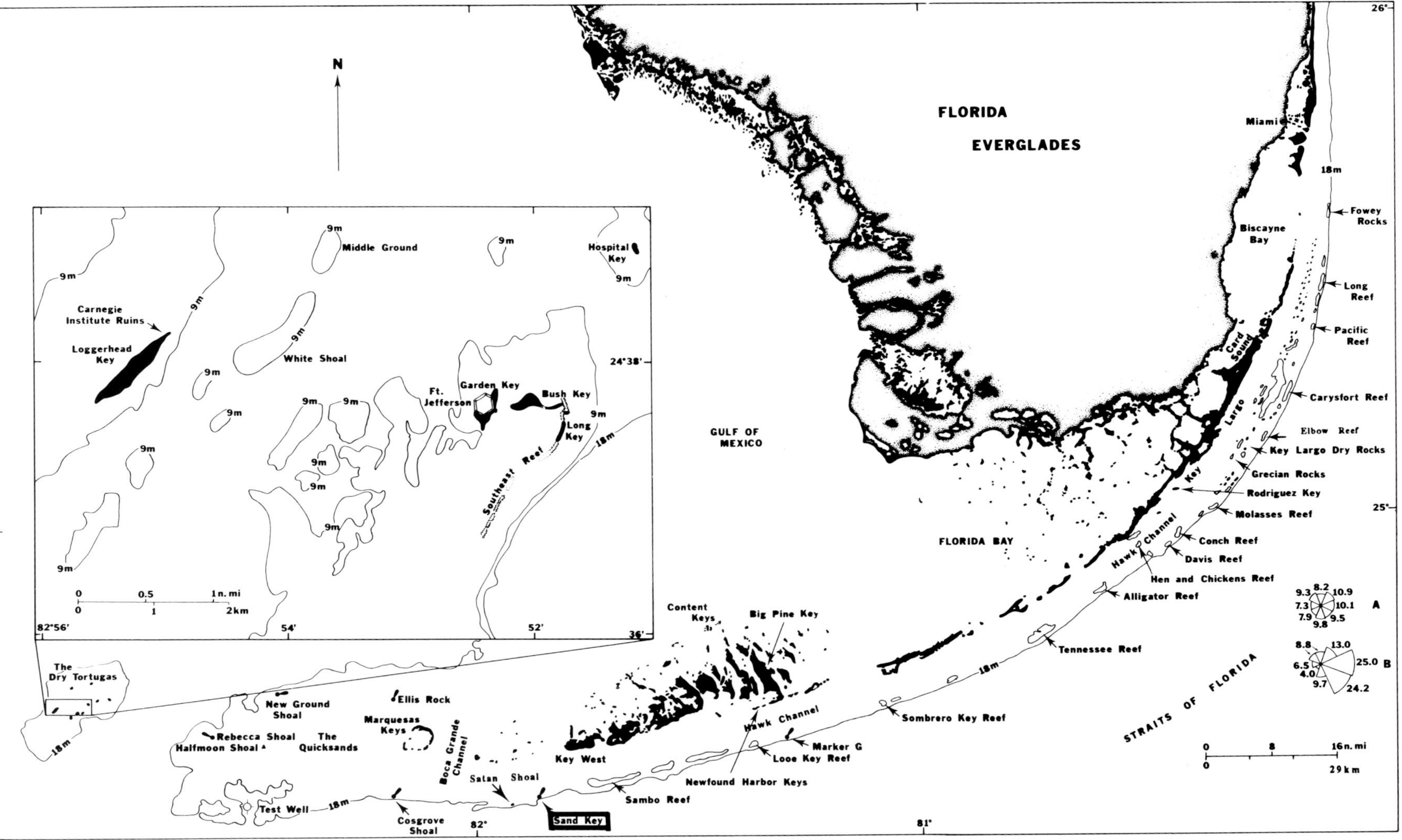

FIG. 5.—Map of south Florida showing Pleistocene upper and lower Florida Keys (in black), Florida Bay and the Dry Tortugas (from Shinn and others, 1989). The linear Keys represent a coral reef and the perpendicular lower Keys, beginning with Big Pine Key, were deposited as ooid-sand shoals. The Florida reef tract is the belt of reefs and sediment seaward (east) of the Keys, and extends from north of Miami to the vicinity of the "test well" at the lower lefthand corner of the map. Rose diagram (A) shows mean of average hourly wind velocities; (B) shows percentage of total hours wind blows from each direction. Arrow by Sand Key (circled) points to location of stratigraphic section shown in Figure 10.

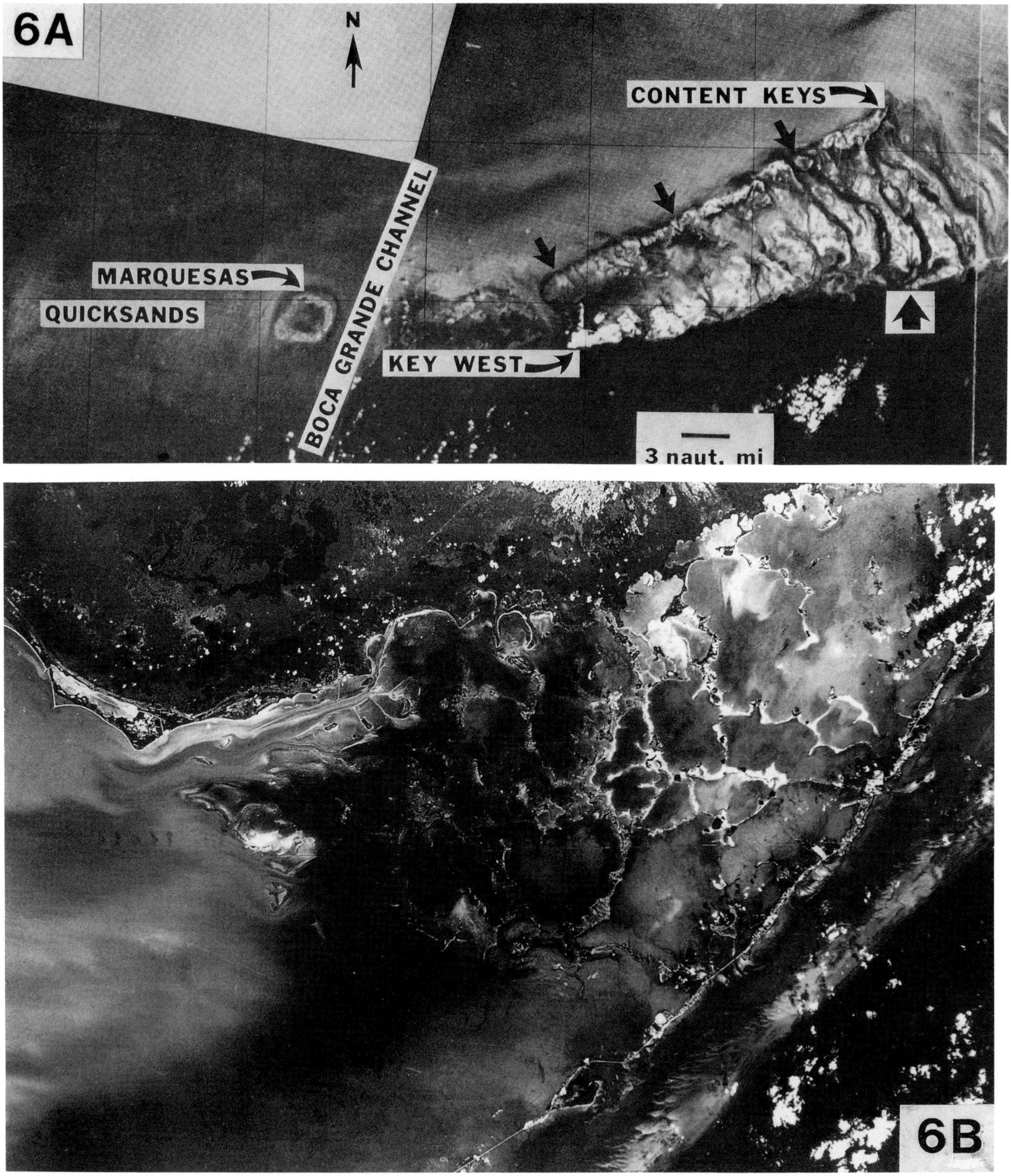

FIG. 6.—(A) Landsat image of lower Keys and Marquesas/Quicksands area west of Key West. Large arrow indicates southern extension of the exposed Pleistocene coral reef trend. Three small arrows show location of Pleistocene oolite-beach deposits that developed against the perpendicular oolite island of the lower Keys (from Shinn and others, 1989). (B) Spot satellite image of Florida Bay and portion of Florida Keys. Note anastomosing mud banks separating sediment-free areas, locally called "lakes," and the strip of reefs and white carbonate sands east of the Keys that characterize the 7 to 8 km-wide reef tract.

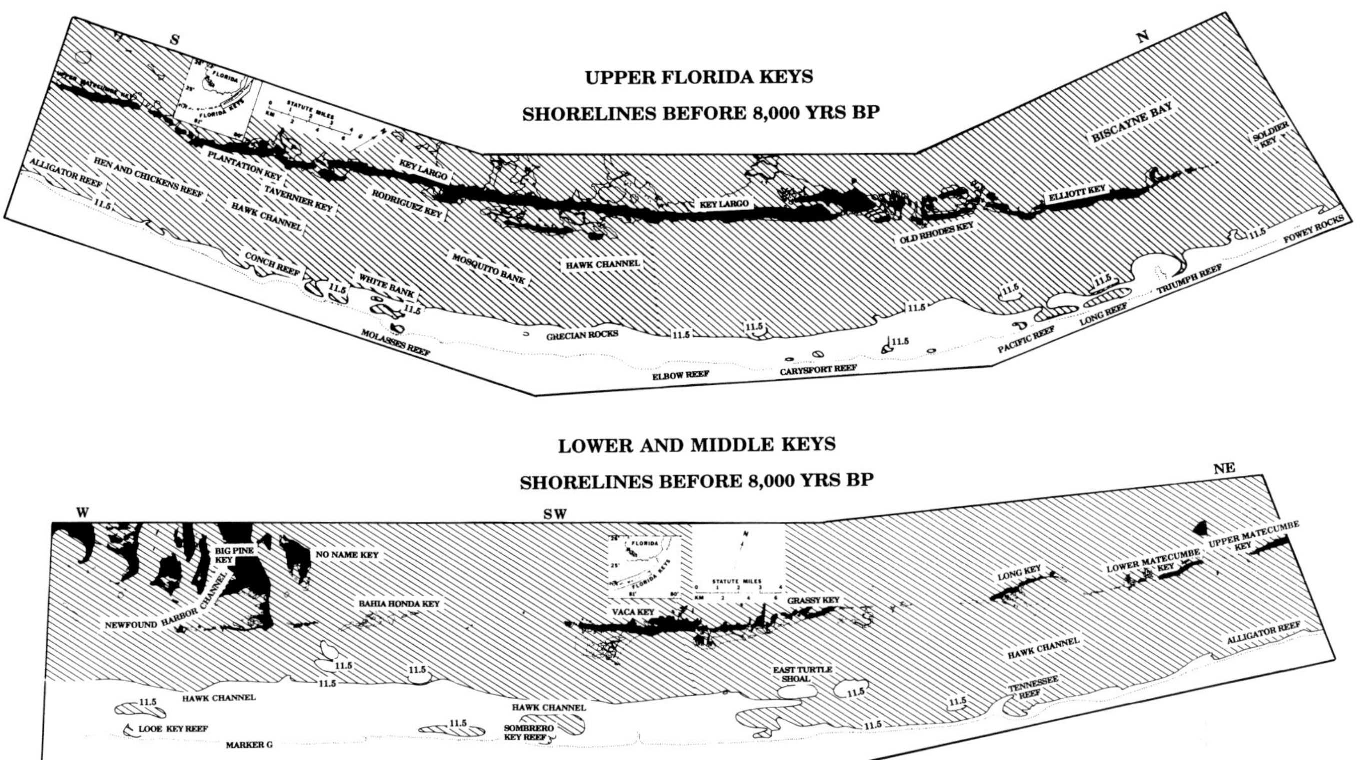

FIG. 7.—Maps showing approximate location of the shoreline when sea level was 11.5 m below present between 10 and 8 ka, (former land area shaded; present land black). Note that modern reefs rest on what were isolated offshore islands (from Lidz and Shinn, 1991).

Widely scattered heads and thickets of *Acropora cervicornis* grow on the outliers and, except in one area, have not accumulated *in situ* to form Holocene reefs. Apparently, periodic storm and hurricane waves sweep coral and accumulated carbonate sediment into the backreef troughs.

Coastal Processes

Wind and waves have a dominating influence on reef development and carbonate-sand distribution. In the lower Keys

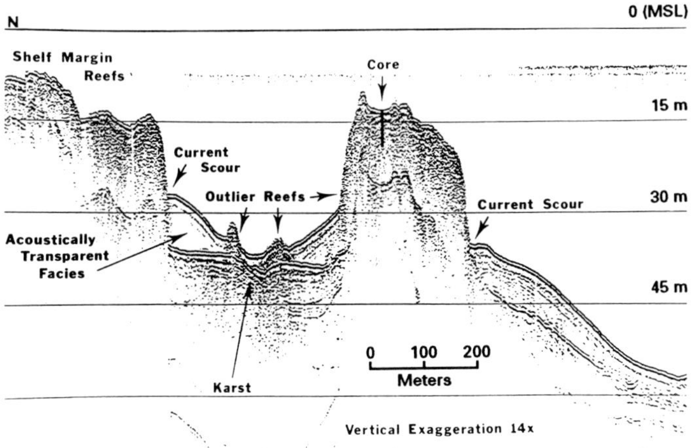

FIG. 8.—High-resolution seismic line across Pleistocene outlier reef off Sand Key Reef (Holocene) southwest of Key West (see Fig. 5 for location; from Lidz and others, 1991).

these processes are predominantly parallel to the platform margin, causing carbonate sands to be transported parallel to the reef trend. In places, reefs have been smothered by sediment, such as on the seaward side of Looe Key Reef (Shinn, 1981; Lidz and others, 1985). Aerial photos show many areas where fan-like deposits of carbonate sand have spilled seaward through depressions between reefs. Hurricanes are the most likely cause of these fans. One large fan, 1 km north of French Reef (Fig. 9A), is known to have been caused by Hurricane Donna (Ball and others, 1967). Aerial photography shows that hurricanes also form blowouts in the sea-grass meadows that grow in the lee of many reefs (Ball and others, 1967). Sand-filled blowouts, much like their counterparts on land, have been shown to migrate and form distinctive sedimentary sequences (Wanless, 1981).

Cross sections of typical bank reefs (Fig. 10), based on underwater core drilling and seismic profiling, show their transgressive nature (Shinn and others, 1989). Transgression results from a combination of hurricanes and the rise of Holocene sea level. During hurricanes, corals and carbonate sediment are periodically transported over and into the backreef area. Ball and others (1967) stated that the average Florida reef is subjected to a hurricane about every 6 to 7 years.

Sediment Accumulation

Carbonate-sediment accumulations are generally thickest near coral reefs and accumulations up to 12 m thick have

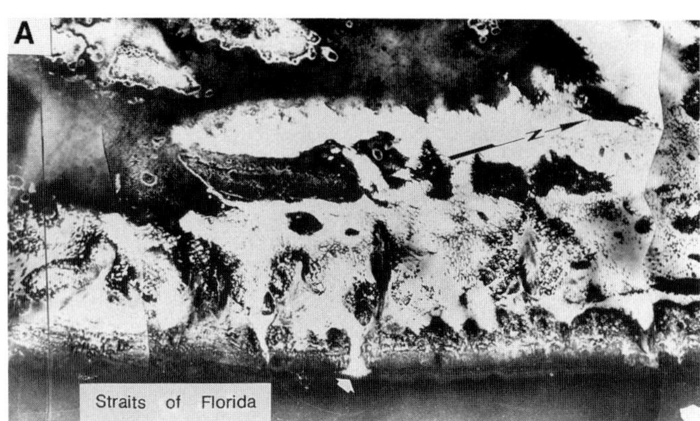

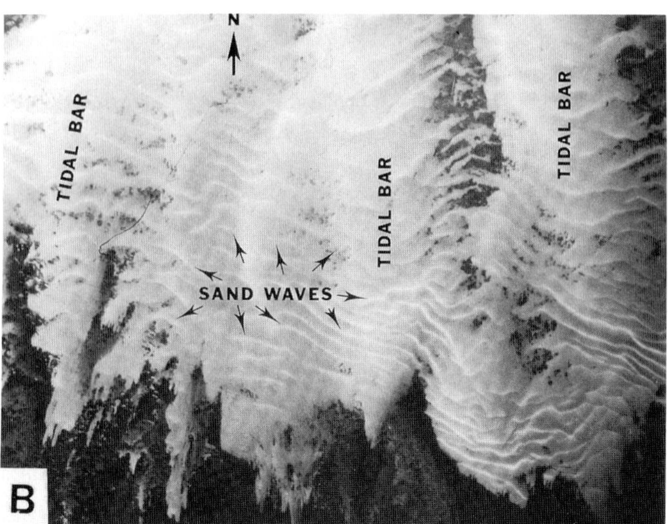

FIG. 9.—(A) Aerial photograph of the area approximately 6 to 8 km west of the Marquesas Keys showing east-west-oriented sand waves overlying less distinct tidal bars. Sand waves are <1 m below sea level at low tide. Distance from top to bottom is approximately 1 km (from Shinn and others, 1990). (B) Vertical aerial photograph of reef tract margin off Key Largo showing seaward-directed carbonate-sand fan formed during Hurricane Donna (arrow).

been located in bedrock lows and troughs landward of some platform-margin bank reefs (Lidz and others, 1985). Carbonate sand forms platform-margin banks up to 8 km wide, 5 to 8 m thick, and as much as 50 km long along the seaward margin (Enos, 1977).

Landward of the carbonate sand and bank-reef accumulation is a bedrock trough, Hawk Channel (Fig. 5), that parallels the entire reef tract. The trough deepens to the south and west, a trend that has been interpreted as resulting from westward Pleistocene tilting of the underlying platform (Shinn and others, 1989). Wackestone and packstone accumulations up to 5 m thick (Fig. 10) occur in the trough (Enos, 1977). The accumulation thins and wedges out landward against the Pleistocene islands and is buried on its seaward side beneath the outer belt of carbonate sand, which is transgressing landward (Fig. 10). The Pleistocene limestone is usually bare of sediment from the shoreline out to depths of 3 to 4 m, except in localized areas of nearshore

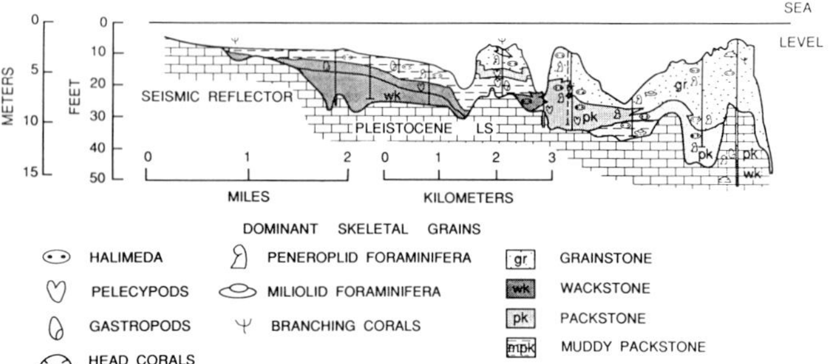

FIG. 10.—Cross section of the Florida reef tract based on seismic profiling and sediment coring (modified from Enos, 1977). Grainstone facies of the outer sand belt overlies packstone, indicating transgression and upward increase in grain size and sorting.

bank accumulation (Turmel and Swanson, 1976; Enos, 1977).

West of Key West, the Quicksands, a 12-m-thick accumulation of carbonate mud and sand, extends 28 km westward over the submerged extension of the lower Keys oolitic limestone (Fig. 5; Shinn and others, 1990). Sand on the Quicksands is composed mainly of *Halimeda* plates heaped into 1- to 2-m-high submarine dunes by north–south reversing tidal currents (Fig. 9*b*). Coral reefs north (New Ground Shoal) and south (the western termination of the Florida reef tract) of the Quicksands are poorly developed. The Tortugas reef and sediment accumulations to the west rest on a separate Pleistocene topographic feature.

Florida Bay

Florida Bay is the roughly triangular area nested between the south Florida mainland and the Pleistocene Florida Keys (Figs. 5 and 6*b*). The western side, beginning at Cape Sable, the southernmost area influenced by terrigenous sediment, is bounded by massisve mud banks. The banks are composed of lime mud and wackstone fronted on their exposed margins with molluscan grainstone and the finger coral, *Porites* sp. All the banks are underlain by a gently sloping Pleistocene limestone surface that is karstic and generally coated with calcrete. The surface of the limestone is about 1 m below sea level in the northeast and north portion of the bay and gently slopes to the west and southwest, where its surface lies 4 to 6 m below sea level, except for rare outliers.

Florida Bay is best known for its linear anastamosing mud banks and mud islands (Fig. 6*b*). There is evidence that mud banks began as Holocene storm levees that formed along the northern shore of Florida Bay (Cottrell, 1989). As sea level rose, the linear features became stranded, eventually drowned, and continued mud production and deposition led to subtidal mudbank development. Migration of the banks, for the most part, removed evidence of their supratidal origin, except where evidence of levee deposition remains beneath larger banks in central Florida Bay (Wanless and Taggett, 1989).

Florida Bay contains 237 mud islands whose history is intimately associated with mud banks (Enos and Perkins, 1977). The islands have accumulated by storm deposition of mud on their surfaces and have, for the most part, kept pace with rising sea level during the past 3,000 to 4,000 years. Their supratidal surfaces, exposed to alternating evaporation, rain and marine flooding, produce diagenetic sedimentary features, and chemically reactive pore waters that lead to precipitation of Mg-calcite, protodolomite and poorly ordered dolomite (Steinen and others, 1977; Swart and others, 1989). Many mud islands rest on a basal freshwater peat and or freshwater lime mud (Scholl, 1964; Davies and Cohen, 1989) that formed a few thousand years ago when the area was swampy, similar to the present Everglades to the north. Coring reveals evidence of island migration, though it is not nearly as pronounced as that of neighboring mud banks (Enos and Perkins, 1979; Enos, 1989). Mangrove peats are accumulating today and make up a significant portion of many mud islands.

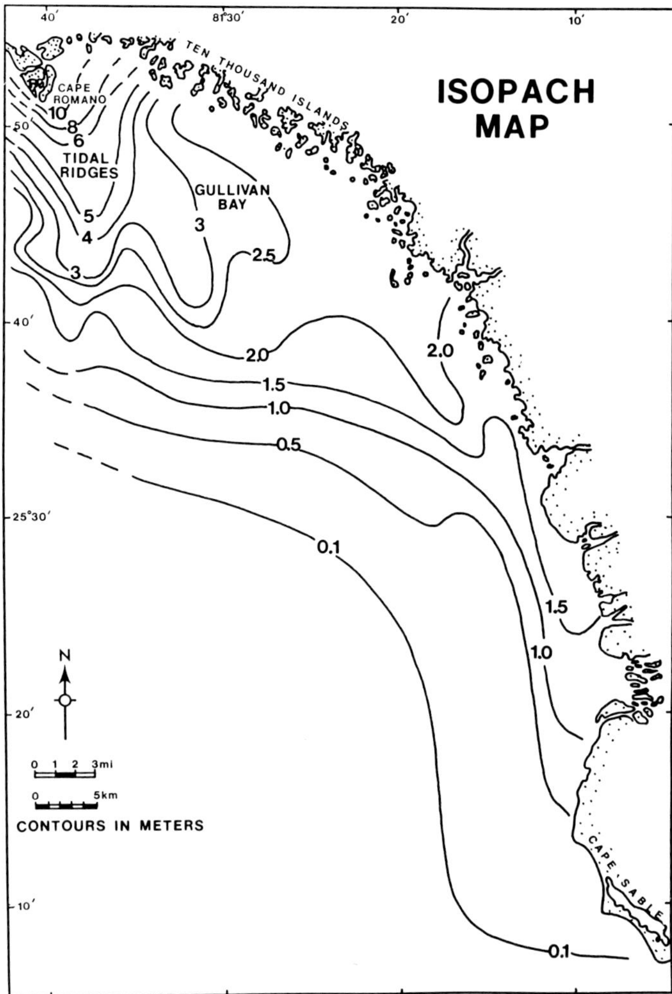

FIG. 11.—Isopach map of Holocene inner-shelf sediments off southwest Florida.

SOUTHWEST FLORIDA MANGROVE COAST

The coast of peninsular Florida between Cape Sable on the north side of Florida Bay and Cape Romano at the southern end of the west-central barrier coast (Fig. 1) is dominated by mangroves. This coastal segment is tide dominated due to the broad, gently sloping continental shelf and the very low mean annual wave height (10 cm) coupled with the broadly embayed coastal configuration. The mangrove coast is contained between two wave-dominated cuspate forelands and can conveniently be divided into two sections: (1) the northern portion, commonly called the Ten Thousand Islands, and (2) the southern portion, which is adjacent to the Everglades and lacks numerous discrete mangrove islands.

General Geology and Geomorphology

The mangrove coast is adjacent to a uniformly sloping inner shelf, which has an average gradient of 1:3,000 to a depth of 10 m. Cape Romano to the north is a quartz-sand-dominated cuspate foreland that is closely associated with an extensive bank of linear and tide-dominated sand ridges (Davis and Klay, 1989; Davis and others, 1989b). Cape Sable to the south is a carbonate-sand foreland area containing three cuspate accretionary-ridge complexes (Roberts and others, 1977). The Ten Thousand Islands coastal area is dominated by mangrove islands separated by numerous tidal channels. Modest upland runoff enters the coast through rivers that empty behind the mangrove islands. This contrasts with the southern part of the mangrove coastal area, where several small rivers empty directly into the open Gulf (Fig. 11).

The mangrove coast is underlain by Pliocene and Pleistocene carbonates of the Tamiami Formation and the Miami Formation (Enos and Perkins, 1977). Overlying the carbonates is a thin, well-sorted, Pleistocene quartz sand that is clayey and rooted locally (Parkinson, 1989). This unit is interpreted as a terrestrial sand sheet of at least partly eolian origin (Davis and Klay, 1989).

Present Morphology and Processes

The modern coastal morphology is the result of about 3,000 years of very slowly rising sea level and a limited sediment supply along with extremely low physical-energy conditions along the coast. Spring-tidal ranges in the north around Gullivan Bay are about 1.4 m and drop to 1.1 m at Cape Sable. Although this range is small over the area, there is a very different tidal influence from north to south. Tidal currents in the linear sand ridges and to the east near Cape Romano are up to 70 cm/sec, but they diminish greatly to the southeast with values of only 17 cm/sec a few kilometers offshore of the central portion of the area (Davis and others, 1989a). Wave energies are low throughout but increase toward the south due to the exposure to strong northerly winds associated with winter frontal passage. The northern portion of the area is in the lee of the coast and thus protected from these winds. Cape Sable owes its wave-dominated configuration to a combination of the northerly winds and the prevailing southwesterly winds.

The Ten Thousand Islands area displays an unusual arrangement of mangrove islands. Adjacent to the irregular mainland coast is a relatively open area called the "chain of bays" (Parkinson, 1989). The relatively open-water coastal environment is wide to the north and narrow to the south. To the north the Ten Thousand Islands are discrete mangrove islands separated by numerous tidal channels, whereas to the south the equivalent area is composed of continuous mangrove stands with only a few streams from the mainland interrupting the coast (Fig. 12). The marked difference in geomorphology is due to a combination of tidal influence and fine-grained sediment availability. The northern area is one of relatively strong tidal currents and a dearth of land-derived mud, whereas the southern portion has sluggish tidal currents and receives a higher rate of fine terrigenous sediment due to the presence of several streams that drain the Big Cypress Swamp and the Everglades (Fig. 12). The difference in geomorphology is responsible for the distribution of fine sediments along the present outer coast. Mud is relatively abundant south of the Ten Thousand Islands where it is carried directly to the open coast, whereas it is only about half as abundant in a corresponding area to the north. In the northern area the supply is generally lower and most mud is trapped in the bays and mangrove swamps and does not reach the coast.

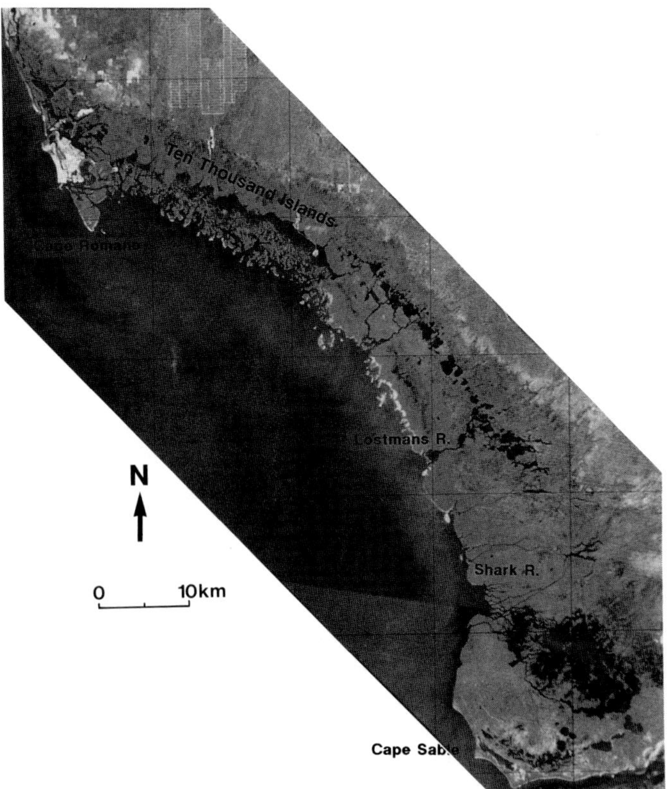

FIG. 12.—Satellite image of the southwest Florida coast showing the Ten Thousand Islands and adjacent areas. The linear trends represent old Quaternary shorelines.

Holocene Stratigraphy

The slow rise of sea level and accompanying transgression during the period of about 7 to 3 ka was accompanied by the accumulation of a distinct Holocene shallow-marine sequence. The sequence extends out onto the inner shelf and is thin overall. It shows a marked thinning south of the Ten Thousand Islands, where it is confined to the present coastal zone (Fig. 11).

The base of the Holocene is marked by an unconformity that locally has coarse carbonate rudites above both the older carbonates and the Pleistocene terrigenous sands. High-resolution seismic data confirm a karstic surface with local relief of a few meters. A peat unit is present but discontinuous just above the basal lag gravels from about 20 km offshore beneath the linear sand ridges to the coastal area where it rises and thickens (Parkinson, 1989). At an offshore location, the peat has been dated at 6,470 + 120 yrs from a depth of 8 m below present sea level (Davis and Klay, 1989), whereas Parkinson (1989) dated what is presumably the same peat at 4,000 to 3,500 yrs in the Ten Thousand Islands area at a depth of 2.5 m below present sea level. These data indicate a shoreline transgression rate of about 6 to 7 m/yr.

There is substantial documentation of a seaward progradation of the coast that is coincident with the slowing of sea level at about 3 ka. The Cape Sable area has apparently prograded up to 8 km (Enos and Perkins, 1979) and achieved its present location at about 1.5 to 1.2 ka (Roberts and others, 1977). Vermitid gastropod reefs began to develop at about 3 ka along much of the coastal region and their growth kept pace with slowly rising sea level (Shier, 1969). The vermitid reefs provided the topographically high areas for mangrove colonization, which eventually led to the domination of the coast by mangrove islands (Shier, 1969; Parkinson, 1989). The combination of the presence of these islands and the sediment accumulation that they foster has resulted in a 2-m-thick progradational sequence (Fig. 13) during the very slowly rising sea level of the late Holocene.

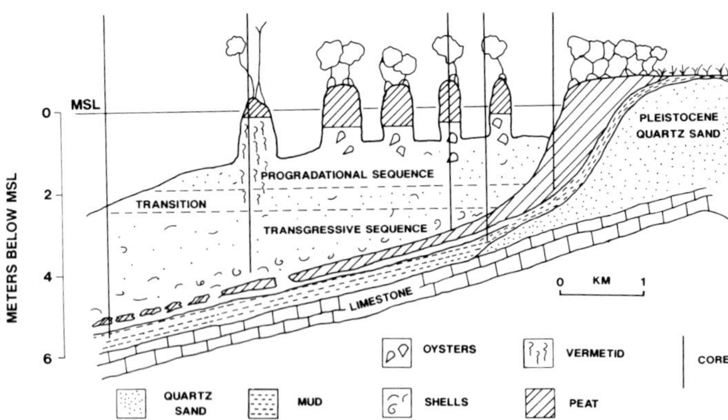

FIG. 13.—Idealized stratigraphic cross section representing the Ten Thousand Islands area showing late Holocene progradation after the initial transgression (after Parkinson, 1989).

WEST-CENTRAL BARRIER SYSTEM

The barrier-inlet system on the west-central peninsula of Florida has the most diverse morphology of any barrier system in the world. The system extends for about 300 km and includes 29 barrier islands and 30 tidal inlets. Included are long, narrow, wave-dominated barriers of both spit and upward-shoaling origins and mixed-energy drumstick barriers of a range of sizes and shapes. Spring-tidal range is just under 1 m and mean annual wave height is about 25 cm at the shore. Barriers extend from Anclote Key on the north to Kice Island at Cape Romano (Fig. 1).

The coastal reach is reasonably straight and oriented west of north throughout most of its extent. There is one major dislocation about two-thirds of the way to the south at Sanibel Island. Subsurface data from coring and seismic surveys indicate that this change in the coast is bedrock influenced (Evans and others, 1989). Two subtle coastal headlands are present; one of Pleistocene exposures in the northern part of the barrier system (central Sand Key; Fig. 14) and one of Miocene strata in the central part of the coast (near Venice Inlet; Fig. 14). The coastal bays protected by the barriers and served by the inlets range greatly in size and shape. Consequently, tidal prisms, inlet size, morphology, and stability are influenced by bays. There is considerable control of barrier islands by the antecedent geology of Neogene and Pleistocene units (Davis and Kuhn, 1985; Evans and others, 1985; Davis and others, 1989b; Evans and others, 1989). At least in the northern portion of the barrier system, the underlying carbonate strata are karstified.

Modern Processes and Geomorphology

The coastal-barrier system is one of mixed tidal and wave energy. Some individual barriers are indicative of more wave than tidal influence, such as Anclote Key, Casey Key and Captiva Island (Fig. 14). Others indicate substantial tidal influence, such as Caladesi Island, Siesta Key, Lovers Key and Marco Island (Fig. 14). The morphology of the barriers is the result of the interplay between these coastal processes. Slight modifications to either tidal or wave parameters can cause a significant shift in the morphology of the barrier-inlet system, especially along a low-energy coast. Wave refraction across ebb-tidal deltas is a major factor effecting these changes.

Hurricanes are fairly common along the barrier coast and have caused important changes during historical time; it is likely that similar changes took place throughout the Holocene. Relatively stable tidal inlets such as Hurricane Pass (1921), Johns Pass (1848) and Redfish Pass (1921) are the result of hurricanes breaching barrier islands in combination with stabilization by pre-Holocene strata. Data from the past century for this reach of barriers show four inlets being opened, 10 being closed and five that were both opened and closed (Fig. 14).

Another important feature of the barrier system is the development of three new barrier islands during a 25-year period beginning in the early 1960s; Three-Rooker Bar north of Honeymoon Island, and North and South Bunces Keys adjacent to Bunces Pass (Fig. 14). Each is the result of upward shoaling without benefit of an event such as a storm. All occur in an area of sediment abundance such as inlet-related shoals or barrier spits and all have a similar morphology. As soon as supratidal conditions persisted, colonization by opportunistic vegetation took place, followed shortly by coppice mounds and then dunes that have achieved elevations of >1 m above the back beach. The wave-dominated barriers are 1.5 to 3.0 km long and have recurved spits at each end (Fig. 15). Further extension is presently prohibited by tidal channels at each end. It is likely that several of the older barriers in this system had similar origins.

Holocene Stratigraphy

The barriers and related coastal environments along the west-central Florida peninsula were developed during the latter portion of the Holocene. The oldest supratidal accumulations on any of the barriers have been dated at near 3,000 yrs (Stapor and others, 1988), a date that is persistent in many barriers along this coast. Nearshore subtidal Holocene sediments beneath the present barrier islands date at 4,500 to 4,200 yrs. Nearby back-barrier deposits have been dated at 4,500 to 3,000 yrs (Evans and others, 1985).

The stratigraphic data base for this coast is spotty. There is good control in the northern third of the area and little to the south. Multiple beach-ridge systems have been analyzed and dated extensively in the southern area (Stapor and others, 1988). One can only compare by analogy, which seems to be appropriate due to the apparent similarities throughout this coastal reach.

The general condition is one of a sediment-starved coast with virtually all Holocene terrigenous sediments having been supplied through reworking during the transgression of the past 7,000 years and continuing to the present. The Holocene sediment prism is thickest under the barriers and thins offshore, so that at depths of 5 to 6 m there is little or no quartz sand overlying the pre-Holocene. There is also a thinning shoreward with the marine Holocene pinching out at the mainland coast. The total Holocene sequence at any given location along this coast is typically less than 7 to 8 m and may be as little as 3 m (Fig. 16). In most areas the Holocene rests unconformably and directly on pre-Pleistocene strata; however, there are locations where a thin, clayey sand Pleistocene unit is present (Davis and Kuhn, 1985; Davis and others, 1989a).

The basal Holocene unit is an organic-rich quartz sand with local peat. The mangrove peat has been dated at 4,500 yrs from a depth of 5 m below present sea level under Anclote Key, the northernmost barrier in the system. The uppermost portion of a similar organic-rich quartz sand has been dated at 1,850 yrs at 2.2 m below sea level.

Above the basal Holocene organic-rich quartz sand, the sequence begins to reflect a wave influence with cleaner, shelly quartz sands on the seaward side and muddy, less-sorted facies on the landward side under what is now a barrier. These are interpreted as shoals, probably subtidal initially, but becoming intertidal and eventually supratidal (Davis and Kuhn, 1985). As sea-level rise slowed between 3.5 and 3.0 ka, waves were able to mold the reworked terrigenous sediment, even though overall wave energy on this shallow and gently sloping inner shelf was quite low.

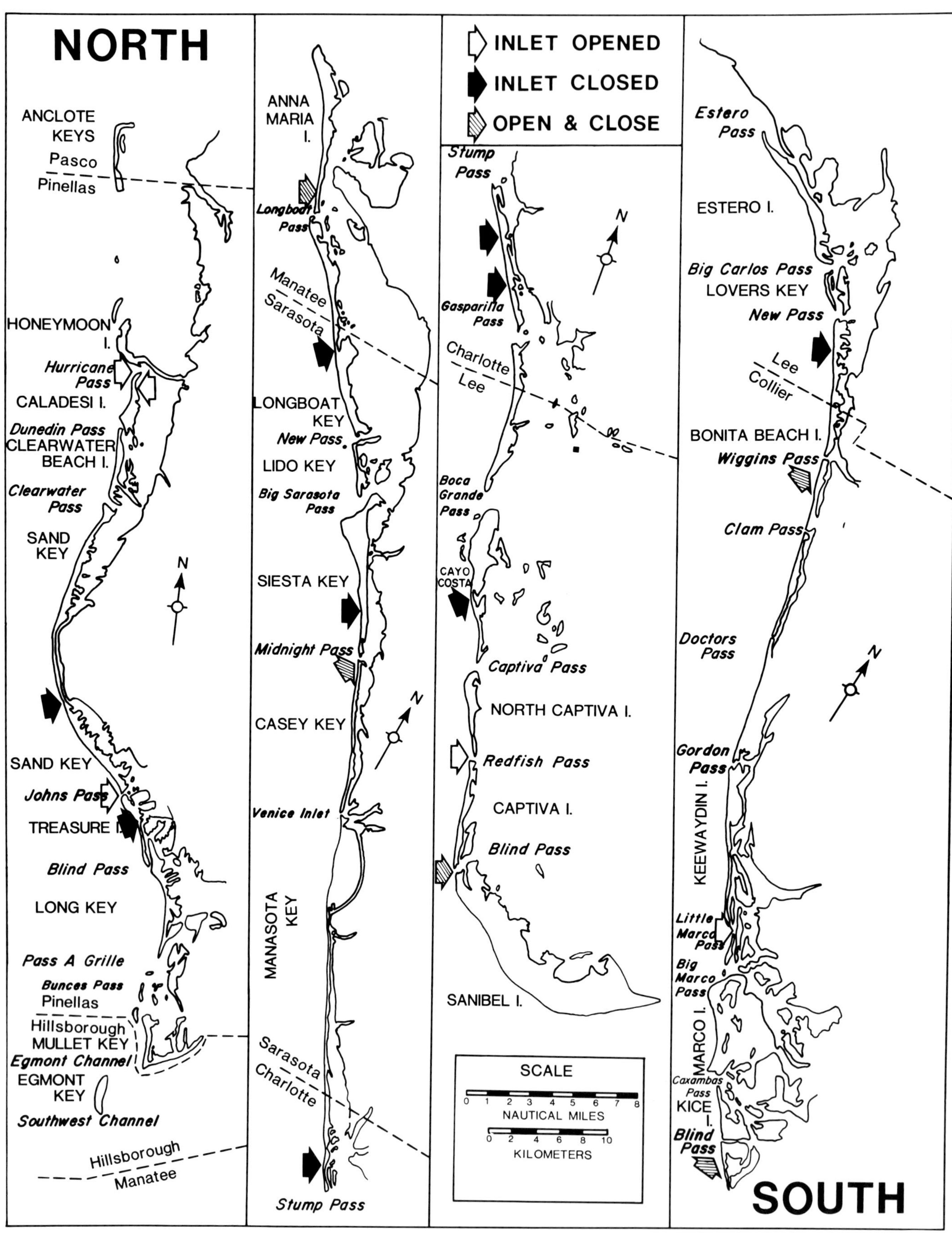
NORTH
INLET OPENED
INLET CLOSED
OPEN & CLOSE
ANCLOTE KEYS
Pasco
Pinellas
HONEYMOON I.
Hurricane Pass
CALADESI I.
Dunedin Pass
CLEARWATER BEACH I.
Clearwater Pass
SAND KEY
SAND KEY
Johns Pass
TREASURE I.
Blind Pass
LONG KEY
Pass A Grille
Bunces Pass
Pinellas
Hillsborough
MULLET KEY
Egmont Channel
EGMONT KEY
Southwest Channel
Hillsborough
Manatee
ANNA MARIA I.
Longboat Pass
Manatee
Sarasota
LONGBOAT KEY
New Pass
LIDO KEY
Big Sarasota Pass
SIESTA KEY
Midnight Pass
CASEY KEY
Venice Inlet
MANASOTA KEY
Sarasota
Charlotte
Stump Pass
Stump Pass
Gasparilla Pass
Charlotte
Lee
Boca Grande Pass
CAYO COSTA
Captiva Pass
NORTH CAPTIVA I.
Redfish Pass
CAPTIVA I.
Blind Pass
SANIBEL I.
SCALE
0 1 2 3 4 5 6 7 8
NAUTICAL MILES
0 2 4 6 8 10
KILOMETERS
Estero Pass
ESTERO I.
Big Carlos Pass
LOVERS KEY
New Pass
Lee
Collier
BONITA BEACH I.
Wiggins Pass
Clam Pass
Doctors Pass
Gordon Pass
KEEWAYDIN I.
Little Marco Pass
Big Marco Pass
MARCO I.
Caxambas Pass
KICE I.
Blind Pass
SOUTH
N

FIG. 15.—Oblique aerial photo at Bunces Pass showing two recently formed barrier islands: North Bunces Key (left), which initially emerged in 1962, and South Bunces Key (right), which emerged in 1977.

Some researchers (e.g., Evans and others, 1985) have suggested that these barriers formed at some distance offshore and migrated to the break in slope of the bedrock where stability was achieved. They do not provide data on the origin of the barrier, and vibracoring data along with the analysis of earlier aerial photos (Hine and others, 1987a) has shown an absence of stratigraphic data to substantiate significant landward migration. Undoubtedly, there was modest migration due to overwashing during the early stages of barrier development, as is seen in historically developed barriers in the area (Davis and Hine, 1989).

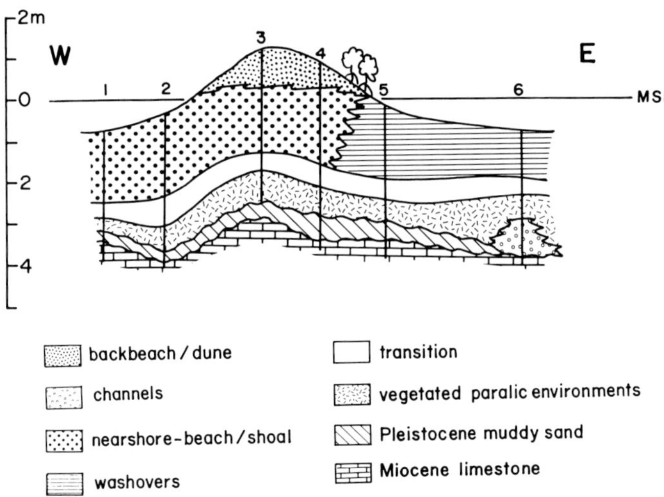

FIG. 16.—Stratigraphic cross section of Anclote Key (see Fig. 14 for location) showing typical stratigraphy of late Holocene barrier island on this coast (modified from Davis and Kuhn, 1985).

The locations along and adjacent to the islands where sediment was relatively abundant became emergent and developed prograding beach-ridge systems relatively early. The 3.0-ka ridges have been identified from Siesta Key, Gasparilla Islands, Cayo Costa, Sanibel Island and Marco Island (Stapor and others, 1988). Younger beach-ridge sets have also been identified on the islands (Fig. 17). The

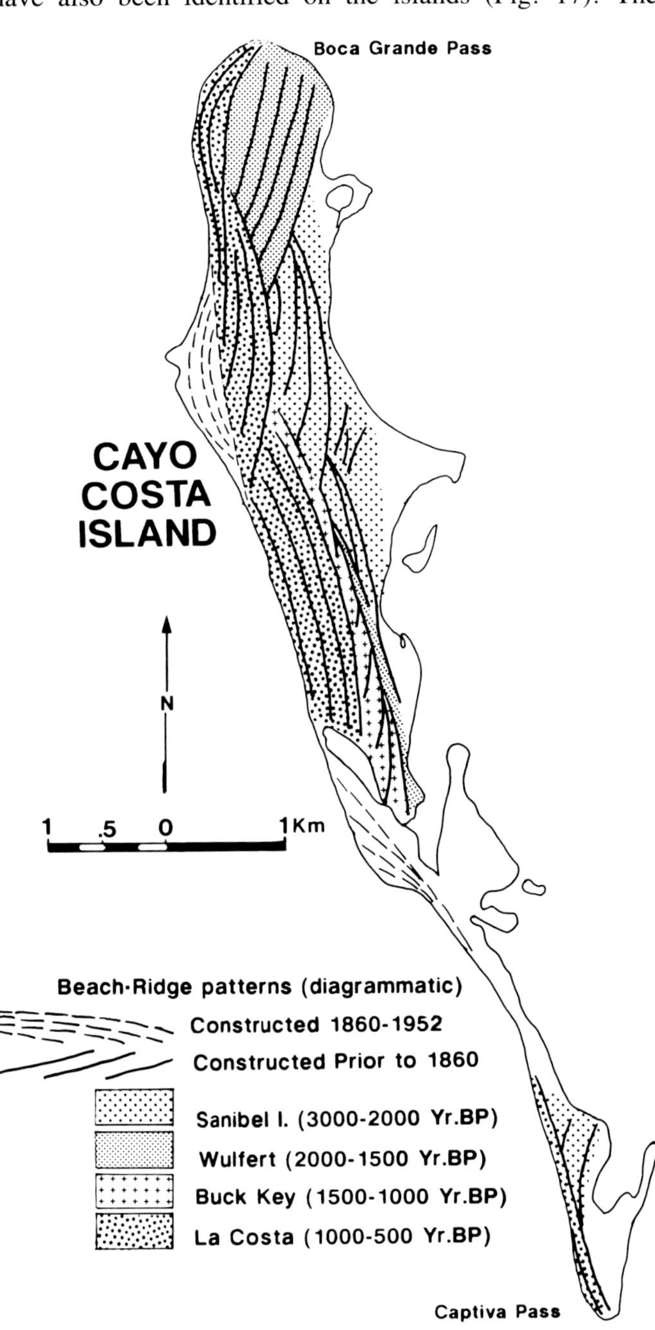

FIG. 17.—Map of Cayo Costa Island (see Fig. 14 for location) showing numerous patterns of beach ridges (modified from Stapor and others, 1988).

FIG. 14.—Islands and inlets that comprise the west-central Florida barrier system. Several inlets have been opened and closed during the past century (from Davis and Gibeaut, 1990).

younger islands are generally narrow with a single beach-dune ridge and have been dated at about 1,500 to 2,000 yrs (Davis and Kuhn, 1985; Stapor and Mathews, 1980; Stapor and others, 1988).

Once the islands became emergent, they were frequently overwashed until beach-foredune ridges became sufficiently elevated to prevent this process. The islands where sediment was abundant built prograding ridge systems (Figs. 14 and 17), whereas those in areas of little sediment were extended along shore and remained narrow (Fig. 14). Associated tidal inlets also were a means for capturing considerable sediment brought to the ebb-tidal deltas via littoral drift. Although some interpretations utilize a transgressive history (e.g., Evans and others, 1985), the barriers in this system probably formed close to their present position. If they had formed some distance offshore and migrated landward, the record of the migration has been destroyed.

MARSHY COAST OF THE BIG BEND AREA

A coastline dominated by a marine to brackish marsh-plant community extends for 350 km from the Appalachicola River Delta to Anclote Key (Fig. 1). This coastline, commonly referred to as the Big Bend Coast, is morphologically complex due to variations in underlying limestone bedrock topography, and the presence of actively discharging freshwater springs, large oyster bioherms, a modern river delta, and possible paleoshorelines. Two key factors have led to development of the plant-dominated coast: (1) the regional lack of siliciclastic sands and muds, and (2) the regional low wave-energy flux.

The shoreline constitutes approximately 25 percent of the entire Florida coastline and along with the inner-shelf system is an appropriate modern example of an epicontinental marine setting that was more prevalent during the high sea-level stands of the Cretaceous.

General Setting

Within the southern Big Bend area, quartz sands were concentrated along the Brooksville Ridge–Ocala Upland (a limestone topographic high a few kilometers inland from the marshy coast) and became thinner/disappeared to the west (Fig. 18).

Holocene sea-level rise has flooded a low-gradient, karstified, bare-rock surface of the ancient carbonate bank where well-cemented, shallow-marine skeletal limestones of Eocene/Oligocene age are exposed (Fig. 18). The quartz sands, as mentioned earlier, lie farther east. The lack of ancient large streams flowing across peninsular Florida in the area of the modern marsh coast is also a factor explaining the dearth of quartz sands. The Suwannee River is the only coastal-plain river that discharges along this coast. Other rivers emanate from springs fed by the Floridan Aquifer and travel only a few kilometers across the coastal plain before reaching the Gulf of Mexico. The Suwannee probably is a relatively geologically young stream formed when the Okeefenokee Swamp developed in south Georgia during the late Neogene/Quaternary (Carver and others, 1986). Also, it is a stream of relatively low water and sand-size sediment discharge due to its drainage basin being dominated by karstic limestone terrain.

Coastal Processes

The Big Bend area has historically been considered a low wave-energy coast. Although not a zero-energy coast, data from wave gauges and observations clearly show that it has a wave climate that is low in comparison to that of the Texas coast or even that of the east coast of Florida, with a mean annual wave height of 10 cm (Tanner, 1960). Frequent extratropical storm passages during the winter and infrequent tropical storms during the summer/fall do provide a local storm surge and sea state that are capable of flooding and eroding the marsh (Hine and others, 1987b).

The general low wave-energy environment is due to the extremely broad (>150 km) shelf width and very low, seaward-dipping gradient (1:5000) inherited from the ancient underlying carbonate platform. Additionally, the dominant extratropical storm winds are from the North/Northwest and blow alongshore or offshore.

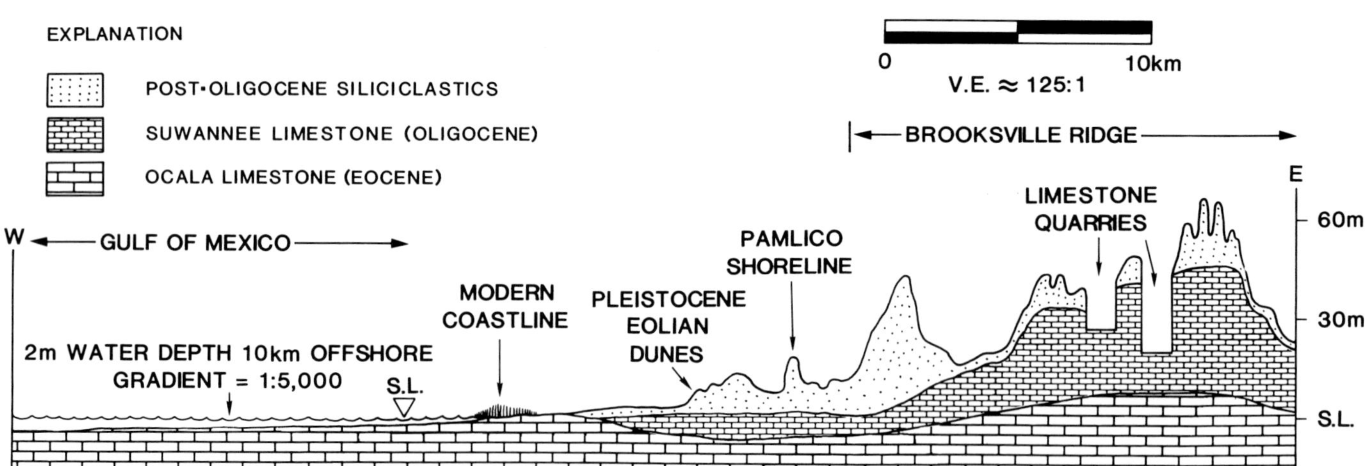

FIG. 18.—Stratigraphic cross section from the Brooksville Ridge across the coastal zone to the Gulf of Mexico (from Hine and others, 1988).

Karstification/Freshwater Discharge

Eocene and Oligocene limestones are exposed and lie within the shallow subsurface of the southern portion of the marsh coast. The karst topography created by dissolution processes is primarily responsible for influencing local coastal morphology, sedimentation and resulting stratigraphy. Two basic karst processes occur here: (1) surface dissolution due to downwelling of acid pore waters from overlying marsh sediments, and (2) regional dissolution, primarily subterranean dissolution, and subsequent collapse due to mixing-zone undersaturuation (Back and Hanshaw, 1970; Plummer, 1975; Back and others, 1986; Fig. 19). The combination of these two factors has produced three easily recognizable horizontal scales of surficial topography. The smallest scale (centimeters to a few meters) is relatively less important than medium- (tens to hundreds of meters) or large-scale (kilometers) karst-induced coastal features. Medium-scale features are rectilinear tidal creeks occupying enlarged joint patterns as well as rock-cored hammocks forming marsh islands (Fig. 20, south of the mouth of the Homosassa River on Fig. 21). Large-scale features are broad, shallow depressions in the bedrock forming shelf embayments, elevated rocky areas between embayments forming marsh-island archipelagoes, and linear channel structures etched in the bedrock by laterally moving spring-discharge events.

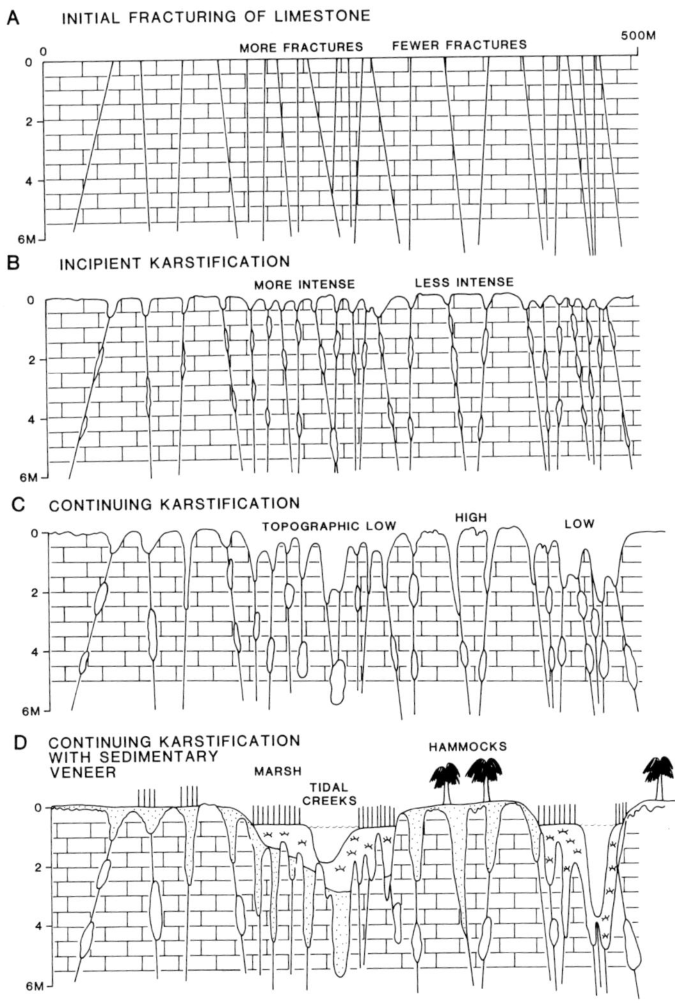

FIG. 19.—Sequential cross sections showing development of karstic area such as that beneath the marshy coastline (from Hine and others, 1988).

Coastal Morphology

The southern 65 km of the 350-km-long marsh coast has been studied in some detail (Hine and Belknap, 1986; Hine and others, 1988a, 1988) and has been divided into four morphologic sectors: (1) berm ridge, (2) marsh peninsula, (3) marsh archipelago, and (4) shelf embayment (Fig. 21). The berm-ridge shoreline is essentially a slowly eroding *Juncus roemerianus* marsh that supports a narrow but distinct sandy beach and berm ridge consisting of skeletal and quartz sands. During storms, the sands are washed over the berm ridge onto the marsh surface, thus creating a transgressive berm-ridge morphology. The marsh is punctuated by tidal creeks, which are bedrock controlled as they connect circular ponds created by sinkholes. The marsh-peninsula shoreline is more digitate and consists of points of land extending out into the Gulf that are tied to bedrock highs. During sea-level rise or storms, the marsh peninsulas retreat from their bedrock anchors and are ultimately stabilized by new bedrock highs located farther landward.

The marsh archipelago is dominated by numerous marsh islands that have formed on a flooded, elevated, topographically irregular bedrock surface. There are three marsh archipelago systems with each bordered by topographically lower shelf embayments (Figs. 21 and 22). The marsh creeks occupy dissolved rectilinear bedrock fractures, or they have connected a series of sinkholes. The marsh islands or hammocks are underlain by localized bedrock highs or nubs. Because of the elevated bedrock surface, the marsh stratigraphic veneer is thin and discontinuous. The limestone bedrock crops out extensively. Relatively thick (7 m) stratigraphic sections are found only in deep sinkholes. The holes are generally filled with basal clean, light-colored Pleistocene quartz sands followed by rooted, organic-rich, dark-colored, fine-grained quartz sands from the modern marsh.

As sea level rises, the marsh hammocks, originally surrounded by marsh grasses, become encircled by enlarging tidal creeks. Eventually, the hammocks become marsh islands having an inner vegetation core of less salt-tolerant trees and shrubs and an outer marsh-grass fringe.

The shelf embayments are broad, sometimes indistinct, indentions into the marsh coast and are morphologically and genetically related to freshwater-discharging springs. The combination of strong tidal flows, exposed bare-rock substrate, and reduced salinities has allowed the development of oyster (*Crassostrea virginica*) bioherms that extend laterally for many kilometers (Bahr and Lanier, 1981; Fig. 4). The bioherms create open-shelf estuarine systems that form distinct depositional basins, as shown by their sediments and stratigraphic relations (Fig. 23).

As sea level rises, the marsh hammocks submerge, the marsh coast retreats, and the rocky highs become new hard substrate for oyster-bioherm growth. The outer bioherms

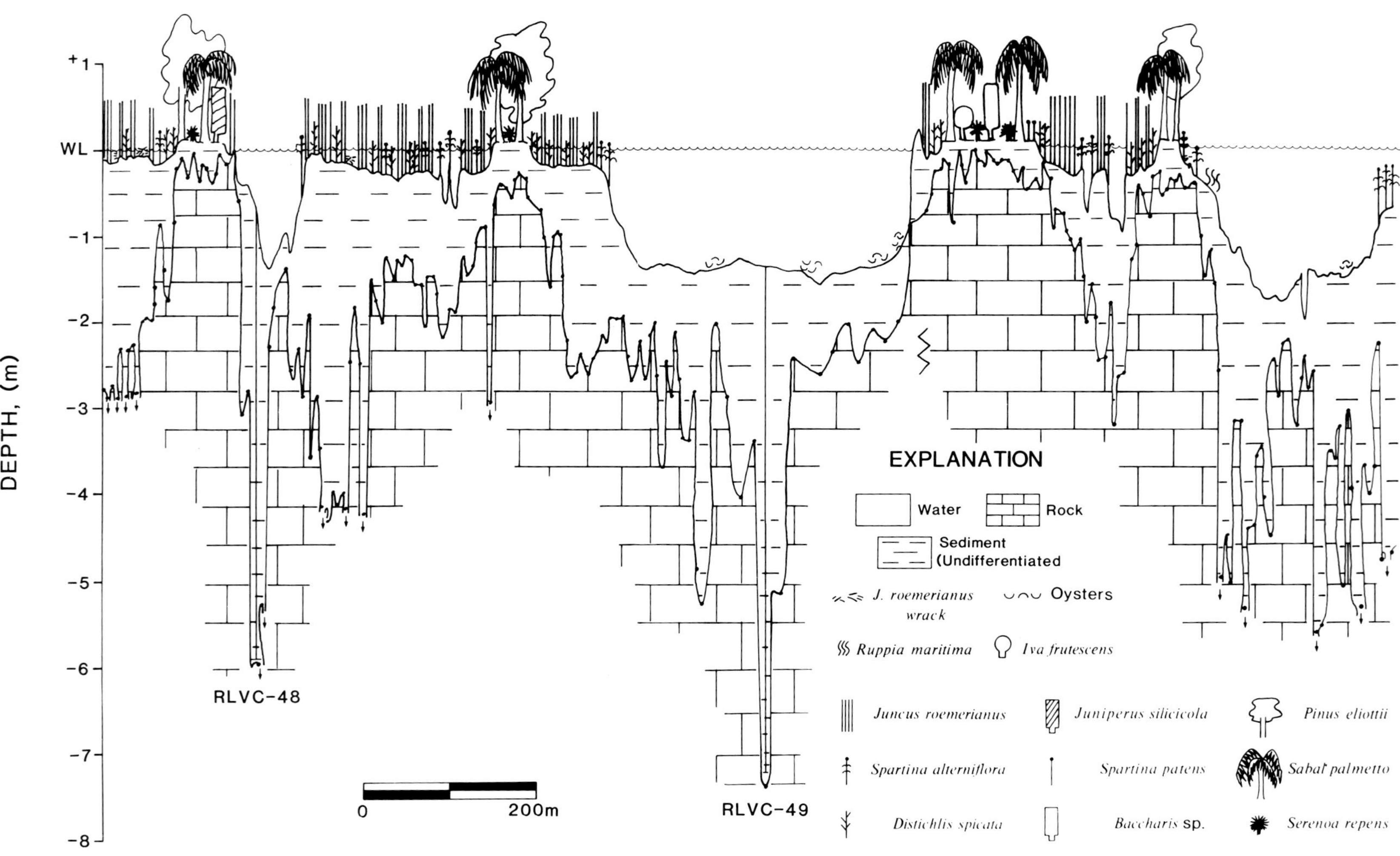

FIG. 20.—Actual stratigraphic cross section showing relations between underlying karstic limestone surface and surface environments (from Hine and others, 1988).

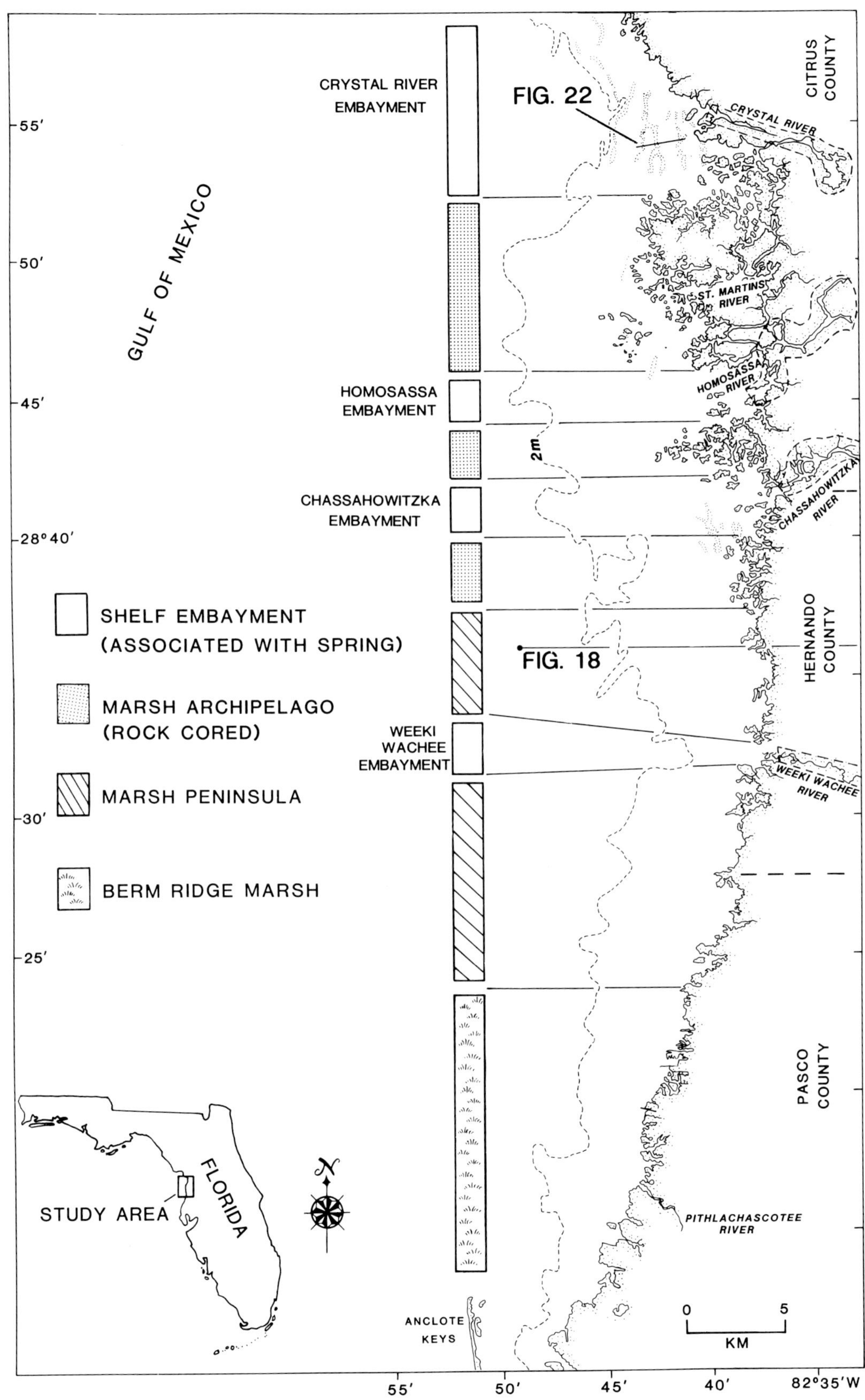

FIG. 21.—Map of marshy coast showing distribution of major morphologic elements (from Hine and others, 1988).

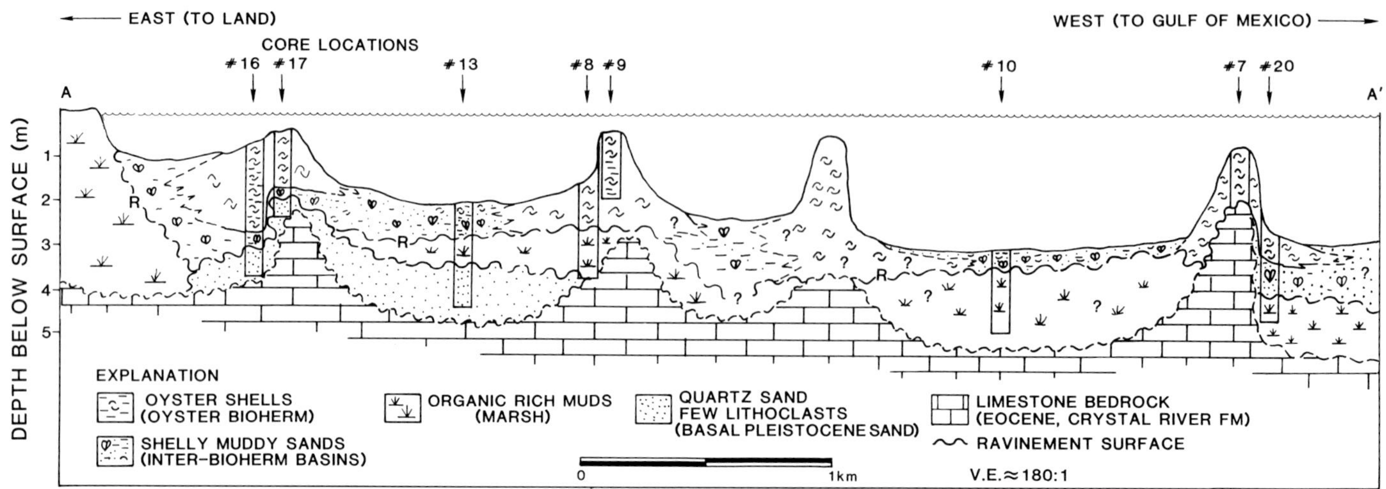

FIG. 22.—Cross section showing stratigraphy associated with a shelf embayment and a marsh archipelago (from Hine and others, 1988).

become exposed to higher salinities, thus allowing oyster predators to decimate the benthic molluscan community. In addition, increased exposure to a higher wave state allows the oyster bioherms to become physically dispersed.

SUMMARY

The Florida peninsula contains a diverse variety of coastal regimes, although it is exclusively within the microtidal range. Additionally, there is wide variation in adjacent shelf morphology, in sediment availability, in sediment composition, and in dominant coastal processes. The result is a fascinating coastal system that has developed during only the past few thousand years. Given its geographic extent, it may well be the most diverse coast in the world.

Included in the spectrum of coastal segments are: a wave-dominated barrier system (east coast), a rimmed carbonate-platform margin (the Keys), tide-dominated vegetated paralic systems (southwest and Big Bend areas), and a mixed-energy barrier system (Gulf central). These diverse suites of morphologies and related stratigraphic sequences have developed largely during the past 3,000 years under limited sediment availability, except on the east coast. The tidal range is rather similar throughout; however, the wave climate varies and thus provides a combination of coastal processes that, although having an overall low-energy climate, produces a wide range of coastal morphologies.

The balance between sea-level change, sediment availability and coastal processes is a delicate one. If a modest shift in any one of these should occur, it is likely that considerable coastal response would take place. The prognosis for increased rise in sea level along with ever-increasing human influence on the coast gives cause for concern about considerable change in the future for all of these coastal environments.

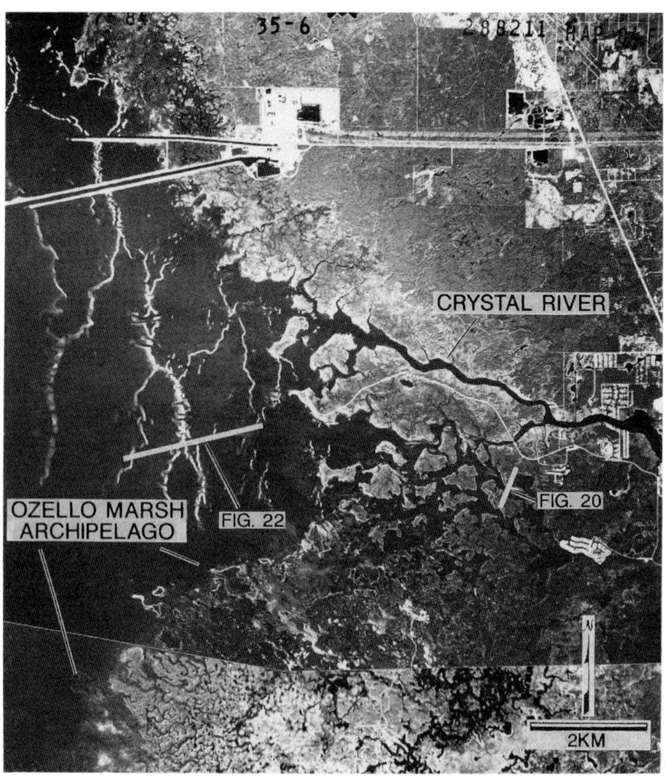

FIG. 23.—Aerial photo of the marshy coast estuary in the vicinity of Crystal River (after Hine and others, 1988).

REFERENCES

ALT, D., AND BROOKS, H. K., 1965, Age of Florida marine terraces: Journal of Geology, p. 406–4111.

BACK, W. B. AND HANSHAW, B. B., 1970, Comparison of chemical hydrogeology of the carbonate peninsulas of Florida and Yucatan: Journal Hydrology, v. 10, p. 330–368.

BACK, W. B., HANSHAW, B. B., HERMAN, J. S., AND VAN DRIEL, J. N., 1986, Differential dissolution of a Pleistocene reef in the ground-water mixing zone of coastal Yucatan, Mexico: Geology, v. 14, p. 137–140.

BAHR, L. M., AND LANIER, W. P., 1981, The ecology of intertidal oyster reefs of the South Atlantic coast: a community profile: Office of Biological Services, U.S. Fish and Wildlife Service, Washington D.C., FWS/OBS-81/15, 105 p.

BALL, M. M., SHINN, E. A., AND STOCKMAN, K. W., 1967, The geologic effects of Hurricane Donna in south Florida: Journal of Geology, v. 75, p. 583–597.

BROOKS, H. K., 1972, Geology of Cape Canaveral, space age geology: Southeastern Geology Society, 16th Field Conference, Tallahassee, Florida, p. 335–44.

BULLEN, R. P., 1975, Implications for some Florida deposits and their archeological contents: Florida Anthropologist, v. 21, p. 14–16.

CARVER, R. E., BROOK, G. A., HYATT, R. A., 1986, Trail Ridge and Okefenokee Swamp, *in* Leathery, T. L., ed., Geological Society of America, Centenial Field Guide Volume 6, p. 331–334.

CHEN, C. S., 1965, The regional lithostratigraphic analysis of Paleocene and Eocene rocks of Florida: Florida Geological Survey Bulletin, No. 45, 105 p.

COTTRELL, D. J., 1989, Holocene evolution of the northeastern coast and islands of Florida Bay (abst.) Bulletin of Marine Science, v. 44, p. 516.

DAVIES, T. D., AND COHEN, A. D. 1989, Composition and significance of the peat deposits of Florida Bay: Bulletin of Marine Science, v. 44, p. 387–391.

DAVIS, R. A., JR., AND GIBEAUT, J. C., 1990, Historical morphodynamics of inlets in Florida—models for coastal zone planning: Florida Sea Grant College Program, Technical Paper 55, 811 p.

DAVIS, R. A., AND HAYES, M. O., 1984, What is a wave-dominated coast? Marine Geology, v. 60, p. 313–329.

DAVIS, R. A., AND HINE, A. C., 1989, Quaternary geology and sedimentology of the barrier island and marshy coast, west-central Florida, U.S.A: 28th International Geological Congress, Field Trip Guidebook T375, American Geophys. Union, Washington, D.C., 38 p.

DAVIS, R. A., AND KLAY, J. M., 1989, Origin and development of Quaternary terrigenous inner shelf sequences, southwest Florida: Trans., Gulf Coast Association of Geologic Society, v. 39, p. 341–347.

DAVIS, R. A., AND KUHN, B. J., 1985, Origin and development of Anclote Key, west-peninsular Florida: Marine Geology, v. 63. 153–171.

DAVIS, R. A., JEWELL, P., AND SUSSKO, R. J., 1989a, Inner continental shelf off southwest Florida, *in* Morton, R. E., and Nummedal, D., eds., GCSSEPM Foundation, 7th Annual Research Conference, p. 53–61.

DAVIS, R. A., KNOWLES, S. C., AND BLAND, M. P., 1989b, Role of hurricanes in the Holocene stratigraphy of estuaries: examples from the Gulf Coast of Florida: Journal of Sedimentation Petrology, v. 59, p. 1052–1061.

ENOS, P., 1977, Quaternary depositional framework of south Florida, part I: carbonate sediment accumulations of the south Florida shelf margin: Geological Society of America Memoir 147, p. 1–130.

ENOS, P., 1989, Islands in the bay–a key habitat of Florida Bay: Bulletin of Marine Science, v. 44, no. 1, 365–386.

ENOS, P., AND PERKINS, R. D., 1974, Quaternary sedimentation in south Florida: Geological Society of American Memoir 147, 198 p.

ENOS, P., AND PERKINS, R. D., 1977, Quaternary sedimentation in south Florida: Geological Society of American Memoir 147, 130 p.

ENOS, P., AND PERKINS, R. D., 1979, Evolution of Florida Bay from island stratigraphy: Geological Society of American Bulletin, v. 90, p. 59–83.

EVANS, M. W., HINE, A. C., AND BELKNAP, D. F., 1989, Quaternary stratigraphy of Charlotte Harbor estuary-lagoon system, southwest Florida: implications of the carbonate-siliclastic transition: Marine Geology, v. 88, p. 319–348.

EVANS, M. W., HINE, A. C., BELKNAP, D. F., AND DAVIS, R. A., 1985, Bedrock controls on barrier island development: west-central Florida coast: Marine Geology, v. 63, p. 263–283.

FIELD, M. E., AND DUANE, D. B., 1974, Geomorphology and sediments of the inner continental shelf, Cape Canaveral, Florida: U.S. Army Corps of Engineers Technology Memo 42, 87 p.

GODFREY, P. J., 1976, Barrier islands of the east coast: Oceanus, v. 19, p. 27–40.

HAYES, M. O., 1979, Barrier island morphology as a function of tidal and wave regime, *in* Leatherman, S. L., ed., Barrier Islands from the Gulf of St. Lawrence to the Gulf of Mexico: New York, Academic Press, p. 1–27.

HEALY, H. L., 1975, Terraces and shorelines of Florida: Florida Bureau of Geology Map Series 71, Tallahassee, FL.

HINE, A. C., 1977, Lily Bank, Bahamas: history of an active oolite sand shoal: Journal of Sedimentation Petrology, v. 47, p. 1554–1582.

HINE, A. C., AND BELKNAP, D. F., 1986, Recent geological history and modern sedimentary processes of the Pasco, Hernando and Citrus County coastline: west-central Florida: Gainesville, University of Florida, Florida Sea Grant College Publishers, 79, 160 p.

HINE, A. C., BELKNAP, D. F., HUTTON, J. G., OSKING, E. B., AND EVANS, M. W., 1988, Recent geological history and modern sedimentary processes along an incipient, low-energy, epicontinental-sea coastline: northwest Florida: Journal of Sedimentation Petrology, v. 58, p. 567–579.

HINE, A. C., EVANS, M. W., DAVIS, R. A., AND BELKNAP, D. F., 1987a, Depositional response to seagrass mortality along a low-energy, barrier island coast: west-central Florida: Journal of Sedimentation Petrology, v. 57, p. 431–439.

HINE, A. C., EVANS, M. W., MEARNS, D. L., AND BELKNAP, D. F., 1987b, Effect of Hurricane Elena on Florida's marsh-dominated coast: Pasco, Hernando, and Citrus counties: Gainesville, University of Florida, Final Report to Florida Sea Grant College, 33 p.

HOFFMEISTER, J. E., AND MULTER, H. G., 1968, Geology and origin of the Florida Keys: Geological Society of America Bulletin, v. 79, p. 1487–1502.

HOFFMEISTER, J. E., STOCKMAN, K. W., AND MULTER, H. G., 1967, Miami Limestone of Florida and its Recent Bahamian counterpart: Geological Society of America Bulletin, v. 78, p. 175–190.

JONES, C. P., AND MEHTA, A. J., 1978, Ponce de Leon Inlet: Gainesville, FL: Florida Sea Grant Program Glossary of Inlets Report 6, 57 p.

KINDINGER, J. L., 1986, Geomorphology and tidal-belt depositional model of lower Florida Keys (abst.): AAPG Bulletin, v. 70, p. 607.

KIRTLEY, D. W., AND TANNER, W. F., 1968, Sabellariid worms: builders of a major reef type: Journal of Sedimentary Petrology, v. 38, p. 73–79.

KOFOED, J. W., 1963, Coastal development in Volusia and Brevard Counties, Florida: Bulletin of Marine Science of the Gulf and Caribbean, v. 13, p. 1–10.

LIDZ, B. H., HINE, A. C., SHINN, E. A., AND KINDINGER, J. L., 1991, Multiple outer-reef tracts along the South Florida bank margin: outlier reefs, a new windward-margin model: Geology, v. 19, p. 115–118.

LIDZ, B. H., ROBBIN, D. M., AND SHINN, E. A., 1985, Holocene carbonate sedimentary petrology and facies accumulation, Looe Key National Marine Sanctuary, Florida: Bulletin of Marine Science, v. 36, p. 672–700.

LIDZ, B. H., AND SHINN, E. A., 1991, Paleoshorelines, reefs and a rising sea: south Florida, U.S.A.: Journal of Coastal Res, v. 7, p. 203–229.

LIGHTY, R. G., 1977, Relict shelf-edge Holocene coral reef, southeast coast of Florida: Proceedings, Third International Coral Reef Symposium, v. 2, Miami, Fla., p. 215–221.

MARINO, J. N., AND MEHTA, A. J., 1986, Sediment volumes around Florida's east coast tidal inlets: COEL, University of Florida (UF/COEL-86/001), Gainesville, FL, 71 p.

MEHTA, A. J., AND BROOKS, H. K., 1973, Mosquito Lagoon barrier beach study: Shore and Beach, v. 41, p. 27–34.

MCKINNEY, M. L., 1984, Suwannee channel of the Paleogene coastal plain: support for the "carbonate suppression model" of basin formation: Geology, v. 12, p. 343–345.

NUMMEDAL, D., OERTEL, G. F., HUBBARD, D. K., AND HINE, A. C., 1977, Tidal inlet variability–Cape Hatteras to Cape Canaveral: Proceedings, Coastal Sediments '77, ASCE, Charleston, South Carolina, p. 543–562.

OSMOND, J. K., MAY, J. P., AND TANNER, W. F., 1970, Age of the Cape Kennedy barrier and lagoon complex: Journal of Geophysical Research, v. 75, p. 468–479.

PARKINSON, R. W., 1989, Decelerating Holocene sea-level rise and its influence on southwest Florida coastal evolution: a transgressive/regressive stratigraphy: Journal of Sedimentation Petrology, v. 59, p. 960–972.

PERKINS, R. D., 1977, Depositional framework of Pleistocene rocks in south Florida, *in* Enos, P., and Perkins, R. D., eds., Quaternary Sedimentation in South Florida, Part II: Geological Society of America Memoir 147, p. 131–198.

PINET, P. R., AND POPENOE, P., 1985, A scenario of Mesozoic-Cenozoic ocean circulation over the Blake Plateau and its environs: Geological Society of America Bulletin, v. 96, p. 618–626.

PLUMMER, L. N., 1975, Mixing of sea water with calcium carbonate ground water: Geological Society of America Memoir 142, p. 216–236.

PRICE, W. A., 1954, Shoreline and coasts of the Gulf of Mexico: U.S. Fish and Wildlife Service Fishery Bulletin 89, v. 55, p. 39–65.

PURI, H. S., AND VERNON, R. O., 1964, Summary of the geology of Florida: a guidebook to the classic exposures: Florida Geological Survey, Special Publication 5, Tallahassee, FL, 312 p.

ROBBIN, D. M., 1981, Subaerial $CaCO_3$ crust–a tool for defining sea-level changes and timing reef initiation: Proceedings, Fourth International Coral Reef Symposium, v. I, Manila, p. 575–579.

ROBBIN, D. M., 1984, A new Holocene sea-level curve for the upper Florida Keys and Florida reef tract, *in* Gleason, P. J., ed., Environments of South Florida, Present and Past, II, p. 437–458.

ROBBIN D. M., AND STIPP, J. J., 1979, Depositional rate of laminated soilstone crusts, Florida Keys: Journals of Sedimentary Petrology, v. 49, p. 175–180.

ROBERTS, H. H., WHELAN, R., AND SMITH, W. G., 1977, Holocene sedimentation at Cape Sable, South Florida: Sedimentation Geology, v. 18, p. 25–60.

SANFORD, S., 1909, The topography and geology of southern Florida: Florida Geological Survey Second Annual Report, p. 175–231.

SCHOLL, D. W., 1964, Recent sedimentary record in mangrove swamps and rise in sea level over the southwestern coast of Florida: Marine Geology, part 1, v. 1, p. 344–366; part 2, v. 2, p. 343–364.

SCHOLL, D. W., CRAIGHEAD, F. C., SR. AND STUIVER, M., 1969, Florida submergence curve revised: its relation to sedimentation rates: Science, v. 163, p. 562–564.

SCHOLL, D. W., AND STUIVER, M., 1967, Recent submergence of south Florida: a comparison with adjacent coasts and other eustatic data: Geological Society of America Bulletin, v. 78, p. 437–454.

SHIER, D. E., 1969, Vermetid reefs and coastal development in the Ten Thousand Islands, southwest Florida: Geological Society of America Bulletin, v. 80, p. 485–508.

SHINN, E. A., HUDSON, J. H., ROBBIN, D. M., AND LIDZ, B., 1981, Spurs and grooves revisited—construction versus erosion, Looe Key Reef, Florida: Proceedings, 4th International Coral Reef Symposium, Manila, Philippines, v. 1, p. 475–483.

SHINN, E. A., LIDZ, B. H., HALLEY, R. B., HUDSON, J. H., AND KINDINGER, J. L., 1989, Reefs of Florida and the Dry Tortugas: International Geological Congress, Field Trip Guidebook T176, American Geophysical Union, Washington, D.C., 53 p.

SHINN, E. A., LIDZ, B. H., AND HOLMES, C. W., 1990, High-energy carbonate-sand accumulation, the Quicksands, Southwest Florida Keys: Journal of Sedimentary Petrology, v. 60, no. 6, p. 952–967.

STAPOR, F. W., AND MATHEWS, T. D., 1980, C-14 chronology of Holocene barrier islands, Lee County, Florida: a preliminary report, *in* Tanner, W. F., ed., Shorelines Past and Present: Department of Geology, Florida State University, Tallahassee, FL p. 47–67.

STAPOR, F. W., MATHEWS, T. D., AND LINDFORS-KEARNS, F. E., 1988, Episodic barrier island growth in southwest Florida: a response to fluctuating Holocene sea level? *in* Maurrasse, J-M. R., ed., Miami Geological Society Memoir 3, p. 149–202.

STAPOR, F. W., AND MAY, J. P., 1983, The cellular nature of littoral drift along the northeast Florida coast: Marine Geology, v. 51, p. 217–237.

STAUBLE, D. K., AND DA COSTA, S. L., 1987, Evaluation of backshore protection techniques: Coastal Zone '87, Waterways Division, ASCE, p. 3233–3247.

STAUBLE, D. K., AND MCNEILL, D. F., 1985, Coastal geology and the occurrence of beachrock: central Florida Atlantic coast: Field Guide, 1985 Geological Society of America Annual Meeting, Orlando, 2 parts, 41 p.

STEINEN, R. P., HALLEY, R. B., AND VIDELOCK, S. L., 1977, Holocene dolomite locality in Florida Bay (abst.): AAPG Bulletin, v. 61, p. 83.

SWART, P. K., BERLER, D., MCNEILL, D., GUZIKOWSKI, M., HARRISON, S. A., AND DEDICK, E., 1989, Interstitial water geochemistry and carbonate diagenesis in the sub-surface of a Holocene mud island in Florida Bay: Bulletin of Marine Science, v. 44, p. 490–514.

SWIFT, D. J. P., KOFOED, J. W., SAULSBURY, F. P., AND SEARS, P., 1972, Holocene evolution of the shelf surface, central and southern Atlantic coast of North America, in Swift, D. J. P., Duane, D. B., and Pilkey, O. H., eds., Shelf Sediment Transport: Process and Pattern: Stroudsburg, Pennsylvania, Dowden, Hutchinson, and Ross, p. 499–574.

TANNER, W. F., 1960, Florida coastal classification: Transport, Gulf Coast Association of Geologic Society, v. 10, p. 259–266.

TURMEL, R. J., AND SWANSON, R. G., 1976, The development of Rodriguez Bank, a Holocene mudbank in the Florida reef tract: Journal of Sedimentary Petrology, v. 46, no. 3, p. 497–518.

VERNON, R. O., 1951, Geology of Citrus and Levy Counties: Florida Geological Survey Bulletin 33, 255 p.

WALTON, T. L., 1976, Littoral drift estimates along the coastline of Florida Gainesville, FL: Florida Sea Grant Program Report 13, 41 p.

WANLESS, H. R., 1981, Fining-upward sedimentary sequences generated in seagrass beds: Journal of Sedimentary Petrology, v. 51, no. 2, p. 445–0454.

WANLESS, H. R., 1982, Editorial: Sea level is rising—so what? Journal of Sedimentary Petrology, v. 52, p. 1051–1054.

WANLESS, H. R., AND TAGETT, M. G., 1989, Origin, growth and evolution of carbonate mudbanks in Florida Bay: Bulletin of Marine Science, v. 44 p. 454–489.

WHITE, W. A., 1958, Some geomorphic features of central peninsular Florida: Florida Geological Survey, Geological Bulletin 41, Tallahassee, FL, 92 p.

WHITE, W. A., 1970, The geomorphology of the Florida peninsula: Florida Geological Survey, Geological Bulletin 51, Tallahassee, FL, 164 p.

HYDROLOGY OF METEORIC DIAGENESIS: EFFECT OF PLEISTOCENE STRATIGRAPHY ON FRESHWATER LENSES OF BIG PINE KEY, FLORIDA

H. LEONARD VACHER, MICHAEL J. WIGHTMAN, AND MARK T. STEWART

Department of Geology, University of South Florida, Tampa FL, 33620

ABSTRACT: The Florida Keys offer an outstanding setting to study the diagenetic environment of one interglacial (the present) superimposed on limestones that formed during earlier interglacials. It has been shown by others that the surficial Miami Formation (oolitic facies) and Key Largo Formation (reef facies) of Big Pine Key were deposited during the last interglacial (unit Q5 of the local time stratigraphy; Q = Quaternary) and are underlain by Key Largo Formation of earlier interglacials (units Q4 and Q3). Mapping of the freshwater/saltwater interface beneath the lenses at Big Pine Key shows that the lenses are considerably foreshortened near the base of the Q5 unit. Dupuit-Ghyben-Herzberg (DGH) modeling of the northern lens indicates an order-of-magnitude contrast in hydraulic conductivity between the Q5 and pre-Q5 limestones, in addition to an order-of-magnitude contrast in hydraulic conductivity between the Q5 and modern analogs. Differences of such magnitude attest to the significance of secondary permeability in these Pleistocene limestones.

DGH flow-net analysis shows that the distribution of freshwater discharge, pore-volume flushing rate, and residence time along flowlines are all significantly affected by the presence of the buried, higher permeability limestones. The case study, therefore, illustrates how the development of secondary porosity and permeability during early meteoric diagenesis feeds back as a stratigraphic control on groundwater flow during continued meteoric exposure.

INTRODUCTION

Environments where carbonate deposits are altered penecontemporaneously with sedimentation include the marine-phreatic, meteoric-vadose, meteoric-phreatic and freshwater/saltwater mixing zones (Purdy, 1968; Matthews, 1974; Longman, 1980; Halley, 1984; Moore, 1989). The significance of the meteoric-phreatic zone to such early diagenetic processes as mineralogic stabilization, cementation, and development of secondary porosity has been documented by many workers in the context of both recent settings and ancient rocks (Land, 1970; Steinen and Matthews, 1973; Steinen and others, 1978; Halley and Harris, 1979; Craig, 1988; Budd, 1988a; Vacher and others, 1990; Budd and Vacher, 1991). Geochemical processes accompanying the circulation of fresh ground water through the meteoric-phreatic zone are examined by Plummer and others (1976), Steinen and others (1978) and Budd (1988b). In this paper, we present a hydrogeologic case study of Big Pine Key, Florida, that shows that one of the consequences of early diagenesis, namely the development of secondary permeability, affects the environment of further diagenesis. That is, the thickness and flux of fresh ground water are fundamentally controlled by the distribution of permeability, which undergoes large-scale changes during meteoric diagenesis.

GEOLOGIC SETTING AND STRATIGRAPHY

The Florida Keys (Fig. 1) are composed of two Pleistocene formations: the Miami and Key Largo Formations (Hoffmeister and Multer, 1968). The Key Largo Formation is exposed in the narrow middle and upper Keys, which are oriented parallel to the shelf margin. The Miami Formation is exposed in the larger lower Keys, which are oriented approximately normal to the shelf margin. The orientation reflects the origin of the deposits (Kindinger, 1986; Shinn and others, 1989): the Key Largo as a shallow-water complex of reefs and associated carbonate sands deposited along a topographic break in slope (Harrison and Coniglio, 1985), and the Miami Formation as an oolitic, tidal-bar sand belt (Ball, 1967; Hoffmeister and others, 1967).

Big Pine Key (Fig. 1) is the easternmost of the lower Keys. Its stratigraphy (Fig. 2) has been worked out by Coniglio and Harrison (1983) from cores of nine wells that were also used for a water-resources investigation (Hanson, 1980). Time-stratigraphic units, which are defined for south Florida by Perkins (1977), cut across the lithostratigraphy defined by the Miami and Key Largo Formations. The time stratigraphy is based on the occurrence of calcrete-coated discontinuity surfaces and calcretes (Perkins, 1977). There are five time-stratigraphic units, Q1 (oldest) through Q5 (youngest), representing successive interglacial stages (Perkins, 1977). The outcropping Q5 unit consists of a facies intergradation of Miami Formation and Key Largo Formation (Coniglio and Harrison, 1983; Kindinger, 1986), which is limited to the southeastern tip of the island (Fig. 2). In the main part of the island, where the freshwater lenses of this study occur, Q5-age Miami limestone is underlain by the Key Largo Formation of units Q4 and Q3 (Coniglio and Harrison, 1983). The Q4/Q3 boundary, which is the major discontinuity in south Florida (Perkins, 1977), lies close to the Miami/Key Largo contact (Fig. 2) and is marked by an especially thick, tight calcrete (Shinn, pers. commun., 1990).

The age of the time-stratigraphic units in Big Pine Key can be inferred from geochronological studies in the south Florida region. U-series dates on corals from exposures of the Key Largo Formation in the upper Keys and ooids from exposures of the Miami Formation in the Miami area are in the range of 120,000 to 140,000 yrs (Broecker and Thurber, 1965; Osmond and others, 1965), thus establishing correlation of the surficial unit, Q5, with the last major interglacial, substage 5e. Amino-acid racemization (AAR) ratios on *Mercenaria* from various south Florida localities (Mitterer, 1974, 1975) define 3 AAR groups, which Mitterer (1975) and Perkins (1977) correlate with the Q5, Q4 and Q3 time-stratigraphic units. Using the AAR ratio and U-series age of Q5 as a calibration, Mitterer (1975) estimates ages of 170,000 to 191,000 yrs for Q4, and 212,000 to more than 223,000 yrs for Q3. Using Mitterer's new technique of calculating ages from AAR ratios (Mitterer and

Quaternary Coasts of the United States: Marine and Lacustrine Systems, SEPM Special Publication No. 48

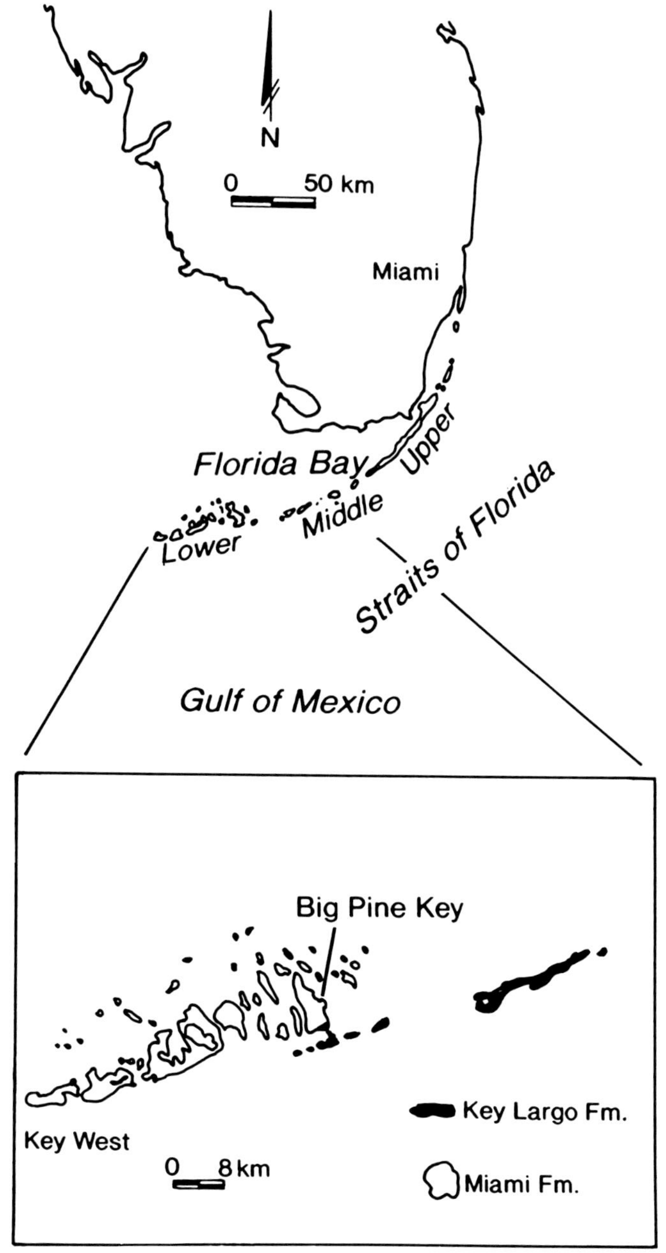

FIG. 1.—Maps showing location of the lower Florida Keys, Big Pine Key, and the Key Largo and Miami Formations.

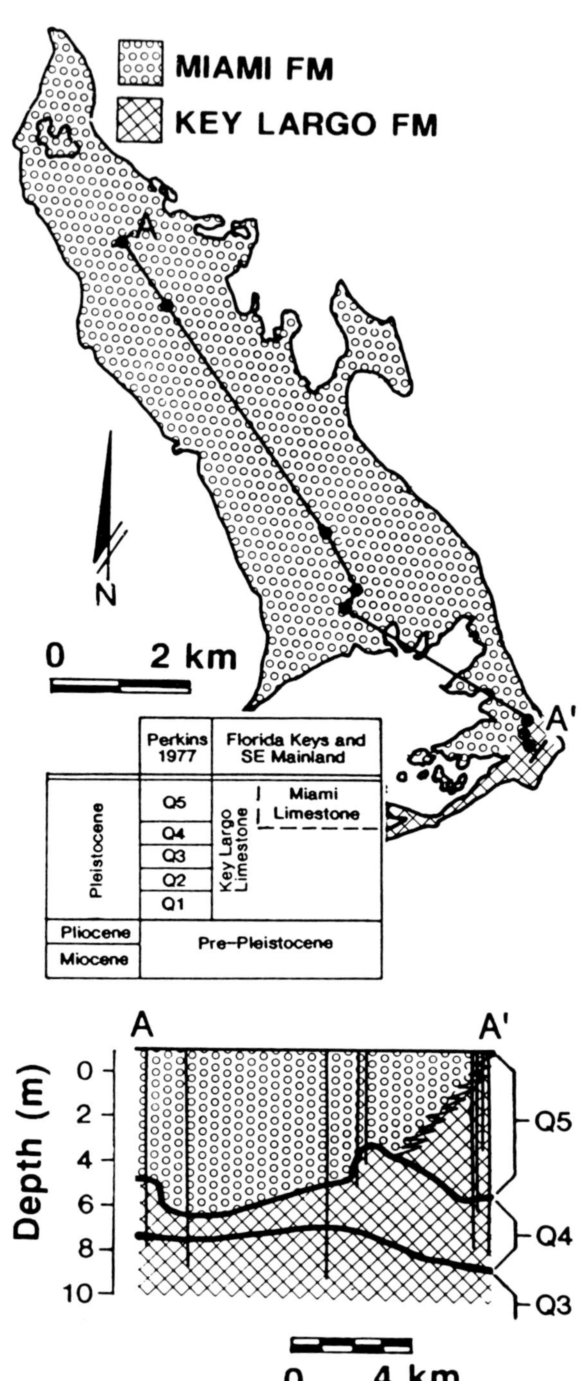

FIG. 2.— Summary of the stratigraphy of Big Pine Key. The Miami Formation (oolitic facies) occurs at the surface of most of the island and is limited to the uppermost, Q5, time-stratigraphic unit. Q4-age Key Largo Formation (reef facies) underlies the Miami Formation (oolite facies) except at the southeastern end of the island, where the Q5 unit includes the Key Largo Formation. Adapted from Coniglio and Harrison (1983) and Harrison and Coniglio (1985).

Kriausakul, 1989), we obtain somewhat larger age estimates: 184,000 to 216,000 yrs for Q4, and 245,000 to more than 260,000 yrs for Q3.

Coniglio and Harrison (1983) describe the petrography of the Big Pine Key limestones. According to their study: (1) the Q5 unit consists of aragonite and low-magnesium calcite, and, in contrast, the sediments of the Q4 and Q3 units are essentially entirely stabilized to low-magnesium calcite; (2) matrix porosity, although extremely variable, does not differ between the three units; and (3) the matrix porosity of Q5 is mostly primary, whereas that of the underlying units is nearly all secondary in the form of dissolutional vugs and molds. Coniglio and Harrison (1983, p. 146) note a trend in the nature of the porosity, in which "the stabilization of sediment mineralogy from aragonite to calcite is accompanied by a distinct change in the porosity fabric from primary to secondary.'' Suchtribution of porosity from primary interstices to secondary vugs can be expected to be paralleled by an increase in bulk permeability (hydraulic conductivity) because individual pores would be larger, though less numerous. If the Q4 and Q3 units are indeed substantially more permeable than the overlying Q5-age Miami Formation, then the geometry of the freshwater lenses in Big Pine Key should be affected (Vacher, 1988).

GEOMETRY OF THE FRESHWATER LENSES

The water-resources study of Hanson (1980) is based on 22 shallow observation wells where downhole variation of salinity was monitored at monthly intervals (6/76 to 4/77) by conductance probe and variable-depth water sampling. The measurements permitted Hanson (1980) to draw maps of the Cl^- variation in slices through the island at depths of 1.5, 3.0 and 4.6 m (5, 10 and 15 ft, respectively). Two lenses were defined (Fig. 3). They tend to expand and shrink laterally in the wet (June–Sept) and dry (Oct–May) seasons.

Wightman (1990) mapped the thickness of the freshwater column during March and August, 1987, by electromagnetic profiling (surface geophysics) using techniques developed by Stewart (1982, 1988, 1990). The surveys involved ground-conductivity readings at 20-m intervals along transects totalling 10.3 and 14.4 km (March and August, respectively). The readings are converted to thickness of fresh water (depth to ''interface'') using a three-layer model with known conductivity and thickness (ground elevation) of the unsaturated zone, fluid conductivity of the fresh water, and fluid conductivity and thickness (infinite) of the salt water (Stewart, 1988, 1990). Comparison of the results with data from Hanson (1980) and use of the technique in other areas (Stewart and others, 1983) suggest that the ''interface'' thus located is in the upper part of the transition zone, commonly between the 2,000 and 4,000-mg/l isochlors.

Figure 4 shows the thickness of the two lenses during the dry and wet seasons of 1987. As shown by these maps, the seasonal expansion and contraction of the area of lenses as documented by the slice maps of Hanson (1980) are not

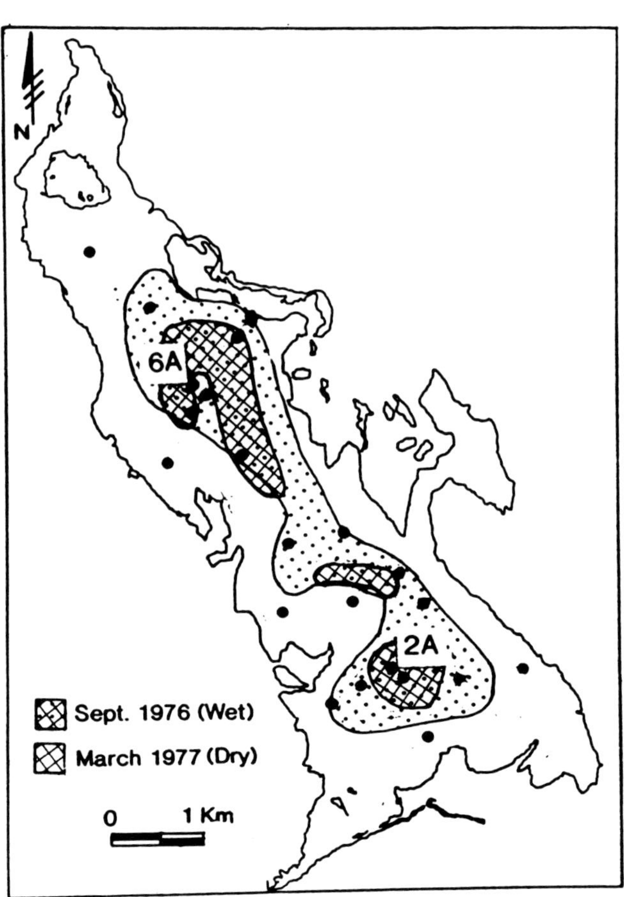

FIG. 3.—Map showing seasonal variation in areal extent of freshwater lenses. The boundaries are the wet-season and dry-season contours of 500-mg/l Cl^- at 1.5-m depth below water table as mapped by Hanson (1980). Adapted from Coniglio and Harrison (1983); data from Hanson (1980).

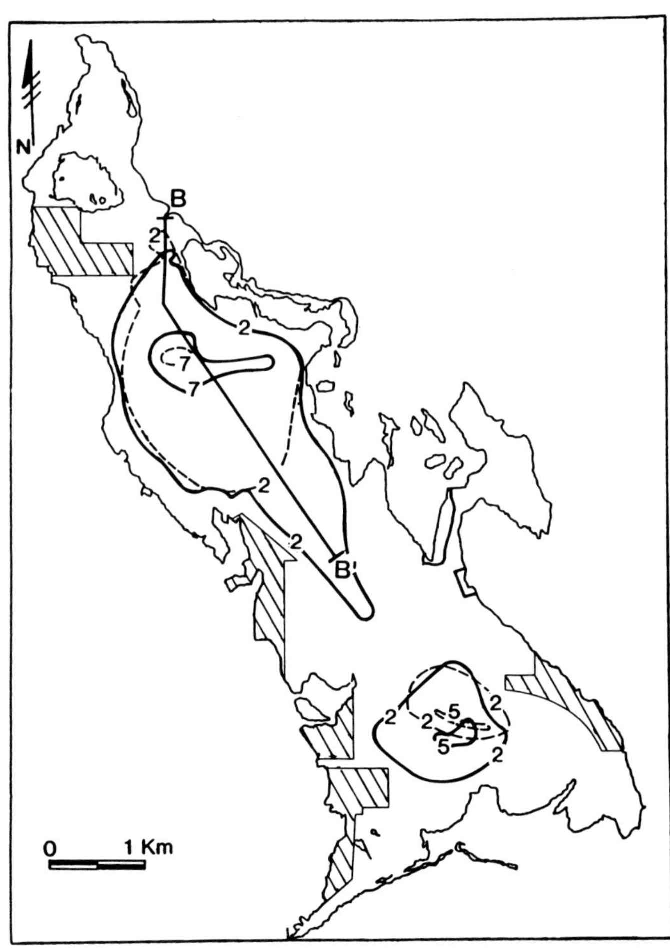

FIG. 4.—Map showing areal extent of freshwater lenses at depths of 2, 5 and 7 m during wet season (solid contours) and dry season (dashed), 1987, by electromagnetic profiling (Wightman, 1990). Cross-hatching indicates areas of finger canals, which clearly influence the area of the lenses.

paralleled by significant thickening and thinning of the lenses.

RELATION BETWEEN STRATIGRAPHY AND THICKNESS OF FRESH GROUND WATER

Observations

Two wells of the Hanson (1980) study penetrate into the salt water beneath the northern lens. Downhole variation of salinity in these wells shows a correspondence between the base of the fresh ground water and the base of the Miami Formation (Fig. 5).

The correspondence between the base of the lens and the location of the geologic contact can be seen in more detail in transects of the geophysical survey. Figure 6 shows one example: a 4.5-km transect along the axis of the northern lens.

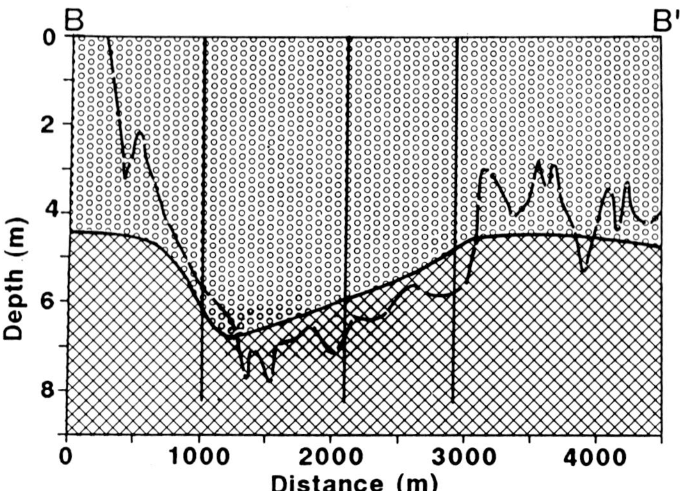

FIG. 6.—Cross section showing correlation of the interface determined by electromagnetic profiling (Wightman, 1990) and the Miami/Key Largo contact. The latter is from contour map of depth to Key Larto in Hanson (1980). Location of cross section is shown in Figure 4. Lithology patterns are as in Figure 2.

Interpretation

The distribution of hydraulic conductivity in an island exerts a fundamental control on the shape on the island's Ghyben-Herzberg lens (Vacher, 1988). The effect of hydraulic conductivity variations can be evaluated by use of Dupuit-Ghyben-Herzberg (DGH) analysis (Vacher, 1988). Figure 7 shows two types of geologic control; in both, one hydrogeologic unit is 10 times more permeable than the other. In the first type of island, there is a lateral geologic variation such that the lens is thinned in the more permeable island sector; Bermuda is an example of this type of island (Vacher, 1978). In the second type of island, the root of the lens is foreshortened in the buried, more permeable hydrogeologic unit; from Figures 5 and 6, we conclude that Big Pine Key is an example of this type of island. Other cases of vertically foreshortened lenses occur in the larger islands of the Bahamas (Cant and Weech, 1986).

Equations from DGH analysis can be used to estimate the magnitude of the contrast in hydraulic conductivity. The technique (Vacher, 1978) involves fitting type curves to the observed configuration of the interface. For the fit used here, a type curve was developed from an analytical model for a circular, two-layer island (Wightman, 1990) and matched to the transect of the northern lens where its areal geometry is most circular. The best fit (Wightman, 1990) for this transect was obtained for $R/K1 = 2 \times 10^{-3}$ and $R/K2 = 1.7 \times 10^{-4}$, where R is recharge, and $K1$ and $K2$ are the hydraulic conductivities of the two hydrogeologic units. As hydraulic conductivity varies directly with intrinsic permeability (K of 1 m/day corresponds to an intrinsic permeability of 1.2 d for fresh water at 20°C), these figures imply a 12-fold permeability contrast between the Q5 layer and the underlying limestones.

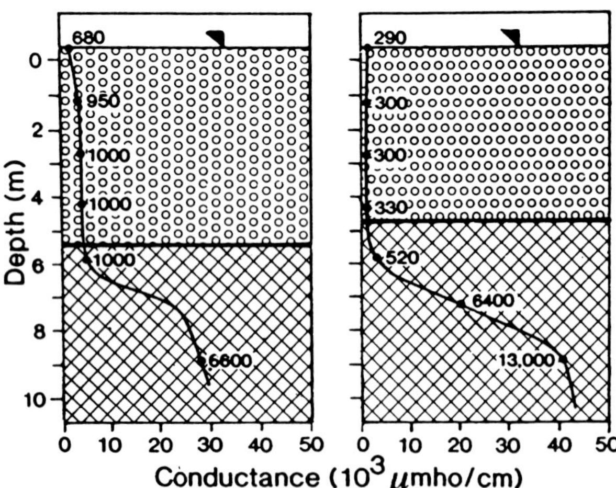

FIG. 5.—Graphs showing conductance vs. depth at wells 6A (left) and 2A (right; modified from Hanson, 1980). Location of wells is shown in Figure 3. Numbers on graphs refer to chlorinity in mg/l. Lithology patterns are as in Figure 2. Note correlation of the Miami/Key Largo contact and the start of the transition zone.

The ratio of Cl^- in the rainfall to that of the freshest ground water in the lens can be used to estimate the island's recharge (Vacher and Ayers, 1980). Using data for Big Pine Key, recharge appears to be about 0.24 m/yr (20 percent of the rainfall). With this value, estimates of the hydraulic conductivities are 120 m/day (140 d) for the upper (Q5) layer and 1,400 m/day (1,700 d) for the underlying (Q4 and Q3) limestones.

In many ways, the hydrogeology of Big Pine Key is like that of Bermuda–even quantitatively. In the area of Bermuda's principal freshwater lens, the upper saturated zone consists of two major hydrogeologic units (Vacher, 1978; Rowe, 1984). The younger unit consists of the Rocky Bay Formation (Vacher and others, 1989), which, like Q5, formed during substage 5e (Harmon and others, 1983; Vacher and Hearty, 1989); in the upper saturated zone, this younger unit has a hydraulic conductivity of about 90 m/day (as compared to 120 m/day for the Q5 unit of Big Pine Key). The older hydrogeologic unit of Bermuda consists of the Belmont Formation and the upper member of the Town Hill Formation (Vacher and others, 1989). The older hydrogeologic unit is like the deeper layer at Big Pine Key (Q4 and

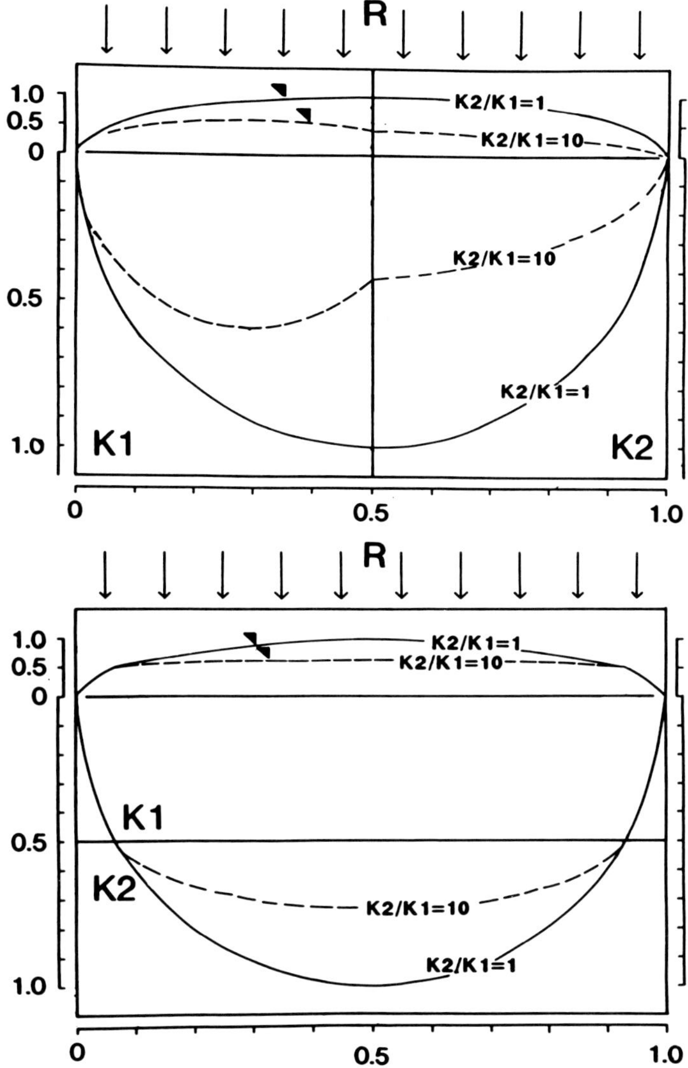

FIG. 7.—Type curves showing the shape of strip-island lenses predicted from Dupuit-Ghyben-Herzberg analysis. For both sets of type curves, the contrast in hydraulic conductivity between geologic units is 10 (dashed) and, for purposes of comparison, one (solid). The vertical axes are dimensionless, defined as the elevation of the water table, or the depth of the interface, relative to the corresponding parameter in the lens where K2/K1 = 1. In the upper case, the two units occur as adjoining strips, with the geologic contact occurring along the central axis of the strip island. In the lower case, the two units are directly superposed, with the geologic contact lying at a depth one-half the maximum depth of the lens that would occur if the deeper, more permeable units were not present. Adapted from Vacher (1988).

Q3) in that: the age of the Belmont is about 200,000 yrs (Harmon and others, 1983; Vacher and Hearty, 1989); AAR ratios suggest that the age of the upper Town Hill is about 300,000 yrs (Hearty, pers. commun., 1991); and the Belmont/Town Hill discontinuity is marked by one of the most prominent paleosols on the island. In the upper saturated zone, this older hydrogeologic unit of Bermuda has a hydraulic conductivity on the order of 1,000 m/day (as compared to 1,400 m/day for the pre-Q5 limestones of Big Pine Key). Indeed, the chief difference between the hydrostratigraphy of Bermuda and that of Big Pine Key is that there is a lateral succession of units in Bermuda, and a vertical succession of units in Big Pine Key.

By the way the estimates of hydraulic conductivity are derived in this paper, they clearly refer to large volumes of rock. They cannot be considered precise measurements. It is difficult to evaluate the uncertainties in the estimates, but we suspect that, like the results of most aquifer tests, the derived values are probably "good" to one significant figure. Even with this lack of precision, two facts stand out. First, there is an order-of-magnitude contrast in the hydraulic conductivity (permeability) between the Late Pleistocene Q5 unit and the underlying Pleistocene limestones. Second, the hydraulic conductivity of the Q5 grainstones is an order of magnitude larger than the hydraulic conductivity of recent depositional analogs (Enos and Sawatsky, 1981; Budd, 1984). These contrasts testify to the significance of the redistribution of porosity that occurs during meteoric diagenesis (Coniglio and Harrison, 1983).

HYDRAULICS OF THE NORTHERN LENS

The presence of a deeper, more permeable layer near the base of the lens profoundly affects the distribution of groundwater flux. Figure 8 shows streamlines in a circular island of radius 1 km for both one-layer and two-layer cases. The size of the island and the values of recharge and hydraulic conductivities are patterned after those of the northern lens of Big Pine Key. The location of the 20 streamlines is determined by techniques of DGH flow-net analysis developed by Vacher and others (1990) for homogeneous, infinite-strip islands and modified by Wightman (1990) for circular and multiple-layer islands. As shown in Figure 8, the lens is thinned in the two-layer case; the streamlines are crowded into the deeper unit; and the streamlines are refracted across the contact. Similar patterns occur in regional aquifers (Freeze and Witherspoon, 1967).

Although the discharge in each of the streamtubes of Figure 8 is the same, there is a difference in the width of the streamtubes at the water table because the island is not an infinite strip. That is, each streamtube (considered three-dimensionally) carries the recharge of an annulus 1/21 of the area of the circular island; annulus widths decrease with increasing distance from the center of the island.

Given a flow net of the island, velocities and residence times can be calculated (Vacher and others, 1990). The spot numbers on Figure 8 show representative flushing and interstitial velocities, assuming that effective porosity is 20 percent. Assuming any value for effective porosity is problematic given the scatter of values found by Coniglio and Harrison (1983). We choose 20 percent because a range extending to a factor of two on either side of this value spans the possibilities (Coniglio and Harrison, 1983). Given that the pore volumes as calculated in Figure 8 vary inversely with effective porosity, the values in the figure can vary by a factor of two because of the value assumed for effective porosity. The distribution of values, however, is controlled by the permeability variation, which is reflected by the crowding of the streamlines.

The presence of the buried, high-permeability unit affects the time that it takes for a parcel of recharge to travel a

QUATERNARY EVOLUTION OF THE APALACHICOLA COAST, NORTHEASTERN GULF OF MEXICO

ERVIN G. OTVOS
Gulf Coast Research Laboratory, P.O. Box 7000, Ocean Springs, Mississippi 39564-7000

ABSTRACT: Pre-existing topography strongly influenced Sangamonian transgression in the Apalachicola area. Late Pleistocene neritic and estuarine Biloxi and Gulfport barrier-complex deposits indicate that interglacial sea level rose from at least −37 m, relative to present sea level, to above +3 m. The Biloxi Formation provides a Gulf of Mexico-wide stratigraphic marker. Subsequent regression first was accompanied by river-channel incision at a level slightly lower than interglacial- and much higher than full glacial-erosion levels. Large eolian dunes formed over Gulfport barrier surfaces, reflecting the regional extent of a Wisconsinan wind system.

Antecedent Pleistocene topography also greatly influenced Holocene sedimentation and associated landforms. The late Holocene history of St. Vincent and "Little St. George" Islands and St. Joseph barrier spit is characterized by strandplain progradation. Unlike St. George Island, the St. Joseph barrier spit did not form through integration of emerging island cores. Multiple erosional episodes characterize the Quaternary barriers. Conclusive field evidence is lacking for Late Holocene sea-level fluctuations. Massive dilution by Apalachicola River runoff and resuspension/homogenization of the bay deposits by occasional hurricanes tend to diminish lateral and vertical salinity gradients in the record of the late Holocene sedimentary cycle. Unlike certain central Gulf Coast areas, contrasts in vertical salinity between Late Holocene lagoonal and neritic deposits in the Apalachicola are weak to nonexistent.

INTRODUCTION

The Apalachicola coast of northwest Florida (Fig. 1) forms a triangular-shaped bulge approximately 72 km wide. Its tectonic framework is the Apalachicola Embayment, a Mesozoic-Neogene structure downwarped along a south-southwest-plunging axis (Schmidt, 1984). Upper Neogene carbonates and siliciclastics underlie Quaternary deposits. Westward, the late Neogene carbonates are gradually replaced by siliciclastic deposits.

An extended period of erosion by the ancestral drainage network preceded accumulation of Sangamonian deposits that are unusually thick in the southwest area. Stream channels cut through highstand deposits during subsequent Wisconsinan lower sea-level stages. Mainland and island barriers formed. After Holocene transgression reached its maximum distance inland, sediments of a regressive hemicycle were deposited in bays and lagoons.

STUDY METHODS

Rotary-drill core samples, obtained by split spoon from 46 holes (Figs. 1 and 2), were utilized in the lithologic and fossil study of the mainland sequences investigated (Otvos, 1985, 1990). Additional cores came from the bay system (Otvos, 1985). Study of exposures, field samples, aerial photos and radiocarbon dates supplemented drill-sample investigations. Cores 45 cm in length were taken at every 1 m and analyzed at close intervals for granulometry and microfossil content. The inferred four salinity categories (oligohaline–lower mesohaline; polyhaline–lower euhaline; mesohaline; and euhaline) were based on foraminiferal assemblages and specific abundance (Table I, in Otvos, 1990).

GEOLOGIC EVOLUTION

Late Neogene Deposits

Alluvial Pleistocene sediments first blanketed a moderately dissected late Neogene terrain (Table 1; Fig. 2), which is underlain by deposits that formed in nearshore and inner-shelf environments. These include limestones, limey marls and semiconsolidated limey biocalcarenites and calcilutites and, to a lesser extent, siliciclastic sediments (Otvos, 1990).

Previously, the carbonates were assigned to the Upper Miocene Choctawhatchee Formation, or Stage. Later, Akers (1972), Schmidt and Clark (1980) and Huddlestun (1984) identified Lower and Lower-to-Middle Pliocene units over the Miocene. Pliocene-Pleistocene planktonic *Globigerina riveroae* (N18-19) and *Globigerinoides ruber* (Blow foraminiferal Zones N18-23) occur in neritic calcareous sediments (Otvos, 1991b). Contrary to Schmidt's suggestion (1984), the Pliocene Jackson Bluff Formation (Zones N18-19) does not extend into the Pleistocene in a time-transgressive fashion. Upper Miocene deposits in drill hole #16 (Lanark Village; Fig. 2; Otvos, 1990), at only 6 m, reflect the erosional thinning of the overlying Pliocene unit.

The uppermost Neogene units include higher concentrations of siliciclastic sediments, deposited in nearshore inshore facies (Table 1). They include 3- to 12-m-thick quartz-sand, silt, and lime-cemented quartz-sandstone intervals. Lithologic distinctions in cores between Neogene and overlying Pleistocene siliciclastics allowed detailed reconstruction, for the first time, of the pre-late Quaternary land surface (Fig. 2).

Greenish- and olive-gray sediments of stiff consistency characterize the distinct Neogene siliciclastics. Occasional dark yellowish-orange and brown colors occur near the top, formed in the pre-Pleistocene weathering horizon. Fossiliferous siliciclastics grade upward into fossil-free silts and quartz sands (drill holes #6, 12, 14b, c, 18; Fig. 3A, B, and Otvos, 1990), signifying increasingly brackish coastal, even alluvial conditions during the late Middle Pliocene regressive stage.

Pre-Sangamonian Quaternary Surface Morphology

Following an extended period of non-deposition and stream erosion, late Pleistocene fluvial deposits covered the Neogene land surface that sloped gently toward the southwest and south. In the southwest, this surface was detected to at least 48 m below sea level. A shallow buried valley occurs between drill holes #22 and 26 (Figs. 2, 3C). Only a Pleistocene veneer overlies the Neogene to the northeast (drill holes #5 and #6). Figure 2 refutes the suggestion by Schnable and Goodell (1968, p. 34) that "during much of

Quaternary Coasts of the United States: Marine and Lacustrine Systems, SEPM Special Publication No. 48

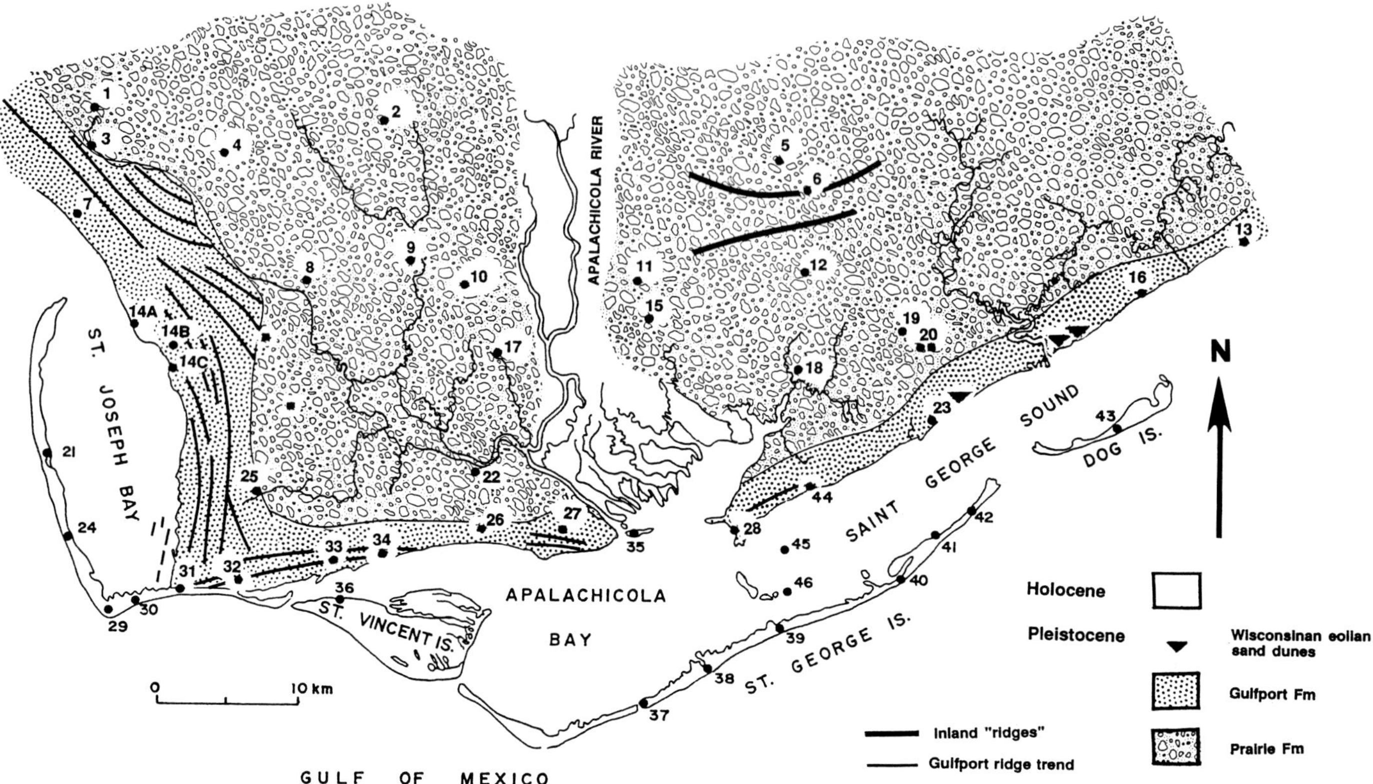

FIG. 1.—Drillhole locations and surface geology, Apalachicola Coast. Dashed lines in St. Joseph Bay mark submerged Gulfport ridges. Two "inland ridges" at drill hole #6 (Brenneman and Tanner, 1958).

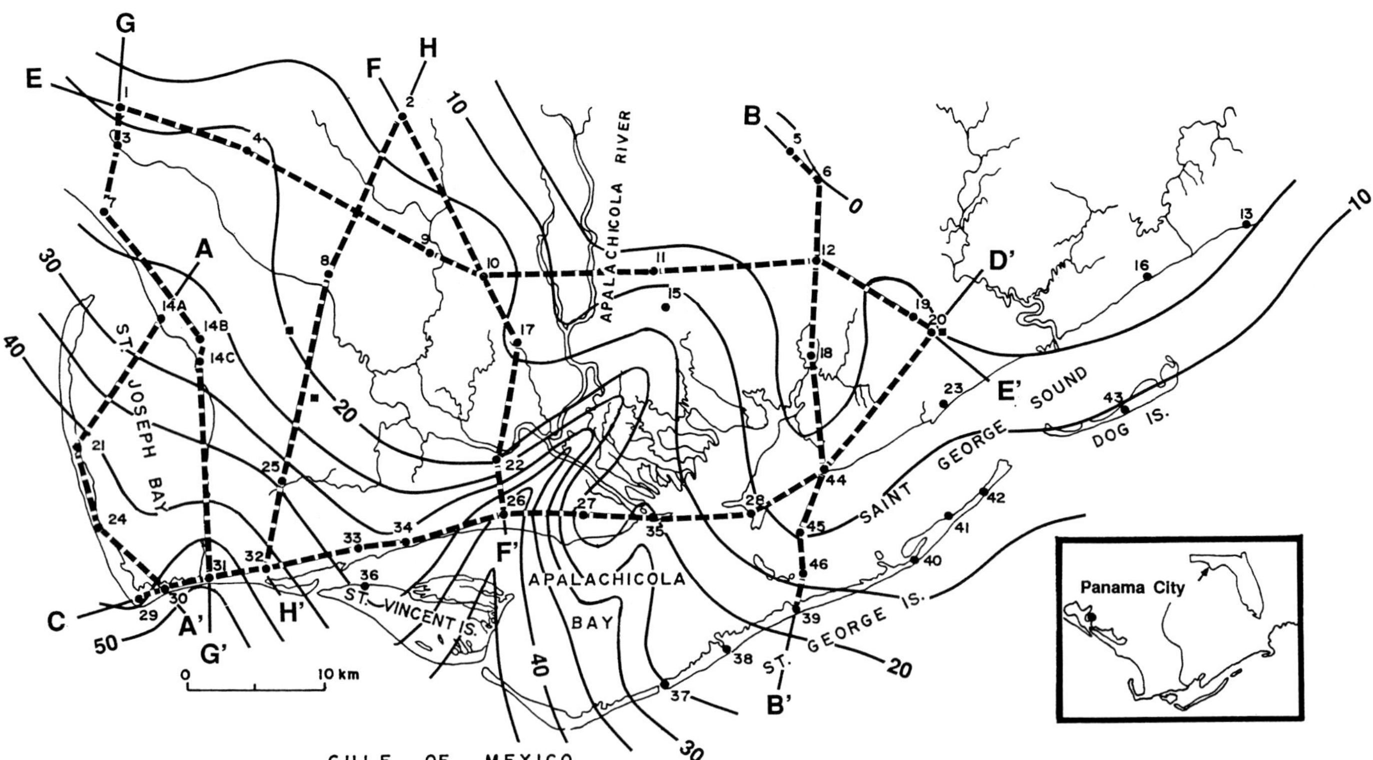

FIG. 2.—Structural contour map on Neogene surface (subsea depths in meters) and cross-section positions.

TABLE 1.—GENERALIZED STRATIGRAPHIC TABLE, UPPER MIOCENE THROUGH QUATERNARY.

	HOLOCENE	Transgressive and regressive (inshore) hemicycles. Mainland and island barriers. Lagoonal-bay, deltaic and floodplain deposits.	
UPPER PLEISTOCENE	WISCONSINAN GLACIAL	Eolian inland dune ridges (Blue Mt., Carrabelle area)	
UPPER PLEISTOCENE	SANGAMONIAN INTERGLACIAL	Prairie Fm. (alluvial) Biloxi Fm. (neritic-to-estuarine deposits) Undifferentiated early and pre-Sangamonian alluvial deposits	Gulfport Fm. (barrier complex)
PLIOCENE	UPPER	Citronelle Fm. (uplands)	
PLIOCENE	MIDDLE		Undifferentiated alluvial and marine siliciclastics
PLIOCENE	LOWER	Perdido Key Fm. (AL-FL border area)	Jackson Bluff Fm. –Intracoastal Fm.
MIOCENE	UPPER	AL-FL border and Pensacola area: Miocene coarse clastics Pensacola Clay (Fm.?)	Choctawhatchee Fm./Stage (=? part of Intracoastal Fm.)

Pleistocene time,'' Upper Miocene high bluffs may have rimmed the Apalachicola Valley on the east.

Late Pleistocene Sedimentary Cycle

Sangamonian sea level reached approximately +7.5 m along the Gulf Coast. Neritic to nearshore/inshore deposits of the Biloxi Formation (Otvos, 1975, 1991a) mark the transgressive stage. A wide barrier strandplain complex (Gulfport Formation) prograded from the highstand shoreline (Otvos, 1985, 1991a, b). Floodplain aggradation led to accumulation of the Prairie Formation and formed the coastal Prairie surface (Fig. 2) landward of the Gulfport strandplain. The Prairie occasionally is interbedded with highly brackish Biloxi facies.

Biloxi Formation.—

An unusually thick Biloxi interval in drill holes #29 and 30 (Fig. 3C) indicated that the Sangamonian sea level rose at least from 37 m, possibly even 48 m below current sea level (Otvos, 1990), which is the greatest known depth of the Biloxi along the Gulf Coast.

At several locations (e.g., drill hole #27; Fig. 2), the formation overlies Neogene siliciclastic beds. The Biloxi, in turn, is overlain by and interfingers with regressive Prairie alluvium. On the mainland, the Biloxi grades upward into Gulfport sands. East of the Mississippi River, the Biloxi regression was not interrupted by the transgression that occurred at certain Texas localities. Due to the dominantly sandy-sediment source and in contrast with its more silty-clayey composition elsewhere, the Biloxi contains little mud. Whereas molluscans and microfossils often are abundant, post-depositional leaching frequently eliminated calcareous fossils from this unit (for example, near Apalachicola; Otvos, 1990). Stratigraphic assignment becomes difficult in deposits from which even nonleachable (siliceous, agglutinated, and chitinous) microfossils are absent.

A fairly complete Biloxi sediment cycle was documented in drill hole #30 (Table 2, Fig. 3C). Only in this hole are brackish deposits overlain by open-marine units that in turn are covered by a low-salinity regressive Biloxi interval. In other drill holes, either the upper or the lower brackish unit is missing. Due to large-scale freshwater influx from the mainland and post-depositional erosion, high-salinity marine intervals are absent from most drill holes. Two Biloxi facies were distinguished:

Shelf Facies.—With one exception (drill hole #39; central St. George Island), the neritic facies (1–6 m) is restricted to the St. Joseph Bay area. Generally, it occurs in thin intervals (drill holes #14C, 21, 24, 30, 31, and 39; Otvos, 1990, p. 24). In certain units, the miliolid content approaches one-third of the total foraminiferal population (*Quinqueloculina lamarckiana, Q. tenagos, Q. seminulum, Triloculina brevidentata, T. trigonula,* and so forth). *Rosalina columbiensis, Brizalina striatula,* and Cribroelphidium poeyanum were among the inner- to mid-shelf species. Species diversity was great. The freshening influence of the mainland runoff is reflected by persistent species with brackish affinities, primarily *Ammonia beccarii, Elphidium incertum mexicanum, E. galvestonense* and *Nonion depressulum matagordanum*. Fine sands and muddy fine sands characterized the brackish facies; slightly granular sands indicated proximity of stream channel mouths.

Nearshore/inshore Brackish Facies.—Most of the Biloxi was deposited in mesohaline–polyhaline and lower polyhaline environents, with brackish foraminiferal taxa predominant (Otvos, 1990). These facies often occur at shallow depths in trenches and drill holes (#19, 20; Fig. 3D). The enclosing sediments range from granular medium sands to muddy sands and muds.

Molluscan species that presently occupy nearby bay, inlet, and other inshore environments were common and include *Anadara transversa, Argopecten irradians, Busycon contrarium, Crassostrea virginica, Chione cancellata, Dinocardium robustum, Mercenaria campechiensis, Oliva sayana, Polinices duplicatus,* and others. Several of them also inhabit shallow nearshore marine environments exposed to mainland runoff. The lowest salinity biotopes contain *Ammotium salsum,* chitinous foraminiferal tests, and diatoms.

Prairie Formation.—

The dominantly sandy fossil-free unit that formed in coalescing coastal flood plains consists of moderately and poorly sorted gray fine sands, slightly granular medium sands, coarse silty fine sands, muddy fine sands, and muds (Otvos, 1990). Yellowish-brown, orange and brownish-gray colors characterize the oxidized upper interval. Pockets of peat occur occasionally.

Prairie deposits overlie the Neogene sediments and interfinger with transgressive Biloxi units. Representing the concluding phase of the Sangamonian regressive hemicycle, the Prairie covers Biloxi deposits and the landward flank of the Gulfport barrier. It forms the ground surface inland of the Gulfport belt. Except where stream and gully erosion intervened, it slopes evenly Gulf-ward from an inland elevation of +9 m.

In a few drill holes, 15 to 26 m of fossil-free post-Neogene siliciclastics underlie the Biloxi and/or Gulfport (Fig. 3A–C). The lowest intervals in these thick alluvial sequences may predate the Prairie.

In contrast to the two to four Pleistocene coast-parallel terraces west of Mobile Bay (Otvos, 1991b), the Prairie is the only Pleistocene coastal surface in the Florida Panhandle. The presence of pre-Sangamonian Pleistocene and Pliocene littoral units and landforms east of the bay inland (e.g., Donoghue and Tanner, this volume) was refuted by

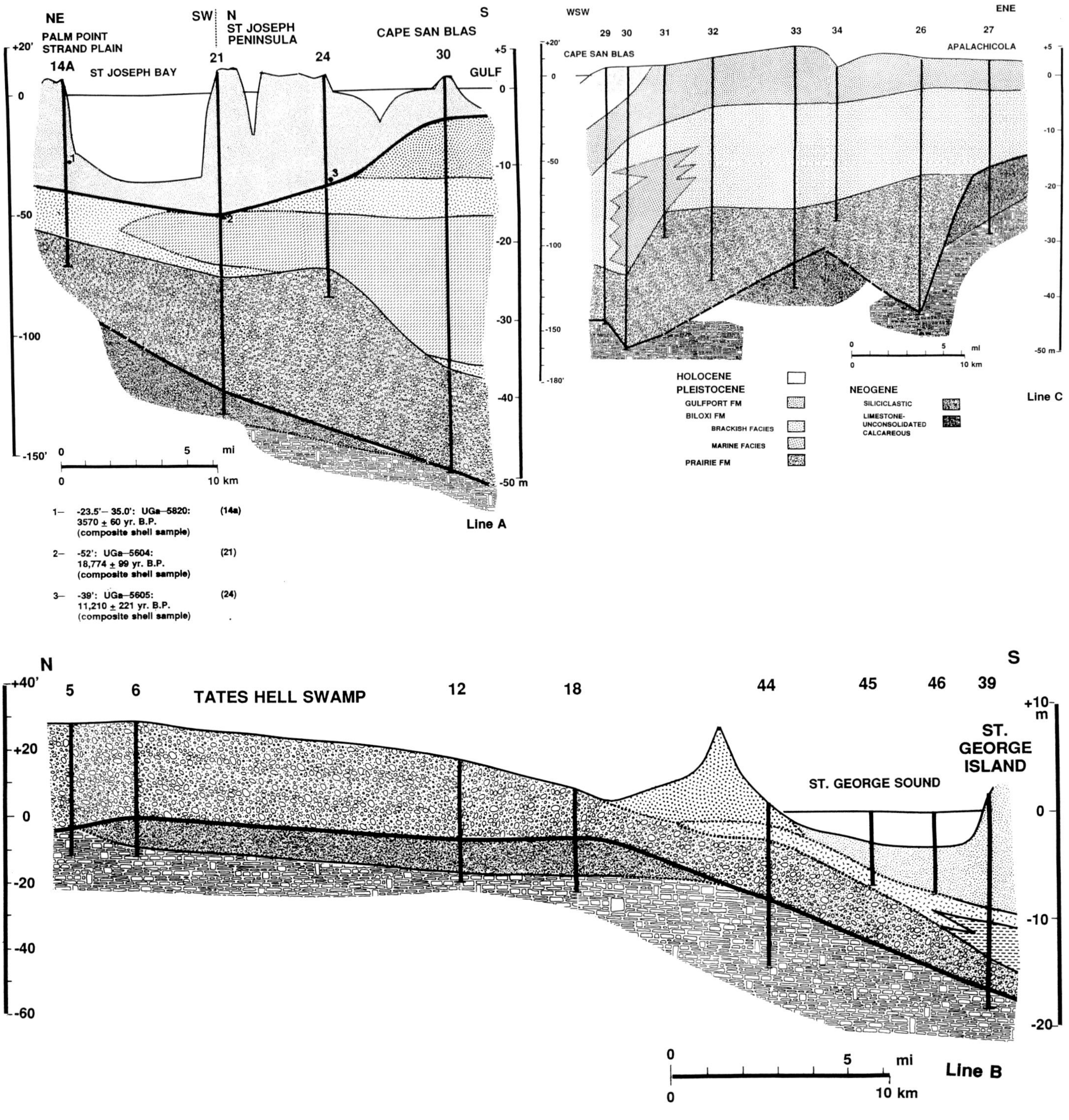

NE
PALM POINT STRAND PLAIN
SW | N
ST JOSEPH PENINSULA
S
CAPE SAN BLAS
14A 21 24 30
ST JOSEPH BAY
GULF
+20' 0 -50 -100 -150'
+5 0 -10 -20 -30 -40 -50 m
0 5 mi
0 10 km
Line A

1— -23.5'– 35.0': UGa–5820: 3570 ± 60 yr. B.P. (composite shell sample) (14a)
2— -52': UGa–5604: 18,774 ± 99 yr. B.P. (composite shell sample) (21)
3— -39': UGa–5605: 11,210 ± 221 yr. B.P. (composite shell sample) (24)

WSW ENE
29 30 31 32 33 34 26 27
CAPE SAN BLAS
APALACHICOLA
+20' 0 -50 -100 -150 -180'
+5 0 -10 -20 -30 -40 -50 m
0 5 mi
0 10 km
Line C

HOLOCENE
PLEISTOCENE
GULFPORT FM
BILOXI FM
BRACKISH FACIES
MARINE FACIES
PRAIRIE FM
NEOGENE
SILICICLASTIC
LIMESTONE-UNCONSOLIDATED CALCAREOUS

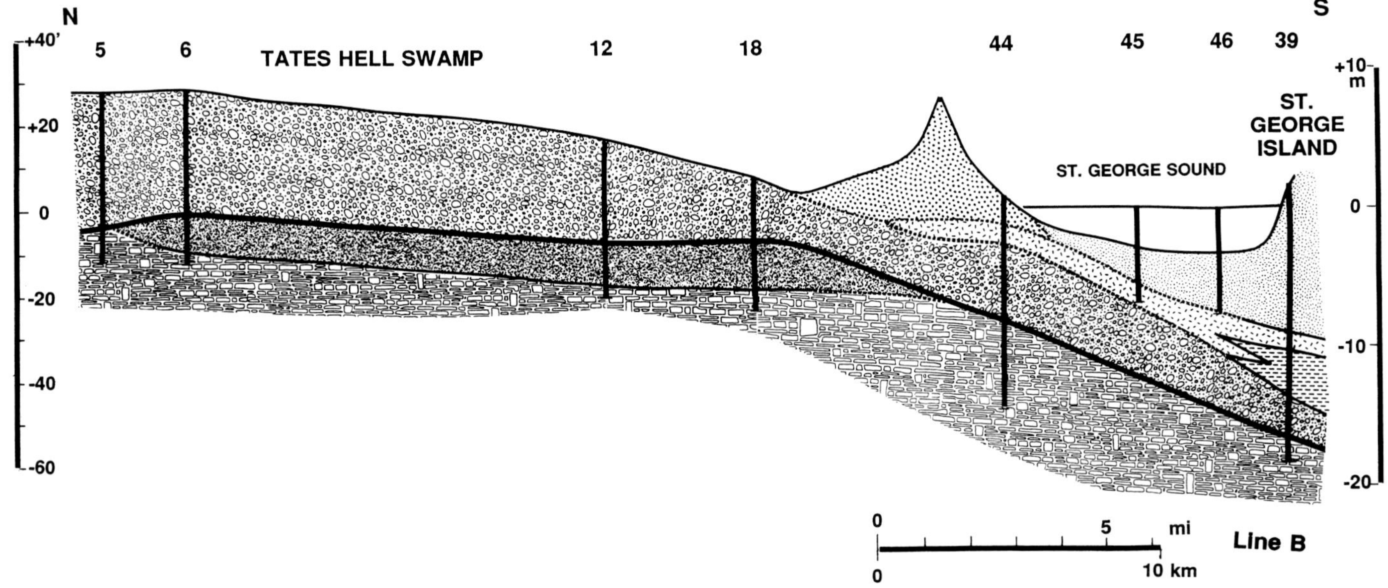

FIG. 3.—Apalachicola Coast geologic cross sections. Locations shown in Figure 2. (A) Line A-A′ across St. Joseph Bay and Peninsula to Cape San Blas. (B) Line B-B′ across Tates Hell Swamp and St. George Sound to St. George Island. (C) Line C-C′ from Cape San Blas to Apalachicola. (D) Line D-D′ from Apalachicola across Apalachicola across Apalachicola River to St. George Island. (E) Line E-E′ from Overstreet and Cerser Swamp across Apalachicola River to Tates Hell Swamp. Rest of lines given in Otvos (1990).

field and sedimentary evidence. All too often, erosion-carved and/or fault-defined interfluve ridges are declared shoreline features in the literature. Extensive erosion, associated with the still active regional uplift, had eliminated all pre-Sangamonian terrace/barrier units (Otvos, 1991a).

Gulfport Formation.—

Doering (1956) was first to recognize the broad coastal barrier at Apalachicola. The barrier deposits (Table 1; Fig. 2) correlate with the Gulfport Formation, described on the

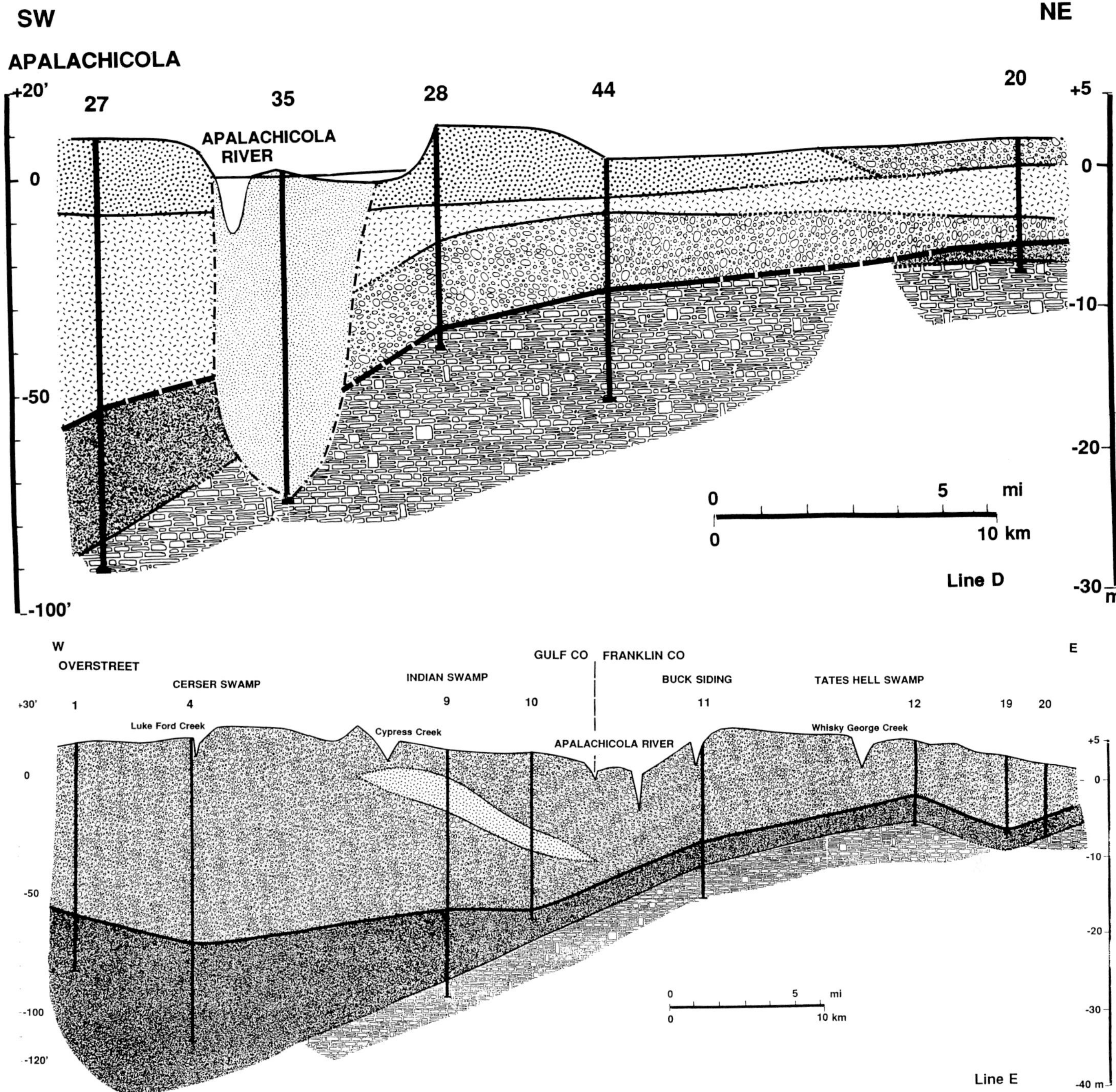

FIG. 3.—Continued.

Mississippi-Alabama coast (Otvos, 1985). A complex of shallow subtidal/shoreface to intertidal deposits, capped probably by eolian sands, underlies the Gulfport strandplain. The formation, 4.5 to 9 m thick, consists of well- to moderately well-sorted, fine, rarely very fine sands, usually of pale yellowish-brown, or yellowish-brown color. Downward-percolating carbon-rich humic-soil acids created dusky yellowish-brown, weakly humate-cemented, dark sandstone horizons at shallow depths.

Parallel-laminated and current-rippled cross-stratified nearshore shallow subtidal sands with occasional cut-and-fill structures are exposed in the southeast bank of the Gulf County Canal at Port St. Joe, about 2.7 m above present sea level.

As many as 30 gently curving subparallel ridges of subdued topography form the 10.5- to 12-km-wide strandplain that rises 4 to 6 m above sea level. Minor age differences do exist between adjacent strandplain sectors: north of drill

TABLE 2.—LOG SUMMARY DRILL HOLE #30 (Figs. 2, 3C).

Age/Fm	Elevation (ft) MSL	Description
Holocene	+5.5 to −12.0	Fine sand, moderately well to poorly sorted, fossiliferous. Brackish nearshore to shoreface with overlying intertidal/dune facies.
Pleistocene		
Gulfport Fm.	−12.0 to −31.0	Fine sand, well to moderately well sorted, dusky yellowish brown (10 YR 2/2)* to pale yellowish brown (10 YR 6/2); humate impregnated. Barrier dune/intertidal facies (end of regressive hemicycle).
Biloxi Fm.	−31.0 to −117.6	
	−31.0 to −58.0	Muddy fine sand, sand, light olive gray (5Y 6/1), most intervals fossiliferous.
	−44.4 to −53.0	Moderately brackish facies (start of regressive hemicycle).
	−53.0 to −113.0	Marine (end of transgressive hemicycle).
	−58.0 to −117.6	Medium sand, very poorly to moderately well sorted, dense to very dense, medium light grey and light grey
	−113.0 to −117.6	Moderately brackish.
Prairie Fm and/or undifferentiated alluvium	−117.6 to −133.0	Mud, fine, sandy, olive-gray, with plant fragments but no brackish-marine fossils.
	−133.0 to −166.0	Slightly granular medium sand, poorly sorted, very dense, white (N9), medium light gray (N6) with molluscans and foraminifera (downhole contaminants).
Neogene	−160.6 to −160.9	Slightly granular, muddy coarse calcarenite with calcareous cement. Very poorly sorted, white (N9), with molluscan fragments. Neritic Neogene taxa include *Asterigerina miocenica*, *Globulina gibba*, and so forth.

*Color codes here and in text follow U.S. Geological Survey Rock-Color Chart.

holes #31 and 32, an older, north–northeast—south–southwest-oriented barrier segment is truncated by an east–northeast—west–southwest-trending Gulfport sector (Fig. 1).

The ridge-plain topography is absent east of Green Point and near the town of Apalachicola. Locally, intensive Wisconsinan wind erosion, combined with scant vegetative cover, may have been the cause (Otvos, 1990).

Relict Stream Channels (Late Sangamonian-Early Wisconsinan?)

Sizable meander traces, imprinted in the Prairie land surface west of Apalachicola River, mark the highest stream aggradation level at the peak of the Sangamonian transgression, before sea level (and with it, erosional base level) began to decline. The channel floors occur 1.2 to 2.4 m lower than the Prairie summit surface, still 1.8 to 6 m above the adjacent broad Apalachicola Valley flood plain. Small, intermittent, underfit creeks (Cypress Creek, Indian Bayou, and Depot Creek) inherited the relict channels (Fig. 4). Meander remnants, formed on the Prairie land surface just before Sangamonian aggradation ended, are common on the Prairie coastal surface (Otvos, 1991a,b).

During subsequent lower sea-level stages, at the time of the Late Sangamonian-Early Wisconsinan regression, Apalachicola River distributary flow excavated the present Lake Wimico Basin and the Jackson River channel (Fig. 4). Excess flood waters from the Depot Creek trunk channel discharged into an ephemeral dead-end distributary. The wide, partly braided, partly meandering relict-channel segment is located in a strandplain swale, downstream from the relict-trunk channel. The channel ends abruptly against a barrier ridge. In the absence of an outlet channel, water apparently dissipated Gulf-ward through the sandy substrate (Fig. 4).

The elevation of the flood channel indicates stream incision prior to the late Wisconsinan. Deeper incision followed in full glacial times (about 22 to 16 ka). At Apalachicola, the channel was cut 22 m below sea level (Schnable and Goodell, 1968). By about 10 ka, the Gulf of Mexico again rose above −22 m and gradually inundated valley floors on the mainland.

Wisconsinan Dune Hills

Four large, steep sand ridges overlie the Gulfport in the Royal Bluff–Carrabelle area (Fig. 2). They consist of slightly oxidized, grayish-orange and light gray sands, moderately well sorted, medium and fine grained (0.22–0.29 mm median diameter). The largest dune hill, 1.1 km long, rises 12 m above its base. A northeast-southwest-axis orientation characterizes the well-preserved parabolic dunes that are concave toward the southwest.

Individual dune hills, round and elongated narrow ridges, also occur at a few northeast Gulf locations between the Apalachicola Coast and Gulf Shores, Alabama. They are somewhat more common between Miramar Beach and Grayton Beach on the central Florida Panhandle. In the absence of marine beaches (prior to their return in the late Holocene), wind-eroded inland surfaces are considered plausible dune-sand sources.

Floodplain-supplied sands constructed dunes along the lower Mississippi River, in the Florida Parishes of Louisiana (Saucier, 1978; Otvos, 1991a), and on the south Atlantic Coastal Plain (Carver and Brook, 1989). Current humid-subtropical conditions preclude the formation of floodplain-fringing dunes. Except for presently semi-arid Texas coast sectors, where Pleistocene barriers underwent significant wind erosion, dune accumulation and eolian reworking is restricted to beach backshore areas on the Gulf.

No significant flood plains existed near the Carrabelle and other northeast-Gulf dune fields of the Wisconsinan. Deflation of sparsely vegetated Gulfport and Prairie surfaces therefore may have constructed dunes during periods of diminished precipitation.

Northeast-paleowind bearings are inferred from the east-northeast and northeast dune orientation of Carrabelle area parabolic dunes, which conforms with azimuths displayed by Georgia coastal-plain ridges 320 km to the northeast. The Georgia-Carolina dunes were considered Wisconsinan, and as "not necessarily" formed under arid conditions (Carver and Brook, 1989).

A Wisconsinan Marine Episode?

Following Schnable and Goodell (1968), Donoghue and Tanner (this volume, their fig. 4) identified two fining-upward sediment sequences in a few drill holes on the mainland shore and the islands. Subdivision was based on finite radiocarbon dates derived from plant matter. The authors considered the "lower sequence" as Sangamonian and the "upper" as a mid-Wisconsinan marine units. Although their dates (34-23 ka) partly coincide with certain relatively ice-free Middle and Late Wisconsinan periods in the glaciated north, the eustatic sea levels stood much lower than the high sea level the authors implied for their "Upper Sequence" (Otvos, 1991a, b). In the North Atlantic, the 32- to 28-ka sea level was at approximately −80 m (Amos and Miller, 1990). Schnable's dates apparently were artifacts of contamination and the associated highly permeable sandy sediments belong to leached, presently fossil-free Biloxi deposits (Otvos, 1990).

Based on data of Brenneman and Tanner (1958), Schnable and Goodell (1968) also mapped a 6- and 9-m-elevation Pleistocene shoreline in the swampy "Tates Hell" area, 16 to 20 km inland from the present shore. Donoghue and Tanner (this volume) again referred to 10-km-long "ridges," located 5 km apart, as "immature Pleistocene barrier-island and sand-ridge" trends, 1 to 4 m above the adjacent sand plain (Fig. 1). The only elevated features are abandoned logging railroad embankments that parallel map contours. Natural topographic ridges are absent. Drill hole #6 (Fig. 2), on one "ridge" location, penetrated only fossil-free, poorly to very poorly sorted Prairie floodplain sands and silty sands (Otvos, 1990).

HOLOCENE

Transgressive Phase—Identification of the Ravinement Diastem

The late Pleistocene antecedent topography influenced the pattern of subsequent Holocene transgression and barrier development. An upward facies shift from brackish inshore/nearshore to higher salinity inner-shelf deposits occurs in drill holes #21, 24, and 29 (Fig. 2; Otvos, 1990). The earliest low-salinity units often have not been preserved. In the Apalachicola estuary (drill hole #35, Fig. 2; Otvos, 1985), salinities were higher during the late Pleistocene than later. Lagoon and bay sediments, deposited after barrier emergence, represent brackish units of the regressive hemicycle. Drill holes #14a, 29, and 35 (Fig. 2) document this upward reduction in salinity.

Hurricane-induced wave and current scouring was shown to have lowered the shallow lagoon floor by a maximum of 3 m (Isphording and others, 1987). This process tends to eliminate and homogenize the Holocene sediment record. Additionally, the heavy Apalachicola River runoff has diminished vertical and lateral salinity gradients and contrast between estuarine and inner-shelf sediments. These are better expressed in central Gulf coastal sequences (Otvos, 1991a).

Ravinement surfaces represent erosional boundaries, formed between inshore (fluvial, estuarine-lagoonal, back-barrier) and marine-shoreface inner-shelf deposits by the transgressing surf zone. The earliest transgressive deposits had been nearly completely eliminated in the Apalachicola area. In addition, the diminished litho- and biofacies contrast between inshore and inner-shelf sediment units in most instances makes recognition of marine-ravinement surfaces within the lower Holocene island- and lagoonal-sediment sequence impossible.

Effects of antecedent topography were most pronounced in southeastern St. Joseph Bay, where rising Gulf waters gradually covered the seaward margin of the Sangamonian strandplain. Elliptical shoals and elongated marsh islands developed over submerged Gulfport ridge crests that parallel the mainland barrier arc (Figs. 2, 5, 6A). Formation and segmentation of large shoal-retreat massifs off Cape San Blas, St. George (Fig. 7B), and Alligator Harbor also occurred during the transgression.

Southern Mainland Strandplain and St. Joseph Barrier Spit

Development of the mainland strandplain and the St. Joseph Peninsula barrier spit was preceded by erosional retreat of the Gulfport shoreline at McNeils and northeast of Cape San Blas. Drift divergence in two opposite directions resulted in sand transport from an eroding central coastal sector on the south shore (Fig. 7A–D). Next, gently curving east-west ridges developed along the Gulfport strandplain sector that later was partly submerged (Figs. 6A, B). Well- to very well-sorted, pale yellowish-brown, grayish-orange and very pale orange (10YR 7/4, 10YR 7/2, 10YR 5/4) sands form steep ridges in the 0.6- to 1.0-km-wide strandplain. The same hues, encountered elsewhere on the Gulf in Late Holocene sands only in certain older portions of Mississippi barrier islands (e.g., Horn and East Ship), also occur in St. Joseph Spit (Peninsula), Indian Peninsula, and St. Vincent Island strandplains (Figs. 2, 7). Locally, intensive weathering, affecting the older Late Holocene ridge sets, may have been responsible for the unusual oxidized colors.

Prograding westward, the newly developing beach ridges wrapped around the tip of Cape San Blas, at the time composed of a corner of the Gulfport strandplain. The Richardson Hammock strandplain area developed next (Fig. 6A). Ancestral Indian Peninsula prograded eastward (Fig. 7A). In contrast to oxidized hues of the older Holocene strandplain generations near Cape San Blas, the 2- to 4-m-high, closely spaced and dune-capped younger ridge sets on the rest of St. Joseph Peninsula (Fig. 5) consist of well-sorted white sands.

By trimming the older ridge generations on St. Joseph Spit, wave erosion supplied sand to new beach-ridge generations that continued to form in a downdrift direction (Figs. 6B, C; 7B, C). No indications exist for any pre-existing islets, attached to and incorporated into the spit in the course of its northward progradation (Figs. 3A, 7B–D). Ridges of the 7- to 10-m-high, steep dune trend along the western shore contrast with the much smaller ridges (Fig. 5) of the progradational strandplain generations. The wide, massive ridges form a continuous line behind the Gulf foreshore.

In the course of its growth, the spit prograded over nearshore and shelf deposits that thicken northward from 2 to 5.5 m to 12.6 m (Fig. 3A; Otvos, 1990). Poorly sorted, muddy, clayey fine sands and fine sands containing benthic foraminiferal species (*Rosalina columbiensis, Nonion de-*

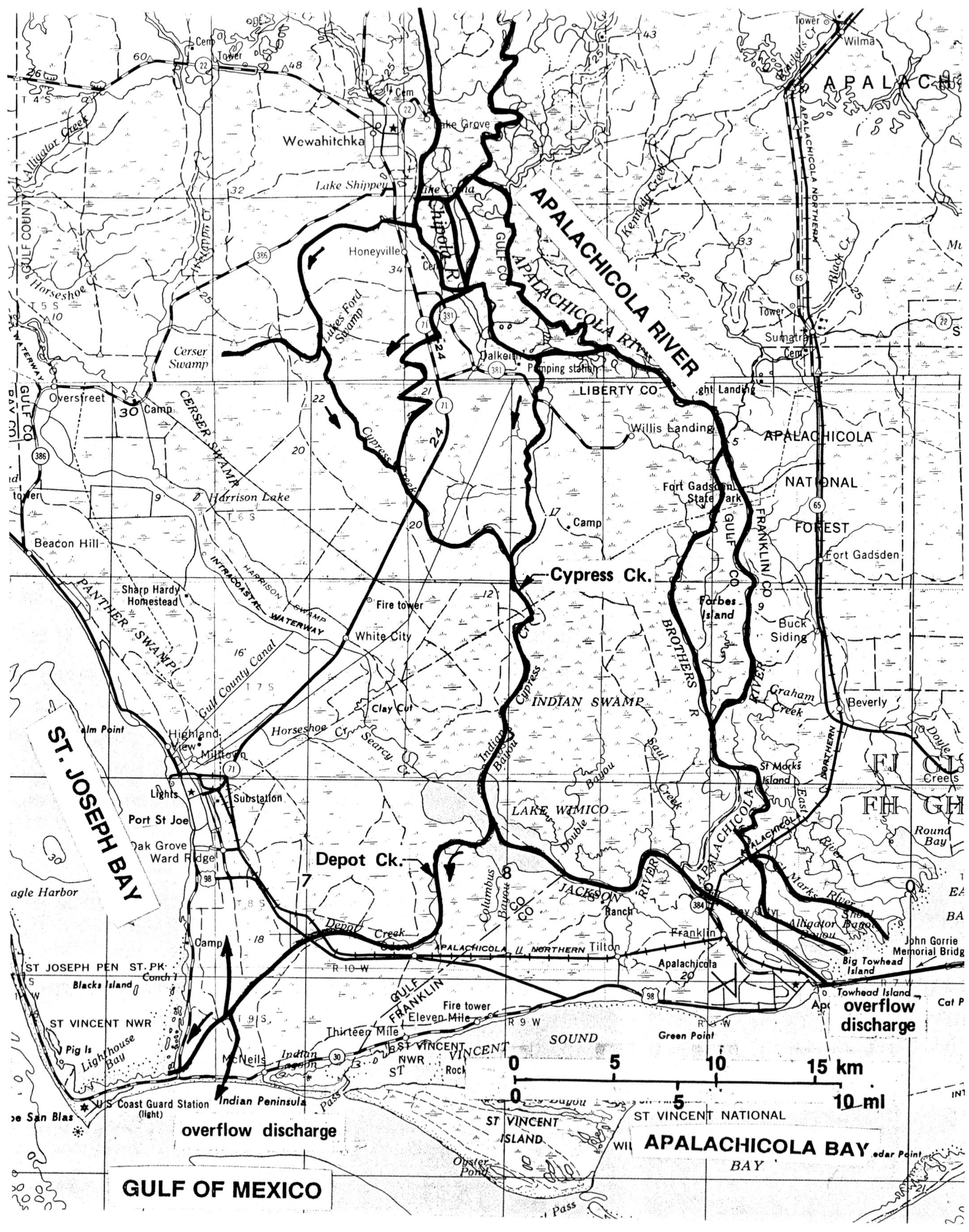
APALACHICOLA RIVER
Cypress Ck.
Depot Ck.
ST. JOSEPH BAY
overflow discharge
overflow discharge
GULF OF MEXICO
APALACHICOLA BAY
0 5 10 15 km
0 5 10 ml
Wewahitchka
Lake Grove
Lake Shippey
Honeyville
Chipola R
Lakes Ford Swamp
Cerser Swamp
Dalkeith
Pumping station
LIBERTY CO
Willis Landing
APALACHICOLA NATIONAL FOREST
Fort Gadsden State Park
Fort Gadsden
Overstreet
Beacon Hill
Harrison Lake
Sharp Hardy Homestead
PANTHER SWAMP
CERSER SWAMP
White City
Fire tower
Gulf County Canal
INDIAN SWAMP
BROTHERS R
Forbes Island
Buck Siding
Graham Creek
Beverly
Highland View
Port St Joe
Oak Grove
Ward Ridge
Substation
LAKE WIMICO
JACKSON RIVER
Franklin
Tilton
Apalachicola
ST JOSEPH PEN ST. PK
ST VINCENT NWR
Indian Peninsula
Thirteen Mile
Eleven Mile
ST VINCENT ISLAND
ST VINCENT NATIONAL
Big Towhead Island
Towhead Island
John Gorrie Memorial Bridge
Coast Guard Station

FIG. 5.—Roadcuts through St. Joseph Spit area Late Holocene strandplain ridges (relict foredunes). (A) East-west ridge complex, northest of Cape San Blas Plantation. (B) Just south of Peninsula State Park gate.

pressulum matagordanum, Cribroelphidium poeyanum, and *Hanzawaia strattoni*) and miliolid forms accumulated on the inner-shelf floor. Age-diagnostic late Quaternary *Gephyrocapsa oceanica* and *G. caribbeanica* nannoplankton species also occurred in this facies (core hole #21, Fig. 2). Upward, the neritic deposits grade into better sorted sandy barrier-platform sediments (shoreface-subtidal facies), deposited in lower polyhaline-mesohaline and euhaline (about 25–32 ‰) environments.

Palm Point Strandplain

The small isolated (0.4-km wide) strandplain, known as Palm Point, adjacent to the western Gulfport strandplain sector (Fig. 7C, D) is underlain by an unusually thick (12 m) Holocene sequence. The steepness of the Pleistocene surface that the transgression encountered (Fig. 3A) is comparable to that of the present Gulf floor off the St. Joseph Peninsula. The lower transgressive sequence that underlies the closely spaced strandplain ridges is a brackish to slightly brackish, nearshore unit of great species diversity.

Estuarine brackish foraminifera (mainly *Ammonia beccarii, Elphidium galvestonense,* and *E. incertum mexicanum*) dominate the upper, regressive sequence. A 3.57-ka radiocarbon date from a composite shell sample, taken from the 7.0- to 10.5-m-interval below sea level, suggests that the overlying strandplain formed quite recently (Otvos, 1990). Strandplain progradation ended by the time growth of the St. Joseph Spit significantly reduced wave energies and drift-transport rates along the new bay's east shore.

Crooked Island—Raffield Barrier Spits

The strandplain complex opposite Tyndall Air Force Base, (Fig. 7A, northwest area) is a very unusual double barrier. Two parallel barrier spits are separated by a sizable lagoon. Emergence of an oblique nearshore bar, tied to the mainland shore southeast and updrift from the Raffield barrier, may have initiated formation of Crooked Island, the younger barrier that blocked further sand supply to Raffield Spit.

BARRIER-ISLAND DEVELOPMENT: SUMMARY

St. Vincent, St. George Island, and Dog Island northeast of St. George formed during slow rise of the Late Holocene sea level. Strandplain-ridge configurations on St. Vincent, "Little St. George" (Fig. 7B–D), northern St. George (at Gap Island), and Dog Island suggest that, as in central Gulf Coast barrier islands (Otvos, 1981, 1985), embryonic-island cores did emerge from shoals. If the Indian Peninsula barrier spit originally extended northeast of the future St. Vincent Island, its fragmented eastern end or shoal continuation may have similarly become a nucleus for the future island (Fig. 7A,B).

Island progradation from the "Gap Island" and Little St. George Island nuclei (Otvos, 1985) through progradational linking with other smaller islands and shoals eventually resulted in formation of the present St. George Island (Fig. 7C, D). Its emergence and resulting isolation of Apalachicola Bay are documented in an abrupt vertical change in foraminiferal spectra in Apalachicola Bay drill hole #10 (Otvos, 1985). Upward in the section, open nearshore conditions gave way to lower salinities. Shore erosion and storm-overwash activity later reshaped the island, destroying most of the original strandplain topography. As in several other investigated Gulf barrier islands (Otvos, 1981), no evidence (such as relict lagoons) has yet been found for ancestral landward-migrating, transgressive or composite barrier-island categories in the late Holocene. However,

←

FIG. 4.—Sangamonian (-Early Wisconsinan?) Apalachicola River distributary system incised in late Pleistocene deposits.

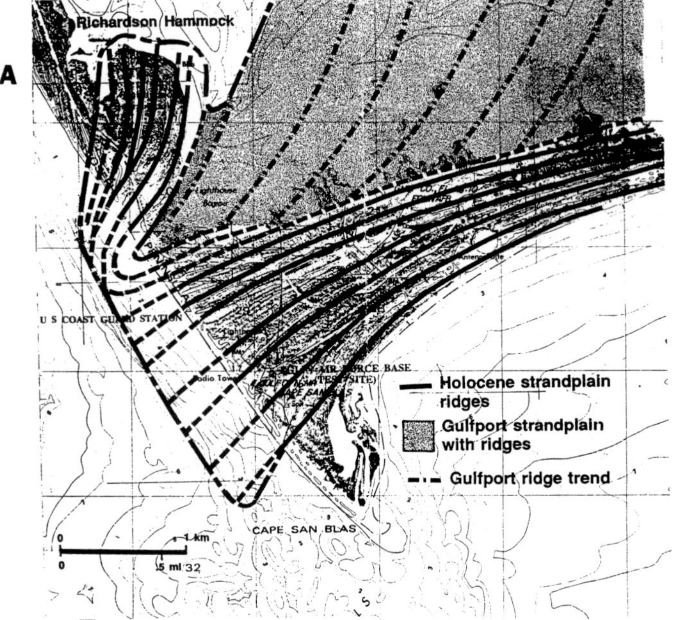

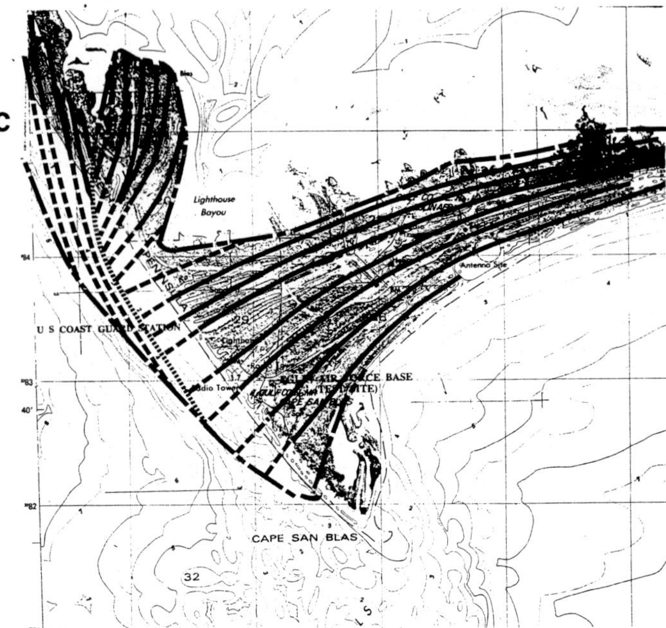

FIG. 6.—(A–C) Generalized early-evolution stages of south St. Joseph barrier area.

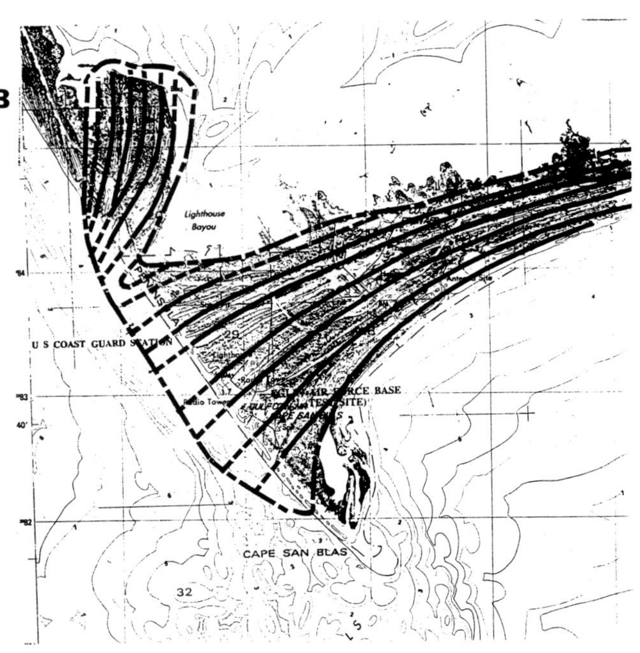

transgressive barriers may have existed during earlier stages of the transgression.

St. Vincent Island

Each of the several beach-ridge generations on this spectacular strandplain island consists of numerous beach ridges. They reflect intermittent episodes of shore erosion and drift reversal in the course of overall island progradation. The earliest, northern strandplain-ridge sets are low, whereas in the south, ridge elevations reach 3 to 6 m. Differences in ridge elevations reflect various factors that influenced dune formation. These include available sand stockpiles on- and offshore, beach width, dune vegetation, strandplain-progradation rates (length of time and degree of ridge exposure to active intertidal sand areas), wind-regime parameters, effects of storm erosion and other influences.

As on other strandplains, the slightly eroded, somewhat subdued beach ridges also represent relict foredunes. As on St. Joseph Spit, high precipitation dunes, created by forest growths, occur along certain shore segments.

Eustatic sea-level rise (current rate approximately 10–12 cm/century; Gornitz and Lebedeff, 1987) may have contributed to the consistently higher elevations of younger ridges. Rainwash, biogenic erosion and compactional subsidence combined to reduce ridge heights. The same processes also occurred in the 5-m-thick, compactible "soupy" mud unit, reported by Stapor (pers. commun., 1987) from his Mallards Slough drill core.

Floor elevations of marsh-bottomed inter-ridge swales rose as they became filled by organic and slopewash matter. Compaction-induced marine inundation of inter-ridge swales and ridges was widespread in northern and eastern island areas. Due to diminishing nearshore bottom-sand sources (Otvos, 1991a), the intensive late Holocene Gulf-wide strandplain growth has slowed or ceased altogether.

Ridge Summits and Sea Levels

Gulf strandplain ridge-summit and swale-floor elevation values in numerous papers have been used to infer significant Holocene sea-level variations during the past three to four millenia (Tanner and others, 1989; Donoghue and Tanner, this volume, to name two). The St. Vincent strandplain ridges, claimed as "storm surf" or "swash-built" beach ridges, actually are not wave-constructed features but steep relict foredunes (Fig. 5) underlain by inter- and subtidal

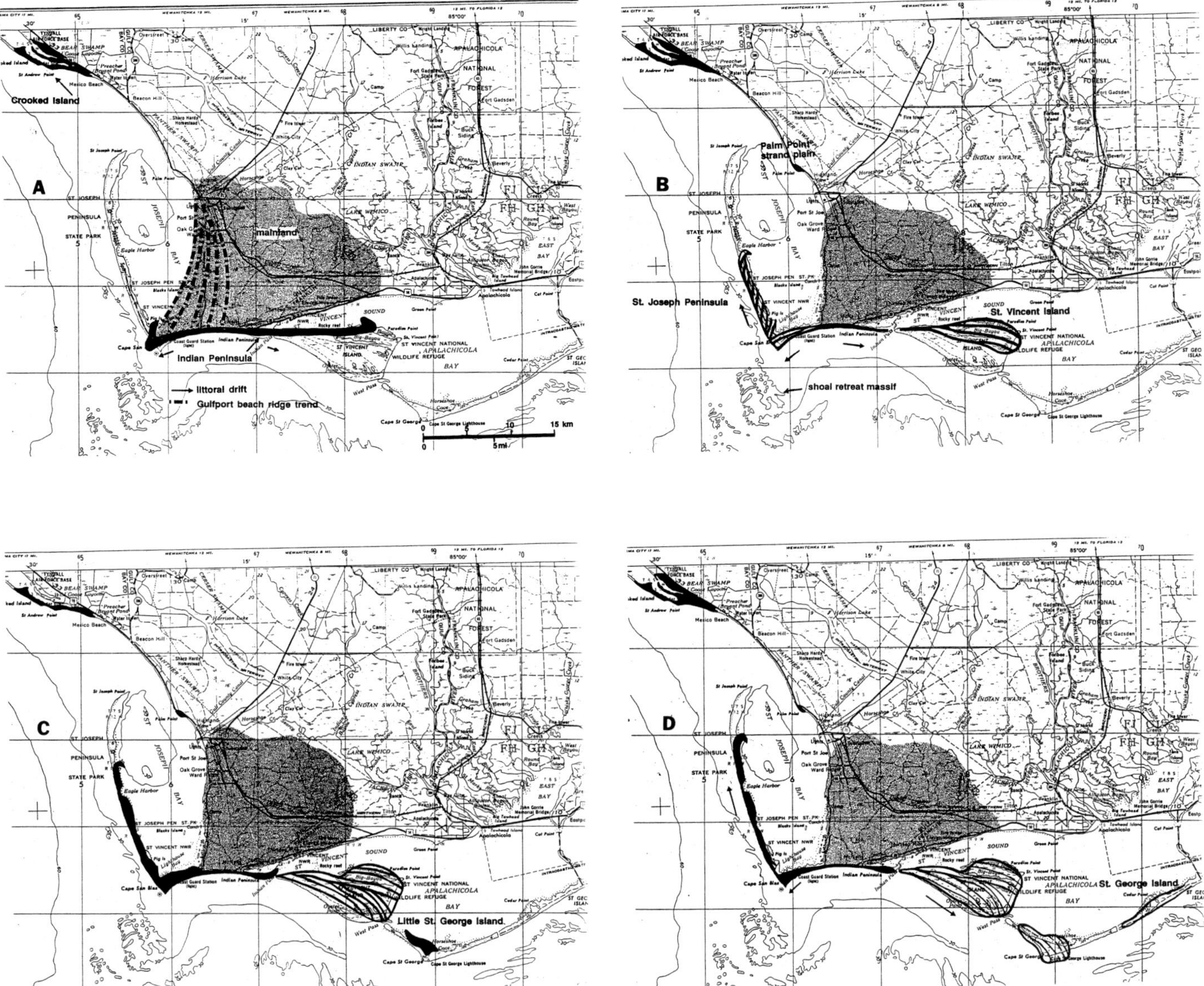

FIG. 7.—(A–D)—Generalized Late Holocene evoluation stages of southwest Apalachicola Coast area. Nearshore arrows indicate littoral-drift directions.

beach sands. Tanner and coworkers credited recurring storm-surge activity with building the largest (5- to 10-m-high). Referring to elevations of allegedly ''swash-built'' ridges and to the position of an Indian archeological site, they suggested Gulf-level fluctuations within the +2-m to −2- to −4-m range.

Instead of constructing ''beach ridges,'' storm waves actually erode and flatten pre-existing sand ridges, leaving level ''storm beaches'' behind. Only gravel-boulder ramparts are known to form during storm-wave activity. Normal swash and surf action on prograding shores results in low ridges, not comparable in dimension, configuration and slope angles to foredunes (Otvos, 1981). Compared with foredunes, swash-built flat-topped berms generally are much lower, narrower and bounded by gentle slopes. Thus, no conclusive evidence yet supports Late Holocene sea-level fluctuations on the Gulf of Mexico.

ACKNOWLEDGMENTS

Drillhole samples were donated by the Florida Bureau of Geology, Florida State University, and Ardaman Engineering, Tallahassee. The thorough and constructive reviews by Drs. Jules R. DuBar, Nicholas K. Coch, John Wehmiller and Charles Fletcher were greatly appreciated. Thanks are due to The Mississippi Mineral Resources Institute and its Director, Dr. J. R. Woolsey, for support of our regional investigations. Radiocarbon dates were provided by the Center for Applied Isotope Studies, University of Georgia. Mr. Wade E. Howat assisted in every phase of the original study and its documentation.

REFERENCES

AKERS, W. H., 1972, Planktonic foraminifera and biostratigraphy of some Neogene formations, northern Florida and Atlantic coastal plain: Tulane Studies in Geology and Paleontology, v. 9, 140 p.

AMOS, C. A., AND MILLER, A. A. L., 1990, The Quaternary stratigraphy of southwest Sable Island Bank, eastern Canada: Geological Society of America Bulletin, v. 102, p. 915–934.

BRENNEMAN, L., AND TANNER, W. F., 1958, Possible abandoned barrier islands in panhandle Florida: Journal of Sedimentary Petrology, v. 28, p. 342–344.

CARVER, R. E., AND BROOK, G. A., 1989, Late Pleistocene paleowind directions, Atlantic Coastal Plain, U.S.A.: Paleogeography, Paleoclimatology and Paleoecology, v. 74, p. 205–216.

DOERING, R. A., 1956, Review of Quaternary surface formations of Gulf Coast regions: American Association of Petroleum Geologists Bulletin, v. 40, p. 1816–1862.

GORNITZ, V., AND LEBEDEFF, S., 1987, Global sea-level changes during the past century, *in*, Nummedal, D., Pilkey, O. H., and Howard, J. D., eds., Sea-Level Fluctuations and Coastal Evolution. SEPM Special Publication 41, p. 3–16.

HUDDLESTUN, P. F., 1984, The Neogene stratigraphy of the central Florida Panhandle: Unpublished Ph.D. Dissertation, Florida State University, Tallahassee, 208 p.

ISPHORDING, W. C., IMSAND, D., AND FLOWERS, G. C., 1987, Storm-related rejuvenation of a northern Gulf of Mexico estuary: Transactions, Gulf Coast Association of Geological Societies, v. 37, p. 357–370.

OTVOS, E. G., 1975, Late Pleistocene transgressive unit (Biloxi Formation), northern Gulf Coast: American Association of Petroleum Geologists Bulletin, v. 59, p. 148–154.

OTVOS, E. G., 1981, Barrier island formation through nearshore aggradation-stratigraphic and field evidence: Marine Geology, v. 43, p. 195–243.

OTVOS, E. G., 1985, Barrier island genesis—question of alternatives for the Apalachicola Coast, northeastern Gulf of Mexico: Journal Coastal Research, v. 1, p. 267–278.

OTVOS, E. G., 1990, Subsurface Evaluation of Mississippi Coastal Sediment Units; Comparison With Apalachicola Area Quaternary Sequence: Final Report: Mississippi Mineral Resources Institute, MMRI-90-IF/U.S. Bureau of Mines G-1194128, 74 p.

OTVOS, E. G., 1991a, Northeastern Gulf Coast Quaternary, in Morrison, R. B., ed., The Geology of North America. Quaternary Nonglacial Geology: Conterminous United States: The Geological Society of America, Decade of North American Geology, v. K-2, p. 588–595, 605–610.

OTVOS, E. G., 1991b, Houston Ridge, Southwest Louisiana—end link in the Late Pleistocene Ingleside barrier chain? Prairie Formation newly defined: Southeastern Geology, v. 31, p. 1–14.

SAUCIER, R. T., 1978, Sand dunes and related eolian features of the lower Mississippi Valley: Geoscience and Man, v. 19, p. 23–40.

SCHMIDT, W., 1984, Neogene stratigraphy and geologic history of the Apalachicola Embayment, Florida: Florida Geological Survey Bulletin, No. 58, 146 p.

SCHMIDT, W., AND CLARK, M. W., 1980, Geology of Bay County, Florida: Florida Bureau of Geology Bulletin, No. 57, 96 p.

SCHNABLE, J. E., AND GOODELL, H. G., 1968, Pleistocene-Recent stratigraphy, evolution, and development of the Apalachicola Coast, Florida: Geological Society of America Special Paper No. 112, 72 p.

TANNER, W. F., DEMIRPOLAT, S., STAPOR, F. W., AND ALVAREZ, L., 1989, The Gulf of Mexico Late Holocene sea level curve: Transactions, Gulf Coast Association of Geological Societies, v. 39, p. 553–562.

QUATERNARY TERRACES AND SHORELINES OF THE PANHANDLE FLORIDA REGION

JOSEPH F. DONOGHUE AND WILLIAM F. TANNER
Department of Geology, Florida State University, Tallahassee, Florida 32306

ABSTRACT: The northeastern Gulf of Mexico coast retains a relatively undisturbed record of paleoshoreline deposits dating from the late Tertiary to the present. The excellent preservation of these features results from long-term conditions of low to moderate wave action, a wide and low-gradient platform, and tectonic stability. The general lowering of sea level from Pliocene time to the present is evident in the decrease in elevation of terrace features with age. The episodic sea-level swings of the late Quaternary and Holocene are recorded in both submerged and raised shoreline features preserved in coastal and nearshore areas. Highstand shoreline features are identified on the basis of morphology, elevation, texture and radiometric dates. Lowstand sequences are identified through seismic data, borehole lithology, and nearshore morphology. The modern coastal zone includes an extensive barrier-lagoon system, major shoals and a large lobate delta actively prograding into an estuary. Many of the present coastal features have direct analogs in both the elevated and submerged paleoshoreline deposits.

INTRODUCTION

The 400-km-long Gulf of Mexico coast of northwest Florida provides a unique laboratory for observing the effects of long-term sea-level change on a stable platform. Preserved in its elevated terraces, beach-ridge plains and inner-shelf sediments is a record of sea-level fluctuations and evolution of the regional sediment pool throughout the Quaternary.

The record of highstands is decipherable in a series of elevated paleoshoreline-deposits, dating from late Tertiary to late Holocene time, standing at elevations from +80 m to +1.5 m. The modern coastal zone is a complex of large shoals, barrier islands and spits, in which beach-ridge sequences preserve a history of late Holocene coastal evolution. At a few locations in the nearshore region, where wave energy is very low, mid- to late Holocene lowstand deposits are preserved in depths between −2 m and −12 m (Tanner and Bates, 1965). At greater water depths, sedimentary evidence of Quaternary regressions and transgressions is recorded in the fluvial deltaic and marine sequences observed in subbottom seismic records of the inner continental shelf.

The northeastern Gulf of Mexico coast (Fig. 1) retains a long and detailed sedimentologic record of geologic events, resulting from a number of factors. These include (1) long-term tectonic stability; (2) low to moderate wave energy; (3) a generally concave-seaward coastline; (4) a mild and preserving climate; (5) a wide and low-gradient shelf, in many places sloping as little as 0.3 m/km; and (6) the absence of a long fetch for wave growth.

As a consequence of these conditions, virtually undisturbed examples of most major sedimentary subenvironments are present in the surface and subsurface deposits of this region. Such environments include: *fluvial deltaic* (including channel, interdistributary, estuarine and delta-front environments): *coastal* (including moderate-energy beach, low-energy beach, cuspate foreland, barrier, lagoon, tidal-inlet, tidal-marsh, dune, and beach-ridge-plain environments); and *inner shelf* (including shoal and transgressive marine deposits).

PRE-PLEISTOCENE SHORELINES: HIGH TERRACES

Evidence for Highstands

The geology of the northeastern margin of the Gulf of Mexico basin has been dominated by the Apalachicola River throughout Neogene and Quaternary time. Riggs (1980) described the Miocene paleogeography of the region as controlled by a paleo-Apalachicola Delta, located near the Florida-Georgia border and prograding southward into the Apalachicola Embayment. Schmidt (1984) noted that the progressively younger units of the embayment exhibit diminishing relief, implying infilling from mid-Miocene to the present. Kofoed and Gorsline (1963) and Gorsline (1963) described relict, deltaic sediments extending eastward from Panama City to the Ochlockonee River (Fig. 1) and inland for 80 km. They attributed the cuspate shape of the regional shoreline to delta building by the Apalachicola River during late Tertiary and Pleistocene time, with subsequent modification by waves and longshore currents.

The Gulf Coastal Plain of the Florida Panhandle contains terraces and other shoreline indicators, which range from approximately 80 m above mean sea level (MSL) down to submarine depths, and which cover an interval of millions of years (Fig. 1). As with similar paleoshoreline features of the Atlantic Coastal Plain, the dating of many of these paleoshorelines is subject to great uncertainty. Dating schemes and nomenclature have frequently been revised over the past few decades. Very few of the terraces can be traced long distances either eastward or westward. Therefore, terrace nomenclature, developed in other areas, is not easily transferred to the Gulf Coastal Plain.

A summary of previous stratigraphic schemes for the elevated terraces of the eastern part of the Gulf Coastal Plain was provided by Winker and Howard (1977a), as well as by Healy (1975). In their reinterpretation, Winker and Howard maintained that the coast east of Mobile Bay, unlike the region to the west, has been sufficiently stable to preserve late Tertiary and Quaternary highstand deposits. They found evidence of three major episodes of Pliocene(?) and Pleistocene submergence and were able to correlate, on the basis of elevation range and state of preservation, terrace deposits of the Gulf Coastal Plain with those of the southeastern Atlantic Coastal Plain, which had previously been described (Winker and Howard, 1977b).

Based on analysis of the Gulf Coastal Plain terrace deposits, Winker and Howard (1977a) concluded that, during periods of submergence, Gulf coastal morphology was similar to that of the present. Beaches prograded near the mouths of the major rivers, such as the Apalachicola, while erosional scarps formed elsewhere. The highest terraces they

Quaternary Coasts of the United States: Marine and Lacustrine Systems, SEPM Special Publication No. 48
 ISBN 0-918985-98-6

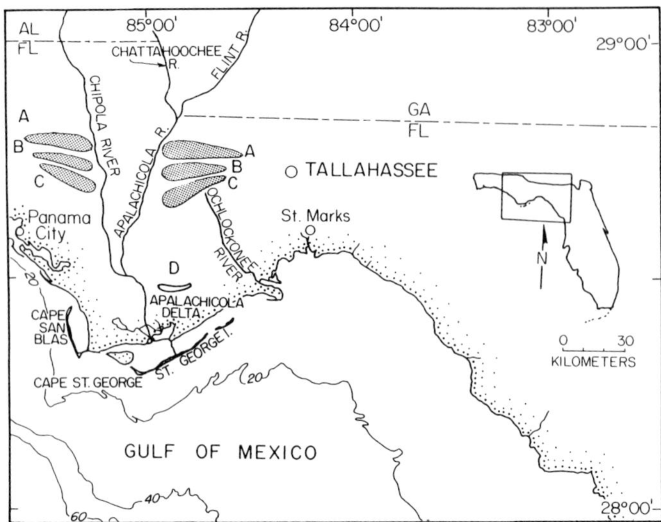

FIG. 1.—Pre-Pleistocene and Pleistocene shorelines of the Apalachicola River coastal region. Shown are outlines of outcrop areas of Pliocene (?) and older terrace sequences at +80 m (labeled A), +50 m (labeled B), and +35 m (labeled C; Tanner, 1966a). All three terraces are considered to be equivalent to the Gadsden Sequence of Winker and Howard (1977a). Location of the +9-m and +6-m pre-Sangamonian (?) terrace (labeled D) is also indicated. Bathymetric contours in meters.

encountered were designated the Gadsden Sequence (Fig. 1). Actually a series of shoreline features at varying elevations above +30 m, the Gadsden Sequence was determined to be roughly correlatable with the Trail Ridge Sequence (Force and Ridoe, 1989) of the southeastern United States, of probable early to mid-Pliocene time.

Winker and Howard (1977a) additionally concluded that a tectonic hinge line occurs south of this sequence. The result has been greater uplift and warping of the older terrace deposits north of the line than for the Pleistocene terraces farther south. The lower terraces, below +30 m, were assigned Quaternary ages.

The highest and oldest paleoshoreline features in the Apalachicola region are apparent barrier-and-lagoon sets at approximately +80, +50 and +35 m elevations (Fig. 1, Tanner, 1966a). At each level there are two old shorelines: one east of, and the other west of, the present Apalachicola River, which flows southward from the Alabama-Georgia-Florida boundary (where the Flint River combines with the Chattahoochee to form the Apalachicola). Extensions of the three old barriers west of the Apalachicola have not been preserved very well due to extensive karst development, but east of the river they are obvious on topographic maps, on aerial photographs and in the field.

Each old barrier is oriented east-west and has its maximum width adjacent to the modern river valley. The east-west length is roughly 20 km. A general description of these paleoshorelines was given by Gremillion and others (1964), except that the barrier deposits were tentatively described by them as Pleistocene. They matched multiple transverse profiles of the terraces with a similar profile across a modern barrier island to the south, St. George Island, and identified the old position of sea level according to the match, resulting in the paleoshoreline positions of +80, +50 and +35 m.

The classic approach to dating elevated marine terraces assigned the highest terrace to the oldest interglacial, the next highest to the second interglacial, and so forth, down to present sea level. In general, this is the method that has been applied in the past for dating Florida terrace deposits (Cooke, 1931, 1932, 1945; MacNeil, 1949; Puri and Vernon, 1959; Hoyt, 1969). In that scheme, the ages of the high terraces in the Apalachicola region would be: +80 m, Aftonian; +50 m, Yarmouthian; +35 m, Sangamonian.

No reliable dates have yet been obtained from the individual high terraces, but the placing of the Plio-Pleistocene boundary at or below +30 m in this region by Winker and Howard (1977a) implies that all of the three high terraces are at least Pliocene contradicting the traditional scheme. Alt and Brooks (1965), on the basis of isolated faunal evidence, assigned an upper Miocene age to the uppermost terrace. Intermediate terraces, down to +15-m elevation, were assigned Pliocene or possible Pleistocene ages. They concluded that the highest unequivocally Pleistocene terrace in Florida was the +8 to 9-m terrace. Tanner (1985), using extrapolation in the absence of reliable dates, estimated that the oldest terrace, at +80 m, was Miocene and the next lower terrace, at +50 m, was early Pliocene. Additional corroboration for a pre-Quaternary age for all of the terraces above +30 m comes from the eustatic record obtained from sequence stratigraphy. Haq and others (1987) indicate that global sea level during the Quaternary did not exceed +20 m, and that sea level has fallen a net 80 m since early Pliocene time. More data are needed, however, to firmly establish the age of the three oldest shoreline sequences of the eastern Gulf Coastal Plain.

Further evidence, which casts into doubt the traditional Quaternary dating scheme for the high terraces, will be presented below. The evidence includes the presence of Sangamonian-age dated materials only 1 to 2 m above modern MSL, and of dated probable Yarmouthian materials at +6 to +9 m.

Sediments of the High Terraces

The sediments of the three terraces east of the Apalachicola River have been studied in detail by a number of investigators (summarized in Tanner, 1982). They are mostly sand and contain no shells. The coarsest material is in the oldest (highest) terrace, close to the river; the sediment becomes finer to the east, southeast and south (to the east: away from the river; to the south; with time). The fining to the east, at any given level, is taken to be a reflection of general dispersion of sediments in the coastal zone, rather than a well-developed eastward-directed shore-parallel current, because (a) similar terraces are located west of the river, where they form a mirror image of the eastern pattern, and (b) the modern coast does not provide evidence for a well-developed west-to-east transport system.

The average mean grain size for the sediments of the +50-m surface, identified as an old barrier, is 1.711 phi, and the average standard deviation is 0.767. The average skewness is 0.282, and the average kurtosis, 4.296. These numbers (Tanner, 1988b) indicate a low- to moderate-energy beach with important complications: additional (perhaps later) wind work, or proximity to an alluvial river (true), or settling from storm conditions on the shoreface, or some combination of these.

In addition to sand, there is a significant amount of silt and clay in the samples obtained: 41 samples had 5 percent or less of fines; 70 samples had 5 to 20 percent fines; and four samples had 20 to 25 percent fines. These values, alone, indicate proximity to a major river mouth.

Grain-size probability plots exhibit the surf "break" (a distinctive kink in the coarser half of the distribution), which indicates a beach environment (Tanner, 1966b). A typical distribution is shown in Figure 3. The fact that the surf break is not located near the tip of the coarse tail and is accompanied by a distinctive fine tail indicates that wave energy was low to moderate, not much different from modern energy level, but quite unlike the waves on the Atlantic coast.

The variability of sample means, and of sample standard deviations, indicates an environment of variable energy, such as a swash zone or coastal-plain stream (Tanner, 1988b). Both of these appear to be correct. The skewness and kurtosis of the grain-size distribution indicate a beach, plus either a settling environment or later eolian reworking, both of which may be correct. The terrace surfaces are far too old, however, by any estimate, to show details of possible beach ridges or original dunes.

There is no evidence from the high terraces for the presence of river distributaries of any kind. On the other hand, the delta front was cuspate in form, with a single barrier, backed by a lagoon, on each side of the river mouth. Therefore–as far as the existing evidence indicates–this delta was not like the modern bird's-foot delta of the Mississippi River.

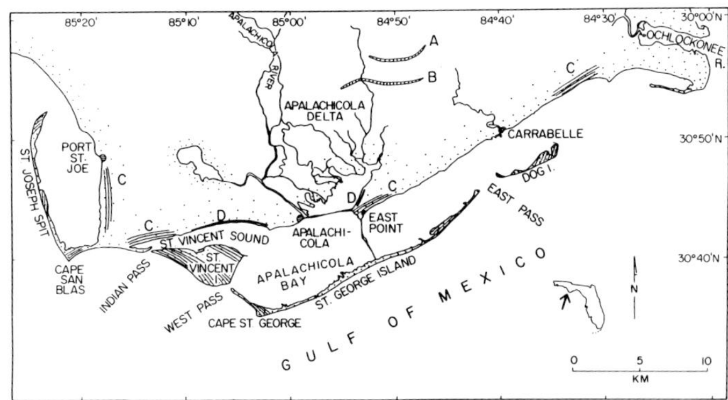

FIG. 2.—Lower Apalachicola River region, showing late Quaternary and Holocene shorelines. Shown are pre-Sangamonian(?) +9-m terrace (labeled A); Pre-Sangamonian(?) +6-m terrace (labeled B; Brenneman and Tanner, 1958; Schnable, 1966); Sangamonian +2-m terrace (labeled C; Von Drehle, 1973); and mid-Holocene +1.5-m terrace (labeled D; Stapor, 1973, 1975). Recent beach ridges on the modern barrier islands and spits are also indicated (Schnable, 1966).

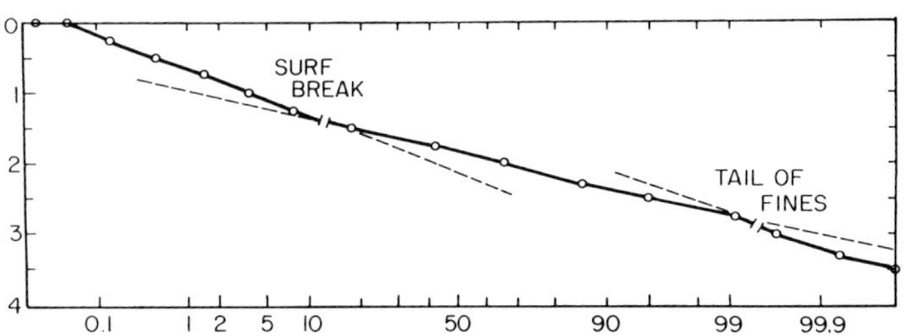

FIG. 3.—Probability plot of a representative sample showing the surf "break," indicating low- to moderate-energy beach and/or shallow-near-shore environment of deposition. Vertical axis shows grain size, on the phi scale (coarse at the top); horizontal axis shows probability scale, marked in percentage. The sample has three obvious segments. Six data points define the coarsest segment; six more define the central segment; the junction between is the surf "break." The geometry is easy to see if the reader holds the edge of the paper almost at eye level, and then looks along the heavy line. The surf "break" differs from the fluvial coarse tail in that the surf "break" inflection is much gentler. Data are from sample 46 of Brenneman (1957).

PLEISTOCENE SHORELINES

Pre-Sangamonian

Winker and Howard's (1977a) second terrace, at elevations between +10 and +30 m, known as the Wakulla Sequence, was correlated by them with the Effingham Sequence of the southeastern Atlantic Coastal Plain. They tentatively assigned it an early Pleistocene age. In the Apalachicola region the next lower shoreline feature to the south, below the three high terraces, is a long, low, narrow, convex-seaward sand body at +9 m and a second similar one nearby at +6 m. The terraces are located approximately 20 km northeast of the front of the present Apalachicola Delta (Fig. 2; Brenneman and Tanner, 1958).

These two topographically subtle features are observed in Tate's Hell, an extensive swamp. Brenneman (1957) sampled both sand bodies and found no shells. The southern (+6 m) sand body has a mean grain diameter of 2.016 phi, a standard deviation of 1.094 (poor sorting for what appears to be an old beach), skewness of 0.194 (not a mature beach), kurtosis of 3.264 (moderate to high wave energy), and a tail of fines (4.0 phi and finer) of 7.2 percent. Almost half the samples showed, on grain-size probability plots, the distinctive surf-zone feature known as the surf "break." It was concluded that each of the two long, narrow sand bodies was an immature barrier island, that is, not extensively reworked (Tanner, 1966b).

Maxwell (1971a) obtained an ionium-disequilibrium date of about 135,000 years on secondary ironstone in the sand of the north (+9 m) terrace. He had previously shown (Maxwell, 1971b) that secondary ironstone in a body of river sand does not become a closed system, for ionium dating purposes, as soon as it is deposited. Rather, it must acquire enough iron-oxide cement over a long period of time to close it. Only the interiors of the thoroughly cemented parts are datable. The lag time between deposition and closure was determined by him, in his Alabama study area, to be about 200,000 years. Unpublished work since then, in a nearby area, showed that the lag time may be much longer than that, but probably not shorter. The same logic apparently can be applied to the two narrow sand bodies at +9-m and +6-m elevation in the Apalachicola area, which therefore may be about 335,000 years old, or older.

They may be as young as Yarmouthian, but clearly predate the Sangamonian.

There are no other known sea-level indicators of the Pleistocene, but pre-Sangamonian, in the region. The +6-m and +9-m sand bodies are very subtle features, and under other circumstances might have been destroyed easily in 100,000 or so years of erosion. Their preservation is due to the fact that they are located on a featureless plain having a seaward slope so gentle (0.3 m/km) that no efficient drainage system has been developed on it.

Otvos (this volume) states that the case has not been made for the origin of these sand bodies. The textural and age data argue in favor of a pre-Sangamonian barrier origin. In addition, these sand bodies have been field-mapped as discrete units, texturally distinct from the surrounding lowlands (Schmidt, 1978; MacNeill, 1949).

Large, obvious coastal features, similar to the +50-m terrace barrier, do not appear between +2 m and +35 m. This covers most of Pleistocene time. It is interesting that most of the Pleistocene in the region is not represented by any visible features.

Sangamonian

Winker and Howard's (1977a) lowest Pleistocene terrace, designated the Escambia Sequence, lies at elevations below +10 m. They correlated it with the Chatham Sequence of the southeastern Atlantic Coastal Plain and assigned a Sangamonian age to the deposits, based in part on U-Th dates of 145,000 to 110,000 yrs reported by Osmond and others (1965) for the Miami Oolite, which is found at a similar elevation in southeast Florida, and by Osmond and others (1970) for equivalent deposits at Cape Canaveral, in northeast Florida.

The Silver Bluff locality studied by Osmond and others (1965) is on the Miami mainland several kilometers west of the modern barrier island. It is therefore taken to be late Sangamonian in age, that is, later than deposition of the oolitic limestone, but pre-Wisconsinan. Whether or not that date is correct, it is clear that there is an important 145- to 110-ka shoreline at about +2 m in South Florida. This date has been extended to many beach ridges in the Apalachicola River region at about +1.5 m to +2 m, in the vicinities of the towns of Carrabelle, East Point, Apalachicola, and Port St. Joe, Florida (Fig. 2).

A set of quartz-sand beach ridges in and near the town of Port St. Joe, Florida, was studied in detail by Von Drehle (1973). Each of these ridges may be a coalesced set of swash-built beach ridges. The Port St. Joe examples are 10 to 15 times as wide as typical modern swash-built ridges, and approximately the same width as a set of modern swash-built ridges.

The important facts about the ridges are as follows: (a) if they are in fact coalesced sets, they should be Sangamonian or older; (b) they overlie and contain well-developed humate, a secondary organic accumulation (Von Drehle, 1973); and (c) they represent ocean-beach development. Coalescence is a slow process, perhaps accomplished largely by slope wash, and has not affected any Holocene beach ridges known to the present authors. Humate growth is very slow; humate has not been found in this part of the world from any Holocene sand body, and must require much more than 5,000 to 10,000 years to form (it does appear underneath one or more of the high terraces, but they are considerably older). From these lines of evidence, it is inferred that the ridges are Sangamonian or older.

The low-angle cross-bedding in the ridges is of the beach type, and marine-shell molds have been found near the base. The mean grain diameter is 2.269 phi, the standard deviation is 0.321, the skewness is −0.242, kurtosis is 4.21, and the tail of fines (4 phi and finer) is 0.04 percent (Von Drehle, 1973). These values are characteristic of a thoroughly reworked, low- to moderate-energy beach far from a river mouth.

Similar quartz-sand beach ridges line the mainland shore both east and west of the town of Apalachicola, Florida (Fig. 2). They have mean grain diameter of 2.2 phi, standard deviation of 0.5 phi units, skewness of 0.02 and kurtosis of 3.6 (Wei, 1985). These data likewise indicate a well-worked beach sand deposited under moderate-energy surf conditions.

The presumably Sangamonian ridges occupy strips more than 2 km wide. Modern beach-ridge plains, built under similar conditions, typically prograde at about 1 m per year. Hence, a width of 2 km requires approximately 2,000 years to form, and 3 km requires about 3,000 years. Two inferences can be drawn: (1) the necessary span of time (more than 2,000 years), prior to construction of mid to late Holocene beach ridges, does not appear to be available during any mid-Holocene highstand, and therefore must be sought in Sangamon time, or earlier; and (2) the fact that these mainland ridges seem to have been built in less than 3,000 years appears to mean that, in Sangamonian time, sea level did not occupy its maximum high position for more than about 25 centuries.

A textural comparison can be made of the various pre-Holocene sands along a north-south traverse extending southward from the +50-m surface (taken as more or less representative of the three highest terraces), across the +6-m feature, to the presumably Sangamonian sands at about +2 m. The data (in phi units) are as shown in Table 1.

The tabulation excludes the Sangamonian ridges near Port St. Joe (Fig. 2) as being too far west of the transect, hence too far from the Apalachicola River. They show much better sorting (0.321), and clearly negative skewness (indicating a mature beach). The history shown in these selected data is one of fining of grain size with time, improvement in sorting, skewness approaching zero (where it clearly indicates beach conditions), and kurtosis staying largely in the low to moderate wave-energy range, which describes modern conditions.

TABLE 1. —TEXTURAL DATA, PRE-HOLOCENE SHORELINE DEPOSITS OF THE APALACHICOLA REGION (PHI UNITS).

Shoreline	Mean	Std. Dev.	Skewness	Kurtosis
+50-m surface	1.711	0.767	0.282	4.296
+6-m surface	2.016	1.094	0.194	3.264
+2-m surface	2.2	0.5	0.02	3.6

Wisconsinan

During the most recent major regression, which began about ka, sea level stood at approximately −100 m (Dillon and Oldale, 1978; Hopley, 1982). More than half the northeastern Gulf of Mexico shelf was subaerially exposed. The extreme sea-level lowering resulted in fluvial dissection of the nearshore marine and fluvial deltaic sediments that had been deposited on the present inner shelf and coastal area during Sangamonian and earlier glacial and interglacial times. The Wisconsinan erosional episode is evident in high-resolution subbottom seismic records, which reveal a dissected surface at the top of the Tertiary limestone bedrock of the continental shelf (Fig. 4). This erosional unconformity lies buried at depths of 25 to 40 m below MSL on the inner shelf.

Lying unconformably upon the limestone are two seismic sequences, each varying in thickness from 0 to 30 m. The sequences are interpreted as two pulses of fluvial deltaic estuarine deposition, which resulted from episodic sea-level rise during late Wisconsinan/Holocene time. A large body of evidence indicates that sea-level rise in the Gulf of Mexico throughout that time has been episodic rather than monotonic, punctuated by stillstands and perhaps regressions (Curray, 1960; Ballard and Uchupi, 1970; Frazier, 1974; Thomas and Anderson, 1988).

Curray's (1960) sea-level data for the northern gulf indicate a minor sea-level fall, followed by a continued rise during the period 9 to 7 ka. Frazier's (1974) Gulf of Mexico sea-level curve shows a prominent stillstand, which he associated with a glacial advance, occurring during the period 10 to 7.5 ka, at which time sea level stood at approximately −16 m MSL. The upper seismic unit observed in the northeastern Gulf of Mexico records typically lies at such depths and may date from that stillstand. In the Cape St. George Shoal area, as well as in the neighboring Cape San Blas Shoal (Fig. 1), progradational deltaic deposits are included within the sequence as identified by vibracore and seismic evidence. These deposits are interpreted as being the result of progradation of a paleo-Apalachicola delta during that time.

Various lines of evidence provide further clues to the origin of the two seismic units (Donoghue, 1989a). Each unit includes, in places, large-scale seaward-dipping cross-beds, associated with delta building by the Apalachicola River. Each is bounded by erosional surfaces showing evidence of fluvial dissection. Furthermore, sediment textural data from vibracores collected on the inner shelf demonstrated that the sands are predominantly of fluvial origin, with the fluvial character becoming more pronounced with depth in the cores (Donoghue, 1989a).

Additional information on the origin of the two sediment sequences overlying the limestone bedrock of the inner shelf is provided by Schnable (1966), who sampled a series of eighteen boreholes on the barrier islands and mainland shore in the region of the Apalachicola River mouth. He identified two late Pleistocene eustatic couplets, that is, the deposits of two complete eustatic cycles, separated by an unconformity. Each of the couplets was described as fining upward, with coarse fluvial sediments at the base and finer estuarine marine deposits at the top. These two upland lithologic units are correlatable with the two seismic sequences observed offshore.

HOLOCENE SHORELINES

Mid-Holocene Highstand

A wave-cut scarp in loose sand at about 1.5 m above MSL has been described from many places in Florida, in-

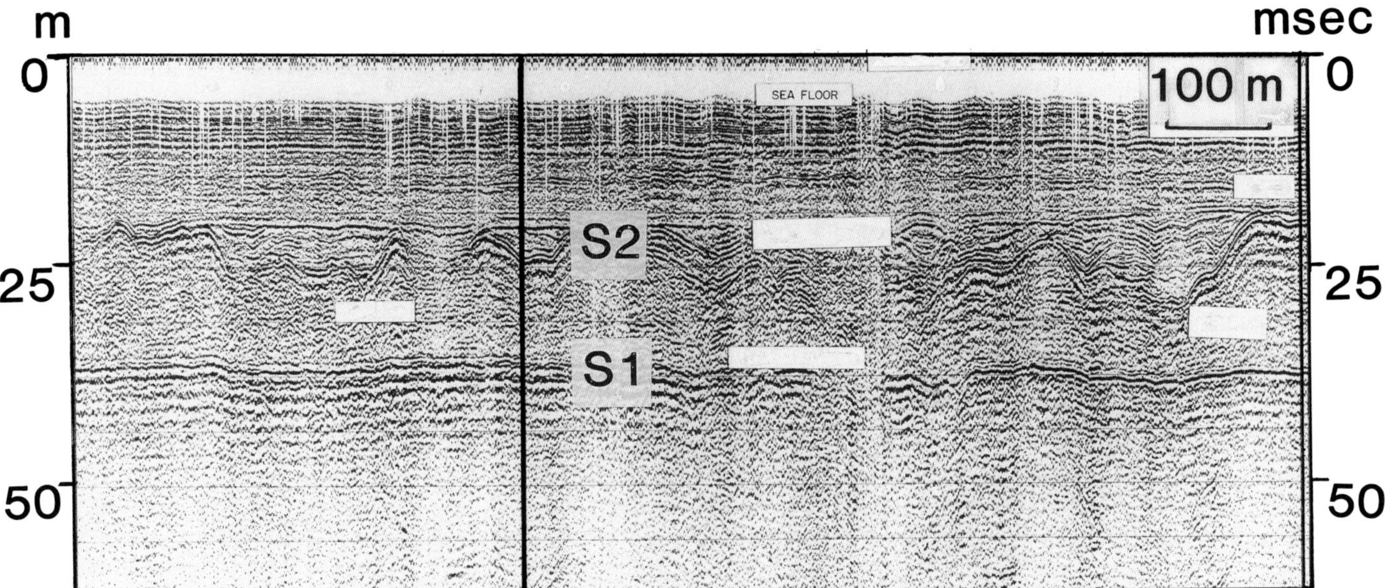

FIG. 4.—Seismic-reflection profile from Apalachicola Bay, southwest of modern Apalachicola River mouth (see Fig. 2 for location) showing two Wisconsinan erosional surfaces. Lower surface (S1) marks the top of the limestone bedrock. Upper surface (S2) separates the two major Wisconsinan seismic sequences observed throughout the region. Scale is in two-way travel time (m/sec) at right and approximate depth (m) at left.

cluding as far west as Pensacola (Stapor, 1973). It is well defined at several places close to the town of Apalachicola (Fig. 2).

Although the scarp is cut in loose sand, it is still standing at the angle of repose of that sand and therefore cannot be more than 10 to 5 ka. However, Indian kitchen middens, of about 3.5 to 3.0 ka, are found on the seaward side of the scarp (Stapor, 1973), placing a younger limit on its age. Furthermore, at least some of those middens are seaward of what was a lagoon between midden and scarp and are associated with beach ridges that were becoming higher with time: that is, resulting from a rising sea level. Because this rise in sea level must have taken several centuries, the scarp must be older than 4.5 to 4.0 ka. These limiting dates place it at about 6 to 5 ka. The scarp is thought to represent the "mid-Holocene high" position of sea level and pre-dates all other known subaerial Holocene features in the area. Its great extent (along hundreds of kilometers of shoreline) precludes the possibility of storm effects or uplift (Stapor, 1973).

How does one distinguish between the Sangamonian shoreline described earlier and evidence for a mid-Holocene highstand at a similar elevation? Along the west and Panhandle coasts of Florida, a wave-cut cliff in loose sand, still standing at the angle of repose, with its toe about 1.5 m above modern MSL is taken as mid-Holocene in age for reasons stated above. Still other features standing at 2 to 4 m above modern MSL are interpreted to be of Sangamonian age, especially if they include a sand body overlying humate, or involve very wide coalesced beach ridges with extremely gentle side slopes, as described earlier. These tests do not work in every case, but the critical point is that there is indeed evidence for both Sangamonian and mid-Holocene highstands, regardless of the degree of success in separating them at any one locality.

Possible Mid-Holocene Lowstands

Along the northwest Florida coast, wave energy decreases toward the east. In the low-energy region of the coast east of St. Marks, Florida (Fig. 1), energy conditions since the mid-Holocene have been sufficiently low to allow preservation of one, and perhaps many, lowstand shorelines. Bathymetric surveys and air photographs (Tanner and Bates, 1965) indicate the presence of a wave-cut scarp at a depth of −2 m, approximately 2 km seaward of the present shoreline. Textural data from near this scarp confirm a former beach environment. Farther offshore, bathymetric surveys suggest other possible shorelines at −7 m, −10 m and possibly −12 m. Based on their depth and their degree of preservation, all of these submerged shorelines are inferred to be of mid-Holocene age.

Late Holocene History

Nearshore region.—

In the nearshore subsurface sediments of the Apalachicola region, the mid- to late Holocene is represented in places by a transgressive sand deposit, typically 5 m or less in thickness (Donoghue, 1980a). It immediately overlies the two lower and thicker seismic sequences, which were interpreted as late Wisconsinan to early Holocene fluvial deltaic estuarine deposits. The transgressive sand is interpreted as being a result of the final phase of sea-level rise during mid- to late Holocene time.

Apalachicola Delta and Estuary.—

The Apalachicola River (Fig. 1) is the fourth largest in the northern Gulf of Mexico basin. Fluvial sediments deposited by the Apalachicola during lowstands are the major source of the quartz sand that dominates the northeastern Gulf of Mexico shelf and shorelines (Doyle and Sparks, 1980). During the present highstand, the river continues to deliver approximately one million tons of sediment per year to the river mouth. However, virtually all of this sediment is presently being retained in the delta and estuary (Donoghue, 1988; Kofoed and Gorsline, 1963). The result has been rapid infilling of the estuary, Apalachicola Bay (Fig. 2), at rates averaging 8 mm/yr (Bedosky, 1987). The modern Apalachicola Delta is likewise migrating rapidly into the bay, increasing the rate at which the estuary fills. With sedimentation rates greatly exceeding the rate of sea-level rise, the estuary is in the final stages of infilling (Tanner, 1966a; Brooks, 1973).

Since at least Pliocene time, the river has continued to migrate southward in response to the general lowering of sea level, filling its lower valley with fluvial deltaic sediments. The Apalachicola Delta is presently prograding southeastward into Apalachicola Bay (Fig. 2) at rates in excess of 2 m/yr (Donoghue, 1989b; Bedosky, 1987). Archaeological evidence indicates that this rapid migration has converted former estuarine environments to freshwater environments within the past 2,000 years (White, 1991; Donoghue, 1989a).

Shallow seismic data from Apalachicola Bay indicate that the river front, in addition to migrating southward over the long term, has also migrated eastward during the mid- to late Holocene (Donoghue, 1989a). The present position of the Apalachicola Delta is only its most recent locale. Ample sedimentologic and geomorphologic evidence exists throughout the region for older delta positions in late Quaternary time (Schnable, 1966; Stapor, 1973).

St. Vincent Island Beach-Ridge Plain.—

Each ridge in a beach-ridge plain gives the geographic position of the beach and the vertical position of sea level at the time of deposition (Tanner, 1988a). Most beach ridges in ridge sets around the Gulf of Mexico were deposited at intervals of 30 to 60 years. Therefore, an orderly sequence of some 100 ridges should include a few as old as 3,000 to 6,000 years. In fact, no Gulf of Mexico beach ridges more than 3,000 to 3,500 years old are now known. For comparison, the beach-ridge set on Isla del Carmen on the southern side of the Gulf of Mexico (in Campeche, Mexico) includes one rather high ridge having a radiocarbon date of 2,330 ±95 yrs (Thom, 1969). This date was taken from the younger part of a coalesced system that splays out along its axis into a number of ridges, indicating that the older part must be approximately 2,600 to 2,800 years old. This appears to be the date of an impor tant highstand of sea level, and the first such stand in the Gulf of Mexico in late Holocene time.

Beach ridges on the islands on the lower west coast of Florida date back to 3 ka (Stapor and others 1988). The youngest radiocarbon-dated shells in each collection were taken to represent the true age, because older shells (back to 7,600 B.P.) were thought to have been reworked from earlier deposits. Youngest ages in oldest ridges from five barrier island localities, were as follows: (1) La Costa Island, 2,985 B.P.; (2) La Costa Island, 2,920 B.P.; (3) Sanibel Island, 3,070 B.P.; (4) Sanibel Island, 2,670 B.P.; (5) Marco Island, 2,770 B.P. These numbers, drawn from an array of 290 dates, imply the same highstand as on Isla del Carmen.

Stapor and others (1988) chose to lump their results in large categories (e.g., 3–2 ka). Their conclusions are therefore not as sharp as the data warrant, but the impressive radiocarbon data set and the excellent geographic coverage should leave no doubt that the oldest ridges in their study area are about 3 ka.

Cheniers (largely made of shell and shell hash) and beach ridges are well developed in southwestern Louisiana. Radiocarbon dates reported for these features extend back to 3 ka (Byrne and others, 1959; Gould and McFarlane 1959). This appears to represent the same highstand found elsewhere around the Gulf of Mexico.

St. Vincent Island is located across the lagoon from, and southwest of, the mainland town of Apalachicola, Florida (Fig. 2). The oldest dates on that island come from an Indian kitchen midden of about 3.5 to 3.0 ka. The oldest beach ridges are landward of, and slightly older than, the kitchen midden, hence at least 3 ka. The ridges represent a rising sea level, coeval with the highstand found in the previously mentioned regions of the Gulf.

Other beach-ridge plains that have been studied along the coasts of the Gulf of Mexico do not go so far back in time. It is therefore thought that an important sea-level rise took place in the centuries up to, and including 3 ka. This rise must have post-dated the "mid-Holocene high" 6 to 5 ka for reasons given above.

On a well-developed beach-ridge plain, such as on St. Vincent Island, the ridges occur in sets of 5 to 18. Individual sets stand either high or low. The vertical difference between sets is typically 1 to 2 m. Each set represents some two to six centuries. The key is set position, not ridge position. A high set cannot indicate a storm lasting for centuries, nor can the alternation of high and low sets be the result of tectonism. Instead, the beach-ridge sequence provides a sea-level history for the last 3,000 years, with rises and falls typically of 1 to 2 m (Tanner and others, 1989). An alternative explanation for high ridges is that they are relict foredunes (Otvos, this volume). Textural evidence and geometric regularity, however, indicate that these ridges are not eolian in origin. Furthermore, archeological evidence from a number of sites indicates multiple rises and falls of sea level on the island over the past few millennia (Braley, 1982; Stapor, 1975), in agreement with the beach-ridge data.

Other islands and peninsulas in the area were initiated much more recently than 3 ka, some of them as late as about 1 ka. Many of them developed from island nuclei, which apparently formed on pre-existing shoals (Tanner, 1990). Therefore, a mid- to late Holocene history of the region would appear as a series of maps, each one showing more and larger islands and peninsulas than the previous one.

Modern Apalachicola Barrier Chain.—

No dates older than 3.5 to 3.0 ka have been obtained from any of the barrier islands of the Apalachicola region of the northeastern Gulf of Mexico (Stapor, 1973, 1975). Their geomorphologic history has therefore been brief, although complex. Modern rates of shoreline migration and nearshore bathymetric change are extremely rapid, given that this is a low- to moderate-energy coast.

Investigation of the historic-chart record (Donoghue and others, 1990) reveals that, over at least the past century, the islands generally have been erosional on both sides, with the greatest amount of shoreline retreat on southeast-facing beaches. Southeast is the dominant wave direction for both fairweather and storm waves.

The easternmost barrier, Dog Island, (Fig. 2), faces into the southeast along virtually its entire length. As a consequence, it is erosional for most of its length. In places the retreat rate of Dog Island's gulf shoreline has averaged as much as 2.5 m/yr over the period 1856 to 1979. The island is also eroding along most of its bay side, thus becoming narrower along almost its entire length, while extending at both ends (Donoghue and others, 1990).

The eastern part of St. George Island, facing southeastward, has been retreating on average more than 1 m/yr over the period 1856 to 1979. The westernmost part of St. George faces southwest. Because southwesterly waves in this region tend to be fairweather waves rather than storm waves, this stretch of coast is constructional rather than erosional. The shoreline on the beaches west of Cape St. George (Fig. 2) have been advancing approximately 2 m/yr over the same period. Similar to the shorelines of Dog Island, the Apalachicola Bay shorelines have been eroding, on average a little less than 1 m/yr over the >120-yr period.

The two prominent capes–Cape St. George and Cape San Blas (Fig. 2)–although generally anchored to two transverse shoals, are migrating alongshore extremely rapidly. Cape St. George migrated westward at a rate of approximately 8 m/yr from 1856 to 1943. The migration rate of Cape San Blas, near the western edge of the barrier rim, is even more rapid. The cape has been migrating eastward, at rates averaging between 10 and 16 m/yr over the period 1868 to 1976, with some points on the shoreline shifting by more than 1 km (Orhan, 1989). Considering this long-term drift rate, Cape San Blas is one of the most mobile coastal features in the United States.

Examination of changes in nearshore bathymetry over the past century indicates that, although the barrier islands are rapidly migrating, they are in approximate sedimentologic equilibrium (Donoghue and others, 1990). Only negligible amounts of sediment have been lost to the nearshore area. Over the past century, major changes have occurred locally, however. Examples include deepening and ebb-delta growth at the two western passes, West Pass and Indian Pass (Fig. 2). In addition, there have been major changes in the elevation of the bottom–both erosion and deposition–off Cape St. George, reflecting the extreme rate of migration of the shoreline there.

The barrier-island rim around the Apalachicola Delta appears to be nearly complete, with two narrow natural passes (at the western end) and two wide natural passes (at the eastern end). The wide passes to the east probably cannot be reduced in size much in the near future, because of the low wave-energy conditions there, coupled with relatively great distance from the supply of new sediment, near the Apalachicola Delta front.

Tidal currents in the lagoon, Apalachicola Bay, move from east to west. Outflow near the western end is maintaining one deep, narrow inlet (West Pass, immediately east of St. Vincent Island) and one narrow, barred inlet (Indian Pass, west of St. Vincent Island). It does not appear to be possible that the western inlet cross section can be reduced significantly.

The Apalachicola barrier island rim appears to be nearly fully enclosed now. Its near-term evolution, barring a major change in sea level, should consist primarily of landward migration and fragmentation of narrow barrier islands, such as Dog Island and St. George Island, and narrowing of the only wide barrier island (St. Vincent). That phase in the Quaternary evolution of this region of the northeastern Gulf of Mexico coast should end when there is no longer a barrier island in the present chain.

ACKNOWLEDGMENTS

The authors are grateful for constructive reviews by John B. Anderson and Richard A. Davis. This paper is a contribution of the International Geological Correlation Program (IGCP) Project 274.

REFERENCES

ALT, D., AND BROOKS, H. K., 1965, Age of the Florida marine terraces: Journal of Geology, v. 73, p. 406–411.

BALLARD, R., AND UCHUPI, E., 1970, Morphology and Quaternary history of the continental shelf of the Gulf Coast of the United States: Bulletin of Marine Science, v. 20, p. 547–559.

BEDOSKY, S., 1987, Recent sediment history of Apalachicola Bay, Florida: Unpublished M. S. Thesis, Florida State University, Tallahassee, 235 p.

BRALEY, C. O., 1982, Archeological testing and evaluation of the Paradise Point site, St. Vincent National Wildlife Refuge, Franklin County, Florida: Unpublished Report to Interagency Archeological Services Division, National Park Service, by Southeastern Wildlife Services, Athens, Georgia, 102 p.

BRENNEMAN, L., 1957, Preliminary sedimentary study of certain sand bodies in the Apalachicola delta: Unpublished M.S. Thesis, Florida State University, Tallahassee, 53 p.

BRENNEMAN, L., AND TANNER, W. F., 1958, Possible abandoned barrier islands in panhandle Florida: Journal of Sedimentary Petrology, v. 28, p. 342–344.

BROOKS, K., 1973, Geological oceanography, *in* Jones, J., Ring, R., Rinkel, M., and Smith, R., eds., A Summary of Knowledge of the Eastern Gulf of Mexico: Florida Institute of Oceanography, St. Petersburg, p. IIE-1 to IIE-27.

BYRNE, J. V., LE ROY, D. O., AND RILEY, C. M., 1959, The chenier plain and its stratigraphy, southwestern Louisiana: Transactions, Gulf Coast Association of Geological Societies, v. 9, p. 237–260.

COOKE, C. W., 1931, Seven coastal terraces in the southeastern states: Washington Academy of Sciences Journal, v. 21, p. 503–513.

COOKE, C. W., 1932, tentative correlation of American glacial chronology with the marine time scale: Washington Academy of Sciences Journal, v. 22, p. 310–313.

COOKE, C. W., 1945, Geology of Florida: Florida Geological Survey Bulletin 29, 342 p.

CURRAY, J. P., 1960, Sediments and history of the Holocene transgression, continental shelf, Northwest Gulf of Mexico, *in* Sherard, F. P., Phleger, F. B., and Vanandel, T. H., eds., Recent Sediments, Northwest Gulf of Mexico: American Association of Petroleum Geologists, Tulsa, p. 221–266.

DILLON, W. AND OLDALE, R., 1978, Late Quaternary sea level curve, reinterpretation based on glaciotectonic influence: Geology, v. 6, p. 56–80.

DONOGHUE, J. F., 1988, Evaluation of sediment loading processes in the Apalachicola Bay estuary: National Oceanographic and Atmospheric Administration Technical Memorandum Series No. NOS-MEMD-17, 72 p.

DONOGHUE, J. F., 1989a, Sedimentary environments of the inner continental shelf, northeastern Gulf of Mexico: Transactions, Gulf Coast Association of Geological Societies, v. 39, p. 355–364.

DONOGHUE, J. F., 1989b, Modern and ancient valleys of the Apalachicola River and estuary, northwest Florida, *in* Tanner, W. F., ed., Coastal Sediment Mobility: Proceedings, 8th Symposium on Coastal Sedimentology, Tallahassee, Florida, p. 231–248.

DONOGHUE, J. F., DEMIRPOLAT, S., AND TANNER, W. F., 1990, Recent shoreline changes, Northeastern Gulf of Mexico, *in* Tanner, W. F., ed., Coastal Sediments and Processes: Proceedings, 9th Symposium on Coastal Sedimentology, p. 51–66.

DOYLE, L., AND SPARKS, T., 1980, Sediments of the Mississippi, Alabama, and Florida (MAFLA) continental shelf: Journal of Sedimentary Petrology, v. 50, p. 905–915.

FORCE, E. R., AND RICH, F. J., 1989, Geologic evolution of the Trail Ridge eolian heavy-mineral sand and underlying peat, northern Florida: U.S. Geological Survey Professional Paper 1499, 16 p.

FRAZIER, D. E., 1974, Depositional episodes, their relationship to the Quaternary stratigraphic framework in the northwestern portion of the Gulf Basin: Texas Bureau of Economic Geology Circular 74-1, 28 p.

GOULD, H. R., AND MCFARLAN, E., 1959, Geologic history of the chenier plain, southwestern Louisiana: Transactions, Gulf Coast Association of Geological Societies, v. 9, p. 261–272.

GORSLINE, D. S., 1963, Oceanography of Apalachicola Bay, Florida, *in* Clements, T., Stevenson, R., and Halmos, D., eds., Essays in Marine Geology in Honor of K. O. Emery: University of Southern California Press, Los Angeles, p. 69–96.

GREMILLION, L. R., TANNER, W. F., AND HUDDLESTUN, P., 1964, Barrier and lagoon sets on high terraces in the Florida Panhandle: Southeastern Geology, v. 6, p. 31–36.

HAQ, B., HARDENBOL, J., AND VAIL, P., 1987, Chronology of fluctuating sea levels since the Triassic: Science, v. 235, p. 1156–1167.

HEALY, H. G., 1975, Terraces and shorelines of Florida: Florida Bureau of Geology Map Series No. 71.

HOPLEY, D., 1982, THe geomorphology of the Great Barrier Reef: Quaternary development of coral reefs: John Wiley, New York, 453 p.

HOYT, J. H., 1969, Late Cenozoic structural movements, northern Florida: Transactions, Gulf Coast Association of Geological Societies, v. 19, p. 1–9.

KOFOED, J., AND GORSLINE, D., 1963, Sedimentary environments in Apalachicola Bay and vicinity: Journal of Sedimentary Petrology, v. 33, p. 205–223.

MACNEIL, F. S., 1949, Pleistocene shorelines in Florida and Georgia: U.S. Geological Survey Professional Paper 221-F, p. 95–106.

MAXWELL, R. W., 1971a, Preliminary ionium date from marine terrace, Florida: Coastal Research Notes, v. 5, p. 9–10.

MAXWELL, R. W., 1971b, Origin and chronology of Alabama River terraces: Transactions, Gulf Coast Association of Geological Societies, v. 21 p. 83–95.

ORHAN, H., 1989, Coastal changes on St. Joseph Peninsula, Florida, 1868–1983, *in* Tanner, W. F., ed., Coastal Sediment Mobility: Proceedings, 8th Symposium on Coastal Sedimentology, p. 47–66.

OSMOND, J. K., CARPENTER, J. R., AND WINDOM, H. I., 1965, $^{230}Th/^{234}U$ age of the Pleistocene corals and oolites of Florida: Journal of Geophysical Research, v. 70, p. 1843–1847.

OSMOND, J. K., MAY, J. P., AND TANNER, W. F., 1970, Age of Cape Kennedy barrier-and-lagoon complex: Journal of Geophysical Research, v. 75, p. 469–479.

PURI, H., AND VERNON, R., 1959, Summary of the Geology of Florida and a Guidebook to the Classic Exposures: Florida Bureau of Geology Special Publication No. 5, 255 p.

RIGGS, S., 1980, Intraclast and pellet phosphorite sedimentation in the Miocene of Florida: Journal of the Geological Society of London, v. 37, p. 285–314.

SCHMIDT, W., 1978, Environmental Geology Series, Apalachicola Sheet: Map Series no. 84, Florida Bureau of Geology, Tallahassee, Florida.

SCHMIDT, W., 1984, Neogene stratigraphy and geologic history of the Apalachicola embayment, Florida: Florida Bureau of Geology Bulletin no. 58, 146 p.

SCHNABLE, J., 1966, The evolution and development of part of the northwest Florida coast: Unpublished Ph.D. Dissertation, Florida State University, Tallahassee, 231 p.

STAPOR, F. W., Jr., 1973, Coastal sand budgets and Holocene beach ridge plain development, northwest Florida. Unpublished Ph.D. Dissertation, Florida State University, Tallahassee, 219 p.

STAPOR, F. W., Jr., 1975, Holocene beach ridge plain development, northwest Florida: Zeitschrift fur Geomorphologie, v. 22, p. 116–144.

STAPOR, F. W., Jr., MATHEWS, T. D., AND LINDFORS-KEARNS, F. E., 1988, Episodic barrier island growth in southwest Florida: a response to fluctuating Holocene sea level?: Miami Geological Society Memoir No. 3, p. 149–202.

TANNER, W. F., 1966a, Late Cenozoic history and coastal morphology of the Apalachicola River region, western Florida, *in* Shirley, M. L., and Ragsdale, J. A., eds., Deltas in their Geologic Framework: Houston Geological Society, Houston, p. 84–97.

TANNER, W. F., 1966b, The surf "break": key to paleogeography?: Sedimentology, v. 7, p. 203–210.

TANNER, W. F., 1982, High marine terraces of Mio-Pliocene age, Florida Panhandle, *in* Scott, T. M., and Upchurch, S. B., eds., Miocene of the Southeastern United States: Florida Bureau of Geology, Special Publication No. 25, Tallahassee, Florida, p. 200–209.

TANNER, W. F., 1985, Late Cenozoic sea level history in the Southeastern United States: TERQUA Symposium Series, Lincoln, Nebraska, v. 1, p. 3–8.

TANNER, W. F., 1988a, Beach ridge data and sea level history from the Americas: Journal of Coastal Research, v. 4, p. 81–91.

TANNER, W. F., 1988b, Paleogeographic inferences from suite statistics: Late Pennsylvanian and Early Permian strata in central Oklahoma: Shale Shaker, v. 38, p. 62–66.

TANNER, W. F., 1990, Origin of barrier islands on sandy coasts: Transactions, Gulf Coast Association of Geological Societies, v. 50, p. 819–823.

TANNER, W. F., AND BATES, J. D., 1965, Submerged beach on a zero-energy coast: Southeastern Geology, v. 7, p. 19–24.

TANNER, W. F., DEMIRPOLAT, S., STAPOR, F. W., AND ALVAREZ, L., 1989, The "Gulf of Mexico" Late Holocene sea level curve: Transactions, Gulf Coast Association of Geological Societies, v. 39, p. 553–562.

THOM, B. G., 1969, Problems of the development of Isla del Carmen, Campeche, Mexico: Zeitschrift für Geomorphologie (new series), v. 13 p. 406–413.

THOMAS, M. A., AND ANDERSON, J. B., 1988, The effect and mechanism of episodic sea level events: the record preserved within late Wisconsinan-Holocene incised valley-fill sequences: Transactions, Gulf Coast Association of Geological Societies, v. 38, p. 399–406.

VON DREHLE, W., 1973, A sedimentary investigation of the large, linear sand bodies exposed in the Gulf County, Florida, Canal: Unpublished M.S. Thesis, Florida State University, Tallahassee, 119 p.

WEI, W., 1985, A sedimentological investigation of the beach ridges along the mainland coast of Gulf and Franklin Counties, Florida: Unpublished M.S. Thesis, Florida State University, Tallahassee, 133 p.

WHITE, N. M., 1991, Testing remote shell mounds of the lower Apalachicola valley, northwest Florida: Florida Anthropologist, v. 44, p. 17–29.

WINKER, C. D., AND HOWARD, J. D., 1977a, Plio-Pleistocene paleogeography of the Florida Gulf Coast interpreted from relic shorelines: Transactions, Gulf Coast Association of Geological Societies, v. 27, p. 409–420.

WINKER, C. D., AND HOWARD, J. D., 1977b, Correlation of tectonically deformed shorelines on the southern Atlantic coastal plain: Geology, v. 5, p. 123–127.

SEDIMENT CHARACTERISTICS AND SEAFLOOR TOPOGRAPHY OF A PALIMPSEST SHELF, MISSISSIPPI-ALABAMA CONTINENTAL SHELF

STEVEN J. PARKER* AND ALBERT W. SHULTZ**

Department of Geology, The University of Alabama, Tuscaloosa, Alabama 35487

AND

WILLIAM W. SCHROEDER

Marine Science Program, The University of Alabama, Dauphin Island, Alabama 36528

ABSTRACT: The sedimentary facies and seafloor topography of the Mississippi-Alabama continental shelf are a product of late Pleistocene-Holocene regression and transgression. This region lies between the fine-grained facies associated with Holocene Mississippi River delta deposition and the carbonate facies on the western Florida shelf. Sediments in the shallow subsurface of the inner shelf were deposited by fluvial and coastal systems that developed on the shelf during Pleistocene sea-level fluctuations. Shoreface retreat associated with Holocene transgression allowed marine and coastal processes to modify these deposits, resulting in the present sediment distribution. Recent sedimentation on the inner shelf in this region has been minor and restricted primarily to a nearshore fine-grained facies.

The Southeast Banks area (SEBA), located on the Mississippi-Alabama continental shelf approximately 23 km south of Morgan Peninsula in water depths of 17 to 27 m, was selected for study based on the irregular seafloor topography and unusual bottom sediments present at Southeast Banks fishing ground, a site well known to local fisherman. Three sedimentary facies were identified in the Southeast Banks area: (1) a sand facies, (2) a shell gravel and sand facies, and (3) a mud, shell gravel and sand facies. Hardbottoms, which consist of shell debris as well as rubble and large fragments of carbonate-cemented sandstone and coquina, occur in association with the sedimentary facies at Southeast Banks.

The sea floor in the SEBA is characterized by a series of northwest-trending ridges and troughs. Lengths and widths of these ridges are highly variable. Additionally, variations in sediment textures are coincident with changes in seafloor topography. Ridge morphology, orientation, and sediment textures associated with the ridges indicate that these features developed as shoreface-connected ridges through marine reworking of older sediments and topographic features. In the present environment, ridge features and bottom sediments are disturbed primarily by periodic high-energy storm events.

INTRODUCTION

Sediments of the Mississippi-Alabama continental shelf generally consist of quartz sand, carbonate sand and gravel, and mud (Ludwick, 1964; Upshaw and others, 1966; Doyle and Sparks, 1980). Previous investigations of the surficial sediments in this area were based on a relatively small number of samples spread over a large area, and as a result were not adequate to identify variations in the sediments over short distances.

Recent studies (Schroeder and others, 1988a; Shultz and others, 1990) suggest that sediment textures on the Alabama inner continental shelf are much more complex than previously described. Shultz and others (1990) observed sediments varying from shell gravel to mud over sampling intervals of less than 1 km. They also noted that variations between sand, shell hash, and mud are associated with bathymetric change in the area. Schroeder and others (1988a) identified hardbottoms (rock outcrops, indurated sediments, and shell gravel) in several areas on the Alabama inner shelf.

A few investigators have described the seafloor topography in the northern Gulf of Mexico (Hyne and Goodell, 1967; Pyle and others, 1975). However, they proposed different origins for the mechanisms that produce the irregular topography. Pyle and others (1975) identified giant- to large-scale bedforms on the Mississippi-Alabama inner shelf. They proposed that these bedforms likely formed during high-energy storm events. Hyne and Goodell (1967) suggested a fluvial origin for the submarine geomorphology offshore of northwest Florida.

The site selected for this study is an area where highly variable sediment textures and irregular seafloor topography are known to exist. The site was chosen for a detailed examination of their relation. The investigation examines the areal distribution of the sedimentary facies as defined by textural and compositional characteristics and the evolution of seafloor topography in the Southeast Banks area. Granulometric analyses, sidescan sonar data, clay mineralogy, and carbonate content provide the basis for identifying the sedimentary facies. The relations between the sedimentary facies, hardbottom areas, and seafloor topography will aid in understanding the complex sedimentological processes responsible for the distribution of sediments on the Mississippi-Alabama continental shelf. In addition, these relations will be useful in understanding the Holocene history of the Alabama shelf.

SETTING

The Mississippi-Alabama continental shelf is a broad, nearly flat region extending from the Mississippi River delta complex on the west to De Soto Canyon on the east (Fig. 1). The outer edge of the shelf is marked by a fairly gradual increase in slope over a distance of several kilometers, but is arbitrarily taken as the 80-m isobath. The shelf is approximately 122 km wide south of Mississippi and narrows to approximately 44 km at the head of De Soto Canyon, with an average width of approximately 95 km. Gradients on the shelf average around 1 m/km.

Seafloor topography and sediment distribution of the Mississippi-Alabama shelf are the result of deltaic progradation, late Pleistocene regression accompanied by erosion

*Present Address: Geological Survey of Alabama, P.O. Box O, Tuscaloosa, Alabama 35486

**Present Address: Conoco Inc., 1000 South Pine, P.O. Box 1267, Ponca City, Oklahoma 74603

Quaternary Coasts of the United States: Marine and Lacustrine Systems, SEPM Special Publication No. 48

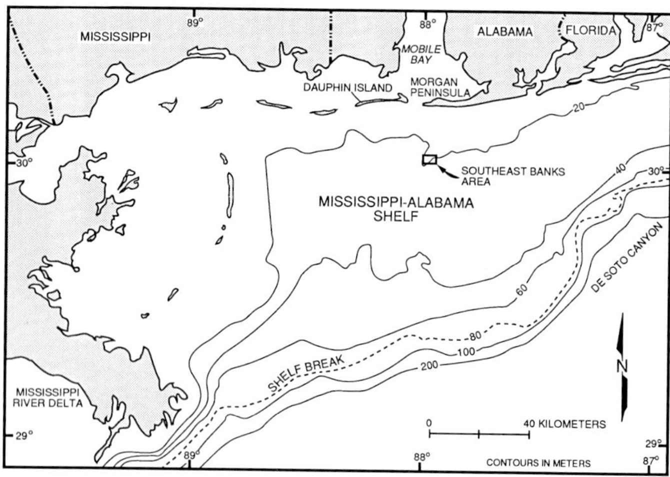

FIG. 1.—Location of the Mississippi-Alabama shelf and the Southeast Banks area (SEBA).

of the exposed shelf by ancient fluvial systems, and reworking by coastal processes associated with Pleistocene-Holocene sea-level rise (Ludwick, 1964; Kindinger, 1988). Sediments underlying the thin Holocene sediment cover consist of fluvial sands and gravels that were deposited during the last sea-level lowstand ending about 18 to 15 ka (Smith, 1986; Kindinger, 1988). Marine, estuarine, and fluvial systems probably prograded seaward and at successively lower elevations from the present Mississippi and Alabama coastline until sea level reached a position approximately 125 m below present sea level (Smith, 1986). Subsequent sea-level rise allowed marine processes to rework sediments. Sedimentation associated with this transgression was minor, allowing relict sediments to remain exposed on the shelf.

Activity of the St. Bernard lobe of the Mississippi River delta caused deposition of a thin veneer of Holocene mud on the extreme western part of the Mississippi-Alabama shelf (Kindinger, 1988). East of Mobile Bay, the shelf is characterized by a ridge-and-trough topography. These features generally trend northwest-southeast and exhibit relief around 2 to 3 m.

Much of the terrigenous material supplied by rivers to the Mississippi-Alabama coastal areas is presently being trapped in bays and estuaries; only small amounts of this material reach the shelf (Doyle and Sparks, 1980). Ryan (1969) estimated that an annual average of 4.3×10^9 kg of suspended sediment and an unknown amount of bedload are transported into Mobile Bay, and estimated that only approximately 30 percent of this amount (1.3×10^9 kg) passes through the estuary into the Gulf of Mexico. Dinnell and others (1990), using sediment plume analysis, estimated that approximately 9.1×10^8 to 1.8×10^9 kg of sediment are transported to the shelf annually.

The area chosen for this study lies approximately 25 km south of Morgan Peninsula on the Alabama inner continental shelf in water depths of 17 to 28 m. The area is known as the Southeast Banks area (SEBA) and includes the Southeast Banks fishing ground, a site well known to local fisherman. The SEBA encompasses nearly 22 km^2 and measures 4.0 km north to south and 5.4 km east to west.

METHODS

The sea floor in the SEBA was surveyed using high-resolution (100 kHz) sidescan sonar, sonic depth profiling, underwater video, grab sampling, and dredging. Sidescan sonar records were collected during May, 1987. Grab and dredge sampling was conducted during the period of January, 1987 to February, 1988. Absence of major storms during this time reduces the likelihood of significant change in bottom sediments or bathymetry during the data-gathering period.

Approximately 85 km of high-resolution sidescan sonar and fathometer records were collected along 21 north-south survey lines. Loran-C navigation was used to establish the positions of these lines; lines were spaced approximately 0.27 km apart. Sidescan sonar records in 300-m widths (150 m on each side of vessel) overlapped 10.5 percent to form a mosaic of the entire area. A bathymetric map was produced from approximately 500 data points picked from fathometer profiles.

Because there are many variables controlling the tonal shades produced on a sonagram, direct interpretation of bed materials from sidescan sonar data alone may be unreliable. Sidescan sonagrams were correlated with bottom sediments by collecting 94 bottom samples along four sampling transects using a Shipek grab sampler (Fig. 2). Additional samples were taken using a Capetown dredge around the hard-bottom sites, located in the southeastern part of the study area. Underwater video coverage is along four transects in the southeastern part of the study area.

Grab samples were analyzed in the laboratory for grain size and carbonate content, and selected samples were analyzed for clay mineralogy. Grain-size distributions were determined by standard sieve and hydrometer methods (Lewis, 1984). The carbonate content of bulk samples was determined by loss through dissolution of the pulverized sample

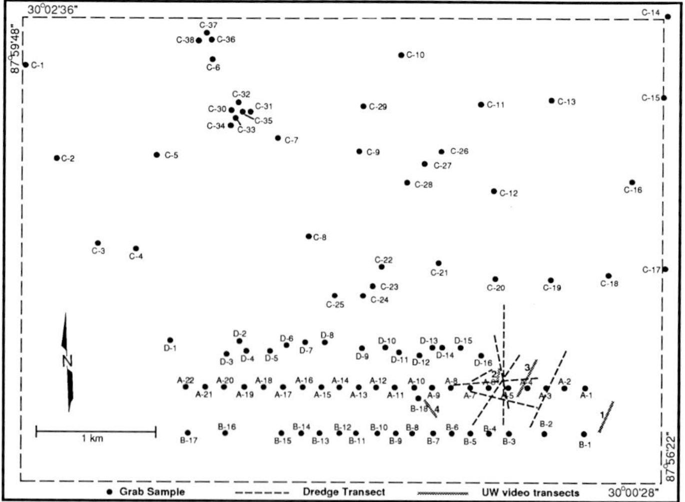

FIG. 2.—Location of grab samples, dredge transects, and underwater video transects in the SEBA.

in a pH 5 buffer solution of sodium acetate and acetic acid.

Samples analyzed for clay mineralogy by X-ray diffractometry were selected on the basis of percentage of clay, geographic location, and water depth. Clay samples were analyzed by X-ray diffractometry using oriented clay slides of particle sizes ≤ 2 μm. Methods of identification and quantification of clay minerals were according to Schultz (1964).

SEAFLOOR TOPOGRAPHY OF THE SOUTHEAST BANKS AREA

The sea floor in the SEBA is characterized by several north- to northwest-trending ridges (Fig. 3). Water depths range from a minimum of approximately 16.8 m on top of the ridges to a maximum of about 22.5 m in the deepest trough. Water depth increases abruptly along a relatively steep slope in the southeastern part of the study area. The slope dips to the southeast and the gradient along the slope is approximately 4 m over a distance of 0.75 km. The greatest water depths occur southeast of this slope and reach a maximum of 28 m. The rest of the study area slopes gently north to south at approximately 1 m/km.

The characteristics of the seafloor ridges in the SEBA, including relief, length, width, and spacing, are highly variable. Ridge heights range from 1 to 4 m with an average height of 2 m. Ridges are spaced an average of 0.5 km and range in length from 0.31 to 2.65 km. Some ridges exhibit relatively straight crestlines; however, some display curved or sinuous crests. Ridges are symmetric or asymmetric and typically show rounded to peaked crests separated by pronounced V-shaped troughs. Steeper slopes generally occur on the landward or northeastern sides of the asymmetrical ridges. In the shallower western and northern parts of the study area, ridges are more continuous and exhibit more northerly trends than elsewhere. Ridges in the deeper southeastern part of the study area are more erratic and exhibit less continuity and form smaller angles with the shoreline. Ridges become terminated at their southeastern ends by the northeast-trending slope. Southeast of this slope, where water depths are significantly deeper, ridges are nonexistent within the study area.

In some cases, smaller bedforms appear to be superimposed on larger, continuous ridges. These bedforms, observed on the sidescan sonagram mosaic, are typically oriented obliquely to the large sand ridges. Small bedforms, apparent from the underwater video coverage, occur in coarse shell gravels and are on the order of 20 to 30 cm high and exhibit wavelengths of around 1 to 2 m.

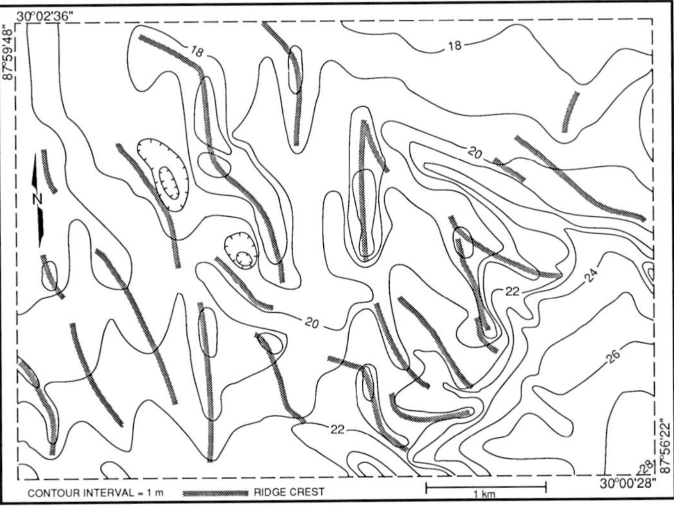

FIG. 3.—Bathymetric map showing location of ridge crests in the SEBA.

SEDIMENTARY FACIES

Introduction

The sedimentary facies in the SEBA were identified using both sidescan sonar and bottom sampling data. Tonal patterns on the sonagram were groundtruthed by bottom sampling. The facies were determined by separation of the tones represented on the sonagram and based on the grain-size data from the bottom samples. The two methods were compared to test the accuracy of the sonagram. In all but a few samples, the tone on the sonagram was consistent with the textural results from the bottom sampling data. In most cases, the inconsistencies were a result of the limited resolution of the sonagram, and were resolved using a larger scale sonagram.

The sonagram can be divided into three dominant shades or tones (Fig. 4). Solid white areas (square and rectangular blocks) are areas of no data. TONE-1 is a white to light gray shade representing a weak to moderate acoustical backscatter (Fig. 4). This reflection is typical of medium-grained to fine-grained sand and silt (Pyle and others, 1975; Williams, 1982). TONE-1 covers the majority of the study area. TONE-2 is a solid black shade occurring in subparallel bands and irregular patches throughout the study area (Fig. 4). This reflection indicates a strong acoustic backscatter typical of gravel to medium sand. TONE-3 is a medium to dark gray region in the southeastern corner of the study area (Fig. 4). The reflection may represent gravel- to silt-size sediment (Pyle and others, 1975; Williams, 1982). These tones were compared with the grain size data from the bottom samples. Based on the results, three sedimentary facies were identified: (1) sand facies, (2) shell gravel and sand facies, and (3) mud, shell gravel, and sand facies. Constituents of the facies are listed in order of importance with the dominant constituent listed last.

Sand Facies

The sand facies, which corresponds with TONE-1 on the sidescan sonagram, covers the majority of the study area (Fig. 5). It typically occurs along the crest and the seaward or southwestern slopes of the ridges and throughout much of the flat areas between ridges. Quartz sand is the dominant constituent of this facies with percent quartz averaging 92.7 percent. Carbonate sand and gravel, silt, and clay are present in minor amounts (Table 1). Quartz sand is generally fine- to medium-grained, subangular to subrounded, and moderately sorted. The coarsest quartz grains generally are around 0.5 phi (0.71 mm) to 0.0 phi (1.0 mm). Mean grain size is 1.88 phi and the average standard deviation is

FIG. 4.—Sidescan sonagram mosaic of the SEBA.

0.74 phi. Sand-size particles constitute an average of 97.4 percent and range from 90.6 to 99.4 percent of samples within the facies (Table 1).

Percent carbonate ranges from 0.4 to 22.5 percent and averages 6.0 percent. Gravel content averages 1.9 percent and ranges from 0.1 to 8.7 percent. The highest amounts of carbonate material in the facies generally occur in samples adjacent to the shell gravel and sand facies. These samples indicate that the boundaries between the two facies may be gradational; however, most of the samples indicate the boundaries are sharp. Between samples D-5 and D-6 (Fig. 2), there is an increase in percent carbonate from 4.7 percent to 93.8 percent.

Silt and clay percentages are negligible, both averaging less than 1 percent. Fine material was resuspended as the underwater video camera came in contact with the bottom, indicating that silt and clay may occur as a thin veneer throughout the study area. Slightly higher silt and clay percentages occur in the northwest-trending troughs.

Shell Gravel and Sand Facies

The shell gravel and sand facies corresponds with TONE-2 on the sidescan sonagram (Fig. 5). It appears black on the sonagram due to the intense accoustical backscatter as a result of shell gravel present in the facies. The shell gravel and sand facies occurs in subparallel bands along the landward or northeastern slopes of the ridges in the study area. To the southeast, these bands become less defined and the facies occurs more as irregular patches. These patches are interconnected and cover an extensive area along the boundary between the mud, shell gravel, and sand facies (Fig. 5).

Sand percentages average 61.8 percent and range from 20.9 to 89.3 percent. Quartz sand comprises an average of 43.6 percent of the sand fraction within the facies; much less than the sand facies. Silt and clay content is minor, with averages less than 1 percent. The average particle size is very coarse sand and mean grain sizes average −0.15 phi (1.11 mm). The facies is poorly sorted; average standard deviation is 1.61 phi (Table 1).

Gravel, comprised exclusively of carbonate material, averages approximately 37.0 percent and ranges from 10.1 percent to 77.0 percent (Table 1). Carbonate content is slightly higher ranging from 19.1 percent to 93.8 percent and averaging 54.0 percent, which indicates that a large part of the sand fraction is carbonate. The majority of car-

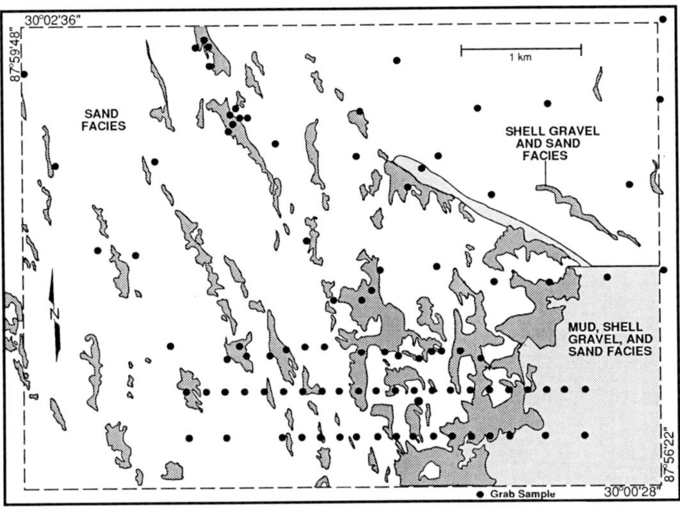

FIG. 5.—Sedimentary facies identified in the SEBA.

TABLE 1.—STATISTICAL, TEXTURAL, AND COMPOSITIONAL CHARACTERISTICS OF THE SEDIMENTARY FACIES

Parameter	Sand Facies	Shell Gravel and Sand Facies	Mud, Shell Gravel, and Sand Facies
No. of samples	68	16	9
Statistics			
Mean grain size (ϕ)			
Average	1.88	−0.13	2.45
Range	1.18–2.36	−1.67–1.2	1.99–3.23
Std. deviation (ϕ)			
Average	0.74	1.61	1.42
Range	0.46–1.24	1.33–2.04	0.67–1.96
Texture			
Percent gravel			
Average	1.9	38.5	6.3
Range	0.1–8.7	10.1–77.0	0.2–14.1
Percent sand			
Average	97.4	60.2	82.2
Range	90.6–99.4	20.9–89.3	62.8–97.6
Percent silt			
Average	0.3	0.5	6.9
Range	0.0–.9	0.0–1.8	.5–25.6
Percent clay			
Average	0.5	0.8	4.7
Range	0.1–1.3	0.3–1.6	1.1–10.6
Composition			
Percent quartz sand			
Average	92.7	43.6	74.4
Range	58.2–99.1	4.1–80.3	54.8–95.1
Percent carbonate			
Average	6.0	55.1	18.3
Range	0.4–22.5	19.10–93.8	3.0–29.8

bonate material recovered in the grab and dredge samples from the SEBA consists of whole and broken fragments of mollusks (bivalves and gastropods), echinoderms, worm tubes, solitary corals, and barnacles. Bivalve shells comprise most of the carbonate fraction. Some of the common shells include cockle shells (*Dinocardium robustum*), calico scallops (*Argopecten gibbus*), and calico clams (*Macrocallista maculata*). Florida fighting conchs (*Strombus alatus*) are among the most common gastropods. Relict shells comprise some of the carbonate material and include oyster shells (*Crassostrea virginica*), blood arks (*Anadara ovalis*), and hardshell clams (*Mercenaria mercenaria*). These shells are typically black or brown and appear worn and abraded.

The extensive shell pavements found in association with this facies have been termed by Schroeder and others (1988a, p. 535) as "hardbottoms," which they define as ". . . a generic term that describes any seafloor feature or deposit with a hard or indurated surface." The term hardbottom was used rather than hardground because these features occur primarily among clastic sediments, are not extensive, and include shell pavements, which are not lithified. Indurated sediments consisting of rubble and large fragments of rock also occur in association with the shell gravel and sand facies. Dredging and underwater video confirmed the presence of rock rubble at a site known as Southeast Banks fishing ground in the southeastern part of the study area. Although dredging and video were not used throughout the area (Fig. 2), fathometer records do not suggest that the bottom roughness associated with rock materials at the Southeast Banks site is widespread.

Hardbottom rocks are primarily carbonate-cemented sandstones, mudstones, and coquina. Sideritic sandstones have been found in the study area but are rare. The hardbottom rocks are composed mainly of subangular quartz sand and shell debris with minor amounts of feldspar, mica, and glauconite (Schroeder and others, 1988a). Mollusk shells, foraminifera tests, and echinoderm fragments are common constituents in these rocks. Aragonite, high Mg-calcite, and pseudodolomite are the dominant cements and probably formed by methane oxidation in shallowly buried sediments (Howard, 1990). Similar materials have been reported from other coastal sites in the Gulf of Mexico (Roberts and Whelan, 1975; Kocurko, 1986; Weiss and Wilkinson, 1988).

Rocks generally range in color from buff to yellowish gray, dark gray, and dark brown. Many of the sandstones are well cemented, whereas shelly sandstones and coquina are porous and friable. Extensive borings by the bivalve *Lithophaga* and the sponge *Cliona* are present in many of the rocks. Encrusting epifauna, including soft corals (*Leptogorgia virgulata* and *Lophogorgia hebes*), barnacles, bryozoa, sponges, and serpulid worms are present in various amounts on the rocks in the study area (Schroeder and others, 1988b).

Hardbottom rocks occur as scattered rubble consisting of rounded, abraded pieces and irregular slabs or outcrops of rock as much as 1 to 2 m in diameter. Large slabs are typically covered with abundant soft corals and other encrusters. In some cases, rocks are completely covered with a loose veneer of fine sand, shell, or mud; however, they often have attached epifauna, which indicates they had exposed surfaces at one time.

Relief associated with these large slabs is generally 20 to 25 cm; however, some rocks exhibit little or no relief. Smaller, abraded rocks generally exhibit little or no faunal encrustation. Lack of epifauna may be due to reworking or because rocks were completely buried. Furthermore, encrustations appear on both the exposed side and the underside of small- to moderate-size slabs (0.5 m in diameter), which may indicate episodic disruption of the slabs. This episodic movement of larger hardbottom rocks may result from intense wave energy produced during hurricanes or strong winter storms. Undercutting by fish and other fauna is also apparent, allowing encrustation to form on the undersides of hardbottom slabs.

Mud, Shell Gravel, and Sand Facies

The mud, shell gravel, and sand facies occurs primarily in the southeastern part of the study area and corresponds with TONE-3 on the sidescan sonagram (Fig. 5). Although sand is the dominant constituent of the facies, considerable percentages of shell gravel and mud also are present. The tonal shade representing this facies results from a moderate to strong backscatter. Sediment textures associated with this tone generally indicate a mixture of gravel, sand, silt, and clay. Sediment samples collected in this facies reflect this wide range of grain sizes.

Sand dominates the facies with percentages ranging from 62.8 percent to 97.9 percent and averaging 83.7 percent. Quartz sand comprises an average of 74.4 percent of the sand fraction. Carbonate content ranges from 3.0 to 29.8 percent and averages 18.3 percent. A considerable portion of the carbonate material is shell gravel, which averages 5.7 percent and ranges from 0.2 percent to 14.1 percent. Silt and clay values are significantly higher than those for the other two facies. The greatest silt and clay percentages are from sample C-27 (see Fig. 2), which is located in a large northwest-trending trough in the northeast to central part of the study area. Silt content ranges from 0.6 percent to 25.6 percent and averages 6.3 percent. Clay percentages average 4.3 percent and range from 1.1 percent to 10.6 percent. The average grain size of the facies is fine sand (2.45 phi), and the sorting is poor with an average standard deviation of 1.42 phi (Table 1).

Because of the significant clay content in the facies, X-ray analysis was run on selected samples to determine the composition of the clay. The dominant clay minerals in the facies are kaolinite, montmorillonite, and illite. Broad smectite peaks suggest that mixed-layer clays may be present; however, no specific diffraction peaks for regular mixed-layer clays were identified. No chlorite peaks were identified in any of the samples analyzed. Kaolinite is the predominant clay, averaging 44.5 percent of the clay fraction and ranging from 37.1 percent to 50.1 percent. Montmorillonite and illite occur in almost equal amounts. Percentage of montmorillonite ranges from 12.1 to 46.5 and averages 27.4 percent. Illite (averaging 29.2 percent and ranging from 18.7 to 45.8 percent) is present in all samples.

Kaolinite percentages remain fairly consistent throughout the samples tested. The highest kaolinite percentage (49.4) is associated with the stiff, cohesive, dark gray clay. Large amounts of this clay were collected within the facies, which may indicate an extensive clay layer exposed at the surface. Variations in the illite and montmorillonite percentages are much more noticeable. Sample C-18 (Fig. 2) has the highest recorded montmorillonite percentage (45.5) and A-1 (Fig. 2) has the highest recorded illite percentage (45.7).

The clay-mineral suite observed in the SEBA indicates provenance is primarily from the Mobile River system with possible influence from the Mississippi River system. The Mobile River system is characterized by a clay-mineral suite consisting of 40 to 50 percent kaolinite, 40 to 50 percent montmorillonite, and 0 to 5 percent illite (Griffin, 1962). The Apalachicola River system to the east contains a clay-mineral suite of 70 to 80 percent kaolinite, 0 to 20 percent montmorillonite, 0 to 5 percent illite, and 0 to 10 percent vermiculite. Clay-mineral composition for the Mississippi River system is 60 to 80 percent montmorillonite, 10 to 20 percent kaolinite, and 20 to 30 percent illite (Griffin, 1962). Based on this information, the Mobile River system appears to be the dominant contributor of clays to the SEBA. However, the high-illite content in the SEBA may also reflect an influence of the Mississippi River on this part of the continental shelf. The low percentages of montmorillonite indicate an influence of the Apalachicola and Mobile River systems. However, montmorillonite is the dominant clay in Mobile Bay, which may reflect the tendency for montmorillonite to flocculate and become trapped in coastal bays (Isphording and Lamb, 1980). *In situ* alteration of the clays may also cause clay percentages in the Southeast Banks to vary from the percentages of the source.

DISCUSSION

Few reports have been published relating sediment distribution and seafloor topography on the Mississippi-Alabama continental shelf. Previous reports by Hyne and Goodell (1967) and Pyle and others (1975) described shelf sediments and topographic features in the vicinity of the study area. These studies identified both relict features and active bedforms; however, they proposed different mechanisms by which these bedforms and the resulting sediment distribution evolved.

Hyne and Goodell (1967) conducted a survey of the ridge-and-trough topography off the coast of Choctawhatchee Bay, Florida, east of the study area. They observed that the ridges and troughs are oriented roughly 70° to the strandline and exhibit a maximum relief of 9 m. The authors considered a sand-wave origin for the ridge-and-trough topography unlikely because of dissimilarities between the sedimentary characteristics of this area and those for sand waves in other areas. They noted that ridge-and-trough topography appears similar to subparallel stream-erosion patterns developed on the Pleistocene and Recent marine terraces of the coastal plain in the area. Therefore, they proposed that during late Wisconsin regression, barriers were deposited on a marine terrace which were subsequently dissected by fluvial action, forming the ridge-and-trough topography. In addition, they proposed that this topography was only slightly modified by Holocene transgression, which resulted in the present asymmetrical shapes and sediment characteristics.

Pyle and others (1975) recorded and described the general distribution of bedforms on the continental shelf in the northeastern Gulf of Mexico. On the Mississippi-Alabama shelf, they identified bedforms including giant sand waves (wavelength [λ] > 30 m, Ripple Index [wavelength/amplitude] = 30–100), large sand waves (λ = 1–30 m, R.I. = 15–30), low-relief swells (λ = a few hundred to 1,000 m, R.I. = much greater than 100), and irregular hummocky topography. Pyle and others (1975) concluded that the hydraulic regime under which these bedforms originate is quite complex. Giant sand waves have been associated with current speeds greater than 51 cm/sec (Stride and Chesterman, 1973). Kenyon and Stride (1970) identified large to giant bedforms associated with currents ranging from 64 to 124 cm/sec. Pyle and others (1975) estimated that currents on the order of 50 to 100 cm/sec were required to form the majority of large and giant sand waves in this area. Further, they proposed that giant bedforms were produced under extreme storm conditions, and large to small sand waves result from less severe conditions induced by frontal passages.

Although there are limited data explaining the mechanisms that produce the seafloor topography and sediment distribution in the northern Gulf of Mexico, large bedforms and associated sediments have been discussed in several papers for other continental shelves around the world. The mechanisms by which large-scale bedforms are produced on continental shelves have been under some debate. Theories by Duane and others (1972), Swift and Field (1981), Parker and others (1982), and Hoogendoorn and Dalrymple (1986) propose that large-scale bedforms form in response to storm-generated currents. Each of these studies also observed that sediment distribution is directly related to the formation of the bedforms.

In their study of nearshore ridges on the Atlantic shelf of the United States, Duane and others (1972) argued that ridges are not degraded barriers because the ridges intersect the coast at oblique angles. They suggested that ridges formed by storm-generated helical flow along retreating shorefaces during sea-level rise.

Swift and Field (1981), in their analysis of ridges on the north Atlantic shelf, developed a model that suggests that ridge formation is a response to erosional shoreface retreat driven by downwelling storm flows. Their model shows that as storm flow passes over the ridge, the upcurrent slopes are eroded and the downcurrent slopes are aggraded. They observed an evolutionary sequence from shoreface ridges through nearshore ridges to offshore ridges and that changes in morphology and sedimentary patterns occur in successively seaward ridges. These changes indicate that ridges occurring in deeper water are less affected by storm flows of the present hydraulic regime. They noted that sediment distribution is related to seafloor topography in each of the ridge sets, which exhibit coarse-grained sediment consisting of shell hash, on the upcurrent or landward flanks, and fine sand on the downcurrent or seaward slopes.

In a study of the Argentine shelf, Parker and others (1982) noted that Swift and Field's (1981) model does not accurately account for the oblique orientation of ridges. They suggested Huthnance's (1982) model is most suitable for large-scale bedforms (sand ridges) because it requires an oblique orientation of the bedform with respect to flow. Parker and others (1982) also observed sediment patterns similar to those described by Swift and Field (1981) with fine sand occurring on the seaward slopes of ridges and shell hash on the landward slopes.

Hoogendoorn and Dalrymple (1986) observed that the ridges on the Scotian shelf formed by obliquely onshore geostrophic flow elevated by high-energy storms. They noted that the difference between these ridges and the United States east coast ridges is due to a more intense storm climate in the Canadian east coast area and to obliquely offshore geostrophic flow of the United States east coast. They too observed shell hash on the upcurrent or landward slopes of ridges and fine sand on the downcurrent or seaward slope.

Stubblefield and others (1984) suggest that ridges on the New Jersey continental shelf not only form by post-transgressive mechanisms but also represent relict features such as degraded barriers from the last marine transgression. Nearshore coast-oblique ridges, similar to those described by Swift and Field (1981), formed as post-transgressive features in response to the storm-dominated hydraulic regime in the area. However, mid-shelf coast-parallel ridges originated as degraded barriers drowned during the last marine transgression. Seaward of the mid-shelf ridges, the authors identified several outer-shelf coast-oblique ridges that exhibit characteristics similar to those of the nearshore ridges. They noted marked differences in the orientation, wavelength, height, and width between the nearshore and outer-shelf coast-oblique ridges and the mid-shelf coast-parallel ridges. Based on these data, along with vibracore and surficial-sediment data, they concluded that the mid-shelf coast-parallel ridges and the outer-shelf coast-oblique ridges that intersect the mid-shelf ridges are an early Holocene analog to the nearshore coast-oblique ridges and the present coastline.

The origin of large-scale bedforms and associated sediment patterns on the Mississippi-Alabama shelf is not clearly understood. However, seafloor topography and sediment distribution occurring in the SEBA exhibit characteristics similar to those previously discussed. Based on these studies, possible origins for the ridges in the SEBA include: (1) a relict topography formed by fluvial processes, (2) active bedforms created by storm-generated waves and currents of the present hydraulic regime, or (3) remnant shoreface-connected ridges formed along a retreating shoreline during Holocene sea-level rise.

Hyne and Goodell's (1967) theory suggesting that this topography is of fluvial origin is highly unlikely considering the regional extent and uniformity of the ridges and troughs. Bathymetric maps of the northern Gulf of Mexico indicate a fairly continuous ridge-and-trough topography from Mobile Bay, Alabama, to Cape San Blas, Florida. In addition, stream patterns along the northeastern Gulf Coast are dendritic and do not appear to follow a northwest-southeast trend.

Based on the information from Pyle and others (1975), the bedforms found in the SEBA are classified as low-relief swells and giant and large sand waves. They proposed that these bedforms form under extreme storm conditions. However, average current speeds for normal conditions within the study area are typically less than 10 cm/sec and rarely exceed 30 cm/sec (Dinnell, 1988). These speeds are much less than the current speeds estimated by Pyle and others (1975) for forming large-scale bedforms. In addition, tidal current and wave activity are relatively minor in the area and are not likely to produce the energy required to form these features. Unfortunately, data on the current velocities during storms are not available; however, current speeds would be expected to be much higher than average during the passage of hurricanes, tropical storms, and intense winter storms.

The nearshore ridges described by Duane and others, (1972), Swift and Field (1981), Parker and others (1982), and Stubblefield and others (1984) commonly show an oblique orientation to the shoreline, a steeper slope on the landward or upcurrent side, coarsest sediments on the upcurrent or landward flank, and finer sediments on the downcurrent or seaward flank. Sediments within the troughs may vary from shell hash to mud. Similar characteristics are exhibited by the ridges found in the SEBA. These ridges commonly form oblique angles with the present shoreline ranging from 26 to 81°. Their position on the shelf, along with their morphologies and orientations suggest they were formed by similar mechanisms as those described on other continental shelves (Swift and Field, 1981; Parker and others, 1982; Stubblefield and others, 1984). These ridges also appear to belong to an extensive group of ridges that occurs at approximately the same depth. Sediment patterns across the ridges in the SEBA also exhibit a similar distribution. Coarse sediment consisting of shell hash occurs along the landward or upcurrent flanks, whereas fine sand occurs along the seaward or downcurrent flank. The ridges in the SEBA differ in size compared to those found on other shelves. SEBA ridges are generally smaller, which may reflect a less intense storm climate in the Gulf of Mexico.

Sediment characteristics in the SEBA also provide evidence that ridges developed as shoreface-connected ridges. Relict shells constitute a large portion of the shell population and include *Crassostrea virginica* (oyster), *Anadara ovalis* (blood ark), and *Mercenaria mercenaria* (hardshell clam). These organisms require estuarine conditions and indicate the presence of a remnant bay, sound, or lagoon in the SEBA. Reworking of the sediments during shoreface retreat resulted in formation of shoreface-connected ridges and deposition of these shells on their landward slopes.

The occurrence of hardbottom rocks as rounded cobbles suggests that these rocks were reworked as the shoreface migrated northward. Recent studies (Schroeder and others, 1988a; Howard, 1990) suggest that the formation of the hardbottom rocks occurred during late Pleistocene and Holocene and that cementation may have occurred in the shallow subsurface of a lagoon or bay environment. However, large fragments of hardbottom rock have apparently withstood reworking by coastal processes during lower sea levels. The fact these rocks are exposed at the surface suggests that Pleistocene-Holocene sea-level rise was fairly rapid and accumulation of Holocene sediments has been relatively minor on this part of the shelf.

As sea level continued to rise, ridges in the SEBA were drowned and were less affected by nearshore storm-generated flow. The ridges presently undergo only slight modification primarily during the passage of major storms. Further analysis of other ridges in this area along with studies of the hydrodynamics of the Mississippi-Alabama shelf are necessary to determine adequately the origin of the seafloor topography and sediment distribution on the Mississippi-Alabama shelf.

SUMMARY

The sediments in the SEBA can be divided into three major facies: (1) sand facies, (2) shell gravel and sand facies, and (3) mud, shell gravel, and sand facies. Previous studies in this area indicate a single extensive sand facies; however, use of a closely spaced sampling grid and side-scan sonar allows separation of these facies on a much smaller scale. The sand facies dominates the study area and consists primarily of fine- to medium-grained, moderately sorted quartz sand. This facies is comparable with the sand facies described in previous literature for the area. The shell gravel and sand facies consists mainly of quartz sand with high percentages of carbonate sand and gravel. It occurs in subparallel, linear bands, or as irregular patches and is generally sharply bounded by the sand facies. The mud, shell gravel, and sand facies is located in the deepest part of the study area and is distinguished from the other facies by a considerably higher silt and clay content.

The distribution of sediments in the Southeast Banks area is associated with the local topography of the sea floor. Seafloor topography is characterized by numerous northwest-trending ridges and troughs. The ridges are generally oblique to the present shoreline and exhibit a variety of morphologies. Shell gravel typically occurs on the landward slopes and fine to medium sand occurs on the seaward slopes.

Hardbottom areas consisting of shell hash and indurated sediments are common on the sea floor in the SEBA. Shell-hash areas are extensive and may contain greater than 70 percent gravel-size material. Hardbottom rocks consist of carbonate-cemented sandstone, mudstone, and coquina. The rocks are commonly encrusted with soft corals, bryozoa, sponges, barnacles, and serpulid worms. Evidence of reworking is apparent in the rounded pieces; however, large slabs appear to be *in situ*.

The seafloor topography and sedimentary characteristics of the SEBA exhibit evidence of shoreface retreat during Holocene transgression. The bedforms in the area most likely formed as shoreface-connected ridges during lower sea levels and were detached from the shoreface by subsequent sea-level rise. Relict shells along with the petrologic data of the hardbottom rocks provide evidence that a bay or sound environment existed in the area during periods of lower sea levels. The relict sediments were reworked by coastal processes during the formation of shoreface-connected ridges, resulting in the present sediment distribution. Modern sedimentation on the inner shelf is minor allowing relict sediments and bedforms to be exposed. High-energy, storm-generated waves and currents of the modern hydraulic regime are responsible for minor, periodic reworking and redistribution of bottom sediments.

ACKNOWLEDGMENTS

Research for this project was supported in part by the National Oceanographic and Atmospheric Administration Office of Sea Grant, Department of Commerce, under Grant No. NA85AA-D-SG005 (Projects R/ER-19-PD and R/ER-19), the Mississippi-Alabama Sea Grant Consortium, The University of Alabama, the Naval Ocean Research and Development Activity, Stennis Space Center, Mississippi, and the Marine Environmental Sciences Consortium, Dauphin Island, Alabama. The authors thank Robert W. Frey and Frank W. Stapor for their careful review and constructive critique of the manuscript.

REFERENCES

DINNELL, S. P., 1988, Circulation and sediment dispersal on the Louisiana-Mississippi-Alabama continental shelf: Unpublished Ph.D. Dissertation, Louisiana State University, Baton Rouge, Louisiana, 173 p.

DINNELL, S. P., SCHROEDER, W. W., AND WISEMAN, W. J., Jr., 1990, Estuarine-shelf exchange using Landsat images of discharge plumes: Journal of Coastal Research, v. 6, p. 789–799.

DOYLE, L. J., AND SPARKS, T. N., 1980, Sediments on the Mississippi, Alabama, and Florida (MAFLA) continental shelf: Journal of Sedimentary Petrology, v. 50, p. 905–915.

DUANE, D. B., FIELD, M. E., MEISBURGER, E. P., SWIFT, D. J. P., AND WILLIAMS, S. J., 1972, Linear shoals on the Atlantic inner shelf, Florida to Long Island, *in* Swift, D. J. P., Duane, D. B., and Pilkey, O. H., eds., Shelf Sediment Transport: Process and Pattern: Dowden, Hutchinson, and Ross, Stroudsburg, Pennsylvania, p. 447–449.

GRIFFIN, G. M., 1962, Regional clay-mineral facies–Products of weathering intensity and current distribution in the northeastern Gulf of Mexico: Geological Society of America Bulletin, v. 73, p. 737–768.

HOOGENDOORN, E. L., AND DALRYLMPLE, R. W., 1986, Morphology, lateral migration and internal structures of shoreface-connected ridges, Sable Island Bank, Nova Scotia, Canada: Geology, v. 14, p. 400–403.

HOWARD, R. O., Jr., 1990, Petrology of hardbottom rocks, Mississippi-Alabama-Florida continental shelf: Unpublished M.S. Thesis, University of Alabama, Tuscaloosa, Alabama, 121 p.

HUTHNANCE, J. M., 1982, On one mechanism forming linear sand banks: Estuarine, Coastal and Shelf Science, v. 14, p. 79–99.

HYNE, N. J., AND GOODELL, H. G., 1967, Origin of the sediments and submarine geomorphology of the inner continental shelf off Choctawhatchee Bay, Florida: Marine Geology, v. 5, p. 299–313.

ISPHORDING, W. C., AND LAMB, G. M., 1980, The sediments of Mobile Bay: Dauphin Island Sea Lab Technical Report No. 80–002, 31 p.

KENYON, N. H., AND STRIDE, A. H., 1970, The tide swept continental shelf sediments between the Shetland Isles and France: Sedimentology, v. 14, p. 159–173.

KINDINGER, J. L., 1988, Seismic stratigraphy of the Mississippi-Alabama shelf and upper continental slope: Marine Geology, v. 83, 79–94.

KOCURKO, M. J., 1986, Interaction of organic matter and crystallization of high magnesium calcite, South Louisiana, *in* Gautier, D. L., ed., Roles of Organic Matter in Sediment Diagenesis: Society of Economic Paleontologists and Mineralogists Special Publication 38, p. 13–21.

LEWIS, D. W., 1984, Practical Sedimentology: Van Nostrand Reinhold Company, Inc., New York, 229 p.

LUDWICK, J. C., 1964, Sediments in northeastern Gulf of Mexico, *in* Miller R. L., ed., Papers in Marine Geology: Macmillan Co., New York, p. 204–238.

PARKER, GERARDO, LANFREDI, N. W., AND SWIFT, D. J. P., 1982, Seafloor response to flow in a southern hemisphere sandridge field: Argentine inner shelf: Sedimentary Geology, v. 33, p. 195–216.

PYLE, T. E., HENRY, V. J., MCCARTHY, J. C., GILE, R. T., AND NEURAUTER, T. W., 1975, Baseline monitoring studies, Mississippi, Alabama, Florida, outer continental shelf, 1975–1976, v. V. Geophysical Investigations for Biolithologic Mapping of the MAFLA-OCS Lease Area: Bureau of Land Management, Washington, D.C., BLM/ST-78/34, 267 p.

ROBERTS, H. H., AND WHELAN, T., III, 1975, Methane-derived carbonate cements in barrier and beach sands of a subtropical delta complex: Geochimica et Cosmochimica Acta, v. 39, p. 146–156.

RYAN, J. J., 1969, A sedimentologic study of Mobile Bay, Alabama: Unpublished M.S. Thesis Florida State University, Tallahassee, Florida, 110 p.

SCHROEDER, W. W., DARDEAU, M. R., DINDO, J. J., FLEISCHER, PETER, HECK, K. L., Jr., AND SHULTZ, A. W., 1988b, Geological and biological aspects of hardbottom environments on the MAFLA shelf, northern Gulf of Mexico, *in* Proceedings, Oceans '88 Conference: v. 1, p. 17–21.

SCHROEDER, W. W., SHULTZ, A. W., AND DINDO, J. J., 1988a, Inner-shelf hardbottom areas, northeastern Gulf of Mexico: Transactions, Gulf Coast Association of Geological Societies, v. 38, p. 535–541.

SCHULTZ, L. G., 1964, Quantitative interpretation of mineralogical composition from X-ray and chemical data for the Pierre Shale: U.S. Geological Survey Professional Paper 391-C, 51 p.

SHULTZ, A. W., SCHROEDER, W. W., AND ABSTON, J. R., 1990, Alongshore and offshore variations in Alabama inner-shelf sediments, *in* Tanner, W. F., ed., Coastal Sediments and Processes: Proceedings, 9th Symposium on Coastal Sedimentology, Geology Department, Florida State University, Tallahassee, Florida, p. 141–152.

SMITH, W. E., 1986, Geomorphology of Coastal Baldwin County, Alabama: Bulletin 132, Geological Survey of Alabama, 133 p.

STRIDE, A. H., AND CHESTERMAN, W. D., 1973, Sedimentation by nontidal currents around northern Denmark: Marine Geology, v. 15, p. 53–58.

STUBBLEFIELD, W. L., MCGRAIL, D. W., AND KERSEY, D. G., 1984, Recognition of transgressive and post-transgressive sand ridges on the New Jersey continental shelf, *in* Tillman, R. W., and Siemers, C. T., eds., Siliciclastic Shelf Sediments: Society of Economic Paleontologists and Mineralogists Special Publication 34, p. 1–23.

SWIFT, D. J. P., AND FIELD, M. E., 1981, Evolution of a classic sand ridge field: Maryland sector, North American inner shelf: Sedimentology, v. 28, p. 461–481.

UPSHAW, C. F., CREATH, W. B., AND BROOKS, F. L., 1966, Sediments and microfauna off the coasts of Mississippi and adjacent states: Bulletin 106, Mississippi Geological Survey, 127 p.

WEISS, C. P., AND WILKINSON, B. H., 1988, Holocene cementation along the central Texas coast: Journal of Sedimentary Petrology, v. 58, p. 468–478.

WILLIAMS, S. J., 1982, Use of high resolution seismic reflection and sidescan sonar equipment for offshore surveys: Coastal Engineering Technical Aid 82-5, U.S. Army Corps of Engineers, Coastal Engineering Research Center, Fort Belvoir, Virginia, 22 p.

QUATERNARY EVOLUTION OF THE EAST TEXAS COAST AND CONTINENTAL SHELF

JOHN B. ANDERSON, MARK A. THOMAS[1], FERNANDO P. SIRINGAN, AND WENDY C. SMYTH
Department of Geology and Geophysics, Rice University, P.O. Box 1892, Houston, Texas 77251

ABSTRACT: A 2,700-km high-resolution seismic-reflection data set, acquired in recent years, has helped resolve some old problems concerning the age of Quaternary formations along the east Texas coast, and has resulted in mapping of the Trinity/Sabine incised valley. The seismic data were used in conjunction with oil company platform borings and recently acquired sediment cores to examine the stratigraphy of the incised-valley fill and to map the distribution of sand bodies on the shelf.

The Trinity/Sabine valley has experienced at least two episodes of incision and infilling. The earliest incision occurred during $\delta^{18}O$ substage 5d and the latest reincision occurred during $\delta^{18}O$ stage 2. The late Wisconsinan-Holocene transgressive deposits that now fill the valley include the following facies (from bottom to top): fluvial; upper estuary/bayhead delta; middle estuary; lower estuary/tidal inlet/tidal delta; and offshore marine deposits. Backstepping parasequences, indicating an episodic rise in sea level, characterize the valley fill.

Sandbody formation and preservation on the shelf also has been influenced strongly by the episodic nature of the late Wisconsinan-Holocene sea-level rise. Sabine Bank, the largest of the sand bodies and the only one studied in detail, is a reworked coastal lithosome that rests on the ravinement surface. Inner-shelf muds contain few discrete storm beds. Relatively thick (<75 cm) amalgamated storm deposits are restricted to sand banks and the incised valley.

The modern Sabine Lake and Galveston Bay estuaries formed initially by flooding of the Sabine and Trinity valleys approximately 8 ka. The subsequent flooding event, which inundated the broad, shallow meander portions of the valleys, occurred approximately 4 ka and appears to have been rapid.

Extant coastal systems of the study area incorporate a wide range of environments, including barriers, strandplains, chenier plains, and tidal inlets. The systems formed predominantly during the stillstand of the past 3,500 years. Galveston Island and Bolivar Peninsula were derived from offshore sand sources. Progradation of the coastal barriers ceased with the exhaustion of the sand supply.

INTRODUCTION

Pioneering research into the Quaternary evolution of the east Texas-Louisiana continental shelf and associated coastal zone began with Fisk and his students nearly 50 years ago. Since then, many geologists have contributed to our understanding of the area; the majority of the studies has been concentrated onshore. Recently, results from an aggressive research program on the Louisiana shelf have been published widely (e.g., Suter and Berryhill, 1985; Suter and Penland, 1985; Suter and others, 1987), whereas research on the east Texas shelf has been less consistent, poorly coordinated, and conducted in scattered locations with different objectives.

Quaternary deposits of the east Texas continental shelf and adjacent coasts and estuaries were examined using high-resolution seismic profiling (3.5-kHz subbottom profiler, boomer, and small water gun) and sediment coring. Abdulah and Anderson (1991) and Bartek and others (1991) discuss similar investigations of the continental shelf offshore of the Brazos and Colorado rivers.

The east Texas continental shelf occupies an intermediate position between the strongly river-dominated Louisiana shelf and the strongly wave-dominated south Texas shelf. Examination of the area permits an evaluation of the interplay between sediment supply and eustasy and their roles in shaping sedimentary packages, at outcrop scale and at seismic stratigraphic scale. The northern Gulf of Mexico continental margin is stable tectonically (Martin, 1978). Long-term (Quaternary) subsidence rates are approximately 0.1 mm/yr or even less (Paine, 1991; Winker, 1979). Still, accommodation space has developed rapidly enough for thick Quaternary sequences to have accumulated on the continental shelf.

Our work has concentrated on the continental shelf between Sabine Pass and Galveston Island and includes Sabine Lake and the Galveston Bay complex (Galveston Bay, Trinity Bay, East Bay, and West Bay). Nearly 2,700 km of high-resolution seismic-reflection data and approximately 300 sediment cores (vibracores, piston cores, and gravity cores) have been collected (Fig. 1A). In addition, the data set includes descriptions from more than 100 platform borings in the area.

QUATERNARY STRATIGRAPHY

Morton and Price (1987) describe the Quaternary geology of the Texas shelf in great detail. Prior to now, the majority of the work done on the Quaternary stratigraphy of the area has been concentrated onshore, where fluvial and fluvial deltaic deposits dominate the stratigraphic record.

Fisk (1944) recognized that incised valleys exist beneath Texas and Louisiana bays and estuaries and that these probably extend offshore. He proposed that the valleys formed during late Pleistocene sea-level lowstands. Bernard and others (1962) suggested that the incised valleys were backfilled with sediment during the ensuing sea-level rise. A number of different investigations support the concept that most modern bays of the Texas-Louisiana shelf occupy old fluvial valleys (Fagg, 1957; Kane, 1959; Parker, 1959; Shepard and Moore, 1960; Behrens, 1963; Rehkemper, 1969; Byrne, 1975; Wilkinson and Byrne, 1977; Morton and McGowen, 1980). More recent studies have mapped the valleys on the continental shelf (Nelson and Bray, 1970; Suter and Berryhill, 1985; Suter and Penland, 1985; Pearson and others, 1986; Suter and others, 1987; Thomas and Anderson, 1988). These studies have led to the recognition of a complex system of late Quaternary incised valleys, deltas, and estuaries on the shelf.

---

[1]Present address: Shell Western Exploration and Production, Gulf Coast Division, Houston, Texas 77210-4252

Quaternary Coasts of the United States: Marine and Lacustrine Systems, SEPM Special Publication No. 48

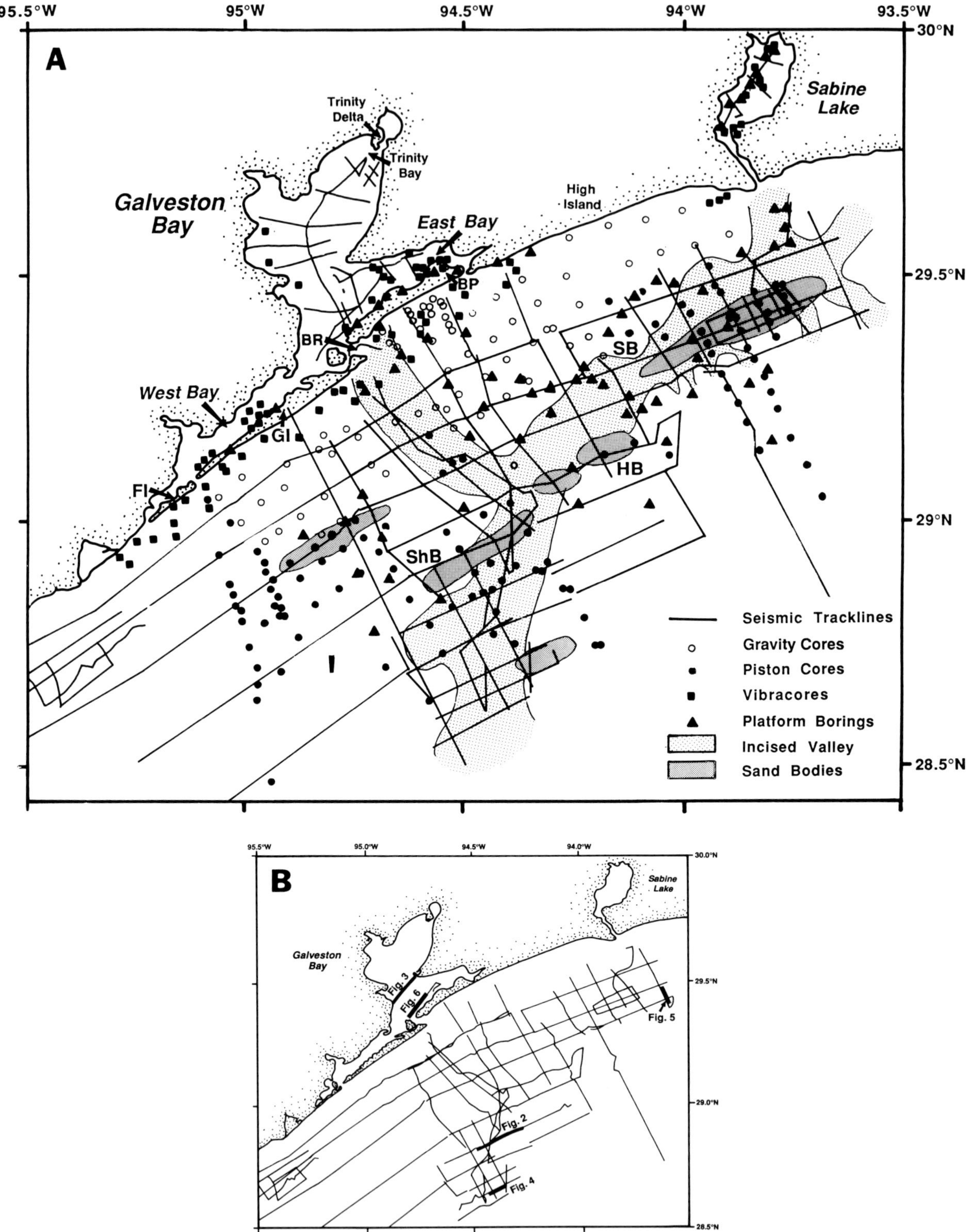

FIG. 1.—(A) Map of the study area showing locations of seismic lines, sediment cores, and platform boring locations. Also shown are locations of Trinity/Sabine incised valley and shelf sand banks. SB = Sabine Bank, HB = Heald Bank, ShB = Shepard Bank, BR = Bolivar Roads, BP = Bolivar Peninsula, GI = Galveston Island, and FI = Follets Island (modified from Anderson and others, 1990). (B) Map of seismic-survey tracklines. Bold line segments indicate locations of profiles presented in this paper.

Beaumont Formation

The most widespread Quaternary deposit by far, both onshore and offshore, is the Beaumont Formation. Hayes and Kennedy (1903) first named the formation and Barton (1930) later described it as a late Pleistocene delta plain. The deposit is believed to be equivalent to the Williana, Bentley, Montgomery, and Prairie terraces of Louisiana (Fisk, 1938, 1944; Bernard, 1950), although these correlations remain controversial (McFarland and LeRoy, 1988). Winker (1979), Coleman and Roberts (1988), and Thomas (1990) conducted seismic stratigraphic investigations of the Quaternary of offshore Texas and Louisiana and concluded that deposition of the Beaumont Formation occurred during $\delta^{18}O$ substage 5e.

The Beaumont Formation consists of clays of various colors and often contains limonite streaks, carbonaceous material, lenses of brown silt and sand, and ferric oxide and carbonate concretions. The surface commonly is mottled and oxidized and is distinguished easily from overlying Holocene deposits. Thickness ranges from 25 to 75 m in east Texas (Bernard, 1950) and from 40 to 225 m in south Texas (Price, 1934). Within the study area, individual deltas probably were similar in size to the modern Trinity delta (Aronow, 1971).

Ingleside Shoreline Trend

Price (1933) first mapped the Ingleside Shoreline Trend. Its age is disputed, but the trend generally is thought to represent Sangamon highstand shoreline deposits (Winker, 1979; Morton and Price, 1987). The deposits occur intermittently from southwest Louisiana to Tamaulipas, Mexico (Price, 1958) and consist of sandy sediments up to 25 m thick that locally contain nearshore marine fossils and accretionary-beach ridges (Morton and Price, 1987).

Winker (1979) noted that the Ingleside interfingers updip with the Beaumont Formation, forming a beach-shoreface sequence that thins seaward. Based on a seismic stratigraphic analysis, Thomas (1990) assigned Ingleside deposition to $\delta^{18}O$ substage 5d. Thomas (1990) showed that the $\delta^{18}O$ substage 5e condensed section extends several kilometers inland of the Ingleside Trend, which supports Graf's (1966) interpretation that, in the Galveston Bay area, the trend is a beach ridge and chenier plain adjacent to a deltaic headland of Trinity River.

Deweyville Formation

Barton (1930) first noticed the well-preserved meander scars of the Deweyville Formation, and noted they were much larger than either modern or Beaumont meanders. Later, Bernard (1950) identified Deweyville terraces along the Neches and Sabine rivers. Deweyville and Beaumont terraces along the Trinity River valley differ in elevation, gradient, channel geometry, preservation of geomorphic features, and sediment texture. Sandy, unconsolidated deposits of various colors generally characterize the terraces.

Deweyville deposits yield ages that range from 9 ka to 36 ka (Bernard and LeBlanc, 1965; Slaughter, 1965; Gagliano and Thom, 1967; Saucier, 1968; 1981; Aronow, 1971; Smith and Saucier, 1971; Otvos, 1980). This range led Aten (1983) and Pearson and others (1986) to suggest a Holocene age. However, Thomas (1990) used high-resolution seismic records to show that significant incision followed Deweyville deposition, separating this depositional event from the Holocene sea-level rise. Thomas (1990) concluded that Deweyville deposition was associated with a higher order sea-level cycle that occurred during $\delta^{18}O$ substage 5c. The younger (Holocene) dates reported by Pearson and others (1986) and Aten (1983) are thought to be from peats that overlie Deweyville terrace deposits (Smyth, 1991).

RESULTS

Trinity/Sabine Incised Valley System

The most prominent late Quaternary feature is the Trinity/Sabine incised valley (Fig. 1A). Initial incision of the valley occurred during $\delta^{18}O$ substage 5d, based on seismic stratigraphic analysis (Thomas, 1990). During the $\delta^{18}O$ stage 2 lowstand, the valley was reincised; the later incision was the deepest (ranging from −35 to −40 m offshore) and left remnants of older Deweyville fluvial terraces along the valley flanks (Figs. 2, 3, 4).

During the late Wisconsinan-Holocene sea-level rise, the Trinity/Sabine valley backfilled completely with fluvial, estuarine and marine sediments (Fig. 2). Seismic records and cores from the valley show that deposition was influenced strongly by the episodic nature of the sea-level rise (Thomas and Anderson, 1988; 1989; Thomas, 1990; Anderson and Thomas, 1991).

Valley-fill facies.—

Fluvial sands are confined to the incised valley and exhibit average thicknesses of 6 to 12 m, based on borings (Thomas, 1990). They are not well imaged seismically. A zone of high-amplitude, discontinuous reflectors (organic muds and peats) separates the fluvial facies from the overlying bay/estuarine facies (Figs. 2, 4). The upper surface exhibits a few meters of relief. The deposits often are referred to as the *bayline* and represent base-level encroachment into the valley.

Horizontal, subparallel reflectors characterize the upper-bay seismic facies (Figs. 2, 4). The upper surface has relief of a few meters. Borings through the facies encountered muds with abundant plant debris, muddy sands, roots, silt, and sand lenses. Distributary-channel fill is rare. The upper bay also includes the bayhead delta facies, which displays subtle, low-angle clinoforms in seismic records, indicating transport down the bay axis. Strike lines show that the deltas consist of stacked lenticular lobes. Organic muds with dispersed wood fragments constitute sediments of the bayhead delta.

In some areas the mid-bay/estuarine facies and/or lower-bay facies (tidal-inlet/flood-tidal delta) directly overlie the fluvial facies within the valley, indicating rapid flooding (e.g., Fig. 4). Laterally continuous, parallel to subparallel reflectors characterize the seismic expression. Sediment cores penetrated muds with shell layers predominantly comprised of *Mulinia* and occasionally thin sand or silt layers, interpreted to be storm deposits. Oyster reefs occur in the mid-

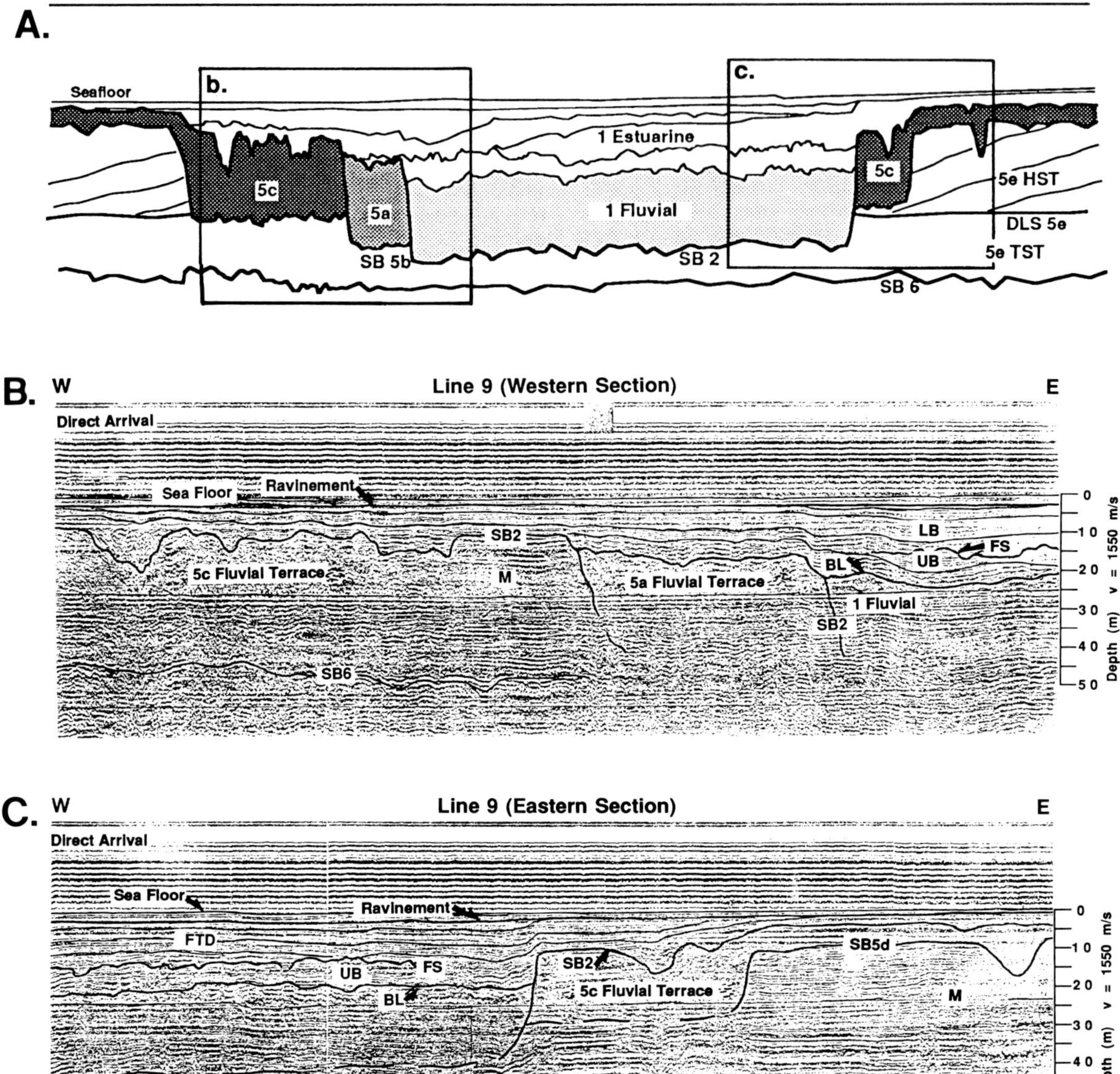

FIG. 2.—Line drawing of an interpreted seismic line (line 9) across the Trinity/Sabine incised valley (A) and portions of the original seismic trackline (B and C). Locations of (B) and (C) are indicated by boxes on (A; from Thomas, 1990). This record illustrates several episodes of incision and infilling ($\delta^{18}O$ substage 5d to $\delta^{18}O$ stage 1). The base of the valley seldom is imaged on seismic records. Note the chaotic to discontinuous and wavy reflectors of the fluvial unit and the irregular but fairly continuous bayline (BL). Subparallel reflectors and an irregular upper surface interpreted as a flooding surface (FS) characterize the upper bay (UB) facies. The lower-bay (LB) and flood-tidal-delta facies (FTD) occur above the upper-bay facies. The flood-tidal-delta facies is characterized by long, low-angle clinoforms indicating transport up the bay axis. SB-Sequence Boundary, TST = transgressive systems tract, DLS = downlap surface, and HST = highstand systems tract (from Thomas, 1990). M = multiple. See Figure 1B for location of trackline.

bay facies of Galveston Bay, although none were observed in the offshore mid-bay facies.

The lower-bay facies includes the tidal-inlet and flood-tidal-delta deposits and is the most easily distinguished seismic facies. Long, low-angle clinoforms, indicating transport up the bay axis, characterize the flood-tidal-delta seismic facies (Fig. 4). The cored deposits include laminated mud, interlaminated mud and sand, and sand. The tidal-inlet facies is characterized by sigmoidal clinoform reflectors; which accrete laterally across the valley (Fig. 4). The basal contact is erosional and may display up to 9 m of relief. Muddy sands with shell layers and a diverse marine and estuarine fauna characterize the tidal-inlet facies.

The ravinement surface separates the estuarine facies from the overlying marine facies (Fig. 2, 4, 5). The surface shows minor relief in the form of small gullies, approximately 1 m deep and less than 30 m wide, possibly representing return-flow storm channels (Thomas, 1990). Good control on the thickness of the marine section shows it to be, on average, less than 1 m thick outside the valley. The marine

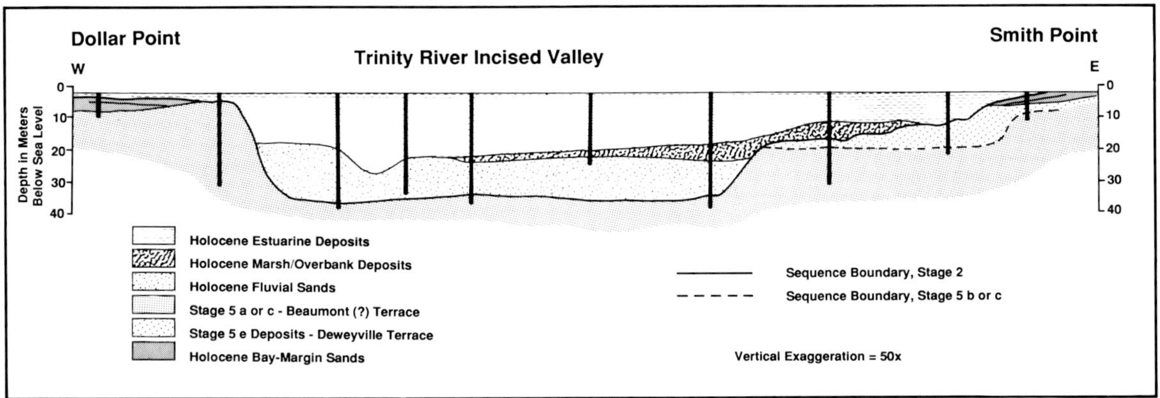

FIG. 3.—Cross section of Galveston Bay showing various stages of valley incision and fill. This section is based on interpretation of high-resolution seismic-reflection profiles and platform borings (from Smyth, 1991). See Figure 1B for line location.

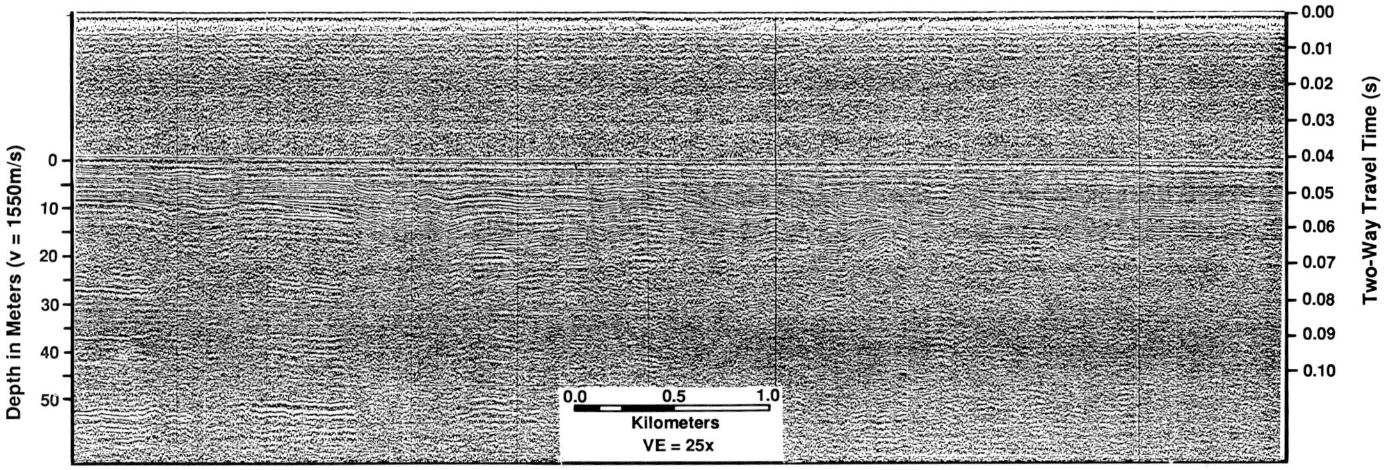

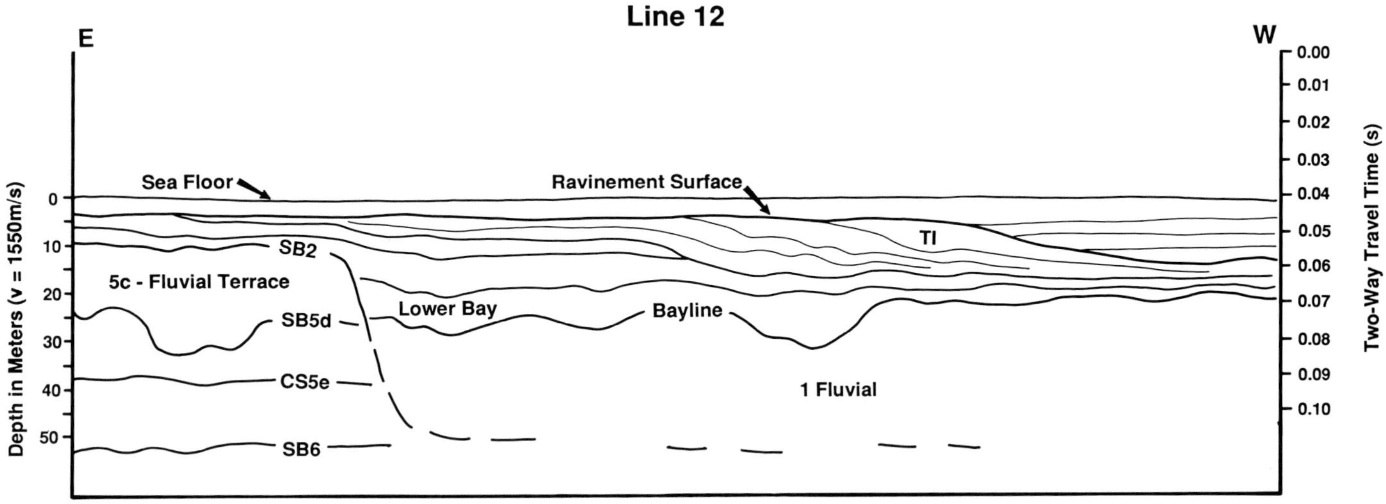

FIG. 4.—Seismic profile 12 across the Trinity/Sabine valley. This line illustrates typical seismic facies. Sigmoidal clinoform reflectors that accrete laterally across the valley characterize the tidal inlet (TI) seismic facies. SB2 = $\delta^{18}O$ stage 2 sequence boundary (from Thomas, 1990). CS = condensed section. See Figure 1B for location of trackline.

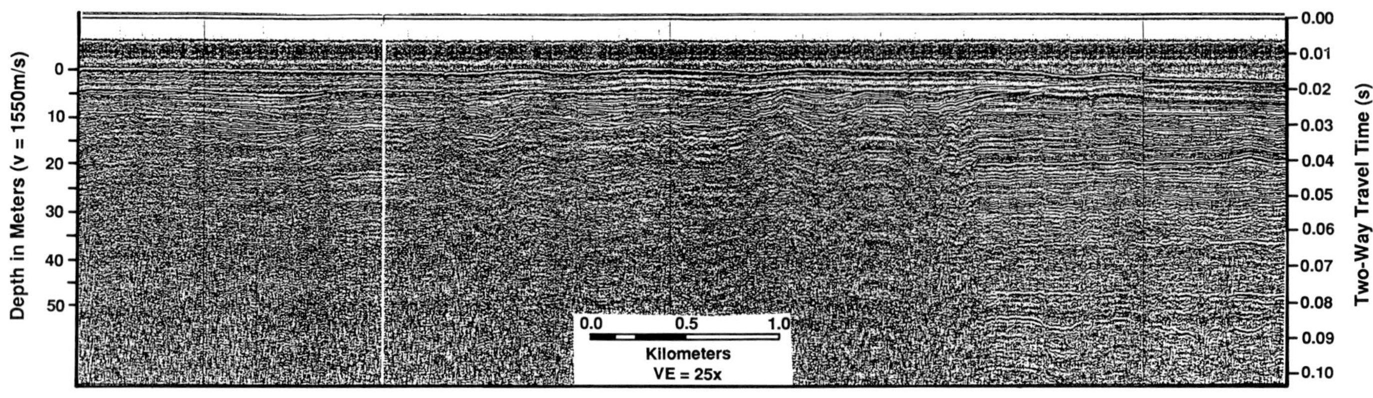

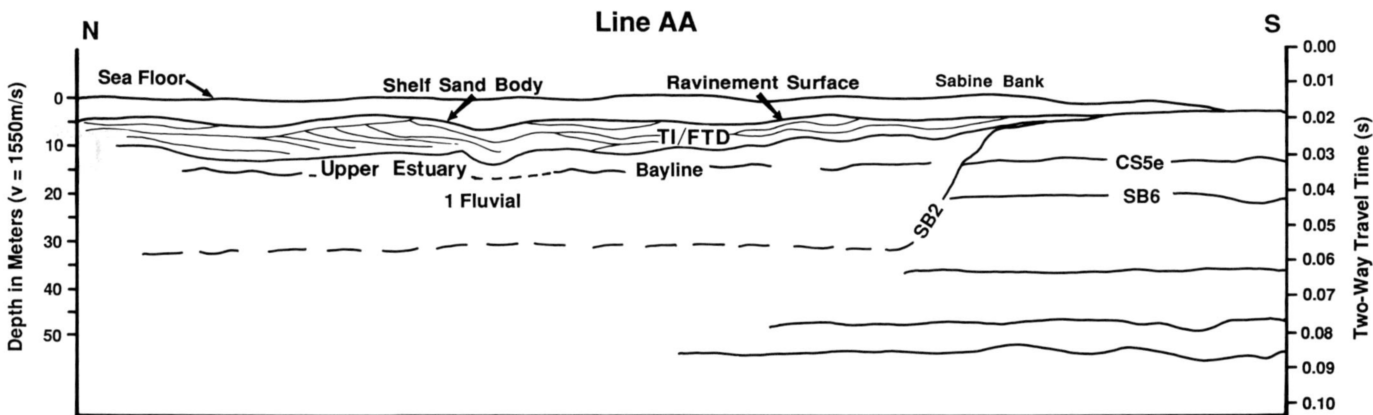

FIG. 5.—Seismic profile AA across Sabine Bank and above an incised valley. Note that the bank is above the ravinement surface. An interpreted tidal-inlet facies exists below the bank (from Thomas, 1990). TI/FTD = tidal-inlet/flood-tidal delta, CS = condensed section, SB = Sequence Boundary. See Figure 1B for trackline location.

facies consists mainly of greenish-gray (glauconitic), sandy, silty muds with marine fossils and abundant burrows.

Valley-Fill Stratigraphy.—

Thomas (1990) mapped the facies within the Trinity/Sabine valley for different time segments of the Holocene. The facies are bounded by intermediate flooding surfaces that lie between the bayline and ravinement surfaces. The intermediate flooding surfaces represent transgressions across a surface that was at sea level during periods of relative stillstand (Thomas and Anderson, 1989). Seismic reflections of these surfaces approximate time lines. Updip, flooding surfaces merge with the bayline and converge with one another. Ideally, during a sea-level stillstand, the components of a complete spectrum of valley-fill environments coexist, and sediments fill a wedge-shaped space resulting from a prior rise in sea level. A sea-level rise is manifested as a landward shift of the bayline and the ravinement surface. Intermediate flooding surfaces bound packages of sediment identified as parasequences (VanWagoner and others, 1988). Thomas's (1990) work led to the observation that backstepping parasequences characterize valley fill. This interpretation supports an episodic sea-level rise (Shepard and Moore, 1960; Nelson and Bray, 1970; Frazier, 1974; Anderson and Thomas, 1991). Studies of the Louisiana shelf provide additional evidence of the episodic nature of Holocene sea-level rise (Penland and others, 1988). Ongoing work on the Texas shelf focuses on gathering radiocarbon dates on key surfaces to constrain the timing of sea-level events.

Rehkemper (1969) conducted a detailed study of the history of Galveston Bay based on borings and seismic (sparker) data, but the quality of the seismic data generally was not suited for detailed facies and stratigraphic analyses. More recently, Smyth (1991) and Anderson and others (1991a) conducted detailed studies using uniboom data coupled with Rehkemper's descriptions of borings and radiometric dates.

Seismic records from the bay show that the most recent incision ($\delta^{18}O$ stage 2) left remnants of older valley fill, including Deweyville fluvial terraces and Ingleside coastal-barrier deposits, along the valley flanks (Figs. 2, 3). The deeper portions (−20 m) of the valley flooded about 8 ka. Evidence for simultaneous flooding of the lower and upper reaches of the bay, in conjunction with the occurrence of middle estuarine deposits resting directly on peats as evidenced from sediment cores, indicates that the initial flooding of the bay was rapid (Smyth, 1991). Approximately 4 ka another rapid flooding event occurred, inundating valley depths shallower than 6 m. This event produced the latest flooding surface recognizable on seismic records from the bay (Smyth, 1991). Sediment cores taken along the flanks of the incised valley sampled middle-bay deposits resting on the Pleistocene surface; a gradational deepening-upward interval is lacking. An intermediate flooding surface is recognized offshore; preserved tidal-inlet and related lithofacies occur beneath Sabine Bank and immediately offshore of Bolivar Peninsula approximately 10 m below sea level.

Regressive conditions were established in Galveston Bay by 2.5 ka. Since that time little change has occurred, except in the Trinity bayhead delta, which has occupied five different lobes in the past 2,600 years (Aten, 1983).

Inner Continental Shelf Sediments.—

Core coverage on the lower shoreface and inner shelf is quite good (Fig. 1A). The great majority of the cores penetrated through Holocene sediments (generally <1 m thick) and into the underlying Pleistocene units; exceptions are those cores taken on banks and over the incised valleys (Siringan and Anderson, in prep.). The Holocene deposits consist mainly of dark, greenish-gray, bioturbated sandy muds. The inner-shelf muds offshore of Galveston Island and Bolivar Peninsula contain few discrete storm beds. Relatively thick (<75 cm) amalgamated storm deposits occur only in sand banks and over the incised valley (Siringan and Anderson, 1991).

Several large sand bodies exist within the study area (Fig. 1A). The largest is Sabine Bank, which exhibits 7.5 m of relief. Seismic records show that the bank rests on the ravinement surface (Fig. 5), indicating transport onshore during the Holocene transgression (Thomas, 1990; Siringan and Anderson, 1991). Thus far, attempts to core through the bank have failed, but a single platform boring from the bank penetrated 4 m of sand. Vibracores from the bank, up to 2 m long, penetrated amalgamated, graded shell to sand units, interpreted as storm deposits, within an overall coarsening-upward sequence. Piston cores acquired seaward of the bank penetrated thin, bioturbated sands resting directly on the Holocene-Pleistocene (ravinement) surface. Piston cores shoreward of the bank penetrated marine muds overlying estuarine/lagoonal clays. Seismic records show units interpreted to be tidal-inlet/tidal-delta deposits beneath Sabine Bank (Fig. 5). The occurrence of these deposits beneath Sabine Bank may be due to the oblique (toward the southwest) landward reworking of the bank (Siringan and Anderson, 1991).

Preservation potential of shelf-sand bodies appears to be linked closely to rates of sea-level rise, as deduced from the Trinity/Sabine valley-fill stratigraphy (Thomas, 1990). Their formation requires relatively prolonged stillstands. If a slow rise follows the stillstand, the sand bodies are reworked, and little trace of their existence remains. If a rapid rise follows the stillstand, preservation potential is increased.

Extant Coastal Systems of the East Texas Gulf Coast

Extant coastal systems along the east Texas Gulf Coast are believed to have evolved sometime during the past 3,500 years, or during the latest sea-level stillstand (Gould and McFarlan, 1959; Bernard and others, 1970; Cole and Anderson, 1982; Smyth, 1991). The topography of the Pleistocene surface largely controls the distribution of barriers and tidal inlets; tidal inlets occur over incised valleys, whereas highly erosional strandplains are associated with deltaic headlands (Bernard and others, 1959; McGowen and others, 1977; Morton, 1977; Wilkinson and Basse, 1978; Price and Parker, 1979; Morton and McGowen, 1980).

Galveston-Bolivar Barrier Complex.—

Based on facies relations and radiocarbon dates, Bernard and others (1959, 1970) interpreted Galveston Island as having formed from a small submerged bar that emerged and grew seaward by beach accretion and southwestward by spit accretion. They interpreted two sections across Galveston Island as offlap sequences. Their western cross section shows Brazos River prodelta muds overlying the Pleistocene-Holocene unconformity, which, in turn, is overlain by barrier-island and shoreface sands. Morton (1979) reinterpreted this section as indicative of a transgressive segment of the island.

Cole and Anderson (1982) acquired sediment cores along four transects across Galveston Island and conducted detailed grain-size and mineralogical analyses on the sediments. Their mineralogical data show that Galveston Island is comprised of approximately equal amounts of sand from three fluvial sources—Trinity-Sabine, Mississippi, and Brazos rivers—with little temporal variation in the relative abundances of sands from the different sources. The barrier-island sands show remarkable grain-size constancy within the subaerial to upper-shoreface interval. The mixed mineralogical assemblage and constant grain size indicate that Galveston Island was derived from an offshore source (Cole and Anderson, 1982) and was not supplied directly by fluvial sources acting in conjunction with longshore currents. Progradation of the island ceased when the offshore sand supply was depleted.

LeBlanc and Hodgson (1959) suggested that Bolivar Peninsula formed by spit accretion. On the basis of morphologic evidence and on Cole and Anderson's (1982) mineralogical data, Eyer (1984) interpreted Bolivar Peninsula as an emergent barrier bar that later was breached by tidal inlets and smaller surge channels. The bars eventually grew and coalesced through spit, inlet, and beach-ridge accretion after sea level reached a stillstand. Vibracores through the main part of the peninsula penetrated well-sorted sands of uniform grain size downcore that contained a mixed heavy-mineral assemblage indicative of an offshore source (Cole and Anderson, 1982).

The overall sandbody geometry (defined by cores from the lobate features on the back side of Bolivar Peninsula) and lithology are consistent with both a flood-tidal-delta and washover-fan origin (Siringan and Anderson, 1991). The faunas in these cores are a low-diversity assemblage and the tidally influenced sediments are of fairly uniform thickness (averaging 1 m). A series of boreholes along Bolivar Peninsula and seismic lines from the same vicinity show—no incised fluvial channels or deeply carved tidal channels associated with the lobate features (Siringan and Anderson,

1991). High-resolution seismic data collected behind the western half of Bolivar Peninsula show southwestward prograding clinoforms (Fig. 6). The combined data indicate that the peninsula was initially a thin, narrow, and low feature, formed predominantly by southwestward spit progradation (Siringan and Anderson, 1991). Storm overwash channels later breached the peninsula. The channels may have served as ephemeral tidal inlets activated during storms and hurricanes. Widening of the peninsula through accretion made it less susceptible to breaching.

Bolivar Roads Tidal Inlet.—

Bolivar Roads is the largest tidal inlet (3 km wide with a natural channel depth of approximately 9 m) on the Texas coast. The natural channel presently is maintained artificially by jetties and by dredging to a minimum depth of 11.5 m. Prior to the construction of the jetties. Bolivar Roads had a well-developed ebb-tidal delta (Morton, 1977; Eyer, 1984). Morton (1977), Mason (1981), and Eyer (1984) document the changes associated with jetty construction.

Seismic profiles on the bay side of Bolivar Roads (Fig. 6) show a well-developed, 10-m-thick flood-tidal-delta sequence extending approximately 5 km bayward. East-west (strike-oriented) seismic lines across this feature record channel cut and fill. Trough-like channel geometry on the seaward side broadens and shallows on the bayward side (Fig. 6). In dip section, both ebb- and flood-directed clinoforms (1°–0.5°) are evident. Westward growth of Bolivar Peninsula is indicated by southwestward-dipping accretionary clinoforms, interpreted as spit-tidal-inlet facies, east of the flood-tidal delta (Fig. 6). A core (TGB-A, from Rehkemper, 1969) taken behind the western half of Bolivar yielded radiocarbon ages of 3,112±240 yrs and 2,375±180 yrs at a depth of 9.3 m, near the base of the flood-tidal delta (Siringan and Anderson, 1991).

Strandplain Chenier.—

Byrne and others (1959) have documented that the southwestern Louisiana chenier plain (including the Sabine Pass area) is underlain by a seaward-thickening wedge of Holocene sediments (7 to 8 m thick) resting on top of the Holocene-Pleistocene unconformity. The lower portion of the sediment wedge consists of coastal-marsh deposits grading upward and seaward into nearshore and offshore marine sediments. The upper section grades upward and seaward into nearshore marine to mudflat and coastal-marsh deposits. Byrne and others (1959) and Gould and McFarlan (1959) infer that the lower transgressive unit corresponds to the latest postglacial rise that ended when sea level reached its present stillstand. The upper offlapping sequence reflects the progradation of the chenier plain during the past 3,000 to 2,800 years of sea-level stillstand (Gould and McFarlan, 1959).

Howe and others (1935) were the first to suggest that the clay deposits of the chenier plain are depositional features, whereas the chenier ridges are erosional features caused by pulses in sediment supply. Generally, it is believed that the variation in sediment supply along Louisiana's west coast was due to the shifting of the Mississippi River mouth, and to a lesser degree, to changes in local sediment supply (Howe and others, 1935; Gould and McFarlan, 1959). Using radiocarbon dates of the cheniers and Mississippi River del-

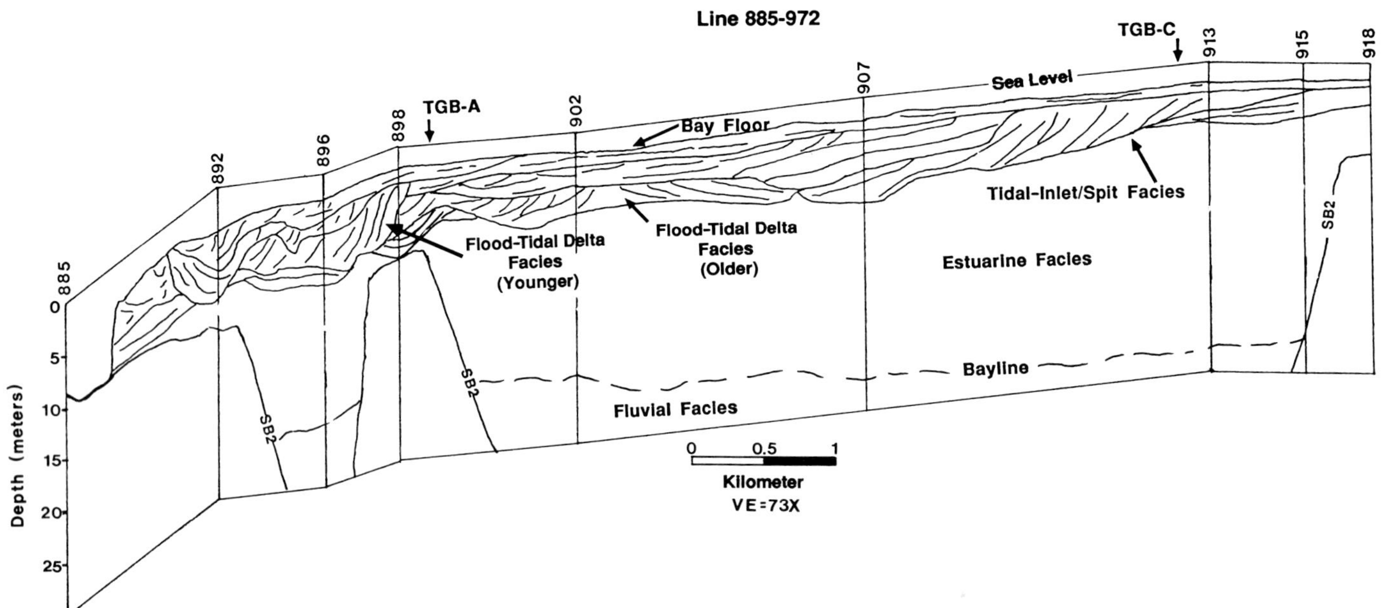

FIG. 6.—Line drawing interpretation of a segment of Williams' seismic profile 885–972 behind Bolivar Peninsula. Southwestward-dipping clinoforms on the right end of the line, interpreted as spit progradation, reflect the southwestward growth of Bolivar Peninsula. At left are channel structures of the present Bolivar Roads flood-tidal delta. The channels become broader and shallower toward the bay. Remnants of an older flood-tidal delta occur to the right of the extant lithosome. Boreholes TGB-A and TGB-C, from Rehkemper (1969), aided the interpretations (modified from Siringan and Anderson, 1991). See Figure 1B for location of trackline.

taic lobes, Gould and McFarlan (1959) demonstrated these more localized effects on sedimentation.

The Sabine River incised valley exerted an important influence on the development of the present configuration of the Sabine Pass chenier plain. As sea level rose during the last transgression, the incised valley was flooded and the estuary formed. Sand spits rapidly developed and restricted the mouth of the estuary (Siringan and Anderson, 1991; Anderson and others, 1991b). Cores from Sabine Lake provide evidence for the rapid establishment of brackish conditions in the lake (Anderson and others, 1991b). Radiocarbon analyses of *Rangia* shells resting just above the Pleistocene surface at approximately −5 m yielded ages of 3,890±120 yrs and 3,440±140 yrs. This period is believed to be the time of the most widespread flooding in the bay (Anderson and others, 1991b).

SUMMARY

The incised Trinity and Sabine river valleys are the dominant late Quaternary features on the east Texas continental shelf. Incision and infilling of the valleys have occurred during more than one glacial eustatic cycle. The episodic nature of the Holocene sea-level rise has exerted a profound influence on the facies architecture of the valley fill and on the formation and preservation of sand banks on the inner shelf. Coastal lithosomes are formed during very slow sea-level rises/relative stillstands and may be preserved by rapid rises and overstepping.

Galveston Bay and Sabine Lake occupy the Trinity and Sabine incised valleys, respectively. The deeper portions (−20 m) of the valleys flooded approximately 8 ka. The initial flooding of the bay was rapid. Approximately 4 ka, another rapid flooding event occurred, inundating the valleys to depths of 6 m below present sea level. Evolution of modern coastal lithosomes has occurred over the past 3,500 years. A thin veneer of sediments covers the inner continental shelf off the north Texas coast. The inner-shelf muds contain very few discrete storm beds. Thick, amalgamated storm deposits are restricted to sand banks and regions over the incised valley. Sabine Bank, the largest sand body within the study area, is a reworked coastal lithosome that has been transported obliquely to the southwest.

ACKNOWLEDGMENTS

Research was made possible through grants from the National Science Foundation (Grant OCE-8908320), the Petroleum Research Fund (PRF 23494-AC8), and by an Industrial Associates Group that includes the following company participants: Amoco, Arco, British Petroleum, Exxon, Shell, Unocal, and Union Pacific. We thank Rufus LeBlanc and Jeffrey Paine for their very helpful comments on the earlier version of this paper and Stephanie Staples Shipp for her technical assistance.

REFERENCES

ABDULAH, K. C., AND ANDERSON, J. B., 1991, Eustatic controls on the evolution of the Pleistocene Brazos-Colorado Deltas, Texas: Gulf Coast Section, Society of Economic Paleontologists and Mineralogists Foundation 12th Annual Research Conference, Program and Abstracts, p. 1–7.

ANDERSON, J. B., SIRINGAN, F. P., TAVIANI, M., AND LAWRENCE, J., 1991b, Origin and evolution of Sabine Lake, Texas-Louisiana: Transactions, Gulf Coast Association of Geological Societies, v. 41, p. 12–16.

ANDERSON, J. B., SIRINGAN, F. P., AND THOMAS, M. A., 1990, Sequence stratigraphy of the late Pleistocene-Holocene Trinity/Sabine valley system: relationship to the distribution of sand bodies within the transgressive systems tract: Gulf Coast Section, Society of Economic Paleontologists and Mineralogists Foundation 12th Annual Research Conference, Program and Abstracts, p. 15–20.

ANDERSON, J. B., SIRINGAN, F. P., SMYTH, W. C., AND THOMAS, M. A., 1991a, Episodic nature of Holocene sea level rise and the evolution of Galveston Bay: Gulf Coast Section, Society of Economic Paleontologists and Mineralogists Foundation 12th Annual Research Conference, Program and Abstracts, p. 8–14.

ANDERSON, J. B., AND THOMAS, M. A., 1991, Marine ice sheet decoupling as a mechanism for rapid, episodic sea level change: the record of such events and their influence on sedimentation: Sedimentary Geology, v. 70, p. 87–104.

ARONOW, S., 1971, Quaternary geology, *in* Wesselman, J. B., ed., Ground-Water Resources of Chambers and Jefferson Counties, Texas: Texas Water Development Board, Report 133, p. 34–53.

ATEN, L. E., 1983, Indians of the Upper Texas Coast: Academic Press, New York, 370 p.

BARTEK, L. R., ANDERSON, J. B., AND ABDULAH, K. C., 1991, A case study of the preservation potential of the coastal lithosomes of small deltaic systems: examples from the Holocene of the Texas continental shelf: Gulf Coast Section, Society of Economic Paleontologists and Mineralogists Foundation 12th Annual Research Conference, Program and Abstracts, p. 15–25.

BARTON, D. C., 1930, Deltaic coastal plain of southeastern Texas: Geological Society of America Bulletin, v. 41, p. 359–382.

BEHRENS, E. W., 1963, Pleistocene river valleys in Aransas and Baffin Bays: University of Texas Publication, Institute of Marine Sciences, v. 9, p. 7–18.

BERNARD, H. A., 1950, Quaternary geology of southeast Texas: Unpublished Ph.D. Dissertation, Louisiana State University, Baton Rouge, 164 p.

BERNARD, H. A. AND LEBLANC, R. J., 1965, Résumé of the Quaternary geology of the northwestern Gulf of Mexico province, *in* Wright, H. E., and Frey, D. G., eds., The Quaternary of the United States: Princeton University Press, Princeton, p. 137–185.

BERNARD, H. A., LEBLANC, R. J., AND MAJOR, C. F., 1962, Recent and Pleistocene Geology of Southeast Texas: Geology of the Gulf Coast and Central Texas and Guidebook of Excursions, Houston Geological Society, Houston, p. 175–224.

BERNARD, H. A., MAJOR, C. F., AND PARROT, B. S., 1959, The Galveston barrier island and environs: a model for predicting reservoir occurrence and trend: Transactions, Gulf Coast Association of Geological Societies, v. 9, p. 221–224.

BERNARD, H. A., MAJOR, C. F., PARROT, B. S., AND LEBLANC, R. J., 1970, Recent Sediments of Southeast Texas: a Field Guide to the Brazos Alluvial and Deltaic Plains and the Galveston Barrier Island Complex: Bureau of Economic Geology, University of Texas at Austin, Guidebook 11, 132 p.

BYRNE, J. R., 1975, Holocene depositional history of Lavaca Bay, central Texas Gulf Coast: Unpublished Ph.D. Dissertation, University of Texas at Austin, 149 p.

BYRNE, J. V., LEROY, D. O., AND RILEY, C. M., 1959, The chenier plain and its stratigraphy, southwestern Louisiana: Transactions, Gulf Coast Association of Geological Societies, v. 9, p. 237–260.

COLE, M. L., AND ANDERSON, J. B., 1982, Detailed grain size and heavy mineralogy of sands of the northeastern Texas Gulf Coast: implications with regard to coastal barrier development: Transactions, Gulf Coast Association of Geological Societies, v. 32, p. 555–563.

COLEMAN, J. M., AND ROBERTS, H. H., 1988, Sedimentary development of the Louisiana continental shelf related to sea level cycles: Geomarine Letters, v. 8, p. 63–119.

EYER, A. D., 1984, Geomorphology and morphologic development of Bolivar Roads Inlet and Bolivar Peninsula, Texas: Unpublished M.S. Thesis, University of Houston, Houston, 261 p.

FAGG, D. B., 1957, The recent marine sediments and Pleistocene surface of Matagorda Bay, Texas: Transactions, Gulf Coast Association of Geological Societies, v. 19, p. 239–261.

PART IV
PACIFIC COASTAL SYSTEMS

HOLOCENE CORAL REEF ON KAUAI, HAWAII: EVIDENCE FOR A SEA-LEVEL HIGHSTAND IN THE CENTRAL PACIFIC

ANTHONY T. JONES

Department of Oceanography, School of Ocean and Earth Science and Technology, University of Hawaii, Honolulu, Hawaii 96822 USA

ABSTRACT: Evidence of a late Holocene sea-level highstand has been found on the island of Kauai, Hawaii. A fossil reef complex in growth position is exposed approximately 500 m inland in the Hanalei River estuary at a maximum elevation of +1.8 m above present sea level. A fossil reef flat is also exposed in the adjacent Waioli Stream. Radiocarbon ages (uncalibrated) of corals within the reef complex range from 4.2 to 3.23 ka. The coral species identified imply a shallow-water environment. Although buried by fluvial sediments, the paleoshoreline should be several tens of meters landward of the study site. The elevation and age of the Kauai reef support recent geophysical models of the last deglaciation and indicate an emergence approximating that in French Polynesia and the Cook Islands. It is proposed that as the relative sea level retreated to its present position, the Hanalei River built a delta that buried the fossil reef with unconsolidated alluvium.

INTRODUCTION

Sea-level fluctuations over the last few glacial cycles during the Pleistocene have profoundly influenced the evolution of coastal systems on low-latitude volcanic islands. The volcanic islands of Hawaii offer a unique opportunity to study sea-level history and paleoceanographic environments in the north-central Pacific. To date, unequivocal evidence for a Holocene sea-level highstand in the Hawaiian Islands has not been obtained. New data from a fossil coral reef on Kauai provide evidence for a paleoshoreline that implies a late Holocene relative sea-level highstand in the north-central Pacific.

Geologic Setting

The Hawaiian Islands are a linear chain of volcanic islands and coral atolls in the north-central Pacific. Built over a stationary "hot spot," the islands move with the Pacific plate, resulting in progressively older islands, atolls, and seamounts in a northwest direction. The islands are located in the tropics and subtropics at latitudes of 19°N to 29°N, where vertical coral reef growth ranges from approximately 14 mm/yr in the southern part of the chain to 1 mm/yr for northern atolls (Grigg, 1982). Although isolated from continental and other shallow-water environments, reef corals have been present continuously in the Hawaiian-Emperor chain for the last 34 million years (Grigg, 1988).

The island of Kauai, dated at 5.0 Ma, is the oldest of the major Hawaiian Islands. Kauai has well-developed beaches, large nearshore-sand deposits, extensively eroded valleys, and large river systems. One such river system, the Hanalei River, occasionally floods the valley plain, as evidenced by extensive alluvium deposits. The river feeds into the almost semicircular Hanalei Bay, which opens to the north (Fig. 1). The bay is protected by two headlands: Puu Poa Point to the east and Makanoa Point to the west.

Being the wettest island in the Hawaiian chain, Kauai is not an ideal setting for preservation of exposed marine carbonates due to dissolution and erosion. Geologic maps of Kauai show no emergent marine deposits (Hinds, 1930; Stearns, 1946; Macdonald and others, 1960) in contrast to the other Hawaiian islands, where marine deposits (raised reefs and marine fossiliferous conglomerates) have been mapped (Stearns and Vaksvik, 1935; Stearns, 1940; Stearns and Macdonald, 1942; Stearns and Macdonald, 1946; Stearns and Macdonald, 1947; Stearns, 1947). However, Hinds (1930) reports the presence of emergent littoral deposits on Kauai and provides a list of marine species.

Hawaii Paleoshoreline Studies

Paleoshorelines in the Hawaiian Islands have been of interest to geologists ever since early workers studied emergent fossil reefs on the island of Oahu (Agassiz, 1889; Dana, 1890; Hitchcock, 1900, 1906; Wentworth and Palmer, 1925; Ostergaard, 1928; Pollock, 1928a, b). During mapping surveys of the Hawaiian Islands, Harold T. Stearns and collaborators noted marine fossils on every island (Oahu: Stearns and Vaksvik, 1935; Lanai: Stearns, 1940; Maui: Stearns and Macdonald, 1942; Hawaii: Stearns and Macdonald, 1946; Molokai: Stearns and Macdonald, 1947; Niihau: Stearns, 1947). Stearns formulated hypotheses as to the cause of these deposits and their chronology (see Stearns, 1935a, b, 1938, 1941, 1945, 1961, 1974). A 1978 synopsis of his work presents a speculative chronology of the paleoshorelines (Stearns, 1978).

Emergent fossil reefs, in particular the Waimanalo stand on Oahu, have played an important role in the development of a eustatic sea-level history for the last interglacial period at about 125 ka (Veeh, 1966; Ku and others, 1974; Easton and Ku, 1981; Brückner and Radtke, 1989). However, Moore and Moore (1984, 1988) have re-interpreted the marine fossil deposits at particularly high elevations (>50 m) in the southern Hawaiian Islands as resulting from giant waves (tsunamis), and not from eustatic sea-level changes.

Recently, the deeply submerged reef terraces seaward of the island of Hawaii have gained attention. Subsidence caused by lithospheric loading has drowned coral reefs during periods of rapid eustatic sea-level rise. Observations of these drowned reefs from dredged material and during manned submersible dives have led to the development of a chronology for the submerged terraces and an assessment of the local rates of crustal subsidence (Campbell, 1984, 1986; Moore and Fornari, 1984; Szabo and Moore, 1986; Moore and Campbell, 1987; Moore and others, 1990a; Moore and others, 1990b; Jones, 1991; Ludwig and others, 1991).

The Holocene Record

Records from Oahu suggest that there was no significant Holocene highstand on Oahu. When Easton and Olson (1976)

Quaternary Coasts of the United States: Marine and Lacustrine Systems, SEPM Special Publication No. 48

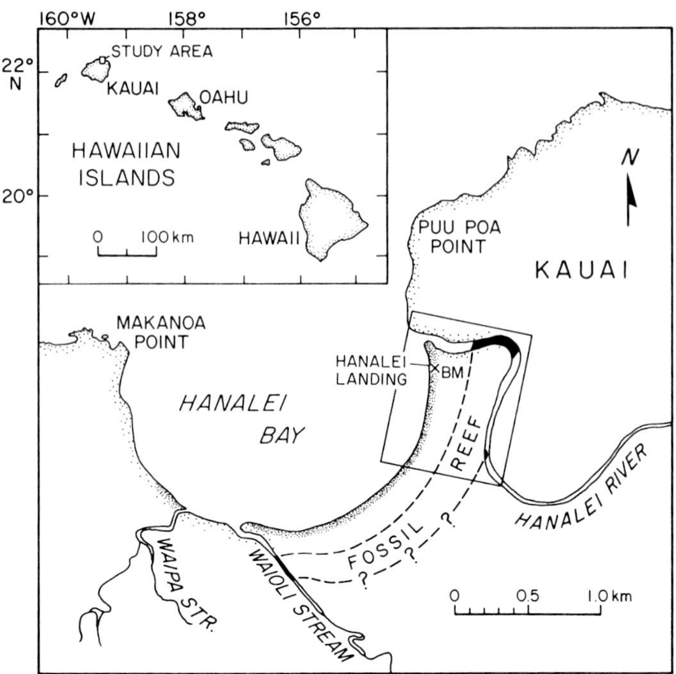

FIG. 1.—Location of fossil reef in Hanalei River estuary, north Kauai. Extent of fossil reef is shown. Box indicates coverage of Figure 2.

drilled the coralgal reef in Hanauma Bay, Oahu, they concluded from ^{14}C chronology that sea level had not risen above present level on Oahu since 7 ka. In addition, after surveying the low benches on Oahu, Bryan (1989) concluded that the wave-cut benches do not provide evidence for a Holocene sea-level stand significantly above present level.

Stearns (1977) argued for a Holocene highstand ("Kapapa stand") on the basis of a dated coral (3.485±0.16 ka) from an outcrop of beach rock in Hanauma Bay ranging in elevation from +1.5 to +2.1 m above present sea level. In noting the elevation (about +2 m) of the conspicuous wave-cut terraces and notches around Oahu, Stearns correlated the elevation of the Kapapa stand with the age and elevation of benches elsewhere in the Pacific. However, Easton (1977) dismissed these arguments because of the transported nature of the coral; he argued that the coral sample had been thrown up by storm surge from the center of the Hanauma Bay reef flat (dated at 3–4 ka) and emphasized the difficulty in correlating elevation of terraces and notches across ocean basins. Hence, geologists have questioned the existence of a Holocene highstand in the Hawaiian Islands.

The Hanalei River Fossil Reef

Approximately 500 m^2 of a fossil coral reef have been uncovered by lateral migration of the river channel at the lower end of the Hanalei River. A smaller reef section about 700 m upstream has also been exposed. The reef framework is typically upright, erect, and largely intact. Inspection of potholes and recesses in the reef reveal the framework is at least 1.5 m thick, an estimate that agrees with those provided by local landowner records of water-well drilling (G. Wilcox, pers. commun., 1990). The dominant reef coral in the Hanalei River exposure, *Porites compressa*, forms a nearly monospecific stand of upright, fingerlike, branching colonies, some greater than 1 m in diameter. Maragos (1977) documents *Porites compressa* in wave-protected areas, where it is spatially competitive and abundant in the modern marine environment of Hawaii. This species accounts for a large portion of the reef material found on Oahu (Pollock, 1928a). The large solitary coral, *Fungia scutaria*, a common member of the modern reef community generally restricted to the upper 3 m of water depth on the reef flat, was also collected at the Hanalei River sites.

The adjacent Waioli Stream, approximately 2 km southwest of the Hanalei River estuary, contains a reef-flat exposure of about 150 m^2, where samples of *Pocillopora meandrina* and *Fungia scutaria* were collected. *Pocillopora meandrina* is a dominant coral growing in Hawaiian waters on hard substratum in depths less than 10 m (Maragos, 1977; R. Grigg, pers. commun., 1991). No reef-framework material was observed during field work in the sediment-filled Waipa Stream.

METHODS AND RESULTS

Fossil corals exposed in the Hanalei River bed were discovered during a visual census of estuarine fish. Field surveys of the Hanalei River estuary and the adjacent Waioli Stream were undertaken to map the geographic extent of the fossil reef. Multiple elevation measurements (mean sea-level datum) were determined using a digital altimeter, specified by the manufacter to have a ±0.3-m precision. Altimeter elevations were then calibrated to a nearby U.S. Coast and Geodetic Survey tidal benchmark at the Hanalei Landing. Tidal range (MLW-MHW) in the Hanalei Bay is 0.4 m (National Ocean Survey, 1989).

Radiocarbon Dating

Corals *(Porites compressa)* preserved in growth position were sampled at sites shown in Figure 2. The outer surfaces were scraped to remove encrusted estuarine bivalves, calcareous serpulid tubes, and filamentous algae. Subsequent examination by X-ray diffraction for signs of diagenetic alteration showed that all samples were greater than 98 percent aragonite. Samples (>5 g) were etched in weak acid to remove outer layers and were ^{14}C dated using benzene synthesis. Ages determined by ^{14}C are based on a ^{14}C half-life of 5,568 years and are reported as conventional ^{14}C years before 1950 (Table 1). Ages ranged from 4.2 to 3.23 ka and are reported without δ^{13}C corrections due to the compensating offset of isotopic fractionation and the surface-water effect.

DISCUSSION

In situ reef corals provide a useful estimate of the minimum elevation of sea level and are readily dated (e.g., Hopley, 1986). Their stenohaline characteristic ensures that coral samples represent growth in sea waters of normal salinity (about 35‰). The presence of *Porites compressa, Pocillopora meandrina*, and *Fungia scutaria* in the Hanalei River exposure implies an ancient shallow-water environ-

FIG. 2.—Aerial photograph of Hanalei River estuary with radiocarbon-sample sites indicated. Aerial photograph from U.S. Army Corps of Engineers (1983).

TABLE 1.—RADIOCARBON AGES OF CORALS FROM HANALEI RIVER ESTUARY, KAUAI

Lab No.	Field No.	$\delta^{13}C$	Conventional ^{14}C Age	Calibrated Age*
Beta-41951	Kau-1	−2.3	4200±70	4620±100
Beta-41952	Kau-2	−2.7	3670±80	3900±110
Beta-41953	Kau-3	−0.7	3230±80	3390±90
Beta-41954	Kau-4	−0.4	4070±80	4480±120

Radiocarbon ages reported as years before 1950 with one standard deviation error.

*Calibrated with CALIB of Stuiver and Reimer (1987) using a reservoir ΔR of 115±50.

ment, possibly analogous to that of modern Kaneohe Bay on the windward side of Oahu, where *Porites compressa* comprises more than 80 percent of the coral cover (Maragos, 1972). In Kaneohe Bay, some *Porites compressa* colonies are exposed at extreme low tides. Even though it is not restricted to the near surface, *Porites compressa* often forms monospecific reefs in these shallow reaches tens to hundreds of meters from the present shoreline. The relation of the Hanalei fossil reef to its ancient strandline at the Hanalei River exposure could not be determined because of burial by fluvial sediments, but the paleoshoreline would be expected to be several tens or hundreds of meters inland from the sample site.

Dates of fossil corals from Kauai are rare. A 10-fathom-deep (about 18 m) reef terrace on Kaheko Reef, off northeast Kauai, was sampled in the 1960s. A ^{14}C age of 8.37±0.25 ka (Hubbs and others, 1965; Inman and Veeh, 1966) was confirmed by a U-series ($^{230}Th/^{234}U$) date of 8±1 ka (Veeh, 1966). The Kaheko reef dates agree with data from recent Holocene sea-level curves from Barbados (Fairbanks, 1989), although a late Holocene emergence is not reported from the Caribbean. Two other ^{14}C dates from Kauai have been reported. Beach rock from Oomano Point at an approximate 1-m elevation was ^{14}C dated at 1.6±0.16 ka and cemented "sandstone" from Kauhou Valley at an elevation of +0.9 to +1.8 m was ^{14}C dated at 15±0.6 ka (Hubbs and others, 1965). The sandstone may have been subaerially transported and deposited during the rapid eustatic rise following the last glacial lowstand (at 18 ka), whereas the beach rock may have been deposited after the late Holocene emergence.

The ^{14}C-dated corals from the Hanalei River exposure presented here indicate that relative sea level stood at least +1.8 m above present on Kauai at 4.2 ka. Since then, relative sea level has dropped to its present level. Whether the retreat was smooth or oscillatory cannot be determined from the fossil reef data. A similar Holocene highstand has been well documented for islands in Micronesia (Buddemeier and others, 1975), the South Pacific (French Polynesia: Pirazzoli and Montaggioni, 1988; Cook Islands: Woodroffe and others, 1990b), and the central Indian Ocean (Woodroofe and others, 1990a).

Preservation of the Hanalei River fossil reef probably resulted from delivery of sediment by the river or periodic flooding, which built a delta that prograded over the exposed fossil reef. As the shoreline shifted seaward, the reef structure remained buried. The reefs, then, would have been protected from erosion and extensive subaerial diagenesis. Finally, during recent time, the Hanalei River at the seaward bend removed the overburden, thus exposing the reef structure.

Implications

The Holocene highstand on Kauai agrees in amplitude and timing with a high-resolution global model of the last deglaciation (Fig. 3). This recently developed model (ICE-3G of Tushingham and Peltier, 1991) considers ice/ocean/solid-earth interactions in such a way as to preserve an equal potential gravitational field and account for glacial isostatic adjustment. The model was constrained by ^{14}C-controlled relative sea-level curves from sites that were ice covered. The model was tested against geologically controlled glacial-retreat isochrons, oxygen isotope data from deep-sea sedimentary cores, and the elevation of coral terraces, in order to constrain net sea-level rise and mass of ice melted. The model was later validated with relative sea-level curves from ice-free sites (Tushingham and Peltier, 1992).

The geophysical model predicts a Holocene rise in the central Pacific of 2 m at 5 ka (Tushingham, 1989). Data from Easton and Olson (1976) generally lie below the predicted sea-level curve, with the exception of 7-ka material at about 10 m (see Stearns, 1977, and Easton, 1977, for

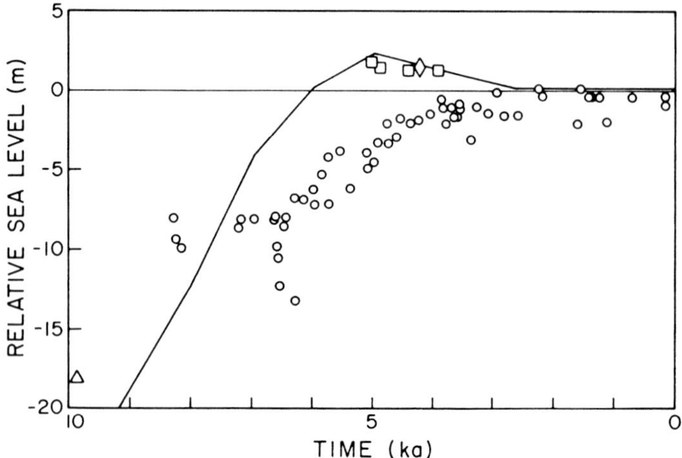

FIG. 3.—Comparison of the observed data from Oahu (○ Easton and Olson, 1976; ◇ Stearns, 1977) and Kauai (△ Inman and Veeh, 1966; □ this report) with the predicted relative sea-level history for Oahu using the ICE-3G melting chronology (Tushingham, 1989).

discussion). No highstand record is indicated in Easton and Olson's data set from Hanauma Bay. The absence may be accounted for by the lack of preservation in Hanauma Bay, Oahu. The fringing reef in this semi-protected crater would not be preserved by a prograding delta, as interpreted for the case of the Hanalei River, but instead may have been planed off by wave action and bioerosion. There is some indication of planing of the reef in their published cross section (Easton and Olson 1976, fig. 3). Stearns' (1977) datum falls near the predicted curve, although the coral fragment was transported from the reef.

The ICE-3G model, however, does not consider any vertical adjustments due to local crustal loading at the central Pacific "hot spot." Lithospheric flexure resulting from crustal loading should be confined to a radius of 400 km around the hot spot (Watt and ten Brink, 1989). Oahu and Kauai are 350 km and 500 km from the hot spot, respectively. Therefore, Kauai may be considered beyond the influence of this flexure. Although the isostatic adjustment of the flexure cannot be ruled out as a factor for Oahu, arguments that Oahu can be considered stationary seem valid (Veeh, 1966; Ku and others, 1974; Moore, 1987). Therefore, the age and elevation of the Kauai fossil reef reported here provide the first solid evidence for a Holocene sea-level highstand in the Hawaiian Islands.

Sea-level change induced by anthropogenically driven global-climate change requires serious consideration. The growing consensus of reports on a widespread mid-Holocene highstand are important as are the data on the duration of the last interglacial period (about 12 ka) and the rapid fall in sea level (20 cm/100 yr) at the conclusion of the last interglacial (Chen and others, 1991). Finally, the thickness and extent of the Hanalei River fossil reef are potentially useful paleoclimatic records for central Pacific Ocean conditions during the Holocene.

CONCLUSIONS

A fossil shallow-water coral reef discovered on Kauai at +1.8 m above present sea level and ^{14}C dated at 4.2 ka establishes a Holocene relative sea-level highstand in Hawaii and the central Pacific Ocean from 4.2 to 3.23 ka. These results support recent high-resolution geophysical models of late Pleistocene deglaciation.

Sea-level history in the Hawaiian Islands, especially during the Holocene, is not well understood. Fossil reefs in the Hanalei River may prove to be an excellent location for extraction of a record of Holocene transgression using modern drilling and seismic techniques and new high-precision methods based on mass-spectrometric uranium-series dating.

ACKNOWLEDGMENTS

I thank Gordon Smith, who directed my attention to the fossil corals in the Hanalei River. E. Wingert provided advice on measuring elevation in a river and supplied a digital altimeter for the field survey. W. B. Bryan, D. C. Cox, W. H. Easton, C. H. Fletcher, R. W. Grigg, F. T. Mackenzie, J. G. Moore, and K. Zaiger commented on early manuscripts. The U.S. Army Corps of Engineers Pacific Ocean Division permitted use of the aerial photograph in Figure 2. Manoa Mapworks prepared the aerial photograph, and Brooks Bays of the School of Ocean and Earth Science and Technology Illustration drafted the figures. Radiocarbon analyses were performed by Beta Analytical Laboratories in Miami, Florida. Research on Kauai was partially funded by the State of Hawaii, the Hawaii Undersea Research Laboratory, the Geological Society of America, and a Harold T. Stearns Fellowship.

REFERENCES

AGASSIZ, A., 1889, The coral reefs of the Hawaiian Islands: Bulletin of the Museum of Comparative Zoology, Harvard, v. 17, p. 121–170.

BUDDEMEIER, R. W., SMITH, S. V., AND KINZIE, R. A., 1975, Holocene windward reef-flat history, Eniwetak Atoll: Geological Society of America Bulletin, v. 86, p. 1581–1584.

BRYAN, W. B., 1989, Pacific Holocene sealevel variation: evidence from windward Oahu (abst.): EOS, Transactions, American Geophysical Union, v. 70, p. 1370.

BRÜCKNER, H., AND RADTKE, U., 1989, Fossile strande und korallenbanke auf Oahu, Hawaii: Essener Geographische Arbeiten, v. 17, p. 291–308.

CAMPBELL, J. F., 1984, Rapid subsidence of Kohala Volcano and its effect on coral reef growth: Geo-Marine Letters, v. 4, p. 31–36.

CAMPBELL, J. F., 1986, Subsidence rate for the southeastern Hawaiian Islands determined from submerged terraces: Geo-Marine Letters, v. 6, p. 137–146.

CHEN, J. H., CURRAN, H. A., WHITE, B., AND WASSERBURG, G. J., 1991, Precise chronology of the last interglacial period: ^{234}U-^{230}Th data from fossil coral reefs in the Bahamas: Geological Society of America Bulletin, v. 103, p. 82–97.

DANA, J. D., 1890, Coral and Coral Islands, 3rd edition: Dodd, Mead, and Company, New York, 440 p.

EASTON, W. H., 1977, Radiocarbon profile of Hanauma Reef, Oahu: reply: Geological Society of America Bulletin, v. 88, p. 1535–1536.

EASTON, W. H., AND KU, T. L., 1981, $^{230}Th/^{234}U$ dates of Pleistocene deposits on Oahu: Bulletin of Marine Science, v. 31, p. 552–557.

EASTON, W. H., AND OLSON, E. A., 1976, Radiocarbon profile of Hanauma Reef, Oahu, Hawaii: Geological Society of America Bulletin, v. 87, p. 711–719.

FAIRBANKS, R. G., 1989, A 17,000-year glacio-eustatic sea level record: influence of glacial melting rates on the Younger Dryas event and deep-ocean circulation: Nature, v. 342, p. 637–642.

GRIGG, R. W., 1982, Darwin Point: a threshold for atoll formation: Coral Reefs, v. 1, p. 1–6.

GRIGG, R. W., 1988, Paleoceanography of coral reefs in the Hawaiian-Emperor Chain: Science, v. 240, p. 1737–1743.

HINDS, N. E. A., 1930, The geology of Kauai and Niihau: Bernice P. Bishop Museum Bulletin, v. 71, 103 p.

HITCHCOCK, C. D., 1900, The geology of Oahu: Geological Society of America Bulletin, v. 11, p. 15–60.

HITCHCOCK, C. D., 1906, The geology of Diamond Head: Geological Society of America Bulletin, v. 17, p. 469–496.

HOPLEY, D., 1986, Corals and reefs as indicators of paleo-sea levels with special reference to the Great Barrier Reef, *in* van de Plassche, O., ed., Sea-Level Research: A Manual for the Collection and Evaluation of Data: Gulliard Printers Ltd., Great Yarmouth, England, p. 195–228.

HUBBS, C. L., BIEN, G. S., AND SUESS, H. E., 1965, La Jolla natural radiocarbon measurements, part IV: Radiocarbon, v. 7, p. 66–117.

INMAN, D. L., AND VEEH, H. H., 1966, Dating the 10-fathom terrace off Hawaii (abst.): EOS, Transactions, American Geophysical Union, v. 47, p. 125.

JONES, A. T., 1991, Drowned reefs in the Alenuihaha Channel: evidence 0for island subsidence and low stands of Quaternary sea-level (abst.): Pacific Science, v. 44, p. 92.

KU, T.-L., KIMMEL, M. A., EASTON, W. H., AND O'NEIL, T. J., 1974, Eustatic sea level 120,000 years ago on Oahu, Hawaii: Science, v. 183, p. 959–962.

LUDWIG, K. R., SZABO, B. J., MOORE, J. G., AND SIMMONS, K. R., 1991, Crustal subsidence rates off Hawaii determined from $^{234}U/^{238}U$ ages of drowned coral reefs: Geology, v. 19, p. 171–174.

MACDONALD, G. A., DAVIS, D. A., AND COX, D. C., 1960, Geology and ground-water resources of the island of Kauai, Hawaii: Hawaii Division of Hydrography Bulletin, v. 13, p. 1–212.

MARAGOS, J. E., 1972, A study of the ecology of Hawaiian reef coral: Unpublished Ph.D. Dissertation, University of Hawaii, Honolulu, 292 p.

MARAGOS, J. E., 1977, Order Scleractinian, *in* DEVANEY, D. M., AND ELDREDGE, L. G., eds., Reef and Shore Fauna of Hawaii: Bernice P. Bishop Museum 64, p. 158–241.

MOORE, G. W., AND MOORE, J. G., 1988, Large scale bedforms in boulder gravel produced by giant waves in Hawaii: Geological Society of America 229, p. 101–110.

MOORE, J. G., 1987, Subsidence of the Hawaiian Ridge: U.S. Geological Survey Professional Paper 1350, p. 85–100.

MOORE, J. G., AND CAMPBELL, J. F., 1987, Age of tilted reefs, Hawaii: Journal of Geophysical Research, v. 92, p. 2641–2636.

MOORE, J. G., CLAGUE, D. A., LUDWIG, K. R., AND MARK, R. K., 1990a, Subsidence and volcanism of the Haleakala Ridge, Hawaii: Journal of Volcanology and Geothermal Research, v. 42, p. 273–284.

MOORE, J. G., AND FORNARI, D. J., 1984, Drowned reefs as indicators of the rate of subsidence of the island of Hawaii: Journal of Geology, v. 92, p. 752–759.

MOORE, J. G., NORMARK, W. R., AND SZABO, B. J., 1990b, Reef growth and volcanism on the submarine southwest rift zone of Mauna Loa, Hawaii: Bulletin of Volcanology, v. 52, p. 375–380.

MOORE, J. G., AND MOORE, G. W., 1984, Deposit from a giant wave on the Island of Lanai, Hawaii: Science, v. 226, p. 1312–1315.

National Ocean Survey, 1989, Tide Tables, High and Low Water Predictions, West Coast of North and South America, including the Hawaiian Islands: U.S. Department of Commerce, National Oceanic and Atmospheric Administration, Rockville, Maryland, 191, p.

OSTERGAARD, J. M., 1928, Fossil marine mollusks of Oahu: Bernice P. Bishop Museum Bulletin, v. 51, 32 p.

PIRAZZOLI, P. A., AND MONTAGGIONI, L. F., 1988, Holocene sea-level changes in French Polynesia: Palaeogeography, Palaeoclimatology, Palaeoecology, v. 68, p. 153–175.

POLLOCK, J. B., 1928a, Fringing and fossil coral reefs of Oahu: Bernice P. Bishop Museum Bulletin, v. 55, 56 p.

POLLOCK, J. B., 1928b, The amount of the geologically recent negative shift of strand line on Oahu: Washington Academy of Science Journal, v. 18, p. 53–59.

STEARNS, H. T., 1935a, Shore benches on the island of Oahu, Hawaii: Geological Society of America Bulletin, v. 46, p. 1467–1482.

STEARNS, H. T., 1935b, Pleistocene shore lines on the islands of Oahu and Maui, Hawaii: Geological Society of America Bulletin, v. 46, p. 1927–1956.

STEARNS, H. T., 1938, Ancient shore lines on the island of Lanai, Hawaii: Geological Society of America Bulletin, v. 49, p. 615–628.

STEARNS, H. T., 1940, Geology and groundwater resources of Lanai and Kahoolawe, Hawaii: Hawaii Division of Hydrography Bulletin, v. 6, 177 p.

STEARNS, H. T., 1941, Shore benches on north Pacific Islands: Geological Society of America Bulletin, v. 52, p. 773–780.

STEARNS, H. T., 1945, Eustatic shore lines in the Pacific: Geological Society of America Bulletin, v. 56, p. 1071–1078.

STEARNS, H. T., 1946, Geology of the Hawaiian Islands: Hawaii Division of Hydrography Bulletin, v. 8, 106 p.

STEARNS, H. T., 1947, Geology and ground-water resources of the island of Niihau, Hawaii: Hawaii Division of Hydrography Bulletin, v. 12, 51 p.

STEARNS, H. T., 1961, Eustatic shorelines on Pacific Islands: Zeitschrift Fur Geomorphologie, Supplement 3, p. 1–16.

STEARNS, H. T., 1974, Submerged shorelines and shelves in the Hawaiian Islands and a revision of some of the eustatic emerged shorelines: Geological Society of America Bulletin, v. 85, p. 795–804.

STEARNS, H. T., 1977, Radiocarbon profile of Hanauma Reef, Oahu: discussion: Geological Society of America Bulletin, v. 88, p. 1535.

STEARNS, H. T., 1978, Quaternary shorelines in the Hawaiian Islands: Bernice P. Bishop Museum Bulletin, v. 237, 57 p.

STEARNS, H. T., AND MACDONALD, G. A., 1942, Geology and ground-water resources of the island of Maui, Hawaii: Hawaii Division of Hydrography Bulletin, v. 7, 344 p.

STEARNS, H. T., AND MACDONALD, G. A., 1946, Geology and ground-water resources of the island of Hawaii: Hawaii Division of Hydrography Bulletin, v. 9, 363 p.

STEARNS, H. T., AND MACDONALD, G. A., 1947, Geology and ground-water resources of the island of Molokai, Hawaii: Hawaii Division of Hydrography Bulletin, v. 11, 113 p.

STEARNS, H. T. AND VAKSVIK, K. N., 1935, Geology and ground-water resources of the island of Oahu, Hawaii: Hawaii Division of Hydrography Bulletin, v. 1, 479 p.

STUIVER, M., AND REIMER, P. J., 1987, User's guide to the Programs CALIB and DISPLAY 2.1: Quaternary Isotope Laboratory, University of Washington, Seattle, 13 p.

SZABO, B. J., AND MOORE, J. G., 1986, Age of −360-m reef terrace, Hawaii, and the rate of late Pleistocene subsidence of the island: Geology, v. 14, p. 967–968.

TUSHINGHAM, A. M., 1989, A global model of late Pleistocene deglaciation: implications for Earth structure and sea level changes: Unpublished Ph.D. Dissertation, University of Toronto, 206 p.

TUSHINGHAM, A. M., AND PELTIER, W. R., 1991, ICE-3G: a new global model of late Pleistocene deglaciation based upon geophysical predictions of post-glacial relative sea level change: Journal of Geophysical Research, v. 96, p. 4497–4523.

TUSHINGHAM, A. M., AND PELTIER, W. R., 1992, Validation of the ICE-3G model of Würm-Winconsin deglaciation using a global data base of relative sea level histories: Journal of Geophysical Research, v. 97, p. 3285–3304.

U.S. Army Corp of Engineers, 1983, Kaua'i Coastal Resource Atlas: Honolulu, Hawaii, Manoa Mapworks, 279 p.

VEEH, H. H., 1966, Th^{230}/U^{238} and U^{234}/U^{238} ages of Pleistocene high sea level stand: Journal of Geophysical Research, v. 71, p. 3379–3386.

WATT, A. B., AND ten BRINK, U.S., 1989, Crustal structure, flexure, and subsidence history of the Hawaiian Islands: Journal of Geophysical Research, v. 94, p. 10473–10500.

WENTWORTH, C. K., AND PALMER, H. S., 1925, Eustatic bench of islands of the North Pacific: Geological Society of America Bulletin, v. 36, p. 521–544.

WOODROFFE, C., MCLEAN, R., POLACH, H., AND WALLENSKY, E., 1990a, Sea level and coral atolls: Late Holocene emergence in the Indian Ocean: Geology, v. 18, p. 62–66.

WOODROFFE, C. D., STODDARD, D. R., SPENCER, T., SCOFFIN, T. P., AND TUDHOPE, A. W., 1990b, Holocene emergence in the Cook Islands, South Pacific: Coral Reefs, v. 9, p. 31–39.

HOLOCENE SEDIMENTARY FRAMEWORK OF GRAYS HARBOR BASIN, WASHINGTON, USA

CURT D. PETERSON
Geology Department, Portland State University, Portland, Oregon 97207
AND
JAMES B. PHIPPS
Grays Harbor College, Aberdeen, Washington 98520

ABSTRACT: High-resolution seismic profiling (300 km) and drill coring (17 sites to depths of 60 m below present sea level) in Grays Harbor, Washington, record the Holocene marine transgression and associated filling of the drowned-river basin (present surface area 290 km^2) in the convergent Cascadia margin. Seismic-reflection records are used to map the morphology of the ancestral Chehalis-Humptulips river valley (axial channel depth 60–70 m below present sea level) and to produce an isopach map of the basin fill (total volume 8.7 km^3). Drill coring establishes general textural sequences of sand and mud coarsening upward to sand and gravel in lower-bay reaches, and gravelly sand and sandy gravel fining upward to sand and mud in upper-bay reaches. Radiocarbon dating of core-sample wood, carbonate shells, and peaty muds yields a deposit depth-age curve in Grays Harbor beginning at 10,760±90 yrs (uncalibrated age in radiocarbon years before present, RCYBP) from 57 m depth below present sea level. Average basin sedimentation rates decrease from 1.2 cm/yr (10.5–7.5 ka) to 0.1 cm/yr (5.5–0 ka), following declining rates of eustatic sea-level rise in middle Holocene time. This trend corresponds to a four-fold decrease in average rates of basin-fill accumulation from 2.0×10^6 m^3/yr (8.5–7.75 ka) to 0.5×10^6 m^3/yr (5.5–0 ka), implying substantial river-sediment bypassing and/or diminished marine-sediment influx in the Grays Harbor basin during the latter half of the Holocene marine transgression.

INTRODUCTION

Relatively few studies of early Holocene coastal stratigraphies have been conducted along convergent-margin coasts, even though such coasts represent a significant proportion of the world's shorelines (Inman and Nordstrom, 1971). The lack of such studies might, in part, reflect the generally poor preservation of transgressive-sequences in shallow shelves and narrow coastal plains of uplifting margins, such as the Cascadia margin of the United States Pacific Northwest. However, Holocene transgressive sequences in this margin are locally preserved in the drowned-river valleys cut below the level of the uplifted coastal plain (Glenn, 1978; Peterson and others, 1984a). The sedimentary sequences preserved within these incised valleys potentially record: (1) the local onset and rate of marine transgression, (2) the partitioning of fine and coarse sediments across the coastal interface, and (3) the stratigraphic development of coastal basins during the transition from high-gradient river valleys (prior to marine transgression) to sediment-filled estuarine basins (during high sea-level stands).

In this paper we present initial results from seismic profiling and drill coring of the Holocene fill in the Grays Harbor basin of southwest Washington. Grays Harbor is the northernmost of three relatively large estuaries; the other two are Willapa Bay and the Columbia River, located in the central Cascadia margin of the northwestern United States (Fig. 1). These three tidal basins owe their large size to a series of protective barrier spits; the spit sediments are apparently derived from the Columbia River (Cooper, 1958; Ballard, 1964; Rankin, 1983). Grays Harbor is a tide-dominated estuary that covers some 223 km^2 in surface area (Scheidegger and Phipps, 1976; Peterson and others, 1984b). Whereas late Holocene coastal deposits in southwest Washington have been recently investigated in tidal basins (Clifton and Phillips, 1980; Atwater, 1987; Sherwood and Creager, 1990) and the continental shelf (Nittrouer, 1978; Sternberg, 1986), the corresponding early Holocene deposits have not previously been examined.

In this study we (1) map the geometry of the ancestral, lower Grays Harbor basin, (2) document the textural facies sequences developed during the filling of the drowned river-valley basin, and (3) estimate rates of basin net sedimentation and bulk sediment accumulation during the Holocene transgression. Finally, the overall depositional evolution of the Grays Harbor basin is related to ancestral-valley morphology, abundant sediment supply from fluvial and marine sources, and high depositional energy in the convergent Cascadia margin of the northeast Pacific Ocean.

BACKGROUND

The present Grays Harbor basin occupies the lower 50 km of the Chehalis River, which originates in the Cascade volcanic arc and in the southwest Washington Coast Range (Barrick, 1976; Wells and others, 1984). Several major tributaries join the lower Chehalis River valley from the north, including the Satsop, Wynoochee, Wishkah, Hoquiam, and Humptulips Rivers (Fig. 1). These high-gradient tributaries flow off the southern boundary of the Olympic Coast Range, an uplifted accretionary complex. The wave-cut platform of the inner continental shelf in the vicinity of Grays Harbor is thought to be covered by a relatively thin transgressive sand sheet, some 10 m in thickness (Nittrouer, 1978). The outermost shelf is intersected by several submarine canyons, including the prominent Astoria Canyon west of the Columbia River mouth (McManus, 1964; Duncan and Kulm, 1970).

The Grays Harbor basin is exposed to relatively high-energy deposition resulting from a 3-m diurnal tidal range, producing channel-current velocities up to 2.5 m/s, intrabasin wind waves up to 1 m in height, shoaling over shallow tidal flats, and high ocean-wave energy, which produces winter-storm surf of 6 to 7 m, significant wave height, seaward of the tidal inlet (Glancy, 1971; Loehr and Ellinger, 1974; Barrick, 1976). Although Grays Harbor is relatively shallow (over 60 percent of its surface area is intertidal), deeper tidal channels (3 to 15 m depth) locally dissect the shallow tidal flats (Fig. 2).

Finally, late Holocene records of episodic coastal uplift and subsidence events (1–2 m vertical displacement) in Grays

Quaternary Coasts of the United States: Marine and Lacustrine Systems, SEPM Special Publication No. 48

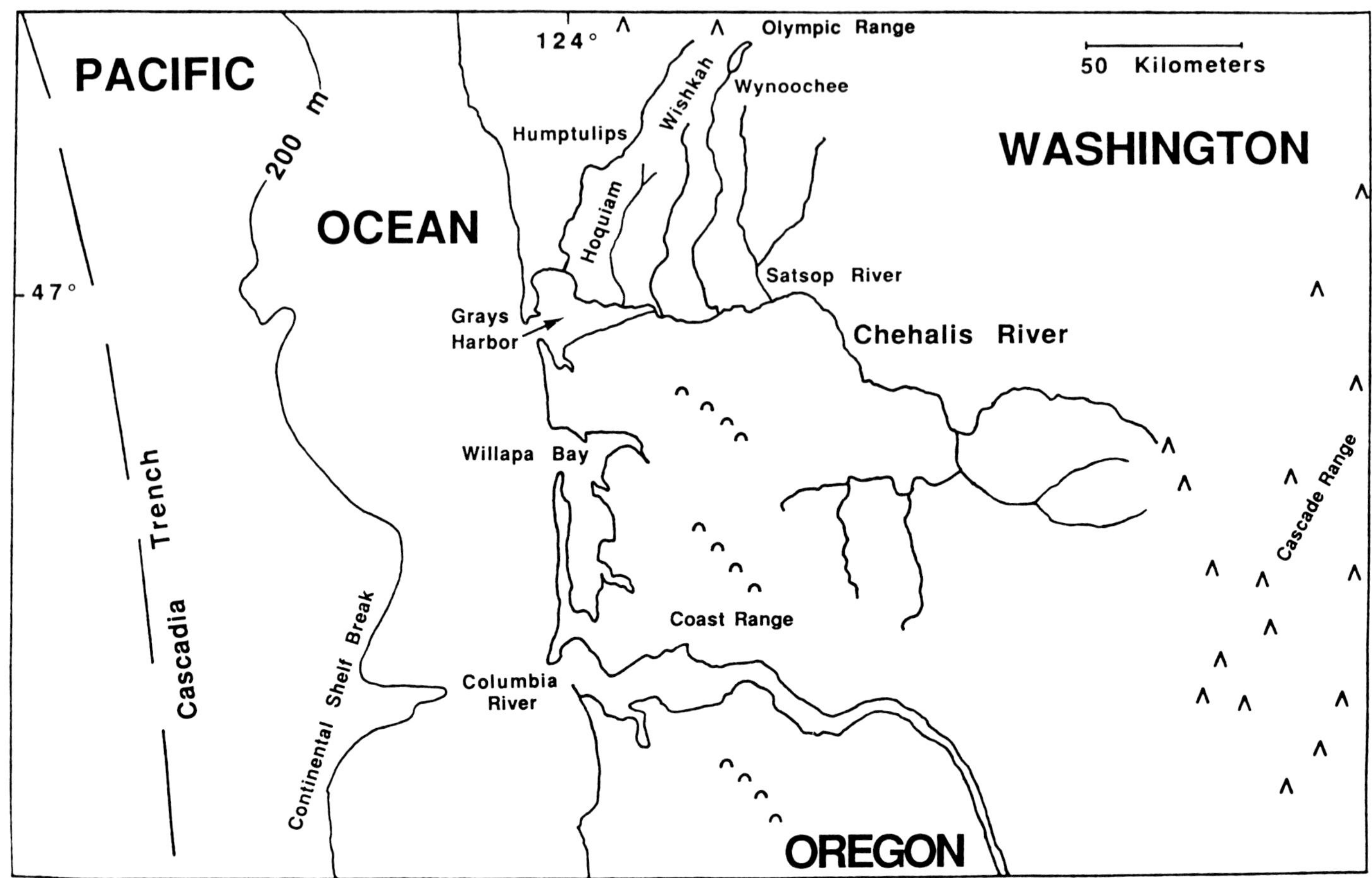

FIG. 1.—Location map of Grays Harbor (solid arrow) and major tributaries with headwaters in the Cascade and Olympic Ranges (open triangles) and the lower-elevation Coast Range (open half circles). The continental-shelf break (200-m depth contour) and buried Cascadia trench (dashed line) are also shown. The Astoria Canyon is shown as an indentation of the shelf break seaward of the Columbia River mouth.

Harbor and other central Cascadia margin bays indicate active tectonic deformation in the subduction-zone margin (Atwater, 1987; Darienzo and Peterson, 1990). The possible effects of such short-term tectonic events (500-year average recurrence interval) on basin depositional evolution are not known. However, an abrupt 1.5 m rise in sea level (possible coseismic-subsidence event) could increase the modern tidal prism (6.0×10^8 m^3; Barrick, 1976) by 50 percent in the shallow Grays Harbor estuary.

METHODS

Approximately 300 km of seismic-reflection trackline were run within the Grays Harbor basin and in the adjacent inner continental shelf in 1985 (Fig. 2). Seismic-reflection profiles were obtained with a uniboom system, producing a 900 J broad-spectrum pulse at a 0.25-second repetition rate. The return signal was received with a 12-element hydrophone, filtered to include 500 to 15,000 Hz, and recorded on an EPC 3200 analog recorder. Anomalous signal attenuation upriver of the Hoquiam River confluence prohibited seismic profiling of the deeper basin fill (below 30 to 40 m subsurface) in the upper-bay reaches. Pre-Holocene acoustic contact depths established at drill sites 1, 2, 3, 4, 13, and 14 (Fig. 2) were compared to time sections of adjacent seismic lines to determine the average seismic velocities (1,600 m/s) for the Holocene fill. Navigation was accomplished by Loran C in the inner shelf (accuracy 100–300 m), and Mini-Ranger, bathymetric charts, and aerial photographs within the estuary (accuracy 10 to 100 m).

Drill-core samples (generally 50–75 cm long × 7.5 cm in diameter) were taken from 12 sites in 1986 (rotary mud drilling sites 1–9 and 15–17), and from five sites previously drilled in 1960 (cable-tool drilling sites 10–14; Bunker, pers. commun. 1991). Extended split-barrel or shelby-tube cores were generally taken at 1.5- or 3-m intervals and at apparent lithologic transitions detected during the drilling. Drilling was terminated at pre-Holocene contacts (sites 1–5, 7, 9, 13, and 14), depth of drill-core refusal in basal gravels/cobbles (sites 8, 15, 16, and 17), or first encounter with gravel in exposed open-bay reaches (sites 6, 10, 11, and 12).

Core samples were analyzed for sediment texture, by wet and dry sieving, and deposit age, by radiocarbon dating of shells, wood, and peats. All radiocarbon analyses were performed by a commercial laboratory in 1986 and 1987, using standard pretreatment leaches, $^{12}C/^{13}C$ adjustments, and ^{14}C-reservoir calibrations after Stuiver and Reimer (1986). Detailed reports on Grays Harbor core-sample analyses are

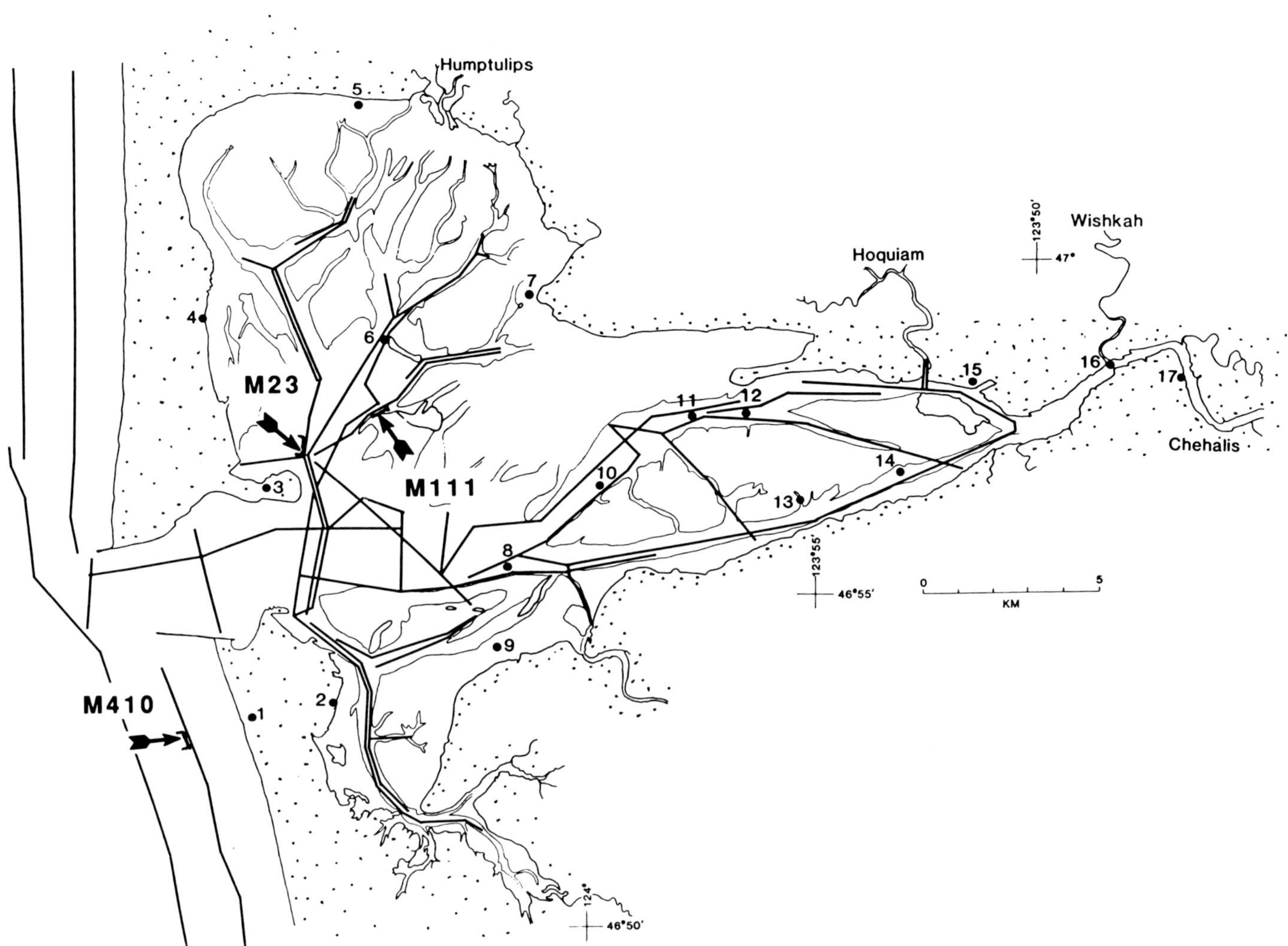

FIG. 2.—Base map of Grays Harbor study area. Shown are mean low low water contours of modern bay bathymetry (thin solid lines); seismic track lines (thick solid lines); selected seismic-profile locations M111, M23, and M410 (arrows); and drill-core sites (solid circles numbered 1–17). The shorelines of the modern bay (stippled) include the barrier spits and the narrow flood plain at the base of upland terraces surrounding the eastern margins of the estuary.

in preparation, including sediment physical properties, source mineralogy, sedimentary structures, and fossil assemblages.

RESULTS

Pre-Holocene River Valley

Seismic records from the lower Grays Harbor basin establish the ancestral-valley bottom (pre-Holocene contact) with clearly defined reflectors or reflector packages. For example, seismic profile M111 from the central-basin area (Fig. 2) shows a distinctive basal reflector at about 30 m below present sea level (Fig. 3A). By comparison, seismic profile M23 located just inside the present tidal inlet (Fig. 2) defines a strong acoustic contact at a depth of 56 to 58 m (Fig. 3B). Several crossing profiles and an adjacent drill hole (site 3) confirm that this contact represents the local depth of axial channel downcutting. The seismic records from profile M23 and many other profiles inshore of the Grays Harbor barrier spits are characterized by strong horizontal reflections, particularly in the top 30 to 40 m of basin fill.

Seismic profiles seaward of the tidal inlet (e.g., profile M410) show deeper basal reflectors at depths of about 70 m (Fig. 3C). The offshore basal reflectors shallow to about 10 m subsurface (i.e., thickness of unconsolidated shelf sediments over the wave-cut platform just north and south of the present Grays Harbor estuary is approximately 10 m). In contrast to the inshore profiles, the offshore profiles are generally characterized by a lack of recognizable horizontal reflections at any depth in the Holocene basin fill.

Shallow submarine terraces around the interior margin of the Grays Harbor basin (core sites 5, 7, 9, 13, and 14; Fig. 2) are cut into weathered deposits of silt, sand and/or gravel that rim the modern bay shoreline. Moore (1965) reports that some of these deposits are of Pleistocene glacial-drift and outwash origin. By comparison, the deeper axes of the ancestral-river valley are cut into semiconsolidated, un-

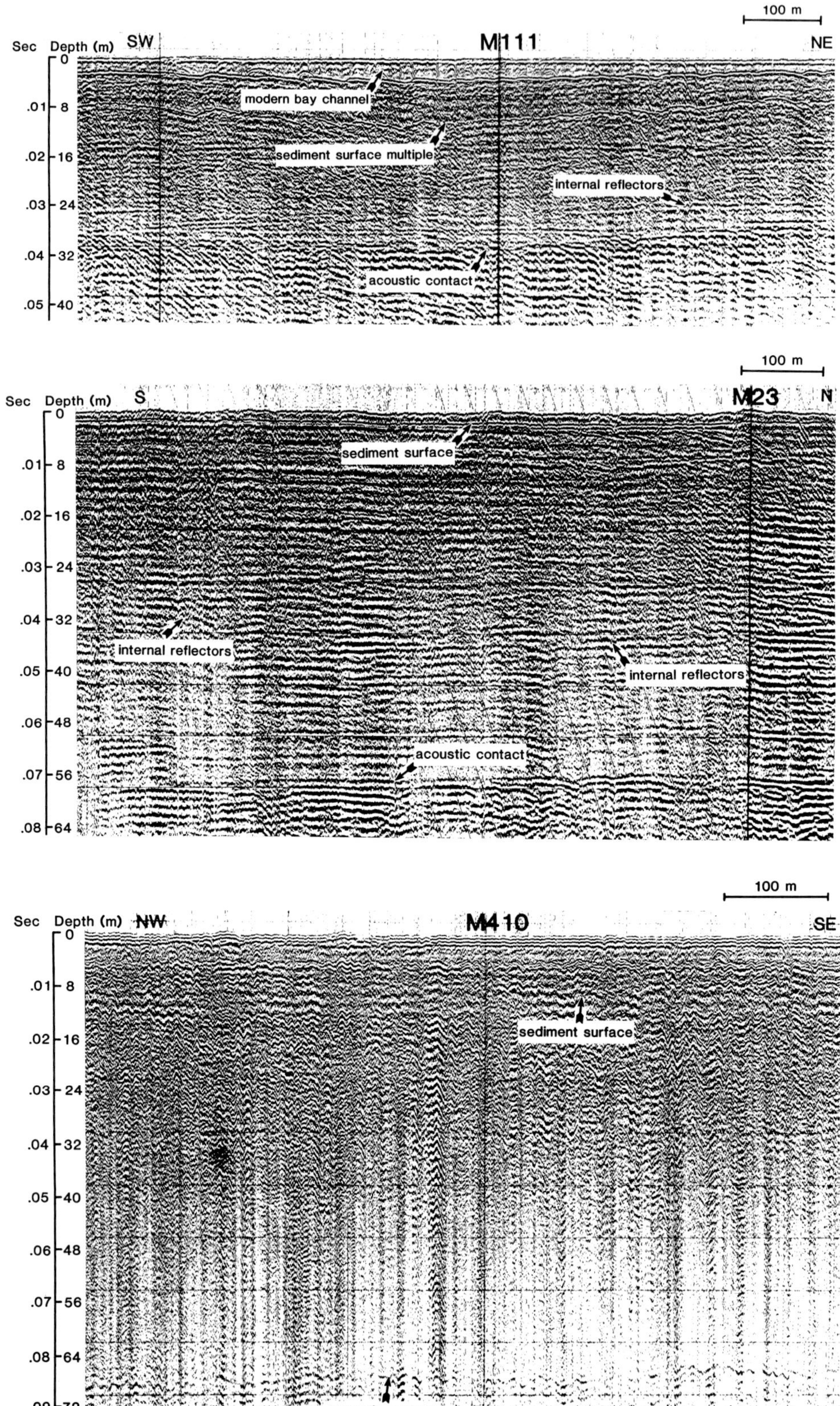

FIG. 3.—Three seismic profiles (A: profile M111; B: profile M23; C; profile M410) showing basal acoustic contacts used to establish ancestral-valley morphology in the Grays Harbor basin. Two-way travel time and depth below present sea level are also shown. In profile M111 a small modern tidal channel and reflection multiple (sediment-surface multiple) are shown. Horizontal reflections and/or reflection multiples (internal reflectors) are shown in the basin fill of profiles M111 and M23. No horizontal internal reflections are observed in the offshore profile M410. Depth of modern sediment surface (sediment surface) in profile M410 is confirmed from water depth (shelf bathymetry) at the profile location.

weathered silts, as observed at 55 to 60 m depth in the bottom cores of drill sites 1, 2, and 3. The age and origin of these pre-Holocene dewatered silts are unknown. Both the weathered Pleistocene sediments and the dewatered silts produce strong acoustic contacts with the overlying Holocene fill.

Figure 4 shows a contour map of the ancestral-valley bottom constructed from seismic profiles throughout the lower-middle basin (below the Hoquiam River confluence) and the adjacent offshore region. The contours of the isopach map represent the total thickness of the Holocene basin fill, i.e., depth to the pre-Holocene contact below present sea level, in the area covered by seismic profiling and drill coring. Large features of the ancestral basin floor are easily resolved using the 10-m contour interval. Such features include two narrow axial valleys (the Chehalis and Humptulips river valleys), which join to form a broad axial valley under the present bay mouth. This broad valley bottom, 10 to 15 km wide, deepens offshore to 70 m below present sea level at a distance of about 5 km west of the axial-valley confluence.

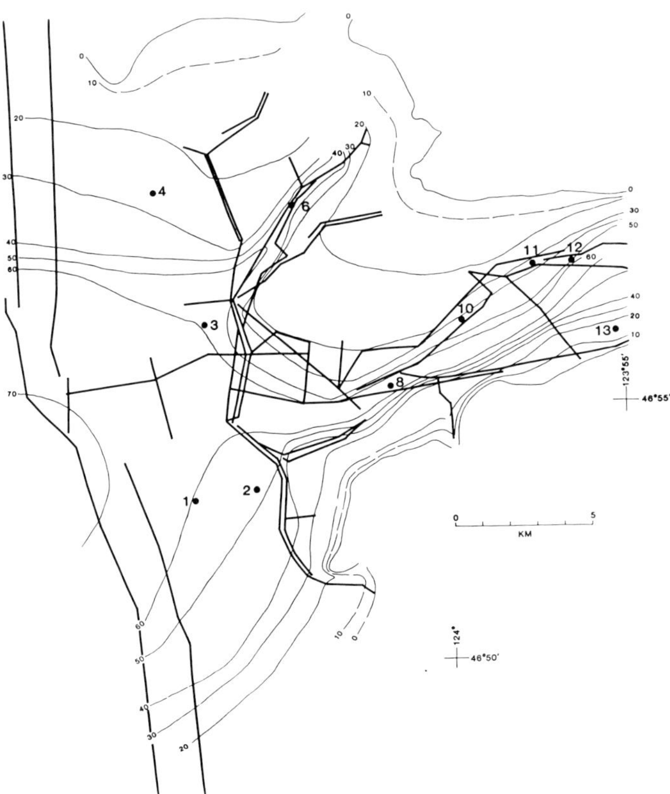

FIG. 4.—Contour map of the ancestral-valley bottom (pre-Holocene contacts) showing Holocene fill isopach in the lower-bay reaches and innermost shelf area of the Grays Harbor basin. Depth contours of the ancestral-basin floor (thin lines) are shown at 10-m intervals from 0 to 70 m depth below modern sea level. Dashed contour lines indicate extrapolation between deeper contours and modern inshore shoreline. Seismic tracklines (bold lines) and drill sites (solid circles) show control points and serve as reference points to modern basin morphology as outlined in Figure 2.

Sediment-Texture Composition

The Holocene fill of the Grays Harbor basin is dominated by coarse-grained sediments throughout its length and depth. For example, muddy sand to sandy gravel characterizes about 80 percent of the 290 analyzed core samples from the 17 drill-core sites (Fig. 5A, B). The sand-rich deposits are typically cross-bedded with minor bioturbation. In contrast, the mud-rich deposits are generally laminated with minor to major bioturbation. Invertebrate shells and wood debris are locally abundant in gravelly channel-lag deposits. Peaty deposits were rarely observed, except in the uppermost bay sites 15, 16, and 17, where variably rooted muds sampled above a 35-m depth (Fig. 5B) suggest episodic marsh or floodplain deposition.

Generally, the mud-rich deposits (>50 percent silt and clay) are restricted to the top 35 m of basin fill in the upper-bay sites (8–17) and to the bottom 35 m of the basin fill in the lower-bay sites (1–6). In contrast, gravelly deposits are located in the bottom 35 m of basin fill in the upper-bay sites (8 and 10–17) and in the top 35 m of basin fill in the most seaward sites (1–4). It is possible that a thin (<10 m) layer of gravel might also underlie the bay fill above the deepest part of the ancestral axial valley (thalwag) apparently located between core sites 2 and 3 (Fig. 4). Nevertheless, the lack of continuity of gravel deposits across the length of the basin in either the top 30 m or bottom 30 m of the basin fill (Fig. 5A, B) suggests two independent gravel sources, e.g., fluvial and marine sources. The mud is presumed to be largely derived from the river sources, whereas the sand fraction is expected to be supplied from both the fluvial and marine sides (Scheidegger and Phipps, 1976).

Alternating vertical sequences of sand, sandy-mud, and sandy-gravel layers occur in various orders of superposition in the top 30 m of the lower-bay sites 2, 3, 4, 6, and 8 (Fig. 5A). The thickness of the alternating layers (3–7 m), together with their lack of correlation between core sites, suggests a possible channel cut-and-fill origin (Clifton and Phillips, 1980). On the other hand, the middle-bay core sites (10–14) demonstrate somewhat less variation in sediment texture, e.g., sand to sandy mud, above 30-m subsurface core depth (Fig. 5B). Thick basal sequences of gravelly sand to sandy gravel underlying the sand and mud deposits of the middle upper-bay reaches (Fig. 5B) might account for the anomalous acoustic-signal attenuation observed in these areas.

Deposit Depth-Age Curve

Radiocarbon ages of 28 wood, shell, and peaty samples (Table 1) are used to construct a deposit depth-age curve for the Grays Harbor basin fill (Fig. 6). The deepest sample dated in the Grays Harbor basin (57 m at site 1) has an uncalibrated radiocarbon age of 10,760±90 yrs, roughly establishing the onset of Holocene marine transgression in the central-basin region. Clusters of similar sample dates (within ±1,000 yrs) from selected subsurface depths (at approximately 30, 20, 10, and 0–5 m) are found at widely spaced core sites (Fig. 6). These trends suggest that basin vertical filling was roughly uniform along the basin length.

TABLE 1.—SEDIMENT-SAMPLE AGES FROM GRAYS HARBOR

Site-Sample	Beta#	Depth (m) Below MSL	Material Dated	RCYBP*±1 S.D.	Calibrated Age**±2 S.D.
15-03	20296	02.0	Wood	330±100	0–540
07-01	20308	01.5	Wood	620±70	520–690
04-01	20293	01.0	Peat	830±60	670–920
02-04	20279	05.0	Shell	1360±70	1160–1390
15-06	20297	06.5	Wood	3190±230	2849–3979
05-06	20309	05.0	Peaty	3380±80	3459–3839
03-11	20287	14.5	Shell	3570±80	3653–4119
01-09	20957	11.5	Shell	4020±120	4149–4849
17-04	20529	03.0	Peat	4120±80	4419–4859
08-02	20301	12.5	Shell	4740±100	5143–5725
06-03	20528	12.5	Shell	5080±90	5649–6160
17-07	20530	11.0	Peaty	5540±80	6189–6523
15-10	20527	12.5	Peaty	5770±140	6299–6889
01-22	20284	31.0	Shell	6040±90	6729–7179
03-14	20288	19.0	Wood	6170±80	6853–7260
06-06	20306	21.0	Shell	6440±110	7169–7499
03-16	20291	22.0	Shell	7320±390	7429–9009
03-21	20289	29.5	Shell	7350±110	7929–8379
17-12	20531	20.0	Peaty	7530±80	8129–8500
15-14	20299	13.5	Peaty	7340±140	
03-30	20290	43.0	Shell	8730±100	
17-16	20532	29.0	Peaty	7930±120	8430–9093
02-20	20280	29.5	Shell	7990±90	8569–9093
15-24	20300	34.0	Wood	8050±110	8592–9159
02-25	20281	37.0	Shell	8940±100	
02-35	20282	52.5	Shell	9700±130	
08-14	20304	47.5	Shell	10110±270	
01-39	20286	57.0	Wood	10760±90	

*Radiocarbon ages in radiocarbon years before present (RCYBP) adjusted by $^{13}C/^{12}C$ analyses for shell material, including±one unit standard deviation error.
**Calibrated ages from ^{14}C reservoir fluctuations for samples younger than 2 ka (Stuiver and Reimer, 1986), including±two units of standard deviation error. Peaty = 10–50 wt. % organics, Peat = >50-wt. % organics

In other words, the net sedimentation rates were similar for corresponding depth intervals throughout the study basin. In contrast, net sedimentation rates were quite variable among different depth intervals (Fig. 6). For example, the earliest period of Holocene deposition (10.5–7.5 ka) is characterized by rapid sedimentation (1.2 cm/yr). Filling later slowed dramatically, particularly after 5.5 ka (0.1 cm/yr) following declining rates of eustatic sea-level rise in middle late Holocene time (Clark and Lingle, 1979).

Aside from the radiocarbon-age uncertainties, the small-scale variability of deposit depth-age relations (Fig. 6) probably reflects differences in local deposit elevations during a given time interval. For example, the dated peaty cores, typically deposited at intertidal to supratidal elevations, are found to be generally 3 to 10 m shallower than dated shell or wood fragments of corresponding ages (Table 1; Fig. 6). Some of the dated shell and wood samples are associated with channel-lag sediments, confirming their deposition at lower intertidal to subtidal elevations. Post-depositional settling might account for only a minor part of the deposit depth-age variability, as the predominance of sand and gravel in the basin fill should diminish the effects of local sediment compaction. Finally, peat and peaty samples (Table 1) provide an approximation of relative sea-level position. The present depths of some of the peaty samples are possibly several meters lower than actual sea-level positions of corresponding age. Such errors could result from the low paleotidal levels (lower intertidal) of some slightly rooted samples (Darienzo and Peterson, 1990) and the possible compaction of underlying muddy deposits in some basin-margin areas.

DISCUSSION

Ancestral River-Valley Morphology and Basin Fill

Both the morphology and fill material of the Grays Harbor basin reflect the very narrow coastal plain between the Coast Range foothills and the continental-shelf platform in the central Cascadia margin. Over a seaward distance of less than 10 km, the narrow tributary valleys converge and then abruptly widen to form a broad axial valley under the present bay mouth (Fig. 4). The valley presumably continues farther offshore as a widening and shallowing basin cut below the seaward-dipping shelf platform. The abrupt transition of the narrow river valleys to the broad axial valley is shown in a computer-drawn three dimensional projection of the ancestral basin floor (Fig. 7). The projection illustrates the general narrowing and steepening of the ancestral-valley side slopes with increasing proximity to the Coast Range foothills. The foothills apparently restricted the lateral movement of the ancestral Chehalis River during the pre-Holocene glacial period of river-valley downcutting.

Sand and muddy sand deposits largely fill the broad lower basin below and shoreward of the present barrier spits (Fig. 5A: sites 1–4). Alternating gravel, sand and mud layers in the lower-bay core sites (2–8) produce high-density contrasts that likely account for the numerous horizontal reflections and/or reflection multiples in the inshore seismic profiles (Fig. 3A, B). By comparison, the most seaward drill site (1) on the outer edge of the present barrier spit (Fig. 2) is dominated by sand and gravelly sand, showing little density contrast and correspondingly few basin-fill reflectors in the offshore seismic profiles (Fig. 3C).

In the uppermost bay reaches east of the Hoquiam River confluence, geotechnical and water-well borehole logs (Eddy, 1966; industry borehole logs, 1962–1986) indicate a continuation of the narrow Chehalis River valley (axial channel 50–60 m depth below present sea level) well upriver of site 17. (Fig. 2). The narrow river valley probably served initially as a conduit, and later as a reservoir, for sand and gravel (Fig. 5B: sites 15–17). The coarse-grained sediments were probably derived from the steep slopes of the Olympic Range, Coast Range, and Cascade Range, and possibly from reworked glacial debris left behind by the retreating Puget Lobe of the Cordilleran ice sheet (Thorson, 1980). Sandy gravel deposits in the drowned Chehalis River valley shallow from −35 m to modern sea level between site 17 and the Wynoochee River confluence (Figs. 1, 2, 5B; Eddy, 1966). The shallowing-upriver wedge of sandy gravels presumably followed the landward retreating head of tide water in middle late Holocene time, and argues against direct deposition from Puget Lobe meltwater streams of pre-Holocene age (Thorson, 1980).

Finally, the onset of marine transgression recorded in Grays Harbor (at approximately −55 m, about 11 to 10 ka) is not substantially different (±10 m) from sea-level positions of similar age reported from some passive margins (Dillion and Odale, 1978). The short-term vertical tectonic displacements reported for Grays Harbor (Atwater, 1987) have not

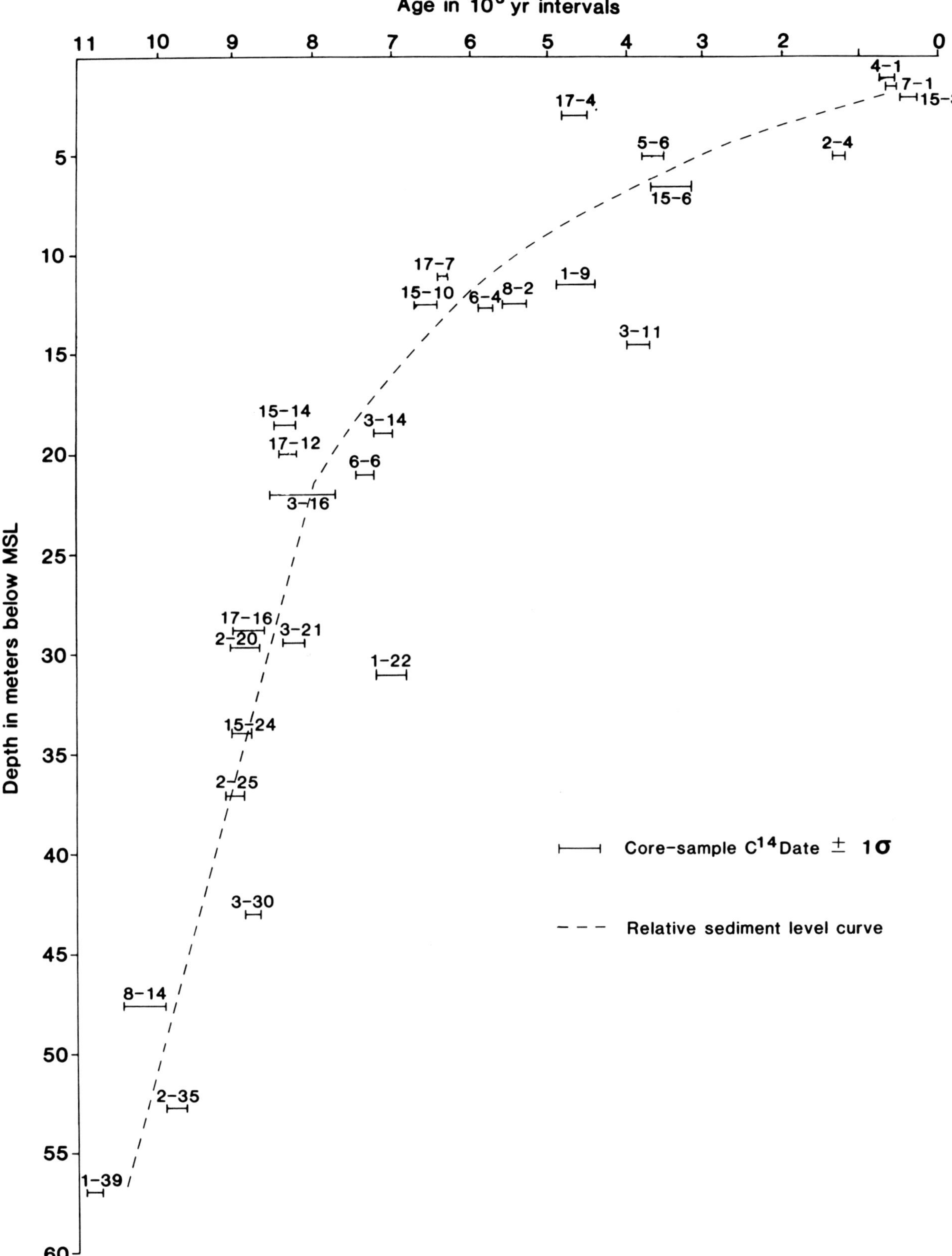

FIG. 6.—Sample-calibrated radiocarbon age (with ±1 standard deviation error) versus depth below modern sea level in the basin fill of Grays Harbor. Table 1 shows sample-calibrated radiocarbon age with ±2 standard deviation error. The average sample depth for a given age range represents the approximate deposit surface depth (dashed line) during deposition. Lateral channel migration has possibly lowered some shell and wood fragments in channel-lag deposits to greater depths than surrounding tidal flats, particularly in late Holocene time.

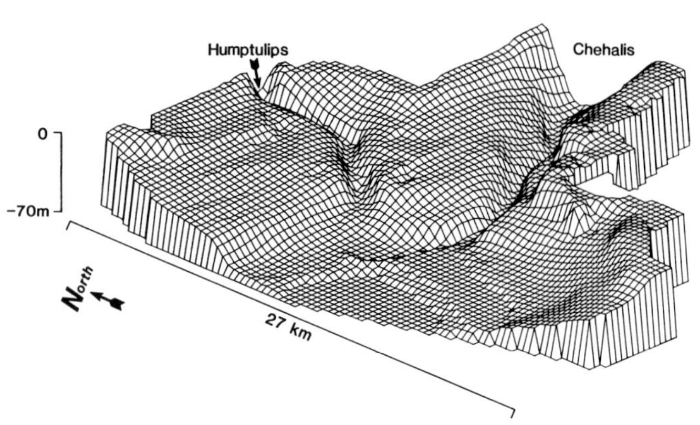

FIG. 7.—Three-dimensional-map projection of the lower Grays Harbor ancestral-valley floor (pre-Holocene contacts) constructed from the 10-m-contour map data (Figure 4) from modern sea level (0 m) to the axial-valley bottom (maximum 70 m depth below present sea level). The map projection shows the abrupt transition from steep-sided, narrow river valleys at the foot of the Coast Range to a broad axial valley cut into the shallow-shelf platform.

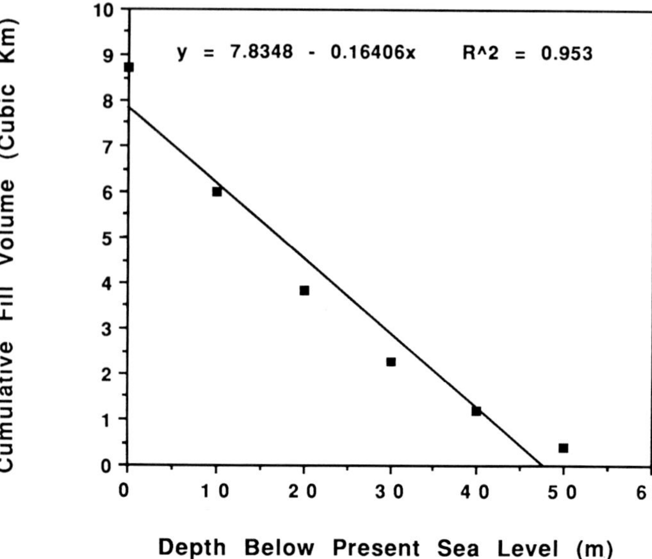

FIG. 8.—Basin-fill cumulative volume (solid diamonds) as a function of depth below modern sea level in the Grays Harbor basin. The relation between depth and cumulative volume fill is nearly linear, e.g., a linear correlation fit of $R^2 = 0.95$.

resulted in any apparent long-term net uplift or subsidence in the Grays Harbor area. These results are consistent with much longer records of coastal-terrace deformation in Willapa Bay (Kennedy, 1978) that indicate very low net-uplift rates (<0.5 mm/yr) for the southwest Washington coast since about 80 ka.

Sediment-Volume Accumulation Rates

The volume of Holocene fill in the Grays Harbor basin was calculated for six depth horizons (at 0, 10, 20, 30, 40, and 50 m below present sea level) from the isopach map (Fig. 4) including an extrapolation of the narrow Chehalis River valley upriver to site 17 (Table 2). The six contour intervals are assigned ages (Table 2) from the average deposit depth-age curve shown in Figure 6. The cumulative volume of basin fill, plotted as a function of depth below present sea level (Fig. 8), increases almost linearly with decreasing depth and is estimated to total about 8.7 km^3.

By comparison, the relations between average deposit age and depth are strongly nonlinear (Fig. 9), suggesting changing fill-accumulation (volume) rates during the Holocene transgression. Indeed, estimates of basin-fill accumulation rate (Table 2; Fig. 10) range from 0.8×10^6 m^3/yr (10.5 to 10 ka) to 2.0×10^6 m^3/yr (8.5 to 7.75 ka) to 0.5×10^6 m^3/yr (5.5 to 0 ka). If deposits from the deepest fill intervals of the upper-bay reaches, i.e., 40 to 60 m depth below present sea level, are older than assumed (Fig. 9), then basin-fill accumulation rates for the earliest Holocene period would be even lower than those shown in Figure 10. The early

TABLE 2.—ACCUMULATION RATES OF GRAYS HARBOR BASIN FILL

Depth Interval (m)	Surface Area (km^2)	Basin-Fill Age (ka)	Basin-Fill Cumulative Volume (km^3)	Basin-Fill Volume Accumulation Rate ($\times 10^6$ $m^3 yr^{-1}$)
0	290	0.0	8.70	0.49
10	251	5.50	6.00	0.97
20	185	7.75	3.82	2.05
30	123	8.50	2.28	1.45
40	94	9.25	1.19	1.04
50	61	10.00	0.41	0.82
60	21	10.50		

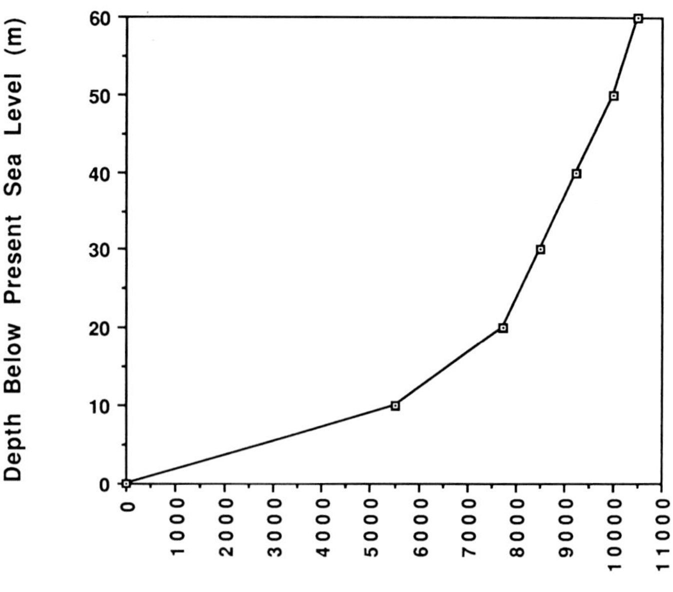

FIG. 9.—Average sediment age (from Fig. 6) as a function of depth below modern sea level in Grays Harbor basin. The relation between deposit depth and age is strongly nonlinear in Grays Harbor.

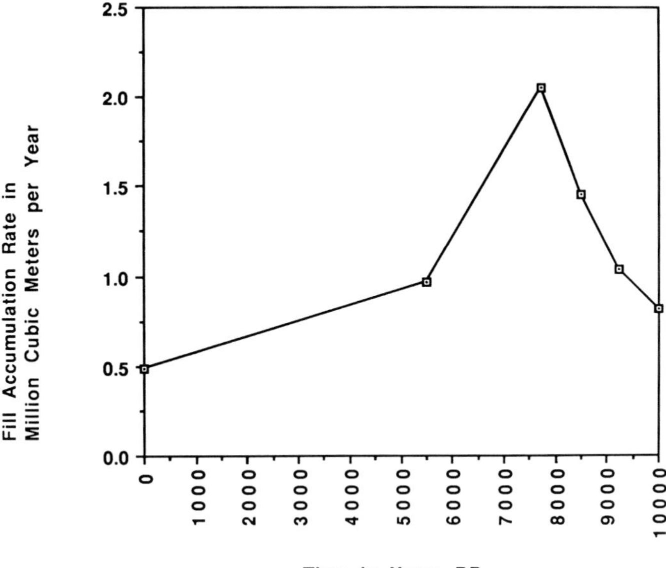

FIG. 10.—Basin-fill accumulation rate (volume fill per year) as a function of time (10–0 ka) during the period of Holocene marine transgression in the Grays Harbor basin. A four-fold decrease in fill-accumulation rate after 8 ka implies river-sediment bypassing or diminished marine-sediment influx in Grays Harbor in late Holocene time.

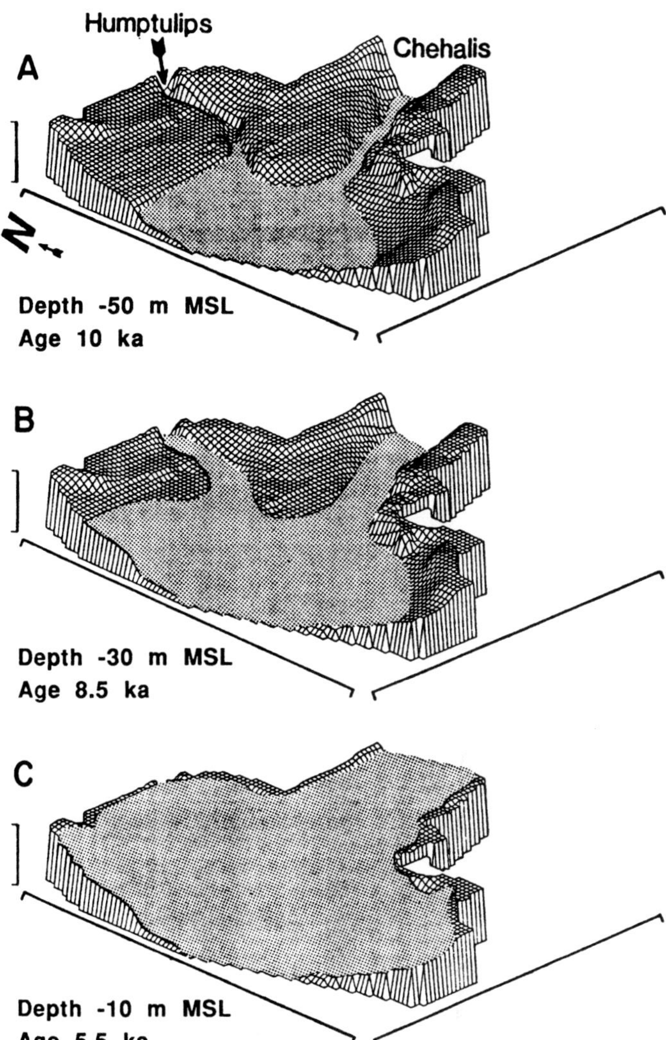

FIG. 11.—Diagrams of deposit level in Grays Harbor basin at three depths of −50 m at 10 ka (upper diagram), −30 m at 8.5 ka (middle diagram) and −10 m at 5.5 ka (lower diagram) relative to present inshore-shoreline level. Vertical scale total 70 m depth, and horizontal scales roughly 25 km in length and width (after Fig. 7).

period of basin filling might have been characterized by substantial bypassing of fine-grained river sediments through the narrow upper-bay reaches (Fig. 11A).

The broadening of the basin with decreasing depth below present sea level (Fig. 11B) provided an increasingly effective catchment as shown by increasing accumulation rates that peaked between 8 and 7 ka (Fig. 10). However, accumulation rates dropped dramatically after this period even though the basin continued to increase in surface area (Fig. 11C). Heusser and others (1980) have reported only minor changes in climatic conditions in the study area since 8 ka, which implies relatively constant sediment production in adjacent drainages. We must conclude that the corresponding four-fold decrease in basin-fill accumulation rate in middle late Holocene time represents terrestrial-sediment bypassing (Kulm and Byrne, 1966) and/or diminished influx of marine sediments to Grays Harbor. Either condition would imply a vertical filling of the basin to shallow subtidal or possibly intertidal depths throughout the late Holocene (5.5–0 ka), as observed elsewhere in the Cascadia margin (Peterson and others, 1984a,b).

Basin Depositional Evolution

Two vertical trends of sediment texture are observed in the Grays Harbor basin fill. Deposits generally coarsen upward in the lower-bay reaches (Fig. 5A, sites 1–3) and fine upward in the upper-bay reaches (Fig. 5B, sites 15–17). These trends are likely due to the landward migration of the tidal-inlet-barrier system in the lower-bay reaches, and to the protected estuarine-channel system in the upper-bay reaches. Under combined conditions of marine transgression and vertical filling of the basin the high-energy tidal-inlet sands and gravels overtopped the protected bay sands and muds in the lower-bay reaches, whereas the protected bay sands and muds overtopped the river gravels in the constricted upper-bay reaches. Significantly, the high-energy tidal-inlet facies of the lower bay (approximately 0–35 m below present sea level) has yet to reach the middle-bay reaches (Fig. 5A,B, sites 8–14). The landward migration of this facies is thus constrained to less than 10 km between sites 1 and 8 (Fig. 2) since early Holocene time (Fig. 6). Similarly, gravels at −35 m in the upper-bay reaches (Fig. 5B, sites 15–17) rise to 0-m depth downstream of the Wynoochee River confluence, a landward migration distance of less than 10 km (Fig. 1) since early Holocene time.

In apparent agreement with the small transgressive distances of the textural-facies boundaries, the Grays Harbor sedimentation rates demonstrate a relatively stationary depocenter throughout the middle to late Holocene transgression. For example, basin net-sedimentation rates (cm/yr) at corresponding depth-age intervals are relatively uniform at different core sites along the length of the basin (Fig. 6). The stable system stands in stark contrast to some passive-margin systems, which show Holocene depocenters undergoing substantial landward retreat, as in Delaware Bay (Kraft, 1971; Knebel and others, 1988; Fletcher and others, 1990), or lateral migration, as in the Mississippi Delta (Fraizer, 1967; Penland and others, 1985). The roughly synchronous vertical accretion of the Grays Harbor basin fill probably results from a high-gradient, deeply incised ancestral-river valley, a sediment supply from both river and marine sources sufficient to keep up with the Holocene sea-level rise, and a very high-energy environment that caused sediment to bypass the shallowing basin and to be removed from the tidal-inlet and barrier shoreface areas. Some or all of these conditions were possibly present during the Holocene transgression in other mid-latitude convergent margin coasts around the Pacific rim.

CONCLUSIONS

1. Seismic profiling of the ancestral Chehalis River valley indicates axial-valley downcutting to 60-m depth under the present barrier-spit shorelines of Grays Harbor. The ancestral-valley morphology is characterized by an abrupt transition between steep-sided river valleys at the foot of the Coast Range to a broad axial valley cut well below the surrounding shelf platform. The morphology of the Grays Harbor basin is a product of the very narrow coastal plain of the convergent Cascadia margin.
2. The Grays Harbor basin fill is dominated by relatively coarse-grained sediments, including sand and gravel, although sandy mud and mud are locally abundant in some core intervals. Coarsening-upward sequences occur in the lower-bay reaches, whereas fining-upward sequences occur in the upper-bay reaches. These sequences suggest combined landward migration and vertical accretion of high-energy tidal-inlet deposits over protected-bay deposits in the lower-bay reaches, and lower energy protected-bay deposits over river gravels in the upper-bay reaches. However, the landward-migration distances are relatively small, e.g., less than 10 km since early Holocene time.
3. The average sedimentation rate (cm/yr) in Grays Harbor decreased by about one order of magnitude from the early Holocene (10–8 ka) to the late Holocene (5–0 ka). By comparison, the basin-fill accumulation rate (m^3/yr) peaked between 8 and 7 ka, before decreasing in the late Holocene to one quarter of the peak rate. Both trends follow declining rates of eustatic sea-level rise in middle late Holocene time, reported from many other continental margins. The trends imply river-sediment bypassing and/or diminished marine-sediment influx in Grays Harbor, likely as a result of shallowing-basin bathymetry during middle to late Holocene time.

ACKNOWLEDGMENTS

We thank Dave Hogg for assisting with seismic profiling, Margaret Mumford for performing laboratory analyses and drafting, and Doann Hamilton for assisting with drafting for this report. Support for this study was provided by the National Science Foundation under grant OCE-8403295 and by the College of Research, Oregon State University. Drill cores taken in 1986 are archived in the refrigerated core repository at Oregon State University, Corvallis, Oregon. Core-sample archival at the College of Oceanography Core Lab, Oregon State University is supported under NSF grant OCE-8800548. All radiocarbon analyses reported here were performed by Beta Analytic Inc. The U.S. Geological Survey, Branch of Pacific-Arctic Marine Geology, and the U.S. Army Corps of Engineers, Seattle District, provided the equipment and vessel time, respectively, for the seismic profiling of the Grays Harbor basin. We thank Kenner Drilling Inc. for undertaking the core-drilling operations in the exposed reaches of Grays Harbor. Finally, we thank LaVerne Kulm and Jerry Glenn for providing many valuable suggestions for the improvement of this paper.

REFERENCES

Atwater, B. F., 1987, Evidence for great Holocene earthquakes along the outer coast of Washington State: Science, v. 36, p. 942–944.

Ballard, R. L., 1964, Distribution of beach sediment near the Columbia River: University of Washington, Department of Oceanography, Technical Report No. 98, 82 p.

Barrick, R. C., 1976, Hydrodynamics of Grays Harbor Estuary, Washington: Appendix A in Maintenance Dredging and the Environment of Grays Harbor Washington: U.S. Army Corps of Engineers, Seattle District, Seattle, Washington, 95 p.

Clark, J. A., and Lingle, C. S., 1979, Predicted relative sea level changes (18,000 years B.P. to present) caused by late-glacial retreat of the Antarctic ice sheet: Quaternary Research, v. 11, p. 279–298.

Clifton, H. E., and Phillips, R. L., 1980, Lateral trends and vertical sequences in estuarine sediments, Willapa Bay, Washington *in* Field, M. A., Bouma, A. H., Colburn, I. P., Douglas, R. G., and Ingle, J. C., eds., Fourth Pacific Coast Paleogeography Symposium, Pacific Section, Society of Economic Paleontologists and Mineralogists, p. 55–71.

Cooper, W. S., 1958, Coastal sand dunes of Oregon and Washington: Geological Society of America Memoir 72, 169 p.

Darienzo, M. E., and Peterson, C. D., 1990, Episodic tectonic subsidence of late Holocene salt marshes, northern Oregon, central Cascadia margin: Tectonics, v. 9, p. 1–22.

Dillon, W. P., and Odale, R. O., 1978, Late Quaternary sea-level curve: reinterpretation based on glaciotectonic influence: Geology, v. 6, p. 56–60.

Duncan, J. R., and Kulm L. D., 1970, Mineralogy, provenance and dispersal history of late Quaternary deep-sea sands in Cascadia Basin and Blanco Fracture Zone off Oregon: Journal of Sedimentary Petrology, v. 40, p. 874–887.

Eddy, P., 1966, Geologic Map of the Lower Chehalis River Valley, Grays Harbor County: State of Washington Department of Conservation, Division of Water Resources.

Fletcher, C. H., III, Knebel, H. J., and Kraft, J. C., 1990, Holocene evolution of an estuarine coast and tidal wetlands: Geological Society of America Bulletin, v. 102, p. 283–297.

Folk, R. L., 1980, Petrology of Sedimentary Rocks: Hemhills, Austin, Texas, 185 p.

Fraizer, D. E., 1967, Deltaic deposits of the Mississippi River: their development and chronology. Transactions, Gulf Coast Association of Geological Societies, p. 287–311.

Glancy, P. A., 1971, Sediment transport by streams in the Chehalis River basin, Washington, Oct. 1961 to Sep. 1965, Tacoma, Washington: U.S. Geological Survey Water-Supply Paper 1798-H, 53 p.

Glenn, J. L., 1978, Sediment sources and Holocene sedimentation history in Tillamook Bay, Oregon: data and preliminary interpretations. U.S. Geological Survey Water Resources Division Open-File Report 78–680, 64 p.

HEUSSER, C. J., HEUSSER, L. E., AND STREETER, S. S., 1980, Quaternary temperatures and precipitation for the northwest coast of North America: Nature, v. 286, p. 702–704.

INMAN, D. L., AND NORDSTROM, C. E., 1971, On the tectonic and morphologic classification of coasts: Journal of Geology, v. 79, p. 1–21.

KENNEDY, G. L., 1978, Pleistocene paleoecology, zoogeography and geochronology of marine invertebrate faunas of the Pacific Northwest Coast (San Francisco Bay to Puget Sound): Unpublished Ph.D. Dissertation: University of California at Davis, 824 p.

KNEBEL, H. J., FLETCHER, C. H., III, AND KRAFT, J. C., 1988, Late Wisconsinan-Holocene paleogeography of Delaware Bay; A large coastal plain estuary: Marine Geology, v. 83, p. 115–133.

KRAFT, J. C., 1971, Sedimentary facies patterns and geologic history of a Holocene marine transgression: Geological Society of America Bulletin, v. 82, p. 2131–2158.

KULM, L. D., AND BYRNE, J. V., 1966, Sedimentary response to hydrography in an Oregon estuary: Marine Geology, v. 4, p. 85–118.

LOEHR, L. C., AND ELLINGER, E. L., 1974, A description of the oceanographic environment of the Washington coast and the Strait of Juan de Fuca: University of Washington, Department of Oceanography for the Oceanographic Institute of Washington, Seattle, Washington, 146 p.

MCMANUS, D. A., 1964, Major bathymetric features near the Coast of Oregon and Washington, and Vancouver Island: Northwest Science, v. 38, p. 65–82.

MOORE, J. L., 1965, Surficial geology of the southwestern Olympic Peninsula. Unpublished M.S. Thesis, University of Washington, Seattle, 63 p.

NITTROUER, C. A., 1978, The process of detrital sediment accumulation in a continental shelf environment: an examination of the Washington shelf: Unpublished Ph.D. Dissertation, University of Washington, Seattle, 243 p.

PENLAND, S., SUTER, J. R., AND BOYD, R., 1985, Barrier island arcs along abandoned Mississippi River deltas: Marine Geology, v. 63, p. 197–233.

PETERSON, C. D., SCHEIDEGGER K., KOMAR, P., AND NIEM, W., 1984b, Sediment composition and hydrography in six high-gradient estuaries of the northwestern United States: Journal of Sedimentary Petrology, v. 54, p. 86–97.

PETERSON, C. D., SCHEIDEGGER, K. F., AND SCHRADER, H. J., 1984a, Holocene depositional evolution of a small active-margin estuary of the northwestern United States: Marine Geology, v. 59, p. 51–83.

RANKIN, D. K., 1983, Holocene geologic history of the Clatsop Plains foredune ridge complex: Unpublished M.S. Thesis, Portland State University, Portland, Oregon, 176 p.

SCHEIDEGGER, K. F., AND PHIPPS, J. B., 1976, Dispersal patterns of sand in Grays Harbor estuary, Washington: Journal of Sedimentary Petrology, v. 46, p. 163–166.

SHERWOOD, C. R., AND CREAGER, J. S., 1990, Sedimentary geology of the Columbia River estuary: Progress in Oceanography, v. 25, p. 15–79.

STERNBERG, R. W., 1986, Transport and accumulation of river-derived sediment on the Washington continental shelf, USA: Journal of the Geological Society of London, v. 143, p. 945–956.

STUIVER, M. AND REIMER, P. J., 1986, A computer program for radiocarbon age calibration: Radiocarbon, v. 28, p. 1022–1030.

THORSON, R. M., 1980, Ice-sheet glaciation of the Puget Lowland, Washington, during the Vashon Stade (late Pleistocene): Quaternary Research, v. 13, p. 303–321.

WELLS, R. E., ENGEBRETSON, D. C., SNAVELY, P. D., Jr., AND COE, R. S., 1984, Cenozoic plate motions and the volcano-tectonic evolution of western Oregon and Washington: Tectonics, v. 3, p. 275–294.

HOLOCENE TIDAL-MARSH STRATIGRAPHY IN SOUTH-CENTRAL OREGON–EVIDENCE FOR LOCALIZED SUDDEN SUBMERGENCE IN THE CASCADIA SUBDUCTION ZONE

ALAN R. NELSON

U.S. Geological Survey, MS 966, P.O. Box 25046, Denver, Colorado 80225

ABSTRACT: Protected tidal inlets in four estuaries along 100 km of the south-central Oregon coast contain tidal-marsh stratigraphic sequences that suggest different styles of relative sea-level rise during the late Holocene. The south-central coast lies 70 to 90 km east of the leading edge of the overriding North America plate in the central part of the Cascadia subduction zone. Localized differences in the style of submergence along this actively deforming coast imply differences in the rate and extent of coastal subsidence or uplift during subduction of the Juan de Fuca plate beneath the North America plate. Thick tidal-marsh peat in the northern part of the study area records gradual submergence (1.0–1.6 mm/yr since 2.0–2.5 ka) that rules out large (>0.5 m), sudden changes in relative sea level, such as have occurred along other subduction-zone coasts during some historic great earthquakes (magnitude >8). In the south part of the study area, changes in intertidal lithofacies across abrupt transgressive and regressive overlap boundaries are consistent with localized coseismic subsidence and uplift of about 0.5 to 1.0 m above faults and folds in the accretionary wedge of the North America plate. Most coseismic deformation on these local structures probably coincides with great subduction earthquakes, but some localized deformation may reflect much smaller earthquakes that occur independently of plate-boundary events.

INTRODUCTION

Vertical tectonic movements have shaped the Holocene evolution of many tectonically active coasts (Lajoie, 1986; Berryman 1987). Much research on Holocene sea levels was prompted by a search for global patterns of sea-level change along coasts having low rates of uplift or subsidence (Kidson, 1982). In contrast, sea-level studies of tectonically active coasts offer hope of identifying tectonically induced changes in sea level and in measuring rates of tectonism (e.g., Bloom, 1980; Newman and others, 1987; Taylor and others, 1990).

The highest rates of relative sea-level (RSL) change on tectonically active coasts result from instantaneous vertical movements during large earthquakes (Peltier, 1987). Study of such coseismic sea-level changes is one aspect of coastal paleoseismology, a branch of neotectonics that seeks to determine the location, magnitude (M), and recurrence (frequency) of individual prehistoric earthquakes. The objectives of coastal paleoseismology studies differ from those of many other Holocene sea-level studies, which focus on correlating gradual changes in RSL over large regions through leveling and radiocarbon dating of basal organic-rich sediment (e.g., Jelgersma, 1961; Newman and others, 1980; Shennan, 1987; 1989; Belknap and others, 1989; van de Plassche, 1990). On many tectonically active coasts, the historic earthquake record is too short to show the future hazard from great earthquakes and accompanying tsunamis (e.g., Vita-Finzi, 1986; Nishenko, 1989). Paleoseismology studies help define such hazards.

No great (M > 8) earthquakes have occurred in the Cascadia subduction zone (Fig. 1A) during the short period of historic settlement (<200 yr) in the Pacific Northwest, but the possibility of such earthquakes has been widely discussed (e.g., Heaton and Hartzell, 1987; West and McCrumb, 1988; Spence, 1989; Adams, 1990). Heaton and Hartzell (1987), Atwater (1987), and Rogers (1988) suggested that the stress accumulating on the interface between the Juan de Fuca and North America plates may be released in earthquakes as large as M 9. During M > 9 earthquakes in Chile and Alaska in the early 1960s, zones 60 to 180 km wide and many hundreds of kilometers long suddenly subsided 0.5 to 2.5 m due to tectonic release of stress in the overriding plate (Plafker, 1969; Plafker and Savage, 1970). These examples show that great plate-interface earthquakes can produce extensive zones of coastal subsidence located arcward of regions of coseismic uplift and thrust faulting in the accretionary wedge of the overriding plate (Plafker, 1972). If the Cascadia subduction zone produces earthquakes as large as M 9, long stretches of the Pacific Northwest coast may have repeatedly subsided as much as 2.5 m. The rise of regional sea level on this coastline during the late Holocene (e.g., Clark and others, 1978; Peterson and others, 1984, fig. 6) should promote preservation of a stratigraphic record of sea-level changes. Thus, if meter-scale coseismic-subsidence events occurred repeatedly, evidence of this deformation should be widespread along the Pacific Northwest coast.

Paleoseismic studies of tidal-wetland stratigraphy offer direct evidence that great Holocene earthquakes probably occurred in the Pacific Northwest. The submergence and burial of peaty marsh and forest soils in muddy estuarine sequences in southwest Washington (Atwater, 1987; Atwater and Yamaguchi, 1991) and northern Oregon (Grant, 1989; Darienzo and Peterson, 1990) seems too widespread >100 km long), too large (>1 m submergence), and too sudden (<10 yr) to be attributed to any process other than coseismic subsidence. Similarities in the stratigraphic sequences in southern Washington and northern Oregon suggest that the sudden submergence of wetlands was caused by coseismic subsidence of regional extent (>100 km) during great thrust earthquakes on the Cascadia plate interface (Darienzo and Peterson, 1990). If so, RSL in northern Oregon and southern Washington during the late Holocene rose in a series of jerks due to coseismic subsidence of parts of the coast hundreds of kilometers long. Thick late Holocene sequences beneath tidal marshes confirm that net interseismic uplift of the coast was slower than the rate of late Holocene RSL rise (Atwater, 1987; Darienzo and Peterson, 1990).

The south-central Oregon coast (Fig. 1B) lies 70 to 90 km east of the inferred seafloor trace of the Cascadia plate-interface thrust (Fig. 1A). Given its distance arcward of the trace of the thrust, this study area could encompass (1) the western parts of zones of coseismic subsidence during great

Quaternary Coasts of the United States: Marine and Lacustrine Systems, SEPM Special Publication No. 48

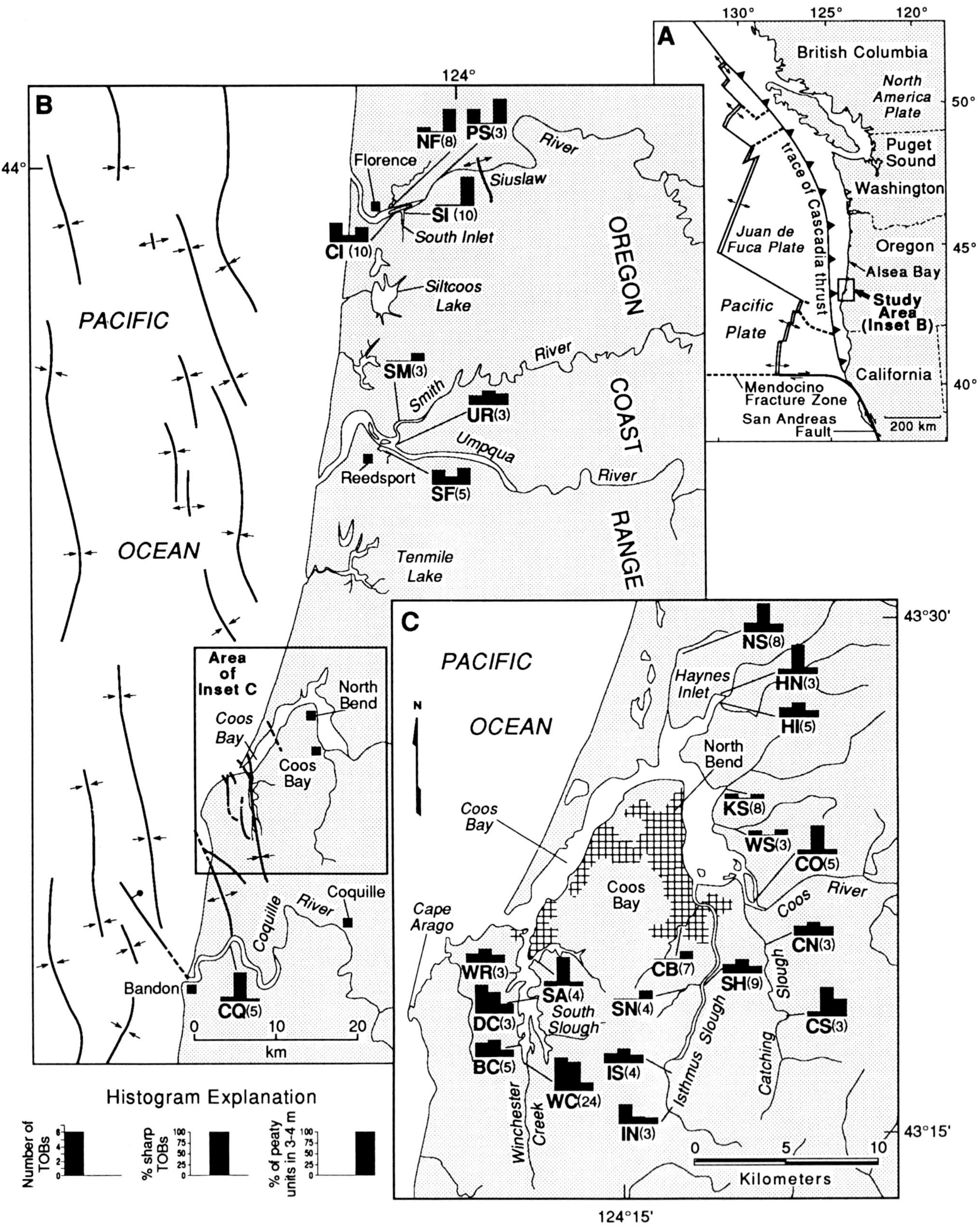
A
130°
125°
120°
British Columbia
North America Plate
50°
Puget Sound
Washington
trace of Cascadia thrust
Juan de Fuca Plate
Oregon
45°
Alsea Bay
Study Area (Inset B)
Pacific Plate
California
40°
Mendocino Fracture Zone
San Andreas Fault
200 km
B
124°
44°
NF(8)
PS(3)
River
Florence
Siuslaw
SI (10)
South Inlet
CI (10)
Siltcoos Lake
OREGON
PACIFIC
SM(3)
River
Smith
UR(3)
COAST
Umpqua
Reedsport
SF(5)
River
OCEAN
Tenmile Lake
RANGE
Area of Inset C
North Bend
Coos Bay
Coos Bay
Coquille
River
Coquille
Bandon
CQ(5)
0
10
20
km
C
43°30'
PACIFIC
OCEAN
N
NS(8)
Haynes Inlet
HN(3)
HI(5)
North Bend
Coos Bay
KS(8)
WS(3)
CO(5)
Coos Bay
Coos
River
Cape Arago
CN(3)
WR(3)
CB(7)
SH(9)
SA(4)
South Slough
SN(4)
Slough
DC(3)
CS(3)
BC(5)
WC(24)
IS(4)
Isthmus Slough
Catching
Winchester Creek
IN(3)
0
5
10
43°15'
Kilometers
124°15'
Histogram Explanation
Number of TOBs
% sharp TOBs
% of peaty units in 3-4 m

plate-interface earthquakes (Atwater, 1988), and/or (2) zones of regional uplift and localized secondary folding and faulting in the accretionary wedge of the overriding plate (McInelly and Kelsey, 1990; Nelson and Personius, 1991; Peterson and Darienzo, 1991). No Holocene, coseismic, surface deformation has been thoroughly documented in the northern part of the study area, but Peterson and Darienzo (1991) inferred as many as 10 coseismic-subsidence events of <1 m (two accompanied by local tsunamis) from the stratigraphic record in the marshes of the Alsea Bay estuary (Fig. 1A). Nelson and Personius (1991) recognized only one active Quaternary structure (east of site PS on the Siuslaw River, Fig. 1B) and no net Holocene deformation of river terraces along the lower Siuslaw and Umpqua Rivers.

In the southern half of the study area, the coast swings west toward the trace of the Cascadia thrust; trends of structures that deform the continental shelf (Fig. 1B; Clarke and others, 1985; Kelsey, 1990, fig. 13) show that this area includes the distal part of the active accretionary wedge of the North America plate. Pleistocene marine terraces in the Coos Bay—Coquille River area also record late Quaternary deformation on northwest-trending folds and flexure-slip and high-angle faults due to crustal shortening in the overriding North America plate (McInelly and Kelsey, 1990; Muhs and others, 1990; Kelsey, 1990). Analogies with deformation during great earthquakes in Chile, Japan, and Alaska indicate that regional uplift of as much as several meters and/or differential movements across folds and faults would be expected in the accretionary wedge during great plate-interface earthquakes (e.g., Plafker, 1969; 1972; Matsuda and others, 1978; Lajoie, 1986; Vita-Finzi, 1986; Berryman and others, 1989). In southern Oregon, late Holocene coseismic deformation has been inferred on an anticline near Cape Blanco, 30 km south of Bandon (Fig. 1B), and on shallow faults and folds in the South Slough area of western Coos Bay (Fig. 1C; Nelson, 1987; Peterson and Darienzo, 1989; McInelly and Kelsey, 1990; Kelsey, 1990). Surface deformation of this type might be reflected as submergence or emergence events of 0.5 to 2 m in the marsh stratigraphic record. Many localized deformation events are probably coincident with plate-interface earthquakes (McInelly and Kelsey, 1990), but shallow earthquakes of moderate magnitude (M 6–7) on local structures (<30 km long) in the overriding plate might also produce local areas of coseismic subsidence or uplift (e.g., Yeats and others, 1981; Lajoie, 1986; Berryman and others, 1989).

A reconnaissance of the late Holocene (<4 ka) stratigraphy at 26 tidal-marsh sites among four estuaries in south-central Oregon (Figs. 1B through 6) was made to determine if a record of sudden, meter-scale submergence events, similar to those in northern Oregon and southern Washington, was widely preserved in estuaries of the southern Cascadia subduction zone. Subtidal sediments (Peterson and others, 1984), marsh vegetation (Jefferson, 1975; Frenkel and Boss, 1988) and modern intertidal microfossil assemblages (Jennings and Nelson, 1992) have been studied in some of the four estuaries, but this is the first stratigraphic study of Holocene intertidal sediments in estuaries south of Alsea Bay. This paper shows how tidal-marsh lithofacies sequences differ along the south-central Oregon coast and argues that these differences are more easily explained by non-tectonic processes or localized coseismic deformation than by zones of regional deformation produced during great plate-interface earthquakes. Characteristics of tidal-marsh sequences are used to infer RSL changes.

EVIDENCE OF RELATIVE SEA-LEVEL CHANGE IN LITHOFACIES SEQUENCES

Tidal-Marsh Lithofacies—Character and Relation to Marsh Elevational Zones

The stratigraphy of the tidal marshes of south-central Oregon estuaries are shown in Figures 2–6 with lithofacies codes similar to those used in fluvial (Miall, 1977) and glacial sedimentology (Eyles and others, 1983). The codes used here incorporate some of the features of the widely used Troels-Smith (1955) system for describing organic-rich sediments (e.g., Birks and Birks, 1980; Shennan, 1986a). The rooted and detrital components of herbaceous and woody plants in peaty lithofacies can be distinguished using the codes, and interpretive modifiers can indicate an inferred origin or depositional environment. The codes and symbols allow abrupt to diffuse lithofacies changes downcore to be concisely indicated in more detail than with simple graph-

FIG. 1.—(A) Major features of the Cascadia subduction zone in the northwestern United States and southwestern Canada, modified from Rogers (1988), Spence (1989), and Wilson (1989). (B) Location of tidal-marsh sites in the Siuslaw River, Umpqua River, and Coquille River estuaries in south-central Oregon in the southern part of the Cascadia zone. (C) Location of tidal-marsh sites in the Coos Bay estuary. Areas of insets (B) and (C) shown in (A) and (B), respectively. (A) The trace of the Cascadia thrust fault (barbed line, barbs point downdip) is placed at the bathymetric boundary between the continental slope and abyssal plain; double lines are spreading ridges, solid lines are strike-slip faults; dashed lines are other faults or political boundaries. (B) Anticlines, synclines, and faults that produce relief on the sea floor or that deform Pleistocene marine or fluvial terraces (modified from Clarke and others, 1985; McInelly and Kelsey, 1990, fig. 13; Nelson and Personius, 1991, fig. 7). Cross-ruled areas in (C) show developed urban areas. For tidal-marsh sites in (B) and (C), histograms summarize characteristics of lithofacies sequences in outcrops or cores along transects perpendicular to the upland-marsh border; characteristics are: (1) number of transgressive overlap boundaries (TOBs), (2) percentage of these TOBs that are abrupt or sharp (<3-mm boundary thickness; see Fig. 2), and (3) percentage of peaty units (lithofacies P, Pg, Pm, Mp, or Mg, see Fig. 2) in upper 3 to 4 m of sediment at each site. Values shown for most sites are averages for several cores <30 m from the upland; number of cores at each site shown in parentheses after site name. Site names from north to south: PS, Palouse Slough; NF, North Fork of the Siuslaw River, SI, South Inlet; CI, Cox Island; SM, Smith River; UR, Umpqua River; SF, Schofield Slough; NS, North Slough; HN, northern Haynes Inlet; HI, Haynes Inlet; KS, Kentuck Slough; WS, Willanch Slough; CO, Coos River; CN, northern Catching Slough; CS, southern Catching Slough; SH, mouth of Shinglehouse Slough; SN, upper part of Shinglehouse Slough; CB, Coalbank Slough; IS, Ithmus Slough; upper part of Ithmus Slough; WR, Warnock Creek; SA, Shana Creek; DC, Day Creek (unpublished data of C.D. Peterson and M.E. Darienzo); BC, Block Creek; WC, Winchester Creek

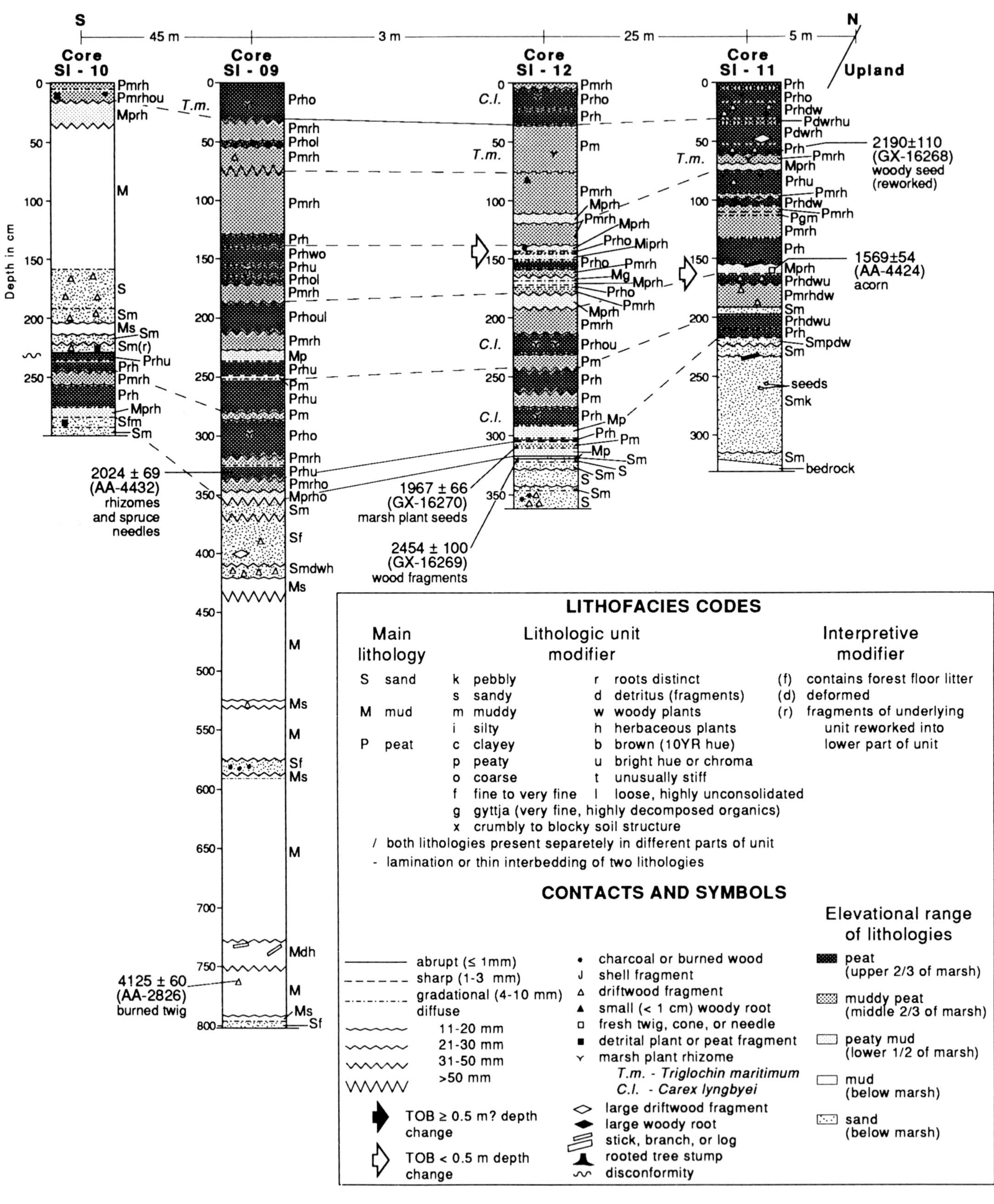
S
N
45 m
3 m
25 m
5 m
Upland
Core SI - 10
Core SI - 09
Core SI - 12
Core SI - 11
Depth in cm
T.m.
C.l.
2024 ± 69
(AA-4432)
rhizomes
and spruce
needles
1967 ± 66
(GX-16270)
marsh plant seeds
2454 ± 100
(GX-16269)
wood fragments
4125 ± 60
(AA-2826)
burned twig
2190±110
(GX-16268)
woody seed
(reworked)
1569±54
(AA-4424)
acorn
seeds
bedrock
LITHOFACIES CODES
Main lithology
S sand
M mud
P peat
Lithologic unit modifier
k pebbly
s sandy
m muddy
i silty
c clayey
p peaty
o coarse
f fine to very fine
g gyttja (very fine, highly decomposed organics)
x crumbly to blocky soil structure
r roots distinct
d detritus (fragments)
w woody plants
h herbaceous plants
b brown (10YR hue)
u bright hue or chroma
t unusually stiff
l loose, highly unconsolidated
/ both lithologies present separetely in different parts of unit
- lamination or thin interbedding of two lithologies
Interpretive modifier
(f) contains forest floor litter
(d) deformed
(r) fragments of underlying unit reworked into lower part of unit
CONTACTS AND SYMBOLS
abrupt (≤ 1mm)
sharp (1-3 mm)
gradational (4-10 mm)
diffuse
11-20 mm
21-30 mm
31-50 mm
>50 mm
TOB ≥ 0.5 m? depth change
TOB < 0.5 m depth change
charcoal or burned wood
shell fragment
driftwood fragment
small (< 1 cm) woody root
fresh twig, cone, or needle
detrital plant or peat fragment
marsh plant rhizome
T.m. - Triglochin maritimum
C.l. - Carex lyngbyei
large driftwood fragment
large woody root
stick, branch, or log
rooted tree stump
disconformity
Elevational range of lithologies
peat (upper 2/3 of marsh)
muddy peat (middle 2/3 of marsh)
peaty mud (lower 1/2 of marsh)
mud (below marsh)
sand (below marsh)

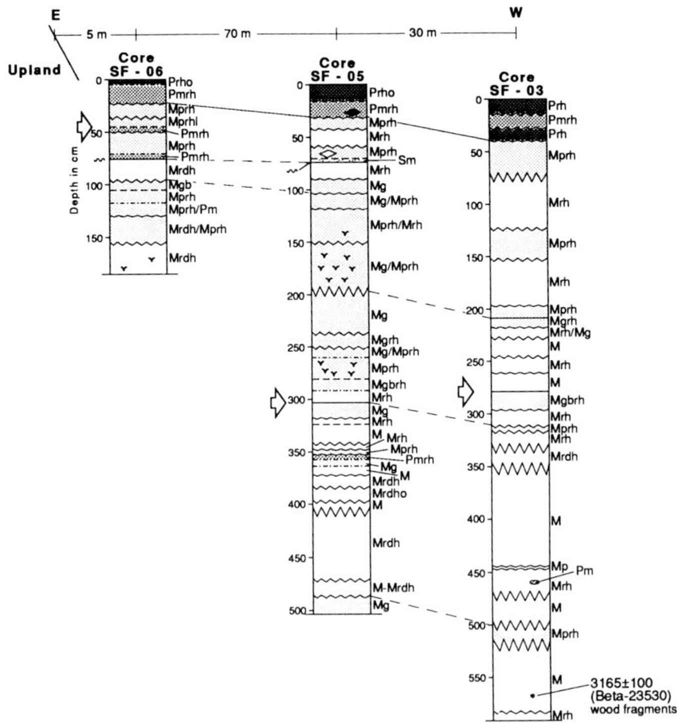

FIG. 3.—Lithofacies sequences, an AMS ^{14}C age, and correlations among cores along part of a transect adjacent to a steep, forested slope in Schofield Slough (a small, freshwater-dominated tributary of the Umpqua River). See Figure 2 caption for explanation of symbols, lithofacies codes, and other diagram features.

ical systems (e.g., Belknap and others, 1989; van de Plassche, 1991). An advantage of letter codes over the graphical Troels-Smith system (e.g., Shennan, 1986a) is that the lithology and environmental interpretation of thin (<10 cm) but highly significant units can be portrayed in the same detail as those of the much thicker units.

Most lithofacies beneath Oregon tidal marshes were deposited in mudflat, low marsh, or high-marsh intertidal environments; only a few units are subtidal sediments or soils formed on steep slopes in the forested upland adjacent to the high marsh. Modern low-marsh and high-marsh environments each occupy about half the 1.2- to 1.5-m vertical range of the marsh, although the composition of plant communities and the degree of zonation vary greatly from marsh to marsh. Dominant members of low-marsh plant communities include *Carex lyngbyei, Triglochin maritimum, Salicornia virginica*, and *Distichlis spicata*. High-marsh communities are more diverse with lower percentages of these same species in addition to *Deschampsia cespitosa, Agrostis alba, Potentilla pacifica* and *Atriplex patula* (Jefferson, 1975; Frenkel and others, 1981; Frenkel and Boss, 1988; Jennings and Nelson, 1992). Mud flats below the edge of the low marsh occasionally bear patches of algae, and *Zostera marina* is common on the floors of tidal channels.

Each intertidal environment has an indicative meaning, i.e., a vertical relation to a reference water level (van de Plassche, 1982; Shennan, 1986b; Chappell, 1987). As in most temperate regions, high-marsh deposits accumulate at or just below mean high water (MHW; Frey and Basan, 1985); therefore, MHW is the reference water level in most marsh stratigraphic studies. In south-central Oregon, estuaries have mean tidal ranges of about 1.6 m. In the middle reaches of the four estuaries studied (Fig. 1B), MHW is about 0.9 m above mean tide level (MTL). The high marsh typically extends from about 0.6 m above to 0.2 m below MHW; the low marsh commonly extends the 0.9 m from near MHW to MTL, and mud flats are usually below MTL (Jennings and Nelson, 1992). These approximate elevational ranges indicate uncertainties of at least 0.5 to 0.8 m for estimates of the former position of MHW (e.g., Frenkel and others, 1981, Table 3) based on lithofacies characteristic of low- and high-marsh environments. Thus, changes in RSL as large as 0.9 m might not produce any distinctive change in intertidal lithofacies, even in the marsh environment. RSL changes of <0.5 m are probably difficult to identify in many lithofacies sequences, although changes as small as 0.1 m may be reflected by lithofacies changes at some marsh sites (e.g., van de Plassche, 1991).

The texture (grain-size distribution), structure, and organic content of lithofacies differ in mudflat, marsh, and upland environments, but these sediment properties do not always distinguish low-marsh and high-marsh environments. Upland forest soils consist of peat or muddy peat with a crumbly subaerial soil structure, typically capped by a layer of forest-floor litter (needles, leaves, cones). Peat with abundant herbaceous roots and without a significant mud component usually indicates the upper two-thirds of the marsh environment (e.g., facies Prho and Prhu, Figs. 2 and 3). "Peat," as used here and elsewhere (e.g., Peterson and Darienzo, 1991), refers to lithofacies that appear to consist of at least half organic material, or for sediment whose structure is dominated by coarse rootlets or fibers;

FIG. 2.—Lithofacies sequences, ^{14}C ages, and probable (solid lines) and possible (dashed lines) correlations of boundaries among cores along north half of transect across mouth of small inlet near mouth of South Inlet in Siuslaw River estuary. Distances between adjacent cores and forested upland shown in meters. Lithofacies codes based on codes of Miall (1977) and Eyles and others (1983), following some features of the descriptive system of Troels-Smith (1955). Observations of modern intertidal zone ranges were used to estimate relative elevation ranges within intertidal zones for common lithologies (shown by shading). (In typical marshes in central Oregon the marsh zone extends from near MTL to 0.0–0.6 m above MHW.) Changes between lithofacies typical of adjacent elevational zones, and changes between non-adjacent zones suggest two types of abrupt transgressive overlap boundaries (TOBs). Abrupt TOBs between non-adjacent lithofacies that may have been produced by 0.5 m or more of sudden submergence shown by filled arrows; unfilled arrows mark abrupt TOBs between lithofacies typical of adjacent zones that probably formed by submergence of $\leq$ 0.5 m. Radiocarbon laboratory numbers and type of sample material are shown for each accelerator mass-spectrometer (AMS) ^{14}C age (age in ^{14}C yr BP as reported by ^{14}C laboratory). Complete ^{14}C data available from author.

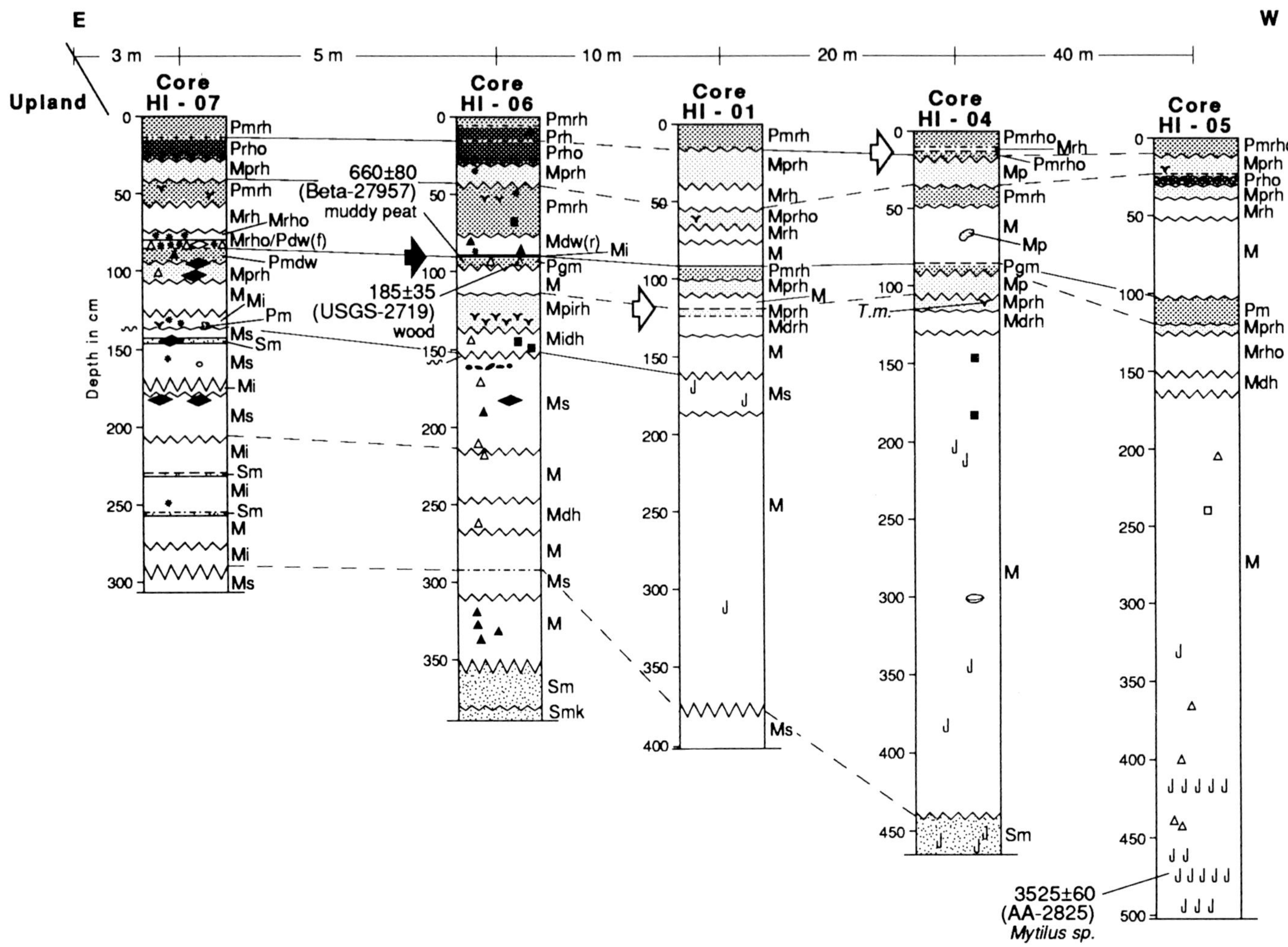

FIG. 4.—Lithofacies sequences, AMS and conventional ¹⁴C ages, and correlations among cores along transect across marsh on south side of Haynes Inlet, the northeastern arm of Coos Bay. See Figure 2 caption for explanation of symbols, lithofacies codes, and other diagram features. Conventional radiocarbon ages (Beta and USGS) come from samples retrieved with an 8-cm-diameter sealed bucket auger. Age of wood fragments within peat is much younger than age on cleaned peat (rootlets and wood fragments removed). Wood fragments may have originated as tree roots or branches from fallen trees injected into the older peat.

some lithofacies of decomposed peat may contain as little as 15 percent organic matter. High-marsh peat deposited adjacent to steep, heavily forested uplands commonly contains woody roots, detrital wood, and litter from nearby overhanging trees (e.g., Prhwdw(f), Figs. 4 and 6). Peat deposited in the middle and lower elevational ranges of the marsh is commonly muddier, has fewer roots and less fibrous structure, and contains less detrital wood (e.g., Pm, Pmrh, Pmrho; Fig. 5) than does peat from the upper parts of the marsh. Peaty mud (Mp) and rooted mud (Mrh) are the most typical lithofacies of the low marsh. Most rootless mud (M) is inferred to have accumulated on intertidal mudflats (Fig. 3). Microfossils and growth-position plant macrofossils, especially the below-ground stems of *Carex lyngbeyi* or *Triglochin maritimum*, offer a more reliable means of distinguishing low-marsh and high-marsh lithofacies than does sediment lithology (e.g., Niering and Warren, 1977; Jennings and Nelson, 1992).

Tendencies of Sea-Level Movement Indicated by Lithofacies Changes

A lithofacies change in a vertical sequence usually indicates a shift to a lower or higher intertidal environment. An upward shift to a lithofacies characteristic of a lower part of the intertidal zone usually reflects a positive tendency of sea-level movement (Streif, 1979; Shennan, 1986b; van de Plassche, 1991); a transgressive overlap boundary (TOB) between two lithofacies marks a positive tendency. Regressive overlap boundaries (ROBs) indicate a negative tendency and a shift to a higher intertidal environment, such as from low marsh to high marsh. TOBs and ROBs do not necessarily indicate sea-level changes because, for example, an ROB may result from rapid aggradation during a period of RSL rise (Shennan, 1982; Davis and Clifton, 1987). In mid-estuary environments, many TOBs and ROBs reflect local changes in sedimentation rates due to the many

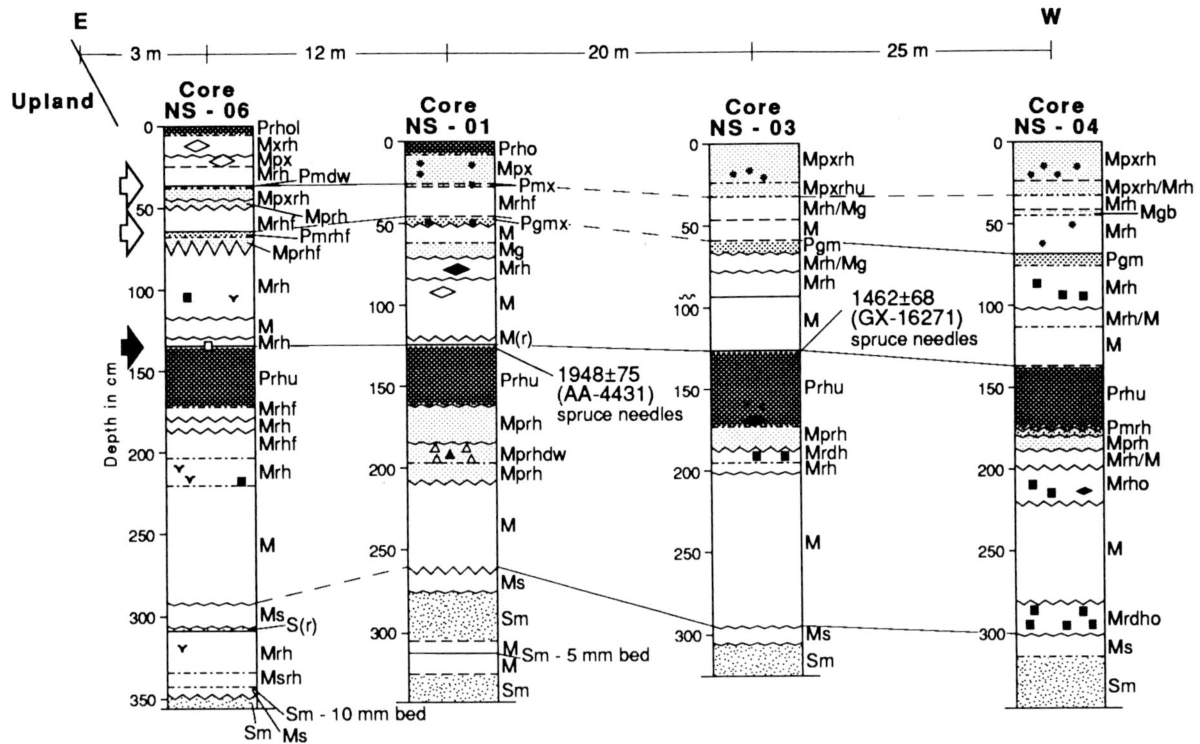

FIG. 5.—Lithofacies sequences, AMS ^{14}C ages, and correlations among cores along transect in inlet on east edge of North Slough, the northernmost arm of Coos Bay estuary. See Figure 2 caption for explanation of symbols, lithofacies codes, and other diagram features. Crumbly soil structure in lithofacies Mpx and Pmx in the upper 0.5 m is probably due to soil oxidation and consolidation during periods when site was completely diked.

highly interrelated factors involved in marsh development, erosion, and/or burial (e.g., Redfield, 1972; Frey and Basan, 1985; Clark and Patterson, 1985; Finkelstein and Hardaway, 1988; Fletcher and others, 1990; Allen, 1990; Peterson and Darienzo, 1991). Nevertheless, if individual TOBs or ROBs can be correlated for tens to hundreds of meters along transects of cores at a site, they can be used as indicators of changes in the rate of RSL rise relative to the local rate of sedimentation or marsh accretion (e.g., Rampino and Sanders, 1981; Streif, 1987; Beale, 1990; van de Plassche, 1991). Only where radiocarbon dating shows many overlap boundaries at widely spaced sites to be synchronous can regionally significant positive and negative tendencies of RSL movement be identified (e.g., Shennan, 1986b; 1987).

Coseismic Significance of Abrupt versus Gradual Overlaps

The sharpness of overlap boundaries is particularly important in paleoseismic studies because subsidence or uplift during large earthquakes should produce abrupt overlap boundaries (e.g., Ovenshine and others, 1976; Atwater, 1987; Darienzo and Peterson, 1990; Plafker, 1990; Nelson and Personius, 1991).

Gradual overlap boundaries (>4 mm, gradational to diffuse on Fig. 2) predominate in marsh sequences at sites that are protected from wave erosion. Few published studies of marsh stratigraphy (most are from passive continental margins) discuss the thickness or character of overlap boundaries, which implies that most boundaries are gradual transitions between lithofacies. Most of the local, regional, and global factors usually invoked to explain overlap boundaries, especially those of more than site significance (e.g., van de Plassche, 1982; 1991; Shennan, 1986b; 1987; Streif, 1987; Fairbridge, 1987; Peltier, 1987; Chappell, 1987), involve processes that operate over decades to hundreds of years. These gradual processes imply gradual facies changes at overlap boundaries in protected estuarine settings (e.g., Behre, 1986, fig. 3). Gradual overlap boundaries could also be produced by slow subsidence or uplift during interseismic accumulation of tectonic strain (Darienzo and Peterson, 1990). However, to the author's knowledge, there is no way along a convergent-margin coast to distinguish overlap boundaries resulting from these gradual, tectonically induced changes in regional uplift rate from those produced by the non-tectonic factors responsible for similar overlap boundaries on the coasts of passive continental margins. Gradual overlap boundaries should be attributed to tectonic processes only at sites where rates of interseismic vertical movement are too high to be the result of other processes that cause sea level to rise or fall (e.g., Newman and others, 1987; Plafker, 1990; Taylor and others, 1990).

Abrupt overlap boundaries (<3 mm; includes sharp to abrupt boundaries of Fig. 2) that are widely found in protected tidal inlets imply sudden events or erosion. In most sea-level studies, abrupt TOBs in protected settings are usually attributed to tidal-channel cutting (e.g., Devoy, 1979,

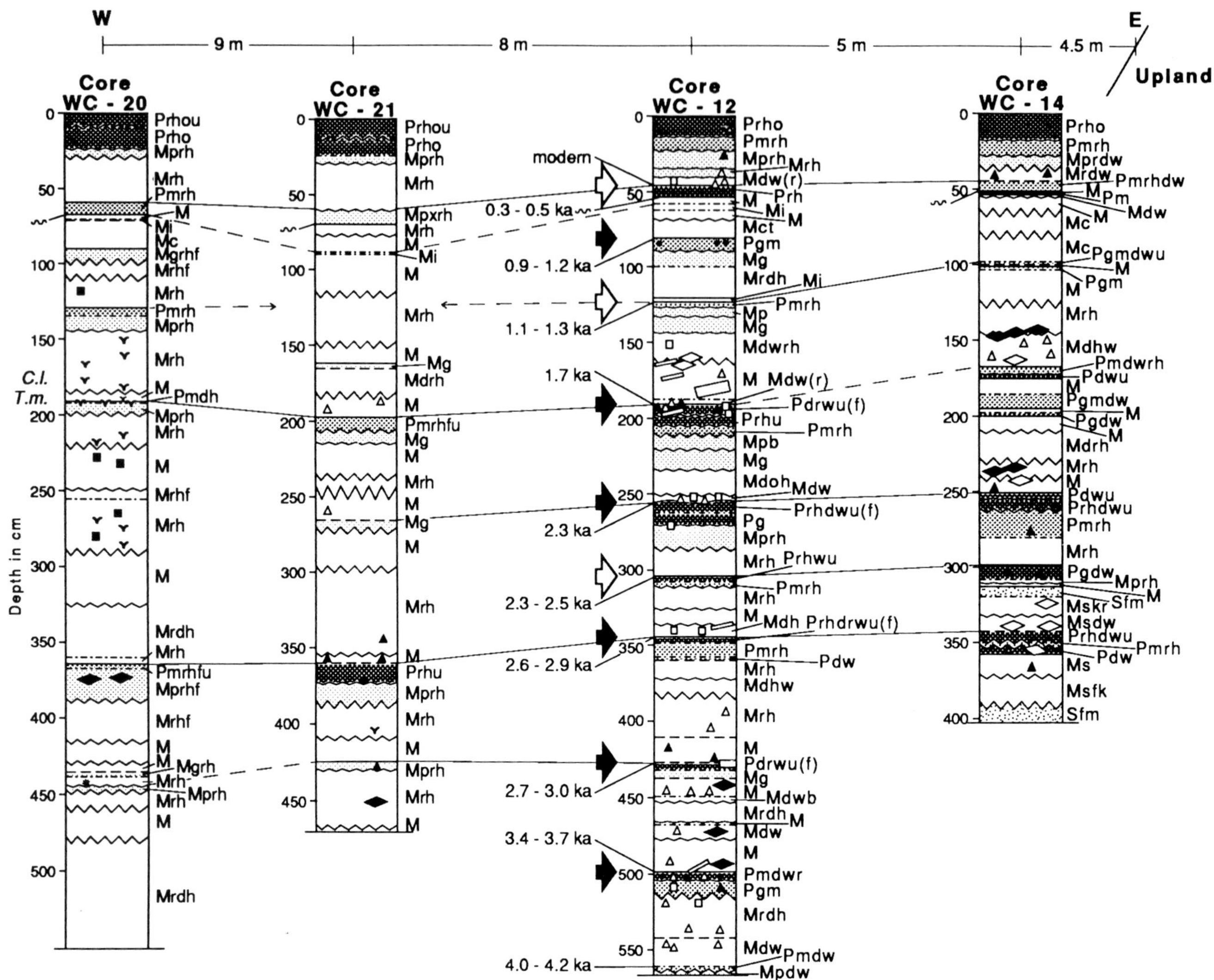

FIG. 6.—Lithofacies sequences, age estimates (in ^{14}C yrs × 1000) and correlations among four of 12 cores along transect across mouth of small inlet adjacent to Winchester Creek (at south end of South Slough arm of Coos Bay). See Figure 2 caption for explanation of symbols, lithofacies codes, and other diagram features. Age estimates for nine abrupt TOBs and one disconformity shown at left of core WC-12 are based on 15 conventional ^{14}C ages and 14 AMS ^{14}C ages from that core. Unpublished diatom- and foraminiferal-assemblage data from core WC-12 suggest that most abrupt TOBs (filled arrows) represent about 0.5 to 1.0 m of rapid submergence.

p. 361; Shennan, 1986b). Abrupt ROBs have been attributed to deposition of floating blocks of peat (Shennan, 1986a), or to slumping of peat blocks into tidal channels (Belknap and others, 1989; Peterson and Darienzo, 1991). Such processes produce abrupt boundaries of limited extent (commonly <10 m) and can be eliminated as potential causes if abrupt boundaries can be traced for many tens of meters across the marsh (Atwater, 1987; Darienzo and Peterson, 1990). Most short-term events that produce estuary-scale submergence, such as river floods, storm tides, and changes in ocean circulation patterns (e.g., Komar and Enfield, 1987), do not last long enough to produce abrupt, extensive TOBs in marsh environments (Peterson and Darienzo, 1991).

Changes in the volume of the tidal prism due to migrating bars or channels are more difficult to dismiss as the cause of abrupt changes in marsh sedimentation (van de Plassche, 1991). Bars can be breached suddenly during floods or storms, and migration of barrier bars at the mouths of major inlets may affect tide levels in large portions of an estuary. For example, the height of maximum high tides in the Alsea Bay estuary (Fig. 1A) increased by as much as 0.7 m due to bay-mouth bar migration induced by the 1983 El Niño Southern Oscillation (Jackson and Rosenfeld, 1987). In this case, however, the higher tides probably raised the elevation of MHW by no more than a few decimeters, and Peterson and Darienzo (1991) found no evidence of marsh burial in the seven years following the 1983 El Niño.

Coseismic subsidence or uplift of intertidal sediments can result in abrupt TOBs and ROBs of on-site scale or regional extent (e.g., Lawson, 1908, p. 82, 85; Wright and Mella, 1963, p. 1368; Ovenshine and others, 1976; Plafker, 1969; 1990). However, the 0.5 to 0.9-m elevational ranges of intertidal environments in south-central Oregon indicate that vertical Earth movements need to be >0.5 m to produce overlap boundaries that can be traced for many tens of meters. Where delicate fossils (such as the rooted bases of marsh-plant leaves abruptly buried by intertidal mud) indicate TOBs were produced in less than a few years, evidence for a coseismic-subsidence origin is particularly strong (Atwater and Yamaguchi, 1991). Abrupt ROBs produced by emergence may be difficult to distinguish from local erosional disconformities. An erosional origin is unlikely, however, if an abrupt, horizontal ROB extends at least tens of meters through intertidal deposits. Evidence of soil formation should be common in sediment beneath an ROB formed in response to significant emergence from the intertidal zone (Streif, 1987; van de Plassche, 1990).

Inferring the Style of Submergence from Lithofacies Sequences

Styles of late Holocene submergence in south-central Oregon can be inferred from lithofacies sequences in tidal-marsh deposits. The relative thickness of peat and mud lithofacies and the number and character of TOBs and ROBs in marsh sequences suggest differing RSL histories along the south-central coast. Figures 1B and 1C show average values of three of these parameters for cores within 5 to 30 m of the upland at most sites. In many temperate tidal marshes, peat accumulation has kept pace with the slow rise of sea level during the late Holocene (e.g., Redfield, 1972; Rampino and Sanders, 1981; Atwater and Belknap, 1980; Reed, 1988; van de Plassche and others, 1989; Beale, 1990). Thus, thick peat and muddy peat lithofacies suggest significant periods of little or no change in the slow rate of RSL rise (Streif, 1987; van de Plassche, 1991). Rapid rises, if they occurred while peaty units were being deposited, were probably small ($<$0.5 m). Sequences of buried marsh soils that have upper TOBs or lower ROBs bounding their peaty surface horizons (A or O horizons) suggest multiple, rapid RSL changes. Where local processes, such as channel cutting or preservation of slump blocks, can be ruled out, abrupt TOBs and ROBs suggest sudden submergence or sudden emergence, respectively. The lack of evidence of upland-soil development in intertidal sequences (except at diked sites; e.g., Fig. 4) suggests that significant ($>$0.5 m) falls in RSL were not common. Lithofacies in subtidal and lower intertidal environments vary too little with water depth to record late Holocene sea-level fluctuations.

LATE HOLOCENE TIDAL-MARSH STRATIGRAPHY IN SOUTH-CENTRAL OREGON

Field Methods

A 1-m-long, 2.5-cm-diameter, half-cylinder gouge corer was used to take reconnaissance cores in the marshes of the four estuaries. Core transects at each site (Figs. 1B and 1C) consisted of 3 to 12 cores to depths of 2 to 7 m, spaced 3 to 40 m apart. At most sites, relative elevations between cores were determined with an Abney hand level and rod. The zonation of plant communities in central Oregon marshes (Frenkel and others, 1981; Jennings and Nelson, 1992) and the plant assemblages at most core sites suggest that the ends of core transects nearest the upland in Figures 2 through 6 are within ±0.2 m of MHW. Core segments were split and described in the field using standard methods (e.g., Belknap and others, 1989; van de Plassche, 1991), and peat, shell, charcoal, and plant macrofossil samples were collected for conventional and accelerator mass-spectrometer (AMS)^{14}C analysis (Figs. 2–6).

The easiest sites to core and interpret were in the middle reaches of the four estuaries. Extensive coastal-dune fields cover much of the coast in south-central Oregon (Cooper, 1958) and sandy sediment, which is difficult to penetrate with a hand corer, is predominant in the marshes in the seaward parts of the estuaries (e.g., Peterson and Darienzo, 1991). Because estuarine sand can be deposited at intertidal or subtidal depths by fluvial or tidal currents at widely varying rates, the depositional environment of sandy deposits is also more difficult to interpret in small-diameter cores than is that of the interbedded peat and mud sequences typical of the middle parts of the estuaries. Likewise, genetic interpretation of TOBs and ROBs near the freshwater-dominated heads of the estuaries is less certain than interpretation of TOBs and ROBs in the middle reaches, because brackish-water peat and mud are commonly difficult to distinguish from freshwater peat and fluvially deposited mud. Most of the cores were taken in undiked marshes in order to avoid the desiccated, compacted deposits typical of diked areas.

Siuslaw River Estuary

Many marshes in the middle reaches of the Siuslaw River estuary are underlain by 2 to 4 m of tidal-marsh peat, as shown by transects of cores at four sites in this part of the estuary. In the North Fork of the Siuslaw drainage, about 1.3 km from its confluence with the main Siuslaw River (site NF, Fig. 1B), the thickness of peat and peaty mud in the upper 4 m of cores in a transect 300 m long varies from 2 m to 4.3 m. Similarly, on Cox Island in the main river (site CI, Fig. 1B), peat is the dominant lithology in the upper 4 m of 10 cores along a north-south transect. Boundaries between lithofacies in cores on both transects are mostly gradational or diffuse. The few abrupt TOBs and ROBs found in some cores near tidal channels could not be correlated; the abrupt overlap boundaries were probably cut by laterally migrating channels.

The most detailed transect in the Siuslaw estuary (Fig. 2) crosses the mouth of a small inlet on the east side of South Inlet, a 2-km-long, north-south-trending arm of the estuary on the south side of the Siuslaw River (site SI, Fig. 1B). Three AMS ^{14}C ages from cores SI-09 and SI-12 show that peat began forming at this site about 2.5 to 2.0 ka. The peaty lithofacies show reddish-brown color (7.5YR hues), high organic content, coarse fibrous texture, and occasional identifiable rhizomes and stems with attached leaf bases of *Triglochin maritimum* and cf. *Carex lyngbeyi*. These properties imply that marsh vegetation has continuously occupied the fringes of the inlet since 2.0 ka. The boundaries between units are mostly gradational to diffuse; exceptions are the contacts of sand beds deposited in channels near the center of the inlet (core SI-10), or on the sides of the inlet, perhaps during large storm tides. Small fluctuations in the rate of submergence may have occurred during deposition of the peaty units, but the changes appear to have been slow (gradual boundaries) and of smaller amplitude than the elevation range of high-marsh environments in this inlet (<0.7 m; Jennings and Nelson, 1992, fig. 4A). Many of the changes (e.g., from peat to muddy peat) could have been produced by changes in local sedimentation rates due to (1) small changes in the position of tidal channels, (2) biologically induced changes in the density or species composition of marsh-plant communities, or (3) changes in tide levels of <30 cm during periods of a few decades. No abrupt TOBs were found that suggest sudden-submergence events of >1 m, comparable to those inferred by Atwater (1987) and Darienzo and Peterson (1990) from marshes in northern Oregon and southern Washington.

The upper 4 m of cores from the upper parts of the Siuslaw estuary (e.g., site PS, Fig. 1B) are dominated by peat and peaty mud, interbedded with thin beds of mud. Gradual boundaries between most lithofacies in the interbedded sequences suggest slow environmental changes during the last few thousand years in the upper part of the estuary. Continuous peat deposition at more seaward sites precludes sudden, large (>0.5 m), regional changes in RSL as the cause of these interbedded-lithofacies sequences. Perhaps the interbedded sequences reflect local changes in sedimentation rates resulting from variations in river-channel positions or river-flood frequencies combined with small changes in tide levels (e.g., Nelson and Personius, 1991, fig. 4).

Umpqua River Estuary

Cores from the marshes of the Umpqua River estuary show a lower percentage of peaty units in the upper 4 m of sediment than do cores from the Siuslaw estuary. At the Umpqua River site (UR, Fig. 1B) the two abrupt TOBs within the peaty upper 2 m of the section were probably cut by migrating tidal channels, perhaps during a river flood. Peaty units in a small marsh fringing the Smith River (site SM, Fig. 1B) are restricted to the upper 1.2 m, are <0.2 m thick, and have gradational boundaries. West of the confluence of the Umpqua with the Smith River, most marsh sediment is too sandy to core easily below 2 m depth; peaty units are restricted to the upper 1 m of cores.

Most of the boundaries in the cores from Schofield Slough (site SF, Fig. 1B), 5 km up a small tributary drainage (Schofield Slough) from the Umpqua River, are gradual; only one abrupt TOB is preserved in each of the cores (Fig. 3). The stratigraphy at site SF is probably representative of the freshwater-dominated upper parts of the estuary. Vegetation in this marsh consists almost entirely of the brackish to freshwater bulrush *Scirpus acutus*. Peaty lithofacies beneath this marsh (Pm, Mp, and Mg, Fig. 3) have a softer, soupier consistency than peaty units in more saline marshes, probably due to the framework created by coarse, intertwined, easily decomposed roots and rhizomes of *S. acutus*. As in most marshes fringing estuaries in Oregon, peaty lithofacies thin away from the upland (e.g., Darienzo and Peterson, 1990). A ^{14}C age from a fragment of burned wood at 5.6 m depth in core SF-03 is younger than ages from samples at similar depths at most other sites. The young age indicates significant compaction of the soft, peaty lithofacies. The percentage of peaty lithofacies in cores from site SF suggests that marshes persisted at the site for thousands of years.

Evidence from the limited number of Umpqua River marsh cores (16) does not preclude sudden rises and falls of RSL during the past 3 ka, but gradual boundaries suggest most changes were slow. Furthermore, with the possible exception of the abrupt ROB at 75 cm depth in cores SF-06 and SF-05 (marked by a disconformity symbol on Fig. 3), lithofacies changes are not consistent with large (>1 m) changes in tide levels. As in the Siuslaw estuary sequences, lithofacies sequences from the marshes of the middle reaches of the Umpqua estuary suggest a slow rise of RSL for most of the past 2 to 3 ka.

Eastern Part of Coos Bay Estuary

The stratigraphy of cores from 13 sites in fringing marshes in the middle and upper parts of the eastern Coos Bay estuary (Fig. 1C) differs from that of the Siuslaw cores. Most Coos Bay cores have a relatively low percentage of peat in the upper 4 m, most have 1 to 2 TOBs in the upper 1.5 m, and at least one of the TOBs is abrupt (e.g., site HI, Fig. 4). Lithofacies sequences at some sites (HI, CN, SH, CB, SN, IS, Fig. 1C) do not differ greatly from those at two of the Umpqua sites (UR, SM), but most Coos Bay sites have less peat and more abrupt TOBs than do cores from the two northern estuaries. Peaty units thin and become less peaty away from the upland, and abrupt TOBs at the tops of peaty units become less distinct (e.g., Fig. 4). Few cores more than 25 m from the upland contain abrupt TOBs that suggest RSL changes significantly >0.5 m. The low percentage of peaty lithofacies below 1.5 m depth shows that, at most sites, modern marshes are more extensive than were any marshes of the past few thousand years. The concentration of peaty units in the upper 1.5 m suggests a general regression during the past 1.0 to 1.5 ka (age estimate based on sediment thickness and ^{14}C ages at eight sites in Coos Bay), perhaps due to decreasing rates of submergence, increasing sedimentation rates, or both.

In the northernmost arm of Coos Bay (site NS, Fig. 1C), the stratigraphy in a small inlet on the south side of North Slough differs from that of other sites in eastern Coos Bay. Here a thick peat is common at a depth of 1.25 to 1.75 m (Fig. 5). The peat is a uniform, reddish-brown, fibrous, high-marsh peat, very similar to the peat at the South Inlet site at the Siuslaw River. The thickness, uniformity, and extent of the peat across the inlet suggest very slow RSL movements for at least a few hundred years during peat deposition. AMS ages on spruce needles (*Picea sitchensis*) at the abrupt TOB at the top of the peat indicate that an extensive high marsh in the North Slough inlet was rapidly submerged about 2.0 to 1.4 ka.

The abruptness of extensive TOBs and ROBs bounding the peaty lithofacies that form the surface horizons (A or O horizons) of buried marsh soils in the upper 1.5 m of cores from eastern Coos Bay differs from site to site. Differences in the abruptness of overlap boundaries implies differences in tendencies and rates of RSL movement. The most organic-rich surface horizons of buried soils at <1.5 m depth near uplands at more than half the sites (NS, HN, HI, CO, CN, CS, IS) have abrupt TOBs at upper boundaries and gradual ROBs at lower boundaries–the same type of contacts described from buried soils in tidal-wetland sequences in northern Oregon and southern Washington (Atwater, 1987; Darienzo and Peterson, 1990). Overlap boundaries on peaty lithofacies above and below the most organic-rich soils in Coos Bay are gradual. The stratigraphy at such sites (e.g., Haynes Inlet, Fig. 4) suggests a rapid-submergence event during a period of slow RSL rise. Upward changes from middle- or high-marsh peat to intertidal mud across the abrupt TOB at the top of the soil suggest about 0.5 to 1.5 m of rapid submergence. Gradual lithofacies changes above and below the organic-rich soil suggest small fluctuations in RSL or in sedimentation rates. Multiple conventional radiocarbon ages on different fractions of organic materials from the surface horizons of peaty soils at sites HI (Fig. 1), CO, and SH range over many hundreds of years (e.g., Fig. 4), so these ages cannot show whether or not the peaty soils at different sites developed and were submerged at about the same time (Grant and others, 1989; Nelson and Personius, 1991). At most sites, submergence probably occurred about 0.6 to 0.2 ka (^{14}C yrs).

At other sites in Coos Bay (KS, WS, IN), TOBs at the tops of peaty lithofacies that form the most organic-rich surface horizons of shallow soils are more gradual. The TOB at the top of the muddy peat at 0.9 m depth at site SH is gradational with the overlying mud and becomes more diffuse toward the upland; 5 m from the upland the upper 1 m of the section is entirely peat and muddy peat. Thus, at some sites the rate of RSL rise during the past 1,500 to 1,000 ^{14}C yrs apparently increased and decreased gradually; rapid submergence did not occur at all Coos Bay sites.

The contrasting RSL tendencies inferred from the different types of overlap boundaries in Coos Bay sequences are more consistent with recurrent coseismic deformation on local faults and folds of the accretionary wedge (Fig. 1B) during a gradual RSL rise than with a scenario of regional zones of subsidence produced during great plate-interface earthquakes. Although evidence of a rapid submergence event in the past 1,500 years is common in eastern Coos Bay, gradual upper TOBs at some sites suggest slow submergence. Very abrupt lower ROBs in the upper 2 m at a few other sites (SH, IS, CS) indicate possible local emergence. Because marsh lithofacies are not commonly found below 1.5 m in eastern Coos Bay, few inferences about sea-level history before 1.5 to 1.0 ka can be made. Available radiocarbon ages are not precise enough to determine whether any submergence or emergence events are synchronous with inferred coseismic-subsidence events in northern Oregon and southern Washington (Grant and others, 1989). Tsunami deposits might also be expected along the wave-exposed shores of an estuary characterized by repeated great earthquakes (e.g., Wright and Mella, 1963; Atwater, 1987). However, no landward-thinning sand beds capping peaty soils, similar to those thought to have been deposited by tsunamis farther north, (Atwater, 1987; Reinhart and Bourgeois, 1989; Darienzo and Peterson, 1990; Woodward and others, 1990; Peterson and Darienzo, 1991) were found in Coos Bay.

South Slough Arm of Western Coos Bay Estuary

Along the north-trending South Slough arm of western Coos Bay (Fig. 1C), late Pleistocene folding and faulting of marine terraces is well documented near tidal-marsh sites (McInelly and Kelsey, 1990). Holocene movement on some of these structures has been inferred (Peterson and Darienzo, 1989; McInelly and Kelsey, 1990; Nelson and Personius, 1991), making the link between coseismic subsidence or uplift and abrupt TOBs or ROBs in marsh sediments stronger in South Slough than elsewhere in south-central Oregon.

As in eastern Coos Bay, almost all marsh sequences in South Slough have a low percentage of peaty lithofacies and a buried peaty soil in the upper 1 m. Abundant tidal sand in sequences in the northern part of South Slough (e.g., sites WR and SA, Fig. 1C) prevented coring deeper than 2 to 3 m, but cores in the southern part of the slough reached depths of 4 to 8 m. At most sites throughout the slough, the TOB at the top of the former surface horizon of the peaty soil in the upper 1 m is abrupt and the lower horizon boundary is gradual. The mud (sand at site SA) immediately overlying the surface horizon contains no marsh-plant roots, suggesting that the TOB marks a substantial (>0.5 m) change in water depth. However, at the Winchester Creek (WC) site, a buried, peaty surface horizon at about 0.5 m depth has an abrupt to gradual upper TOB and an abrupt lower ROB (marked by a disconformity symbol on Fig. 6). A modern AMS ^{14}C age (100 percent modern carbon) from spruce needles at the top of this horizon shows that this soil was submerged and buried during this century, apparently following breaching of the dike that formerly prevented tides from reaching the site. The extent of the abrupt ROB at site WC and ages from the base of the overlying peaty horizon suggest the ROB may have been produced by emergence about 0.5 to 0.3 ka. Regardless of the process that produced the ROB, the differing sharpness of overlaps bounding the surface horizons of shallow peaty soils in South Slough suggests that soils at different sites formed in different ways and, therefore, that all peaty soil TOBs and ROBs at 0.5 to 1.0-m depth are not necessarily synchronous.

The more protected inlets in the southern half of South Slough (e.g., sites DC, BC, and WC) contain the most distinctive marsh-lithofacies sequences in the region. The percentage of peat in the upper 3 to 4 m is low, but multiple, thin peaty horizons, most with abrupt upper TOBs and gradual lower ROBs, are interbedded with mud units (e.g., Fig. 6). Nine abrupt TOBs in the past 4,000 ^{14}C yrs are recorded at the WC site, but available ^{14}C ages are not precise enough to show how closely TOBs are spaced in time or whether some TOBs correlate with those identified farther north along the Pacific Northwest coast (e.g., Darienzo and Peterson, 1990; Peterson and Darienzo, 1991). The lithofacies sequences are similar to those described by Atwater (1987), except that (1) peat lithofacies thin and become less peaty closer to the upland than in Washington, (2) buried peaty soils of spruce swamps have not been found,

GRANT, W. C., 1989, More evidence from tidal-marsh stratigraphy for multiple late Holocene subduction earthquakes along the northern Oregon coast: Geological Society of America Abstracts with Programs, v. 21, p. 86.

GRANT, W. C., ATWATER, B. F., CARVER, G. A., DARIENZO, M. E., NELSON, A. R., PETERSON, C. D., AND VICK, G. S., 1989, Radiocarbon dating of late Holocene coastal subsidence above the Cascadia subduction zone—compilation for Washington, Oregon, and northern California (abst.): EOS, Transactions, American Geophysical Union, v. 70, p. 1331.

HEATON, T. H., AND HARTZELL, S. H., 1987, Earthquake hazards on the Cascadia subduction zone: Science, v. 236, p. 162–168.

JACKSON, P. L., AND ROSENFELD, C. L., 1987, Erosional changes at Alsea spit, Waldport, Oregon: Oregon Geology, v. 49, p. 55–59.

JEFFERSON, C. A., 1975, Plant communities and succession in Oregon coastal salt marshes; Unpublished Ph.D. Dissertation, Department of Geography, Oregon State University, Corvallis, 192 p.

JELGERSMA, S., 1961, Holocene sea-level changes in The Netherlands: Meded. Geol. Sticht. Serie C, v. VI, 7, p. 1–100.

JENNINGS, A. E., AND NELSON, A. R., 1992, Foraminiferal assemblage zones in Oregon tidal marshes—relation to marsh floral zones and sea level: Journal of Foraminiferal Research, v. 22, no. 1, p. 13–29.

KELSEY, H. M., 1990, Late Quaternary deformation of marine terraces on the Cascadia Subduction Zone near Cape Blanco, Oregon: Tectonics, v. 9, p. 983–1014.

KIDSON, C., 1982, Sea level changes in the Holocene: Quaternary Science Reviews, v. 1, p. 121–151.

KOMAR, P. D., AND ENFIELD, D. B., 1987, Short-term sea-level changes and coastal erosion, *in* Nummedal, D., Pilkey, O. H., and Howard, J. D., eds., Sea-Level Fluctuation and Coastal Evolution: Society of Economic Paleontologists and Mineralogists Special Publication 41, p. 17–28.

LAJOIE, K. R., 1986, Coastal tectonics, *in* Wallace, R. E., Panel Chairman, Active Tectonics: National Academy Press, Washington, D.C., p. 95–124.

LAWSON, A. C., Chairman, 1908, The California Earthquake of April 18, 1906—Report of the State Earthquake Investigation Commission: Carnegie Institute of Washington, Washington, D.C., 450 p.

MATSUDA, T., OTA, Y., ANDO, M., AND YONEKURA, N., 1978, Fault mechanism and recurrence time of major earthquakes in southern Kanto district, Japan, as deduced from coastal terrace data: Geological Society of America Bulletin, v. 89, p. 1610–1618.

MCINELLY, G. W., AND KELSEY, H. M., 1990, Late Quaternary tectonic deformation in the Cape Arago–Bandon region of coastal Oregon as deduced from wave-cut platforms: Journal of Geophysical Research, v. 95, p. 6699–6714.

MIALL, A. D., 1977, A review of the braided river depositional environment: Earth Science Reviews, v. 13, p. 1–62.

MUHS, D. R., KELSEY, H. M., MILLER, G. H., AND KENNEDY, G. L., WHELAN, J. F., AND MCINELLY, G. W., 1990, Age estimates and uplift rates for late Pleistocene marine terraces: southern Oregon portion of the Cascadia forearc: Journal of Geophysical Research, v. 95, No. B5, p. 6685–6698.

NELSON, A. R., 1987, Apparent gradual rise in relative sea level on the south-central Oregon coast during the late Holocene—implications for the great Cascadia earthquake hypothesis (abst.): EOS, Transactions, American Geophysical Union, v. 68, p. 1240.

NELSON, A. R., AND PERSONIUS, S. F., 1991, Great earthquake potential in Oregon and Washington: an overview of recent coastal geologic studies and their bearing on segmentation of Holocene ruptures in the central Cascadia subduction zone, *in* Rogers, A. M., Kockelman, W. J., Priest, George, and Walsh, T. J., eds., Assessing and Reducing Earthquake Hazards in the Pacific Northwest: U.S. Geological Survey Open-File Report 91-441A, 29 p.

NEWMAN, W. S., CINQUEMANI, L. J., PARDI, R. P., AND MARCUS, L. F., 1980, Holocene deleveling of the United States east coast, *in* Morner, N.-A., ed., Earth Rheology, Isostasy, and Eustasy: John Wiley, New York, p. 449–463.

NEWMAN, W. S., CINQUEMANI, L. J., SPERLING, J. A., AND MARCUS, L., AND PARDI, R. R., 1987, Holocene neotectonics and the Ramapo fault zone sea-level anomaly—a study of varying marine transgression rates in the lower Hudson estuary, New York and New Jersey, *in* Nummedal, D., Pilkey, O. H., and Howard, J. D., eds.; Sea-Level Fluctuation and Coastal Evolution: Society of Economic Paleontologists and Mineralogists Special Publication 41, p. 97–114.

NIERING, W. A., AND WARREN, R. S., 1977, Our dynamic tidal marshes—vegetation changes as revealed by peat analysis: The Connecticut Arboretum Bulletin, no. 22, New London, Connecticut, 12 p.

NISHENKO, S. P., 1989, Circum-Pacific seismic potential 1989–1999: U.S. Geological Survey Open-File Report 89–86, 126 p.

OVENSHINE, A. T., LAWSON, D. E., AND BARTSCH-WINKLER, S. R., 1976, The Placer River silt—an intertidal deposit caused by the 1964 earthquake: U.S. Geological Survey Professional Paper 543-F, 28 p.

PELTIER, W. R., 1987, Mechanisms of relative sea-level change and the geophysical responses to ice-water loading, *in* Devoy, R. J. N., ed., Sea Surface Studies—A Global View: Croom Helm, London, p. 57–94.

PETERSON, C. D., AND DARIENZO, M. E., 1989, Episodic abrupt tectonic subsidence recorded in late Holocene deposits of the South Slough syncline—an on-land expression of shelf fold belt deformation from the southern Cascadia margin: Geological Society of America Abstracts with Programs, v. 21, p. 129.

PETERSON, C. D., AND DARIENZO, M. E., 1991, Discrimination of climatic, oceanic and tectonic forcing of marsh burial events from Alsea Bay, Oregon, USA, *in* Rogers, A. M., Kockelman, W. J., Priest, George, and Walsh, T. J., eds., Assessing and Reducing Earthquake Hazards in the Pacific Northwest: U.S. Geological Survey Open-File Report 91-441C, 53 p.

PETERSON, C. D., SCHEIDEGGER, K. F., AND SCHRADER, K. J., 1984, Holocene depositional evolution of a small active margin estuary of the Northwestern United States: Marine Geology, v. 59, p. 51–83.

PLAFKER, G., 1969, Tectonics of the March 27, 1964 Alaska Earthquake: U.S. Geological Survey Professional Paper 543–I, 94 p.

PLAFKER, G., 1972, Alaskan earthquake of 1964 and Chilean earthquake of 1960—implications for arc tectonics: Journal of Geophysical Research, v. 77, p. 901–925.

PLAFKER, G., 1990, Regional vertical tectonic displacement of shorelines in south-central Alaska during and between great earthquakes: Northwest Science, v. 64, p. 250–258.

PLAFKER, G., AND SAVAGE, J. C., 1970, Mechanism of the Chilean earthquakes of May 21 and 22, 1960: Geological Society of America Bulletin, v. 81, p. 1001–1030.

RAMPINO, M. R., AND SANDERS, J. E., 1981, Episodic growth of Holocene tidal marshes in the northeastern United States—a possible indicator of eustatic sea-level fluctuations: Geology, v. 9, p. 63–67.

REDFIELD, A. C., 1972, Development of a New England salt marsh: Ecological Monographs, v. 42, p. 201–237.

REED, D. J., 1988, Sediment dynamics and deposition in a retreating coastal salt marsh: Estuarine, Coastal, and Shelf Science, v. 26, p. 67–79.

REINHART, M. A., AND BOURGEOIS, J., 1989, Tsunami favored over storm or seiche for sand deposit overlying buried Holocene peat, Willapa Bay, WA (abst.): EOS, Transactions, American Geophysical Union, v. 70, p. 1331.

ROGERS, G. C., 1988, An assessment of the megathrust earthquake potential of the Cascadia subduction zone: Canadian Journal of Earth Sciences, v. 25, p. 844–852.

SHENNAN, I., 1982, Interpretation of Flandrian sea-level data from the Fenland, England: Proceedings of the Geologists Association, v. 83, p. 53–63.

SHENNAN, I., 1986a, Flandrian sea-level changes in the Fenland. I—The geographical setting and evidence of relative sea-level changes: Journal of Quaternary Science, v. 1, p. 119–154.

SHENNAN, I., 1986b, Flandrian sea-level changes in the Fenland. II—Tendencies of sea-level movement, altitudinal changes, and local and regional factors: Journal of Quaternary Science, v. 1, p. 155–180.

SHENNAN, I., 1987, Global analysis and correlation of sea-level data, *in* Devoy, R. J. N., ed., Sea Surface Studies—A Global View: Croom Helm, London, p. 198–230.

SHENNAN, I., 1989, Holocene crustal movements and sea-level changes in Great Britain: Journal of Quaternary Science, v. 4, p. 77–89.

SPENCE, WILLIAM, 1989, Stress origins and earthquake potentials in Cascadia: Journal of Geophysical Research, v. 94, p. 3076–3088.

STREIF, H., 1979, Cyclic formation of coastal deposits and their indications of vertical sea-level changes: Oceanis, v. 5, p. 303–306.

STREIF, H., 1987, Barrier islands, tidal flats, and coastal marshes resulting from a relative rise of sea level in East Frisia on the German North Sea coast, *in* van der Linden, W. J. M., Cloetingh, S. A. P. L., Kaasschieter, J. P. K., van de Graaff, W. J. E., Vandenberghe, J., and van der Gun, J. A. M., eds., Coastal Lowlands: Geology and Geotechnology: Kluwer Academic Publishers, Dordrecht, The Netherlands, p. 213–223.

TAYLOR, F. W., EDWARDS, R. L., WASSERBERG, G. J., AND FROHLICH, C., 1990, Seismic recurrence intervals and timing of aseismic subduction inferred from emerged corals and reefs of the central Vanuatu (New Hebrides) frontal arc: Journal of Geophysical Research, v. 95, p. 393–408.

TROELS-SMITH, J., 1955, Characterization of unconsolidated sediments: Geological Survey of Denmark, Series IV, v. 3, 72 p.

VAN DE PLASSCHE, O. 1982, Sea-level change and water-level movements in the Netherlands during the Holocene: Mededelingen rijks geologische dienst, v. 36–1, 93 p.

VAN DE PLASSCHE, O. 1990, Mid-Holocene sea-level change on the eastern shore of Virginia: Marine Geology, v. 91, p. 149–154.

VAN DE PLASSCHE, O. 1991, Late Holocene sea-level fluctuations on the shore of Connecticut inferred from transgressive and regressive overlap boundaries in salt-marsh deposits, *in* Gayes, P. T., Lewis, R. S., and Bokuniewicz, H. G., eds., Quaternary Coastal Evolution of Southern New England: Journal of Coastal Research, special issue no. 11, p. 159–179.

VAN DE PLASSCHE, O., MOOK, W. G., AND BLOOM, A. L., 1989, Submergence of coastal Connecticut 6000–3000 (^{14}C) years B.P.: Marine Geology, v. 86, p. 349–354.

VITA-FINZI, C., 1986, Recent Earth movements: An Introduction to Neotectonics: Academic Press, London, U.K., 226 p.

WEST, D. O., AND MCCRUMB, D. R., 1988, Coastline uplift in Oregon and Washington and the nature of Cascadia subduction-zone tectonics: Geology, v. 16, p. 169–172.

WILSON, D. S., 1989, Deformation of the so-called Gorda plate: Journal of Geophysical Research, v. 94, p. 3065–3075.

WOODWARD, J., WHITE, J., AND CUMMINGS, R., 1990, Paleoseismicity and the archaeological record: areas of investigation on the northern Oregon coast: Oregon Geology, v. 52, p. 57–65.

WRIGHT, C. AND MELLA, A., 1963, Modifications to the soil pattern of south-central Chile resulting from seismic and associated phenomena during the period May to August 1960: Bulletin of the Seismological Society of America, v. 53, p. 1367–1402.

YEATS, R. S., CLARK, M. N., KELLER, E. A., AND ROCKWELL, T. K., 1981, Active fault hazard in southern California: ground rupture versus seismic shaking: Geological Society of America Bulletin, v. 92, p. 189–196.

THERMOLUMINESCENCE AGES OF ESTUARINE DEPOSITS ASSOCIATED WITH QUATERNARY MARINE TERRACES, SOUTH-CENTRAL CALIFORNIA

GLENN W. BERGER AND KATHRYN L. HANSON

Department of Geology, Western Washington University, Bellingham, Washington 98225; and Geomatrix Consultants, 100 Pine St., Suite 1000, 10th Floor, San Francisco, California 94111

ABSTRACT: Many coastal areas of California and Oregon have emergent Quaternary marine terraces that have been assigned only indirect, approximate ages because of the paucity of suitable datable material for the U-series method, or because of the ≈40-ka limit of the radiocarbon-dating method. We report three thermoluminescence (TL) ages for clayey estuarine deposits from the San Simeon area of south-central California. A weighted mean TL age of 95±13 ka for clayey silt underlying marine deposits of the San Simeon (Q_s) terrace suggests that this terrace correlates to oxygen isotope substage 5c (105 ka), or more likely substage 5a (80 ka). A single-analysis TL age of 230±70 ka for clayey beds in tilted estuarine deposits at San Simeon Bay is consistent with the available stratigraphic and age control for the underlying Careaga Formation (early Pleistocene or Pliocene) and the overlying Q_p marine terrace (≈60 ka). This TL age most probably compares to a time of high sea level at 200±20 ka (equivalent to marine oxygen isotope stage 7). The large analytical error in this TL age is consistent with deposition shortly before or after the ≈200 ka highstand of sea.

INTRODUCTION

Emergent Quaternary marine terraces are present along most of the south-central California coastline from San Simeon to the north to the Santa Maria Valley to the south (Weber, 1983; Lajoie, 1986; Hanson and others, this volume). The terraces are remnants of abandoned wave-cut platforms that are preserved in step-like sequences along uplifted coastlines. In the San Simeon area (Fig. 1), flights of four and five marine terraces have been mapped in detail near the southern onshore reach of the San Simeon fault zone (Hanson and others, this volume).

Determining rates of Quaternary deformation is one of the main reasons for studying deposits associated with such terraces. Direct dating of these deposits is seldom attempted because of the paucity of suitable material (coral is preferred for the U-Th series method but is rarely preserved) or lack of a widely applicable dating method. Rather, the general procedure is to map and correlate local terrace sequences to worldwide paleosea-level data. This "rates-without-dates" approach has clear limitations.

As part of the development of the thermoluminescence (TL) sediment-dating method (Aitken, 1985; Berger, 1988; Forman, 1989), one of us (GWB) has applied TL-dating procedures to coastal sediments from several locations in western North America. The advantage of TL dating is that the burial time of the detrital minerals themselves can be measured, potentially over the time range of 1 to 300 ka. We report TL-age results for three samples from fine-grained estuarine deposits in the San Simeon area (Fig. 1).

PRINCIPLES OF TL DATING

Wintle and Huntley (1980) proposed that, for simplicity, the TL in sediments can be thought of as having a light-sensitive and a light-insensitive component. Several studies (the most detailed of which is that of Berger, 1990) have documented that insolation before burial reduces some or all of the light-sensitive TL in sediments to zero, the effectiveness of zeroing depending upon depositional conditions. After burial, ionizing radiation (mainly alpha, beta, and gamma) from ambient U, Th, and K will replace the electrons in emptied light-sensitive electron traps (particular types of crystal defects) at a constant rate (over the less than 1-Ma age range of TL dating). Subsequent laboratory measurements can extract a light-sensitive TL signal proportional to the burial time.

Wintle and Huntley (1980) suggested that the most suitable method for separating these two TL components in the measurement of the post-burial absorbed dose is their R-gamma partial-bleach method. With this method, controlled laboratory illumination (optical bleaching) or natural sunlight is applied to various subsamples of the sediment to construct a sample-dependent dose response of the bleachable TL signal. Extrapolation of this response to the dose corresponding to zero bleachable TL yields an equivalent-dose (D_E) value. If this D_E reaches a constant value as a function of TL-readout temperature (the "plateau test"; Aitken, 1985), then the plateau value is used in the age equation

$$t = D_E/(\dot{D}_\alpha + \dot{D}_\beta + \dot{D}_\gamma + \dot{D}_{cr}).$$

The alpha, beta, gamma, and cosmic-ray dose-rate components in the denominator can be evaluated as outlined elsewhere (Aitken, 1985; Berger, 1988).

SAMPLE SELECTION AND TREATMENT

Sample Descriptions

Three small (≈100 gm) blocks of clayey sediment were collected at the locations shown in Figure 1. The generalized stratigraphy at the more western site is given in Figure 2. Sample SIM-1a was collected from the center of a 60- to 90-cm-thick bed of clay and silty clay and associated organic-rich lenses that is thought to represent a littoral-sublittoral depositional environment. According to field observations, these sediments fill an older abandoned fluvial channel of Oak Knoll Creek that was probably incised during a previous sea-level lowstand. Aggradation and deposition of the estuarine sediments likely occurred during the rise to the sea-level highstand that produced the San Simeon (Q_s) terrace. Sample SIM-2a was collected about 50 m northwest of SIM-1a in the same clayey bed but has a siltier texture. The San Simeon (Q_s) terrace marine deposits overlying these samples (Fig. 2) have been assigned independent age estimates of ≈80 or ≈105 ka (Hanson and others, this volume). Detrital charcoal from these estuarine

Quaternary Coasts of the United States: Marine and Lacustrine Systems, SEPM Special Publication No. 48

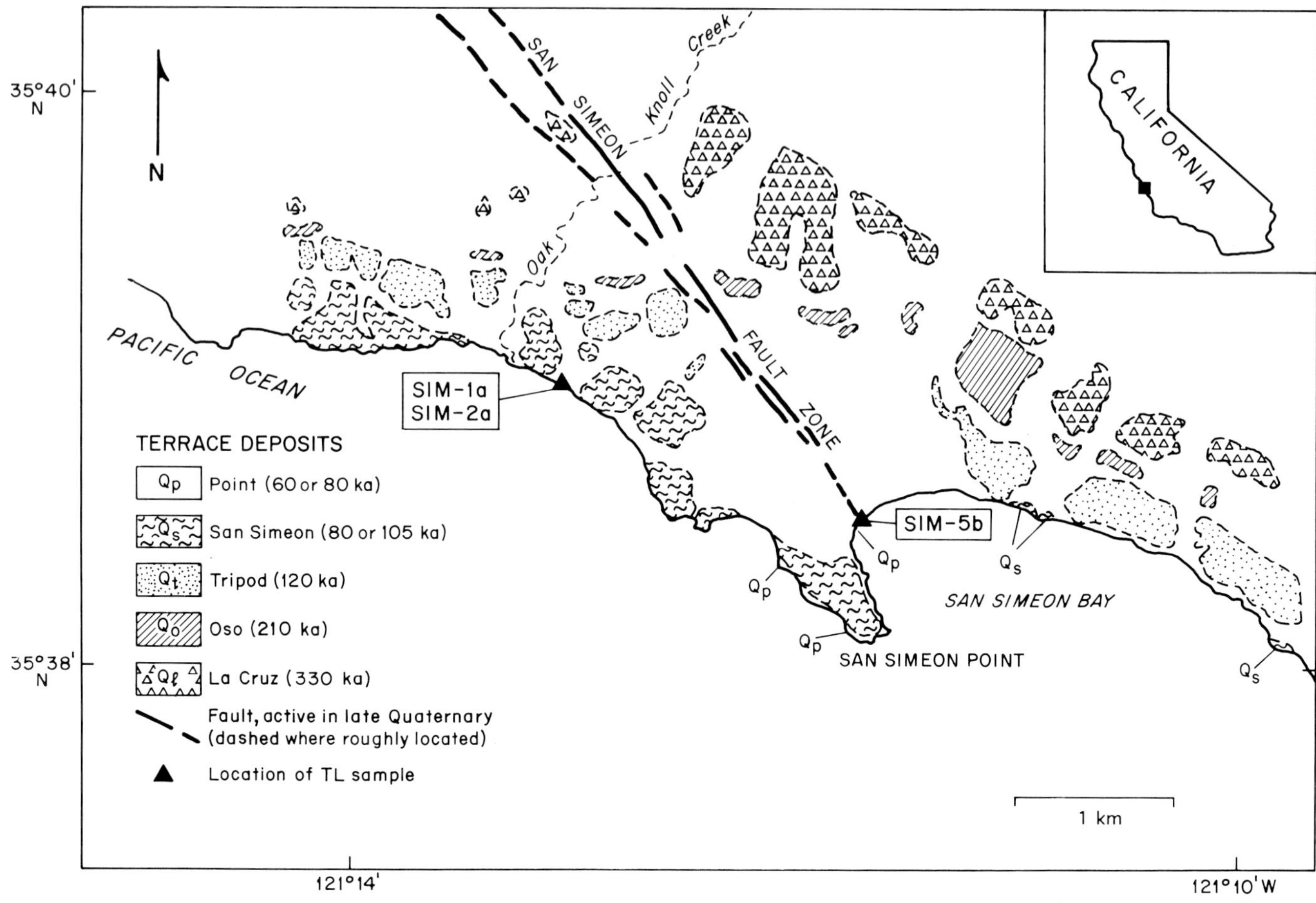

FIG. 1.—Locations of TL samples and marine terraces along the southern onshore reach of the San Simeon fault zone (modified from Fig. 4 of Hanson and others, this volume). Q_p deposits are exposed in only three places in the area illustrated here. TL-sample locations are SIM-1a and SIM-2a, 35°38′58″N, 121°12′ 53.2″W; SIM-5b, 35°38′26.4″N, 121°11′40.1″W.

sediments yielded only a limiting radiocarbon age of >40.5±1.1 ka (Hanson and others, this volume).

The third TL sample (SIM-5b in Fig. 1) was collected from an organic-rich silty clay horizon within an interbedded sequence of silty fine sand, sandy silt, and fossiliferous sand (Fig. 3). These sediments appear to have been laid down in a quiet-water lagoon or an estuarine environment. They unconformably overlap a fossiliferous, shallow marine, pebbly sandstone that Hall (1975) tentatively correlated with the Pliocene and Pleistocene Careaga Formation. Racemic D/L amino-acid racemization ratios have been obtained on mollusc samples from the Careaga Formation (J. F. Wehmiller, pers. commun., 1991). Fauna from the putative Careaga deposits at San Simeon are interpreted by G. Kennedy (Los Angeles County Museum of Natural History, pers. commun., 1986) to be early Quaternary (>500 ka) and possibly Pliocene in age. The estuarine sediments that overlie these beds are in turn overlain by basal marine deposits of the Point (Q_p) terrace (Fig. 3a).

The estuarine deposits show an apparent dip of 20 to 40° to the northeast, suggesting that these sediments were tilted after deposition. In addition, the Careaga beds, the estuarine sediments, and the overlying Q_p terrace wave-cut platform and the associated basal marine deposits are all displaced by at least one and probably multiple fault traces within the San Simeon fault zone.

TL Methods

Samples for TL measurements were collected and handled so as to preclude significant exposure to light. The analytical steps followed in the preparation and treatment of the chosen polymineralic 4- to 11-μm grains are outlined in Berger (1988). Gamma irradiations were performed at Simon Fraser University. Irradiated subsamples were stored for eight days at 75°C to remove unstable laboratory-induced TL (Berger, 1987). An OG570 glass filter in the >570-nm f_1 arrangement of Berger (1990) was used for optical bleaching. The saturating-exponential model of Berger and others (1987) was used for regression and error analyses in the partial-bleach method. Other analytic procedures are mentioned in table footnotes.

RESULTS

Dosimetric data used in the TL-age equation are given in Table 1. Representative TL "glow curves" for sample

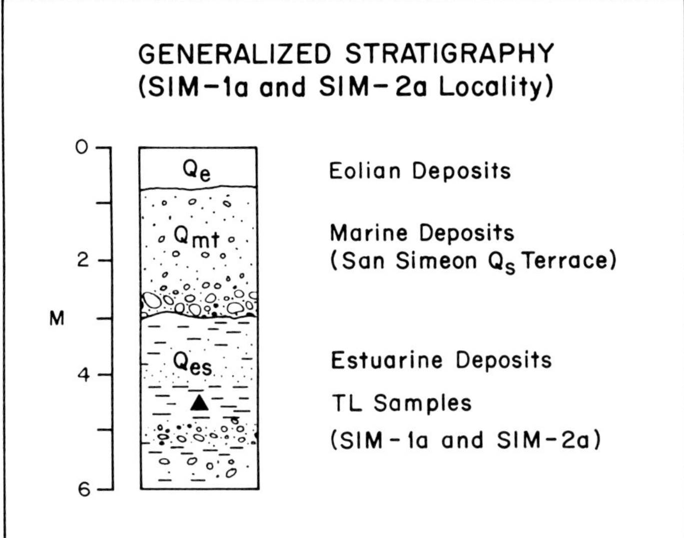

FIG. 2.—Stratigraphic section for sites SIM-1a and SIM-2a of Figure 1. Q_e—fine sand and sandy silt; well sorted. Q_{mt}—sand and pebbly sand overlying basal sandy cobble and boulder gravel; dark yellowish-brown; massive to moderately well-bedded; well-sorted; well-rounded clasts. Q_{es}—interbedded dark olive gray clay, silty clay, and sandy silt; organic-rich lenses, locally containing detrital charcoal and wood fragments; grades laterally into a sequence of predominantly fluvial pebbly sand and sandy gravel.

SIM-1a (Fig. 4) show nothing unusual, but the presence of a bleach-resistant signal at >≈350°C may indicate a substantial quartz component in this 4- to 11-μm polymineral sample (no separate mineralogic examination was made). Similar signals were observed for the two other samples and are discussed later.

An illustration of the partial-bleach method is given in Figure 5 for sample SIM-1a. The inset shows the resulting D_E values for samples SIM-1a and SIM-2a. Similar data for sample SIM-5b are presented in Figure 6. There the required larger extrapolations and more nearly "saturated" TL-growth curves reflect the greater age of the third sample. The results of these partial-bleach experiments are summarized in Table 2. Because samples SIM-1a and SIM-2a are from the same bed, they represent independent duplicate analyses of material of the same age. The agreement to within 1σ of the TL ages for these samples (Table 2) indicates that no significant uncorrected systematic analytic errors distinguish these two results. Thus, a weighted mean TL age of 95±13 ka (Table 2, footnote d) was calculated for these two samples.

DISCUSSION

TL-Age Accuracy

One source of inaccuracy could arise from any failure of the zeroing assumption; however, we believe that this source is insignificant for any estuarine muds older than ≈5 ka. The light-sensitive TL in surface tidal-flat muds from the Arcata Bay lagoon/estuary in northern California (Berger and others, 1991) and in tidal-flat and delta-foreslope muds from the Fraser River delta in southern British Columbia (Berger, 1990; Berger and others, 1990) has been shown to be sufficiently well zeroed to permit accurate TL dating beyond about 5 ka without any zero-point corrections.

Another potential inaccuracy could arise from the presence of a significant quartz-mineral component in the analyzed polymineralic fine-grain fraction, even though feldspar TL is 20 to 50 times more intense per gram than is quartz TL (Berger, 1984). Quartz TL is not important for the samples in Figure 5, where plateaus are observed well below ≈350°C, but could be a problem for sample SIM-5b, for which the plateau develops only above ≈310°C (Fig. 6). If any quartz TL is *relatively* bleach resistant both in nature (assuming deposition only in turbid water; see Berger, 1988, 1990) and under the >570-nm laboratory optical bleaching (Berger, 1988), then we would expect no significant effect on the accuracy of the D_e values in Figure 6.

In addition, our 75°C (8 d) preheat treatment may not have removed completely all laboratory-induced unstable TL from these samples. If so, then these results are minimum ages; however, based on previous experience with other samples, we consider that this is unlikely. Nevertheless, any similar future TL dating could employ additional stability tests to assess this possibility more fully. Finally, our calculated 230-ka age could be an underestimate if the relevant electron traps above ≈300°C in sample SIM-5b have lifetimes of the order 0.5 to 1 Ma. For this reason too, further TL tests on this and similar samples would be useful.

We conclude that the single TL-age estimate of 230±70 ka for sample SIM-5b can be considered accurate, though additional TL tests would provide a useful check.

Geological Interpretation

The three TL ages reported here have some significance because directly datable materials from the San Simeon area are restricted to charcoal for radiocarbon dating, a fragment of mammal bone from marine gravel overlying the Q_p marine platform (Hanson and others, this volume), and these three TL samples. A mean U-series age of 46±2 ka for the bone fragment is consistent with its stratigraphic context but provides only a minimum limiting age for this terrace. Radiocarbon analyses of charcoal from the estuarine sediments in Figure 2 suggest an age of >40.5±1.1 ka for these deposits (Hanson and others, this volume). The weighted mean TL age of 95±13 ka for samples SIM-1a and SIM-2a (Fig. 2) suggests that the San Simeon (Q_s) terrace correlates to oxygen-isotope substage 5c (105 ka) or substage 5a (80 ka) because the estuarine sediments predate this marine terrace stratigraphically. This age interpretation for the Q_s terrace is also consistent with a correlation of the Tripod (Q_t) terrace (Fig. 1) with the 120-ka Cayucos terrace (Hanson and others, this volume). Other arguments by Hanson and colleagues based on terrace altitudinal spacing, geomorphic expression, and comparison of soil properties favor an age of ≈80 ka for the Q_s terrace.

The TL age of 230±70 ka for the estuarine deposits in Figure 3B is consistent with the available stratigraphic and

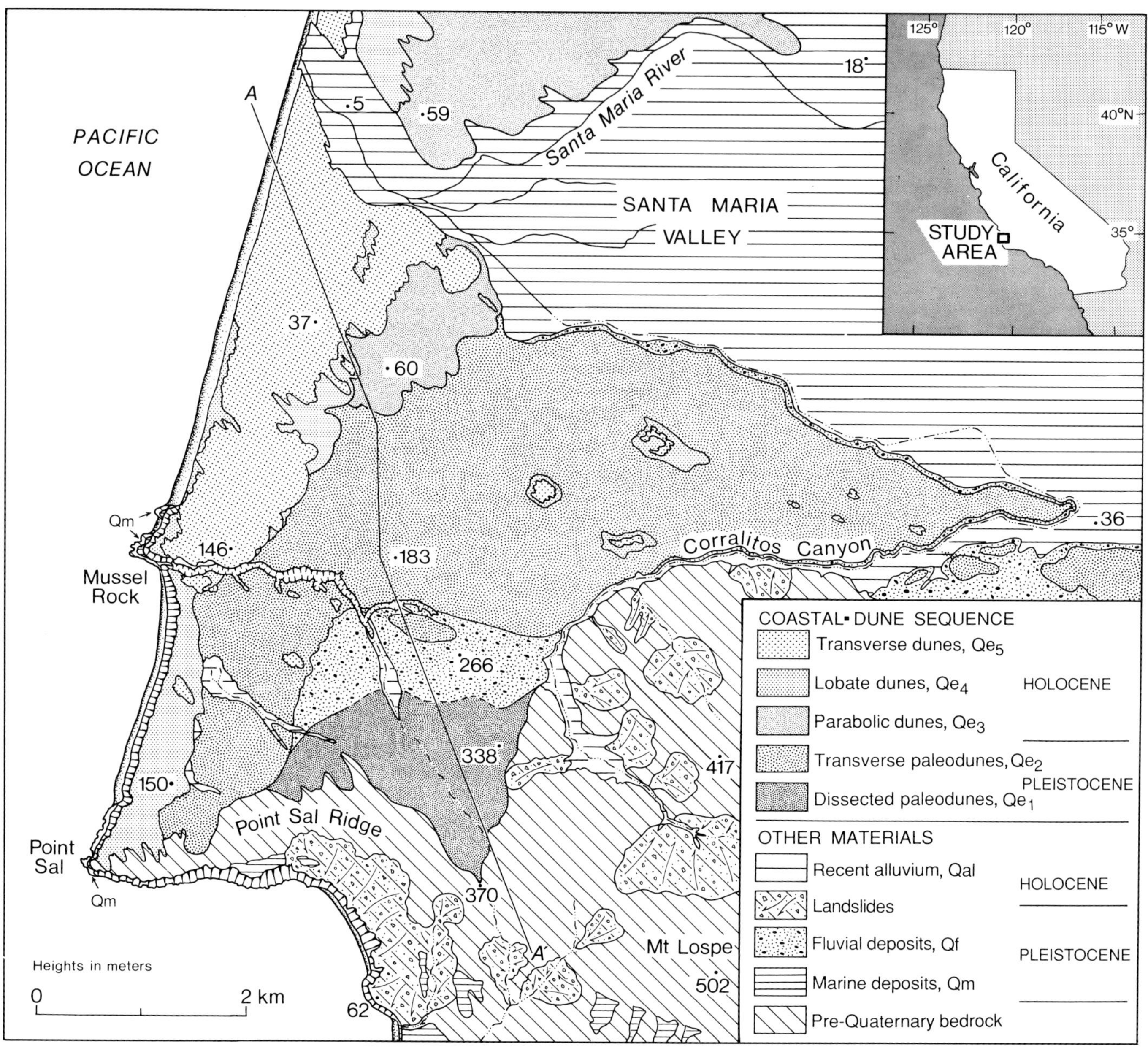

FIG. 1.—The late Quaternary coastal-dune sequence and associated materials near Point Sal, south-central California. Section A–A′ relates to Figure 3.

The late Quaternary deposits exceed 100 m in thickness in and adjacent to Mussel Rock ravine. Two wells 3 km north-northeast of Mussel Rock have revealed 198 m and 207 m of undifferentiated Quaternary deposits that Hall (1982) suggested are Orcutt sand but that may include older Paso Robles materials. The top of the Paso Robles Formation is thought to lie 56 m below sea level 2.5 km east of the Santa Maria River mouth and at sea level near Corralitos Canyon. Important exposures of late Quaternary deposits in Mussel Rock ravine are shown in Figure 2. Although no section reveals a complete late Quaternary sequence, cross-correlations between exposures, together with the surface form of various deposits, suggest the sequence discussed later and shown in Figure 3.

Marine-Terrace Deposits (Qm)

Around Mussel Rock and Point Sal, an irregular bedrock platform 6 to 10 m above mean sea level is overlain by up to 4 m of stratified, quartz-rich beach deposits showing some cross-bedding with individual sets dipping at up to 10°. The deposits contain a 0.5-m-thick basal gravel composed of Tertiary shale chips and blocks derived from nearby buried sea cliffs, overlain by well-rounded cobbles and pebbles in a moderately well-sorted, medium to coarse sandy matrix.

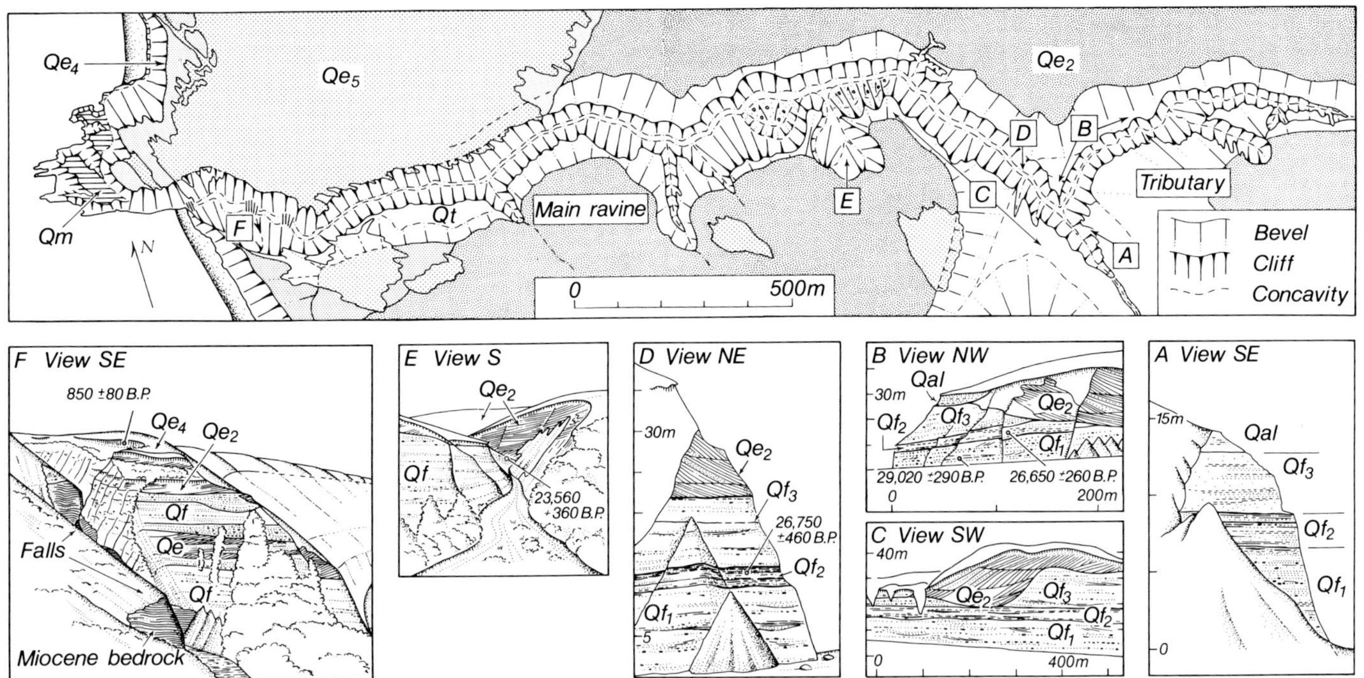

FIG. 2.—Mussel Rock ravine, showing six sections where late Quaternary fluvial and eolian deposits are associated.

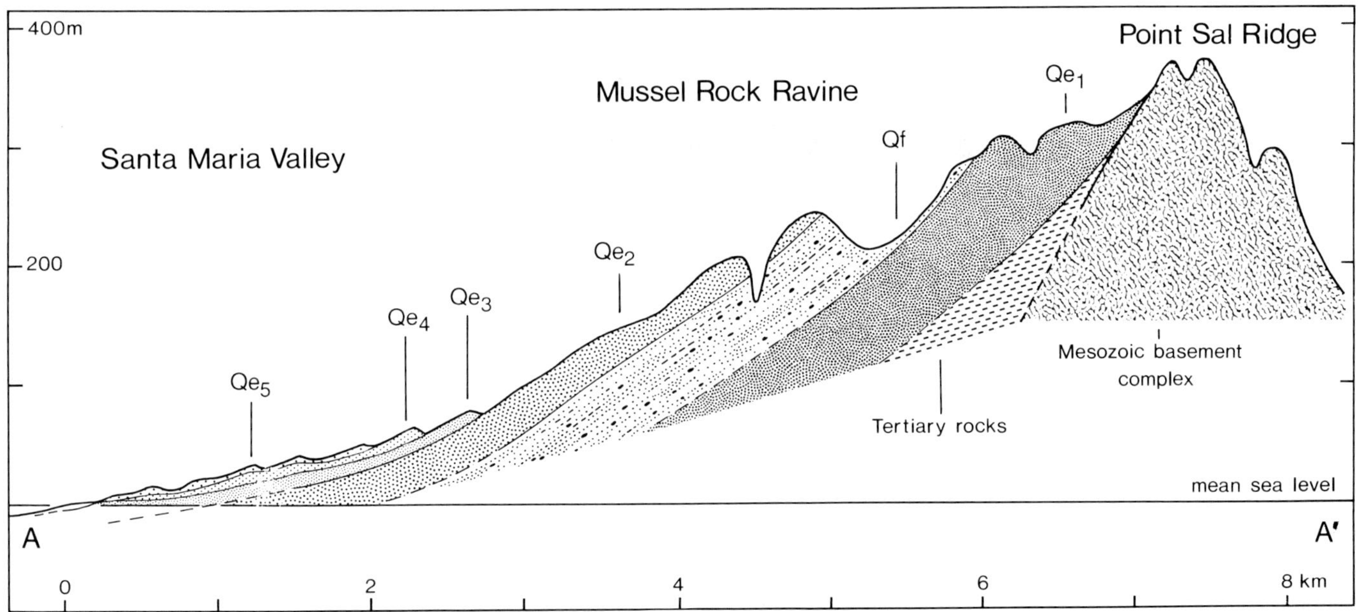

FIG. 3.—Idealized cross section through the late Quaternary depositional sequence between the Santa Maria Valley and the Point Sal Ridge (A–A′ in Figure 1). Near km 2, exploratory wells indicate a 200-m thickness of undifferentiated Quaternary deposits, which probably include the older Paso Robles Formation. The Orcutt Frontal Fault, a seismically active, high-angle, southwest-dipping reverse fault, extends east-west beneath, and probably within the sequence near km 4.

The larger clasts contain pholad borings and include much igneous and metamorphic material. The marine-terrace deposits are overlain by ferruginous sands and gravels of colluvial, fluvial, and eolian origin, but the sequence is largely obscured by recent dune sands.

Dissected Paleodunes (Qe_1)

The oldest dune sand in the area extends as a triangular outcrop for 2 km upslope from the head of Mussel Rock ravine to a maximum elevation of 370 m on the Point Sal

Ridge (Fig. 1). The dissected paleodune deposit is a strongly indurated, moderately well-sorted sand whose original structure has been largely destroyed by weathering and ferricrete development. Iron-coated grains and iron cement give the sand a reddish-yellow (Munsell 7.5YR 7/8) to light reddish-brown (5YR 6/4) appearance, leaching to light gray (10YR 7/1) along joints and root casts. Iron pisoliths up to 2 cm in diameter characterize surface soils, derived colluvium, and sections to depths of 4 m. Although its northern margins are obscure, this unit appears to descend beneath later fluvial units, to which it supplied abundant sandy sediment. Because the morphology of its subcrop is unknown, the total thickness of the unit is uncertain, but it may exceed 50 m and include non-eolian sediments.

Fluvial Deposits (Qf_1, Qf_2, Qf_3)

Fluvial deposits, approaching 50 m in thickness, are well revealed in the walls of Mussel Rock ravine, notably at elevations of 180 to 260 m above sea level near the confluence of main and tributary ravines 2.3 km from the coast (Fig. 2). Despite lateral and vertical variations, these deposits usually reveal two mostly fluvial units separated by a mixed fluvial-paludal unit.

The lower fluvial unit (Qf_1) comprises a well-stratified, moderately sorted to moderately well-sorted medium sand containing subsidiary gravel lenses and thin seams of silt and clay. Sections up to 12 m thick are exposed. Pebbles and small cobbles include Franciscan ophiolite, reddish sandstones and shale chips from Tertiary rocks, and heterogeneous mud balls, all presumably derived from the Point Sal Ridge and Casmalia Hills. Ubiquitous iron pisoliths were probably flushed from the dissected paleodunes (Qe_1) exposed upstream. Individual sand beds range from 2 to 12 cm in thickness. Owing to variable iron oxidation, their color ranges from light gray (10YR 7/2), through yellowish-brown (10YR 5/4) and reddish-brown (5YR 4/4) to red (2.5YR 4/6). Because the present channel gradient steepens more rapidly than the 1° to 2° dip of these beds, the lower fluvial unit is progressively exposed downstream (Fig. 2A–C). This unit reflects higher energy sediment transport than the overlying fluvial materials. Charcoal from a 0- to 60-cm-thick wedge of silty clay near the top of this unit, 2 m above the tributary-ravine floor (Fig. 2B), yielded a ^{14}C age of 29,020±290 yrs b.p. (Beta-31912; errors quoted to one standard deviation).

The mixed fluvial-paludal unit (Qf_2) consists mostly of silt beds, individually 2 to 45 cm thick, separated by moderately sorted medium to fine sand and some dense clay seams 0.5 to 1 cm thick. The exposed unit varies in thickness from 2 to 6 m. Color varies from olive (5Y 5/3) through grayish-brown (2.5Y 5/2) to dark gray (10YR 4/1). The often greenish color and fine texture indicate deposition under partly anaerobic, paludal conditions, probably in ponds blocked by downstream accumulations of eolian sand. Well-sorted, cross-bedded sands of probable eolian origin occur within and above the unit. Charcoal fragments from a 30-cm-thick clayey silt near the top of the unit, 10 m above the tributary-ravine floor 110 m northeast of the confluence (Fig. 2B), yielded a ^{14}C age of 26,650±260 yrs b.p. (Beta-21212). A wood sample from a clayey silt in similar stratigraphic position in the main ravine 200 m north-northwest of the confluence (Fig. 2D) yielded a ^{14}C age of 26,750±460 yrs b.p. (Beta-17456). Analyzed with a Coulter ultrasonic multisizer, the latter deposit revealed 5 percent sand, 67 percent silt, and 28 percent clay by volume, with a mean in the medium-silt range.

The upper fluvial unit (Qf_3) comprises a moderately sorted medium sand up to 14 m thick, with minor silt and clay seams but few pebbles (Fig. 2A–2D). Some well-sorted cross-bedded sands could be eolian in origin, compared with near-horizontal, fining-upward fluvial beds. Color ranges from light gray (5YR 7/1) through brown (5YR 5/4) to reddish-yellow (7.5YR 7/8). A gully south of the main ravine, 700 m north-northwest of the confluence, reveals a 40-cm-thick seam of grayish-brown sandy silt dipping 25°W beneath overlying eolian sand (Fig. 2E). The deposit contains 30 percent sand, 62 percent silt, and 8 percent clay by volume, with a mean in the coarse-silt range. Charred wood from the deposit 15 cm beneath the overlying eolian sand yielded a ^{14}C age of 23,560±360 yrs b.p. (Beta-26458). Near the mouth of Mussel Rock ravine, a 40-m-high section exposes a complex sequence of fluvial sands and gravels, including channel-fill deposits, interbedded with cross-bedded eolian sands (Fig. 2F).

Transverse Paleodunes (Qe_2)

The transverse paleodunes (Qe_2) that overlie the fluvial deposits are the thickest and most extensive surface dunes in the area, inviting comparison with similar stabilized dunes elsewhere along the California coast (Fig. 1). Though subdued, they reveal a series of transverse ridges trending N 45° E near the coast but N 75° E farther inland where the Point Sal Ridge influences near-surface wind patterns. Subdued lobate forms also occur. These paleodunes extend inland to Corralitos Canyon and straddle Mussel Rock ravine, whose beveled upper slopes reflect the erodibility of their poorly consolidated sands (Fig. 2). The ravine exposes up to 20 m of cross-bedded eolian sands with sets dipping up to 30°, and total thickness probably exceeds 30 m (Fig. 2B–E). Averaging this thickness over the dune area south of the Santa Maria River, suggests that the transverse paleodunes may contain as much as 1.3 km^3 of sand. The paleodunes feather out around 200 m above sea level southeast of the tributary ravine and, excepting Mussel Rock ravine and two small valleys farther south, are relatively undissected. Deposits of the transverse paleodunes comprise frosted, well-sorted, quartz-rich, medium sands varying in color from white (5Y 8/1), through yellow (5Y 8/8) and brown (7.5YR 5/2) to reddish-yellow (5YR 6/8). They mostly postdate the mixed fluvial-paludal unit (Qf_2) deposited before 26 ka but at their base interfinger with the upper fluvial unit (Qf_3) that was deposited between about 26 and 23 ka. The transverse paleodunes largely blocked drainage off the Point Sal Ridge, causing later alluvial deposits (*Qal*) to overlie older fluvial deposits (*Qf*) above the confluence (Fig. 2A, B), prior to the cutting of the present ravine.

Parabolic, Lobate, and Active Transverse Dunes (Qe_3, Qe_4, Qe_5)

Parabolic dunes (Qe_3), with long axes trending between N 45° W and N 60°W and crest elevations reaching 60 m above sea level, overlap the transverse paleodunes toward the Santa Maria River (Fig. 1). Toward Mussel Rock ravine, lobate dunes (Qe_4) overlap the parabolic dunes directly onto the transverse paleodunes and their associated soil, extending 1 km inland above the ravine. The dunes also overlie fluvial deposits and paleodunes in a 200- to 600-m-wide belt climbing to an elevation of 150 m above sea level between the ravine and Point Sal, blocking off two small drainage basins en route (Fig. 3). North of the ravine, both lobate and parabolic dunes are mantled by a thin sheet of active dunes (Qe_5) largely devoid of vegetation and whose transverse ridges trend N 45° E in the north and N 60° E toward Mussel Rock; the active dunes are migrating southeast at a rate of 2 to 5 m/yr. (Fig. 4). Using the older dunes as a ramp, these active dunes climb to nearly 150 m above sea level above Mussel Rock ravine. Toward the north, the active dunes are composed largely of sand derived via the beach from the Santa Maria River. Farther south and inland, they comprise much-reworked older sand.

Despite the local absence of reliable dating materials, the parabolic dunes are morphologically similar to dunes that had become stabilized before 3 ka along the coast south of Morro Bay, some 40 km to the northwest (Orme, 1990). Above Mussel Rock, shells of *Mytilus californianus, Protothaca staminea, Acanthina spirata*, and *Ostrea lurida* from a midden within the lobate dunes yielded an uncorrected ^{14}C age of 1,120±80 yrs b.p. (Beta-31913). Approximately 400 m to the southeast, above the south slopes of Mussel Rock ravine, shells of *Mytilus californianus* from a midden at a depth of 1 m, within a 2-m-thick paleosol formed in lobate dunes, yielded an uncorrected ^{14}C age of 850±80 yrs b.p. (Beta-31525).

DISCUSSION AND SUMMARY

The fluvial deposits (Qf_1–Qf_3) are identified with the Orcutt sand (Woodring and Bramlette, 1950) or Orcutt Formation (Worts, 1951). Radiocarbon ages suggest that the fluvial deposits exposed in Mussel Rock ravine occur late in the Orcutt sequence, between 30 and 23 ka. Their lithology indicates erosion of the Point Sal Ridge and Casmalia Hills, whereas their abundant sands and iron pisoliths are largely derived from the strongly indurated eolian sands

FIG. 4.—Coastal dunes extending north past Mussel Rock to the Santa Maria Valley. In the foreground, south of Mussel Rock ravine, lobate dunes mantle transverse paleodunes along the cliff top. North of the ravine, active transverse dunes largely overlap lobate and parabolic dunes (Spence Collection, UCLA, November 1947).

(Qe_1) with ferricrete horizons that emerge from beneath the fluvial deposits on the ridge's higher slopes. Shale-rich fluvial deposits north of Corralitos Canyon have a more distant source within the Santa Maria basin. The greenish silts and gray clays were mostly deposited under paludal conditions, probably in shallow ponds created by downstream accumulations of dune sand. Well-sorted, cross-bedded eolian sands (Qe_2) interfinger with and overlie the higher fluvial and paludal units and at the surface provide a widespread blanket of subdued transverse paleodunes. These are overlain by mostly unconsolidated sands of the stabilized parabolic (Qe_3) and lobate (Qe_4) dunes and by active transverse dunes (Qe_5). Color variations in the deposits, though remarkable, provide no indication of age but reflect weathering changes linked to erratic water-table fluctuations in materials of variable porosity and permeability.

The deposits provide valuable information on late Quaternary events along the south-central California coast. Suggested relations with oxygen isotope stages and sea-level trends are shown in Table 1. The marine-terrace deposits that emerge from beneath the fluvial and eolian complex around Mussel Rock are of uncertain age. However, amino-acid analysis of the bivalve *Protothaca staminea* from the lowest marine terrace around Point Sal suggests correlation with a similar low terrace flanking the San Luis Range, 30 to 40 km to the northwest, which $^{230}Th/^{234}U$ ages place around 85 to 80 ka, or oxygen isotope substage 5a (PG & E, 1988; Clark, 1990; Hanson and others, this volume). Because eustatic sea level was around 5 m below present in substage 5a (Bloom and others, 1974), some tectonic uplift of the Casmalia Hills is indicated. The Mussel Rock marine terrace could be younger, possibly stage 3 at about 50 to 40 ka when eustatic sea level was around 40 m below the present, but this implies a very high rate of tectonic uplift. It could be older, perhaps related to substage 5e around 130 to 120 ka, but its elevation 6–10 m above sea level would then imply no subsequent tectonic uplift, a view inconsistent with other evidence (Clark, 1990).

It is reasonable to place the dissected paleodunes (Qe_1) in stage 4 during the eustatic emergence of the substage 5a coastline, and the fluvial deposits (Qf_1–Qf_3) in an aggradation phase linked to a higher base level during stage 3. During the transition between stages 3 and 2, when sea level fell rapidly and the Santa Maria River and other streams prograded across the emerging continental shelf, abundant sand became available for dune construction and the transverse paleodunes (Qe_2) formed. The paleodunes, whose surface slopes beneath present sea level, probably continued to accumulate during stage 2, the last glacial maximum, but have since been lost or submerged seaward of the present coast.

The bracketing ages of 30 to 23 ka for the fluvial deposits, which are thought to cap a much thicker fluvial sequence, and the ^{14}C age of 23,560±360 yrs b.p. near the base of the transverse paleodunes support the above scenario. Whereas Bard and others (1990) indicated that ^{14}C ages may be systematically younger than U-Th ages calibrated through dendrochronology, with a maximum difference of about 3,500 years at 20 ka, this potential discrepancy does not affect the overall scenario, although it may expand the time frame for its accomplishment.

The parabolic dunes (Qe_3) probably began accumulating during the Flandrian transgression late in stage 2. Whatever their initial shape, it is likely that the parabolic form developed as sand supplies became restricted by rising sea level, until the noses of the parabolas became stabilized some distance downwind. Near Morro Bay, 40 km northwest of Mussel Rock, midden materials beneath the noses of parabolic dunes have yielded ^{14}C ages between 3,080±90 and 4,160±70 yrs b.p., indicating the time around which these migrating dune forms were stabilized (Orme, 1990). The lobate dunes (Qe_4) represent renewed eolian activity of unknown cause, although similar lobate dunes near Morro Bay overlie an extensive fire horizon around 1,730±90 yrs b.p., which was caused either by lightning or early hunting peoples. Lobate dunes around Mussel Rock have yielded uncorrected ^{14}C ages of 1,120±80 and 850±80 yrs b.p. The present transverse dunes were probably activated within the past 200 years.

The depositional sequence is shown schematically in Figure 3. The slope at the base of the later eolian units (Qe_2–Qe_5) is largely explained by deposition on a coastal ramp, with dunes feathering out against the Point Sal Ridge. Such an explanation, however, is less feasible for the fluvial deposits, whose subcrop slope, together with that of the underlying dissected paleodunes (Qe_1), must have been affected by late Quaternary tectonic uplift of the Point Sal Ridge, which Clark (1990) estimated at 0.14–0.17 mm/yr. The subsurface dimensions of these deposits are uncertain, particularly as the sequence has probably been influenced by the Orcutt Frontal Fault, which reaches the coast north of Mussel Rock.

Extensive paleodunes similar to Qe_2 deposits occur elsewhere along the California coast, notably behind Monterey Bay, Morro Bay, and Santa Monica Bay. These dunes were probably also initiated during the transition between oxygen-isotope stages 3 and 2. It is reasonable to assume that the formation of extensive transverse dunes is favored during marine regressions when abundant sand supplies become available as rivers extend their courses across continental shelves mantled with stranded marine sediments. Conversely, during marine transgressions, sand supplies become limited, although the instability of the coastal zone probably maintains vigorous beach-dune interaction. As sea-level rise slows down or ceases, sand supplies are progressively restricted and parabolic dunes tend to form

TABLE 1.—LATE QUATERNARY DEPOSITS NEAR POINT SAL AND SUGGESTED OXYGEN ISOTOPE STAGES AND SEA-LEVEL RELATIONS

Late Quaternary Deposits	Oxygen Isotope Stage	Relative Sea Level
Qe_5 Active transverse dunes		High
Qe_4 Lobate dunes	1	High
Qe_3 Parabolic dunes		Rising to high
Qe_2 Transverse paleodunes	2	Low
Qf_3 Upper fluvial unit		Falling to low
Qf_2 Mixed fluvial-paludal unit	3	High
Qf_1 Lower fluvial unit		High
Qe_1 Dissected paleodunes	4	Low
Qm Marine-terrace deposits	5a	High

downwind as winds shift existing sand inland. Lobate dunes are intermediate between transverse and parabolic dunes.

Thus, it is hypothesized that extensive transverse dunes form during periods of relative sea-level fall and plentiful sand budgets, whereas parabolic dunes form during and after periods of relative sea-level rise and diminished sand budgets. This scenario does not explain why thin transverse dunes are now forming locally north and south of the Santa Maria River mouth, but it is likely that land-use changes within the past 200 years have increased local sediment budgets, both from rivers and from destabilization of older dunes by overgrazing and other activities.

ACKNOWLEDGEMENTS

Reviews of this paper by G. B. Griggs and K. Berryman are gratefully acknowledged. The author also thanks Richard Ford, Lin Xiao-Dong, and Amalie Jo Orme for field and laboratory assistance. The study was supported by a grant from the Academic Senate, University of California, Los Angeles.

REFERENCES

BARD, E., HAMELIN, B., FAIRBANKS, R. G., AND ZINDLER, A., 1990, Calibration of the ^{14}C timescale over the past 30,000 years using mass spectrometric U-Th ages from Barbados corals: Nature, v. 345, p. 405–410.

BLOOM, A. L., BROEKER, W. S., CHAPPELL, J. M., MATTHEWS, R. K., AND MESOLELLA, K. J., 1974, Quaternary sea level fluctuations on a tectonic coast: New $^{230}Th/^{234}U$ dates from the Huon Peninsula, New Guinea: Quaternary Research, v. 4, p. 185–205.

CLARK, D. G., 1990, Late Quaternary tectonic deformation in the Casmalia Range, coastal south-central California, *in* Lettis, W. R., Hanson, K. L., Kelson, K. I., and Wesling, J. R., eds., Neotectonics of South-Central Coastal California: Friends of the Pleistocene, Pacific Cell, Field Trip Guidebook, p. 349–383.

COOPER, W. S., 1967, Coastal Dunes of California: Geological Society of America Memoir 104, 131 p.

FAIRBANKS, H. W., 1896, The geology of Point Sal: University of California Publications in the Geological Sciences, v. 2, p. 1–92.

HALL, C. A., 1982, Pre-Monterey subcrop and structure contour maps, western San Luis Obispo and Santa Barbara counties, south-central California: U.S. Geological Survey Map MF-1384, 28 p.

ORME, A. R., 1990, The instability of Holocene coastal dunes: the case of the Morro dunes, California, *in* Nordstrom, K. F., Psuty, N. P., and Carter, R. W. G., eds., Coastal Dunes: Form and Process: Wiley, New York, p. 317–338.

ORME, A. R., AND TCHAKERIAN, V. P., 1986, Quaternary dunes of the Pacific coast of the Californias, *in* Nickling, W. G., ed., Aeolian Geomorphology: Allen and Unwin, London, p. 149–175.

PG & E (PACIFIC GAS AND ELECTRIC COMPANY), 1988, Final report of the Diablo Canyon long-term seismic program: U.S. Nuclear Regulatory Commission Docket Nos. 50–275, and 50–323 (unpaginated).

WOODRING, W. P., AND BRAMLETTE, M. N., 1950, Geology and paleontology of the Santa Maria district, California: U.S. Geological Survey Professional Paper 222, 185 p.

WORTS, G. F., 1951, Geology and ground-water resources of the Santa Maria valley area, California: U.S. Geological Survey Water Supply Paper 1000, 169 p.

AMINOSTRATIGRAPHY OF SOUTHERN CALIFORNIA QUATERNARY MARINE TERRACES

JOHN F. WEHMILLER
Department of Geology, University of Delaware, Newark, Delaware 19716

ABSTRACT: Amino-acid enantiomeric ratios, or D/L values, in molluscs from central and southern California Quaternary coastal units are compared and summarized using a combination of stratigraphic control, kinetic modeling, and current temperature differences along the coast. Methods for comparison of analytical data obtained by both gas-chromatographic and liquid-chromatographic methods are used to normalize results obtained over nearly 15 years into a common format. Most of the results imply ages that fall within stage 5 of the marine oxygen isotope record. Distinct aminozones representing at least three pre-stage 5 depositional episodes are also recognized. Discrepancies between amino-acid age estimates and those derived from independent evidence are quantified using kinetic modeling, and an anomalous local thermal history is invoked to explain one particular conflict between aminostratigraphic and morphostratigraphic interpretations.

INTRODUCTION

Five papers in this volume employ aminostratigraphy in the analysis of the chronology of Quaternary coastal features in south-central and southern California (Hanson and others; Kennedy and others; Kern and Rockwell; Muhs and others; Rockwell and others). This paper integrates these localized aminostratigraphic sections (based on data obtained by several laboratories and/or analytical methods over a period of nearly 15 years) into a regional framework that incorporates some recent developments in the modeling of amino-acid racemization kinetics (Wehmiller and others, 1988a; Hsu and others, 1989; Mitterer and Kriausakul, 1989). In addition, this approach facilitates comparison of aminostratigraphic results with other, independent chronostratigraphic information.

THE REGIONAL AMINOSTRATIGRAPHIC FRAMEWORK

Figure 1 shows the specific coastal segments of California for which aminostratigraphic data are discussed here. Some of these results and the basic format for their regional comparison and correlation have been presented elsewhere (Wehmiller and others, 1977; Lajoie and others, 1980; Kennedy and others, 1982; Wehmiller, 1982, 1984a, 1990). In these publications, and in related research, D/L data for two kinetically similar amino acids, leucine and isoleucine, have been most frequently reported, although the gas-chromatographic (GC) methods that yield leucine D/L values also yield D/L data for four to seven other amino acids[1] (e.g., Muhs and others, this volume). D-alloisoleucine/L-isoleucine values (hereafter referred to as A/I values), determined by conventional ion-exchange high-pressure liquid chromatography (HPLC), are more commonly utilized by most aminostratigraphic laboratories because of several analytical advantages of the HPLC method. Although leucine and isoleucine are chemically similar, differences in their analytical chemistry and kinetics of racemization (or epimerization, in the L-isoleucine to D-alloisoleucine conversion) require that a normalization procedure exists if the two types of results are to be compared. For leucine and isoleucine data obtained in the same laboratory, on the same mollusc sample, equation 1 (from Wehmiller and others, 1988b) describes the relation between A/I values and D/L leucine values:

$$A/I = 1.2712 \times [(D/L)^{1.4625}] \quad (1)$$

In principle, equation (1) can be used to convert leucine D/L values reported in this volume to the A/I values that are now more commonly measured. Detailed comparisons of results obtained in different laboratories are not easily quantified because of known interlaboratory differences (Wehmiller, 1984b), but these factors are not significant for the questions addressed in this paper.

Presentations of aminostratigraphic data often show results as plots of D/L (A/I) values against current mean annual temperatures (CMAT) within a region of study (Wehmiller, 1982; 1990; Miller and Mangerud, 1985; Hearty and others, 1986; Bowen and Sykes, 1988; Clark and others, 1989). This format incorporates the two basic assumptions of aminostratigraphy: (1) that sites within a study region with the same (or similar) CMAT values have had similar temperature histories, so that differences in D/L values for samples at these sites can be interpreted to represent age differences, and (2) that interregional differences in CMAT represent approximate differences between the effective temperatures (the integrated kinetic effect of the entire Quaternary temperature history) for these regions. Isochrons, lines connecting D/L values for coeval samples at different CMATs, are theoretically smooth functions of temperature (Wehmiller, 1982; Wehmiller and others, 1988a).

Used here as a reference isochron for comparison of leucine and isoleucine data from California coastal terraces is

$$A/I = 0.1709 \times 10^{[0.0242 \times (CMAT)]} \quad (2)$$

(from Wehmiller and others, 1988a).

Equation (2), based partly on data obtained from calibrated terrace localities in central and southern California, relates A/I values in ~120-ka (substage 5e)[2] samples of the bivalve mollusc *Protothaca staminea* to CMAT. Equation (2) is a regression to data from calibrated (isotopically

[1]alanine, valine, glutamic acid, proline, phenylalanine, aspartic acid, and occasionally isoleucine.

[2]In this and previous publications, substage 5e has been "assigned" the age of 120 ka, although it is widely recognized that this substage had finite duration and that calibration samples from this substage might be between 135 and 120 ka in age. Any changes in the age assigned to these samples (and the associated isochron) would produce proportionate changes in the estimated ages of other samples (isochrons) that depend on this calibration.

Quaternary Coasts of the United States: Marine and Lacustrine Systems, SEPM Special Publication No. 48
 ISBN 0-918985-98-6

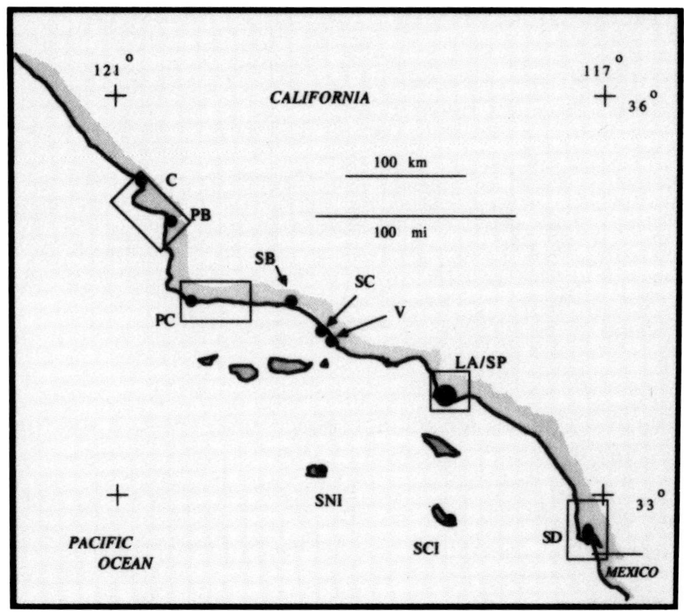

FIG. 1.—Map of central and southern California, showing aminostratigraphic study areas (in boxes) discussed in this volume. C, Cayucos; PB, Pismo Beach; PC, Point Conception; SB, Santa Barbara; SC, Sea Cliff; V, Ventura; LA, Los Angeles; SP, San Pedro; SNI, San Nicolas Island; SCI, San Clemente Island; SD, San Diego. Data for SB, V, SC, SNI, and SCI are cited in one or more of the papers for comparison with aminostratigraphic results in the study areas.

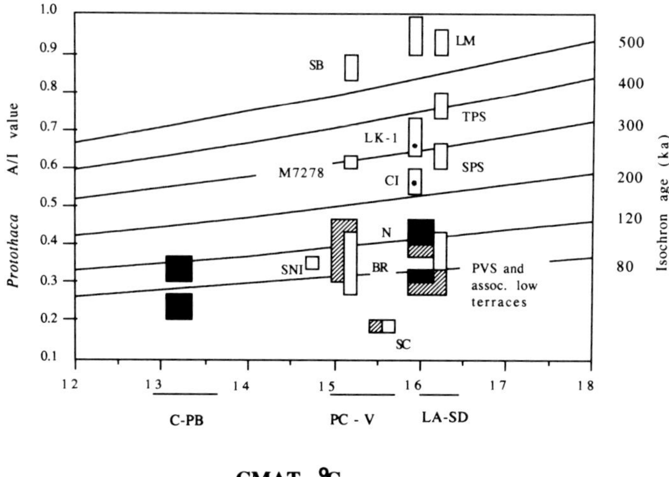

FIG. 2.—Kinetic-model isochrons for isoleucine (A/I values) in *Protothaca staminea*, developed as outlined in text. Abbreviations as in Figure 1, plus: SC, Sea Cliff; BR and N, Bird Rock and Nestor terraces, Point Loma, San Diego; CI and LK-1, Coronado Island and Solona Beach; PVS, SPS, TPS, and LM, Palos Verdes Sand, San Pedro Sand, Timms Point Silt, and Lomita Marl, respectively; SB, Santa Barbara Formation; M7278. Solid or scored boxes denote data actually obtained for *P. staminea*. Open boxes represent *Saxidomus* data converted to "equivalent *Protothaca* values" using equations in Lajoie and others (1980). Pairs of squares for Nestor and Bird Rock terraces, and for C-PB region, depict two regions where aminostratigraphic data conform to local stratigraphic control provided by a paired early to late stage 5 terrace sequence. Scored boxes show that data for Point Conception and Palos Verdes Sand (and San Pedro terrace sites) are broadly correlative with stage 5 calibrations; the San Pedro data, in most cases, conform to an early to late stage 5 sequence that is consistent with local terrace geomorphology. Data for M7278, CI, and LK-1 represent significant conflicts with independent chronostratigraphic information, as discussed in text. Data (from *Epilucina* and *Tegula* samples) for San Nicolas and San Clemente Islands (Fig. 1) are discussed in Muhs and others (this volume) and in Muhs (1983, 1985) but are not plotted here because of difficulty of conversion to equivalent *Protothaca* values. Data for San Nicolas Island from Wehmiller and others (1977), as discussed in Wehmiller (1984a, 1990).

or otherwise—see Kennedy and others, 1982; Wehmiller, 1990) terrace localities between coastal Washington, USA, and Baja California Sur, Mexico (CMAT range of approximately 11°C.).

Equation (2) can serve as the theoretical basis for calculation of other isochrons, *if a suitable kinetic model* is selected and *if the assumption of identical effective temperatures* (for sites within a given CMAT region) is applied. The simplest equation for this purpose is that for parabolic kinetics (Mitterer and Kriausakul, 1989):

$$A/I = k \times [t^{0.5}] \quad (3)$$

Although this parabolic kinetic model yields age estimates that are not identical to those previously derived from southern California racemization data (Wehmiller and others, 1977; Wehmiller and Belknap, 1978), it is used here to develop a generalized aminostratigraphic framework for coastal California. Differences between the parabolic kinetic-model age estimates and those derived from other kinetic models (Wehmiller and others, 1977; Wehmiller and Belknap, 1978) are relatively small (±10 percent) for D/L or A/I values less than 0.70.

Figure 2 shows an array of isochrons for specified ages, using equation (2) as the calibration (120 ka) isochron and using equation (3) to derive the isochrons for other ages. The uncertainty of this modeling approach is difficult to quantify because of the numerous assumptions that are inherent to any kinetic and/or temperature model, but a model such as that shown in Figure 2 is useful because it permits reasonable estimates of age or temperature differences for samples that can be assumed to have "experienced" common environmental (temperature) histories. Examples of these estimates are described later.

Also shown in Figure 2 are the ranges of A/I values—(converted, where necessary, using equation 1)—for the four coastal segments outlined in Figure 1. The data shown in Figure 2 are for *Protothaca staminea*, a frequently analyzed species, although in a few cases data for *Saxidomus* are also incorporated into the figure (using intergeneric conversions from Lajoie and others, 1980) because of the general analytical reliability of this genus (Kennedy and others, 1982). Other amino-acid isochrons and D/L values could be presented in the same format as shown in Figure 2, but, in all but a few cases, the relative chronologic information gained from these plots would not change the conclusions derived from Figure 2. Although each isochron actually represents an age range because of the inherent uncertainties of kinetic modeling, the model shown in Figure 2 agrees well with the few available middle Pleistocene age controls. In particular, data for samples from the Ventura Basin (Wehmiller, 1990, fig. 14) and the Lomita Marl (Ponti and others, 1987) constrain the position of any "middle Pleistocene" isochrons (~800–700 ka) because of radiometric and/or paleomagnetic control in this age range.

DISCUSSION

The results plotted in Figure 2 indicate that the majority of the analyzed California terrace deposits discussed in this volume fall within the range of 120 to 80 ka, equivalent to all of stage 5 of the marine oxygen isotope record (Shackleton and Opdyke, 1973). Paired aminozones representing early and late stage 5 terraces in stratigraphic succession are found in San Diego (Point Loma, Nestor and Bird Rock terraces), San Pedro, and in the Cayucos-Pismo Beach region. Aminozones representing older apparent ages are found in stratigraphically older units or on older (higher) terraces in the San Diego, Los Angeles (San Pedro), Ventura, and Santa Barbara areas (Figs. 1, 2). A distinctly younger aminozone is found in the Sea Cliff area (Figs. 1, 2), where rapid tectonic uplift has exposed a sequence of latest Pleistocene and Holocene emergent terraces (Lajoie and others, 1979).

The results that plot within the stage 5 range are not always consistent with independent stratigraphic controls, and these discrepancies raise questions about the true resolving power of aminostratigraphic methods for the relatively small age differences within stage 5. In addition, in a few instances, the independent evidence yields ages between 120 and 80 ka, but D/L data suggest ages two to three times greater. These discrepancies are discussed in detail in papers elsewhere in this volume, but a few of these cases are mentioned here to provide examples of the use of the isochron models (Fig. 2) in the quantification of apparent conflicts.

Examples of D/L values being much higher than expected, given their geomorphic or stratigraphic position, include the Bird Rock terrace at Solona Beach (site LK-1) and the Bay Point Formation at Coronado Island (site CI); both in the San Diego area (Kern and Rockwell, this volume), and the second terrace at Cañada de Alegria (USGS loc. M7278; Kennedy and others, this volume; Rockwell and others, this volume) east of Point Conception. In each of these cases, racemization data indicate an age of ~300 ka, but geomorphic evidence suggests an age within isotope stage 5. Data for locality M7278 can be explained as a consequence of a probable local anomaly in temperature history, as hydrothermal activity (not noted at the time of sample collection) associated with the nearby South Branch of the Santa Ynez Fault could have caused more extensive racemization of these samples (Rockwell and others, this volume). Results from this locality provoked extensive discussion (see Kennedy and others, and Rockwell and others, this volume) because they were reproducible from several collections but difficult to reconcile with the terrace elevation. Subsequent analyses of two *Saxidomus* samples from M7278 (unpublished data) revealed anomalously low D/L values in the free amino-acid fraction, suggesting an unusual diagenetic history for these samples. If the M7278 terrace is assigned an age of *roughly* 100 ka (as required by its geomorphic position), then the aminostatigraphic results can be explained (using the isochrons in Fig. 2) by an effective temperature about 8°C warmer than for other nearby localities. This estimate does not seem unreasonably large, and the postulated history of these samples compares well with a similar "thermal alteration" record demonstrated by Miller and Hopkins (1980).

No such obvious thermal factors can be cited as explanations for the anomalously high ratios from the two San Diego localities, although there is some concern that the very shallow-burial depths (<0.5 m) for some of the bay deposits (e.g., Coronado Island; Kern and Rockwell, this volume) may have resulted in higher effective temperatures than for some of the open-coast sites in the region. Site LK-1 in Solona Beach remains as a serious unresolved question, because its geomorphic position argues for an 80-ka age (Kern and Rockwell, this volume), whereas all the D/L values (on well-preserved samples of multiple genera) argue for a pre-stage 5 age assignment. Reworking of older shells into a younger deposit is one obvious mechanism that could produce "higher-than-expected" D/L values, but even reworking cannot be easily invoked in the case of site LK-1, because (1) all the analyzed shells were complete (although not articulated) valves, and (2) such an argument would require that 100 percent of the sampled molluscan population had been reworked. Slumping of a discrete block of older terrace sediments onto a younger terrace deposit could produce these observed results, but evidence of such slumping was not apparent at the time of shell collection.

Samples altered by diagenetic leaching or contamination might be expected to exhibit D/L values lower than their "true" values, due to (1) the addition of L amino acids by contamination, or (2) the removal of extensively racemized free amino acids by leaching. Several probable examples of these processes can be found in the California terrace studies. Two terraces in the Cayucos–Pismo Beach region yielded chalky shell fragments that gave "apparent amino-acid age estimates" of late stage 5, or roughly 75±15 ka, but these age estimates are younger than the ages derived from independent evidence (Hanson and others, this volume, table 1, sites W87–89 and G86–100). These particular shells were generally in poor condition (fragmented, chalky), and both qualitative and quantitative analyses of the chromatograms and amino-acid abundances indicated that the samples were of questionable value. Nevertheless, samples from other terraces in the region had similar preservation characteristics and chromatographic appearances and often gave "reasonable" results. Without the independent control, therefore, the results for W87–89 and G86–100 would not have been recognized as anomalous. Samples from USGS locality M7822, near Cañada de Alegria east of Point Conception, apparently also show the effects of leaching. Here, two chalky fragments of *Saxidomus* (usually considered a reliable genus because of the valve's robust nature), yielded D/L values suggesting an early stage 5 (~120 ka) age assignment. These results, however, were questioned because that terrace remnant is significantly higher than the one considered to be the region's 120-ka surface (Rockwell and others, this volume). The chalky nature of the samples was noted at the time of analysis, but the chromatographic quality did not appear to be unusual. This question can only be resolved with additional analyses and more thorough evaluation of the preservation state of the samples in question.

Several *Protothaca staminea* samples (mostly shell fragments) from terraces in the region east of Point Conception yielded D/L results of questionable value in resolving terrace ages within the stage 5 range; fortunately, geomorphic discrimination of the terraces (just as in the Cayucos–Pismo Beach region) permits evaluation of the *Protothaca* results. D/L leucine values ranging from 0.35 to 0.48 are found in *Protothaca* samples from the three stage 5 terraces east of Point Conception (Kennedy and others, this volume; Rockwell and others, this volume), but rarely do the *Protothaca* D/L values conform to the local morphostratigraphic sequence. This range of D/L values is typical for other early to late stage 5 sequences (i.e., the Nestor–Bird Rock terrace pair on Point Loma, near San Diego), but the inconsistency of the D/L results with the terrace sequence indicates that the currently available data for this genus in the Point Conception area have not been successful in resolving age differences *within stage 5*. Conversely, *Saxidomus* data from early and late stage 5 terraces east of Point Conception appear to have resolved these age differences (Kennedy and others, this volume; Rockwell and others, this volume). Limited data for the A/I value in the free amino-acid pool from selected *Protothaca* samples (Rockwell and others, this volume) suggest that this measurement is successful in resolving the slight age differences within stage 5, but further analyses are required to explore this potentially useful approach to high-resolution aminostratigraphic study of late Pleistocene terraces along the North American Pacific coast. Because many of the problematic *Protothaca* samples from the Point Conception area were fragments, it is possible that reworking of older material (within stage 5) has resulted in the mixing of samples of different ages on these closely spaced terraces. Unfortunately, the statistics of the results do not permit a clear evaluation of this possibility.

SUMMARY

The geochronology of marine-terrace and basin deposits along the central and southern California coast is a major component of any study of coastal evolution in the region. Aminostratigraphy has proven to be useful in such efforts, although examples of the ''failure'' of the method can be identified. Kinetic-model isochrons provide a chronostratigraphic framework for comparison of results from different temperature regions, but the inherent uncertainties of the isochrons and potential differences in environmental histories of analyzed samples must be appreciated and constantly re-evaluated. The dominant aminozone from central and southern California represents coastal deposition or terrace formation during the (broadly defined) ''last interglacial'' (stage 5); at least three additional early to middle Pleistocene aminozones are identified in the Ventura, San Pedro/Los Angeles, and San Diego regions. Because of the resolving power of aminostratigraphy, it is probable that each of these aminozones represents more than one interglacial stage of coastal deposition or terrace formation, as inferred from D/L data for higher terraces in the Palos Verdes Hills/San Pedro region.

ACKNOWLEDGMENTS

Critical comments from George Kennedy, Kathryn Hanson, Daniel Muhs, and Thomas Rockwell aided in preparation of this paper. Support for aminostratigraphic research at the University of Delaware from the U.S. Geological Survey, the National Science Foundation, Woodward-Clyde, and Geomatrix Consultants is gratefully acknowledged.

REFERENCES

Bowen, D. Q., and Sykes, G. A., 1988, Correlation of marine events and glaciations on the northeast Atlantic margin: Philosophical Transactions of the Royal Society of London, v. B318, p. 619–635.

Clark, P. U., Nelson, A. R., McCoy, W. D., Miller, B. B., and Barnes, D. K., 1989, Quarternary aminostratigraphy of Mississippi Valley loess: Geological Society of American Bulletin, v. 101, p. 918–926.

Hearty, P. J., Miller, G. H., Stearns, C. E., and Szabo, B. J., 1986, Aminostratigraphy of Quaternary shorelines in the Mediterranean Basin: Geological Society of America Bulletin, v. 97, p. 850–858.

Hsu, J. T., Leonard, E. M., and Wehmiller, J. F., 1989, Aminostratigraphy of Peruvian and Chilean Quaternary marine terraces: Quaternary Science Reviews, v. 8, p. 255–262.

Kennedy, G. L., Lajoie, K. R., and Wehmiller, J. F., 1982, Aminostratigraphy and faunal correlations of late Quaternary marine terraces, Pacific coast, USA: Nature, v. 299, p. 545–547.

Lajoie, K. R., Kern, J. P., Wehmiller, J. F., Kennedy, G. L., Mathieson, S. A., Sarna-Wojcicki, A. M., Yerkes, R. F., and, McCrory, P. A., 1979, Quaternary marine shorelines and crustal deformation, San Diego to Santa Barbara, California, *in* Abbott, P. L., ed., Geological Excursions in the Southern California Area: San Diego State University, San Diego, p. 3–15.

Lajoie, K. R., Wehmiller, J. F., and Kennedy, G. L., 1980, Inter- and intrageneric trends in apparent racemization kinetics of amino acids in Quaternary mollusks, *in* Hare, P. E., Hoering, T. C. and King, K., Jr., eds., Biogeochemistry of Amino Acids: John Wiley, New York, p. 305–340.

Miller, G. H., and Hopkins, D. M., 1980, Degradation of molluscan shell protein by lava-induced transient heat flow, Pribilof Islands, Alaska: implications for amino acid geochronology and radiocarbon dating, *in* Hare, P. E., Hoering, T. C., and King, K., Jr., eds., Biogeochemistry of Amino Acids: John Wiley, New York, p. 445–452.

Miller, G. H., and Mangerud, J., 1985, Aminostratigraphy of European marine interglacial deposits: Quaternary Science Reviews: v. 4, p. 215–278.

Mitterer, R. M., and Kriausakul, N., 1989, Calculation of amino acid racemization ages based on apparent parabolic kinetics: Quaternary Science Reviews, v. 8, p. 353–357.

Muhs, D. R., 1983, Quaternary sea level events on northern San Clemente Island, California: Quaternary Research, v. 20, p. 322–341.

Muhs, D. R., 1985, Amino acid age estimates of marine terraces and sea levels on San Nicolas Island, California: Geology, v. 13, p. 58–61.

Ponti, D. J., Lajoie, K. R., and Appel, S. H., 1987, Aminostratigraphic classification of the ''type'' Quaternary marine formations from the San Pedro–Palos Verdes area, Los Angeles Basin, California: Geological Society of America Abstracts with Programs, v. 19, p. 440–441.

Shackleton, N. J., and Opdyke, N, D., 1973, Oxygen isotope and palaeomagnetic stratigraphy of equatorial Pacific core V28-238: oxygen isotopic temperatures and ice volumes on a 10^5 year and 10^6 year scale: Quaternary Research, v. 3, p. 39–55.

Wehmiller, J. F., 1982, A review of amino acid racemization studies in Quaternary mollusks: stratigraphic and chronologic applications in coastal and interglacial sites, Pacific and Atlantic coasts, United States, United Kingdom, Baffin Island, and tropical islands: Quaternary Science Reviews, v. 1, p. 83–120.

Wehmiller, J. F., 1984a, Relative and absolute dating of Quaternary mollusks with amino acid racemization: evaluation, application, questions, *in* Mahaney, W. C., ed., Quaternary Dating Methods: Elsevier, Amsterdam, p. 171–193.

Wehmiller, J. F., 1984b, Interlaboratory comparison of amino acid enantiomeric ratios in fossil mollusks: Quaternary Research, v. 22, p. 109–120.

Wehmiller, J. F., 1990, Amino acid racemization: applications in chemical taxonomy and chronostratigraphy of Quaternary fossils, *in* Carter, J. G., ed., Skeletal Biomineralization: Patterns, Processes, and Evolutionary Trends, v. 1: Van Nostrand Reinhold, New York, p. 583–608.

Wehmiller, J. F., and Belknap, D. F., 1978, Alternative kinetic models for the interpretation of amino acid enantiomeric ratios in Pleistocene mollusks: examples from California, Washington, and Florida: Quaternary Research, v. 9, p. 330–348.

Wehmiller, J. F., Lajoie, K. R., Kvenvolden, K. A., Peterson, E., Belknap, D. F., Kennedy, G. L., Addicott, W. O., Vedder, J. G., and Wright, R. W., 1977, Correlation and chronology of Pacific Coast marine terraces of continental United States by amino acid stereochemistry: technique evaluation, relative ages, kinetic model ages, and geologic implications: U.S. Geological Survey Open-File Report 77–680, 196 p.

Wehmiller, J. F., Hsu, J. T., and Leonard, E. M., 1988a, Latitudinal isochrons of amino acid enantiomeric ratios in Quaternary molluscs: implications of North and South American Pacific Coast data for aminostratigraphy and effective temperature gradients: Geological Society of America Abstracts with Programs v. 20, p. A53.

Wehmiller, J. F., Belknap, D. F., Boutin, B. S., Mirecki, J. E., Rahaim, S. D., and York, L. L., 1988b, A review of the aminostratigraphy of Quaternary mollusks from United States Atlantic Coastal Plain sites, *in* Easterbrook, D. L., ed., Dating Quaternary Sediments: Geological Society of America Special Paper 227, p. 69–110.

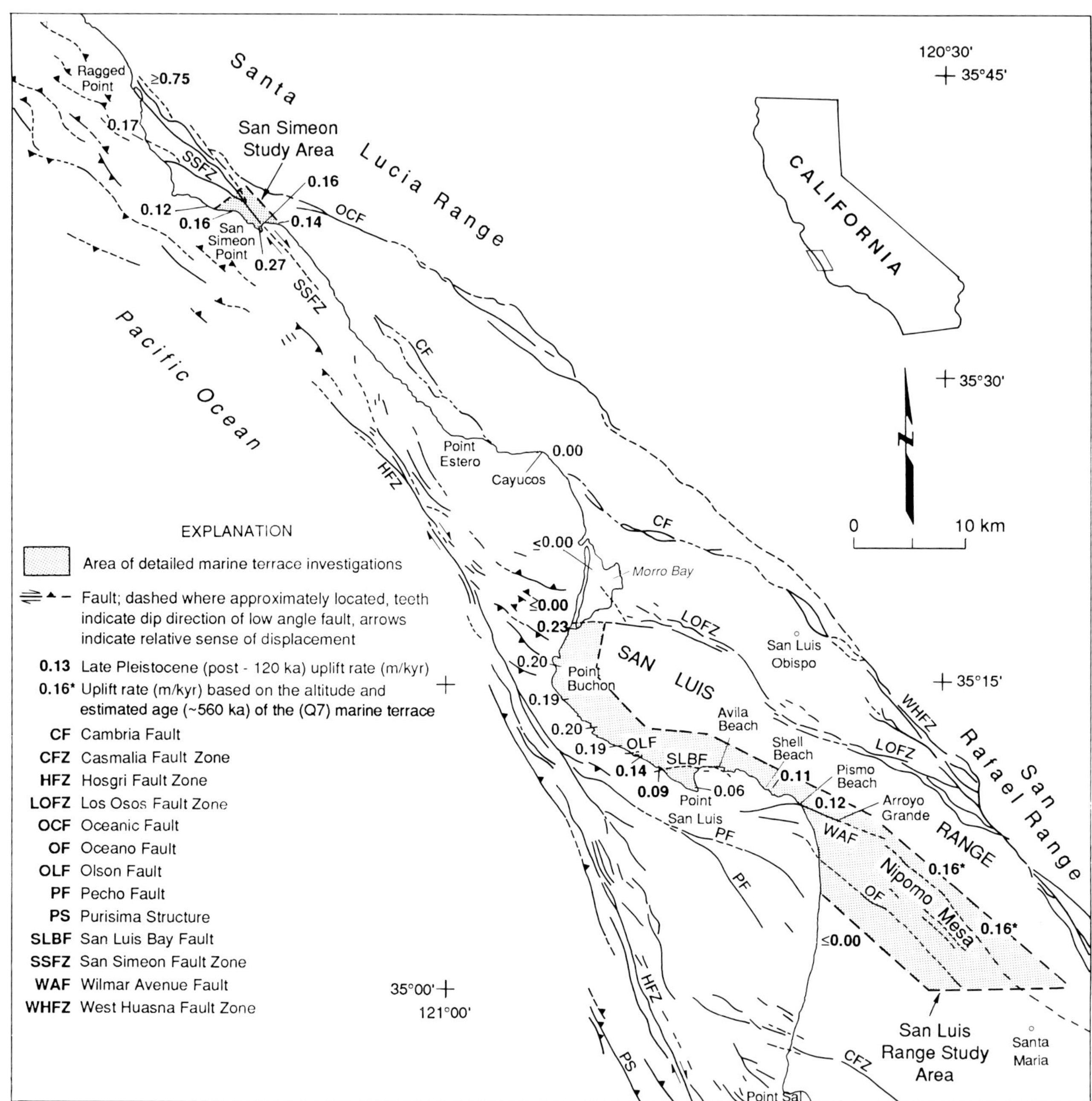

FIG. 1.—Map showing locations of marine-terrace investigations and uplift rates along the south-central coast of California.

Lettis and others, 1990); and (3) assess Pleistocene slip rates for individual faults (Hanson and Lettis, 1990; Lettis and others, 1990; Lettis and Hall, 1990). These investigations indicate that the region north of approximately 35° latitude has undergone late Pleistocene contractional deformation via dip-slip faulting and block uplift (or subsidence).

MARINE-TERRACE DISTRIBUTIONS

Well-preserved flights of emergent marine terraces are present along the margin of the San Luis Range and along the coast between San Simeon and Cayucos (Fig. 1). These terraces provide a means to assess the rate and pattern of deformation in the region. However, emergent marine terraces are absent in the Morro Bay and Santa Maria Valley areas. In addition to the presence of Quaternary fluvial deposits at depths of about 200 m below sea level (PG&E, 1988), the lack of marine terraces in these areas suggests that the regions bordering the San Luis Range are subsiding.

We investigated the distribution of marine terraces throughout the San Simeon and San Luis Range areas.

Through a program of aerial photograph interpretation, aerial reconnaissance, detailed field mapping, seismic-refraction studies, and soil-geomorphology studies, detailed maps of the marine terraces were prepared at scales of 1:24,000, 1:12,000, and 1:2,400 between Morro Bay and the Santa Maria Valley and along the southern onshore reach of the San Simeon fault zone. Where the platforms are mantled by thick (up to 30 m) deposits, additional data on platform altitude were obtained by drilling 271 boreholes. In the coastal region between these two study areas, investigations included reconnaissance mapping of coastal landforms and Quaternary deposits to confirm the presence or absence of marine terraces. In the Cayucos area, data on soil development and age of the first emergent terrace was also obtained for use in calibrating terrace ages and correlating terrace sequences between the San Simeon and San Luis Range study areas.

San Luis Range Area

Prominent marine terraces flank the entire western and southwestern margins of the San Luis Range. The distribution and preservation of marine terraces, however, vary considerably within this coastal region. The lowest two terraces (Q_1 and Q_2) are well preserved and form relatively continuous, broad platforms throughout the region between Morro Bay and Pismo Beach. However, at Point San Luis, the lowest (Q_1) terrace is not preserved, possibly because the uplift rate is not high enough to allow preservation. Marine deposits up to 2 m thick overlying the Q_1 and Q_2 wave-cut platforms are typically overlain by 15 to 30 m of alluvial, colluvial, and eolian sediments that form a prominent terrace surface extending from near the range front to the coast. Altitudes of shoreline angles of the Q_1 terrace range from 14±2 m in the area just north of Point Buchon to 5±1 m just north of the San Luis Bay fault. The Q_2 terrace likewise decreases in altitude to the south from a maximum of 34±1 m in the region just south of Morro Bay to a minimum of 13±1 m in the Point San Luis region. Lower terraces are locally displaced and/or deformed by several reverse faults, including the Los Osos, San Luis Bay, and Wilmar Avenue faults and the inferred Olson fault (Figs. 2 and 3). Based on vertical displacements of terrace strandlines and platforms, these faults have low slip rates ranging from 0.04 to 0.08 mm/yr, depending on the dip of the fault (Kelson and others, 1987; Lettis and others, 1990).

Well-preserved flights of older terraces occur only in the coastal region between Morro Bay and Point Buchon (Montaña de Oro State Park) and in the Pismo Beach area (Fig. 1). Remnants of 12 platforms ranging in altitude from 14±1 m to 247±6 m are present in Montaña de Oro State Park (Fig. 2A). The wave-cut platforms of the higher terraces in this region commonly are well preserved and overlain by a veneer of marine gravel and eolian sand, alluvium, and/or colluvium. Along the steep rugged coastline between Point Buchon and the Olson fault (Fig. 1), however, only a few, small remnants of the higher marine terraces are present. The paucity of higher marine terraces in this area is not attributed to a lack of uplift, but rather to lack of preservation as a result of: (1) seacliff retreat during the 120-ka sea-level highstand, (2) erosion by hillslope processes, and (3) a limited initial extent of the platforms due to lack of development on resistant bedrock.

Higher terrace remnants also are present in the Point San Luis area between the Olson fault and Avila Beach. In general, remnants in this area are preserved as dissected erosional benches that typically are stripped of marine deposits. At least seven higher erosional benches are present in this area. However, lateral correlation of terrace remnants is complicated by deformation associated with the San Luis Bay fault and possibly the Olson fault (Fig. 1).

From Avila Beach south to Pismo Beach, higher terraces are extensively eroded and only locally preserved, probably because of the same factors presented above for poor preservation of higher terraces in the Point Buchon to Olson fault area. In the Pismo Beach area, at least nine terraces range in altitude up to 137±3 m and typically are spaced less than 15 m apart (Fig. 2B). South of Pismo Beach, lower terraces are buried beneath a 60- to 75-m-thick mantle of eolian sand and alluvium in the Arroyo Grande and Nipomo Mesa areas (Fig. 1). Two of the older terraces, at altitudes of approximately 87 and 110 m, are relatively continuous for at least 25 km south of Arroyo Grande Creek to the southern limit of our study area. In contrast to the terraces from Morro Bay to Pismo Beach, which parallel the present coastline, the terraces south of Pismo Beach diverge from the present coastline and extend to the southeast at least 21 km inland, suggesting that much of the Santa Maria Valley was inundated during the middle to late Pleistocene.

San Simeon Area

Along the southern onshore reach of the San Simeon fault zone are four terraces northeast of, and five terraces southwest, of the fault zone (Fig. 4). We assign informal names (modified from Weber, 1983) to each of these terraces, which are from youngest to oldest, the Point (Q_p), San Simeon (Q_s), Tripod (Q_t), Oso (Q_o), and La Cruz (Q_{lc}) terraces. Numerous higher, older terraces present in the area were not mapped in detail during this study. These higher terraces indicate that coastal uplift was occurring prior to development of the La Cruz terrace, probably throughout much of the Quaternary. A profile illustrating the altitudes of shoreline angles on both sides of the fault zone indicates substantial vertical deformation of the terraces (Fig. 5).

The lowest terrace, the Point Terrace, occurs only within and to the southwest of the San Simeon fault zone. The terrace, which was grouped with the next higher terrace by Weber (1983), is a narrow, discontinuous bench around San Simeon Point that has a shoreline angle altitude of 6.4 to 8.5 m. Terrace deposits overlying the wave-cut platform are displaced by the fault zone.

The San Simeon terrace is the lowest emergent terrace present on both sides of the San Simeon fault zone. Northeast of the fault zone, discontinuous, narrow remnants of the terrace have shoreline-angle altitudes of approximately 5 to 7 m. Fluvial erosion and subsequent burial by extensive eolian sand modify and obscure the platform near its intersection with the San Simeon fault zone on the north-

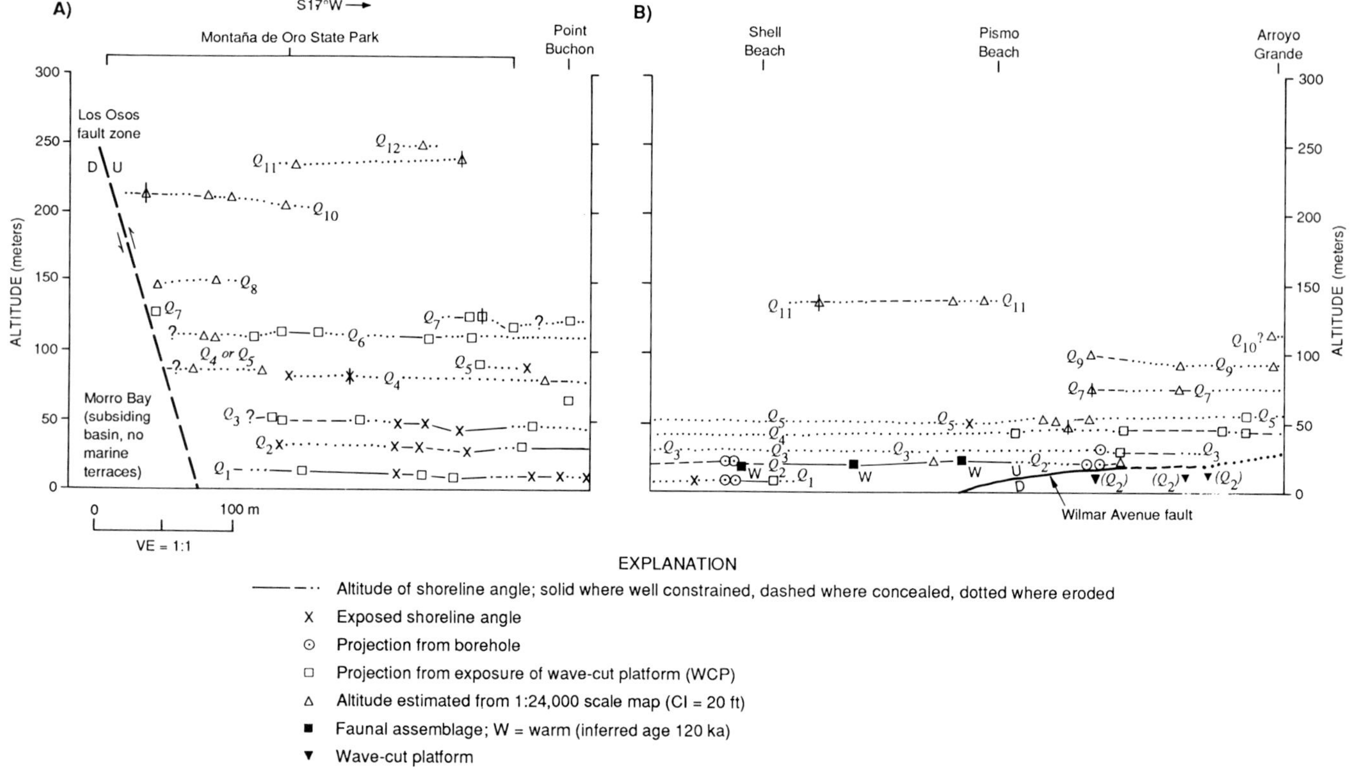

FIG. 2.—Shore-parallel profiles of flights of marine terraces in Montaña de Oro State Park (A) and Pismo Beach (B) areas. Note the greater terrace altitudinal spacing in the Montaña de Oro region, where uplift based on the ~120-ka (Q_2) terrace is ~0.2m/ka versus ~0.1 m/ka in the Pismo Beach area.

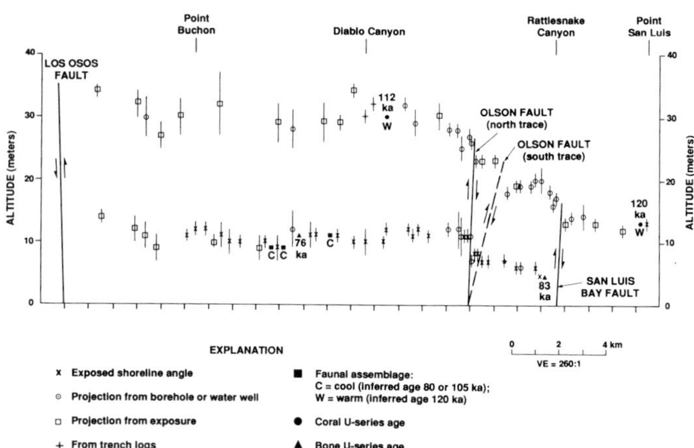

FIG. 3.—Shore-parallel profile of the Q_1 (80 ka) and Q_2 (120 ka) marine terrace-shoreline angles between the Los Osos fault and Point San Luis, illustrating deformation associated with the Los Osos, Olson and San Luis Bay faults.

eastern side of the fault. Within and directly southwest of the fault zone, the San Simeon terrace reaches its highest altitude of 24±1 m. With increasing distance from the fault zone, the altitude of the terrace shoreline angle southwest of the fault progressively decreases to 9±2 m (Fig. 5).

The Tripod terrace is the broadest and best expressed of the lower marine terraces on both sides of the fault zone. Northwest of Airport Creek on the southwestern side of the fault and on the entire northeastern side of the fault, the Tripod terrace is the second lowest emergent terrace. The Tripod terrace is the third highest terrace within and adjacent to the fault zone on the southwest side. Northeast of the fault zone, the terrace maintains a relatively consistent altitude of 23 to 25 m (Fig. 5). Southwest of the fault zone, the terrace occurs at an elevation of 21±1 m. Toward the fault zone, however, the terrace rises from 25 m near Adobe Creek to 38 m in the area bounded by the two primary strands of the fault zone (Fig. 5).

Weber (1983) mapped discontinuous-terrace remnants slightly higher than the Tripod terrace as the Oso terrace in the region north of our study area. Remnants of this terrace are present on both sides of the San Simeon fault zone along its southern onshore reach. The location and altitude (32 to 35 m) of the shoreline angle of the Oso terrace northeast of the fault zone is well constrained by natural exposures of the wave-cut platform and from exploration pits. The altitude of the shoreline angle of the Oso terrace rises from 37±1 m near Adobe Creek, to ≥40±1 m directly west of the fault zone, and to a maximum of 47±2 m within the fault zone.

The highest marine terrace mapped in detail in the San Simeon area during this study is the La Cruz terrace. Northeast of the fault zone, the terrace, which has a shoreline-angle altitude of 53±2 m, forms a broad, well-expressed geomorphic surface south of Oak Knoll Creek (Fig. 4).

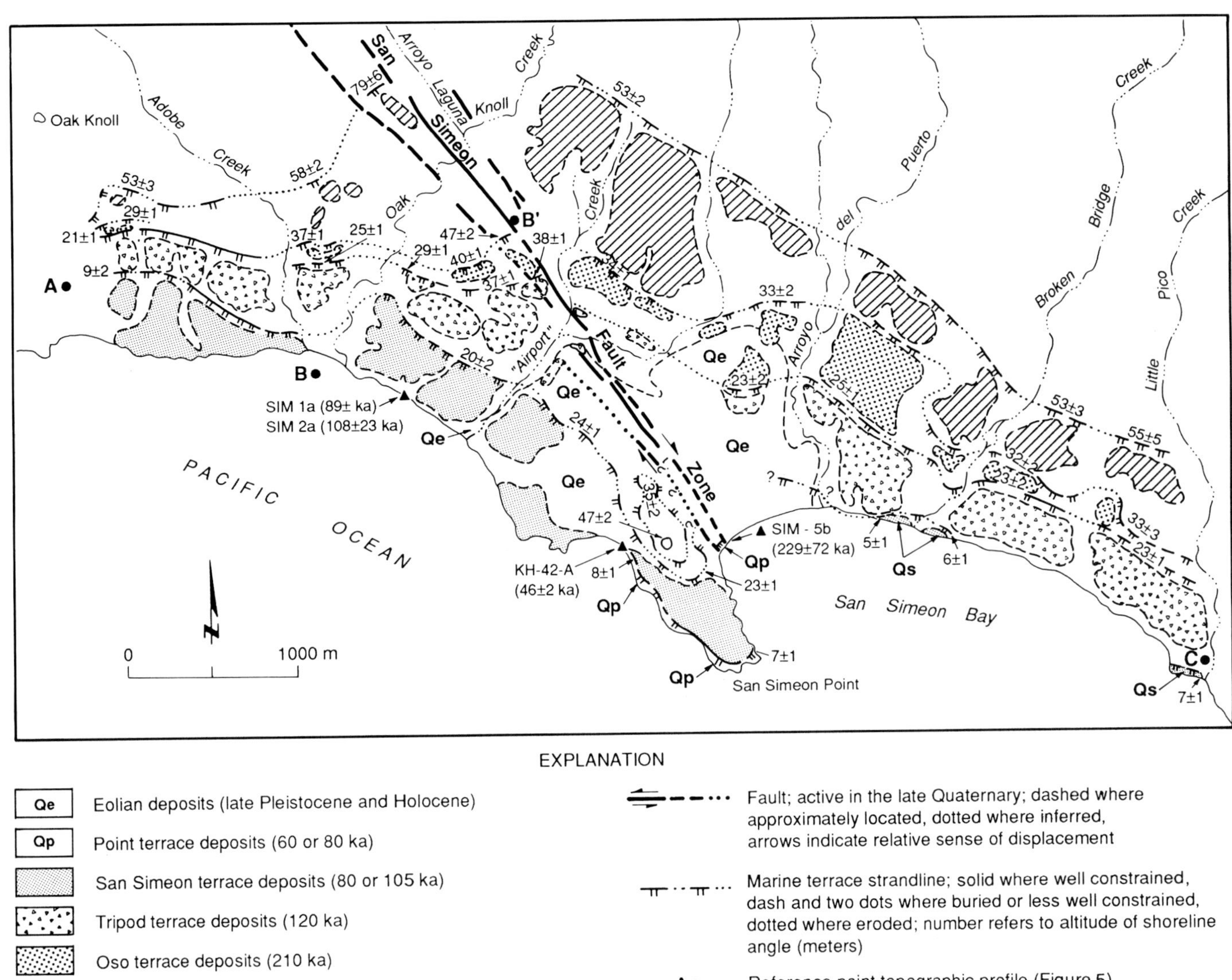

FIG. 4.—Geologic map of marine terraces along the southern onshore reach of the San Simeon fault zone.

Discontinuous remnants of the La Cruz terrace are preserved southwest of and within the San Simeon fault zone. The shoreline angles for these remnants, which show a pattern similar to those of the lower terraces, gradually rise from an altitude of 53±3 m southwest of Adobe Creek to 58±2 m between Adobe Creek and the fault zone. The highest terrace remnant occurs between the two active strands of the fault at a shoreline-angle altitude of 79±6 m (Fig. 5).

ESTIMATED AGES AND CORRELATION OF MARINE TERRACES

San Luis Range Area

The marine-terrace sequence in the San Luis Range area is one of the best-dated marine-terrace sequences along the western North American coastline. The ages of the lower two terraces (terraces Q_1 and Q_2) between Point Buchon and Pismo Beach are constrained by 11 uranium-series ages on coral and vertebrate bone samples, 10 amino-acid racemization values on marine invertebrate shells, and 14 paleoclimatic analyses of invertebrate faunal assemblages (Table 1). Nine samples collected from the Q_2 terrace, which provide the best age control for an individual terrace in the study area, indicate that the Q_2 terrace is ~120 ka.

Six samples of marine- and terrestrial-mammal teeth and bones collected from deposits overlying the lowest marine platform (Q_1) yield uranium-series ages ranging from approximately 49 to 83 ka. We consider these ages to be minimum values, but, because the maximum numerical ages are around 80 ka, a correlation to the 80-ka high sea stand is preferred over a correlation to the 105-ka high sea stand. The 80-ka interpretation is consistent with evidence from other areas along the coast of California and Baja California, Mexico, that indicate the ~105-ka terrace is rarely preserved and that couplets of low, well-preserved, broad ter-

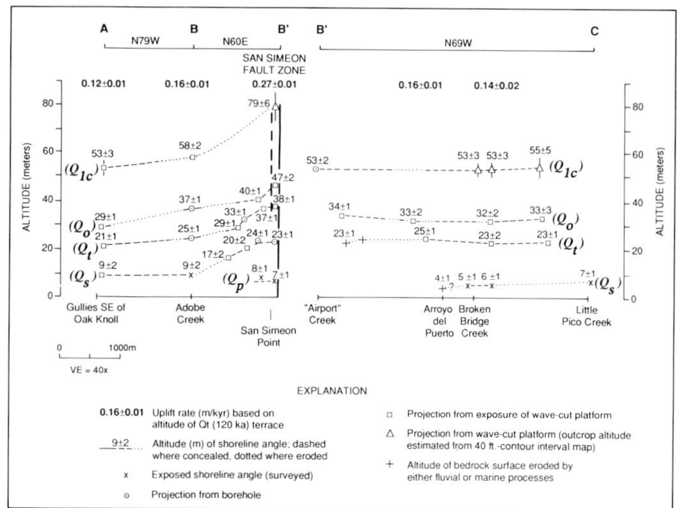

FIG. 5.—Topographic profile illustrating vertical deformation of marine terraces across the southern onshore San Simeon fault zone. Location of profiles and description of map units provided on Figure 4.

races generally correlate to the ~80-ka and ~120-ka global highstands (Lajoie and others, 1979; Muhs and others, 1987, 1988; Rockwell and others, 1989).

There are no uranium-series ages or faunal assemblage data available for independent age determination of the higher terraces. However, using the ages of the lower two terraces (~80 ka and ~120 ka) to calibrate terrace spacing to the paleosea-level data from other dated terraces (see discussion of uplift rates later), we estimate that terraces Q_3 and Q_4 formed during the approximately 210- and 330-ka sea-level highstands, respectively. We speculate that terraces higher than Q_4 are middle to early Pleistocene.

San Simeon Area

Age estimates of marine terraces at San Simeon are based on numerical ages, correlation of one of the terraces to the dated terrace in the Cayucos area, and correlation of terrace altitudinal spacing to sea-level curves. Prior to our study, no datable materials had been found in the marine-terrace deposits. Numerical ages obtained during this study include one uranium-series age on bone, one radiocarbon age on charcoal, and two thermoluminescence ages on estuarine deposits. Uranium-series analysis of a whale bone fragment (sample KH-42-A, Table 1) from deposits overlying the wave-cut platform of the Point (Q_p) terrace provides a minimum age of 46±2 ka for the terrace. Radiocarbon analysis of detrital charcoal from estuarine deposits underlying the wave-cut platform of the next higher San Simeon (Q_s) terrace yields a minimum age of >40,480±1080 ^{14}C yrs. Thermoluminescence analyses of two samples of estuarine silty clay (samples SIM-1a and SIM-2a) from the same locality as the radiocarbon sample provide a weighted mean age of 95±13 ka (Berger and Hanson, this volume), suggesting that the overlying younger Q_s terrace wave-cut platform correlates to the ~105-ka highstand or more likely the ~80-ka highstand.

The numerical ages were obtained from terrace remnants southwest of the San Simeon fault zone. To provide correlation and age control of terraces across the fault zone, we used conventional correlative methods, including:

1. lateral correlation of the second emergent marine terrace (Q_t) northeast of the fault zone to the 120-ka Cayucos terrace (samples C-5e and LACMNH 10731, Table 1; Muhs and others, 1988) as mapped by Weber (1983);
2. comparison of relative soil profile development for terraces on both sides of the fault zone. Results of the soil stratigraphic studies indicate that the second, third, and fourth marine terraces southwest of and within the fault zone correlate to the first, second and third terraces northeast of the fault zone (Rockwell and others, 1987; Rockwell, pers. comm., 1990).
3. comparison of relative altitudinal spacing between terraces and correlation to Quaternary sea-level curves (see discussion of uplift rates later). Results of this analysis support correlations across the fault zone and provide estimates of the ages of the Oso and La Cruz terraces.

Estimated ages of the San Simeon terraces on the basis of numerical- and correlated-age estimates, relative geomorphic preservation, and stratigraphic and topographic position and correlations to terraces in the San Luis Range area are presented in Table 2.

ESTIMATES OF PALEOSEA LEVEL, UPLIFT RATES, AND AGES BASED ON TERRACE ALTITUDINAL SPACING

Altitudinal spacing of marine terraces is dependent on: (1) terrace ages, (2) sea level at the time of platform formation (paleosea level), (3) elapsed time between consecutive sea-level highstands, and (4) rates of uplift. Paleosea level during the ~120-ka highstand is estimated at +2 to 10 m, with a value of +6 m commonly used (see summary in Lajoie, 1986). Paleosea level for other highstands commonly is estimated by assuming an uplift rate based on the present elevation of the ~120-ka terrace using a +6-m paleosea level, and applying this uplift rate and present altitude of another dated terrace to obtain a paleosea level. Using this approach in the San Luis Range area, where we have good altitude and age control for the 120-ka and 80-ka terraces, we estimate paleosea level during the 80-ka high-stand was −4±1 m (relative to present level). This value is considerably closer to modern sea level than to estimates of −15 to −16 m for equivalent terraces in New Guinea, Barbados, and Haiti (Chappell and Shackleton, 1986; Matthews, 1973; Dodge and others, 1983), but is in good agreement with a revised estimate of −7 m in New Guinea (Bloom and Yonekura, 1985) and −5±2 m along the coast of southern California and Baja California, Mexico (Muhs and others, 1988; Rockwell and others, 1989).

Graphical comparison of relative terrace spacing to Quaternary sea-level curves calculated in this manner has been used to infer ages of marine terraces and evaluate temporal and spatial changes in uplift rates (Pillans, 1983; Weber, 1983; Bull, 1985). Our graphical approach for evaluating long-term uplift and terrace ages based on altitudinal spac-

TABLE 1.—SUMMARY OF AGE ESTIMATES FOR FOSSIL SAMPLES BASED ON FAUNAL ASSEMBLAGE, AMINO-ACID, AND URANIUM-SERIES DATA.

Study Sample Number	Location (latitude, longitude)	Terrace (WCP[1] elevation)	Material Dated	Faunal Assemblage[2] (Interpreted Age, ka)	Amino Acid Racemization[3] (Interpreted Age, ka)	U-Series[4] (ka)	Comments[1]
500p	Shell Beach (Shelter Cove) (35° 09' 12.5"N, 120° 39' 48.5"W)	Q_2 (22 m)	shell hash	NES (80 or 105)	75 ± 15	--	Composite sample from two closely spaced sites; shell hash 0.5-1 m thick overlying WCP
SLO-4	Mallagh Landing (35° 10' 30.5"N, 120° 43' 17"W)	Q_1 (~4m)	coral from shell hash	--	--	96 ± 2	$^{234}U/^{238}U$ ratio indicates sample probably assimilated secondary uranium; age is thus considered unreliable
SLO-5	Mallagh Landing (35° 10' 30.5"N, 120° 43' 17"W)	Q_1 (~ 4m)	coral from shell hash	--	--	93 ± 2	$^{234}U/^{238}U$ ratio indicates sample probably assimilated secondary uranium; age is considered unreliable
503p	Shell Beach (Sunset Palisades) (35° 10' 1.5"N, 120° 41' 33.7"W)	Q_1 14 ft (4.3 m)	shell coquina	NES (80 or 105)	--	--	Shell coquina overlying WCP, 10-15 cm thick, moderately consolidated
520p	Mallagh Landing (35° 10' 31"N, 120° 43' 7.6"W)	Q_2 (13-15 m)	shell hash	SES? (120)	125 ± 5	--	Shell hash 0.5 m thick overlying cobbles on WCP
SLO-6	Mallagh Landing (35° 10' 29.7"N, 120° 43' 5.1"W)	Q_2	coral from shell hash	--	--	99 ± 2	Uranium concentration is high; U-series age is not consistent with results for Sample 520p from the same terrace
522p	Pirate's Cove (east of Mallagh Landing) (35° 10' 32.5"N, 120° 42' 40.3"W)	Q_2 (13.4 m)	fossiliferous sand	SES (120)	125 ± 5	--	Fossiliferous sand 0.3 m thick
523p	Pismo Beach (Harbor View Avenue) (35° 08' 44.2"N, 120° 38' 48.2"W)	Q_2 (4 m)	fossiliferous sand	SES (120)	125 ± 5	--	Fossil horizon 0.3-0.6 m thick; collected 2.4 m above WCP on downthrown side of Wilmar Avenue fault
524p	~0.15 km west of mouth of Pecho Creek (35° 10' 44.2"N, 120° 47' 35.7"W)	Q_1 (3 - 3.4 m)	whale rib	--	--	Th 49 ± 1 Pa 51 ± 4	Collected from marine gravel ~ 0.6 m above WCP
525p	Shell Beach (35° 09' 38.5"N, 120° 40' 35.7"W)	Q_2 (16.4 - 18 m)	fossil hash	SES (120)	125 ± 5	--	Fossil hash 1 m thick overlying WCP. Collected from boreholes 110 and 114.
KIK-42	Shell Beach (Shelter Cove) (35° 09' 10"N, 120° 39' 37.6"W)	Q_2 (19 m)	coquina	SES (120)	125 ± 5	--	Coquina 3.5 m thick overlying WCP; WCP is laterally continuous; USGS fossil locality M7498
KIK-44	Pirate's Cove (east of Mallagh Landing) (35° 10' 29.2"N, 120° 42' 33"W)	Q_1 (6.1 m)	fossiliferous pebble conglomerate	NES (80 or 105)	75 ± 15	--	Fossiliferous conglomerate 20 cm thick overlying WCP, at SLA
G86-100	Point San Luis (35° 09' 45.5"N, 120° 45' 58.8"W)	Q_2 (12.2 m)	fossil hash	SES (120)	75 ± 15[5]	--	Amino acid and faunal assemblage results are discordant
W87-10	Point San Luis (35° 09' 38.4"N, 120° 45' 20.4"W)	Q_2 (13.4 m)	coral from fossiliferous sand	SES (120)	--	Th 117 ± 3	Collected from 2.6 m thick marine sandy gravel approximately 1.1 to 3.7 m above WCP
W87-61	Irish Canyon (35° 10' 58.7"N, 120° 47' 59.1"W)	Q_1 (4.9 m)	horse dentary	--	--	Th 62 ± 2 Pa 66 ± 4	Collected from alluvium 30 cm above WCP
W87-62	Irish Canyon (35° 11' 01"N, 120° 48' 3.9"W)	Q_1 (4.9 m)	Bison dentary	--	--	Th 51 ± 2 Pa 43 ± 3	Collected from alluvium 10 cm above WCP
W87-89	San Luis Obispo Bay (35° 10' 13"N, 120° 45' 23.6"W)	Q_2 (at SLA)[1] (12.8 m)	fossiliferous sand	SES (120)	75 ± 15[5]	--	Amino acid and faunal assemblage results discordant
KH-42-A	San Simeon Point (35° 38' 30.6"N, 120° 12' 6.4"W)	Point Terrace Q_p (8.9 m)	mammal bone fragment	--	--	Th 46 ± 2 Pa 49 ± 4	Collected from marine gravel 1.2 m above WCP near SLA
LM-100	~0.6 km NW of the mouth of Crowbar Canyon Creek (35° 13' 38.7"N, 120° 52' 23.9"W)	Q_1 (5.5 m)	large mammal bone fragment	--	--	Th 76 ± 3 Pa 76 ± 4	Collected from marine sand approximately 1.2 m above WCP.
W87-101	Near mouth of Pecho Creek (35° 10' 43.7"N, 120° 47' 31"W)	Q_1 (4.6 m)	marine mammal tooth	--	--	Th 83 ± 3 Pa 109 +57,-25	Sloth, horse, and camel bones collected from same site.
W87-103	Approximately 1 km northwest of the mouth of Deer Creek (35° 11' 34.7"N, 120° 49' 16.4"W)	Q_1 (7 m)	whale rib	--	--	Th 63 ± 2 Pa 58 ± 3	
W87-135 (SLO-7)	Diablo Canyon Power Plant (35° 12' 22.5"N, 120° 50' 54.5"W)	Q_2 (26 m)	coral from fossiliferous sand	SES? (120)	--	Th 112 ± 4	Uranium concentration is high
LM-74	~1.6 km northwest of mouth of Crowbar Canyon Creek (35° 14' 1.9"N, 120° 52' 52.4"W)	Q_1 (5.8 m)	fossiliferous sand	NES (80 or 105)	75 ± 15	--	USGS[5] locality M7280 and LACMNH[5] locality 5640
LM-124	Pecho Ranch, NE of Lion Rock (35° 13' 8.5"N, 120° 51' 59.4"W)	Q_1 (8.5 - 9)	fossiliferous sand	NES (80 or 105)	--	--	USGS[5] locality M7285 and LACMNH[5] locality 5639
C-5e	Cayucos . (35° 26' 51.5"N, 120° 55' 20.6"W)	Cayucos Terrace (4 m)	whale rib	--	--	Th 124 ± 4 Pa 108 +21,-14	Sample from basal marine deposits directly overlying WCP of first emergent terrace
LACMNH #10731	Cayucos (35° 26' 58"N, 120° 54' 47.6"W)	Cayucos Terrace (4 m)	coral from shell hash	SES (120)	125 ± 5	Th 117 ± 3	Sample collected by G. Kennedy

[1]WCP - wave-cut platform; SLA - shoreline angle
[2]Faunal determinations by G.L. Kennedy, Los Angeles County Museum of Natural History, written communication, 1986-1987; NES - northern extralimital species present; SES - southern extralimital species present
[3]Amino acid racemization analyses and interpreted ages of shell samples are from J.F. Wehmiller, Department of Geology, University of Delaware (written communication, 1987)
[4]U-Series ages of coral are from D.R. Muhs, U.S. Geological Survey (written communiction, 1989; personal communication, 1990). U-series ags of bone and teeth are from T.L. Ku, University of Southern California (written communication, 1987).
[5]USGS - U.S. Geological Survey, LACHMNH - Los Angeles County Museum of Natural History

ing of wave-cut platforms is illustrated in Figure 6. Based on published estimated paleosea levels (y-intercept values) for globally recognized highstands at ~80, ~105, ~120, ~210, and ~330 ka, lines are drawn that represent the predicted altitude of each terrace for any given rate of constant uplift. Uncertainties in the estimated altitudes of the paleosea level are shown by an envelope around each line. Paleosea levels for the 80-ka, 105-ka, and 120-ka highstands are based on the California-Japan paleosea level estimates (Muhs and others, 1988; 1990; Machida, 1975), which are supported by local paleosea level data for the 80-ka terrace. Values for the 210-ka and 330-ka highstands are estimated

TABLE 2.—REGIONAL CORRELATION OF MARINE TERRACES

Sea Level Highstand (ka)	San Simeon	Cayucos	San Luis Range
60	*Qp*?	—	
80	*Qp* or *Qs*	—	*Q*1
105	*Qs*?	—	—
120	*Qt*	Cayucos Terrace	*Q*2
210	*Qo*	—	*Q*3
330	Q_{1c}	—	*Q*4

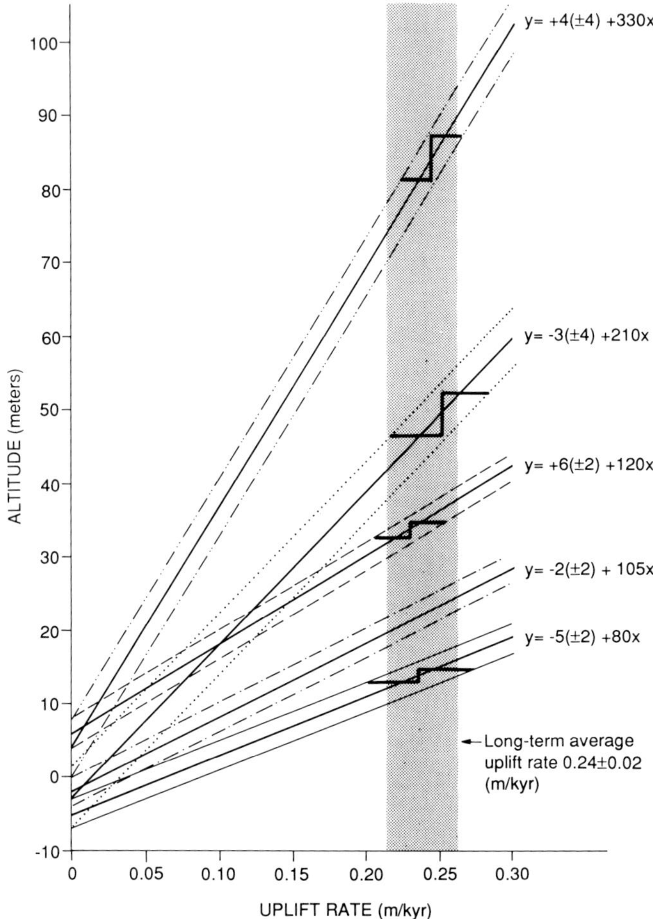

FIG. 6.—Example of a graphic approach to estimate average long-term uplift rate for a flight of marine terraces at Montaña de Oro State Park, San Luis Range area. Solid line represents range of uplift rate, given uncertainties in the present altitude of marine terrace, and the paleosea-level at which it formed. The stippled pattern represents the most likely range for the long-term average uplift rate.

from New Guinea (Merritts and Bull, 1989) and Barbados (Bender and others, 1979) data.

On the basis of independent age control for the 80- and 120-ka terraces and well-constrained terrace-shoreline angles, we have used this graphical approach in our study area to: (1) estimate ages for undated terraces within the lower four to five terraces of the terrace sequence, and (2) assess long-term rates and patterns of uplift. We recognize that interpretations based on such graphic approaches may not be unique, given the assumptions that are used to estimate paleosea levels and the possibility that erosion or reoccupation of platforms and tectonic factors may complicate the highstand record. However, in most areas the observed altitudes of terrace-shoreline angles match reasonably well the predicted altitudes based on assumed correlations to the globally recognized highstands and uniform long-term uplift rates similar to rates calculated for the 120-ka terrace (Fig. 6). This suggests that these age estimates are reasonable and that rates and patterns of uplift throughout most of the study areas, as indicated on Figure 1, may have been relatively uniform for the past approximately 330 ka.

LATE QUATERNARY DEFORMATION

The present morphology of the coastline between the approximate latitudes of San Simeon and Santa Maria consists of prominent headlands associated with uplifting areas and embayments associated with subsiding regions. The morphology suggests that the general pattern of uplift and subsidence indicated by the distribution of late Quaternary marine terraces (Fig. 1) has persisted throughout much or all of the Quaternary. Thus, the morphology of this coastline reflects, to a large degree, the locations and activities of several Quaternary faults.

On the basis of distribution of marine terraces, we interpret that the San Luis Range has undergone late Quaternary uplift as a relatively rigid crustal block with little or no internal deformation of the block (Fig. 3). Quaternary uplift of the San Luis Range at rates of 0.06 to 0.23 m/ka occurred primarily via reverse faulting along the northeastern and southwestern margins of the range (Lettis and others, 1990). The southwest-dipping Los Osos fault zone borders the uplifting range to the northeast. In contrast, deformation along the southwestern margin of the range occurs across a broad zone consisting of several northeast-dipping identified or inferred reverse faults, including the Olson, San Luis Bay, Wilmar Avenue, Oceano, and Pecho faults (Fig. 1). Detailed evaluation of the deformation of marine terraces that flank the western and southwestern parts of the San Luis Range provided a means to recognize these faults and evaluate their style and rate of Quaternary deformation (Kelson and others, 1987; Hanson and others, 1989).

Marine terraces in the San Simeon area are differentally uplifted across the main active trace of the San Simeon fault. Southwest of the fault zone, higher rates of uplift are accommodated, in part, by localized upwarping or tilting (Fig. 5). Graphical analysis of the altitudinal spacing of terraces in the regions on both sides of, and away from, the San Simeon fault zone, however, yield uniform and similar long-term rates of uplift. In the vicinity of Adobe Creek southwest of the fault zone, the long-term average uplift rate is 0.17±0.02 m/ka. Northeast of the fault zone near Broken Bridge Creek, the long-term uplift rate is 0.16±0.01 m/ka. The long-term average uplift rate in the vicinity of the fault based on the 120-ka terrace within the fault zone is estimated to be about 0.27±.01 m/ka. On the basis of the patterns and rates of uplift, the maximum vertical component of slip across the fault zone is on the order of 0.1 mm/

yr. This is a factor of 10 less than the estimated 1 to 3 mm/yr lateral component, indicating that the San Simeon fault zone is predominantly a strike-slip fault.

CONCLUSIONS

Tectonic processes have significantly influenced the evolution of south-central coastal California. Rates and patterns of uplift and subsidence reflect the location and activity of several faults. Correlation of marine terraces in the region using detailed mapping and age-dating investigations in the San Simeon and San Luis Range areas allows assessment of Quaternary uplift and deformation. In the coastal region between Morro Bay and the northeastern margin of the Santa Maria Valley (San Luis Range area), a sequence of at least 12 elevated marine terraces is locally present. The lowest two terraces in this sequence (Q_1 and Q_2) are ~80 ka and ~120 ka, respectively, on the basis of uranium-series ages, amino-acid racemization analyses, paleoclimatic analyses of invertebrate faunal assemblages, and comparison of geomorphic and terrace-altitudinal spacing. Based on interpreted ages and terrace spacing between the lowest two emergent terraces along the coast from Morro Bay to Pismo Beach, we estimate paleosea level during the 80-ka highstand was -4 ± 1 m (relative to present level). This value is in general agreement with other recent estimates from coastal California, Mexico, and Japan, but is significantly higher than estimates from New Guinea and Barbados.

Near San Simeon, flights of four and five terraces, ranging in age from 60 or 80 ka to 330 ka, have been mapped to the northeast and southwest, respectively, of the southern onshore San Simeon fault zone. Estimated ages and correlation of terraces across the San Simeon fault zone are based on lateral correlation of the Tripod (Q_t) terrace to the ~120-ka Cayucos terrace, comparison of soil geomorphology, geomorphic expression and terrace-altitudinal spacing. Estimated late Pleistocene uplift rates are 0.17 ± 0.02 m/ka southwest and 0.16 ± 0.01 m/ka northeast of the fault and approximately 0.24 m/ka for uplifted and warped areas within and adjacent to the fault zone.

ACKNOWLEDGMENTS

The research benefitted substantially from discussions with G. E. Weber, who introduced us to marine terraces in the San Simeon area, and T. K. Rockwell (San Diego State University), who conducted the soil-geomorphology studies in the San Simeon area and assisted in various other dating analyses. We also thank D. R. Muhs (U.S. Geological Survey), G. L. Kennedy (Los Angeles County Museum of Natural History), T. L. Ku (University of Southern California), J. F. Wehmiller (University of Delaware) and G. W. Berger (University of Western Washington) for helpful discussions and analytical results. D. Merritts and G. E. Weber provided helpful comments on an early draft of this paper. Support for this research was provided by Pacific Gas and Electric Company.

REFERENCES

Bender, M. L., Fairbanks, R. G., Taylor, F. W., Matthews, R. K., Goddard, J. G., and Broecker, W. S., 1979, Uranium-series dating of the Pleistocene reef tracts of Barbados, West Indies: Geological Society of America Bulletin, Part I, v. 90, p. 577–594.

Bloom, A. L., and Yonekura, N., 1985, Coastal terraces generated by sea-level change and tectonic uplift, *in* Waldenberg, M. J., ed., Models in Geomorphology: Allen and Unwin, Boston, p. 139–154.

Bull, W. B., 1985, Correlation of flights of global marine terraces, *in* Morisawa, M., and Hack, J., eds., Tectonic Geomorphology: Proceedings, 15th Annual Binghamton Geomorphology Symposium, Allen and Unwin, Boston, p. 129–152.

Chappell, J., and Shackleton, N. J., 1986, Oxygen isotopes and sea level: Nature, v. 324, p. 137–140.

Dodge, R. E., Fairbanks, R. G., Benninger, K. L., and Maurrasse, E., 1983, Pleistocene sea levels from raised coral reefs of Haiti: Science, v. 219, p. 1423–1425.

Hall, C. A., Jr., 1975, San Simeon-Hosgri fault system, coastal California: economic and environmental implications: Science, v. 190, p. 1291–1294.

Hamilton, D. H., 1984, The tectonic boundary of coastal central California: Unpublished Ph.D. Dissertation, Stanford University, California, 290 p.

Hanson, K. L., and Lettis, W. R., 1990, Estimated Pleistocene slip rate for the San Simeon fault zone, south-central coastal California, *in* Lettis, W. R., Hanson, K. L., Kelson, K. I., and Wesling, J. R., eds., Neotectonics of South-Central Coastal California: Friends of the Pleistocene Field Trip Guidebook, Geomatrix Consultants, Inc., San Francisco, p. 191–254.

Hanson, K. L., Lettis, W. R., Wesling, J. R., Kelson, K. I., and Mezger, L., 1989, Correlation and dating of marine terraces, south-central coast California: implications for late Quaternary crustal deformation: 28th International Geological Congress Abstracts, v. 2, p. 2.26–2.27.

Kelson, K. I., Lettis, W. R., Weber, G. E., Kennedy, G. L., and Wehmiller, J. F., 1987, Amount and timing of deformation along the Wilmar Avenue, Pismo, and San Miguelito faults, Pismo Beach, California: Geological Society of America Abstracts with Programs, Cordilleran Section, v. 20, p. 150.

Lajoie, K. R., 1986, Coastal tectonics, *in* Wallace, R. E., ed., Active Tectonics: National Academy Press, Washington, D.C., p. 95–124.

Lajoie, K. R., Weber, G. E., Mathieson, S. A., and Wallace, J., 1979, Quaternary tectonics of coastal Santa Cruz and San Mateo counties, California, as indicated by deformed marine terraces and alluvial deposits, *in* Weber, G. E., Lajoie, K. R., and Griggs, G. B., eds., Coastal Tectonics and Coastal Geologic Hazards in Santa Cruz and San Mateo Counties, California: Geological Society of America Guidebook, Cordilleran Section, San Jose, p. 61–80.

Lettis, W. R., and Hall, N. T., 1990, Evidence for segmentation of the Los Osos fault zone, south-central coastal California, *in* Lettis, W. R., Hanson, K. L., Kelson, K. I., and Wesling, J. R., eds., Neotectonics of South-Central Coastal California: Friends of the Pleistocene Field Trip Guidebook, Geomatrix Consultants, Inc., San Francisco, p. 299–347.

Lettis, W. R., Kelson, K. I., Angell, M. M., Hanson, K. L., Wesling, J. R., and Hall, N. T., 1990, Quaternary deformation of the San Luis Range, San Luis Obispo County, California, *in* Lettis, W. R., Hanson, K. L., Kelson, K. I., and Wesling, J. R., eds., Neotectonics of South-Central Coastal California: Friends of the Pleistocene Field Trip Guidebook, Geomatrix Consultants, Inc., San Francisco, p. 259–290.

Machida, H., 1975, Pleistocene sea level of south Kanto, Japan, analysed by tephrochronology: Bulletin of the Royal Society of New Zealand, v. 13, p. 215–222.

Matthews, R. K., 1973, Relative elevation of late Pleistocene high sea level stands: Barbados uplift rates and their implications: Quaternary Research, v. 3, p. 147–153.

Merritts, D., and Bull, W. B., 1989, Interpreting Quaternary uplift rates at the Mendicino triple junction, northern California, from uplifted marine terraces: Geology, v. 17, p. 1020–1024.

Mezger, E. B., Hanson, K. L., Hall, N. T., and Hunt, T. D., 1987, Evidence for Quaternary faulting in Los Osos Valley, San Luis Obispo County, California: Geological Society of America Abstracts with Programs, v. 19, p. 432.

Muhs, D. R., Kelsey, H. M., Miller, G. H., Kennedy, G. L., Whelan, J. F., and McInelly, G. W., 1990, Age estimates and uplift rates for late Pleistocene marine terraces: southern Oregon portion of Cascadia forearc: Journal of Geophysical Research, v. 95, p. 6685–6698.

MUHS, D. R., KENNEDY, G. L., AND MILLER, G. H., 1987, New uranium-series ages of marine terraces and late Quaternary sea level history, San Nichols Island, California: Geological Society of America Abstracts with Programs, v. 19, p. 780–781.

MUHS, D. R., KENNEDY, G. L., AND ROCKWELL, T., 1988, Uranium-series ages of corals from marine terraces, Pacific Coast of North America: implications for the timing and magnitude of late Pleistocene sea-level changes: American Quaternary Association, Program and Abstracts, p. 140.

PACIFIC GAS AND ELECTRIC COMPANY, 1974, Geology of the southern Coast Ranges and the adjoining offshore continental margin of California, with special reference to the geology in the vicinity of the San Luis Range and Estero Bay: Appendix 2.5D, FSAR, Diablo Canyon Nuclear Power Plant, AEC Docket Nos. 50–275 and 50–323, 112 p.

PACIFIC GAS AND ELECTRIC COMPANY, 1975, Appendix 2.5E, Final Safety Analysis Report for Diablo Canyon Nuclear Power Plant: U.S. Atomic Energy Commission Docket Nos. 50–275 and 50–323, 126 p.

PACIFIC GAS AND ELECTRIC COMPANY, 1988, Final Report of the Diablo Canyon Long Term Seismic Program: U.S. Nuclear Regulatory Commission Docket Nos. 50–275 and 50–323, 669 p.

PILLANS, B., 1983, Upper Quaternary marine terrace chronology and deformation, south Taranak, New Zealand: Geology, v. 11, p. 292–297.

ROCKWELL, T. K., BICKNER, F. R., VAUGHAN, P. R. AND HANSON, K. L., 1987, Applications of soil geomorphology to dating and correlating coastal terrace deposits across the San Simeon fault zone, central California: Geological Society of America Abstracts with Programs, v. 19, p. 444.

ROCKWELL, T. K., MUHS, D. R., KENNEDY, G. L., HATCH, M. E., WILSON, S. H., AND KLINGER, R. E., 1989, Uranium-series ages, faunal correlations and tectonic deformation of marine terraces within the Agua Blanca fault zone at Punta Banda, northern Baja California, Mexico, *in* Abbott, P. L., ed., Geologic Studies in Baja California: Pacific Section, Society of Economic Paleontologists and Mineralogists, p. 1–16.

WEBER, G. E., 1983, Geologic investigation of the marine terraces of the San Simeon region and Pleistocene activity of the San Simeon fault zone, San Luis Obispo County, California: U.S. Geological Survey Technical Report, 66 p.

WEBER, G. E., LETTIS, W. R., AND HANSON, K. L., 1987, Late Pleistocene uplift rates along the central California coast, Cape San Martin to Santa Maria Valley: Geological Society of America Abstracts with Programs, v. 19, p. 462.

AGES AND DEFORMATION OF MARINE TERRACES BETWEEN POINT CONCEPTION AND GAVIOTA, WESTERN TRANSVERSE RANGES, CALIFORNIA

THOMAS K. ROCKWELL
Department of Geological Sciences, San Diego State University, San Diego, California 92182
JEFF NOLAN
G.E. Weber and Assoc., 120 Westgate Drive, Watsonville, California 95076
DONALD L. JOHNSON
Department of Geography, University of Illinois, Urbana, Illinois 61801
AND
ROY H. PATTERSON
Dames and Moore, 6 Hutton Centre Dr., Suite 700, Santa Ana, California 92707

ABSTRACT: Five well-expressed marine abrasion platforms, and up to 10 additional poorly expressed ones, are preserved from Point Conception to Gaviota in the western Transverse Ranges, southern California. The terraces, which range in elevation from 10 to over 300 m, and in age from 80 ka to over 1 Ma, record the long-term history of uplift for this region of rapid tectonic convergence. The lowest four terraces have direct age control and correlate to oxygen isotope stages 5a, 5c, 5e, and 7. The fifth terrace presumably correlates to stage 9. Age control is based on: (1) amino-acid stereochemistry on mollusk shells collected from the lower four terraces; (2) uranium-series analyses of bone and teeth from the first and fourth terraces and their alluvial covers; (3) the assemblages of molluscan fauna associated with the first, second, and third terraces; and (4) inferred long-term rates of uplift based on the ages and elevations of the terraces.

The terraces are folded across the Government Point syncline near Point Conception. The shoreline angle for the 80-ka terrace decreases in elevation by 8 to 10 m across the hinge of the syncline, with respect to its elevation to the east and west. Furthermore, platform gradients are steeper on the north limb, indicating asymmetric synclinal deformation. Deformation of the lowest three terraces indicates average folding rates in the range of 2 to 9° per million years, consistent with the total post-Miocene deformation. Along with the changes in slope and elevation for the terrace platforms and shorelines, the presence of several flexural-slip bedding-plane faults cutting the platforms and their overlying deposits also supports active folding of the syncline.

The South Branch of the Santa Ynez fault displaces all of the terraces that cross it. The vertical component of slip, based on displacement of the shoreline angle, indicates a vertical rate of 0.05 mm/yr. The horizontal component of the slip rate was not determined but must be low due to the lack of significant deflection of the shorelines beyond that of the modern shoreline.

Our data clearly indicate that the westernmost Santa Ynez Mountains are actively folding in association with north-south crustal shortening across the Transverse Ranges. The rates of folding estimated from the rates of terrace deformation, however, suggest that only a small fraction of the total regional shortening is accommodated onshore: the balance is apparently accommodated offshore to the south in the Santa Barbara Channel.

INTRODUCTION

Documentation of the ages and elevations of coastal marine terraces is useful in assessing local and regional rates of vertical crustal deformation in California (Woodring and others, 1946; Birkeland, 1972; Bradley and Griggs, 1976; Kern, 1977; Lajoie and others, 1979; Rockwell and others, 1989). As part of the geoseismic investigations conducted for a proposed liquefied natural gas (LNG) marine terminal and associated facilities near Point Conception, California (Dames and Moore, 1980; 1981), the coastal terraces were mapped in detail from about 3 km north of Point Conception to Gaviota, about 25 km to the east (Fig. 1).

Point Conception is the westernmost onshore extent of the Transverse Ranges structural province of southern California. The Transverse Ranges are an east-west-trending system of folds and thrust faults (Yeats, 1981; Rockwell, 1988; Rockwell and others, 1988) resulting, at least in part, from north-south crustal convergence. Part of this convergence has been attributed to transfer of slip from right-lateral faults of the southern California Borderland to the offshore zone of faulting in central California (Weldon and Humphreys, 1986). Rates of convergence are high: the shortening rate near Ventura is estimated at 17±4 mm/yr (Rockwell, 1983; Rockwell and others, 1984) and 22 mm/yr (Yeats, 1983) with local vertical rates exceeding 10 mm/yr (Yeats, 1976; Rockwell and others, 1988). Most of the structures that accommodate these high rates trend offshore between Ventura and Santa Barbara and project westward to the south of Point Conception. One of the purposes of this study is to assess the onshore rates and styles of deformation in the Point Conception area in order to further constrain the locus of deformation in the westernmost Transverse Ranges.

Marine terraces are gently sloping wave-abrasion platforms that are commonly overlain by a veneer of both marine and nonmarine deposits. In the Point Conception area, the marine deposits typically consist of a thin (30–60 cm) sheet of fine, well-sorted sand, although accumulations up to several meters are locally present. Continental deposits typically consist of fine-grained alluvial-fan and colluvial deposits shed from nearby canyons and mountain slopes, and less common eolian sand and silt. In many cases, several marine-abrasion platforms, corresponding to discrete sea-level highstands, are covered by the most recent depositional sequence, resulting in a single, wide terrace surface.

METHODS

Identification of Marine Terraces

Planar bedrock surfaces interpreted as wave-cut platforms occur throughout the field area. The landward edge of each of these surfaces is defined by an abrupt steepening of slope at the old shoreline angle. The relief of the slope,

Quaternary Coasts of the United States: Marine and Lacustrine Systems, SEPM Special Publication No. 48

or paleosea cliff, behind each shoreline angle varies from precipitous to barely perceptible, depending on conditions that include paleotopography, orientation and competence of bedrock strata, and extent of erosion prior to burial.

Elevated wave-cut platforms are distinguished in the field by relatively planar bedrock surfaces that are directly overlain by marine deposits of well-sorted, massive to low-angle cross-bedded sand with local pebble- to boulder-size, rounded clasts. Relief on abrasion surfaces is usually less than 20 to 30 cm where cut across the soft Sisquoc Formation shale, but may be up to 50 cm or more where cut across the more resistant Monterey Formation. Abrasion platforms are also characterized by the local presence of: (1) marine mollusk (Pholadidae) borings into the platform surface and the directly overlying cobbles and boulders; (2) marine molluscan fauna; and (3) large tar blebs on the bedrock surface or as clasts within the well-sorted sand. The tar blebs come from natural oil seeps in the Santa Barbara Channel and are common on beaches today as well.

Terrace Mapping.—

During site investigations for the LNG facility, several large trenches were excavated that exposed the marine-abrasion platforms. Detailed logs of these surveyed trenches are available and were used to quantify elevations of the platforms (Dames and Moore, 1980). Additionally, more than 500 large-diameter borings through the alluvial core were used to determine depth to the abrasion-platform surface. Combined with the trenching data, the borehole information provided sufficient spatial control on platform elevations to prepare contour maps of the bedrock surface. These maps were used to delineate individual abrasion platforms and to construct detailed cross sections depicting the slopes of the platforms.

East and west of the LNG site, the abrasion platforms were mapped on large-scale aerial photographs (1 in. = 500 ft) and U.S. Geological Survey 7.5-min topographic maps. Large-scale close-interval contour maps (1 in. = 100 ft; 2-ft contour interval) were also used in the eastern part of the study area.

A limited amount of refraction seismic profiling from earlier studies (Converse and others, 1970; Dames and Moore, 1980) was also used to locate shoreline angles in areas where wave-cut platforms are covered by thick alluvial and colluvial deposits. The velocity heterogeneity of the Quaternary sediments along with the common occurrence of low-velocity zones of weathered bedrock limited the application of this technique.

Elevations of platform and seacliff exposures were established by transit-stadia surveys. Closure of survey loops indicate that elevational inaccuracies are on the order of ±30 cm or less in all cases. This level of error is within the uncertainty limits associated with micro-relief on the platforms.

Exposures of the shoreline angles for the low terraces are scarce because much of the coastal plain is covered by thick terrestrial deposits. In many cases, the elevation of the shoreline angle was estimated by surveying several platform exposures in canyon walls and projecting the platform surface upgradient to its intersection with the paleosea cliff. Inaccuracies associated with the shoreline-angle elevation for these cases is dependent on the length of the projection, but is estimated to be under 2 m.

TERRACE PRESERVATION AND CHRONOLOGY

Five well-preserved marine-abrasion platforms are present along the low coastal plain and lowest portions of the adjacent hills between Gaviota and Point Conception (Fig. 1). The platforms are generally cut across the massively bedded and relatively incompetent mudstones of the Sisquoc Formation, although locally they are cut across the more competent, siliceous shales of the Monterey Formation. Higher benches of presumed marine-origin were identified and surveyed up to an elevation of 330 m (Dames and Moore, 1981), but those data are omitted from this study because elevational control was by hand-level surveys, and because most are only poorly preserved notches and benches cut into the resistant Monterey Formation.

The first four emergent platforms are all preserved at the proposed LNG terminal site and were well-exposed in

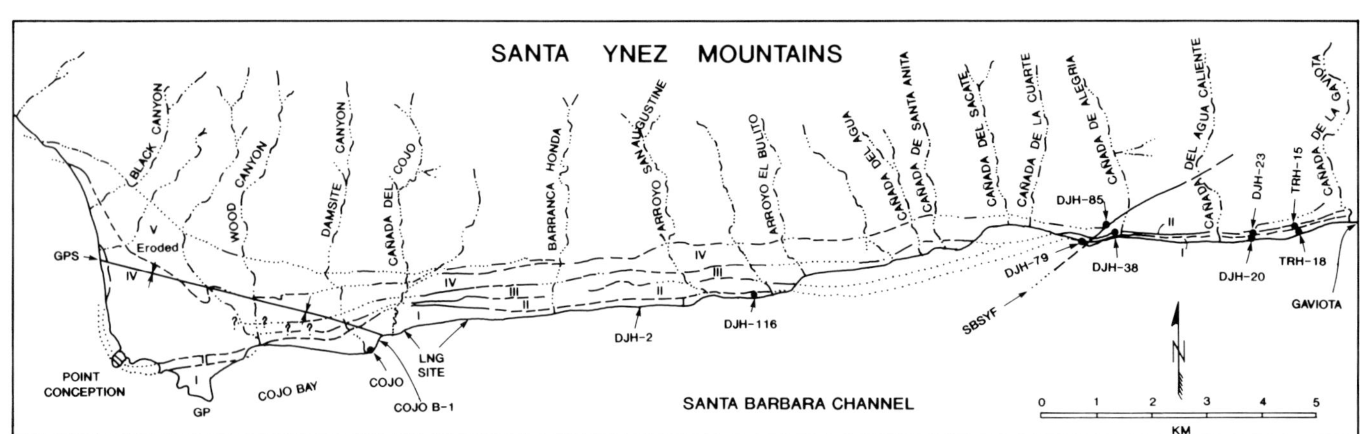

FIG. 1.—Map of marine terraces between Gaviota and 3 km north of Point Conception. Platform numbers refer to those in the text. GP = Government Point, SBSYF = South Branch of Santa Ynez fault. Cojo, DJH, and TRH numbers refer to fossil localities in marine deposits.

trenches and borings (Dames and Moore, 1980). The fifth platform is also preserved in this area, although the abrasion surface is not exposed. The terrace and stratigraphic data from the terminal site investigations are definitive in terms of shoreline-angle elevation and platform slope, and are considered the baseline for extending the terrace mapping to the west and east.

Terrace Dating and Correlation.—

Several methods were employed to date and correlate the abrasion platforms (Dames and Moore, 1980, 1981). U-series dates were obtained on bone, teeth, and shell fragments from the marine and nonmarine deposits overlying the platforms (Table 1). Similarly, where the marine deposits contained fossils, amino-acid stereochemistry analyses were performed to establish the terrace correlations and to estimate terrace ages (Table 2) (Kennedy and Wehmiller, 1986; Kennedy and others, this volume; Wehmiller, this volume). Along with the amino-acid data, the assemblage of molluscan fauna was analyzed for temperature aspect (Kennedy and others, this volume) to assist with the age estimates and terrace correlations (Table 2). Finally, the relative terrace spacing and correlation of terraces to the marine oxygen isotope stages (Bull, 1985) was used to constrain the age estimates, even though uplift rates have varied throughout the study area.

The first emergent terrace, herein named the Cojo terrace, is widely preserved along the coast from Point Conception to Gaviota, and varies in width up to 1 km. Locally, the terrace is missing due to modern seacliff erosion. For example, the terrace is not preserved north of Point Conception for at least 3 km. The shoreline-angle elevation of this terrace varies from about 16 m at Point Conception to around 10 m where it crosses the hinge of the Government Point syncline (Fig. 2). It rises to about 17 m on the north limb of the syncline, decreases to about 11 m west of San Augustine Beach, and then rises to about 17 m to the east before it passes offshore near Arroyo El Bulito (Fig. 1).

Three lines of direct evidence indicate that the Cojo terrace correlates to substage 5a of Shackleton and Opdyke (1973). First, a well-preserved mammal bone, found on the abrasion platform in the marine-sand layer, yielded concordant thorium and protactinium ages of 87 ka (sample DJH-116: Table 1). Second, the molluscan faunal assemblage (Cojo site) contains a strong cool-water aspect with several northern extralimital species (Kennedy and others, this volume). Finally, amino-acid stereochemistry on selected mollusks from this locality yielded ratios consistent with substage 5a, although the data from *Macoma* suggest a possibly even younger age (Table 2).

Other U-series results are consistent with a substage 5a age as well (Table 1). Three bones were collected from sediments overlying the Cojo terrace at or near the proposed LNG site (refer to Dames and Moore, 1980, for precise locations). Two of these, samples AO-3.5 and Cojo B-1, yielded concordant U/Th and U/Pa results, dating the sediment cover at 70 to 60 ka. Another bone, DJH-2, yielded only very marginally concordant results and indicated recent uranium loss. Thus, its age is considered a maximum and is in agreement with the other bone dates.

U-series analyses were also performed on five marine-mollusk shells-collected from the Cojo site (Table 1). Of these, only two (samples Cojo 1Ba and Cojo 2A) produced $^{232}Th/^{234}U$ ratios low enough to indicate a closed system and be considered valid. Both gave late stage 5 ages, consistent with the U-series bone data, but thorium was not analyzed for the latter (Cojo 2A) sample. As a consequence, concordancy could only be checked for the first. The other shell samples all had moderate to high $^{232}Th/^{234}U$ ratios or were discordant, indicating contamination and an open system:

The first emergent terrace on either side of the South Branch Santa Ynez fault (SBSYF) in the eastern part of the study area is correlated to the substage 5a Cojo terrace (Figs. 1 and 2) based on locally developed age control from both sides of the fault. Both the first and second emergent terraces east of the fault have fauna with a distinct cool-water aspect (Kennedy and others, this volume) which, along with their amino-acid chemistry (Table 2), indicates their correlation to substages 5a and 5c, respectively. Furthermore, terrace three, preserved with marine fossils near Arroyo Hondo some 6 km farther east, contains an abundant and distinct warm-water faunal assemblage (Kennedy and Wehmiller, 1986; Kennedy and others, this volume), establishing a correlation to substage 5e.

The amino-acid chemistry (Table 2) agrees well with these interpretations, and supports correlation of the lowest terrace at the SBSYF to the Cojo terrace near Point Conception. The free D-allo/L-iso values are the most convincing set of amino-acid data in Table 2 for establishing these correlations. Both the Cojo terrace near Point Conception and the lowest terraces east and west of the SBSYF all have mean free amino-acid racimization (AAR) ratios of about 0.65 to 0.68 for *Protothaca*, with similar values for *Macoma*. The free ratios for the second terrace are similar but slightly higher, as expected, ranging from 0.67 to 0.71 for *Protothaca*. For comparison, the warm-water Arroyo Hondo 5e terrace yielded a *Protothaca* AAR (free) ratio of 0.76.

The total amino-acid ratios generally agree well with the free amino-acid data. For leucine and valine (Table 2), Cojo *Protothaca* samples yielded mean ratios of 0.38 ± 0.05 and 0.23 ± 0.03 ($n = 18$), respectively. East of the SBSYF, *Protothaca* yielded similar results with localities DJH-20 and TRH-18, having mean values of 0.40 ± 0.05 and 0.25 ± 0.03 ($n = 6$) for leucine and valine, respectively. These values, although very slightly higher, are statistically indistinguishable from one another.

Protothaca fossils recovered from the second terrace east of the fault yielded leucine and valine AAR ratios of 0.39 ± 0.05 and 0.25 ± 0.03, respectively. These values are virtually identical to those from *Protothaca* recovered from the first terrace. However, *Protothaca* samples collected from the warm-water substage 5e locality at Arroyo Hondo also yielded similar AAR ratios of 0.39 ± 0.04 and 0.25 ± 0.01 ($n = 5$) for leucine and valine, respectively. The latter samples, however, were chalky and appeared to have been leached. If the AAR ratios are representative for the stage 5e terrace in this area, then it would appear that *Protothaca* yields inconclusive AAR results when one is attempting to differentiate individual stage 5 substages. Alternatively, if

TABLE 1.—URANIUM-SERIES AGES OF BONES, SHELLS AND A TOOTH COLLECTED FROM THE MARINE AND TERRESTIAL ALLUVIUM OVERLYING THE ABRASION PLATFORMS.

Sample	Material	Terrace Number	Terrace SLA Elev (m)	U/Th* Age (ka)	U/Pa* Age (ka)	232Th/234U* Ratio	Comments
Cojo B-2	Bone	modern alluvium		7±0.4	5.5±0.35		Best estimate of 4±1 ka after correcting for initially present uranium and thorium. Collected from Holocene terrace to Canada del Cojo.
AO-3.5	Bone	overlies 1st		60±2	66±7		Concordant U-series ages support a post-stage 5a age for the alluvium overlying the Cojo terrace.
DJH-2	Bone	overlies 1st		107±6	>124		Marginally concordant U/Th and U/Pa ages. Chemistry indicates recent loss of uranium, therefore, age is a maximum.
Cojo B-1a	Bone	overlies 1st		69.8±3.5	67.9±4.8		Concordant results between separate bone fractions and splits as well as between U/Th and U/Pa. Average age is 68.2±1.8 ka, and agrees with other samples overlying the Cojo terrace. Supports a stage 5a age for the underlying Cojo terrace.
Cojo B-1b	Bone	overlies 1st		65.4±3			
Cojo B-1c	Bone	overlies 1st		69.8±3.9			
Cojo B-1d	Bone	overlies 1st		68.0±2.1			
DJH-79	Bone	1st	27	108±6	>96		Marginally concordant U/Th and U/Pa ages. Best estimate is 80-108 ka.
DJH-116	Bone	1st	17	87±4	87±12		Concordant U/Th and U/Pa ages. Bone found resting on abrasion platform. Indicates stage 5a age for Cojo terrace.
COJO-1A	Mytilus	1st	15	142±13	>142	0.035±0.004	Moderate 232Th/234U ratio, probably age is a maximum and is too high.
COJO-1Ba	Tellina (inner shell)	1st	15	82±4		0.004±0.00	No concordancy check. Low 232Th/234U ratio support stage 5a age.
COJO-1Bb	Tellina (outer shell)	1st	15	178±21	>178	0.010±0.001	Exterior parts of shell of COJO-1Ba.
COJO-2A	Mytilus	1st	15	96±8	90+50/-25	0.107±0.009	High 232Th/234U ratio, therefore this is a maximum age and supports a stage 5a age.
COJO-2B	mixed sp.	1st	15	37±2	84±9	0.017±0.001	Grossly discordant ages; clearly an open system. No age estimate.
BSE-1	Bone	overlies 4th		162±13	157+inf/-32		Concordant thorium and protactinium ages. Supports stage 7 age of terrace 4.
TCL-1	Bone	overlies 4th		149±13	≥125		Concordant with Pa at equilibrian limit. Age indicated by U/Th age.
TCL-4	Bone	overlies 4th		153±11	21±2		Grossly discordant results suggest recent uranium uptake and indicate that the U/Th age is a minimum.
TSE-1	Tooth (enamel)	overlies 4th		164±13	125+79/-30		These two fractions yielded internally consistent and concordant results, within the limits of the Pa method, and indicate an age of 150-180 ka.
TSE-1	Tooth (dentin)	overlies 4th		171±15	125+45/-25		

* Analyses by T. Ku for Dames and Moore; in Dames and Moore (1980).

the samples had been substantially leached, as their appearance would suggest, then the results cannot be compared with the well-preserved fossils collected from the Cojo terrace or from the sites near the SBSYF.

Saxidomus samples, however, apparently do yield reasonable AAR ratios consistent with the established geomorphic relations (Table 2). Based on the *Saxidomus* data, the Cojo terrace correlates to both the lowest terrace at Arroyo Hondo, which also bears cool-water fauna, and to the lowest terrace east of the SBSYF.

Site DJH-79, also on the first terrace and very close to the SBSYF, gave significantly higher ratios for leucine and

TABLE 2.—SUMMARY OF AMINO-ACID RATIOS FOR FOSSIL MOLLUSKS COLLECTED FROM THE STUDY AREA.

Locality	Terrace[1] Number	Faunal Aspect	Type Sample[2]	D-Alloisolucine[3] Free	Total	Leucine[4]	Valine[4]
DJH-20+ TRH-18	1E	Cool	P	0.65±0.01 (n = 2)	0.32±0.03 (n = 2)	0.40±0.05 (n = 6)	0.25±0.03 (n = 6)
			S	—	—	0.42±0.01 (n = 5)	0.26±0.01 (n = 5)
			M	0.70 (n = 1)		0.405±0.02 (n = 2)	0.225±0.02 (n = 2)
Arroyo Hondo	1E	Cool	P	—	—	0.42 (n = 6)	(n = 1) 0.26 (n = 1)
			S	—	—	0.38±0.04 (n = 2)	0.23±0.02 (n = 2)
DJH-23+ TRH-15	2E	Cool	P	0.71±0.01 (n = 3)	0.31±0.03 (n = 3)	0.39±0.05 (n = 4)	0.25±0.03 (n = 4)
DJH-38[5]	2E	Cool	P	0.67 (n = 1)	0.36 (n = 1)	0.64 (n = 1)	0.36 (n = 1)
			S	0.70 (n = 1)	0.48 (n = 1)	0.62±0.03 (n = 4)	0.39±0.02 (n = 5)
Arroyo Hondo	3E	Warm	P	0.76 (n = 1)	0.47 (n = 1)	0.39±0.04[6] (n = 5)	0.25±0.01 (n = 5)
			S	—	—	0.53±0.0 (n = 2)	0.33±0.01 (n = 2)
COJO	1W	Cool	P	0.68±0.04 (n = 7)	0.36±0.07 (n = 4)	0.38±0.05 (n = 17)	0.23±0.03 (n = 18)
			S	—	—	0.35±0.04 (n = 10)	0.23±0.02 (n = 10)
			M	0.64±0.02 (n = 6)	0.25 (n = 1)	0.40±0.02 (n = 13)	0.23±0.02 (n = 12)
DJH-79[5]	1W	n.d.	P	0.68 (n = 1)	0.37 (n = 1)	0.48±0.0 (n = 2)	0.30±0.01 (n = 2)
DJH-85[6]	4W	n.d.	S	—	—	0.51±0.02 (n = 2)	0.34±0.01 (n = 2)
Sea Cliff (45–60 ka)	—	Cold	P	0.58 (n = 1)	0.20 (n = 1)	0.29±0.03 (n = 12)	0.17±0.02 (n = 12)
S.B. Form. 500 ka	—	n.d.	S	1.22 (n = 1)	0.96 (n = 1)	0.79 (n = 1)	0.664 (n = 1)

[1]Terrace numbers refer to the first through fourth terraces east (E) and west (W) of the SYSBF.
[2]Refers to the genera of mollusk analyzed: P = *Protothaca*, S = *Saxidomus*, M = *Macoma*.
[3]Analyzed by P. E. Hare, 1981, *in* Dames and Moore (1981), App. A.5.
[4]Analyzed by J. F. Wehmiller, 1980, *in* Dames and Moore (1981), App. A.1, and unpublished data.
[5]These samples may have been thermally affected. They were collected adjacent to the south branch of the Santa Ynez fault; the surrounding bedrock is hydrothermally altered and an active steam geyser was observed venting out of the fault zone in the spring of 1981.
[6]These samples were all weathered and chalky. The ratios are probably too low.

valine on *Protothaca* of 0.48 (n = 2) and 0.30±0.01 (n = 2), respectively. Similarly, sample site DJH-38 at Alegria Canyon on the second emergent terrace yielded higher than expected values. These samples were collected close to the fault in deposits that exhibit evidence of hydrothermal alteration. During the year following the terrace mapping, steam and hot water vented from the fault in the modern sea cliff for a period of time, with water temperatures up to 70°C (observations by TKR). We interpret the elevated total AAR ratios to be due to this thermal effect. Interestingly, the free AAR ratios were not apparently affected by the elevated temperatures (Table 2).

Based on these data, we correlate the first emergent terrace east and west of the SBSYF, the Cojo terrace, and the lower terrace at Arroyo Hondo to substage 5a (Shackleton and Opdyke, 1973) at about 80 ka (Bloom and others, 1974). The second and third emergent terraces, both near Point Conception and near the SBSYF, respectively, correlate to substage 5c and 5e. It should be noted, however, that the third (substage 5e) terrace is not preserved in the area of the SBSYF, whereas this terrace is otherwise commonly well preserved elsewhere.

The fourth terrace is dated by two methods. At the proposed LNG site, three bones and a tooth were recovered from the terrestrial alluvium overlying the marine section. Two of the bones and the tooth gave concordant U/Th and U/Pa results, within the limits of the protactinium method, and date the section overlying the terrace at 170 to 150 ka (Table 1). The third bone yielded grossly discordant results, suggesting recent uptake of uranium, and therefore a minimum U/Th age of 153 ka. These dates correlate well to the period of sea-level lowering during stage 6, at a time when alluvial sedimentation would be expected on an exposed abrasion platform, and indicate that the underlying terrace correlates to stage 7, estimated at about 230 to 200 ka (Chappell, 1983).

Mollusks were collected from the fourth terrace just west of the SBSYF (loc. DJH-85, Figs. 1, 2). The collection site was poor and the poorly preserved shells of only six species were found, none of which are diagnostic indicators of paleotemperature (Kennedy and others, this volume). Two *Saxidomus* were analyzed for amino acids (Table 2) and yielded ratios that are similar to those of *Saxidomus* collected from the Arroyo Hondo substage 5e terrace. How-

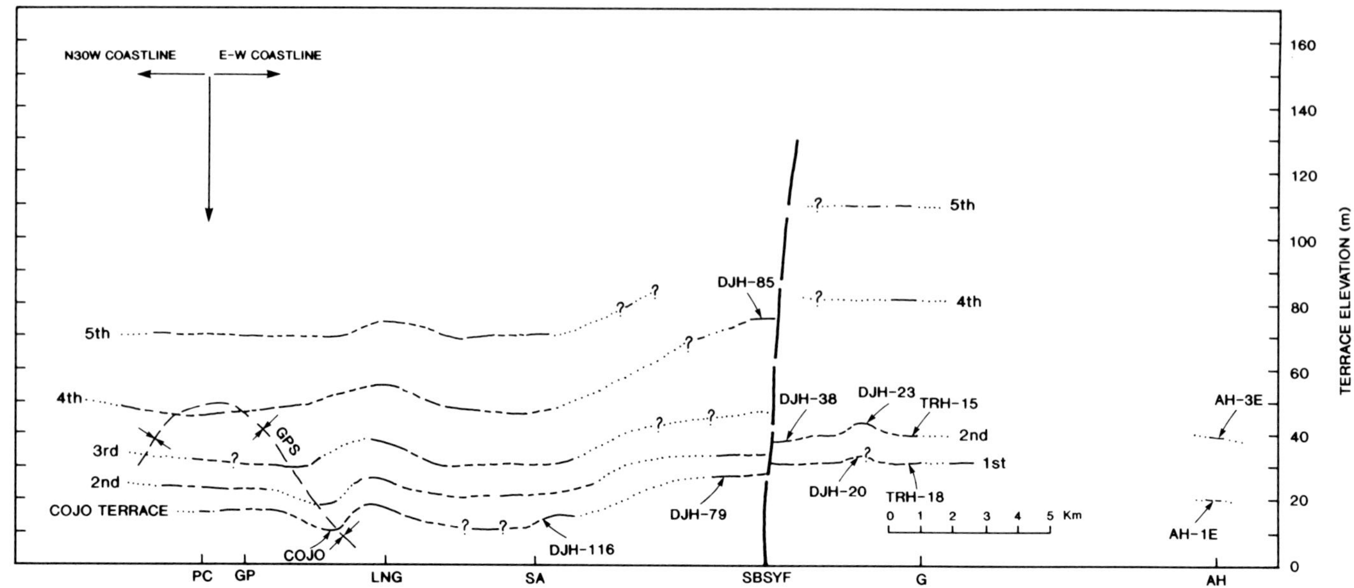

FIG. 2.—Shore-parallel profiles of the elevations of shorelines mapped in this study. PC = Point Conception, GP = Government Point, LNG centers on proposed LNG site, SA = the San Augustine drainage, SBSYF = South Branch of Santa Ynez fault, G = Gaviota, AH = Arroyo Hondo, GPS = Government Point syncline. Cojo, DJH, TRH, and AH numbers refer to fossil localities in marine deposits.

ever, the elevation of the fourth terrace west of the SBSYF is too high for a stage 5e age, unless the uplift rate at that site was several times higher between 120 and 105 ka than is indicated by the long-term average rate. Alternatively, amino acids were leached from the samples, an interpretation that is consistent with their poor state of preservation as noted at the time of their collection. In such a case, the AAR ratios would probably be lower than expected, an explanation that seems more reasonable.

No direct age control is available for the fifth marine terrace. It is correlated to stage 9 at about 340 to 320 ka (Chappell, 1983) based on its relative elevation at the SBSYF and the LNG site.

Higher notches and benches of probable marine-abrasion origin are present up to about 300 m. Many of these benches have concordant elevations from ridge to ridge where not affected by folding or faulting, suggesting once extensive terraces. Cobbles, sometimes with pholad borings, are common at the back edges of many of the benches and appear to be all that remains of the original marine deposits, although one high-level bench (~160 m) near the SBSYF contains fossil mollusks in a marine sand matrix (Dames and Moore, 1981). Based on their elevations and the rate of uplift for the area, we interpret these benches and notches to represent marine-abrasion platforms up to and possibly exceeding 1 Ma in age (Dames and Moore, 1980).

Correlation Using A Constant Uplift Rate.—

The ages and correlations suggested earlier can be tested. If the rate of uplift has been reasonably constant for any one area over the time interval of interest (rates obviously vary from area to area), and assuming that the Cojo terrace is 80 ka (substage 5a) as indicated by virtually all of the data, then a plot of uplift (terrace elevation minus its paleosea level) versus terrace age should define a relatively straight line (Fig. 3). We performed this test using the paleosea-level estimates of Muhs and others (1988), determined from the California and Baja California coast, and those of Bloom and Yonekura (1985), determined from New Guinea. Figure 3 plots only data from the four areas where we have the best elevational control on at least three terraces: Point Conception, the proposed LNG site, and east

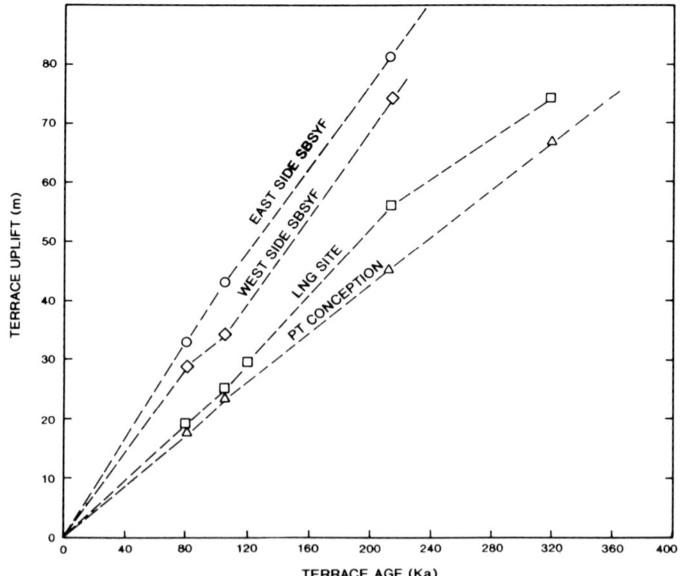

FIG. 3.—Plot of terrace uplift versus age for four transects in the study area. Uplift is present shoreline elevation minus the paleosea level at which it was cut. Paleosea levels used are from Muhs and others (1988) and Bloom and Yonekura (1985).

and west of the SBSYF. In all four areas, the inferred ages of the terraces produce reasonably constant average uplift rates, which suggests that our terrace correlations and age estimates are correct. These interpretations are used later to resolve styles and rates of post-terrace deformation.

TECTONIC DEFORMATION

Long-Period Warping and Variations in Uplift

All five terraces exhibit warping parallel to the coastline between Point Conception and Gaviota (Fig. 2). Part of the warping occurs where the terrace shorelines cross the Government Point syncline or the SBSYF, but elsewhere, warping is apparently associated with differential uplift.

As discussed earlier, the Cojo terrace shoreline angle varies from 16 m at Point Conception to about 10 m at the hinge of the Government Point syncline (Fig. 2). To the east, it rises fairly sharply to nearly 18 m on the western side of the proposed LNG site, but decreases to 15 m on the eastern side, and ultimately decreases to a low of about 11 m some 2 km west of the San Augustine drainage. Eastward from there, it rises to about 16 m near Arroyo El Bulito, where it has been removed by seacliff erosion. The Cojo platform reappears about a 0.5 km west of the SBSYF at an elevation of 27 m, jumps to 31 m as it crosses the fault, and stays at about 30 m eastward toward Gaviota. Farther east, although not mapped in detail, this terrace was observed to decrease in elevation to about 20 m at Arroyo Hondo.

The second (substage 5c), third (substage 5e) and fourth (stage 7) terraces are similarly deformed (Fig. 2). The shoreline angle of the fourth terrace varies in elevation from about 45 m where it crosses the Government Point syncline to about 55 m at the proposed LNG site. At the SBSYF, this terrace rises to 73 m, similar in magnitude (relative to its age) to the first terrace.

A plot of shoreline elevations versus distance north and south of the Government Point synclinal hinge is shown on Figure 4. The four shorelines depicted are, from lowest to highest, the Cojo terrace, and the second, fourth and fifth terraces. There were insufficient data to plot the third terrace due to lack of exposures. In Figure 4, south limb data are shown on the left; north limb data are on the right. Additionally, in several cases the shorelines on each limb are divided into east and west sections, according to the portion of the field area from which they are derived. The west section is for the area principally north of Point Conception.

The observations indicate first that folding appears to be occurring at a faster rate on the north limb than on the south limb of the Government Point syncline. This is consistent with the steeper dips observed in bedrock on the north limb. Second, folding is not proceeding at a constant rate along the fold axis, but rather is more rapid toward the east. Because of this lateral variation, derived fold rates are dependent upon the location along the fold axis. Third, a progressive increase in tilt on successively older shorelines should be observed. Shorelines associated with terraces 4 and 5 exhibit this phenomenon; however, shorelines of the two lowest terraces show greater deformation than the fourth terrace. This phenomenon is attributed to the location of the data points used to construct shoreline-angle geometries for the various platforms. The fourth and fifth terraces of Figure 4 utilize data from a more westerly portion of the fold and therefore reflect the lower rate of folding observed here.

It is not possible to construct a profile such that tilts of the various shorelines can be directly compared on a single cross section of the fold. The tilts on shorelines corresponding to the first two platforms depend on geophysically determined elevations near the hinge of the syncline. Therefore, an error in the geophysical interpretation may also affect the derived fold rates. Two factors that exceed the resolution of the data, but which may also affect determined fold rates, are localized, relatively short wavelength deformation or time variations in the fold rate.

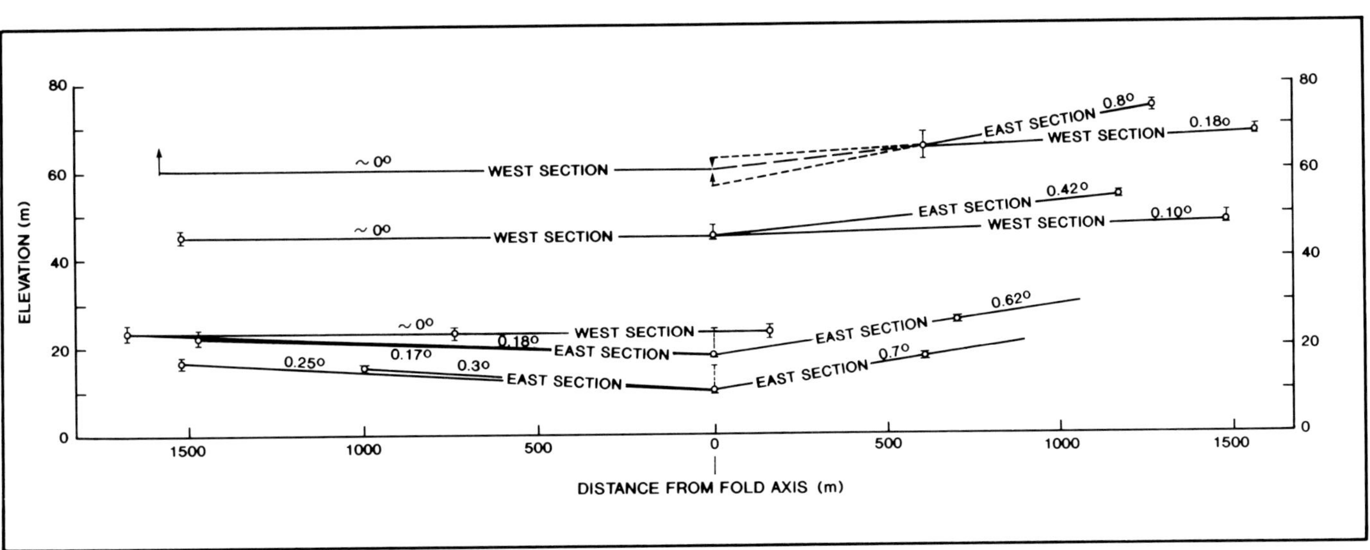

FIG. 4.—Shoreline-angle geometry across the hinge of the Government Point syncline (from Dames and Moore, 1980).

Independent support for the tilts observed on the Cojo terrace shoreline is given by examining the variation in platform gradient. The Cojo platform at the proposed LNG site has an average seaward gradient of about 2.2°, where not affected by faulting. On the south limb of the syncline at Government Point, however, it has a seaward gradient of only 0.9 to 1.0°. The 2.2° gradient of this platform at the LNG site is higher than that of average platform slopes measured elsewhere in California by approximately the amount indicated by shoreline tilt. In addition, the disparity in gradient between the two locations (located on opposing limbs of the syncline) becomes negligibly small if they are adjusted by the amount of tilt indicated by the shorelines.

Fold rates were computed using the shoreline (platform) ages as discussed previously. Two rates were determined for the proposed LNG site area: A maximum rate of 9° per million years (0.16 urads/yr), calculated from the geophysically measured 10-m Cojo terrace-shoreline elevation over the hinge of the fold; and a minimum rate of about 2° per million years (3.6×10^{-2} urads/yr) calculated from tilts observed on the fourth terrace shoreline. These rates are consistent with the post-Miocene deformation of the bedrock.

In addition to folding, bedrock and the local alluvial cover in the area have been displaced by reverse bedding-plane faults (Dames and Moore, 1980). The faults occur on both limbs of the fold and have been active during the period documented for folding. We interpret the faults to reflect flexural-slip folding of the Government Point syncline. Flexural-slip faulting is widely documented in the Transverse Ranges (Yeats and others, 1981; Rockwell and others, 1984; Rockwell, 1988) and elsewhere when layered sedimentary rocks are folded.

DISCUSSION AND CONCLUSIONS

Our data show that the marine terraces between Gaviota and Point Conception clearly are uplifted and warped. Folding across the Government Point syncline has been demonstrated by differences in elevation of individual paleoshorelines across the fold, differences in slope of the Cojo terrace on opposing limbs, and by the occurrence of bedding-plane faulting on both limbs of the fold.

Average rates of uplift for the coastline range from about 0.15–0.23 m/ka near Point Conception and the proposed LNG site to 0.30 m/ka on the east side of the South Branch Santa Ynez fault near Gaviota. These rates are substantially lower than those determined in areas of very active folding to the east in the Ventura Basin (Yeats, 1976; Rockwell and others, 1988), where uplift rates locally exceed 10 mm/yr. The rates are generally higher, however, than in other areas of the California coastline not directly affected by either active faulting or folding. For instance, the average rate of uplift between San Diego and Los Angeles is about 0.10 to 0.14 mm/yr (though with higher local rates adjacent to the Rose Canyon, Newport-Inglewood, and Palos Verdes faults; Kern and Rockwell, this volume; Muhs and others, this volume). North of the Transverse Ranges, the uplift rate varies from near zero to 0.20 mm/yr between the Santa Maria basin and Cayucos (Hanson and others, 1990, and this volume). At San Simeon, the average uplift rate away from the strike-slip San Simeon fault is about 0.15 mm/yr, similar to that near San Diego. These data indicate that the rate of uplift near Point Conception is slightly higher than that for most of the California coast, but substantially lower than in areas of rapid convergence, such as across the Ventura Avenue anticline.

The terraces are displaced by the SBSYF, which exhibits a vertical component of the slip rate of 0.05 mm/yr. The shoreline angles are not recognizably deflected across the fault by more than the modern shoreline, although minor strike slip would be difficult to delineate because of the thick colluvial cover. These data suggest that if the SBSYF is dominantly a strike-slip fault, as suggested previously, then the slip rate on this fault is low.

The Transverse Ranges are characterized by active east-west trending folds and thrust faults associated with north-south crustal shortening (Yeats, 1983; Rockwell and others, 1984; 1988; Rockwell, 1988). This activity accommodates transfer of strike-slip motion between the onshore coastal and offshore borderland faults of southern California, and the coastal and offshore faults of central California (Weldon and Humphreys, 1986). Our data indicate that folding in the westernmost onshore Transverse Ranges has continued throughout the late Quaternary, but with rates substantially less than those of the Ventura area to the east, where high crustal-shortening rates have been demonstrated.

ACKNOWLEDGMENTS

We thank Western LNG Terminal Associates for permission to publish these results, the many Dames and Moore personnel who helped generate data during the course of these studies, and the landowners of Hollister and Bixby Ranches, who granted land access. We also thank Bill Lettis and Phil Kern for their excellent reviews of an earlier draft of this paper.

REFERENCES CITED

Birkeland, P. W., 1972, Late Quaternary eustatic sea-level changes along the Malibu coast, Los Angeles County, California: Journal of Geology, v. 80, p. 432–448.

Bloom, A. L., Broecker, W. S., Chappell, J. M. A., Matthews, R. K., and Mesolella, K. J., 1974, Quaternary sea level fluctuations on a tectonic coast: New $^{230}Th/^{234}U$ dates from the Huon peninsula, New Guinea: Quaternary Research, v. 4, p. 185–205.

Bloom, A. L., and Yonekura, N., 1985, Coastal terraces generated by sea-level change and tectonic uplift, *in* Woldenberg, M. J., ed. Models in Geomorphology. Allen and Unwin, Boston, p. 139–154.

Bradley, W. C., and Griggs, G. B., 1976, Form, genesis, and deformation of central California wave-cut platforms: Geological Society of America Bulletin, v. 87, p. 433–449.

Bull, W. B., 1985, Correlation of flights of global marine terraces, *in* Marisawa, M., and Hack, J., eds., Tectonic Geomorphology. Proceedings, 15th Annual Geomorphology Symposium, State University of New York at Binghampton: George Allen and Unwin, Hemel Hempstead, England, p. 129–152.

Chappell, J., 1983, A revised sea-level record for the last 300,000 years on Papua New Guinea: Search, v. 14, p. 99–101.

Converse, Davis and Associates, and Dixon, S. J., 1970, Geologic feasibility investigation. Phase I study, Hollister Ranch plant siting, Santa Barbara County, California, for Southern California Edison Company, Los Angeles, California: Converse, Davis and Associates Project No. 70-414-AH, Pasadena, California (unpaginated).

DAMES AND MOORE, 1980, Final geoseismic investigation, proposed LNG terminal, Little Cojo Bay, California: for Western LNG Terminal Associates, 6 vols. (unpaginated).

DAMES AND MOORE, 1981, Analysis of data, marine terrace studies and age dating, final geoseimic investigation, proposed LNG Terminal, Little Cojo Bay, California: for Western LNG Terminal Associates, 51 p. plus 5 appendices

HANSON, K. L., WESLING, J. R., LETTIS, W. R., KELSON, K., AND MEZGER, L., 1990, Correlation, ages, and uplift rates of Quaternary, marine terraces: south-central coastal California, *in* Lettis, W. R., Hanson, K. L., Kelson, K. I., and Wesling, J. R., eds., Neotectonics of South-Central Coastal California: Friends of the Pleistocene Guidebook, Pacific Cell, p. 139–190.

KENNEDY, G. L., AND WEHMILLER, J. F., 1986, Paleoclimatic implications of Quaternary marine invertebrate faunas from southwestern Santa Barbara County, California: Western Society of Malacologists, Annual Report, v. 18, p. 22–23.

KERN, J. P., 1977, Origin and history of upper Pleistocene marine terraces, San Diego, California: Geological Society of America Bulletin, v. 88, p. 1553–1566.

LAJOIE, K. P., KERN, J. P., WEHMILLER, J. F, KENNEDY, G. L., MATHIESON, S. A., SARNA-WOJCICKI, A. M., YERKES, R. F., AND MCCRORY, P. F., 1979, Quaternary marine shorelines and crustal deformation, San Diego to Santa Barbara, California, *in* Abbott, P. L., ed., Geologic Excursions in the Southern California Area: Geological Society of America Annual Meeting, Field Trip Guidebook, p. 3–15.

MUHS, D. R., KENNEDY, G. L., AND ROCKWELL, T. K., 1988, Uranium-series ages of corals from marine terraces, Pacific coast of North America: implications for the timing and magnitude of late Pleistocene sea-level changes: American Quaternary Association, Program and Abstracts, 10th Biennial Meeting, p. 140.

ROCKWELL, T. K., 1983, Soil chronology, geology and neotectonics of the north-central Ventura basin, California: Unpublished Ph.D. Dissertation, University of California at Santa Barbara, 424 p.

ROCKWELL, T. K., 1988, Neotectonics of the San Cayetano fault, Transverse Ranges, California: Geological Society of America Bulletin, v. 100, p. 500–513.

ROCKWELL, T. K., KELLER, E. A., CLARK, M. N., AND JOHNSON, D. L., 1984, Chronology and rates of faulting of Ventura River terraces, California: Geological Society of America Bulletin, v. 95, p. 1466–1474.

ROCKWELL, T. K., KELLER, E. A., AND DEMBROFF, G. R., 1988, Quaternary rate of folding of the Ventura Avenue anticline, western Transverse Ranges, southern California: Geological Society of America Bulletin, v. 100, p. 850–858.

ROCKWELL, T. K., MUHS, D. R., KENNEDY, G. L., HATCH, M. E., WILSON, S. H., AND KLINGER, R. E., 1989, Uranium-series ages, faunal correlations and tectonic deformation of marine terraces within the Agua Blanca fault zone at Punta Banda, northern Baja California, Mexico, *in* Abbott, P. L., ed., Geologic Studies in Baja California: Pacific Section, Society of Economic Paleontologists and Mineralogists Book 63, p. 1–16.

SHACKLETON, N. J., AND OPDYKE, N. D., 1973, Oxygen isotope and paleomagnetic stratigraphy of equatorial Pacific core V28-238: Oxygen isotope temperatures and ice volumes on a 10^5-year and 10^6-year scale: Quaternary Research, v. 3, p. 39–55.

WELDON, R., AND HUMPHREYS, E. D., 1986, A kinematic model of southern California: Tectonics, v. 5, no. 1, p. 33–48.

WOODRING, W. P., BRAMLETTER, M. N., AND KEW, W. S. W., 1946, Geology and paleontology of Palos Verdes Hills, California: U. S. Geological Survey Professional Paper 207, 145 p.

YEATS, R. S., 1976, Neogene tectonics of the central Ventura basin, *in* Fritsche, A. E., Best, H. T., Jr., and Wornardt, W. W., eds., The Neogene Symposium: Pacific Section, Society of Economic Paleontologists and Mineralogists, p. 19–32.

YEATS, R. S., 1981, Quaternary flake tectonics of the California Transverse Ranges: Geology, v. 9, p. 16–20.

YEATS, R. S., 1983, Large-scale Quaternary detachments in Ventura basin, southern California: Journal of Geophysical Research, v. 88, p. 569–583.

YEATS, R. S., CLARK, M. N., KELLER, E. A., AND ROCKWELL, T. K., 1981, Active fault hazard in southern California: ground rupture versus seismic shaking: Geological Society of America Bulletin, v. 92, p. 189–196.

PALEOECOLOGY AND PALEOZOOGEOGRAPHY OF LATE PLEISTOCENE MARINE-TERRACE FAUNAS OF SOUTHWESTERN SANTA BARBARA COUNTY, CALIFORNIA

GEORGE L. KENNEDY

Earth Sciences Division, Los Angeles County Museum of Natural History, 900 Exposition Boulevard, Los Angeles, California 90007

JOHN F. WEHMILLER

Department of Geology, University of Delaware, Newark, Delaware 19716

AND

THOMAS K. ROCKWELL

Department of Geological Sciences, San Diego State University, San Diego, California 92182

ABSTRACT: Over 165 species of megainvertebrates, mostly mollusks, from the lowest three (of approximately 17) emergent marine terraces along the southwestern coast of Santa Barbara County, California, provide the basis for both paleoecological and paleoclimatic interpretation. Most of the fossiliferous exposures on the two lowest terraces at Cañada de Alegria, Gaviota, and Arroyo Hondo are situated at or very near to the original shoreline angle (i.e., base of the fossil sea cliff) of their respective terraces, physically constraining any interpretation of their depositional setting. The overall composition of their faunas, however, suggests sources from several near-shore marine habitats, all at intertidal to shallow inner sublittoral depths (0 to 18–27 m). Only the sandy bottom faunas from localities at Cojo Bay differ, representing a position about 700 to 750 m from shore and a water depth of about 15 m.

Composite faunas from localities on the first terrace at Cojo Bay, Gaviota, and Arroyo Hondo, and from the second terrace at Cañada de Alegria contain from four to 12 extralimital northern species each, indicative of slightly cooler water paleoclimatic conditions, comparable to those occurring today in the vicinity of Monterey Bay, central California. The fauna from the third terrace at Arroyo Hondo contains four extralimital southern gastropods, whose zoogeographic ranges suggest a more southerly geographic equivalency and slightly warmer water conditions, comparable to those occurring today between San Diego, California, and Ensenada, Baja California.

On the basis of terrace mapping, uranium-series age estimates of bones, amino-acid racemization and epimerization analyses of bivalve mollusks, long-term uplift studies, and the zoogeographic signatures of the terrace faunas, the three lowest terraces are assigned ages of 85 to 80 ka (Cojo Bay, Gaviota, and lower Arroyo Hondo localities), 105 to 100 ka (Cañada de Alegria second terrace locality), and 130 to 120 ka (Arroyo Hondo upper terrace localities), correlative with dated sea-level highstands recorded elsewhere as reef terraces, and in deep-sea sediments as marine oxygen isotope substages 5a, 5c, and 5e, respectively.

INTRODUCTION

Marine terraces are moderately well developed along the southern California coast between Point Conception and Santa Barbara in southwestern Santa Barbara County. Up to 17 marine terraces may be present along the south slope of the Santa Ynez Mountains (Upson, 1951). These terraces range in elevation from near sea level at the coast, to over 300 m inland, and may range in age from 80 ka to over 1 Ma. Only the five lowest terraces are well expressed and only the four lowest have some sort of direct age control (Rockwell and others, this volume).

Fossil invertebrate remains are exposed at several localities on the low terraces between Point Conception and Santa Barbara, but are represented best by collections obtained east of Gaviota, mainly because limited access has precluded active collecting along the coast west of Gaviota. Recent geologic investigations between Point Conception and Gaviota have revealed several previously unstudied fossiliferous exposures on the three lowest terraces that parallel the coast (Kennedy and Wehmiller, 1986). The purpose of this paper is to describe the paleoecology and zoogeography of these faunas, and to establish a geochronology for them that can be used to correlate them with temporally equivalent terrace deposits on other parts of the California coast, and with worldwide climatic and sea-level fluctuations recorded in the coral reef record of tropical islands (Bloom and others, 1974; Bender and others, 1979; Harmon and others, 1983) and inherent in the oxygen isotope record of the deep sea (Shackleton and Opdyke, 1973).

Previous reports of Pleistocene marine-terrace fossils from southwestern Santa Barbara County are few. Dibblee (1950) briefly mentioned the presence of fossiliferous beach sand 1.6 km (1 mi) west of Gaviota and at Alegria Canyon, and Upson (1951) and Hoskins (1957) cited 14 and 26 taxa, respectively, from the locality west of Gaviota. Kaufman and others (1971), in a study of uranium-series dating of fossil mollusks, cited a "large clam" (possibly *Tresus*) from Gaviota, and four taxa (*Mytilus, Chione* [unlikely, possibly *Protothaca*], "clam," and *Olivella*) from Cojo Canyon. The collections studied herein contain material from each of these localities.

Fossiliferous marine deposits east of Arroyo Hondo also have been described from near Tajiguas railroad siding (Upson, 1951; Hoskins, 1957; Valentine, 1961), at El Capitan State Beach (Upson, 1951; Valentine, 1961), near Naples railroad siding (Hoskins, 1957), and at Isla Vista near Goleta (Oldroyd and Grant, 1931; Hoskins, 1957; Wright, 1972). The closest marine-terrace faunas to the north of Point Conception are from Point Sal and along the southwest flank of the Casmalia Hills in northwestern Santa Barbara County (Fairbanks, 1896; Woodring and Bramlette, 1950 [1951]; Kennedy and others, 1990).

Between Point Conception and Arroyo Hondo, fossil faunas (Table 1) are presently available from 22 collecting stations, of which 15 are treated herein (see Appendix for collection data). These represent several geographic areas (Fig. 1) and temporal periods of deposition (Fig. 2). Location of these fossil deposits are, from west to east: (1) at the east end of Cojo Bay and west of the mouth of Cañada del Cojo (3 collections from 2 localities); (2) on the faulted ridge crest west of Cañada de Alegria (1 locality); (3) along the shoreline angle of the second terrace above the mouth of Cañada de Alegria (2 collections from 1 locality); (4)

Quaternary Coasts of the United States: Marine and Lacustrine Systems, SEPM Special Publication No. 48

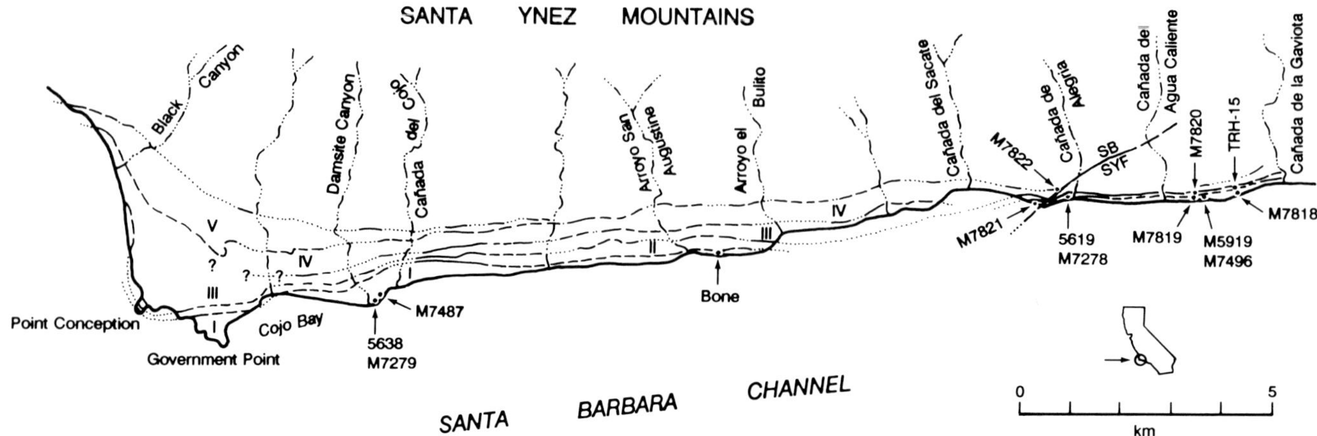

FIG. 1.—Index map showing marine terraces, numbered I through V, and fossil localities between Point Conception and Gaviota, Santa Barbara County, California. SBSYF, South Branch of the Santa Ynez fault. (Modified from Rockwell and others, this volume.)

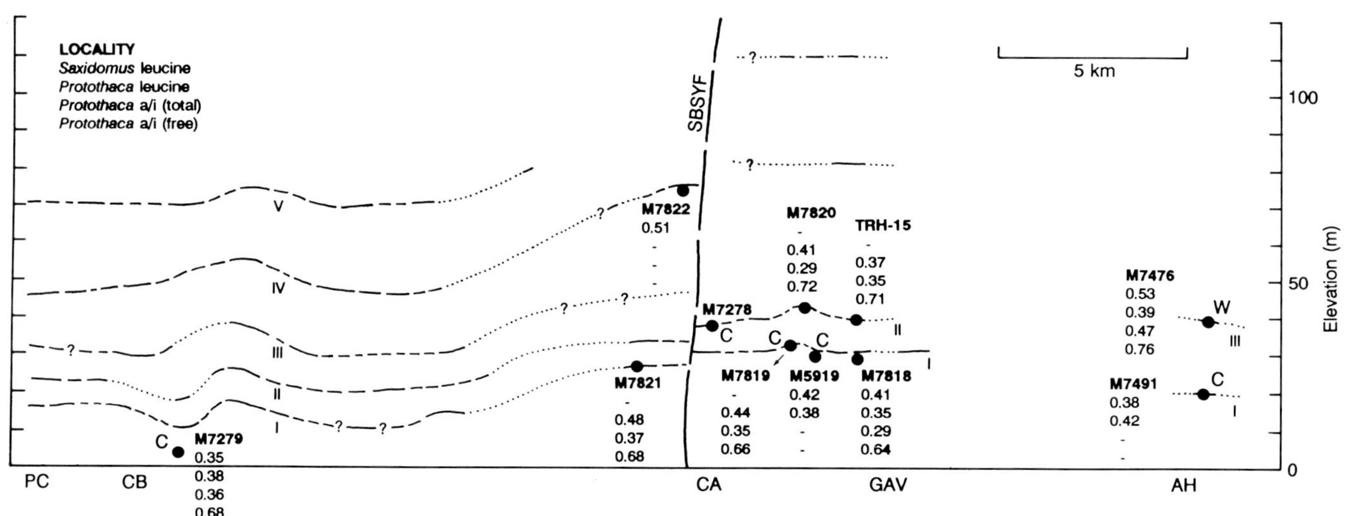

FIG. 2.—Schematic diagram showing the elevation of the inner edges (shoreline angles) of terraces I through V as mapped by Rockwell and others (this volume), fossil localities (solid circles) identified by their institutional numbers (in bold), zoogeographic signature of selected terrace faunas (C, cool; W, warm), and mean values of selected amino-acid D/L values. Up to four values are given for each locality, representing, in order, the *Saxidomus* D/L leucine mean value (total hydrolyzate), the *Protothaca* D/L leucine mean value (total hydrolyzate), and the *Protothaca* D-alloisoleucine/L-isoleucine mean value (total hydrolyzate), and the *Protothaca* D-alloisoleucine/L-isoleucine mean value (free amino acids). D/L values for leucine in *Saxidomus* appear to define two groups, one ranging from 0.35 to 0.43 for localities mapped on the first, late stage 5 terrace and associated with cool-water faunal elements, and the other represented by D/L values of 0.53, representing an early stage 5 terrace and associated with a warm-water faunal element. Results for locality M7822 are equivocal because the shells were chalky and poorly preserved. Results for locality M7278 are not shown, but are discussed in Rockwell and others (this volume); samples from this locality yielded anomalously high D/L values, which is attributed to the effects of geothermal heating (Rockwell and others, this volume). Abbreviations: PC, Point Conception; CB, Cojo Bay; SBSYF, South Branch Santa Ynez fault; CA, Canãda de Alegria; GAV, Gaviota; and AH, Arroyo Hondo.

along the sea cliff 1.3 to 1.56 km west of Gaviota State Beach (3 localities, including that of Hoskins, 1957); and (5) east of Gaviota between Arroyo Hondo and Cañada de la Huerta, on two successive terrace levels (3 localities on the upper terrace, 1 locality on the lower terrace). Several additional fossiliferous localities on the first and second terraces at Cañada de Alegria and eastward to Gaviota (Fig. 1) yielded so few specimens that they are not treated herein, although shells from them were used in dating and correlation studies of the terraces (Patterson and others, 1982; Rockwell and others, this volume, figs. 1, 2).

MARINE TERRACES

The prominent marine terraces that fringe the coast of western Santa Barbara County, from Vandenberg Air Force Base in the north, and southward to Point Conception and eastward to beyond El Capitan State Beach, are shown on the 1:62,500-scale geologic maps of Dibblee (1950, pls. 1, 2), although he does not discriminate the several abrasion platforms that make up this broad, composite terrace. The most thorough study of the marine terraces and their elevational relations is that of Upson (1951), who recognized up to 17 marine terraces on the south flank of the Santa Ynez Mountains, from just west of Gaviota to just east of El Capitan State Beach. The remnants of terrace levels above 60 m are poorly preserved, but most appear to represent surfaces that were once continuous and more extensive.

The terrace mapping of Upson (1951) is complemented to the west by terrace mapping that accompanied the geotechnical investigations necessary for siting a proposed liquid natural gas (LNG) facility near Cojo Bay (Dames and Moore, 1980, 1981; Yerkes and others, 1981; Goodin and others, 1982; Patterson and others, 1982; Rockwell and others, this volume). Figure 1 is a generalized map showing the five lowest emergent marine terraces as mapped from Point Conception to Gaviota (Rockwell and others, this volume). Segments of some of these terraces pinch out, or rather, have been removed by erosion and shoreline retreat during the subsequent sea-level highstands, so that not all of the terraces are present along the entire coastline. Of particular note is the apparent absence of the second terrace at Arroyo Hondo, and the third terrace at Cañada de Alegria, where the South Branch of the Santa Ynez fault intersects the coast. Correlation of some terrace remnants is difficult in areas where the alluvial and colluvial cover is so thick that it obscures the geomorphic relations of the inner margins (i.e., shoreline angles) of the terraces.

The first emergent terrace, named the Cojo Terrace by Rockwell and others (this volume), is nearly 1 km wide at Government Point, about 0.75 km wide at the east end of Cojo Bay, then narrows as it parallels the coast to the east to Arroyo el Bulito, where modern seacliff retreat has removed it (Fig. 1). It appears again just west of Cañada de Alegria and the South Branch of the Santa Ynez fault, and is present eastward to Arroyo Hondo as a very narrow platform remnant.

The second terrace as mapped by Rockwell and others (this volume) is nowhere more than a few hundred meters wide and can be traced from Point Conception eastward to near Arroyo el Bulito, where it has also been removed by erosion and seacliff retreat. It appears again just west of Cañada de Alegria, where it is faulted and vertically displaced by the South Branch of the Santa Ynez fault. Between Cañada de Alegria and Gaviota Creek, it is present only as a very narrow platform, mostly covered by alluvium and colluvium. The second terrace is absent at Arroyo Hondo.

The third terrace is up to 2 km wide at Point Conception, but for most of its extent, is only hundreds of meters wide or less along the coast from Cojo Bay to just west of Cañada del Sacate, where it is truncated and removed by headward erosion of the modern sea cliff. It forms a very narrow platform west of Gaviota Creek, but east of there it widens and forms a broader platform (cf. Upson, 1951). The only fossil localities on the third terrace are in highway road cuts between Arroyo Hondo and Cañada de la Huerta, although reworked shell fragments thought to have been derived from this terrace (as opposed to being kitchen-midden shells) were also found near Molino Canyon.

The fourth terrace is clearly defined near Point Conception and can be traced eastward to the South Branch of the Santa Ynez fault, where it is vertically displaced. Only a single fossil locality (USGS loc. M7822) has been identified on this surface, although exposures east of Gaviota Creek were not examined.

PALEOECOLOGY

The invertebrate faunas from the study area are mainly from sediments that have been deposited directly on a marine-abrasion platform. In areas of the coast where the bedrock lithology is fairly uniform over several kilometers, the geomorphic configuration of the emergent-terrace platforms generally mirrors that found in the modern beach and offshore platform. Thus, more resistant bedrock in a particular area will result in resistant headlands in both the modern and fossil terraces, just as long straight beaches may parallel each other on the modern and fossil platforms.

All of the faunas contain species that were derived from a variety of nearshore marine habitats, from a gravel and boulder lag deposited at the base of the fossil sea cliff, to shelly sands deposited up to 0.75 km offshore. The faunas from Arroyo Hondo and from Cañada de Alegria occur at or within a few meters' distance of the original shoreline angle (i.e., the base of the fossil sea cliff), whereas the Gaviota faunas occur in basal gravels deposited up to 40 m from shore (Hoskins, 1957, and the main fauna from Cojo Bay was probably deposited from 700 to 750 m from shore. The faunas (see Table 1) will be examined in this order.

Cañada de Alegria

Second Terrace.—

Thirty-seven species of mollusks (20 bivalves, 14 gastropods, 1 scaphopod, and 2 or more chitons) and eight species of other invertebrates were collected from the shoreline angle of the second, 38-m terrace exposed above and just west of the mouth of Cañada de Alegria (Table 1; LACMNH loc. 5619, USGS loc. M7278). Almost all specimens are reworked and water-worn fragments that were deposited with littoral or supralittoral gravels at the base of the fossil sea cliff. The most common species are the bivalves *Platyodon cancellatus, Protothaca staminea*, and *Saxidomus gigantea*, all represented entirely by fragments, and the gastropod *Olivella biplicata*. These species suggest that the fauna was derived from a combination of several basic nearshore marine habitats, but mainly from sandy, level bottoms, areas of coarse sand among rocks, or exposed soft-rock bottoms, all at intertidal to shallow inner sublittoral depths (0 to 18–27 m). Most of the remaining species are represented by only a few fragments, but all can be assigned to one of these basic habitat groups.

Mollusks that normally live in sandy sublittoral level-bottom communities include species of *Pecten, Protothaca, Saxidomus, Siliqua, Tellina, Tresus*, and *Olivella*, as well as the echinoid *Dendraster*. *Protothaca*, however, also occurs in coarse sand among rocks in protected areas of the outer coast, a habitat suitable for many of the epifaunal rock dwellers, including the bivalve *Crassadoma*, the gastropods *Acanthina, Ceratostoma, Diodora, Haliotis, Serpulorbis*, and *Tegula*, and barnacles. Several of the epi-

TABLE 1.—CONTINUED

Acanthina spirata (Blainville)	-	5f	-	1,11f	15f	-	5,20f	2,1p	3	1	-	1,3f	-	-
Acmaea mitra Rathke	-	-	-	-	-	-	-	-	-	-	-	-	-	1
Acmaeidae, indet.	-	-	-	-	-	-	-	-	-	-	-	1	-	-
Acteocina sp.	1	-	-	-	-	-	-	-	-	-	-	-	-	-
Alvinia compacta (Carpenter)	-	1	-	-	-	-	-	-	-	-	-	-	-	-
Alvinia filosa (Carpenter)	-	1	-	-	-	-	-	-	-	-	-	-	-	-
Amphissa columbiana Dall	3j	8j,5f	-	-	-	-	2p,2f	-	-	-	-	-	-	-
Antiplanes (Rectiplanes) sp., cf. *A. (R.)* *thalaea* (Dall)	-	1p	-	-	-	-	?1f	-	-	-	-	-	-	-
Astraea undosa (Wood)	-	-	-	-	-	-	-	-	-	-	-	-	-	1po
Astraea (Megastraea) sp., indet.	-	-	-	-	-	-	-	-	-	-	-	1f	-	-
Balcis sp.	-	1	-	-	-	-	-	-	-	-	-	-	-	-
Barleeia sp.	1	6,?1	-	-	-	-	-	-	-	-	-	-	-	-
Bittium eschrichtii (Middendorff)	1	1	-	-	-	-	-	-	-	-	-	-	-	-
Bittium sp., ? *B. rugatum* (Carpenter)	-	6	-	-	-	-	-	-	-	-	-	-	-	-
Bittium sp. or spp.	46	41,16p	-	-	-	-	3	-	-	-	-	-	-	-
Bursa californica (Hinds)	-	-	-	-	-	-	?1f	-	-	1f	-	1p,1f	1f,?2f	1f
Calliostoma canaliculatum (Lightfoot)	1j	1j,6f	-	-	-	-	-	-	-	-	-	-	-	-
Calliostoma ligatum (Gould)	-	1f,?2f,?3j	-	-	-	-	-	-	-	-	-	-	-	-
? *Calliostoma* sp., indet.	-	-	-	-	-	-	1f	-	-	-	-	-	-	-
Ceratostoma foliatum (Gmelin)	-	2f	-	1f	-	-	-	-	-	-	-	-	-	-
Ceratostoma nuttalli (Conrad)	-	-	-	-	-	-	-	-	1	-	-	-	-	-
Cerithiopsis sp.	-	1	-	-	-	-	-	-	-	-	-	-	-	-
Clathromangelia interfossa (Carpenter)	-	1	-	-	-	-	-	-	-	-	-	-	-	-
Clathromangelia nitens (Carpenter)	4	2	-	-	-	-	-	-	-	-	-	-	-	-
Clathurella canfieldi Dall	-	1	-	-	-	-	-	-	-	-	-	-	-	-
"*Collisella*" *scabra* (Gould)	1	3,1f	-	-	-	-	-	-	-	-	-	-	-	-
Conus californicus Reeve	-	1j	-	-	-	-	2,1f	-	1p	1	-	14,2p,12f	2,1f	1,[1]
Crepidula adunca Sowerby	-	4,1f	?1j	-	-	-	3p,4f	1	-	-	-	-	-	-
Crepidula onyx Sowerby	1j	28,10f	-	-	-	-	-	-	-	-	-	-	-	-
Crepidula perforans (Valenciennes)	12,8j	-	-	-	-	-	-	-	-	-	-	-	-	-
Diodora aspera (Rathke)	1j,1f	1	-	1f	3f	-	2f	-	-	1f	-	2f	-	-
Discurria insessa (Hinds)	-	-	-	2f	1f	-	-	-	-	-	-	-	-	-
Epitonium (Nitidiscala) sp.	1j	3j	-	-	-	-	-	-	-	-	-	-	-	-
Fissurella volcano Reeve	-	-	-	-	-	-	-	-	-	-	-	-	1f	-
Fissurellidea bimaculata (Dall)	1j	1f	-	-	-	-	1	-	-	-	-	-	-	-
Fusinus kobelti (Dall)	cf.	1f	-	-	-	-	?	1	-	-	-	-	-	-
Haliotis kamtschatkana Jonas	-	1f	-	-	-	-	-	-	-	-	-	-	-	-

Haliotis sp., cf. *H. rufescens* Swainson	-	-	-	1f	1f,?1f	-	-	-	-	-	-	-	-	-
Hipponix antiquatus (Linnaeus)	-	-	-	-	-	-	2,1p	-	-	-	-	-	-	-
Homalopoma luridum (Dall)	-	-	-	-	-	-	2	-	-	-	-	-	-	-
Iselica sp., cf. *I. ovoidea* (Gould)	1	-	-	-	-	-	-	-	-	-	-	-	-	-
Kelletia kelletii (Forbes)	-	-	-	-	-	-	-	-	-	-	-	2f	-	-
Kurtziella plumbea (Hinds)	7	12,2f	-	-	-	-	-	-	-	-	-	-	-	-
Lacuna marmorata Dall	40	75,1f	-	-	-	-	-	-	-	-	-	-	-	-
Lacuna sp., indet.	-	1	-	-	-	-	-	-	-	-	-	1f	-	-
Lirabuccinum dirum (Reeve)	-	1f	-	-	3f	-	?	2p	-	-	-	?1f	-	-
Lirularia discors McLean	7	5,7f	-	-	-	-	-	-	-	-	-	-	-	-
Littorina scutulata Gould	50+	68,13f	-	-	-	-	-	-	-	-	-	-	-	-
Lottia digitalis (Rathke)	23j,2p	10,3p	-	-	-	-	-	1	-	-	-	-	-	-
Lottia limatula (Carpenter)	1	-	-	-	-	-	-	-	-	-	-	-	-	-
Lottia pelta (Rathke)	7j	2pj	-	-	-	-	-	-	-	-	-	-	-	-
Lottia triangularis (Carpenter)	1j	-	-	-	-	-	-	-	-	-	-	-	-	-
Lottia sp., indet.	7f	4f	-	-	-	-	-	-	-	-	-	?1	-	-
Mitra idae Melvill	-	-	-	-	-	-	-	-	-	-	-	?2f	1f	-
Mitrella carinata (Hinds)	38	36	-	-	-	-	9,5p,4f	8,1f	-	-	-	?1p	-	-
Mitrella gausapata (Gould)	x	20	1	-	-	-	-	-	-	-	-	-	-	-
Mitrella tuberosa (Carpenter)	50+	80	-	-	-	-	1	1	-	-	-	-	-	-
Mitrella spp.	x	68,2f	-	-	-	-	-	-	3f	-	-	-	-	-
Nassarius fossatus (Gould)	1f,>10j	7j,22f	2f	-	-	-	1,1p,13f	1p	-	3f	-	1p,3f	1f	1f
Nassarius mendicus (Gould) [includes *N. m.* var. *cooperi* (Forbes) and *N. m.* var. *indisputabilis* (Oldroyd)]	3,1j,4p	6,11f	-	3f	1f	-	27,13p, >60f	8	1,1p	1p	4f	1p,2f	-	-
Nassarius perpinguis (Hinds)	4,1p	5,5f	1,2f	-	-	-	5,2p	1	-	?1f	-	-	-	-
Natica clausa Broderip and Sowerby	-	1j	-	-	-	-	-	-	-	-	-	-	-	-
Naticidae, indet.	2f	9f	x	4f	-	-	5f	1f	-	1f	2f	16f	-	-
Neptunea tabulata (Baird)	-	1f	-	-	-	-	-	-	-	-	-	-	-	-
Neverita reclusiana (Deshayes)	-	5f	-	-	-	-	1p,1f	-	2f	-	-	1f	1p	-
Nucella emarginata (Deshayes)	1,1j	1,6f	1	-	-	-	?1f	-	-	-	-	-	-	-
Nucella lamellosa (Gmelin)	2f,2j	4j,20f	-	-	-	-	1,2p,18f	-	-	-	?1f	-	-	-
Nucella canaliculata (Duclos)	3j	2j,20f	-	-	-	-	1p,2f	-	-	?1f	-	-	-	-
Nucella sp., indet.	-	-	1f	-	-	-	-	-	-	-	-	-	-	-
Ocenebra beta (Dall)	-	6	-	-	-	-	-	-	-	-	-	-	-	-
Ocenebra foveolata (Hinds)	-	-	-	-	?4f	-	-	1f	-	-	-	-	-	-
Ocenebra sp., ? *O. interfossa* Carpenter	1	2p,2f	-	-	-	-	-	-	-	-	-	-	-	-
Ocenebra lurida (Middendorff)	1	1,1p,5f	1f	-	-	-	-	-	-	-	-	-	-	-

TABLE 1.—CONTINUED

Ocenebra sp., indet.	x	14,13f	1j,1f	4f	2f	-	x	1f	1	-	1p	-	-	
Odostomia (*Evalea*) sp.	-	3	-	-	-	-	-	-	-	-	-	-	-	-
Olivella biplicata (Sowerby)	4,>50j	63,8f	1p,3f	2,num f	1p,27f	x	num	8,1p,2f	43,10p,6f	22,12p	7,5p,15f	35,5p,10f	5,6f	2,6f,[1]
Olivella pycna Berry	100+	176,11p	4	-	1	-	53,9p	2	-	1,1p	-	-	-	
Opalia funiculata (Carpenter)	-	-	-	-	-	-	-	-	-	-	-	1		-
Ophiodermella incisa (Carpenter)	-	4,5f	-	-	-	-	1p,3j,2f	1	-	1	-	-	-	-
Polinices draconis (Dall)	-	-	-	-	-	-	-	-	1	-	-	-	-	-
? *Polinices lewisii* (Gould)	-	-	-	-	-	-	2f	1f	-	-	-	1f	-	-
Propebela tabulata (Carpenter)	6	12,2f,?1	-	-	-	-	-	1	-	-	-	-	-	-
Pteropurpura festiva (Hinds)	-	-	-	-	-	-	-	-	-	-	-	1p,1f	-	-
Serpulorbis squamigerus (Carpenter)	-	-	-	2f	-	-	2f	-	-	1p	-	14f	12f	?2f
Tegula funebralis (Adams)	?1f	1,2j,37f	-	3f	1f,?1f	-	1f	-	-	-	-	1,10f	1,1f	-
Tegula sp., ? *T. brunnea* Philippi	-	2f	-	-	1f	-	-	-	-	-	-	-	-	-
Terebra pedroana Dall	-	-	-	-	-	-	-	-	-	-	-	1p	-	-
Tricolia pulloides (Carpenter)	-	1	-	-	-	-	-	-	-	-	-	-	-	-
Tricolia sp., indet.	-	3	-	-	-	-	-	-	-	-	-	-	-	-
Trimusculus reticulatus (Sowerby)	1	2	-	-	-	-	-	-	-	-	-	-	-	-
Turbonilla spp. (3+)	8	5	-	-	-	-	-	-	-	-	-	-	-	-
Turritella cooperi Carpenter	2,2j	2,2f	-	-	-	-	1p,1f	-	-	1jf	-	-	-	-
Vermetidae, indet.	x	-	-	-	-	-	x	-	-	-	-	-	-	-
MOLLUSCA: SCAPHOPODA														
Dentalium neohexgonum Sharp & Pilsbry	4p,4f	5,6p	-	1,2f	2p	-	-	-	-	-	-	-	-	-
Dentalium pretiosum Sowerby	1,1j,2f	2,2p	-	-	-	-	1p	-	-	-	-	-	-	-
MOLLUSCA: POLYPLACOPHORA														
Cryptochiton stelleri (Middendorff)	-	-	-	1f	-	-	-	-	-	-	-	-	-	-
Indet. chiton valves	16	23,31f	-	?1f	5f	-	x	-	-	-	-	1,4p	-	-
ANNELIDA: POLYCHAETA														
Dodecaceria sp.	-	-	-	-	-	-	-	-	-	-	-	-	-	[x]
Polydora sp., ? *P. commensalis* Andrews	x	x	-	x	x	-	6+	x	x	x	1	x	x	-
Spionidae, indet.	x	x	x	x	x	-	x	x	x	x	x	x	x	x
ARTHROPODA: CRUSTACEA (Cirripedia)														
Balanus crenatus Bruguiere	cf.	x,cf.	-	-	-	-	-	-	-	-	-	-	-	-
Balanus glandula Darwin	x	x	-	-	-	-	-	-	-	-	-	-	-	-
Balanus nubilus Darwin	x	1w	-	-	?2w	-	-	?1w	-	-	-	-	-	-
Balanus pacificus Pilsbry	-	x	-	-	-	-	-	-	-	-	-	-	-	-
Balanus sp./spp.	x	x	-	x	-	-	-	x	-	x	-	-	-	-
Balanomorpha, indet.	x	>200w,74o	-	12w	6w,1of	-	-	1f	-	-	1w	10w	1f	-
Concavus sp., cf *C. aquila* Pilsbry	-	-	-	-	x	-	-	-	-	-	-	-	-	-

Cryptolepas sp.	-	1w	-	-	-	-	-	-	-	?1w	-	-	-	-
Megabalanus californicus (Pilsbry)	cf.	8w	-	-	-	-	-	-	-	-	-	-	1w	-
Pollicipes polymerus Sowerby	-	6w	-	-	-	-	-	-	-	-	-	-	-	-
Semibalanus cariosus (Pallas)	x	>8w	12w,2pw	-	-	-	28w	-	1w	?4w	-	-	-	-
Solidobalanus hesperius (Pilsbry)	x	x	-	-	-	-	-	-	-	-	-	-	-	-
Tetraclita rubescens Darwin	-	-	-	-	-	-	-	-	-	-	-	4w	-	?1w
ARTHROPODA: CRUSTACEA (Malacostraca)														
Brachyuran chelae	80	59,2f	-	-	1	-	-	-	-	-	-	-	-	-
ECHINODERMATA: ECHINOIDEA														
? *Dendraster excentricus* (Eschscholtz)	-	1f	-	1f	-	-	1f	-	-	-	-	2f	-	-
? *Strongylocentrotus* sp. or spp. (spines)	100+	40	-	2	-	-	-	-	-	-	-	-	-	-
CHORDATA: PISCES														
Microgadus proximus (Girard) (otoliths)	-	2	-	-	-	-	-	-	-	-	-	-	-	-
Teleosteii, indet. (otolith)	-	1	-	-	-	-	-	-	-	-	-	-	-	-
CHORDATA: MAMMALIA														
Enhydra lutris (Linnaeus) (milk tooth, dP^3)	-	-	-	-	-	1	-	-	-	-	-	-	-	-

Abbreviations: cf., compares with; f, fragment; j, juvenile; m, mesoplax; num, numerous; o, operculum (gastropods) or opercular plates (barnacles); p, partial specimen; w, wall plates (barnacles); x, present; ?, identification questionable; >, greater than.

faunal species also live on macroscopic marine algae, present along the coast today as well as in the late Pleistocene (Lindberg, 1980). The soft-rock assemblage at Cañada de Alegria is dominated by infaunal boring or nestling bivalves, such as *Chaceia, Parapholas, Penitella, Petricola, Platyodon*, and possibly *Zirfaea*.

Fourth Terrace.—

A meager fauna of eight species of mollusks (7 bivalves, and 1 gastropod), only four of which are identifiable to species, was recovered from pockets of indurated sand on the fourth(?), 74-m terrace above Cañada de Alegria on the western, downthrown side of the South Branch Santa Ynez fault (Table 1; USGS loc. M7822). Normally such a depauperate fauna would not warrant particular attention, but the age of this fauna (see Rockwell and others, and Wehmiller, this volume) provides an independent check on the anomalously older age previously cited for the next lower, second terrace (Yerkes and others, 1981, fig. 15A; Kennedy and others, 1982, fig. 2).

No single species was particularly abundant in the small collection, although 14 fragments of *Platyodon cancellatus* were recovered. The remaining species are represented by one to seven fragments each. *Platyodon* typically lives in burrows in soft sedimentary rocks in more or less exposed areas along the open coast (Fitch, 1953).

Arroyo Hondo

First Terrace.—

Thirty-three species of mollusks (17 bivalves, 16 gastropods) and six species of other invertebrates were recovered from the lower, 20-m terrace exposed in the sea cliff 8 to 9 km east of Gaviota, between Arroyo Hondo and Cañada de la Huerta (Table 1; LACMNH loc. 8234, USGS loc. M7491). The terrace platform here is very narrow, and the fossil locality is only a few meters from the probable shoreline angle. The fossils thus represent detrital shells preserved in the littoral marine gravel and boulder lag deposited close to the base of the fossil sea cliff.

The three abundant species in the fauna (*Protothaca staminea, Platyodon cancellatus*, and *Olivella biplicata*) all occur today in intertidal to shallow inner sublittoral depths (0 to 18–27 m). The assemblage represents a mixture of material derived from protected sandy bottoms (*Olivella, Tresus*), rocky or boulder-covered bottoms with intervening pockets of coarse sand (*Protothaca*), or exposed soft-rock bottoms with epifaunal as well as infaunal boring species (*Platyodon*, and the pholadids *Penitella* spp. and *Parapholas californica*). The remaining species are all uncommon and mostly represented by fewer than five specimens each; all fall into the habitat categories suggested by the three most abundant species. Most of the gastropods are represented by single specimens.

Third Terrace.—

At least 42 species of mollusks (16 bivalves, 25 gastropods, and 1 or more chitons) and seven species of other invertebrates were collected from three localities on the third, 38-m terrace between Arroyo Hondo and Cañada de la Huerta (Table 1; LACMNH loc. 8235, USGS locs. M7476, M7489, M7490). (The second terrace is absent at Arroyo Hondo.) Most of the fossils were obtained from USGS loc. M7476, a highway road cut that very closely parallels the base of the fossil sea cliff. Locality M7489 is east of M7476, in Cañada de la Huerta, and locality M7490 (since destroyed by highway widening) is west and south, representing a position about 20 m from the old sea cliff. The physical position of the fossiliferous exposures relative to the terrace shoreline angle suggests that these faunas were deposited in among the gravel lag very near the base of the sea cliff, and probably within the intertidal zone.

The most common species in the fauna, *Tresus nuttallii*, is represented entirely by broken and wave-worn fragments. *Olivella biplicata, Protothaca staminea, Mytilus californianus, Conus californicus*, and *Penitella* spp. are also relatively common, but most species are represented by very few specimens. The assemblage suggests, as do others from this coast, derivation from a variety of nearshore marine habitats, all at intertidal to shallow inner sublittoral depths (0 to 18–27 m). *Tresus nuttallii*, for example, typically lives in sandy bottoms just offshore, a suitable habitat also for both *Olivella* and *Conus*. Most of the other gastropods are represented by single specimens or only a few fragments. Bivalve species are represented almost entirely by worn fragments, indicative of considerable reworking and post-mortem transport. *Mytilus californianus* represents an intertidal rocky shore habitat, and an infaunal (endolithic) community is indicated by species of *Penitella, Chaceia ovoidea* and *Platyodon cancellatus*. *Protothaca staminea* lives in sand, often in rocky areas or on boulder-covered bottoms.

Gaviota

Invertebrate fossils are present on both the first and second terraces exposed in the sea cliff and in road cuts west of Gaviota State Beach, but are only common enough on the first terrace to warrant discussion. Upson (1951) and Hoskins (1957) reported 14 and 26 taxa, respectively, from similar localities on the first terrace. Upson's fossils are housed at the U.S. National Museum of Natural History (USGS collections) and were not examined for this study, but he does not list any species that were not found in this study from the first terrace. The collection of Hoskins (1957) has been re-examined and is listed in Table 1 (USGS loc. M5919).

First Terrace.—

At least 66 species of mollusks (30 bivalves, 34 gastropods, 1 scaphopod, and 1 or more chitons), six species of other invertebrates, and one marine mammal (sea otter) are represented in the collections from the first, 24 to 27-m terrace exposed in the sea cliff west of Gaviota State Beach (Table 1; USGS locs. M5919, M7496, and M7819). Bivalve specimens typically are found only as broken and wave-worn fragments, and are more abundant toward the eastern end of the terrace outcrop. Gastropods, particularly *Olivella* and some of the less common epifaunal species, generally have remained intact, especially toward the western end of the exposure (USGS loc. M7819), where the shells may have been protected from wave action by a projecting headland at the north end of the fossil beach, or have been derived from that area.

Species that are relatively common toward either end of the first terrace exposure in the sea cliff are *Platyodon cancellatus* (east), *Olivella pycna* and *O. biplicata* (west), *Nassarius mendicus* (west), *Penitella* spp. (west), and *Protothaca staminea* and *Tresus nuttallii* (both east). These species indicate that the fauna as a whole was derived from a variety of nearshore habitats, probably all at intertidal to shallow inner sublittoral depths (0 to 18–27 m). The most common species, *Platyodon cancellatus*, lives in burrows in soft sedimentary rocks in more or less exposed areas of the coast. The pholadids (mostly unidentified pieces of *Penitella*) are most common near the west end of the outcrop, suggesting they may have been derived from intertidal to sublittoral exposures of bedrock at this point. *Tresus nuttallii* lives in sheltered sandy bottoms offshore to 30-m depths, which is also a suitable habitat for the two common species of *Olivella*. *Protothaca* typically inhabits coarse sand, and is often found associated with rocky or boulder-covered areas. *Nassarius mendicus*, also common at the west end of the old beach, is often found along the open coast at low tide as well as at sublittoral depths.

Cojo Bay

Richly fossiliferous marine sands exposed in the sea cliff at the east end of Cojo Bay have yielded the greatest diversity of organisms (more than 125 species) of any terrace deposit in the study area. However, relative abundance and diversity figures may be biased because of the ease of screening and collecting material on the outcrop, particularly in comparison with samples collected from some of the basal conglomerates. At least 105 species of mollusks (35 bivalves, 65 gastropods, 2 scaphopods, and several species of chitons), nine or more species of barnacles, 10 species of other invertebrates, and two fish (based on otoliths) were collected from two localities on the first, 16-m terrace at Cojo Bay (Table 1, LACMNH loc. 5638, USGS locs. M7279 and M7487). Most of these taxa are from USGS loc. M7279, but one bivalve, and one coral and four gastropod species were found only at localities M7487 and 5638, respectively (Table 1).

The fossiliferous marine sands overlying the terrace platform contain only scattered pebbles, in contrast to the basal conglomerates at many of the nearshore localities, and indicate a greater distance from the original shoreline. The main collecting site is approximately 700 to 750 m from the shoreline angle of this terrace as mapped by Rockwell and others (this volume). Elevational differences between the outcrop and shoreline angle suggest a water depth of about 15 m, in agreement with the bathymetry suggested by many of the fossil species.

Because of the large sample size, many of the less common species are found in greater abundance here than are common species from some of the less fossiliferous localities (cf. Table 1). Particularly abundant species or groups of species include some of the smaller gastropods (*Olivella pycna* and *O. biplicata*, *Mitrella tuberosa* and indeterminate *M.* spp., *Lacuna marmorata*, and *Littorina scutulata*), the bivalve *Tellina bodegensis*, barnacle wall plates, and malacostracan chelae ("crab claws").

Composition of the fauna suggests that it was derived from a variety of nearshore habitats. *Littorina scutulata*, for example, is common in the upper part of the intertidal zone, and the numerous barnacles are common in intertidal and adlittoral depths (0 to 9 m) along rocky shores. *Lacuna marmorata* is typically a rocky intertidal species, but is sometimes associated with eel grass, whereas species of *Mitrella* are chiefly sublittoral, and often found below kelp canopies. Species of *Olivella* are often found in protected sandy areas of the open coast at low tide, and are also abundant in shallow water offshore along exposed sandy beaches. The only common bivalve, *Tellina bodegensis*, typically occurs in sandy bottoms intertidally and offshore. Additional microhabitats are also represented by some of the less abundant species, including a variety of mobile and attached epifaunal species, as well as infaunal boring and nestling species.

PALEOZOOGEOGRAPHY

The overriding controlling factor in the distribution of living species in the coastal waters of the eastern Pacific is temperature and its effect on reproductive capability. Because the coastline of North America is dominantly a north-south one, the distribution of temperatures is mainly a reflection of latitudinal position as one moves from the equator to the poles. The distribution of isotherms along the coast, however, is not simply one of impinging on the coast in a regular ladder-like fashion, but is dynamically complex and varies with the season (cf. Robinson, 1973; Lynn, 1967), as well as being influenced by current systems, coastal physiography, upwelling, and short-term cyclic phenomena, such as El Niño Southern-Oscillation (ENSO) events. Why certain temperature ranges affect the distribution of species the way they do is unknown, but a significant number of species on any given coastline will have geographically similar ranges that terminate at or near a provincial boundary that coincides with a limited region of greater temperature change (Valentine, 1966). Probably the most striking provincial boundary in the eastern Pacific, and perhaps the one that is best understood, is in the region of Point Conception, where the cooler water fauna of the Oregonian Province, driven southward by the California Current, is juxtaposed with the warmer water fauna of the Californian Province, carried northward by the Southern California Countercurrent.

The extralimital occurrence of fossil species at localities outside their modern range limits is inferred to be the result of global warming or cooling that affects the distribution, but not necessarily the pattern, of marine isotherms. It is the presence of extralimital species in the fossil record that allows one to make paleozoogeographic inferences, although, because of the dynamic complexities of the marine hydrosphere, it is preferable to couch any zoogeographic-range shifts in terms of geography rather than attempting to hypothesize temperature changes on a numerical basis.

The cool-water aspect of terrace faunas on the Santa Barbara and Ventura County coasts has been recognized for many years (cf. comments in Addicott, 1964). Various hypotheses have been proposed to explain the zoogeographic discrepancy between the cool-water faunas here and the warmer water ones from the Los Angeles Basin, including that of a land barrier separating the two basins of deposition. A barrier to marine organisms is also a bridge for terrestrial organisms and, as such, was also hypothesized by vertebrate paleontologists in an effort to explain the presence of fossil mammoth bones on the offshore northern Channel Islands during the Pleistocene. The land-bridge hypothesis appears to be no longer defensible based on both biologic and geologic evidence (Johnson, 1978; Junger and Johnson, 1980; Wenner and Johnson, 1980).

The perceived zoogeographic "anomaly" of cool-water and warm-water faunas in juxtaposition remained unexplained until the advent of amino-acid racemization studies (e.g., Wehmiller and others, 1977), when dating coastal Pleistocene faunas began in earnest. By the middle to late 1970s, it was apparent that the warm-water and cool-water faunas were not simply geographically disparate examples of contemporaneous conditions, but represented temporally different sea-level events (highstands) that could be related to sea-level stands recorded on other coasts and to climatic episodes documented by isotopic fluctuations in foraminifera from deep-sea cores (Kennedy, 1982a). Thus, most of the difference in temperature (or zoogeographic) aspect of invertebrate faunas from emergent marine terraces along the coast of California could be explained by superimposing the effects of a eustatically fluctuating sea level onto a coastline that was undergoing local and/or regional uplift such that terraces of differing ages (with faunas of different zoogeographic aspect) were uplifted to different levels above modern sea level (Kern, 1977; Kennedy, 1978; Lajoie and others, 1979). In tectonically stable areas, the lowest emergent terrace is usually that formed about 125 ka, when sea level was approximately 6 m higher than it is today (Ku and others, 1974; Harmon and others, 1983), and oceanic-surface waters slightly warmer, at least seasonally, than they are today off the California coast (Kennedy, 1978, 1982b, 1988; Kennedy and others, 1982). In areas that are undergoing tectonic uplift, such as much of the Santa Barbara County coast (Lajoie and others, 1979; Rockwell and others, this volume), the lowest emergent terrace usually dates to a later sea-level highstand that occurred about 85 to 80 ka, when sea level was perhaps 5 to 7 m lower than it is today (Muhs and others, 1988; Rockwell and others, 1989), and the coastal-surface waters of the eastern Pacific were slightly cooler (Kennedy, 1978, 1982b; Kennedy and others, 1982; Kennedy and Wehmiller, 1986). Only in areas of very rapid tectonic uplift, such as at Goleta and near Ventura, California, are emergent terraces found that correlate to interstadial sea levels of oxygen isotope stage 3 (Wehmiller and others, 1977; Lajoie and others, 1979), when sea level was much lower (Bloom and others, 1974) and oceanic-surface waters much cooler than they are today (cf. Addicott and Greene, 1974; Aharon, 1983; Aharon and Chappell, 1986). A further consequence of very rapid uplift is the formation of Holocene emergent marine terraces that have faunas that are zoogeographically similar to those of today.

The invertebrate faunas from emergent terraces along the Santa Barbara coast can be characterized by whether they indicate marine paleoclimatic conditions that were either warmer or cooler than those occurring today along the same coastline. Some of the depauperate faunas from localities between Cañada de Alegria and Gaviota do not contain enough species to provide an adequate paleoclimatic signature, and thus are not treated herein. The definition of whether a fauna indicates "warmer water" or "cooler water" conditions than today is based on a comparison of the modern zoogeographic ranges of the fossil species with the latitude of the fossil locality. Species whose modern ranges are limited to areas only north of the fossil locality are referred to as extralimital northern species; their presence farther south would indicate a southerly shift of marine isotherms and cooler water conditions. Conversely, those species whose modern zoogeographic ranges are limited to areas south of the reference locality are referred to as extralimital southern species. Their existence in fossil localities north of their present range is indicative of a northward isothermal shift and at least seasonally warmer water conditions. The modern latitudinal (or zoogeographic) equivalence of a fossil fauna can be refined even more by species whose modern ranges may encompass the latitude of the fossil locality, but which terminate near the range end points of the extralimital species' ranges. Thus, for example, the geographic overlap of ranges of extralimital northern species that may range today only as far south as Monterey Bay, and generally southward-ranging, nonextralimital species that today range only as far north as Monterey Bay, define a very narrow latitudinal or geographic area of equivalency. Although this system of analysis works well on many parts of the Pacific coast, there are examples of faunas whose composite zoogeographic ranges are not entirely concordant, a phenomenon that may be due to a greater seasonality (warmer summers, cooler winters) as a result of precessional forcing of paleoclimatic cycles (Kennedy, 1988).

Cool-Water Faunas.—

The composite faunas (Table 1) from Cojo Bay, Cañada de Alegria (second terrace), Gaviota (first terrace), and Arroyo Hondo (first terrace), although varying considerably in species diversity and numerical abundance of material, are all zoogeographically similar and indicate marine temperatures no warmer than those occurring today in the vicinity of Monterey Bay, central California. At least 13 extralimital northern species (Table 2) are present in these cool-water faunas. Two bivalve species, *Penitella turnerae* and *Saxidomus gigantea*, are present in all of them. The remaining 11 species occur in one to three faunas apiece.

The geographic equivalency of these fossil faunas is also limited by approximately 17 species whose terminal northern occurrences also lie along the central California coast, most between Monterey Bay and Bodega Bay (Table 3). Only *Acanthina spirata* is present in each of the cool-water faunas, but 12 species were found at Cojo Bay, seven at Cañada de Alegria, 14 at Gaviota, and eight at Arroyo Hondo. In conjunction with the extralimital species ranges,

TABLE 2.—EXTRALIMITAL SPECIES

	Localities and terrace levels*					
Northern species	Cojo (I)	Alegria (II)	Gaviota (I)	Arroyo Hondo (I)	Arroyo Hondo (III)	Southern Range Endpoint (Reference**)
Mollusca: Bivalvia						
Entodesma saxicola (Baird)	X	—	—	—	—	Shell Beach (2)
Macoma expansa Carpenter	X	—	—	—	—	Oceano (3)
Penitella turnerae Evans & Fisher	X	X	X	X	—	Point Sal (1, 2)
Saxidomus gigantea (Deshayes)	X	X	X	X	—	Monterey (4)
Mollusca: Gastropoda						
Amphissa columbiana Dall	X	—	X	—	—	Point Arguello (4)
Lirabuccinum dirum (Reeve)	X	X	X	—	—	Franklin Point, San Mateo Co. (4)
Nucella canaliculata (Duclos)	X	—	X	X	—	Cayucos (5)
Nucella lamellosa (Gmelin)	X	—	X	—	—	Santa Cruz (4)
Ophiodermella incisa (Carpenter)	X	—	X	X	—	Monterey (?) (1)
Propebela tabulata (Carpenter)	X	—	X	—	—	Monterey(?) (6)
Mollusca: Polyplacophora						
Cryptochiton stelleri (Middendorff)	—	X	—	—	—	San Nicolas Island (7), but rare south of Monterey; in Holocene middens in southern Calif. and nothern Baja California as far south as Punta Baja (8)
Arthropoda: Crustacea (Cirripedia)						
Semibalanus cariosus (Pallas)	X	—	X	?	—	Morro Bay (9)
Solidobalanus hesperius (Pilsbry)	X	—	—	—	—	Monterey Bay (9)
Southern species						Northern range endpoint
Mollusca: Gastropoda						
Acanthina lugubris (Sowerby)	—	—	—	—	X	San Diego (10), or just north of Ensenada, Baja California (8)
Opalia funiculata (Carpenter)	—	—	—	—	X	Refugio State Beach (11)
Pteropurpura festiva (Hinds)	—	—	—	—	X	Santa Barbara (12)
Terebra pedroana Dall	—	—	—	—	X	Oxnard (1), or Santa Monica (13)

*Terrace levels: I, first terrace; II, second terrace; III, third terrace.
**References: 1, Herein, LACM Recent mollusk collection; 2, Jablonski and Valentine (1990); 3, Coan (1971); 4, Kennedy (1978); 5, Abbott and Haderlie (1980a); 6, Dall (1921); 7, Abbott and Haderlie (1980b); 8, Rockwell and others (1989); 9, Newman and Abbott (1980); 10, Abbott (1974); 11, DuShane (1979); 12, McLean (1969); 13, Bratcher and Cernohorsky (1987).

these species sharply delimit any possible modern geographic equivalency for the fossil faunas to that around Monterey Bay, central California. Based on current (1950–1962) mean monthly temperatures for Monterey Bay and the coastal region east of Point Conception (Lynn, 1967), the local marine conditions may have been up to 1°C cooler in the coolest month (April) and 1.5° to 2°C cooler in the warmest month (October/November) during the sea-level highstands of middle and late stage 5 (105 to 80 ka), although assignment of temperature changes from geographic-range shifts is equivocal at best.

Warm-Water Faunas.—

In addition to the more numerous cool-water faunas discussed earlier, faunas from two areas indicate marine climatic conditions either as warm as, or warmer than, those occurring along the coast today (Kennedy and Wehmiller, 1986). The first is a deposit, probably representing an Indian kitchen midden, that has been reworked by landsliding into Cañada del Agua Caliente (Fig. 1) and is discussed by Kennedy and Wehmiller (1986); the second, at Arroyo Hondo, represents an unequivocal marine-terrace deposit.

The composite fauna from localities on the third, 38-m terrace between Arroyo Hondo and Cañada de la Huerta (Table 1) contains four gastropods that have extralimital southern ranges (Table 2). Three of these, *Opalia funiculata, Pteropurpura festiva*, and *Terebra pedroana* range today only as far north as Refugio State Beach, Santa Barbara, and Oxnard, respectively, none of which is significantly south (or east) of its fossil occurrence. *Acanthina lugubris*, however, ranges today only slightly north of Ensenada in northwestern Baja California (Rockwell and others, 1989, table 3), or possibly to San Diego (Abbott, 1974), and is thus extralimital by some 300 to 400 km.

Several other species in this fauna also have geographic ranges that help place limits on the modern latitudinal equivalency of this fauna (Table 3). *Clinocardium nuttalli*, for instance, is known today only as far south as San Diego, California, and five species (Table 3) range only as far south as Santo Tomas (one), Camalu (three), or Isla San Martin (one), all in northern Baja California. Zoogeographically, the fauna here is probably equivalent to that found today between San Diego, California, and Ensenada, Baja California. Based on current (1950–1962) mean monthly temperatures for the coast east of Point Conception and for the region between San Diego and Ensenada (Lynn, 1967), the study area may have been up to 2° to 2.5°C warmer in the coolest month (April near Point Conception, February in San Diego), and 1.5° to 3°C warmer in the warmest month (November for the Point Conception area, late September/early October for San Diego) during the interglacial sea-level highstand of oxygen isotope substage 5e, approximately 125 ka.

AGE AND CORRELATION

Pleistocene marine-terrace faunas can be dated by several methods, but only uranium-series disequilibria analyses of

TABLE 3.—SELECTED NONEXTRALIMITAL SPECIES USEFUL IN PALEOZOOGEOGRAPHIC INTERPRETATION

Species	Localities and terrace levels*					Northern (N) or Southern (S) Range endpoint (Reference**)
	Cojo (I)	Alegria (II)	Gaviota (I)	Arroyo Hondo (I)	Arroyo Hondo (III)	
Mollusca: Bivalvia						
Chaceia ovoidea (Gould)	X	X	X	—	—	N: Santa Cruz (2)
Clinocardium nuttalli (Conrad)	X	X	X	X	?	S: San Diego (3)
Leptopecten latiauratus (Conrad)	X	—	—	—	—	N: Point Reyes (2)
Lucinisca nuttalli (Conrad)	X	X	X	—	—	N: Monterey (2)
Nuculana taphria (Dall)	X	—	X	—	—	N: Monterey (4)
Parapholas californica (Conrad)	—	X	X	X	—	N: Bodega Bay (2)
Platyodon cancellatus (Conrad)	X	X	X	X	X	S: Camalu, Baja California (1)
Tivela stultorum (Mawe)	—	—	X	X	X	N: Half Moon Bay (2)
Mollusca: Gastropoda						
Acanthina punctulata (Sowerby)	—	—	—	—	X	S: Santo Tomas, Baja California (2)
Acanthina spirata (Blainville)	X	X	X	X	X	N: Anchor Bay (1) S: Camalu, Baja California (2)
Acmaea mitra Rathke	—	—	—	—	X	S: Isla San Martin, Baja California (2)
Bursa californica (Hinds)	—	—	?	X	—	N: Monterey Bay (2)
Ceratostoma nuttalli (Conrad)	—	—	X	—	—	N: Point Conception (2)
Conus californicus Reeve	X	—	X	X	X	N: Farallon Islands (2)
Diodora aspera (Rathke)	X	X	X	X	X	S: Camalu, Baja California (2)
Fusinus kobelti (Dall)	X	—	X	—	—	N: Monterey (1), but mainly south of Santa Barbara (2)
Nassarius perpinguis (Hinds)	X	—	X	?	—	N: Point Reyes (2)
Neverita reclusiana (Deshayes)	X	—	X	—	X	N: Crescent City, but rare north of Mugu Lagoon (5)
Ocenebra foveolata (Hinds)	—	?	X	—	—	N: Monterey Bay (2)
Serpulorbis squamigerus (Carpenter)	—	X	X	X	X	N: Monterey(?) (6); Montana de Oro (Kennedy, pers. observation)
Turritella cooperi Carpenter	X	—	X	X	—	N: Monterey Bay (2)
Arthropoda: Crustacea (Cirripedia)						
Balanus crenatus Bruguiere	X	—	—	—	—	S: Santa Barbara (7)
Balanus pacificus Pilsbry	X	—	—	—	—	N: Monterey Bay (7)
Concavus sp. cf. *C. aquila* Pilsbry	—	X	—	—	—	N: San Francisco (7) S: San Digeo (7)
Megabalanus californicus (Pilsbry)	X	—	—	—	X	N: Humboldt Bay (on buoys), Monterey Bay on mainland coast (7)
Tetraclita rubescens Darwin	—	—	—	—	X	N: San Francisco Bay (7)

*Terrace levels: I, first terrace; II, second terrace; III, third terrace.

**References: 1, Herein, LACM Recent mollusk collection; 2, McLean (1969); 3, Fitch (1953); 4, Jablonski and Valentine (1990); 5, Marincovich (1977); 6, Dall (1921); 7, Newman and Abbott (1980).

coral, and measurements of the extent of amino-acid racemization (and epimerization) in mollusks are routinely applied in efforts to date fossiliferous deposits on the lowest marine terraces. Corals are relatively uncommon in California marine-terrace deposits, and only a single partial specimen from the first terrace at Cojo Bay was recovered, too little for analysis by standard alpha spectrometry. Nevertheless, U-series (^{230}Th/^{234}U) analyses of bone, teeth, and mollusk shells have been performed on materials from the study area (Kaufman and others, 1971; Dames and Moore, 1980; see summary in Rockwell and others, this volume, table 1), although they are frequently unsuitable sample types and commonly yield incorrect ages.

The extent of racemization (and epimerization) of amino acids in the proteinaceous matrix of marine-mollusk shells, particularly bivalves, has been used extensively on the Pacific Coast to date or correlate upper Pleistocene marine-terrace deposits (e.g., Wehmiller and others, 1977; Kennedy, 1978; Kennedy and others, 1982; and reviews of Pacific Coast applications in Wehmiller, 1982, 1984, 1990, this volume, and Muhs, 1991). Although aminostratigraphy is considered a relative-dating technique, so many samples have now been analyzed from the Pacific Coast region (British Columbia, Canada, to Baja California Sur, Mexico) that an extensive, although mainly unpublished, comparative data base exists from which to interpret any new analytical data from localities within this geographic continuum. New U-series dates on corals (Muhs and others, 1988, and in prep.) have continued to support previous age estimates based on amino-acid data, particularly for late Pleistocene terraces that date to the 80 or 125 ka sea-level highstands.

The extent of racemization or epimerization of several amino acids in bivalve mollusks from localities between Cojo Bay and Arroyo Hondo are summarized by Rockwell and others (this volume, table 2) and in condensed form in Figure 2. Figure 2 relates mean enantiomeric ratios (D/L values) for *Protothaca* and *Saxidomus* samples from mapped terrace surfaces (Rockwell and others, this volume) with the zoogeographic signature (warmer or cooler than modern) of the faunas from those localities. Amino-acid data used in Figure 2 are from Dames and Moore (1980, appendicies A.1, A.5) and Kennedy and others (1982, fig. 2). Figure 2 (herein) demonstrates that the total hydrolyzate data for *Protothaca* from these terraces do not satisfactorily conform to the observed terrace relative-age sequence. Free D-alloisoleucine/L-isoleucine data from selected *Protothaca* samples do conform to the relative-age sequence as indicated by field mapping. *Saxidomus* data (for D/L leucine and other amino acids) generally conform to the terrace sequence. All the results, with the exception of those from Cañada de Alegria (USGS locs. M7278 and M7822), are consistent with early to late stage 5 age assignments for these terraces (see Wehmiller, this volume), but higher resolution age distinctions can only be derived from selected results. Rockwell and others (this volume) attribute at least some of this lack of precision to poor preservation and the fragmental nature of many of the analyzed samples. Implications of these observations are discussed further by Rockwell and others (this volume).

First Terrace.—

Concordant thorium (^{230}Th/^{234}U) and protactinium (^{231}Pa/^{235}U) ages of 87,000±4,000 yrs and 87,000±12,000 yrs (Rockwell and others, this volume, table 1) from an unidentified mammal bone in the marine sediments overlying the Cojo Terrace near Arroyo San Augustine (Fig. 1) suggest this terrace is correlative with the 80-ka sea-level highstand, and with substage 5a of the marine oxygen isotope record (Shackleton and Opdyke, 1973). However, bone takes up ^{238}U and ^{235}U secondarily, and any uranium-series age must be considered a minimum one, although it may be close to the true age (D. R. Muhs, pers. commun., 1991). The U-series age is also supported by concordant ^{230}Th/^{234}U and ^{231}Pa/^{235}U ages of between 60,000 and 70,000 yrs (60,000±2,000 and 66,000±7,000; 69,800±3,500 and 67,900±4,800) for two unidentified bones in the alluvial-terrace sediments overlying the Cojo Terrace near Cojo Bay (Rockwell and others, this volume, table 1). Although these post-date the period of marine abrasion and deposition, they serve as minimum ages for this terrace.

Numerous amino-acid analyses (Fig. 2) of *Protothaca* and *Saxidomus* from the first terrace at Cojo Bay, Cañada de Alegria, Gaviota, and Arroyo Hondo generally support the correlation of this terrace with the 80-ka sea-level highstand (see Rockwell and others, and Wehmiller, this volume, for further discussion). Leucine data in *Saxidomus* from Gaviota and Arroyo Hondo show particularly good correlation of this terrace with the 80-ka terrace recorded elsewhere along the California coast (Kennedy and others, 1982, fig. 2).

Further correlation is supported by the cool-water (zoogeographic) signature (Tables 2, 3) of all of the faunas (Table 1) from the first, Cojo Terrace, which, east of the South Branch Santa Ynez fault, can be physically traced in seacliff exposures. Although the cool-water signature does not preclude correlation with the 105- to 100-ka sea-level highstand of substage 5c, correlation with the 130- to 120-ka highstand is reliably excluded.

Second Terrace.—

The age of the second terrace can be constrained by its intermediate position between the first terrace, estimated to be about 80 ka and the third terrace (see later discussion), estimated to be about 130 to 120 ka, suggesting correlation with the intermediate sea-level highstand that occurred between 105 and 100 ka (Shackleton and Opdyke, 1973).

Even under the best of conditions, differences in racemization are such that terraces that differ in age by only 20,000 yrs (for example, terraces formed during substages 5a and 5c, or 5c and 5e) may not be reliably distinguished (Wehmiller, 1982, 1984, 1990). Samples from the first and second terraces at Gaviota have yielded D/L values (Fig. 2) that do not always distinguish these geomorphically discrete terraces (Patterson and others, 1982; Rockwell and others, this volume). The ratios, however, are less than those from the third terrace at Arroyo Hondo, which is correlated to the 125-ka sea-level highstand on the basis of its warm-water faunal element and supporting D/L values in *Saxidomus* (Kennedy and others, 1982; Kennedy and Wehmiller, 1986).

Higher than expected ratios from the second terrace at Cañada de Alegria (LACMNH loc. 5619, USGS loc. M7278) led Yerkes and others (1981, fig. 15A) and Kennedy and others (1982, fig. 2) to infer a greater antiquity to this locality than is suggested herein (see also Rockwell and others, this volume). The cool-water zoogeographic signature of the fauna at Cañada de Alegria mitigates against correlation with substage 5e, but is consistent with sea-levels or sea-surface temperatures suggested by the marine-oxygen isotope record (Shackleton and Opdyke, 1973), and with its geomorphic position between the first, ~80-ka, and third, ~125-ka terraces. If correctly dated, this would represent the only documented 105- to 100-ka terrace fauna on the Pacific Coast of North America (not including the few specimens from the Gaviota second terrace localities), and is particularly important in understanding the paleoclimatic conditions that existed during oxygen isotope substage 5c.

Third Terrace.—

On the basis of altitudinal spacing and local uplift rates, Rockwell and others (this volume) correlate the 38-m terrace segment at Arroyo Hondo with the third terrace, and with the ~125-ka sea-level highstand. This correlation is strongly supported by amino-acid ratios, particularly leucine in *Saxidomus* (Kennedy and others, 1982, fig. 2). The correlation with the ~125-ka highstand is further supported by the warm-water faunal elements at these localities (Tables 2, 3; Kennedy and Wehmiller, 1986), which differ considerably from the cool-water aspect of the faunas from the lower two terraces that are correlated with the ~80- and ~105-ka sea-level highstands.

Fourth Terrace.—

Only a single fossil locality (USGS loc. M7822) has been identified on the fourth terrace. *Saxidomus* from this surface yielded leucine ratios (mean 0.51±0.02, n = 2) suggestive of a ~125-ka age when compared to similar ratios at other localities in southern California (Kennedy and others, 1982, fig. 2). The depauperate fauna here did not yield any zoogeographically indicative species, and it is not possible to validate the 125-ka age assignment on this basis. The altitudinal spacing of terraces at Cañada de Alegria suggested to Rockwell and others (this volume) that this terrace remnant is a continuation of the fourth terrace mapped to the west, and thus may be correlative with one of the sea-level highstands that occurred approximately 220 to 180 ka, as recorded on Barbados and Bermuda (Bender and others, 1979; Harmon and others, 1983). Further collecting

WENNER, A. M., AND JOHNSON, D. L., 1980, Land vertebrates on the California Channel Islands: sweepstakes or bridges?, *in* Power, D. M., ed., The California Islands: Proceedings of a Multidisciplinary Symposium: Santa Barbara Museum of Natural History, Santa Barbara, p. 497–530.

WOODRING, W. P., AND BRAMLETTE, M. N., 1950 [1951], Geology and paleontology of the Santa Maria district, California: U.S. Geological Survey Professional Paper 222, 185 p.

WRIGHT, R. H., 1972, Late Pleistocene marine fauna, Goleta, California: Journal of Paleontology, v. 46, p. 688–695.

YERKES, R. F., GREENE, H. G., TINSLEY, J. C., AND LAJOIE, K. R., 1981, Seismotectonic setting of the Santa Barbara Channel area, southern California: U.S. Geological Survey Miscellaneous Field Studies Map MF-1169, 25 p., 2 map sheets.

APPENDIX: LOCALITY DESCRIPTIONS

All localities are along the coast of southwestern Santa Barbara County, California (Fig. 1). Coordinates are those of the Universal Transverse Mercator grid system (UTM), zone 10 (1927 datum).

LOS ANGELES COUNTY MUSEUM OF NATURAL HISTORY (LACMNH), Invertebrate Paleontology Section, Los Angeles, California.

LACMNH loc. 5619. Pleistocene. Marine sand and pebble-cobble conglomerate exposed along shoreline angle in roadcut above and west of mouth of Canãda de Alegria, on Hollister Ranch ~4.1 km west of pier at Gaviota State Beach. Elevation 38.3 m (125.6 ft; surveyed). UTM coordinates, 3817390m N, 750540m E. Collected by G. L. Kennedy (GK-78-3) and M. L. Werner (MW-103), 10 April 1978. Same locality as USGS loc. M7278.

LACMNH loc. 5638. Pleistocene. Lense of abundantly fossiliferous marine sand with scattered pebbles exposed in sea cliff ~1 m above abrasion platform at east end of Cojo Bay and southwest of proposed LNG site, ~4.39 km east of Point Conception lighthouse. Estimated elevation 3 to 5 m. UTM coordinates, 3814860m N, 736850m E. Collected by G. L. Kennedy and K. R. Lajoie (KRL-51878), 18 May 1978. Same locality as USGS loc. M7279.

LACMNH loc. 8234. Pleistocene. Marine sand and pebble-cobble conglomerate exposed along terrace platform in sea cliff below Southern Pacific Railroad tracks, 15 m on either side of railroad phone box 45, between Arroyo Hondo and Canãda de la Huerta, 8.3 km east of pier at Gaviota State Beach. Elevation 20 m. Same locality as USGS loc. M7491, except collected by G. L. Kennedy (GK-85-12) and T. K. Rockwell, 10 August 1985; Kennedy and D. R. Muhs, 28 March 1990.

LACMNH loc. 8235. Pleistocene. Marine sand and pebble-cobble conglomerate overlying terrace platform exposed in south-facing roadcut on north side of U.S. Highway 101 (east-west section of northbound lanes) between Arroyo Hondo and Cañada de la Huerta, 8.5 km east of pier at Gaviota State Beach. Estimated elevation 35 to 40 m. Same locality as USGS loc. M7476, except collected by G. L. Kennedy (GK-85-13) and T. K. Rockwell, 10 August 1985; Kennedy and D. R. Muhs, 28 March 1990.

LACMNH loc. 12612. Pleistocene. Marine sand and pebble-cobble conglomerate exposed on terrace platform near top of sea cliff and below Southern Pacific Railroad tracks, ~1.56 km west of pier at Gaviota State Beach. Elevation 27 m. Same locality as USGS loc. M7819, except collected by G. L. Kennedy (GK-90-5) and D. R. Muhs, 28 March 1990.

U.S. GEOLOGICAL SURVEY (USGS), Branch of Paleontology and Stratigraphy, Menlo Park, California.

USGS loc. M5919. Pleistocene. Small pod of coarse sand and gravel about 3 ft thick exposed on terrace platform near top of sea cliff, 1.46 km west of pier at Gaviota State Beach. Estimated elevation 24 to 27 m. UTM coordinates, 3817390m N, 753190m E. Collected by C. W. Hoskins (H56-26), 1956. Same terrace platform as at LACMNH loc. 12612 and USGS locs. M7496 and M7819.

USGS loc. M7278. Pleistocene. Marine sand and pebble-cobble conglomerate exposed at terrace shoreline angle above and west of mouth of Cañada de Alegria. Elevation 38.3 m. Same locality as LACMNH loc. 5619, except collected by G. L. Kennedy, 30 January 1980.

USGS loc. M7279. Pleistocene. Abundantly fossiliferous marine sand exposed in sea cliff at east end of Cojo Bay. Same locality as LACMNH loc. 5638, except collected by G. L. Kennedy, J. F. Wehmiller, and K. R. Lajoie, 30 August 1978; and Kennedy, Lajoie, and S. A. Mathieson, 29 January 1980.

USGS loc. M7476. Pleistocene. Marine sand and pebble-cobble conglomerate overlying terrace platform exposed in roadcut on north side of U.S. Highway 101 (east-west section of northbound lanes) between Arroyo Hondo and Cañada de la Huerta, ~8.5 km east of pier at Gaviota State Beach. Estimated elevation 35–40 m. UTM coordinates, 3818320m N, 763140m E. Collected by G. L. Kennedy, J. F. Wehmiller, and K. R. Lajoie, 30 August 1978; Kennedy, Lajoie, and S. A. Mathieson, 31 January 1980; and Kennedy and Mathieson, 25 March 1981. Same locality as LACMNH loc. 8235.

USGS loc. M7487. Pleistocene. Marine sand exposed in sea cliff ~0.42 km west-southwest of mouth of Cañada del Cojo, at east end of Cojo Bay ~4.66 km east of Point Conception lighthouse. Estimated elevation 4 to 6 m. UTM coordinates, 3814965m N, 737100m E. Collected by G. L. Kennedy, S. A. Mathieson, and K. R. Lajoie, 29 January 1980.

USGS loc. M7489. Pleistocene. Terrace abrasion platform exposed in roadcut on north side of U. S. Highway 101 (east-west section of northbound lanes) on east side of Cañada de la Huerta, ~8.7 km east of pier at Gaviota State Beach. Estimated elevation 35 to 40 m. UTM coordinates, 3818290m N, 763320m E. Collected by S. A. Mathieson and K. R. Lajoie, 31 January 1980.

USGS loc. M7490. Pleistocene. Marine sand and pebble-cobble conglomerate on terrace platform exposed in north-facing roadcut of U. S. Highway 101 (east-west section of northbound lanes), between Arroyo Hondo and Cañada de la Huerta, ~8.29 km east of pier at Gaviota State Beach. Estimated elevation 35 to 40 m. UTM coordinates, 3818290m N, 762920m E. Collected by K. R. Lajoie, 31 January 1980; and G. L. Kennedy and S. A. Mathieson, 25 March 1981.

USGS loc. M7491. Pleistocene. Marine sand and pebble-cobble conglomerate exposed along terrace platform in sea cliff below Southern Pacific Railroad tracks, ~15 m on either side of railroad phone box 45, between Arroyo Hondo and Canãda de la Huerta, ~8.3 km east of pier at Gaviota State Beach. Elevation 20 m. UTM coordinates, 3818200m N, 762940m E. Collected by G. L. Kennedy, S. A. Mathieson, and K. R. Lajoie, 31 January 1980; and Kennedy and Mathieson, 25 March 1981. Same locality as LACMNH loc. 8234.

USGS loc. M7496. Pleistocene. Marine sand and pebble-cobble conglomerate exposed along terrace platform near top of sea cliff ~1.32 to 1.40 km west of pier at Gaviota State Beach. Estimated elevation 24 to 27 m. UTM coordinates, 3817390m N, 753240m E. Collected by A. M. Sarna-Wojcicki and K. R. Lajoie (H56-26), 21 December 1977 and (?) 8 July 1978; Lajoie (GV-51778) and G. L. Kennedy, 17 May 1978; and Kennedy, Lajoie, and J. F. Wehmiller, 30 August 1978. Same terrace platform as at LACMNH loc. 12612 and USGS locs. M5919 and M7819.

USGS loc. M7819. Pleistocene. Marine sand and pebble-cobble conglomerate exposed on terrace platform near top of sea cliff and below Southern Pacific Railroad tracks, ~1.56 km west of pier at Gaviota State Beach. Elevation 27 m (87.8 ft; Dames and Moore, 1980, 1981). UTM coordinates, 3817370m N, 753090m E. Collected by G. L. Kennedy (DJH-20) and S. A. Mathieson, 24 and 26 March 1981. Same locality as LACMNH loc. 12612, same terrace platform as USGS locs. M5919 and M7496.

USGS loc. M7822. Pleistocene. Indurated marine sand in depressions in terrace platform exposed in east-facing roadcut along dirt road near ridge crest on west side of Canãda de Alegria, Hollister Ranch, ~4.14 km west of pier at Gaviota State Beach. Locality is on northern, down-thrown side of the South Branch Santa Ynez fault (Rockwell and others, this volume). Estimated elevation ~74 m. Approximate UTM coordinates, 3817520m N, 750490m E. Collected by G. L. Kennedy and S. A. Mathieson, 25 March 1981.

AMINOSTRATIGRAPHY AND OXYGEN ISOTOPE STRATIGRAPHY OF MARINE-TERRACE DEPOSITS, PALOS VERDES HILLS AND SAN PEDRO AREAS, LOS ANGELES COUNTY, CALIFORNIA

DANIEL R. MUHS

U.S. Geological Survey, MS 424, Box 25046, Federal Center, Denver, Colorado 80225

GIFFORD H. MILLER

Center for Geochronological Research, Institute of Arctic and Alpine Research, University of Colorado, Boulder, Colorado 80309

JOSEPH F. WHELAN

U.S. Geological Survey, MS 963, Box 25046, Federal Center, Denver, Colorado 80225

AND

GEORGE L. KENNEDY

Earth Sciences Division, Los Angeles County Museum of Natural History, 900 Exposition Blvd., Los Angeles, California 90007

ABSTRACT: Amino-acid and oxygen isotope data for fossils from terraces of the Palos Verdes Hills and San Pedro areas in Los Angeles County, California, shed new light on the ages of terraces, sea-level history, marine paleotemperatures, and late Quaternary tectonics in this region. Low terraces on the Palos Verdes peninsula correlate with the ~80-ka and ~125-ka sea-level highstands that are also recorded as terraces on other coasts. In San Pedro, the Palos Verdes sand (the deposit on what is mapped as the first terrace by Woodring and others, 1946) was previously thought to be a single deposit; amino-acid, oxygen isotope, U-series, and faunal data indicate that deposits of two ages, representing the 80-ka and 125-ka highstands occur within this unit. Oxygen isotope data show that on open, exposed parts of the Palos Verdes peninsula, ocean waters during the 125-ka highstand were cooler than present (by about 2.3–2.6°C) similar to what has been reported for other exposed coastal areas in California. In contrast, in the protected embayment environment around San Pedro, water temperatures during the 125-ka highstand were as warm or warmer than present. During the 80-ka highstand, water temperatures were significantly cooler than present even in the relatively protected embayment environment of the San Pedro area.

Late Quaternary tectonic-uplift rates can be calculated from terrace ages and elevations. Correlation of the lowest terraces around the Point Fermin area shows that the Cabrillo fault has a late Quaternary vertical-movement rate of 0.20 m/ka, based on the difference in uplift rates on the upthrown and downthrown sides of the fault. Elsewhere in the Palos Verdes Hills–San Pedro area, late Quaternary uplift rates vary from 0.32 m/ka to possibly as high as 0.72 m/ka. These rates, which reflect vertical movement on the Palos Verdes fault, are in broad agreement with estimated Holocene vertical rates of movement determined for offshore portions of the fault.

INTRODUCTION

The marine-terrace sequence in the Palos Verdes Hills–San Pedro area of southern California (Fig. 1), where 13 terraces rise to an elevation of almost 400 m, has long attracted the attention of researchers. Marine terraces in California, like those on many other tectonically active coastlines, record interglacial sea-level highstands superimposed on long-term tectonic uplift. Because of the number of terraces and their elevations, the Palos Verdes Hills area has been of interest for providing a detailed record of these combined processes of sea-level fluctuations and tectonics.

Critical to developing sea-level and tectonic history of a flight of marine terraces is reliable dating of the terrace deposits. Several attempts have been made to date the marine terraces on the Palos Verdes Hills. Uranium-series methods were applied to marine-terrace mollusks by Fanale and Schaeffer (1965), Szabo and Rosholt (1969), Szabo and Vedder (1971), and Kaufman and others (1971). However, it is now known that mollusks take up uranium secondarily and do not always form closed systems with respect to ^{238}U and its long-lived daughter products (Kaufman and others, 1971). The open-system uranium-trend dating method was applied to some of the Palos Verdes Hills terraces (Muhs and others, 1989), but this method is still experimental. One of the earliest feasibility studies for amino-acid geochronology studies was conducted on shells collected from terraces of the Palos Verdes Hills (Mitterer and Hare, 1967). Age estimates for marine-terrace mollusks based on amino-acid racemization were also reported by Wehmiller and others (1977), Muhs (1984), Bryant (1987), and Wehmiller (1990), but these studies were limited both in the number of localities sampled and in the number of shells analyzed at each locality. Thus, much remains to be learned about the ages of terraces in the Palos Verdes Hills–San Pedro area.

This study presents amino-acid and oxygen isotope data for marine-terrace mollusks that shed new light on terrace ages. With these age estimates, it is possible to correlate terrace-forming intervals with sea-level events recorded on other coasts. Late Quaternary sea-level highstands have been recorded on other coasts at ~60 ka, ~80 ka, ~105 ka, ~125 ka, and 190 to 220 ka (Mesolella and others, 1969; Veeh and Chappell, 1970; James and others, 1971; Bloom and others, 1974; Dodge and others, 1983; Harmon and others, 1983; Edwards and others, 1987; Bard and others, 1990; Ku and others, 1990; Muhs and others, 1992). Oxygen isotope data also provide information on ocean paleotemperatures. Ocean paleotemperatures estimated from oxygen isotope data for this area are compared with similar data for other parts of the Pacific Ocean bordering the United States (Valentine and Meade 1961; Muhs and Kyser, 1987; Muhs and others, 1990). In addition to their importance in developing a local sea-level chronology, terrace ages can, when combined with marine-terrace shoreline-angle (inner-edge) elevations and corrected for sea-level fluctuations, yield long-term average rates of tectonic uplift. From these data, we can infer at least minimum late Pleistocene vertical-slip rates for the faults that occur around or on the Palos Verdes Hills area, such as the Palos Verdes Hills and Cabrillo faults. Both of these faults extend offshore to the southeast of the Palos Verdes Hills. Late Pleistocene–Holocene slip rates have been calculated for the offshore portions of these faults

Quaternary Coasts of the United States: Marine and Lacustrine Systems, SEPM Special Publication No. 48

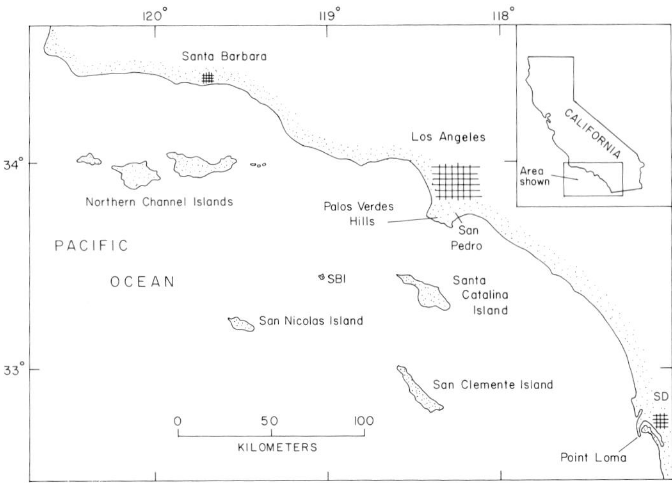

FIG. 1.—Index map of coastal southern California showing location of the Palos Verdes Hills, San Pedro, and other localities referred to in the text. SBI, Santa Barbara Island; SD, San Diego.

by Fischer and others (1987). Age estimates for the marine terraces allow a comparison of slip rates between onshore and offshore portions of the faults.

STRUCTURAL GEOLOGY

The Palos Verdes peninsula–San Pedro area is bounded on its northeast, landward side by the northwest-trending Palos Verdes Hills fault (Fig. 2). This fault, which Nardin and Henyey (1978) interpreted to have experienced dextral strike slip during the Quaternary, extends more than 100 km from the Santa Monica shelf south to near Lasuen Knoll, which is west of the city of San Clemente (Fischer and others, 1987). The only onshore expression of the fault is in the Palos Verdes Hills area. The Palos Verdes Hills fault also has a vertical component of movement, which is reverse, south side up (Yerkes and others, 1965; Nardin and Henyey, 1978; Fischer and others, 1987). The 13 elevated marine terraces on the Palos Verdes Hills are one of the main expressions of Quaternary activity of the Palos Verdes Hills fault, but Fischer and others (1987) documented that late Pleistocene and Holocene offshore sediments are also displaced by the fault. They reported vertical offsets of Holocene shelf sediments, *en echelon* topographic anomalies, and seafloor scarps, and they estimated average Holocene vertical-slip rates of about 0.1 to 0.4 m/ka for the offshore portion of the Palos Verdes Hills fault. Parts of the Palos Verdes Hills crustal block are also displaced by the smaller Cabrillo fault (Fig. 2). Fischer and others (1987) estimated an average Holocene vertical-slip rate of 0.4 to 0.7 m/ka for the offshore portion of the Cabrillo fault.

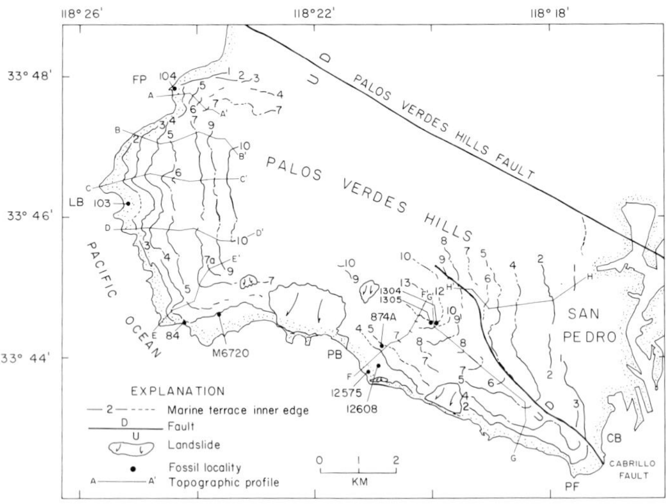

FIG. 2.—Map of the Palos Verdes Hills–San Pedro area showing marine-terrace inner edges, major faults, fossil localities outside of San Pedro, and locations of topographic profiles. Terrace inner edges are taken from Woodring and others (1946), and terrace numbering system is unchanged from those authors. Location of Palos Verdes Hills and Cabrillo faults from Junger and Wagner (1977) and Woodring and others (1946). FP, Flatrock Point; LB, Lunada Bay; PB, Portuguese Bend; PF, Point Fermin; CB, Cabrillo Beach.

MARINE TERRACES

Marine terraces on the Palos Verdes Hills are generally well expressed topographically, although urbanization has significantly modified the natural landscape. Woodring and others (1946) mapped in detail the terraces in both the Palos Verdes Hills (Fig. 2) and San Pedro areas (Fig. 3). Our field observations and interpretation of 1954 (1:20,000) and 1972 (1:30,000) aerial photographs confirm the general landform mapping of these previous workers. Thus, the terraces mapped in Figures 2 and 3 are derived from Woodring and others (1946) and the terrace numbers on those maps are unchanged from their nomenclature. On the basis of new platform-elevation measurements, Bryant (1987) suggested that many more terraces may be present than originally mapped by Woodring and others (1946). However, individual terrace boundaries can be clearly defined only by the location of the inner edge, or shoreline angle, which is the junction of the marine platform and the sea cliff. Bryant's (1987) new elevation measurements appear to be of the marine platforms rather than the shoreline angles, so his hypothesis of additional terraces must await further field study. For the purposes of our study, we have assumed that the map of Woodring and others (1946) is basically correct in terms of landform mapping but not necessarily in terms of lateral correlation of terrace segments.

The best reference point on a marine terrace for paleosea-level or uplift-rate studies is the shoreline angle. In the San Diego area, modern shoreline angles are usually formed within 1 m of present mean sea level (Kern, 1977). Unfortunately, shoreline angles of uplifted marine terraces are rarely exposed in southern California because the sides of canyons that dissect the terraces are usually covered with colluvium. In addition, urbanization has removed or obscured many exposures. The topographically defined inner edges of even well-preserved terraces have elevations that are usually greater than the shoreline-angle elevations because wedges of colluvium or alluvium have accumulated at the bases of the former sea cliffs. The post-emergence

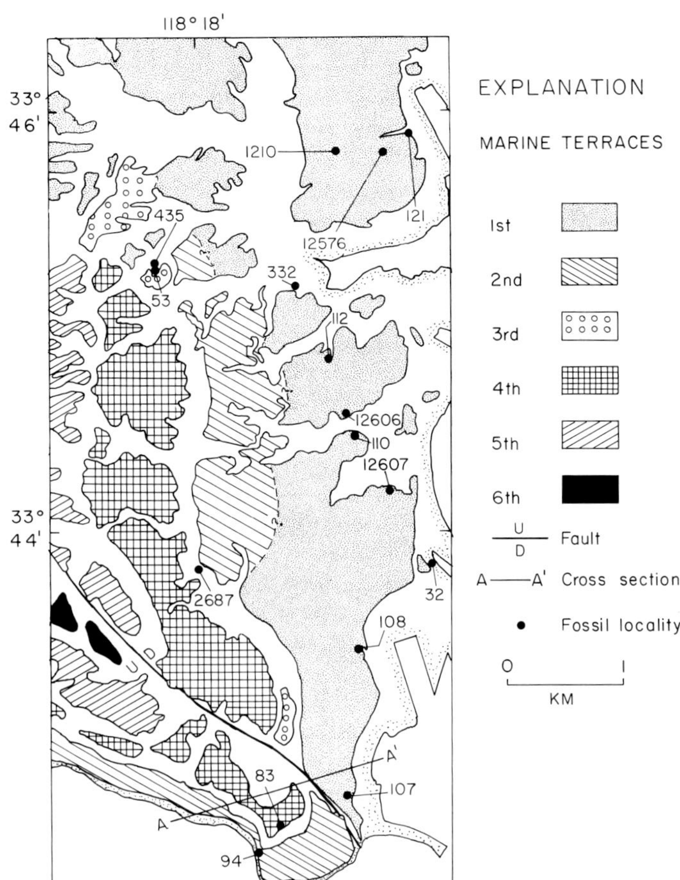

FIG. 3.—Map of marine terraces (modified from Woodring and others, 1946) and fossil localities in the San Pedro area. Inner edge of the first terrace (Palos Verdes sand) is not well expressed at most places and was not mapped in many places by Woodring and others (1946); contacts shown here are approximate. Terrace numbering system is unchanged from Woodring and others (1946). Cross section A–A′ is shown in Figure 10.

terrace-sedimentation history in the Palos Verdes Hills area is highly complex and precludes the use of soils as a reliable mapping and correlation tool.

In order to estimate shoreline-angle elevations on these sediment-covered terraces, we enlarged the shore-normal topographic profiles (locations shown on Fig. 2) constructed by Woodring and others (1946) and projected the outer-platform gradient inland and the maximum slope angle on the sea cliff downward. The intersection of these projections is the approximate position of the shoreline angle, and the elevations are probably accurate to within ±4 m, based on graphical errors alone. Other errors are present because the marine and terrestrial covers of the terraces vary both in thickness and shore-normal extent, but we cannot estimate the magnitude of these errors because there are so few good exposures.

Using these approximate shoreline-angle elevations and estimates of the elevations of the outer-platform edges, we constructed shore-parallel profiles of the terraces in order to examine terrace continuity and to evaluate terrace correlations as mapped by Woodring and others (1946) (Fig. 4). However, correlation of some of the lower terraces, as shown on Figure 4, is based on aminostratigraphic data, as discussed later. Results indicate that even within the rather large uncertainty of the shoreline-angle elevation measurements, the lower terraces can be correlated laterally and are essentially horizontal from topographic profile B north of Lunada Bay southeast to topographic profile F, on the southeast side of the Portuguese Bend landslide. However, elevations of terraces along topographic profiles A, G, and H do not match well with elevations of terraces along profiles B through F (Fig. 4). Profile A, near Flatrock Point, is in an area that has a northwest-trending fault (normal, south side up; described by Riccio and Mills, 1977) that displaces terrace deposits and overlying non-marine deposits. This fault was not previously identified by Woodring and others (1946), and it is possible that the terraces on profile A do not correlate with those on profile B because of displacements by this or other faults. Profiles G and H are on the upthrown and downthrown sides, respectively, of the Cabrillo fault (Fig. 2). It is possible that the terraces in this area have been displaced along the Cabrillo fault, and terrace correlation across the fault based on shoreline-angle elevation has not been attempted. One of our goals was to seek terrace fossils on both sides of the Cabrillo fault that would enable us to make lateral correlations that are not based on shoreline-angle elevation.

URANIUM-SERIES DATING

Corals are the most suitable materials for uranium-series dating, but unfortunately few corals occur in most marine-terrace deposits of the Pacific Coast of North America. Nevertheless, some solitary corals and colonial hydrocorals have been recovered, and in recent years a number of new U-series ages have been generated (Muhs and Szabo, 1982; Rockwell and others, 1989; Muhs and others, 1990, 1992; Muhs, 1992). These recent studies are important to our study

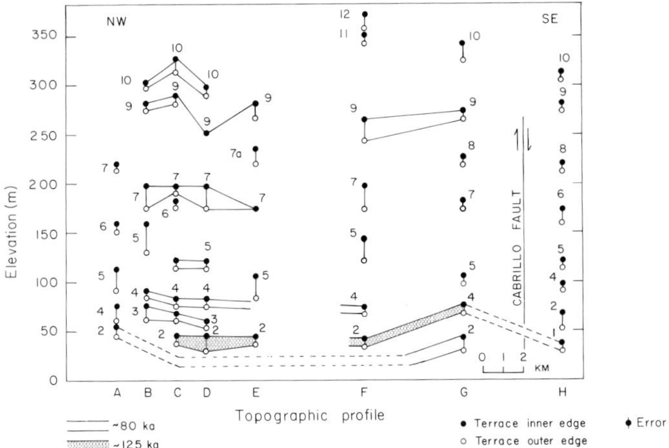

FIG. 4.—Shore-parallel profiles of marine-terrace inner and outer edges in the Palos Verdes Hills and San Pedro. Probable correlations, based on either elevation or amino-acid and stable-isotope data, are shown by solid lines; tentative correlations are shown by dashed lines.

because they provide localities from which amino-acid ratios can be calibrated. Three of these localities, San Clemente Island, San Nicolas Island, and Point Loma near San Diego (Fig. 1), are close to the Palos Verdes Hills and San Pedro. U-series analysis of coral from the lowest two terraces at Point Loma and San Nicolas Island indicate that the terraces are ~80 ka and ~125 ka (Ku and Kern, 1974; Muhs and others, 1992; Muhs, 1992). Muhs and Szabo (1982) reported an age of 127±7 ka for a hydrocoral from the second terrace on San Clemente Island.

It is now known that U-series ages of fossil mollusks are not reliable because these organisms do not take up uranium from seawater during growth and often act as open systems with respect to uranium and its daughter products after death (Kaufman and others, 1971). However, in cases where there are concordant $^{230}Th/^{234}U$ and $^{231}Pa/^{235}U$ ages, these are reliable *minimum* age estimates for fossil mollusks. Szabo and Rosholt (1969) report U-series analyses of fossil mollusks collected from the first terrace in San Pedro near locality 112 (Fig. 3) of Woodring and others (1946). Four of their samples showed concordant $^{230}Th/^{234}U$ and $^{231}Pa/^{235}U$ ages within limits of analytical error, indicating that these four samples have been closed systems with respect to ^{238}U, ^{235}U, and their daughter products since initial uranium uptake. Thus, we can regard the calculated ages as reasonable minimum ages for the shells. The calculated $^{230}Th/^{234}U$ ages range from 92±8 ka to 110±6 ka, and the oldest age is the closest minimum age of the deposit. Thus, deposits of the first terrace in this part of San Pedro are no younger than about 104 ka, based on the analytical uncertainty. The first terrace deposits in northern San Pedro could therefore have been deposited during the highstands recorded on other coasts at ~105 ka or ~125 ka but could not have been deposited during the ~60-ka or ~80-ka highstands.

AMINO-ACID GEOCHRONOLOGY

Principles of Amino-Acid Geochronology

Aminostratigraphy is based on the observation that the protein of living organisms contains only amino acids of the L (lero, or left handed) configuration. Upon death of an organism, amino acids of the L configuration convert to amino acids of the D (dextro, or right handed) configuration, a process referred to as racemization. A similar process, called epimerization, is the conversion of L-isoleucine (Ile) to D-alloisoleucine (aIle). Racemization and epimerization are reversible reactions that result in increased D/L or aIle/Ile ratios through time until an equilibrium ratio of 1.00 to 1.30 (depending on the amino acid) is reached. Thus, in a simplified view, a higher aIle/Ile or D/L ratio in a fossil indicates a relatively greater age. Amino-acid ratios in fossils are also highly dependent on environmental temperature history, genus, and a variety of diagenetic processes (see Miller and Hare, 1980; Wehmiller, 1982; Miller and Brigham-Grette, 1989; and Muhs, 1991, for reviews). The simplest application of amino-acid ratios in geochronological studies is relative age determination and lateral correlation, or "aminostratigraphy" (Miller and Hare, 1980). The main assumption in this approach is that localities studied have had roughly similar temperature histories. Stratigraphic units containing mollusks with amino-acid ratios that cluster around a certain value can be identified as "aminozones" (Nelson, 1982).

In this study, two different methods for amino-acid analyses were employed. Analyses conducted at the University of Colorado used liquid chromatography. The method in use at that laboratory at present is referred to as "method A" and is described in detail by Miller (1985). Liquid-chromatography methods only resolve D-alloisoleucine and L-isoleucine as separate isomers; D and L configurations of other amino acids are not resolved. Analyses conducted at the University of Delaware used gas chromatography. This method does not resolve D-alloisoleucine and L-isoleucine well but does resolve the D and L isomers of several other amino acids. Some of the data obtained by gas chromatography were previously reported by Wehmiller and others (1977), but new data are reported here. Detailed analytical methods for gas chromatography are given by Wehmiller and others (1977).

Amino-Acid Ratios in Mollusks from Dated Terraces on San Nicolas Island

Because of the paucity of corals suitable for uranium-series dating in terrace deposits of the Palos Verdes Hills, we used amino-acid ratios in dated terraces from nearby study areas. As discussed earlier, the two lowest terraces on San Nicolas Island have corals with U-series ages of ~80 ka and ~125 ka. We collected specimens of the gastropod *Tegula* and the bivalve *Epilucina*, two of the most common genera in the Palos Verdes Hills, from the dated San Nicolas Island terraces (Fig. 5) to determine whether either of them could discriminate between 80-and 125-ka deposits. Some of these data were previously reported by Muhs (1985), but most are new.

Results indicate that *Tegula* is a relatively good discriminator of 80-ka and 125-ka deposits (Table 1). Five *Tegula*

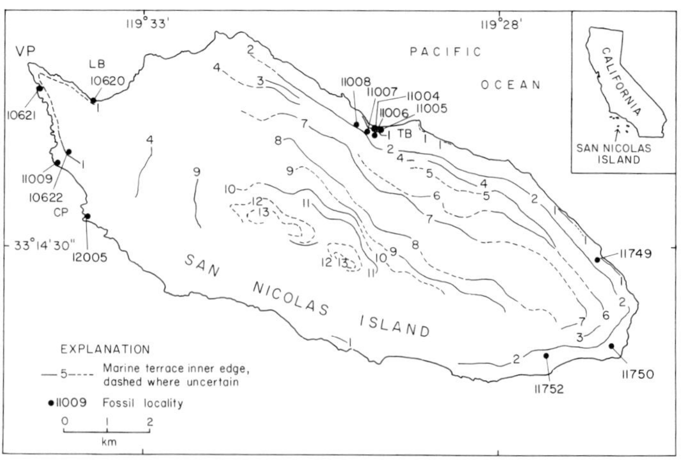

FIG. 5.—Map of inner edges of marine terraces and fossil localities on San Nicolas Island. All inner-edge data are from Vedder and Norris (1963) with the exception of those the first terrace, which was mapped by D. R. Muhs and G. L. Kennedy. VP, Vizcaino Point; CP, Cormorant Point; LB, Laser Bay; TB, Tranquility Beach.

TABLE 1.—RATIOS OF D-ALLOISOLEUCINE TO L-ISOLEUCINE (AILE/ILE) IN *TEGULA* AND *EPILUCINA* FROM U-SERIES-DATED 80-KA AND 125-KA TERRACES ON SAN NICOLAS ISLAND (SNI) AND SAN CLEMENTE ISLAND (SCI) USING LIQUID CHROMATOGRAPHY.

Location[1]	LACMNH Loc.[2]	Terrace	U-series Age (ka)	Genus	AAL[3]	aIle/Ile	Mean±s.d.
SNI, LB	10620	1	80	*Tegula*	5035A	0.36	
					B	0.28	
					C	0.20	
					D	0.29	0.28±0.05
					E	0.29	(n = 5)
SNI, TB	11006	2	125	*Tegula*	3653A	0.50	
					B	0.47	
					C	0.45	
					D	0.35	
					E	0.42	
					5276A	0.46	
					B	0.45	
					C	0.35	
					D	0.37	
					E	0.48	
SNI, VP	10622	2	125	*Tegula*	5038B	0.48	
					C	0.54	
					D	0.56	0.46±0.06
					E	0.52	(n = 14)
SNI, TB	11004	1	80	*Epil.*[4]	3651A	0.33	
					B	0.35	
					C	0.29	
					D	0.34	
					E	0.29	
SNI, CP	12005	1	80	*Epil.*[4]	5039A	0.34	
					B	0.30	
					C	0.43	
					D	0.30	
					E	0.34	
SNI, LB	10620	1	80	*Epil.*[4]	5034A	0.36	
					B	0.39	
					C	0.34	
					D	0.39	
					E	0.34	
SNI, VP	10621	1	80	*Epil.*[4]	5036A	0.40	
					B	0.41	
					C	0.48	
					D	0.40	0.36±0.05
					E	0.40	(n = 20)
SNI, TB	11006	2	125	*Epil.*[4]	3652A	0.34	
					B	0.38	
					C	0.30	
					D	0.34	
					E	0.47	
					5290A	0.45	
					B	0.40	
					C	0.34	
					D	0.37	
					E	0.34	
SNI, VP	10622	2	125	*Epil.*[4]	5037A	0.52	
					B	0.40	
					C	0.47	
					D	0.47	0.41±0.07
					E	0.50	(n = 15)
SCI, EP	10725	2	125	*Tegula*	3650A	0.59	
					B	0.58	
					C	0.60	
					D	0.56	0.59±0.02
					E	0.60	(n = 5)

[1]SNI, San Nicolas Island; LB, Laser Bay; TB, Tranquility Beach; VP, Vizcaino Point; CP, Cormorant Point; SCI, San Clemente Island; EP, Eel Point.
[2]LACMNH, Los Angeles County Museum of Natural History fossil locality.
[3]AAL, Amino Acid Laboratory (University of Colorado) number.
[4]*Epilucina californica*.

individuals from the 80-ka terrace gave a mean aIle/Ile of 0.28±0.05, whereas 14 *Tegula* individuals from the 125-ka terrace gave a mean aIle/Ile of 0.46±0.06. If we omit the four individuals from the relatively shallow terrace deposit at locality 10622, where surface-heating effects may have taken place, the mean for the 10 remaining individuals is 0.43±0.05, still significantly higher than the mean for the shells from the 80-ka terrace. Thus, we are confident that *Tegula* can discriminate between 80-ka and 125-ka deposits.

The results for *Epilucina* indicate that this genus is not effective for discriminating between 80-ka and 125-ka deposits (Table 1). The mean aIle/Ile for 20 individuals from the 80-ka terrace is 0.36±0.05, and the mean for 15 individuals from the 125-ka terrace is 0.41±0.07, which is not significantly different. This difference is reduced even further if we eliminate the five individuals from locality 10622 (again, where the terrace deposit is shallow and surface-heating effects may have taken place); the resultant mean of 10 individuals from the 125-ka terrace is 0.37±0.05, which is almost identical to the ratio for the 80-ka-terrace shells. Muhs (1985) showed that *Epilucina* was effective in discriminating only two broad relative ages among terraces 1, 2, 4, 5, and 10 on San Nicolas Island. We conclude that *Epilucina* may be capable of distinguishing late Quaternary from middle or early Quaternary terraces, but this genus cannot discriminate between 80-ka and 125-ka terraces.

For most of our investigations in the Palos Verdes Hills and San Pedro areas, *Tegula* amino-acid data were used, based on the demonstrated suitability of this genus on San Nicolas Island. *Tegula* occurs primarily in exposed, high-energy, rocky-shore environments, but is largely absent in protected, sandy, quiet-water environments such as embayments. Quiet environments are more common than high-energy environments in San Pedro. For these localities, we rely on the bivalve *Protothaca staminea*, which is common in the terrace deposits.

Aminostratigraphy of Low-Elevation Terrace Deposits in San Pedro and the Palos Verdes Hills

In order to correlate deposits in the Palos Verdes Hills and San Pedro areas with the dated terraces on San Nicolas Island, we sought low terrace sequences that, based on independent evidence, could correspond to the 80-ka and 125-ka highseastands. Kennedy and others (1982) showed that in many parts of the Pacific coast of the United States, 125-ka-terrace deposits commonly contain extraliminal southern species, whereas 80-ka-terrace deposits contain significant numbers of extraliminal northern species. In the Point Fermin area of San Pedro, on the upthrown side of the Cabrillo fault (Figs. 2 and 3), the two lowest terraces are mapped as terrace "2" and terrace "4" by Woodring and others (1946). The fauna of the "fourth" terrace (locality 83, Fig. 3) was described in detail by Valentine (1962) and contains three extralimital southern species, *Bernardina bakeri, Crassinella pacifica*, and *Acanthina lugubris*. In contrast, the fauna on the "second" terrace (locality 94, Fig. 3) was studied by Chase and Chace (1919) and contains a number of extraliminal northern species. Thus, this terrace pair appeared to be a possible 80-ka to 125-ka pair, and accordingly *Tegula* was analyzed from this sequence. The results support our hypothesis when compared with dated deposits on San Nicolas and San Clemente Islands (Table

2 and Fig. 6). In Figure 6, the mean aIle/Ile ratios in *Tegula* are plotted as a function of current mean annual temperature. Mean aIle/Ile ratios should increase with higher integrated thermal histories, which we believe can be proxied crudely by present mean annual air temperatures. The mean aIle/Ile ratio in *Tegula* from the "fourth" terrace at Point Fermin is significantly higher than that for *Tegula* from the "second" terrace; these two means are slighter higher, respectively, than those from the 125-ka and 80-ka terraces on cooler San Nicolas Island. Because San Clemente Island has about the same mean annual air temperature as San Pedro, we would expect similar mean aIle/Ile ratios in the "fourth" terrace at Point Fermin and the 125-ka terrace on San Clemente Island. Our results support this hypothesis, as the the two ratios (0.58±0.05 and 0.59±0.02) are not significantly different (Tables 1 and 2; Fig. 6). We therefore correlate the "fourth" and "second" terraces on Point Fermin with the ~125-ka and ~80-ka highstands of sea level, respectively.

Using this terrace pair at Point Fermin as a starting point, terraces around the Palos Verdes peninsula and in the San Pedro area can be correlated using amino-acid ratios. What has been mapped as the "second" terrace at Flatrock Point (locality 104) has ratios that suggest correlation with the "second" terrace at Point Fermin, but the "second" terrace localities at Lunada Bay (locality 103) and southeast of Portuguese Bend (locality 12575) suggest correlation with the "fourth" terrace at Point Fermin (Table 2 and Fig. 7). On the downthrown side of the Cabrillo fault, a low terrace has been mapped as the "first" terrace by Woodring and others (1946). We collected *Tegula* from locality 108 on this terrace (Fig. 3), a locality that is referred to in the older literature as "Crawfish George's." This locality is significant in that it contains a distinct cool-water fauna, with at least eight extralimital northern species and only three extralimital southern species (Valentine and Meade, 1961). The mean aIle/Ile ratio for *Tegula* from this locality is 0.37±0.05, which is identical to the mean ratio for *Tegula* from the "second" terrace at Point Fermin. The cool-water aspect to the faunas from both localities further reinforces this correlation (Fig. 7). Thus, amino-acid data for *Tegula* indicate that segments of the "second" terrace as mapped by Woodring and others (1946) are not the same age everywhere; some segments are correlative with what has been mapped as the "fourth" terrace, and other segments are correlative with what has been mapped as the "first" terrace.

Comparison of *Protothaca* data from two localities where both *Protothaca* and *Tegula* are present indicates that *Protothaca* can also be used to differentiate and correlate terraces in this area. As discussed earlier, we correlate the "second" terrace at locality 12575 with the ~125-ka highstand on the basis of the *Tegula* data. *Protothaca* shells from this locality have a mean aIle/Ile ratio of 0.35±0.02. At Crawfish George's, *Tegula* data suggest correlation of this deposit with the 80-ka highstand, and *Protothaca* shells have a mean aIle/Ile ratio of 0.28±0.04. Using these two localities for calibration, the age of the lowest mapped terrace within San Pedro can be evaluated. Based on *Protothaca* data, the deposit exposed at 8th and Center Streets

TABLE 2.—RATIOS OF D-ALLOISOLEUCINE TO L-ISOLEUCINE (AILE/ILE) IN *PROTOTHACA* AND *TEGULA* FROM MARINE-TERRACE DEPOSITS IN SAN PEDRO AND THE PALOS VERDES HILLS (PVH), CALIFORNIA, USING LIQUID CHROMATOGRAPHY.

Location[1]	Fossil Loc.[2]	Terrace[3]	Species[4]	AAL[5]	aIle/Ile	Mean±s.d.
San Pedro	12607	1	*P.s.*	6459A	0.22	0.28±0.07
				B	0.26	(n = 5)
				C	0.42	
				D	0.25	
				E	0.25	
San Pedro	WBK 108	1	*P.s.*	5361A	0.27	0.28±0.04
				B	0.29	(n = 5)
				C	0.29	
				D	0.21	
				E	0.32	
San Pedro	12606	1	*P.s.*	5362A	0.40	0.36±0.03
				B	0.35	(n = 4)
				C	0.32	
				D	nd	
				E	0.35	
San Pedro	12576	1	*P.s.*	6456A	0.38	0.34±0.04
				B	0.29	(n = 6)
				C	0.34	
				D	0.32	
				E	0.30	
				F	0.39	
PVH	12575	2	*P.s.*	6457A	0.35	0.35±0.02
				B	0.34	(n = 4)
				C	0.38	
				D	0.32	
PVH	12608	4	*P.s.*	6460A	0.44	0.40±0.04
				B	0.36	(n = 2)
San Pedro	WBK 108	1	*T.f.*	5360A	0.36	0.37±0.05
				B	0.46	(n = 5)
				C	0.32	
				D	0.34	
				E	0.37	
San Pedro	WBK 94	2	*T.f.*	6466A	0.40	0.37±0.08
				B	0.28	(n = 5)
				C	0.52	
				D	0.32	
				E	0.32	
PVH	WBK 104	2	*T.*sp.	6465A	0.44	0.39±0.03
				B	0.39	(n = 5)
				C	0.38	
				D	0.37	
				E	0.35	
PVH	WBK 103	2	*T.*sp.	5359A	0.56	0.47±0.06
				B	0.48	(n = 5)
				C	0.48	
				D	0.38	
				E	0.46	
PVH	12575	2	*T.f.*	6458A	0.56	0.53±0.05
				B	0.48	(n = 5)
				C	0.48	
				D	0.52	
				E	0.60	
San Pedro	WBK 83	4	*T.*sp.	5358A	0.62	0.58±0.05
				B	0.51	(n = 5)
				C	0.62	
				D	0.62	
				E	0.54	
PVH	12608	4	*T.*sp.	6461A	0.66	0.65±0.05
				B	0.59	(n = 5)
				C	0.73	
				D	0.67	
				E	0.62	

[1]PVH, Palos Verdes Hills.
[2]All fossil localities are Los Angeles County Museum of Natural History localities except those preceded by WBK, which are from Woodring and others (1946). It should be noted that although fossils used here were collected from or near localities of Woodring and others (1946), they were collected by the authors, not Woodring and others (1946).
[3]Terrace numbers refer to nomenclature of Woodring and others (1946).
[4]Species: *P.s.*, *Protothaca staminea*; *T.f.*, *Tegula funebralis*; *T.*sp., *Tegula* sp.
[5]AAL, Amino Acid Laboratory (University of Colorado) number.

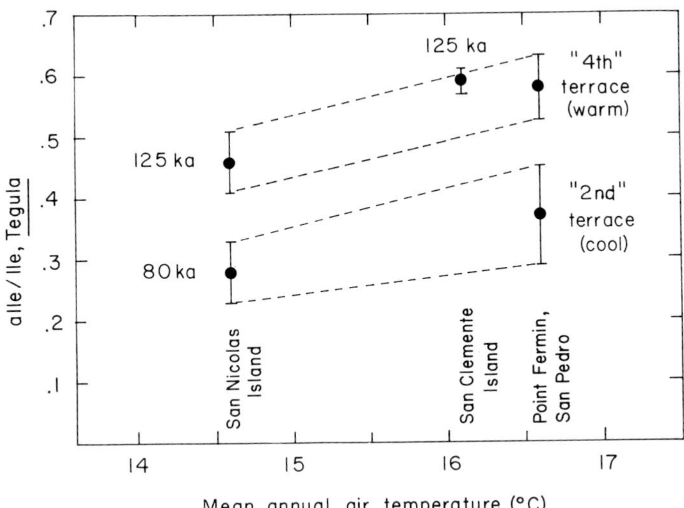

FIG. 6.—Mean alle/Ile ratios in *Tegula* plotted as a function of mean annual air temperature, showing aminostratigraphic correlation of marine terraces in the Point Fermin area with U-series-dated marine terraces on San Nicolas Island and San Clemente Island. "Warm" and "cool" refer to thermal aspects of the terrace faunas.

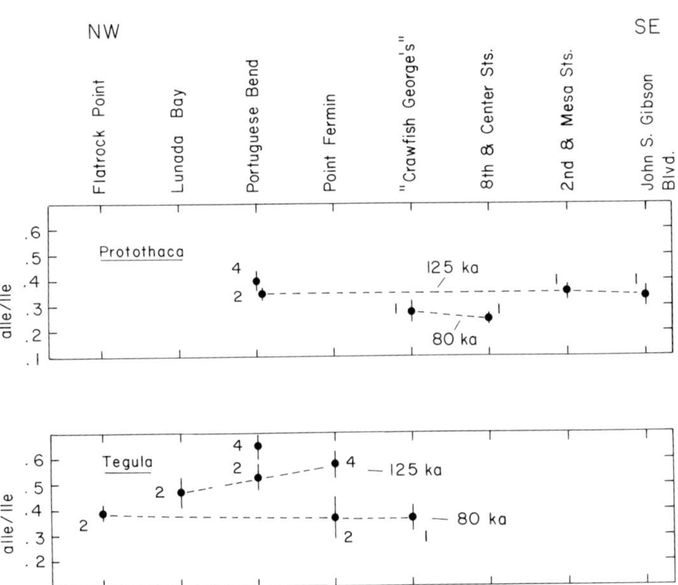

FIG. 7.—Aminostratigraphic correlation of marine terraces in the Palos Verdes Hills–San Pedro area in a northwest-southeast direction, based on alle/Ile ratios in *Protothaca* and *Tegula*. Terrace numbers given are those of Woodring and others (1946).

(locality 12607) correlates with the ~80-ka highstand, and the deposits at 2nd and Mesa Streets (locality 12606) and John S. Gibson Boulevard (locality 12576) correlate with the ~125-ka highstand. Although Woodring and others (1946) mapped a large area within San Pedro as belonging to the "first" terrace (Fig. 3), our data indicate that two ages of deposits are present within this mapping unit. D/L ratios of other amino acids (leucine, glutamic acid, valine, alanine, proline, and phenylalanine) (Table 3) from *Tivela*, *Macoma*, and *Protothaca* support this conclusion.

Locality 121 is near locality 12576 in northern San Pedro (Fig. 3), and shells from the former show higher D/L ratios for most amino acids than those from shells collected from locality 110, near 3rd and Mesa Streets in central San Pedro (Fig. 8). The magnitude of difference as it pertains to age is best shown by comparison with *Protothaca* D/L data reported by Wehmiller and others (1977) for the two low terraces found on Point Loma near San Diego. The older of these two terraces, the Nestor terrace, has a coral U-series age of about 125 ka, and the lower Bird Rock terrace has a coral U-series age of ~80 ka, as discussed earlier. Comparison of D/L ratios in *Protothaca* from the Point Loma terraces with those from San Pedro (Fig. 8) supports the conclusion that the northern San Pedro locality (121) represents a deposit whose age is probably 125 ka, whereas the central San Pedro locality (110) represents a deposit whose age is probably ~80 ka. These data are consistent with the difference in age for the deposits of the "first" terrace based on the alle/Ile data. Such an interpretation is also consistent with the U-series data that indicate a minimum age of ~110 ka for the deposits found at locality 112 in northern San Pedro (Fig. 3). What is problematic about these observations is that our data indicate that the age of deposits found at 3rd and Mesa Streets (locality 110) is ~80 ka, whereas the age of deposits found at 2nd and Mesa Streets (locality 12606) is ~125 ka. Is there a contact between these two deposits of different ages somewhere along the one-city-block distance separating these two localities? Nearly complete urbanization of this area has not allowed us to identify such a contact in the field, and examination of older maps has not revealed any possibilities either. Another possible interpretation is that the contact is farther north than 2nd and Mesa Streets, but some of the shells from the 2nd and Mesa Street locality have been reworked from older, 125-ka deposits. Ponti (1989) came to similar conclusions about the presence of two ages of deposits in San Pedro and thought that the contact between the two units was somewhere between locality 12606 and locality 112 (Fig. 3).

It is interesting to note that Arnold (1903), Woodring (1935, 1957), Woodring and others (1946), Valentine (1961), and Valentine and Meade (1961) all point out that deposits of the first terrace in northern San Pedro have a distinct warm-water fauna, whereas deposits of the first terrace in central and southern San Pedro have a distinct cool-water fauna. These workers propose various mechanisms to explain this geographic zonation, including transportation of the cool-water species from greater depths by storm waves, reworking from older units, and changes in depth or temperature tolerances. Our data indicate that this geographic zonation exists because there are two ages of deposits represented by the "first" terrace: a ~125 ka unit with a warm-water fauna in northern San Pedro and a ~80 ka unit with a cool-water fauna in southern San Pedro.

OXYGEN ISOTOPE STRATIGRAPHY OF LOW-ELEVATION TERRACES

We can check the validity of some aminostratigraphic correlations by the use of oxygen isotope compositions of fossil mollusks. The oxygen isotope composition of a fossil

TABLE 3.—D/L RATIOS IN VARIOUS AMINO ACIDS OF FOSSIL MOLLUSKS FROM SAN PEDRO AND THE PALOS VERDES HILLS, DETERMINED BY GAS CHROMATOGRAPHY.

Loc.[1]	Unit[2]	Species[3]	Sample #[4]	Leucine	Glutamic Acid	Valine	Alanine	Proline	Phenylalanine
WBK 110	1	*S.*	JW76-60	0.54	0.43	0.39	0.75	0.66	0.51
		P.	JW76-63	0.42	0.37	0.24	0.59	0.43	0.51
		M.	JW77-120	0.51	0.42	0.40	0.75	0.67	0.60
		T.	JW76-61	0.47	0.36	0.30	0.71	0.57	0.52
WBK 121	1	*P.s.*	75-54	0.48	0.39	0.32	0.75	0.54	0.56
		M.n.	76-1	0.67	0.55	0.44	0.83	0.73	0.80
		T.s.	76-2	0.49	0.41	0.35	0.77	0.61	0.58
		Ch.u.	75-55	0.49	0.40	0.34	0.83	0.57	0.57
		Ch.u.	75-56	0.51	0.43	0.38	0.85	0.61	0.57
		S.n.	75-52	0.53	0.42	0.37	0.83	0.66	0.58
		S.n.	75-53	0.54	0.42	0.34	0.87	0.67	0.58
LACM 1210	1	*S.n.*	76-16	0.53	0.42	0.34	0.84	0.64	0.54
LACM 2687	2	*Ch.*	JW77-118-1	0.39	0.38	0.35	0.67	0.43	0.46
		Ch.	JW77-118-2	0.40	0.36	0.40	0.68	0.46	0.48
LACM 332	SPS	*S.n.*	76-6	0.60	0.49	0.41	0.93	0.70	0.69
		S.n.	76-7	0.63	0.50	0.48	0.94	0.74	0.67
		P.s.	76-23	0.64	0.59	0.56	0.93	0.71	0.74
		P.s.	76-23a	0.61	0.55	0.51	0.94	0.67	nd
		M.n.	76-59a	0.79	0.71	0.77	0.94	0.85	0.95
WBK 32	TPS	*S.n.*	76-8	0.69	0.54	0.52	0.94	0.80	0.70
		S.n.	76-9	0.75	0.58	0.57	0.93	0.90	0.73
WBK 53a	LM	*S.*	JW79-81	0.82	0.75	0.62	0.95	0.89	0.81
		S.	JW79-82-1	0.80	0.76	0.61	1.02	0.86	nd
SDS 0288	LM	*S.*	JW79-82-2	0.81	0.71	0.71	0.88	nd	0.70
USGS M6720	4	*E.c.*	76-73	0.66	0.59	0.46	0.87	0.67	0.73
WBK 84	4	*T.g.*	75-76	0.67	0.57	0.42	0.87	0.91	0.89
LDGO 874A	5	*T.g.*	75-62	0.84	0.68	0.63	0.93	0.95	1.00
near LACM 1304	12	*T.g.*	75-61	0.94	0.81	0.85	1.02	0.98	1.02

[1]Locality abbreviations: WBK, Woodring and others (1946); LACM, Los Angeles County Museum of Natural History; SDS, San Diego Society of Natural History; USGS, U.S. Geological Survey; LDGO, Lamont-Doherty Geological Observatory. Samples designated with localities from Woodring and others (1946) were collected in recent years by J. F. Wehmiller and/or co-investigators at or near localities of Woodring and others (1946).

[2]Abbreviations for geologic units: 1, 2, 4, 5, and 12 refer to terraces with those designations made by Woodring and others (1946); SPS, San Pedro sand; TPS, Timms Point silt; LM, Lomita marl.

[3]Abbreviations for genus or species: *S.n., Saxidomus nuttalli; S., Saxidomus; P.s., Protothaca staminea; P., Protothaca; M.n., Macoma nasuta; M., Macoma; T.s., Tivela stultorum; T., Tivela; Ch.u., Chione undatella; Ch., Chione; T.g., Tegula gallina; E.c., Epilucina californica.*

[4]Amino-acid laboratory number of John F. Wehmiller, University of Delaware.

mollusk or coral is a function of the isotopic composition and temperature of seawater at the time of shell precipitation. Studies from dated terraces on Barbados, New Guinea, and in California have shown that the oxygen-isotopic compositions of mollusks and corals from 80-ka and 125-ka terraces are significantly different from one another and can be used for terrace correlation (Fairbanks and Matthews, 1978; Aharon, 1983; Muhs and Kyser, 1987). We sampled modern and fossil mollusks from terrace deposits in San Pedro and the Palos Verdes Hills in order to verify our aminostratigraphic correlations and to assess paleowater temperatures at the times of terrace formation. Sampling and laboratory methods followed those of Muhs and Kyser (1987).

In their study of the oxygen-isotopic composition of fossil mollusks from 80-ka and 125-ka terraces in California, Muhs and Kyser (1987) observed that both ages of mollusks are enriched in the heavier isotope compared to that of modern shells, but 80-ka shells have the highest $\delta^{18}O$ values. Mean $\delta^{18}O$ values for fossil *Epilucina* from 125-ka terraces on San Nicolas Island, San Clemente Island, and Point Loma (the Nestor terrace) are 0.43 to 0.48 per mil heavier than those of their modern counterparts, suggesting that on *open, exposed coasts*, water temperatures at 125 ka were cooler than at present. Mean $\delta^{18}O$ values for *Epilucina* fossils from 80-ka terraces on San Nicolas and San Clemente Islands are 0.8 to 1.0 per mil heavier than values in modern shells from those localities.

Epilucina shells were taken from three localities on the "second" terrace on the Palos Verdes Hills (localities 12575, 103, and 104) and modern shells from near these localities. Two of these localities (12575 and 103) have aIle/Ile ratios in *Tegula* that suggest correlation with the ~125-ka highstand, whereas the third locality (104) has aIle/Ile ratios that suggest correlation with the ~80-ka highstand. When the $\delta^{18}O$ values for the fossil shells from localities 12475 and 103, along with values from their modern counterparts, are plotted with the San Nicolas Island, San Clemente Island, and Point Loma data, a correlation with the ~125-ka highstand is implied (Table 4 and Fig. 9, in agreement with the amino-acid data. Data from shells at locality 104, when plotted with those of their modern equivalents, are intermediate between the 80-ka and 125-ka trends (Fig. 9). It is possible that the terrace at locality 104 correlates with the ~105-ka highstand recorded as emergent coral reefs on Barbados and New Guinea (Mesolella and others, 1969; Bloom and others, 1974), but because we have no firmly dated 105-ka terraces elsewhere in California, we cannot test this hypothesis in any rigorous fashion.

In the Point Fermin–Cabrillo Beach area, *Epilucina* shells from the "fourth" terrace (locality 83) that we correlated with the 125-ka highstand and from the first terrace near

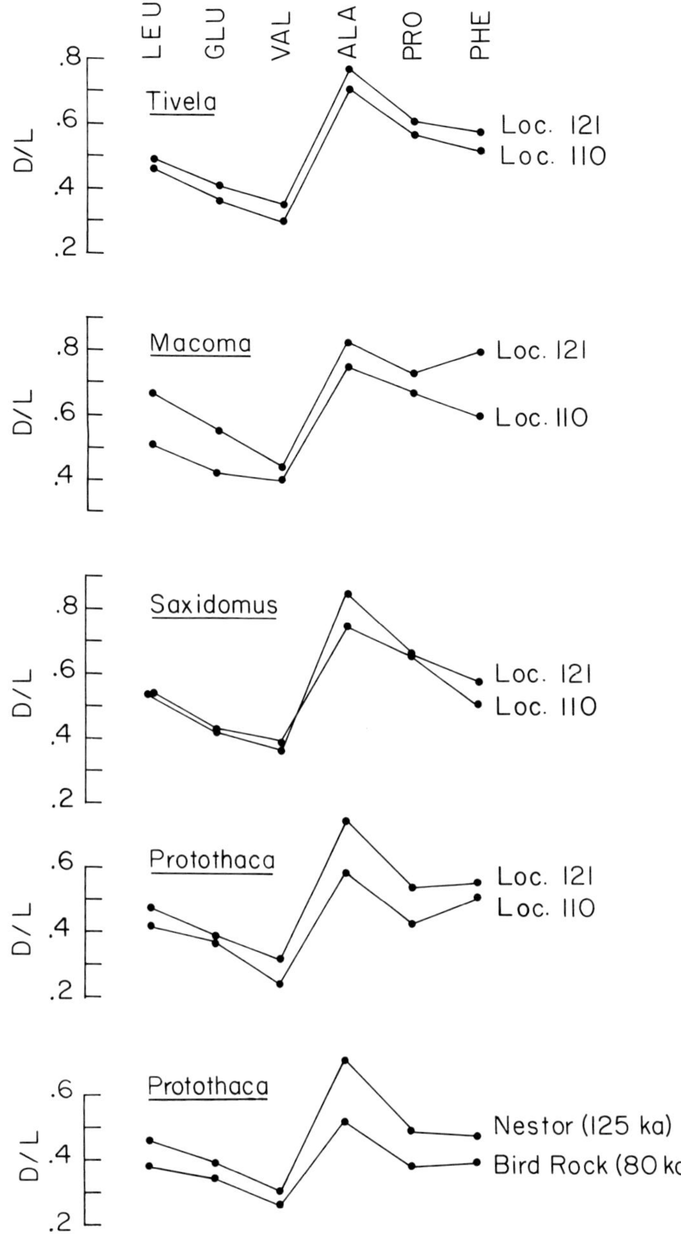

FIG. 8.—D/L ratios in four genera of mollusks for two fossil localities of the first terrace compared to similar ratios for 80-ka and 125-ka terrace fossils from Point Loma near San Diego. Amino-acids: LEU, leucine; GLU, glutamic acid; VAL, valine; ALA, alanine; PRO, proline; PHE, phenylalanine.

Cabrillo Beach (locality 107) that we correlated with the 80-ka highstand were analyzed. Shells from locality 107 (~80 ka) have significantly (0.58 per mil) heavier $\delta^{18}O$ values than shells from locality 83 (~125 ka), which is what we would hypothesize based on their age assignments and their faunal thermal aspects (Table 4 and Fig. 9). Our data for modern shells collected from Cabrillo Beach (south of the breakwater) have values that are not significantly different from those from locality 107 (Table 4), and we suspect that these are fossils that have been reworked onto the

TABLE 4.—STABLE-ISOTOPE COMPOSITION OF MODERN AND FOSSIL MOLLUSKS FROM MARINE TERRACES IN THE PALOS VERDES HILLS (PVH) AND SAN PEDRO, CALIFORNIA.

Location	Terrace	Genus	Fossil Loc.	$\delta^{13}C$(PDB) (‰)	$\delta^{18}O$(PDB) (‰)	$\delta^{18}O$ Mean±s.d. (‰)
PVH	Modern	*Epilucina*	near 12575	2.76	0.48	
				3.65	0.44	
				2.07	0.35	0.42±0.05
PVH	Modern	*Epilucina*	near 103	2.22	0.28	
				1.95	0.36	
				2.57	0.52	0.39±0.10
PVH	Modern	*Epilucina*	near 104	2.53	0.33	
				2.70	0.32	
				2.49	0.22	0.29±0.05
PVH	2	*Epilucina*	12575	1.80	0.84	
				2.40	0.97	
				1.87	1.01	0.94±0.07
PVH	2	*Epilucina*	103	2.28	0.61	
				2.27	0.96	
				2.54	0.94	0.84±0.16
PVH	2	*Epilucina*	104	2.86	0.73	
				2.76	1.13	0.93±0.20
San Pedro	Modern	*Epilucina*	near 107	3.12	0.92	
				2.45	0.87	
				1.52	1.06	0.95±0.08
San Pedro	1	*Epilucina*	107	2.07	1.27	
				1.24	0.97	
				1.18	1.02	1.08±0.13
San Pedro	4	*Epilucina*	83	2.28	0.42	
				2.29	0.51	
				0.87	0.56	0.50±0.06
San Pedro	Modern	*Protothaca*	near 108	0.85	−0.37	
				1.15	−0.25	
				0.82	−0.16	−0.26±0.09
San Pedro	1	*Protothaca*	108	0.88	0.15	
				−0.28	0.23	
				−0.32	0.45	0.28±0.13
San Pedro	1	*Protothaca*	12576	0.91	−0.07	
				−0.05	−0.70	
				0.55	−0.27	−0.35±0.26

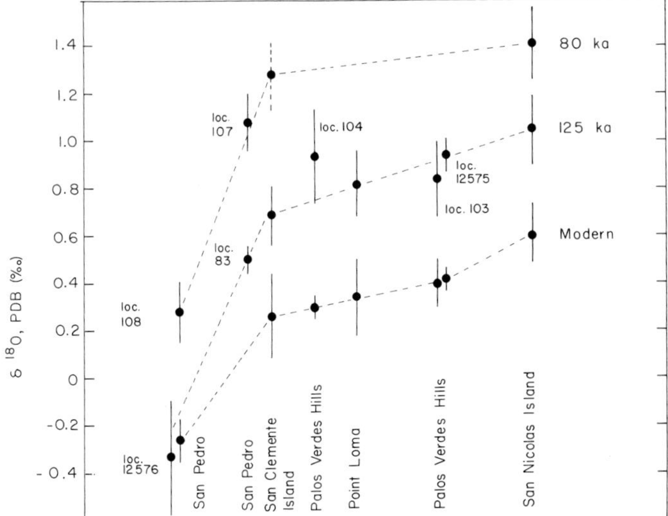

FIG. 9.—Oxygen isotope composition of terrace fossils shown as a function of age for dated terraces on San Clemente Island, Point Loma, and San Nicolas Island and their correlation with fossil localities in the Palos Verdes Hills and San Pedro. All fossils are *Epilucina* with the exception of localities those at 12576 and 108, which are *Protothaca*. Oxygen isotope data for dated terraces are from Muhs and Kyser (1987).

modern beach. The $\delta^{18}O$ composition of the modern shells indicates water temperatures ~1.5°C cooler than at San Nicolas Island, which has a current mean water temperature of ~15°C (Lynn, 1967). Measured water temperatures in the Palos Verdes Hills–San Pedro area are about 16°C (Lynn, 1967).

Within San Pedro, *Epilucina* is not found at most fossil localities, owing to the lack of rocky-shore environments in this area at the time of terrace formation. We therefore analyzed specimens of *Protothaca* that are common in the terrace deposits. Because of possible species effects on oxygen isotope composition, we consider these data independently and compare them to modern *Protothaca* isotopic compositions rather than to the *Epilucina* data. Shells from locality 12576 in northern San Pedro and Crawfish George's in southern San Pedro are estimated to be ~125 ka and ~80 ka, respectively, based on aminostratigraphic correlations. Analyses of modern *Protothaca* shells from Cabrillo Beach, north of the breakwater, indicate that shells from locality 12576 (~125 ka) are not significantly different from the modern shells (Table 4 and Fig. 9), but shells from Crawfish George's (~80 ka) have, as expected, significantly heavier values than either the modern or ~125-ka shells. These data support the aminostratigraphic correlations and are consistent with the generally warm-water aspect of terrace faunas in northern San Pedro and the cool-water aspect of terrace faunas from Crawfish George's (Valentine, 1961; Valentine and Meade, 1961).

Approximate paleotemperature reductions at 125 ka (on the exposed part of the Palos Verdes Hills) and at 80 ka (in San Pedro) can be calculated. In making these calculations, we have assumed that (1) sea level was at −5 m relative to present at 80 ka and at +6 m relative to present at 125 ka and (2) a $\delta^{18}O$ (PDB) shift of 0.11 per mil is equivalent to 10 m of sea level (Fairbanks and Matthews, 1978) and that 0.23 per mil is equivalent to about 1°C. Based on the data from localities 103 and 12575 and modern samples collected near those localities, we infer temperature reductions of 2.3 to 2.6°C at 125 ka on the open, exposed parts of the Palos Verdes Hills. These reductions agree with estimates made for other southern California localities by Muhs and Kyser (1987). As discussed earlier, in the protected parts of San Pedro, water temperatures at 125 ka appear to have been as warm as or warmer than those of the present. However, based on the data from shells collected at Crawfish George's, water temperatures around San Pedro at 80 ka appear to have been about 2.1°C cooler than at present. This temperature reduction is consistent with the cool-water aspect of the terrace fauna (Valentine and Meade, 1961).

One problem that our data and those of Muhs and Kyser (1987) present is the implication of cooler than present paleowater temperatures for some localities at ~125 ka. These observations appear to conflict with observations made earlier of extralimital southern species (i.e., warm-water faunas) present in many terrace deposits that have been correlated with the ~125-ka highstand (cf. Kennedy and others, 1982). It should be noted, however, that the four localities where isotopic data show cooler than present waters at 125 ka (San Nicolas Island, San Clemente Island, Point Loma, and part of the Palos Verdes Hills) are all open, exposed environments that experience strong onshore winds and have the potential for particularly strong upwelling. In contrast, a locality such as 12576 in northern San Pedro (Fig. 3) was in a protected embayment at the time of deposition and would not have been a likely environment for significant upwelling. The oxygen-isotopic composition of shells from the latter environment do not indicate water temperatures that were significantly different from modern water temperatures. Therefore, during the highstand at 125 ka, open coastal areas may have been exposed to stronger onshore winds and thus may have been subject to greater upwelling, which would result in cooler water temperatures on average and could explain the heavier oxygen-isotopic composition in fossils from those localities. Protected embayments, such as those adjacent to northern San Pedro, would not have been subject to significant upwelling, which would explain the similarity to present oxygen isotope temperatures in fossils from such localities. Such a model is consistent with the faunal data as well: deposits from the 125-ka terraces on the open, exposed coasts of San Nicolas Island and Point Loma contain a number of extralimital northern species as well as extralimital southern species (Valentine and Meade, 1961; Vedder and Norris, 1963). In contrast, fossil localities that occur in protected embayments are dominated by extralimital southern species. Kennedy (1988) suggested that during the highstand at 125 ka, greater seasonality may have prevailed on the California coast, possibly as the result of orbital forcing.

Increased frequency or intensity of upwelling as a possible cause for the cooler water temperatures at 125 ka on the open, exposed parts of the California coast may be explained by a recent model presented by Bakun (1990). He suggested that during periods of global warming (such as that envisioned for the 21st century due to greenhouse-gas buildup), continental-margin areas would become warm faster than nearshore ocean waters. This would increase the offshore-onshore pressure gradient, which in turn would increase the strength of alongshore winds. The increased wind strength would enhance the potential for wind-driven upwelling and result in cooler water temperatures along certain coastlines even though the climate on the adjacent continent would be relatively warm. Such a model invites testing for the last interglacial highstand in California at 125 ka and is in part supported by the data presented here.

AMINOSTRATIGRAPHY OF HIGHER MARINE TERRACES AND NON-TERRACED MARINE DEPOSITS

We cannot generate age estimates or attempt lateral correlation for higher marine terraces in the Palos Verdes Hills because of limited data. However, ratios of aIle/Ile for *Tegula* collected from higher terraces given by Muhs (1984) and D/L ratios in this genus collected from the fourth, fifth and twelfth terraces given here (Table 3) indicate that *Tegula* continues to show increasing values over longer time periods. This suggests that should more fossil localities be found, lateral correlation of higher terraces using aminostratigraphy might be possible.

Mean leucine D/L values in *Saxidomus* shells collected from the Palos Verdes sand (from locality 121), the San Pedro sand, the Timms Point silt, and the Lomita marl are

0.53, 0.62, 0.72, and 0.81, respectively (Table 3 and Wehmiller, 1990). These ratios are consistent with the stratigraphic relations among the units as given by Woodring and others (1946). The Palos Verdes sand at locality 121 in northern San Pedro is probably ~125 ka, as discussed previously. The three older marine units appear to be mid-Pleistocene, in agreement with the conclusions of Ponti (1989), who studied these units in considerable detail. If the Lomita marl was deposited during the Pliocene, as suggested by the data of Obradovich (1968), we would expect D/L ratios in *Saxidomus* to be around 0.90 or greater. Wehmiller and others (1977) reported leucine D/L ratios of 0.90 to 0.93 for *Saxidomus* shells from the Fernando and San Diego Formations, which are Pliocene.

IMPLICATIONS FOR LATE QUATERNARY TECTONICS

Marine-terrace data are particularly suitable for studies of Quaternary tectonics because terrace platforms begin as essentially horizontal surfaces (in a shore-parallel sense) that form near sea level. After formation, these platforms can be uplifted, submerged, warped, or faulted depending on the tectonic setting. Along the Pacific Coast of North America, numerous studies have documented tectonic deformation of marine terraces (Birkeland, 1972; Ku and Kern, 1974; Bradley and Griggs, 1976; Kern, 1977; Lajoie and others, 1979, 1982; Muhs and Szabo, 1982; Merritts and Bull, 1989; Rockwell and others, 1989; Hanson and others, 1990; Kelsey, 1990; Lettis and others, 1990; McInelly and Kelsey, 1990; Muhs and others, 1990).

To calculate an uplift rate for a marine terrace, it is necessary to know the age of the terrace, its present elevation (represented by the elevation of the shoreline angle, the closest approximation to mean sea level at the time of terrace formation), and the paleosea level at the time of terrace formation. A number of terrace segments have been correlated here with the well-known 80-ka and 125-ka highstands, based on amino-acid ratios and oxygen isotope composition of the terrace mollusks. The shoreline-angle elevations have been estimated using topographic profiles. Paleosea-level elevations at 80 ka and 125 ka can be estimated using studies from other areas. On tectonically stable coasts distant from plate boundaries, the 125-ka terrace is commonly the only terrace present and is usually 2 to 10 m above present sea level (Ku and others, 1974; Neumann and Moore, 1975; Harmon and others, 1983). Following the lead of other workers (e.g., Bloom and others, 1974), we assume that sea level at 125 ka was about +6 m relative to present. For the 80-ka highstand, paleosea level is usually estimated by calculating an uplift rate for the 125-ka terrace and assuming that this rate is applicable for the 80-ka terrace. The elevation of the 80-ka terrace can then be used to calculate a paleosea level. Unfortunately, estimates of paleosea level derived in this manner differ from coast to coast. The latest estimates for the Huon Peninsula of New Guinea suggest that sea level at 80 ka was −19 m relative to present (Chappell and Shackleton, 1986), whereas estimates from Baja California and southern California suggest that −5 m is more appropriate (Rockwell and others, 1989; Kennedy and others, 1990; Muhs and others, 1988, 1992). For this study, we assume a paleosea level of −5 m relative to present, as that value seems to be most consistent for other localities studied on the Pacific Coast of North America.

A cross section in the area around Point Fermin (Fig. 3) shows how terraces have been displaced by the Cabrillo fault (Fig. 10). The "second" terrace on the upthrown side of the Cabrillo fault has amino-acid ratios that correlate with the "first" terrace on the downthrown side of the fault, but the shoreline-angle elevations of these two terraces are different (~46 m on the upthrown side vs. ~30 m on the downthrown side). Both terraces are correlated with the 80-ka highstand as discussed earlier. Using the assumed −5-m paleosea-level, uplift rates of 0.64 m/ka are derived for the upthrown side of the fault and 0.44 m/ka for the downthrown side of the fault. The difference in uplift rates, 0.20 m/ka, is an approximation of the average rate of vertical movement of the Cabrillo fault over the last 80 ka. This estimate is lower by a factor of two to three relative to Holocene vertical rates of 0.4 to 0.7 m/ka estimated for offshore portions of the Cabrillo fault by Fischer and others (1987). It is possible that the rate of vertical movement varies spatially along this fault or has increased through time.

Our age assignments and estimates of shoreline-angle elevations result in a range of uplift rates for different parts of the Palos Verdes peninsula (Fig. 11). Over most of the westernmost part of the peninsula, from Lunada Bay to just southeast of Portuguese Bend (Fig. 2), the 46-m "second" terrace, with its age assignment of ~125 ka, results in an uplift rate of about 0.32 m/ka. This rate compares favorably with rates derived for other parts of the California and northern Baja California coast that are dominated by what are thought to be strike-slip faults (Ku and Kern, 1974; Kern, 1977; Muhs and Szabo, 1982; Muhs, 1983, 1985; Rockwell and others, 1989; Hanson and others, 1990; Lettis and others, 1990; Muhs and others, 1990). On the upthrown side of the Cabrillo fault, as discussed earlier, the uplift rate is significantly higher, on the order of 0.56 m/ka (if calculated from the ~125-ka "fourth" terrace) to 0.64 m/ka (if calculated from the ~80-ka "second" terrace). On the downthrown side of the Cabrillo fault, the uplift rate based

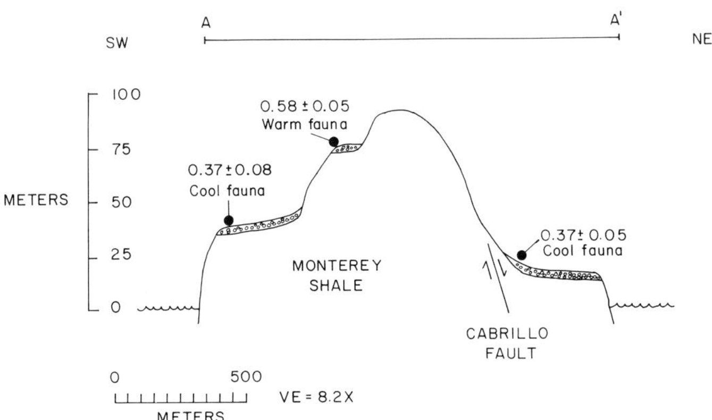

FIG. 10.—Geologic cross section across Point Fermin (location shown in Fig. 3) showing terraces, the Cabrillo fault, thermal aspects of terrace faunas, and alle/Ile ratios in *Tegula*.

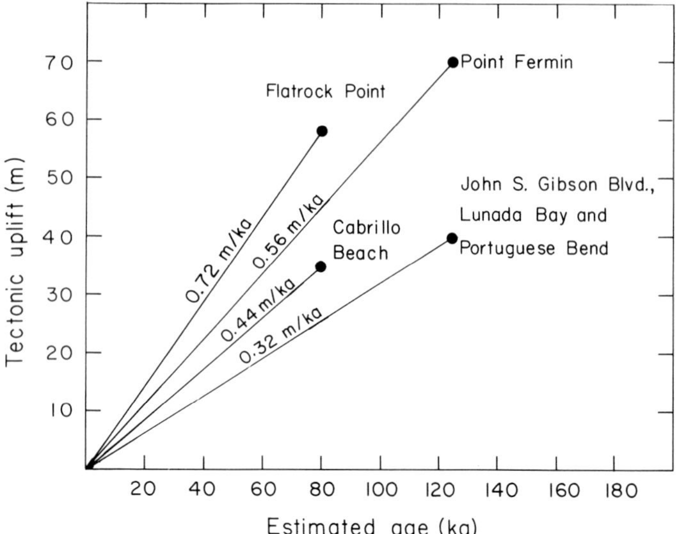

FIG. 11.—Amount of tectonic uplift of marine terraces at various localities around the Palos Verdes peninsula and San Pedro shown as a function of estimated terrace age. Slopes of lines are uplift rates. Rate given for Flatrock Point is a maximum value, based on an 80-ka-age estimate; rate given for the terrace at John S. Gibson Boulevard is a maximum value based on an estimated maximum shoreline-angle elevation of 46 m.

on the 80-ka terrace is lower, on the order of 0.44 m/ka. However, this rate is still higher than those for most other parts of the peninsula. It is difficult to calculate uplift rates based on the 125-ka terrace east of the Cabrillo fault because shoreline angles in this area are neither exposed in the field nor easily estimated from topographic profiles. At locality 12576, the wave-cut platform has an elevation of ~19 m, yielding a minimum uplift rate of around 0.10 m/ka. Based on interpretation of the terrace map by Woodring and others (1946), the maximum possible elevation for the shoreline angle of this terrace is on the order of 46 m, which yields an uplift rate of 0.32 m/ka, similar to the rates for the other parts of the peninsula.

The locality for which we have the greatest uncertainty in uplift rate is the "second" terrace at Flatrock Point, locality 104 (Fig. 2). The terrace platform here is close to the shoreline angle and has an elevation of about 53 m. AIle/Ile ratios in *Tegula* correlate the terrace with the 80-ka highstand (Fig. 7). If this correlation is correct, these data result in an uplift rate of 0.72 m/ka, which is greater than the uplift rates for most of the Palos Verdes Hills and northern San Pedro by more than a factor of two (Fig. 11). On the other hand, stable-isotope data in *Epilucina* indicate that this terrace could correlate with the ~125-ka or ~105-ka highstands which would yield uplift rates of 0.38 m/ka and 0.52 m/ka, respectively, assuming a paleosea level of −2 m at 105 ka (Rockwell and others, 1989). We cannot reject any of the uplift-rate estimates until we have an unambiguous age estimate for this terrace. As pointed out earlier, Riccio and Mills (1977) cited evidence of a previously unreported fault in this area. However, their data showed that the terrace at locality 104 would be on the downthrown side of this fault, which should not yield higher than average uplift rates. Further field investigations in the Flatrock Point area are warranted, both for terrace-age estimates and to evaluate the structural relations in this area.

The estimated late Quaternary uplift rates for the Palos Verdes Hills–San Pedro area, ranging from 0.32 to 0.64 m/ka (and possibly as high as 0.72 m/ka), represent approximate values for the rate of vertical movement of the Palos Verdes Hills fault. If marine deposits whose age and elevations were known existed on the northeast side of the fault, we could calculate a more precise rate of vertical movement, but it is not clear from the terrace map of Woodring and others (1946) where such deposits might be or how they might be correlated. Our uplift rates are in broad agreement with Holocene vertical rates of 0.1 to 0.4 m/ka estimated for offshore portions of the Palos Verdes Hills fault by Fischer and others (1987).

CONCLUSIONS

Studies of the terraces in the Palos Verdes Hills and San Pedro areas (and calibration areas such as San Nicolas Island) yield the following conclusions:

1. Amino-acid ratios in shells from dated terraces on San Nicolas Island show that the gastropod *Tegula* is an excellent discriminator for 80-ka and 125-ka deposits; in contrast, the bivalve *Epilucina* cannot distinguish these two ages of deposits.
2. Amino-acid data from *Tegula* and *Protothaca* and oxygen isotope data from *Epilucina* indicate that most segments of the lowest terrace in the Palos Verdes peninsula (mapped as the "second" terrace by Woodring and others, 1946) are correlative with the 125-ka highstand. However, at Point Fermin (and possibly at Flatrock Point), segments of the "second" terrace are correlative with the ~80-ka highstand.
3. Amino-acid data from *Tegula, Protothaca, Macoma*, and *Tivela* and oxygen isotope data for *Protothaca* indicate that deposits of the "first" terrace in San Pedro actually consist of sediments with two ages: an older (~125 ka) deposit that occurs in northern San Pedro and is characterized by a warm-water fauna, and a younger (~80 ka) deposit that occurs in southern San Pedro and is characterized by a cool-water fauna.
4. Oxygen isotope data from *Epilucina* and *Protothaca* indicate that on open, exposed parts of the Palos Verdes peninsula, ocean temperatures during the 125-ka highstand were cooler than those of the present, similar to what has been observed on other parts of the California coast that have open, exposed environments. Paleotemperature calculations indicate that water temperatures at 125 ka were 2.3 to 2.6°C cooler than the present, possibly as the result of more frequent or more intense upwelling when continental temperatures were actually warmer than present. In contrast, in the protected-embayment type of environment that has characterized the San Pedro area, ocean temperatures during the 125-ka highstand were probably at least as warm as present temperatures. However, even in the relatively protected environment of the San Pedro area, ocean temperatures

during the 80-ka highstand were significantly cooler (~2.1°C) than at present.

5. Correlation of the lowest terraces around Point Fermin indicates that the Cabrillo fault has experienced about 16 m of vertical movement in the last 80 ka, for a vertical-slip rate of 0.20 m/ka. Late Quaternary vertical-movement rates for the Cabrillo fault determined in this study are significantly lower than Holocene rates of vertical movement estimated by other workers for offshore portions of this fault.
6. Uplift rates around the rest of the peninsula vary from 0.32 m/ka to possibly as high as 0.72 m/ka. For most parts of the Palos Verdes peninsula, late Quaternary uplift rates estimated in this study are comparable to Holocene rates of vertical movement estimated by other workers for offshore portions of the Palos Verdes Hills fault.

ACKNOWLEDGMENTS

Most of this work was funded by the Global Change and Climate History Program of the U.S. Geological Survey. Initial work was funded by the Wisconsin Alumni Research Foundation while Muhs was at the University of Wisconsin-Madison. T. R. Rowland, J. N. Rosholt, R. R. Shroba, and R. B. Vaughn assisted with fieldwork at various times, and P. B. Maat prepared some of the samples for oxygen isotope analysis. J. F. Wehmiller kindly provided numerous unpublished gas-chromatographic amino-acid data and helpful comments on their interpretation. The authors appreciate helpful discussions on the geology and aminostratigraphy of the area with D. J. Ponti. Katherine L. Hanson, R. M. Mitterer, A. R. Nelson, B. J. Szabo, M. E. Bryant, and A. L. Bloom provided helpful comments on earlier drafts of this paper.

REFERENCES

AHARON, P., 1983, 140,000 year isotope climate record from raised coral reefs in New Guinea: Nature, v. 304, p. 720–723.

ARNOLD, R., 1903, The paleontology and stratigraphy of the marine Pliocene and Pleistocene of San Pedro, California: Memoirs of the California Academy of Science, v. 3, 420 p.

BAKUN, A., 1990, Global climate change and intensification of coastal ocean upwelling: Science, v. 247, p. 198–201.

BARD, E., HAMELIN, B., AND FAIRBANKS, R. G., 1990, U-Th ages obtained by mass spectrometry in corals from Barbados: sea level during the past 130,000 years: Nature, v. 346, p. 456–458.

BIRKELAND, P. W., 1972, Late Quaternary eustatic sea-level changes along the Malibu coast, Los Angeles County, California: Journal of Geology, v. 80, p. 432–448.

BLOOM, A. L., BROECKER, W. S., CHAPPELL, J. M. A., MATTHEWS, R. K., AND MESOLELLA, K. J., 1974, Quaternary sea level fluctuations on a tectonic coast: new $^{230}Th/^{234}U$ dates from the Huon Peninsula, New Guinea: Quaternary Research, v. 4, p. 185–205.

BRADLEY, W. C., AND GRIGGS, G. B., 1976, Form, genesis, and deformation of central California wave-cut platforms: Geological Society of America Bulletin, v. 87, p. 433–449.

BRYANT, M. E., 1987, Emergent marine terraces and Quaternary tectonics, Palos Verdes Peninsula, California, *in* FISCHER, P. J., and MESA[2] INC., eds., Geology of the Palos Verdes Peninsula and San Pedro Bay: Pacific Section, Society of Economic Paleontologists and Mineralogists and American Association of Petroleum Geologists, Los Angeles, p. 63–78.

CHACE, E. P., AND CHACE, E. M., 1919, An unreported exposure of the San Pedro Pleistocene: Lorquinia, v. 2, p. 1–3.

CHAPPELL, J., AND SHACKLETON, N. J., 1986, Oxygen isotopes and sea level: Nature, v. 324, p. 137–140.

DODGE, R. E., FAIRBANKS, R. G., BENNINGER, L. K., AND MAURRASSE, F., 1983, Pleistocene sea levels from raised coral reefs of Haiti: Science, v. 219, p. 1423–1425.

EDWARDS, R. L., CHEN, J. H., KU, T.-L., AND WASSERBURG, G. J., 1987, Precise timing of the last interglacial period from mass spectrometric determination of thorium-230 in corals: Science, v. 236, p. 1547–1553.

FAIRBANKS, R. G., AND MATTHEWS, R. K., 1978, The marine oxygen isotope record in Pleistocene coral, Barbados, West Indies: Quaternary Research, v. 10, p. 181–196.

FANALE, F. P., AND SCHAEFFER, O. A., 1965, Helium-uranium ratios for Pleistocene and Tertiary fossil aragonites: Science, v. 149, p. 312–317.

FISCHER, P. J., MESA[2] INC., PATTERSON, R. H., DARROW, A. C., RUDAT, J. H., AND SIMILA, G., 1987, The Palos Verdes fault zone: onshore to offshore, *in* Fischer, P. J., and MESA[2] Inc., eds., Geology of the Palos Verdes Peninsula and San Pedro Bay: Pacific Section, Society of Economic Paleontologists and Mineralogists and American Association of Petroleum Geologists, Los Angeles, p. 91–133.

HANSON, K. L., WESLING, J. R., LETTIS, W. R., KELSON, K., AND MEZGER, L., 1990, Correlation, ages, and uplift rates of Quaternary marine terraces: south-central coastal California, *in* Lettis, W. R., Hanson, K. L., Kelson, K. I., and Wesling, J. R., eds., Neotectonics of South-Central Coastal California: Friends of the Pleistocene, Pacific Cell, Field Trip Guidebook, p. 139–190.

HARMON, R. S., MITTERER, R. M., KRIAUSAKUL, N., LAND, L. S., SCHWARCZ, H. P., GARRETT, P., LARSON, G. J., VACHER, H. L., AND ROWE, M., 1983, U-series and amino-acid racemization geochronology of Bermuda: implications for eustatic sea-level fluctuation over the past 250,000 years: Palaeogeography, Palaeoclimatology, Palaeoecology, v. 44, p. 41–70.

JAMES, N. P., MOUNTJOY, E. W., AND OMURA, A., 1971, An early Wisconsin reef terrace at Barbados, West Indies, and its climatic implications: Geological Society of America Bulletin, v. 82, p. 2011–2018.

JUNGER, A., AND WAGNER, H. C., 1977, Geology of the Santa Monica and San Pedro Basins, California continental borderland: U.S. Geological Survey Miscellaneous Field Studies Map MF-820.

KAUFMAN, A., BROECKER, W. S., KU, T.-L., AND THURBER, D. L., 1971, The status of U-series methods of mollusk dating: Geochimica et Cosmochimica Acta, v. 35, p. 1155–1183.

KELSEY, H. M., 1990, Late Quaternary deformation of marine terraces on the Cascadia subduction zone near Cape Blanco, Oregon: Tectonics, v. 9, p. 983–1014.

KENNEDY, G. L., 1988, Zoogeographic discordancy in late Pleistocene northeastern Pacific marine invertebrate distributions explained by astronomical theory of climatic change: Geological Society of America Abstracts with Programs, v. 20, p. A207–A208.

KENNEDY, G. L., LAJOIE, K. R., AND WEHMILLER, J. F., 1982, Aminostratigraphy and faunal correlations of late Quaternary marine terraces, Pacific Coast, USA: Nature, v. 299, p. 545–547.

KENNEDY, G. L., CLARK, D. G., AND WEHMILLER, J. F., 1990, The southern Oregonian province in the late Pleistocene: changing paleoclimates and sea level history based on integrated faunal and aminostratigraphic studies, Casmalia Hills, coastal California: Geological Society of America Abstracts with Programs, v. 22, p. A147.

KERN, J. P., 1977, Origin and history of upper Pleistocene marine terraces, San Diego, California: Geological Society of America Bulletin, v. 88, p. 1553–1566.

KU, T.-L., AND KERN, J. P., 1974, Uranium-series age of the upper Pleistocene Nestor terrace, San Diego, California: Geological Society of America Bulletin, v. 85, p. 1713–1716.

KU, T.-L., IVANOVICH, M., AND LUO, S., 1990, U-series dating of last interglacial high sea stands: Barbados revisited: Quaternary Research, v. 33, p. 129–147.

KU, T.-L., KIMMEL, M. A., EASTON, W. H., AND O'NEIL, T. J., 1974, Eustatic sea level 120,000 years ago on Oahu, Hawaii: Science, v. 183, p. 959–962.

LAJOIE, K. R., SARNA-WOJCICKI, A. M., AND OTA, Y., 1982, Emergent Holocene marine terraces at Ventura and Cape Mendocino, California—indicators of high tectonic uplift rates: Geological Society of America Abstracts with Programs, v. 14, p. 178.

LAJOIE, K. R., KERN, J. P., WEHMILLER, J. F., KENNEDY, G. L., MATHIESON, S. A., SARNA-WOJCICKI, A. M., YERKES, R. F., AND MCCRORY, P. A., 1979, Quaternary marine shorelines and crustal deformation,

San Diego to Santa Barbara, California, *in* Abbott, P. L., ed., Geological Excursions in the Southern California Area: Department of Geological Sciences, San Diego State University, p. 3–15.

LETTIS, W. R., KELSON, K. I., WESLING, J. R., ANGELL, M., HANSON, K. L., AND HALL, N. T., 1990, Quaternary deformation of the San Luis Range, San Luis Obispo County, California, *in* Lettis, W. R., Hanson, K. L., Kelson, K. I., and Wesling, J. R., eds., Neotectonics of South-Central Coastal California: Friends of the Pleistocene, Pacific Cell, Field Trip Guidebook, p. 259–290.

LYNN, R. J., 1967, Seasonal variation of temperature and salinity at 10 meters in the California Current: California Cooperative Oceanic Fisheries Investigations Reports, v. 11, p. 157–186.

MCINELLY, G. W., AND KELSEY, H. M., 1990, Late Quaternary tectonic deformation in the Cape Arago–Bandon region of coastal Oregon as deduced from wave-cut platforms: Journal of Geophysical Research, v. 95, p. 6699–6713.

MERRITTS, D., AND BULL, W. B., 1989, Interpreting Quaternary uplift rates at the Mendocino triple junction, northern California, from uplifted marine terraces: Geology, v. 17, p. 1020–1024.

MESOLELLA, K. J., MATTHEWS, R. K., BROECKER, W. S., AND THURBER, D. L., 1969, The astronomical theory of climatic change: Barbados data: Journal of Geology, v. 77, p. 250–274.

MILLER, G. H., 1985, Aminostratigraphy of Baffin Island shell-bearing deposits, *in* Andrews, J. T., ed., Quaternary Environments: Baffin Island, Baffin Bay, and West Greenland: Allen and Unwin, London, p. 394–427.

MILLER, G. H., AND BRIGHAM-GRETTE, J., 1989, Amino acid geochronology: resolution and precision in carbonate fossils: Quaternary International, v. 1, p. 111–128.

MILLER, G. H., AND HARE, P. E., 1980, Amino acid geochronology: integrity of the carbonate matrix and potential of molluscan fossils, *in* Hare, P. E., Hoering, T. C., and King, K., Jr., eds., Biogeochemistry of Amino Acids: Wiley, New York, p. 415–443.

MITTERER, R. M., AND HARE, P. E., 1967, Diagenesis of amino acids in fossil shells as a potential geochronometer: Geological Society of America Abstracts with Programs, v. 115, p. 152–153.

MUHS, D. R., 1983, Quaternary sea level events on northern San Clemente Island, California: Quaternary Research, v. 20, p. 322–341.

MUHS, D. R., 1984, Aminostratigraphy and kinetic model ages of marine terraces on the Palos Verdes Peninsula, California: American Quaternary Association Eighth Biennial Meeting, Program and Abstracts, p. 89.

MUHS, D. R., 1985, Amino acid age estimates of marine terraces and sea levels on San Nicolas Island, California: Geology, v. 13, p. 58–61.

MUHS, D. R., 1991, Amino acid geochronology of fossil mollusks, *in* Morrison, R. B., ed., Quaternary Non-glacial Geology; Conterminous U.S.: Geological Society of America, Decade of North American Geology, v. K-2, p. 65–68.

MUHS, D. R., 1992, The last interglacial-glacial transition in North America: evidence from uranium-series dating of coastal deposits, *in* Clark, P. U., and Lea, P. D., eds., The Last Interglacial-Glacial Transition in North America: Geological Society of America Special Paper, v. 270, p. 31–52.

MUHS, D. R., AND KYSER, T. K., 1987, Stable isotope compositions of fossil mollusks from southern California: evidence for a cool last interglacial ocean: Geology, v. 15, p. 119–122.

MUHS, D. R., AND SZABO, B. J., 1982, Uranium-series age of the Eel Point terrace, San Clemente Island, California: Geology, v. 10, p. 23–26.

MUHS, D. R., ROCKWELL, T. K., AND KENNEDY, G. L., 1992, Late Quaternary uplift rates of marine terraces on the Pacific Coast of North America, southern Oregon to Baja California Sur: Quaternary International, in press.

MUHS, D. R., ROSHOLT, J. N., AND BUSH, C. A., 1989, The uranium-trend dating method: principles and application for southern California marine terrace deposits: Quaternary International, v. 1, p. 19–34.

MUHS, D. R., KELSEY, H. M., MILLER, G. H., KENNEDY, G. L., WHELAN, J. F., AND MCINELLY, G. W., 1990, Age estimates and uplift rates for late Quaternary marine terraces: southern Oregon portion of the Cascadia forearc: Journal of Geophysical Research, v. 95, p. 6685–6698.

NARDIN, T. R., AND HENYEY, T. L., 1978, Pliocene-Pleistocene diastrophism of Santa Monica and San Pedro shelves, California continental borderland: American Association of Petroleum Geologists Bulletin, v. 62, p. 247–272.

NELSON, A. R., 1982, Aminostratigraphy of Quaternary marine and glaciomarine sediments, Qivitu Peninsula, Baffin Island: Canadian Journal of Earth Sciences, v. 19, p. 945–961.

NEUMANN, A. C., AND MOORE, W. S., 1975, Sea level events and Pleistocene coral ages in the northern Bahamas: Quaternary Research, v. 5, p. 215–224.

OBRADOVICH, J. D., 1968, The potential use of glauconite for late-Cenozoic geochronology, *in* Morrison, R. B., and Wright, H. E., Jr., eds., Means of Correlation of Quaternary Successions: University of Utah Press, Salt Lake City, p. 267–279.

PONTI, D. J., 1989, Aminostratigraphy and chronostratigraphy of Pleistocene marine sediments, southwestern Los Angeles Basin, California: Unpublished Ph.D. Dissertation, University of Colorado, Boulder, 409 p.

RICCIO, J. F., AND MILLS, M. F., 1977, Faulted upper Pleistocene marine terrace, Palos Verdes Hills, California: American Association of Petroleum Geologists Bulletin, v. 61, p. 2001–2004.

ROCKWELL, T. K., MUHS, D. R., KENNEDY, G. L., HATCH, M. E., WILSON, S. H., AND KLINGER, R. E., 1989, Uranium-series ages, faunal correlations and tectonic deformation of marine terraces within the Agua Blanca fault zone at Punta Banda, northern Baja California, Mexico, *in* Abbott, P. L., ed., Geologic Studies in Baja California: Pacific Section, Society of Economic Paleontologists and Mineralogists, Los Angeles, p. 1–16.

SZABO, B. J., AND ROSHOLT, J. N., 1969, Uranium-series dating of Pleistocene molluscan shells from southern California—an open system model: Journal of Geophysical Research, v. 74, p. 3253–3260.

SZABO, B. J., AND VEDDER, J. G., 1971, Uranium-series dating of some Pleistocene marine deposits in southern California: Earth and Planetary Science Letters, v. 11, p. 283–290.

VALENTINE, J. W., 1961, Paleoecologic molluscan geography of the Californian Pleistocene: University of California Publications in the Geological Sciences, v. 34, p. 309–442.

VALENTINE, J. W., 1962, Pleistocene molluscan notes, 4. Older terrace faunas from Palos Verdes Hills, California: Journal of Geology, v. 70, p. 92–101.

VALENTINE, J. W., AND MEADE, R. F., 1961, Californian Pleistocene paleotemperatures: University of California Publications in the Geological Sciences, v. 40, p. 1–45.

VEDDER, J. G., AND NORRIS, R. M., 1963, Geology of San Nicolas Island, California: U.S. Geological Survey Professional Paper 369, 65 p.

VEEH, H. H., AND CHAPPELL, J., 1970, Astronomical theory of climatic change: Support from New Guinea: Science, v. 166, p. 862–865.

WEHMILLER, J. F., 1982, A review of amino acid racemization studies in Quaternary mollusks: stratigraphic and chronologic applications in coastal and interglacial sites, Pacific and Atlantic Coasts, United States, United Kingdom, Baffin Island, and tropical islands: Quaternary Science Reviews, v. 1, p. 83–120.

WEHMILLER, J. F., 1990, Amino acid racemization: applications in chemical taxonomy and chronostratigraphy of Quaternary fossils, *in* Carter, J. B., ed., Skeletal Biomineralization: Patterns, Processes, and Evolutionary Trends: Van Nostrand Reinhold, New York, p. 583–608.

WEHMILLER, J. F., LAJOIE, K. R., KVENVOLDEN, K. A., PETERSON, E., BELKNAP, D. F., KENNEDY, G. L., ADDICOTT, W. O., VEDDER, J. G., AND WRIGHT, R. W., 1977, Correlation and chronology of Pacific Coast marine terrace deposits of continental United States by fossil amino acid stereochemistry—technique evaluation, relative ages, kinetic model ages, and geologic implications: U.S. Geological Survey Open-File Report 77–680, 196 p.

WOODRING, W. P., 1935, Fossils from the marine Pleistocene terraces of the San Pedro Hills, California: American Journal of Science, v. 29, p. 292–305.

WOODRING, W. P., 1957, Marine Pleistocene of California: Geological Society of America Memoir 67, p. 589–597.

WOODRING, W. P., BRAMLETTE, M. N., AND KEW, W. S. W., 1946, Geology and paleontology of Palos Verdes Hills, California: U.S. Geological Survey Professional Paper 207, 145 p.

YERKES, R. F., MCCULLOH, T. H., SCHOELLHAMER, J. E., AND VEDDER, J. G., 1965, Geology of the Los Angeles Basin, California—an introduction: U.S. Geological Survey Professional Paper 420-A, 57 p.

CHRONOLOGY AND DEFORMATION OF QUATERNARY MARINE SHORELINES, SAN DIEGO COUNTY, CALIFORNIA

J. PHILIP KERN AND THOMAS K. ROCKWELL

Department of Geological Sciences, San Diego State University, San Diego, California 92182

ABSTRACT: A series of 16 marine terraces and associated landforms and sedimentary deposits provides a record of Quaternary paleogeographic and tectonic history of coastal San Diego County, California. The width of individual terrace platforms varies dramatically, from several kilometers in places where the platforms are cut into poorly lithified sedimentary strata, to a few meters or tens of meters where the platforms are cut in more resistant rocks. Beach ridges formed of dune sands lie just landward of most of the terrace shorelines. Radiometric ages and amino-acid racemization ratios, together with shoreline elevations, have been used to calculate average rate of uplift and to construct a chronology of the 16 shorelines, which range in age from 80 ka to perhaps as old as 1.29 ma. Shoreline-angle elevations suggest that nearly the entire length of coastal San Diego County has been uplifted at a rate of 0.13 to 0.14 m/ka during the Quaternary. Both higher and lower rates are recorded in areas deformed by the Rose Canyon fault zone. Changes in configuration of successive shorelines indicate that rocks on one side of the fault have been rising through middle and late Quaternary time and that the Rose Canyon fault has been active for at least the past 1 my.

INTRODUCTION

The elevations and geometries of 16 raised marine shorelines, together with evidence from associated sediments, fossils, and faults, have been used to reconstruct the Quaternary geomorphic and tectonic history of coastal San Diego County, California. Both marine and stream terraces are present, and a variety of marine and non-marine sediments occupy the terrace-abrasion platforms and adjacent sedimentary basins. Shoreline geometry has been modified both by regional uplift and by extensive faulting in the right-slip wrench system of the Rose Canyon fault zone. This paper is based on mapping of Quaternary landforms, sediments, and faults in parts of 13 7.5-minute quadrangles that cover the coastal zone from downtown San Diego to north of Oceanside (Fig. 1). The terraces are named for prominent geomorphic or cultural features; these names, shown on Figure 1, are used throughout the paper.

The sediment-masked topographic forms of some of the higher abrasion platforms, which are exceptionally broad and separated by comparatively low sea cliffs, are further obscured by faulting and folding and by destruction or burial of already scant exposures by urban development. As a result, only three terraces in the study area had been described (Hanna, 1926; Hertlein and Grant, 1944; Kennedy, 1975; Kennedy and Peterson, 1975; Kern, 1977) before the work of McCrory and Lajoie (1979). The present study addresses 13 additional terraces discovered in large part through detailed mapping and measurement of elevations of terrace shorelines and abrasion platforms throughout the coastal zone.

Origin or deformation of San Diego area terraces is the subject of studies by Ellis (1919), Palmer (1967), Peterson (1970), Kern (1973a, 1973b, 1977), Ku and Kern (1974), Moore and Kennedy (1975), Lajoie and others (1979), and McCrory and Lajoie (1979). Among the many general works that include substantial reference to terraces in maps or text, the principal ones are those by Hanna (1926), Hertlein and Grant (1944), McEuen and Pinckney (1972), Ziony and Buchanan (1972), Artim and Pinckney (1973), Kennedy (1975), Kennedy and Peterson (1975), and Kennedy and others (1975). Radiometric and amino-acid racemization ages are given in papers by Ku and Kern (1974), Wehmiller and others (1977), Masters and Bada (1977, 1979), Karrow and Bada (1980), and Muhs and others (1988).

FIELD METHODS

Except for precise total-station surveying of several shoreline angles at Point Loma and Torrey Pines, elevations of terrace shorelines and abrasion platforms were determined by leveling with Brunton compass or hand level from bench marks, street intersections, buildings, ridge tops, and other points shown on detailed field maps printed by San Diego City and County. The maps (scale 1:2,400, 200 ft = 1 in.) have a contour interval of 5 ft (1.5 m) and include many spot elevations, especially at street intersections. Their dates of publication are varied but most of the maps used in this study were no more than one to two years old at the time of mapping, which began in 1975 and is ongoing. Measured elevations are judged in most cases to be in error by no more than 1 to 2 m. The relative accuracy in measurement of shoreline-angle elevations not only allows for ready recognition of the separate terraces, but also provides a sound basis for detection and measurement of faults and folds affecting shorelines and abrasion platforms.

Modern shorelines in the San Diego area are within 1 to 2 m of mean sea level. Elevations of emergent shorelines provide measures of sea-level change, tectonic movement, or both. If the original heights are known relative to present sea level, raised shorelines directly record vertical tectonic motions. Unfortunately, the shorelines themselves are not commonly exposed, but abrasion platforms exposed near paleosea cliffs give a minimum elevation generally close to that of corresponding shoreline angles. Similarly, the highest elevation of any body of marine strata, whether on a terrace or not, gives a minimum elevation for sea level at the time of deposition.

SHORELINE GEOMORPHIC FEATURES

Previous lack of recognition of most of the terraces in San Diego County has resulted primarily from the absence of well-developed topographic forms, which are camouflaged by overlying strata of highly variable and commonly substantial thickness. This effect is greatest for the higher,

Quaternary Coasts of the United States: Marine and Lacustrine Systems, SEPM Special Publication No. 48

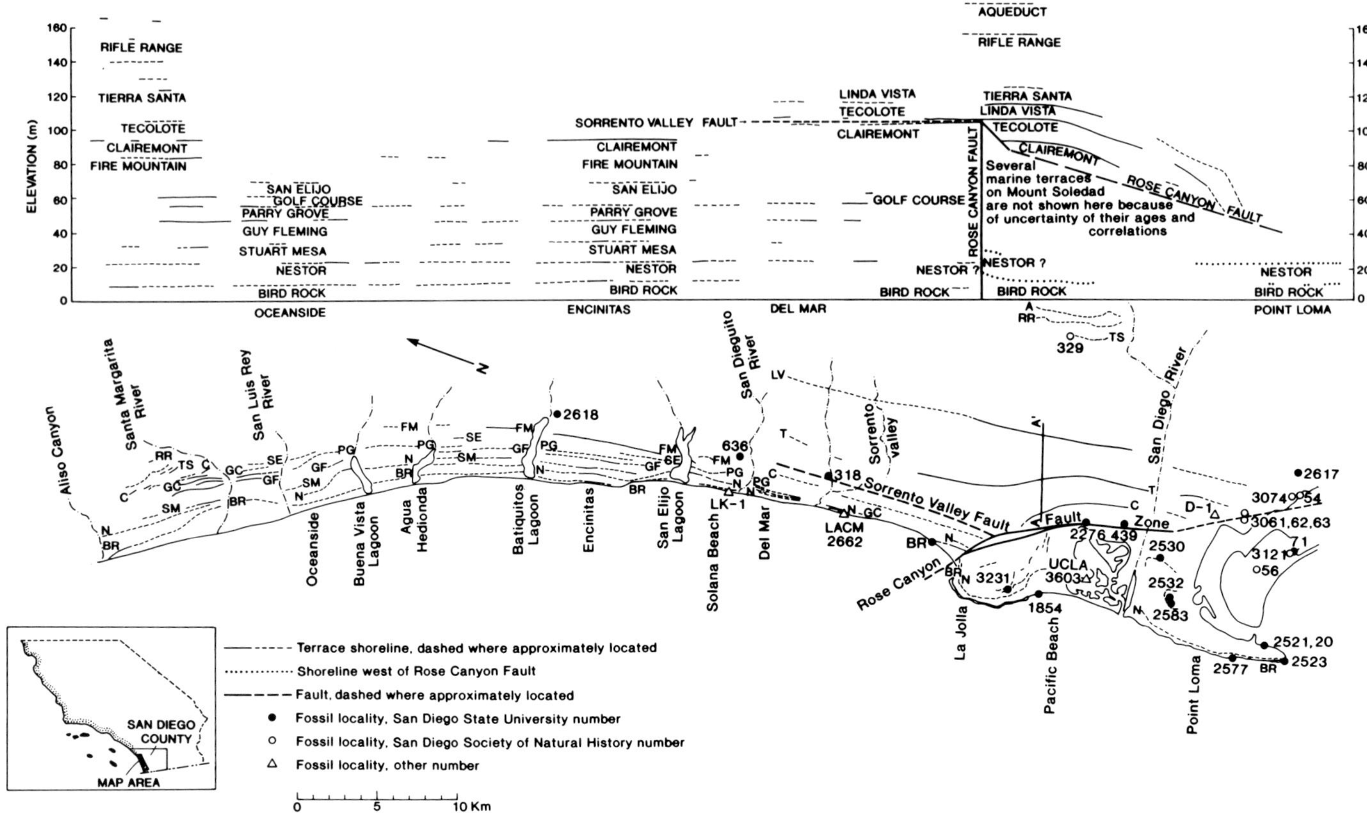

FIG. 1.—Map and profile showing terrace shorelines, major faults, and fossil localities, San Diego to Camp Pendleton. UCLA—University of California, Los Angeles; LACM—Los Angeles County Museum; LK-1—Wehmiller sample.

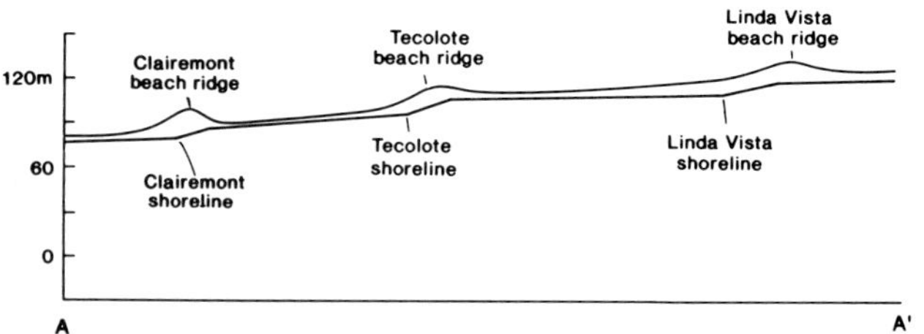

FIG. 2.—Cross section A–A′, showing positions of beach ridges relative to terrace shorelines.

wider platforms, but even many of the narrow near-coastal benches are not clearly revealed in the topography.

General Platform Morphology

The marine-abrasion platforms in this study range in width from a few meters to more than 2 km (Fig. 1). In nearly all cases, a strong correlation exists between terrace width and the relative erodibility of the bedrock in which the terrace is cut. For instance, the Clairemont, Tecolote, Linda Vista, and Tierra Santa terraces combine to form a 12-km-wide plain where they are cut into easily erodible Eocene (Friars, Stadium, Mission Valley, and Scripps Formations) and Pliocene (San Diego Formation) strata at San Diego. The same terraces narrow to only a few tens of meters or less where they are cut into the much more resistant San Onofre Breccia north of Oceanside. Similarly, the Nestor and Bird Rock terraces are widest where cut into the easily erodible Eocene strata and are narrow or absent where cut into resistant Cretaceous (Point Loma and Cabrillo Formations) strata and the Eocene Torrey Sandstone. Approximate seaward dips on the kilometers-wide platforms, where not affected by faults, range between 0.15° and 0.35°.

Beach Ridges

Prominent beach ridges occupy the higher terrace-abrasion platforms landward of the Linda Vista, Tecolote, and Clairemont terrace shorelines. Lower ridges occur on the Tierra Santa platform, which has at least 15, and the Clairemont, which has 2 or more. In many areas the landforms, especially the smaller ones, have been destroyed by urban development and can be seen now only on older topographic maps or aerial photographs.

The beach ridges are closely parallel to the underlying sea cliffs and shorelines, generally just landward of them but in some places slightly seaward. Though the beach ridges on the higher, broader terraces rise prominently above the thinner terrace deposits on the platforms to either side, they are subtle landforms with heights no greater than 15 m and widths up to 1 km.

The beach ridges were thought to have originated as lines of eolian dunes behind and parallel to the shore (Emery,

1950; Peterson, 1970), but in fact they include a variety of marine sedimentary facies. The base of the ridge commonly has imbricated flat-pebble conglomerates interbedded with coarse- to medium-grained sandstones with heavy-mineral laminae. The beach deposits commonly are cross-bedded and channeled and may dip gently seaward. Some ridges contain siltstones or mudstones near their base on the landward side; the sediments probably were deposited in lagoons that formed behind barrier beaches on a coast of very low relief. Higher parts of the ridges commonly consist of clean, well-sorted, fine- to medium-grained sand, in some places in meters-thick sets of apparently eolian cross-strata and in other places in thin planar laminations.

The consistent position of the major beach ridges over the paleosea cliffs indicates that they formed at the time of cutting of the lower terrace of the pair. For instance, the sand ridge that occurs over the paleosea cliff of the Bird Rock terrace and at the seaward edge of the Nestor terrace dates to the time of formation of the Bird Rock terrace. Similarly, the beach ridge at the seaward edge of the Tecolote terrace formed at the same time as the Clairemont terrace.

TERRACE CHRONOLOGY

Marine invertebrate fossils have been collected from 34 localities in sediments associated with at least five of San Diego County's 16 shorelines. The localities have yielded a wide variety of fossils, including coral from both the Bird Rock and Nestor terraces. U-series analysis of the coral (Ku and Kern, 1974; Muhs and others, 1988) yielded ages of about 120 ka (oxygen isotope substage 5e of Shackleton and Opdyke, 1973) for the Nestor terrace at Point Loma and 80 ka (substage 5a) for the Bird Rock terrace, also at Point Loma. The radiometric ages are used as a calibration for the 94 amino-acid racemization (AAR) ratios determined on mollusks from the 34 fossiliferous localities (Table 1).

Fossils from locality LK-1, which geomorphically seems to be on the Bird Rock terrace, appear to have been thermally affected based on much higher than expected ratios. The AAR data from the Bird Rock and Nestor terraces are internally consistent. Clear differences in racemization rate exist depending on the species analyzed (Tables 1 and 2), but the data nevertheless provide a reasonable basis on which to assign ages to other localities that are not associated with a specific platform, such as bay deposits.

The data suggest that some localities may contain mixed populations or ages of taxa. The downtown localities, which are not on a terrace but in bay deposits, appear to range from about 200 ka to perhaps 800 ka, with the differences due to recovery of mollusks from at least three different stratigraphic units. The AAR ratios from the Coronado Island localities suggest that most of these date to about 200 ka, or stage 7, although sample SDSNH-56 may be as young as 120 ka.

To assign ages to the abrasion platforms older than the Nestor terrace, two methods were employed. First, terrace elevations are predicted (Table 3) for known sea-level highstands that have been dated elsewhere (Bloom and others, 1974; Bender and others, 1979; Chappel, 1983). The stage 7 and 9 terraces are assumed to have been cut at about +2 m relative to modern sea level (Harmon and others, 1983). The uplift rate of .013 to .014 m/ka, which is determined from the Nestor terrace, is assumed to have been reasonably constant throughout the late Quaternary. If uplift has not been constant, our age estimates for the older terraces

TABLE 1.—MEAN AMINO-ACID RATIOS FOR FOSSIL MOLLUSKS COLLECTED FROM THE MARINE-SAND DEPOSITS

Locality[1]	Genus[2]	Allo/Iso[3]	D/L Leucine[4]	D/L Valine[4]
Bird Rock Localities (80 ka)				
SDSU-2523	P		0.38±0.01 (2)	0.25±0.02 (2)
SDSU-2618	C	0.30		
SDSU-2618	C		0.38±0.04 (2)	0.31±0.07 (2)
SDSU-318	C	0.36	0.36±0.02 (4)	0.27±0.01 (4)
Scripps Pier	My	0.64		
LK-1[5]	P		0.60±0.01 (2)	0.42±0.01 (2)
LK-1[5]	C	0.55		
LK-1[5]	S		0.64±0.02 (3)	0.45±0.01 (3)
LK-1[5]	T		0.60 (1)	0.42 (1)
Nestor Localities (120 ka)				
SDSU-2577	C	0.36		
	P		0.46±0.02 (7)	0.30±0.02 (7)
SDSU-2520	C		0.44 (1)	0.29 (1)
SDSU-2521	P		0.44 (1)	0.30 (1)
LACM 2662	C	0.34		
LACM 2662	T	0.40	0.44 (1)	0.35 (1)
LACM 2662	My	0.78		
LACM 2662	S		0.47±0.07 (3)	0.31±0.05 (3)
LACM 2662	P		0.47±0.02 (3)	0.35±0.03 (3)
SDSU-1854	C	0.39	0.55 (1)	0.39 (1)
	P		0.51±0.05 (4)	0.37±0.04 (3)
Higher Terraces				
SDSU-3231	C		0.62 (1)	0.55 (1)
Tecolote	C		0.79 (1)	0.78 (1)
SDSU-329	T	1.2	0.99±0.03 (2)	0.99±0.02 (2)
	C		0.93 (1)	0.96 (1)
Coronado Island				
SDSNH-56	T	0.42	0.52±0.04 (2)	0.35±0.01 (2)
SDSNH-3121	T		0.57±0.01 (2)	0.35±0.0 (2)
SDSU-71	T		0.56	0.40
Downtown				
SDSNH-3061	C	0.57	0.69±0.03 (2)	0.48±0.01 (2)
SDSNH-3062	C	0.62	0.66 (1)	0.50 (1)
SDSNH-3063		0.68		
SDSNH-3064	C		0.82±0.03 (2)	0.67±0.01 (2)
SDSNH-3074	C		0.55±0.01 (2)	0.37±0.01 (2)
SDSNH-54	C		0.60 (1)	0.42 (1)
SDSNH-54	O		0.63 (1)	0.60 (1)
SDSU-2617	C		0.84±0.02 (2)	0.82±0.03 (2)
Other Localities				
UCLA-3603	T	0.43	0.59±0.03 (3)	0.38±0.0 (2)
Del Mar NE	C	0.58		
SDSU-2583	O		0.89	0.85
SDSU-2532	O		0.83	0.83
Border Loc.	P		0.48	0.38
Border Loc.	T	0.48		
Border Loc.	C	0.50		
SDSU-636	C	0.59	0.60±0.00 (2)	0.45±0.01 (2)
SDSU-439	C	0.51	0.63±0.02 (2)	0.49±0.01 (2)
SDSU-439	P	0.58		
SDSU-2276	C	0.68±0.02	0.65 (1)	0.52 (1)
SDSU-2530	C		0.76±0.02 (2)	0.69±0.01 (2)
SDSU-2530	O		0.81	0.73
SDSNH-3133	C		0.82±0.01	0.68±0.06 (2)

[1]For localities, refer to Figure 1.

[2]P = *Protothaca*, S = *Saxidomus*, M = *Macoma*, My = *Mytilus*, C = *Chione*, T = *Tivela*, O = *Ostrea*.

[3]Ratio pf D-alloisoleucine to L-isoleucine; from Karrow and Bada (1980) and Masters and Bada (1977).

[4]Analyzed by J. F. Wehmiller; numbers in parentheses indicate number of individuals analyzed (see Wehmiller, this volume).

[5]Samples may have been thermally affected.

TABLE 2.—SUMMARY OF AMINO-ACID RACEMIZATION RESULTS FROM MARINE-ABRASION PLATFORMS IN THE STUDY AREA

Terrace	Age (ka)	Genus of Mollusk Analyzed[1]				
		P	C	T	My	S
Bird Rock	80	0.38	0.37	—	—	—
		0.25	0.28	—	—	—
		—	0.33	—	0.64	—
Nestor	120	0.48	0.49	0.44	—	0.47
		0.33	0.34	0.35	—	0.31
		—	0.36	0.40	0.78	—
Golf Course?	450[2]	—	0.62	—	—	—
		—	0.55	—	—	—
		—	—	—	—	—
Tecolote	800	—	0.79	—	—	—
		—	0.78	—	—	—
		—	—	—	—	—
Tierra Santa	930	—	—	0.99	—	0.93
		—	—	0.99	—	0.96
		—	—	1.20	—	—

[1]Results from several localities are summarized with only the mean values listed here. The first (upper) ratio of each triplet = leucine, the second = valine, the third (lowest) = allo/isoleucine. Refer to Table 1 for all ratios and their errors.
[2]Samples were poorly preserved and may have been substantially leached.

TABLE 4.—PREDICTED AGES OF HIGHER TERRACES USING THE UPLIFT RATE (0.133 m/ka) DETERMINED FOR THE NESTOR TERRACE

Terrace	Average Elevation (m)[1]	Predicted Age (ka)[2]
Stuart Mesa	32	225
Guy Fleming	46	345
Parry Grove	55	413
Golf Course	60	450
San Elijo	68	510
Fire Mountain	84	630
Clairemont	93	698
Tecolote	106	795
Linda Vista	114	855
Tierra Santa	124	930
Unnamed	130	975
Unnamed	140	1050
Rifle Range	155	1160
Aqueduct	172	1290

[1]Elevations are for areas in North County away from active faults. All higher (stage 9 and older) terraces are assumed to have been cut at 0 m relative to modern sea level.
[2]Given an assumed paleosea-level error of ±10 m, all ages have errors of ±75 ka

TABLE 3.—PREDICTED AND ACTUAL ELEVATIONS OF TERRACES FOR THE SAN DIEGO AREA

Terrace	Oxygen Isotope Stage	Inferred Age (ka)[1]	Predicted Terrace Elevation (m)[2]	Actual Terrace Elevation (m)	Discrepancy (m)
Bird Rock	5a	80	9	9–11	0–2
Nestor	5e	120	baseline	22–23	0
Stuart Mesa	7	200–230	29–33	32–34	0–2
Guy Fleming	9	320–340	43–48	46–48	0–3

[1]Ages of the stage 7 and 9 terraces from Chappell (1983).
[2]The 120-ka Nestor terrace and 80-ka Bird Rock terrace are assumed to have been cut at +6 m and −2 m, respectively, relative to modern sea level (Bloom and Yonekura, 1985; Muhs and others, 1988). The stage 7 terrace is assumed to have been cut at +2 m (Harmon and others, 1983), whereas the stage 9 terrace is assumed to have been cut at 0 m relative to modern sea level.

may be in error. The predicted elevations for stage 7 and 9 terraces are 29 to 33 m and 43 to 48 m, respectively. In North County, away from active faults, the Stuart Mesa and Guy Fleming terraces are at about 33 and 47 m, respectively, in close agreement with the predicted elevations. Based on this agreement, and realizing that the paleosea-level estimates for the terraces may be several meters off, we interpret these two terraces as correlating with the two highstands.

In the second method we estimate the ages of the higher terraces (Table 4) by assuming that (1) the uplift rate has been reasonably constant for the coastal area away from areas of faulting throughout the mid- to late Quaternary, and (2) the terraces were cut at or near modern sea level. Given as much as a 10-m error in paleosea level, an age-error estimate of ±75 ka should be applied to the higher terraces.

TERRACE DEFORMATION

Analysis of Figure 1 leads to several conclusions about the long-term geomorphic evolution of the San Diego coastal plain. The entire region is being uplifted at a low but fairly uniform rate of 0.13 to 0.14 m/ka. Slightly higher rates are recorded adjacent to the Rose Canyon fault at Mount Soledad, whereas lower uplift rates and even subsidence are indicated in the downtown area and in San Diego Bay.

The vertical rate of slip across the Rose Canyon fault is low, with the Nestor and Bird Rock terraces vertically displaced by only 10 to 15 m. The strike-slip rate is substantially higher, but the apparent offset of several hundred meters in the coastline is in part due to the primary coastal configuration at the time of terrace cutting. This inference is based on the fact that the Bird Rock terrace has an apparent deflection that is slightly greater than that of the Nestor terrace. Similarly, an older terrace on Mount Soledad appears to be deflected less than the Nestor terrace. This relation is the opposite of what would be expected if all of the shoreline deflection is due to strike slip on the right-lateral Rose Canyon fault. Further work may resolve the details of these relations.

The configuration of the shorelines and beach ridges indicates that Mount Soledad has been rising throughout much of the mid- to late Quaternary. The Tecolote shoreline is bowed strongly toward Mount Soledad, indicating an island or shoaling at time of terrace formation. That a promontory was present when the Clairemont terrace was cut is suggested by the fact that the Clairemont shoreline appears to wrap around Mount Soledad. The Linda Vista terrace does not appear to deflect westward in this area, but the change in average strike of its shoreline north and south of San Diego suggests that a submarine obstruction was probably present when the Linda Vista terrace was cut. From the shoreline configurations, it is clear that the Rose Canyon fault has been active for at least 1 million years, the approximate age of the Linda Vista terrace.

Sometime after cutting of the Clairemont terrace, it and the Tecolote and Linda Vista terraces were warped down into the fault in the downtown area near San Diego Bay (Fig. 1). More recent terrace formation has resulted in the younger terraces cutting across the Clairemont and older terraces at an oblique angle in that area. Many of the buried fossil localities in the downtown area have high AAR ratios and probably are correlative with the older terraces (Table 2). Some may actually be on buried abrasion platforms.

CONCLUSIONS

Sixteen shorelines are preserved as terraces, beach ridges, and other shoreline features along the San Diego County coast. Ages of the shorelines range from 80 ka to perhaps as old as 1.29 ma. Shoreline-angle elevations show that most of the coastal zone has been uplifted uniformly at an average long-term rate of 0.13 to 0.14 m/ka. However, both higher and lower uplift rates have prevailed along the Rose Canyon fault, which is shown by its effects on shoreline configurations to have been active for at least the past million years. The average long-term uplift rate for the San Diego region is similar to those for other areas of coastal California that are dominated by strike-slip tectonics (Lajoie and others, 1979; Ponti, 1989; Hanson and others, 1990), but the higher uplift rate for Mount Soledad in the Rose Canyon fault zone is lower than that at Punta Banda (Rockwell and others, 1989) and Palos Verdes (Muhs and others, this volume). The San Diego rates also are generally much lower than those in the Transverse Ranges (Yeats, 1976; Lajoie and others, 1979; Rockwell and others, 1988; Rockwell and others, this volume), where regional shortening results in substantially higher rates of vertical tectonics.

ACKNOWLEDGMENTS

The extensive mapping on which this paper is based would not have been completed without the encouragement of Ken Lajoie, who also provided maps, aerial photographs, and other materials. John Wehmiller provided more than 100 amino-acid analyses, which form most of the basis for the shoreline chronology presented here. Among the many others who contributed to this study in a variety of ways are Tom Demere, Bob Dowlen, Leonard Eisenberg, Eric Frost, Mike Hart, Monte Marshall, and Bradford Schmalfuss. The authors also thank Gary Carver and Dan Muhs for reviewing the manuscript. Partial support of terrace mapping was provided under U.S. Geological Survey Contract Number 14-08-21898.

REFERENCES

Artim, E. R., and Pinckney, C. J., 1973, La Nacion fault system, San Diego, California: Geological Society of America Bulletin, v. 84, p. 1075–1080.

Bender, M. L., Fairbanks, R. G., Taylor, F. W., Matthews, R. K., Goddard, J. G., and Broecker, W. S., 1979, Uranium-series dating of the Pleistocene reef tracts of Barbados, West Indies: Geological Society of America Bulletin, v. 90, p. 577–594.

Bloom, A. L., Broecker, W. S., Chappel, J. M. A., Matthews, R. K., and Mesolella, K. J., 1974, Quaternary sea-level fluctuations on a tectonic coast: New Th/U dates from the Huon Peninsula, New Guinea: Quaternary Research, v. 4, p. 185–205.

Bloom, A. L. and Yonekura, N., 1985, Coastal terraces generated by sea-level change and tectonic uplift, *in* Woldenberg, M. J., (ed.), Models in Geomorphology: Allen and Unwin, Boston, p. 139–154.

Chappell, J. M. A., 1983, A revised sea-level record for the last 300,000 years on Papua New Guinea: Search, v. 14, p. 99–101.

Ellis, A. J., 1919, Physiography, *in* Ellis, A. J., and Lee, C. H., eds., Geology and Groundwater of the Western Part of San Diego County, California: U.S. Geological Survey Water Supply Paper, v. 446, p. 20–50.

Emery, K. O., 1950, Ironstone concretions and beach ridges of San Diego County, California: Journal of California Mining and Geology, v. 46, p. 213–221.

Hanna, M. A., 1926, Geology of the La Jolla quadrangle, California: University of California Publications in Geological Science, v. 16, p. 187–246.

Hanson, K. L., Wesling, J. R., Lettis, W. R., Kelson, K., and Mezger, L., 1990, Correlation, ages, and uplift rates of Quaternary marine terraces: south-central coastal California, *in* Lettis, W. R., Hanson, K. L., Kelson, K. I., and Wesling, J. R., eds., Neotectonics of South-Central Coastal California: Friends of the Pleistocene Guidebook, Pacific Cell, p. 139–190.

Harmon, R. S., Mitterer, R. M., Kriausakul, N., Land, L. S., Schwarcz, H. P., Garrett, P., Larson, G. J., Vacher, H. L., and Rowe, M., 1983, U-series and amino-acid racemization geochronology of Bermuda: implications for eustatic sea-level fluctuations over the past 250,000 years: Palaeogeography, Palaeoclimatology, Palaeoecology, v. 44, p. 41–70.

Hertlein, L. G., and Grant, U. S., IV, 1944, The geology and paleontology of the marine Pliocene of San Diego, California: San Diego Society of Natural History Memoir, v. 2, Part 1, Geology, 72 p.

Karrow, P. F., and Bada, J. L., 1980, Amino acid racemization dating of Quaternary raised marine terraces in San Diego County, California: Geology, v. 8, p. 200–204.

Kennedy, M. P., 1975, Geology of the San Diego metropolitan area, California: western San Diego metropolitan area: California Division of Mines and Geology Bulletin, v. 200, p. 1–39.

Kennedy, M. P., and Peterson, G. L., 1975, Geology of the San Diego metropolitan area, California: eastern San Diego metropolitan area: California Division of Mines and Geology Bulletin, v. 200, p. 45–56.

Kennedy, M. P., Tan, S. S., Chapman, R. H., and Chace, G. W., 1975, Character and recency of faulting, San Diego area, California: California Division of Mines and Geology Special Report, v. 123, 33 p.

Kern, J. P., 1973a, Origin and history of two upper Pleistocene marine terraces at San Diego, California: Geological Society of America Abstracts with Programs, v. 5, Cordilleran Section, p. 66.

Kern, J. P., 1973b, Late Quaternary deformation of the Nestor terrace on the east side of Point Loma, San Diego, California, *in* Ross, A., and Dowlen, R. J., eds., Studies on the Geology and Geologic Hazards of the Greater San Diego Area, California: Guidebook, May 1973 Field Trip of the San Diego Association of Geologists and Association of Engineering Geologists, p. 43–45.

Kern, J. P., 1977, Origin and history of upper Pleistocene marine terraces, San Diego, California: Geological Society of America Bulletin, v. 88, p. 1553–1566.

Ku, T-L., and Kern, J. P., 1974, Uranium-series age of the upper Pleistocene Nestor terrace, San Diego, California: Geological Society of America Bulletin, v. 85, p. 1713–1716.

Lajoie, K. R., Kern, J. P., Wehmiller, J. F., Kennedy, G. L., Mathieson, S. A., Sarna-Wojcicki, A. M., Yerkes, R. F., and McCrory, P. F., 1979, Quaternary marine shorelines and crustal deformation, San Diego to Santa Barbara, California, *in* Abbott, P. L. ed., Geological Excursions in the Southern California Area: Geological Society of America Annual Meeting, p. 3–15.

Masters, P. M., and Bada, J. L., 1977, Racemization of isoleucine in fossil molluscs from Indian midden and interglacial terraces in southern California: Earth and Planetary Science Letters, v. 37, p. 173–183.

Masters, P. M., and Bada, J. L., 1979, Amino acid racemization dating of fossil shells from southern California, *in* Berger, R., ed., Radiocarbon Dating: Proceedings, International Conference on Radiocarbon Dating, no. 9, p. 757–773.

McCrory, P. F., and Lajoie, K. R., 1979, Marine terrace deformation, San Diego County, California (abst.), *in* Whitten, C. A., ed., Recent Crustal Movements: Tectonophysics, v. 52, no. 1–4, p. 407–408.

McEuen, R. B., and Pinckney, C. J., 1972, Seismic risk in San Diego, California: San Diego Society of Natural History Transactions, v. 17, p. 33–62.

Moore, G. W., and Kennedy, M. P., 1975, Quaternary faults at San Diego Bay, California: U.S. Geological Survey Research, v. 3, p. 589–595.

Muhs, D. R., Kennedy, G. L., and Rockwell, T. K., 1988, Uranium-series ages of corals from marine terraces, Pacific coast of North America: implications for the timing and magnitude of late Pleistocene sea-level changes: American Quaternary Association, Program and Abstracts, 10th Biennial Meeting, p. 140.

PALMER, L. A., 1967, Marine terraces of California, Oregon, and Washington: Unpublished Ph.D. Dissertation, University of California, Los Angeles, 379 p.

PETERSON, G. L., 1970, Quaternary deformation of the San Diego area, southwestern California, *in* Pacific Slope Geology of Northern Baja California and Adjacent Alta California: American Association of Petroleum Geologists, Society of Economic Paleontologists and Mineralogists, and Society of Economic Geologists, Pacific Section Fall Field Trip Guidebook, p. 120–126.

PONTI, D. J., 1989, Aminostratigraphy and chronostratigraphy of Pleistocene marine sediments, southwestern Los Angeles basin, California: Unpublished Ph.D. Dissertation, University of Colorado, Boulder, 409 p.

ROCKWELL, T. K., KELLER, E. A., AND DEMBROFF, G. R., 1988, Quaternary rate of folding of the Ventura Avenue anticline, western Transverse Ranges, southern California: Geological Society of America Bulletin, v. 100, p. 850–858.

ROCKWELL, T. K., MUHS, D. R., KENNEDY, G. L., WILSON, S., HATCH, M. E., AND KLINGER, R., 1989, Uranium-series ages, faunal correlations and tectonic deformation of marine terraces within the Agua Blanca fault zone at Punta Banda, northern Baja California, Mexico, *in* Abbott, P. L., ed., Geologic Studies in Baja California: Society of Economic Paleontologists and Mineralogists Pacific Section Book 63, p. 1–16.

SHACKLETON, N. J., AND OPDYKE, N. D., 1973, Oxygen isotope and paleomagnetic stratigraphy of equatorial Pacific core V28-238: oxygen isotope temperatures and ice volumes on a 10^5 and 10^6 year scale: Quaternary Research, v. 3, p. 39–55.

WEHMILLER, J. F., LAJOIE, K. R., KVENVOLDEN, K. A., PETERSON, E., BELKNAP, D. F., KENNEDY, G. L., ADDICOTT, W. O., VEDDER, J. G., AND WRIGHT, R. W., 1977, Correlation and chronology of Pacific coast marine terrace deposits of continental United States by fossil amino acid stereochemistry—technique evaluation, relative ages, kinetic model ages and geologic implications: U.S. Geological Survey Open-File Report 77–680, 106 p.

YEATS, R. S., 1976, Neogene tectonics of the central Ventura basin, *in* Fritsche, A. E., Ter Best, H., Jr., and Wornardt, W. W., eds., The Neogene Symposium: Society of Economic Paleontologists and Mineralogists, Pacific Section, p. 19–32.

ZIONY, J. I., AND BUCHANAN, J. M., 1972, Preliminary report on recency of faulting in the greater San Diego area, California: (unnumbered) U.S. Geological Survey Open-File Report, 16 p.

PART V
LACUSTRINE COASTAL SYSTEMS

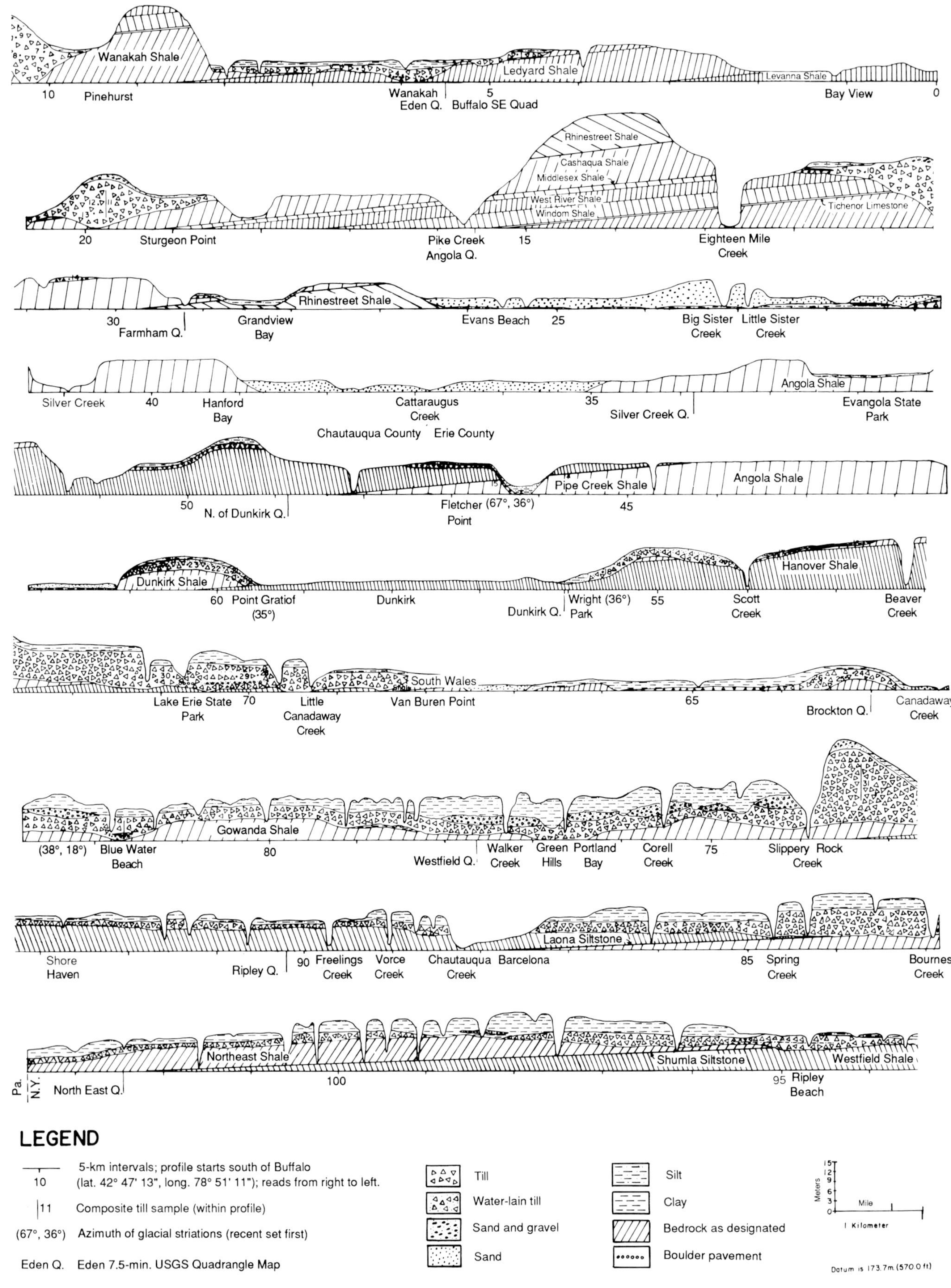

Wanakah Shale
10 Pinehurst
Wanakah
Eden Q.
5
Buffalo SE Quad
Ledyard Shale
Levanna Shale
Bay View
0
Rhinestreet Shale
Cashaqua Shale
Middlesex Shale
West River Shale
Windom Shale
Tichenor Limestone
20 Sturgeon Point
Pike Creek
Angola Q.
15
Eighteen Mile Creek
Rhinestreet Shale
30
Farmham Q.
Grandview Bay
Evans Beach
25
Big Sister Creek
Little Sister Creek
Silver Creek
40
Hanford Bay
Cattaraugus Creek
Chautauqua County
Erie County
35
Silver Creek Q.
Angola Shale
Evangola State Park
50
N. of Dunkirk Q.
Fletcher (67°, 36°) Point
Pipe Creek Shale
45
Angola Shale
Dunkirk Shale
60 Point Gratiof (35°)
Dunkirk
Dunkirk Q.
Wright (36°) Park
55
Scott Creek
Hanover Shale
Beaver Creek
Lake Erie State Park
70
Little Canadaway Creek
Van Buren Point
South Wales
65
Brockton Q.
Canadaway Creek
(38°, 18°)
Blue Water Beach
Gowanda Shale
80
Westfield Q.
Walker Creek
Green Hills
Portland Bay
Corell Creek
75
Slippery Rock Creek
Shore Haven
Ripley Q.
90
Freelings Creek
Vorce Creek
Chautauqua Creek
Barcelona
Laona Siltstone
85
Spring Creek
Bournes Creek
Pa.
N.Y.
North East Q.
Northeast Shale
100
Shumla Siltstone
95
Ripley Beach
Westfield Shale
LEGEND
5-km intervals; profile starts south of Buffalo
10 (lat. 42° 47' 13", long. 78° 51' 11"); reads from right to left.
11 Composite till sample (within profile)
(67°, 36°) Azimuth of glacial striations (recent set first)
Eden Q. Eden 7.5-min. USGS Quadrangle Map
Till
Water-lain till
Sand and gravel
Sand
Silt
Clay
Bedrock as designated
Boulder pavement
Meters
Mile
1 Kilometer
Datum is 173.7m (570.0 ft)

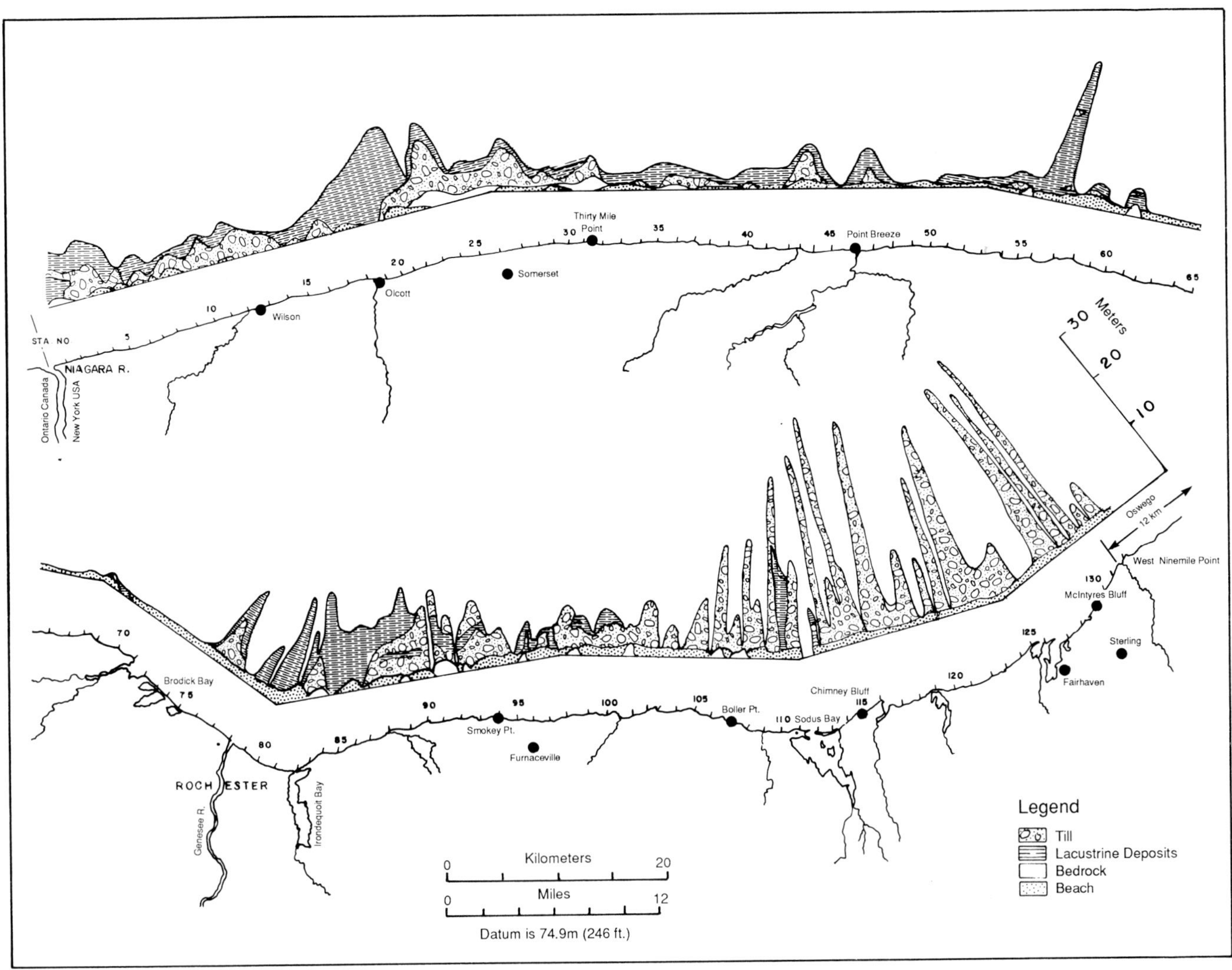

FIG. 3.—Composite stratigraphic section along the south shore of Lake Ontario in New York. From Brennan and Calkin (1984). Reprinted courtesy of New York Sea Grant Institute.

Electric and Gas Corporation. Figure 4 displays the composite stratigraphy for each of the three reaches, including the representative area around Somerset and the surrounding bluffs.

Lower glaciolacustrine beds.—

The oldest unit is a red or, less commonly, gray glaciolacustrine clay and silt as much as 1 thick. The unit has been exposed occasionally in the Somerset bluff (station 26), where nearby test borings (Dames and Moore, 1974; M. Pendleton, pers. commun., 1975) show it underlying a red stony till and overlying the red Queenston bedrock.

Red stony diamicton.—

A very stony, massive, red (2.5 YR 4/2 to 5 YR 3/2) diamicton, interpreted to be lodgment till, lies stratigraphically above the lower glaciolacustrine unit, but more commonly has been observed lying directly on Queenston bedrock. Locally, basal contacts are gradational due to brecciation of the rock.

The matrix is poorly sorted and composed predominantly of sandy silt, ranging to silty sand in the east (Fig. 5). Stones are concentrated near the base, where clasts may make up 50 to 80 percent by weight and reach 50 cm in diameter. Clasts consist predominantly of red-brown micaceous

←

FIG. 2.—Bedrock and surficial geologic stratigraphy along the New York shore of Lake Erie. The above-shore sections are continuous from northeast to southwest if each is read from right to left. From Geier (1980), Geier and Calkin (1983). Reprinted courtesy of New York Sea Grant Institute.

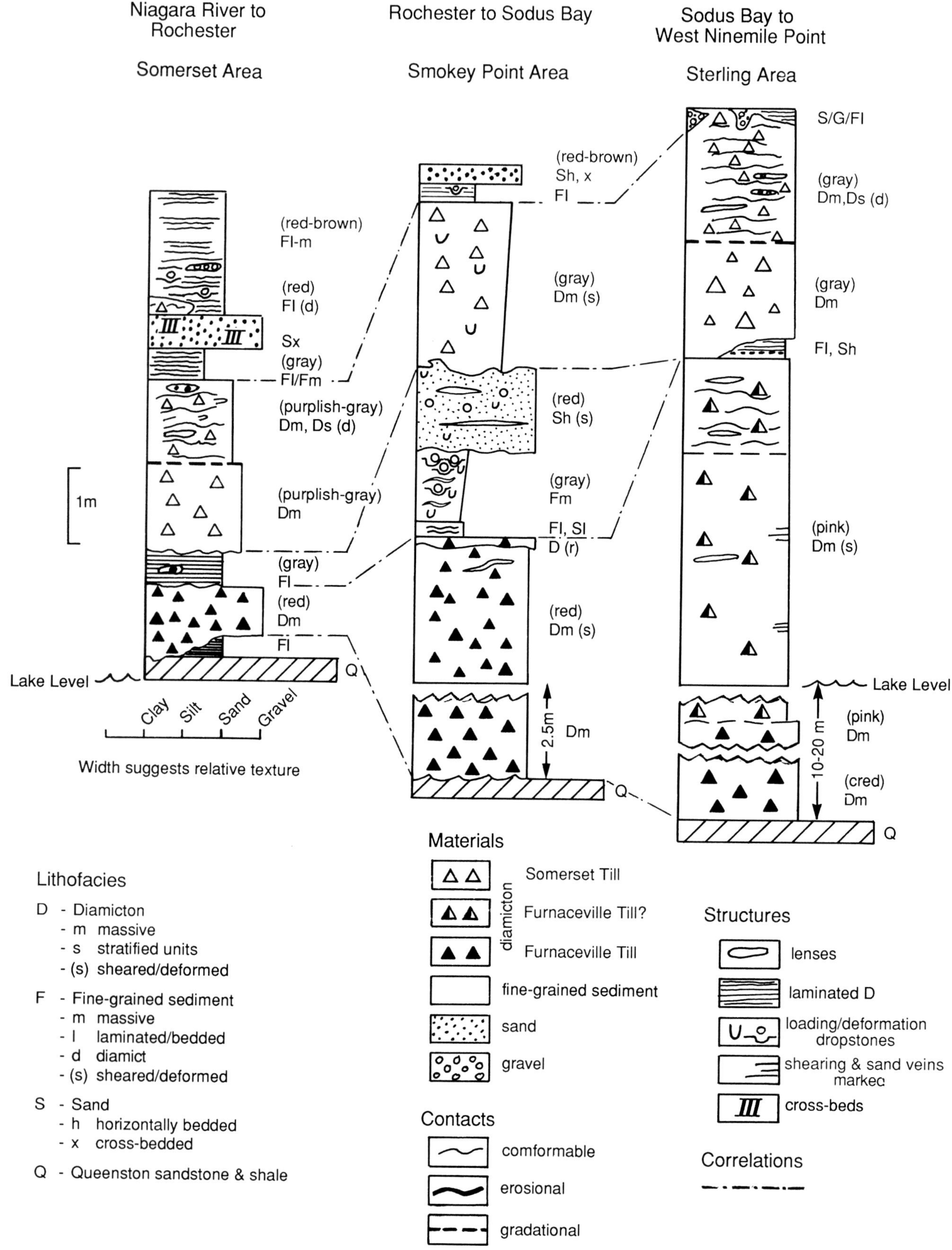
Niagara River to Rochester
Somerset Area
Rochester to Sodus Bay
Smokey Point Area
Sodus Bay to West Ninemile Point
Sterling Area
(red-brown) Fl-m
(red) Fl (d)
Sx
(gray) Fl/Fm
(purplish-gray) Dm, Ds (d)
(purplish-gray) Dm
(gray) Fl
(red) Dm
Fl
Q
1m
Lake Level
Clay
Silt
Sand
Gravel
Width suggests relative texture
(red-brown) Sh, x
Fl
(gray) Dm (s)
(red) Sh (s)
(gray) Fm
Fl, Sl
D (r)
(red) Dm (s)
2.5m
Dm
Q
S/G/Fl
(gray) Dm,Ds (d)
(gray) Dm
Fl, Sh
(pink) Dm (s)
Lake Level
(pink) Dm
10-20 m
(cred) Dm
Q
Materials
Somerset Till
Furnaceville Till?
Furnaceville Till
diamicton
fine-grained sediment
sand
gravel
Contacts
comformable
erosional
gradational
Structures
lenses
laminated D
loading/deformation dropstones
shearing & sand veins markeo
cross-beds
Correlations
Lithofacies
D - Diamicton
- m massive
- s stratified units
- (s) sheared/deformed
F - Fine-grained sediment
- m massive
- l laminated/bedded
- d diamict
- (s) sheared/deformed
S - Sand
- h horizontally bedded
- x cross-bedded
Q - Queenston sandstone & shale

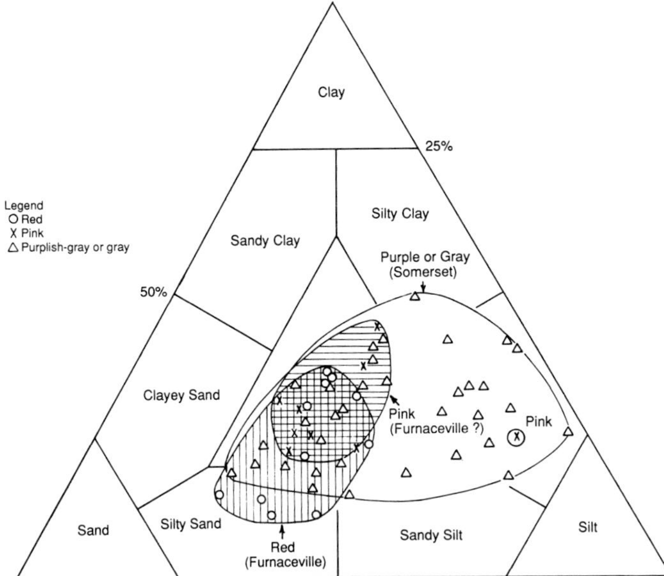

FIG. 5.—Textural composition of till matrices on the south shore of Lake Ontario. Taken from 72 diamicton samples of Brennan and Calkin (1984). Sediment definitions after Shepard (1954).

Queenston sandstone. Near the Niagara River, the percentage of Queenston clasts is equaled by that of gray, fine- to medium-grained micaceous sandstone. Other rock types include medium to dark gray fossiliferous Trenton limestone, tan micaceous-quartz sandstone, white marble, and minor plutonic rocks. The stones are angular to subangular and tabular in all but small pebble sizes, which are better rounded. Particularly at the base, this sediment displays minimal evidence of clast abrasion or sorting by ice during glacial transport. Some evidence of alternative modes of deposition are shown at the upper contacts, as indicated at many exposures by discontinuous lenses of silt and fine sand.

Exposures of the red stony till are scattered west of Rochester and are recorded near stations 26 (where it reaches a maximum thickness of 2 m), 40, 55, and 59 (Fig. 3). The till can be traced in the subsurface 18 km south, almost to the Niagara Escarpment in Niagara County (Smith, 1990).

Intertill glaciolacustrine deposits.—

Either glaciolacustrine deposits or till may overlie the red stony till unit. Where a superjacent diamicton occurs in the bluff, as at stations 20, 88, and 93, the two units may be separated by discontinuous gray or, less frequently, red glaciolacustrine clay to fine sand up to 2.5 m thick. Elsewhere, gray silt cuts into the red till or even incorporates fragments of it (M. Pendleton, pers. commun., 1978).

Purplish-gray silty diamicton (Somerset Till).—

The predominant diamicton of Niagara River-to-Rochester reach is a gray or purplish-gray (5 YR 5/1 to 10 YR 6/2) unit. In view of its apparent origin, the diamicton is named the Somerset Till after the town in Niagara County, New York, where it is exposed in shore bluffs (station 26) and has been studied in test borings for the Somerset Generating Plant. The unit lies directly on the red Queenston Formation in a few stations, where it shows sharp color and textural contacts with the red bedrock (stations 23, 26, and 36), but in most areas it rests on glaciolacustrine deposits or the lower red stony till (station 26).

Two facies of the Somerset Till are distinguished at many localities. The main body of the Somerset Till is a compact, massive, texturally and lithologically homogeneous silty till. The till generally grades upward to a subaqueously deposited diamicton facies that includes bedded and sorted sediment. This facies, in turn, commonly has indistinct upper contacts with the overlying glaciolacustrine sediments that blanket most bluff tops. One or both till facies may be present in any one exposure, and maximum thicknesses reach 3 m or even more in drumlins.

The till as a whole has a sandy-silt matrix, but texture is widely variable (Fig. 5). It is less stony than the underlying till, with clasts averaging about 4 percent of the total mass but ranging up to 50 percent. Slightly higher percentages occur in the lower facies than in the upper. Larger stones, often bullet-shaped with striations, show strong evidence of glacial contact. The lithologic assemblage resembles that of the underlying red stony till but has higher ratios of subrounded greenish-gray to subangular red sandstone or siltstone.

The upper facies of the Somerset Till is 0.5 to 3 m thick and is composed of a massive to laminated diamicton enclosing subhorizontal, variably colored lenses, stringers, or wisp-like beds of sand, silt, or, less commonly, clay (e.g., station 93, where rhythmically bedded sediment is included). Similar stratified diamictons have been described from the north shore of Lake Erie (e.g., Dreimanis, 1982) as "waterlain" tills. Individual sorted and bedded units range up to 0.5 m thick, locally contain small-scale current structures, and are rarely continuous laterally for more than a few meters.

The beds of the upper facies may be greatly wrinkled to grossly tilted due to loading or flowage of saturated sediment. Isoclinal or flow folds (Evenson and others, 1977) occur on a scale of less than 1 m within some individual diamicton units. In places, stones occur along distinct horizons, as in the boulder pavement first described by Gilbert (1898) on both sides of the inlet at Wilson (stations 11 and 13). The boulder-pavement zone appears to lie at the contact between the massive, compact, purplish-gray till (lower

←

FIG. 4.—Composite lithofacies section and stratigraphic correlations for each of three reaches of the Lake Ontario shore in New York: Niagara River–Rochester, Rochester–Sodus Bay, and Sodus Bay–West Ninemile Point. Labels at top of each section indicate areas of the reach where most units are well represented.

facies) and the overlying flow-deposited diamicton (upper facies). Boulders in this zone bear parallel striations, suggesting that they accumulated as a glacial lag deposit during sedimentation bypassing or erosion.

Where deposits of the upper facies lie beneath more than 4 m of lake silts or clay, they remain unweathered and, at first glance, closely resemble the lower compact-till facies. However, because of their greater permeability, they are more easily oxidized and display colors of higher chroma than the compact till.

Ice-marginal deposits.—

Gravelly silt, coarse sand and gravel, pebbly lag deposits, and silty till lenses interpreted as flow till (Hartshorn, 1958; Evenson and others, 1977) occur within the upper glaciolacustrine unit at lateral intervals of less than 3 to 5 km. In places they are clearly separated from the underlying purplish-gray Somerset Till by fine lacustrine deposits and/or an erosional unconformity with a thin pebble-lag gravel. In other areas, deposition clearly involved direct glacier contact.

Ice-marginal deposition is particularly apparent at three locations. East of Olcott Village (station 20), 10 m of cross-bedded sand overlie 6 m of rhythmically bedded gray lake clay and till. At Point Breeze (station 47), purplish-gray Somerset Till on gravelly proglacial drift is itself overlain by 6 m of well-rounded gravel, cross-bedded sand, and lake silt. A high ridge west of Hamlin Beach State Park (station 51) is developed on an ice-contact deposit, perhaps a moulin. Thin gray till (Somerset Till?) with clasts of gray clay underlies cemented and imbricated coarse-cobble gravel, which in turn is overlain by 20 m of cross-bedded lake sand with a few red flow-till lenses within the upper 3 m.

Upper glaciolacustrine deposits.—

Glaciolacustrine deposits averaging 3 to 5 m in thickness overlie the Somerset Till and older sediments. Red or red-brown, oxidized, massive to thin-bedded silt predominates, but rhythmic gray clay and silt units up to 0.08 m are common directly above the Somerset Till. Fine, yellow-brown, oxidized, sandy beds become more important in the upper part. Locally, dropstones range to cobble size. Lower contacts may be diffuse or, in more than 10 percent of the reach, display a pebble or boulder lag on the till or lake-clay interface. Intricate folding and deformation of the bedding are prominent where clay is interbedded with silt and sand. The sediments were deposited primarily in glacial Lake Iroquois, though deposition in the Niagara region had begun earlier in glacial Lake Dawson.

Rochester to Sodus Bay (Stations 80–110)

In the 48-km reach between Rochester and Sodus Bay, the terrane is rolling and markedly dissected by streams lakeward from the sand plain of early Lake Iroquois and a partly wave-modified drumlin and moraine belt (Fig. 3). A representative section for this reach is at Smokey Point, town of Ontario, (stations 93–94) near the Ginna Nuclear Power Station and is described later. In ascending order the units consist of a red stony diamicton, glaciolacustrine deposits, a gray silt diamicton, and thin glaciolacustrine deposits locally (Fig. 4).

Red stony diamicton (Furnaceville Till).—

The oldest exposed Pleistocene unit is a massive, red (5 YR 5/2) stony, sand-silt-clay diamicton (Fig. 5), named the Furnaceville Till after the village of Furnaceville (town of Ontario, Wayne County). The unit is well exposed at Smokey Point (stations 93–94) and displayed in test borings for the Ginna Nuclear Power Station (Dames and Moore, 1965; J. Szymanski, pers. commun., 1973).

The Furnaceville Till lies at the base of bluff exposures or on Queenston bedrock. The till is up to 7 m thick between stations 83 and 103. The upper contact is irregular, displaying evidence of erosion and reworking where the Furnaceville Till is directly overlain by younger till or by glaciolacustrine sediments. The matrix of the Furnaceville Till is silty sand to sand-silt-clay (Fig. 5). Clasts typically make up 5 to 10 percent by volume; glacially shaped stones are common and homogeneously dispersed. However, near bedrock contacts, less-rounded clasts may comprise 30 to 50 percent. More than 50 percent of these clasts consist of subequal proportions of subangular, red Queenston sandstone and greenish-gray Oswego sandstone; 25 percent of the clasts consist of well-rounded, gray, fossiliferous Trenton limestone; the remainder consists of minor crystalline rocks and clasts of red clay derived from an older lake deposit.

Intertill glaciolacustrine deposits.—

At Smokey Point (stations 93–94), a cemented-lag gravel with subrounded pebbles overlies the Furnaceville Till (J. Szymanski, pers. commun., 1973), though several kilometers west of Smokey Point the Furnaceville Till is overlain by up to 7 m of glaciolacustrine sand enclosing red-till clasts. A nearly equal number of bluff sections expose deformed gray or red-silt beds over Furnaceville Till and/or reworked silt with lenses and interbeds of rhythmically bedded clay. Near the upper contact, the gray silts are interlaminated with red sand and contain striated and rounded dropstones that increase in size and abundance toward the upper till contact (Fig. 4).

Gray-silt diamicton.—

Light gray or greenish-gray (10 YR 4/1), rarely purplish-gray (10 YR 6/2; station 109), clayey silt or sand-silt-clay diamicton (Fig. 5) generally overlies sorted glaciolacustrine deposits but locally rests directly on the Furnaceville Till. Up to 5 m thick in bluff exposures, the unit is relatively homogeneous and massive in most exposures but grades laterally upward into subaqueously deposited stratified diamicton and lacustrine clay in several exposures.

The stones of the unit in the vicinity of Smokey Point (Fig. 3) are subrounded to rounded and faceted by glacial transport, but stones are smaller and more sparse than in the underlying Furnaceville Till, or in the type area of the Somerset Till to the west. They average from 3 to 5 percent by volume and generally include, in decreasing order of abundance, red Queenston sandstone, dark Trenton limestone and dolomite, and greenish-gray Oswego sandstone.

In one location near station 93, the light-gray, contorted, subaqueously deposited till is unconformably overlain by stratified diamicton. This is interpreted by Salomon (1974, her station 15) as representing a distinct, though minor, readvance.

Upper glaciolacustrine deposits.—

Glaciolacustrine deposits are generally thin or entirely absent in bluffs that cut into drumlins in the Rochester-to-Sodus Bay reach, but they are present at the surface in interdrumlin lows as red-brown fine sand and silty sand deformed over finer units. The units occur with strong current structures and glacially worked dropstones.

Sodus Bay to West Ninemile Point (Stations 110–132)

Large northeast-oriented drumlins separated by narrow wetlands characterize the terrane between Sodus Bay and West Ninemile Point (Fig. 3), a distance of 35 km. Many drumlins appear wave-flattened above 122-m elevation but are less modified below this level because water depths were greater in the eastern end than in the western end of Lake Iroquois (Fairchild, 1929; Gillette, 1940; Muller and Prest, 1985; Francek, 1990). Drumlin crests at about 122 m above sea level, or lower would have been submerged at least 10 m even during the early low (Newfane) phase of Lake Iroquois. Bluffs between McIntyres Bluff and Moon Bay, in the town of Sterling, Cayuga County, (Fig. 3; stations 129–131) afford representative stratigraphic exposures and abundant subsurface data (Dames and Moore, 1973).

Probable Furnaceville correlatives.—

Between Boller Point (station 107) and station 134, the bedrock surface and red stony till descend below lake level and are not exposed east of station 109 until the Queenston shale crops out near station 132 west of Oswego. However, borings near McIntyres Bluff (station 130) reveal that a red clayey to sandy-silt till with varying amounts of fine to coarse gravel discontinuously overlies bedrock 10 to 12 m below mean lake level. The till reaches a maximum thickness of 7 m above the underlying bedrock and grades upward to a somewhat sandier, thick, pink basal till (Fig. 4; C. Bandoian, pers. commun., 1973).

A pink sand-silt-clay till (e.g., 10 YR 6/2 to 5 YR 4/3) replaces the typical red stony Furnaceville Till on the Queenston bedrock at the base of drumlins west of Sodus Bay (stations 106, 107). Two superimposed facies, a lower massive facies and an upper partly stratified facies, form the major part of exposed portions of the wave-truncated drumlins and substantial parts of interdrumlin exposures eastward through McIntyres Bluff (station 129). The unit is particularly well exposed in Lake Bluff drumlin east of Sodus Bay (station 113) and in the well-known 40-m-high Chimney Bluff drumlin (station 114; Slater, 1929).

The lower pink facies of the till is a homogeneous basal deposit with general textural and lithologic homogeneity in the cores of drumlins. However, even in this compact facies a few irregular horizons of gray sandy-silt till occur. The matrix consists of silty sand or sand-silt-clay (Fig. 5), with 2 to 10 percent of the clasts being glacially striated, polished, and faceted stones. In decreasing abundance are greenish-gray Oswego sandstone, red Queenston sandstone, dark Trenton limestone, and granitic gneiss. Percentages of pebbles and cobbles of gray Oswego sandstone increase eastward toward Oswego.

Ten to 20 percent of the drumlin exposures consist of subaqueous-flow diamicton and sorted units. The sorted units are subhorizontal to mildly convex upward normal to the long axes of the drumlins. Till in the drumlins displays distinct fissility on micro- and macroscales and includes cross-cutting silt or clay-rich veins a few centimeters thick that have generally been related to subglacial lodgment or stacking (Muller, 1974; 1983a). Details of the dipartite structure, the fabric, and the origin of these drumlins of north-central New York have been discussed by Fairchild (1907; 1929), Slater (1929), and Gillette (1940).

Gray drumlin diamicton.—

In some drumlin and interdrumlin bluffs between Sterling and West Ninemile Point (stations 128–132), younger gray silty sand or sand-silt-clay till, weathering to grayish-brown, forms an 8- to 9-m cap over the lower, thicker, more homogeneous facies of pink till. Here, as to the west, a generally homogeneous basal facies of gray till grades upward irregularly to subaqueous, reworked, stratified diamicton or to subaqueous-flow facies forming a gray concentric shell on the drumlins (Dames and Moore, 1973).

Till stones are dominated by gray Oswego sandstone with much less Queenston sandstone and Trenton limestone. Igneous and metamorphic rocks comprise up to 10 percent of the gray till.

Contacts within and between the lower pink drumlin till and the overlying gray drumlin till are relatively distinct. At station 127, for example, 10 percent of the clasts are red sandstone in the gray drumlin till, as compared to 25 percent in the underlying pink drumlin till. The gray till has a slightly sandier matrix (C. Bandoian, pers. commun. 1973), is somewhat less resistant to erosion, and at station 129 is separated from the lower pink till by a 0.3-m-thick horizon of stratified sediment. These relations argue for differences in source terrane and suggest that the lower pink drumlin till and the upper gray drumlin till are distinct stratigraphic units related to separate episodes of glaciation.

Upper glaciolacustrine deposits.—

Drumlin surfaces generally lack continuous lacustrine cover. However, red and gray glaciolacustrine fine sand, silt, and clay are spread locally over till on drumlin flanks and in interdrumlin flats to thicknesses of as much as 4 m. Lower drumlins in the Sterling area (station 129) display coarse to fine gravel as well as sand in channels cut into the gray till (C. Bandoian, pers. commun., 1973).

At Lake Bluff (station 114) and Chimney Bluff (station 115) near Sodus Bay, gray, poorly bedded, and slightly pebbly sorted deposits of lacustrine sand, silt, and clay overlap pink till on drumlin flanks. Undulating and irregular bedding suggests that these deposits may have been partly washed from the higher drumlin surfaces during glacial Lake Iroquois time. They have the finer textures, lower bulk densities (<2.0 gm/cc), and secondary structures characteristic of the gray-silt till and subaqueous-till facies in the Rochester-to-Sodus-Bay reach and may be correlative with them.

Correlation and Lateral Trends Along the Ontario Shore

Pre-Furnaceville drift.—

Red or, less commonly, gray lacustrine clay and silt lie on Queenston bedrock below the lower diamicton in the Somerset area (near station 26) and are locally recorded in a few other sites in the Niagara-to-Rochester reach. These materials are presumably vestiges of sediments deposited in the ice-marginal lake into which the ice sheet spread prior to deposition of the overlying diamicton.

Furnaceville Till.—

The oldest Pleistocene unit that is widely exposed in the Ontario shore bluffs is red stony diamicton. In the Rochester-to-Sodus Bay reach, this lower diamicton, as it exists at Smokey Point near Furnaceville (stations 93, 94), is designated the Furnaceville Till. Where the lower contact is visible, Furnaceville Till generally overlies the Queenston Formation, but it may locally lie on glaciolacustrine and even colluvial deposits.

Correlated with the type Furnaceville Till on the basis of similarity in properties and stratigraphic position is the red stony till exposed in the bluffs of the Niagara-to-Rochester reach (Fig. 5).

In the Sodus Bay-to-Oswego reach, till considered to be correlative with the Furnaceville Till exists in typical red stony facies, but also as the pink till characteristically prominent in drumlin cores. Outcrops of typical red stony Furnaceville Till pass below lake level near the Oswego county line. However, borings near McIntyres Bluff (station 127) reveal a discontinuous but apparently correlative red clayey to sandy-silt till that overlies bedrock 10 to 12 m below mean lake level. Near Sterling, a similar red till exists below the pink drumlin diamicton of bluff exposures. Data from the borings and bluff exposures, which attest to overall maturity and transitional lower contacts of the drumlin, suggest that the pink drumlin till is a thicker and more glacially altered facies of the Furnaceville Till.

Analysis of 37 till-matrix samples yielded an average sand:silt:clay ratio of 45:32:23, an average bulk density of 2.24 g/cc, and an average moisture content of 11 percent (Brennan and Calkin, 1984). The Furnaceville Till is distinguished from the upper (Somerset) till sheet along most of the shore by its stratigraphic position (Fig. 4), red color, high percentages of red Queenston clasts, generally sandier matrix, and compactness, as well as by its lack of significant water-lain facies. These characteristics reflect the Queenston bedrock as the dominant source terrane. Indeed, caution in correlation is warranted because red stony till may exist as a facies of stratigraphically unrelated units where they are similarly dominated by assimilation of underlying Queenston bedrock.

Somerset Till.—

The upper diamicton in bluff exposures (station 26) and test borings near the Somerset Generating Plant is a purplish-gray silty unit designated as Somerset Till. The unit rests directly on Queenston bedrock, on the Furnaceville Till or its pink drumlin facies, or on glaciolacustrine deposits. It is widely overlain by postglacial lacustrine beds but also exists as the surface sediment.

Units correlative with purple or gray Somerset Till in its type area include the more clayey gray till in the reach between Rochester and Sodus Bay and the gray till that composes the outer portions of drumlins east of Sodus Bay (Fig. 4). Two facies characterize the Somerset Till, particularly in the Niagara-to-Rochester reach but elsewhere as well: a lower, rather massive, and homogeneous unit widely overlain by a facies marked by features associated with deposition by subaqueous flow, possibly subaqueous basal meltout till (e.g., Dreimanis, 1982).

Incorporation of underlying sediments must have contributed extensively to the generally finer texture and lighter chroma of the Somerset Till as compared to the underlying Furnaceville Till. This assertion is supported by textural studies that show that neither the silt nor sand fraction in the purple to gray till can be closely related to rock types represented in the very coarse fraction, as was the case for the lower till (Salomon, 1974). Analysis of till-matrix samples from 35 sites yielded an average sand:silt:clay ratio of 28:43:29, an average bulk density of 2.09 g/cc, and an average moisture content of 14 percent (Brennan and Calkin, 1984).

The Somerset Till is distinguished from the lower red Furnaceville Till by its stratigraphic position (Fig. 4), color, generally lower stone content, deficiency in red clasts, finer texture (Fig. 5), and the characteristic occurrence of water-laid facies over a massive component.

Principal-component analysis of till matrices and coarse-sand lithology in 19 samples of the lower red till and 81 samples from the upper purple or gray till (middle till of Salomon, 1974) show sedimentologic trends suggested by other studies of bluff materials. Salomon's data (1974) show an increase in the proportions of green siltstone and metamorphic and igneous rocks in sand-size grains northeastward to the outcrop of green Oswego sandstone and toward the Adirondack Mountains. Conversely, the proportion of red siltstone, as well as the finer silt- and clay-grain sizes, increases westward (i.e., away from sand sources) and southward (i.e., away from the Oswego and into the Queenston exposure zone).

Upper glaciolacustrine and ice-contact deposits.—

Glaciolacustrine sand, silt, and clay are ubiquitous along the bluff tops except over crests of drumlins and are generally ascribed to deposition in glacial Lake Iroquois. However, striking differences in the thickness and character of these deposits were noted west of Rochester. At least 20 of the more than 100 sections measured contain cross-bedded sand, cobble gravel, and diamicton lenses showing evidence of flow and stony pods in the lacustrine deposits. Fluvial-gravel units also occur in the Sterling area. The evidence is indicative of a minor glacier readvance that formed the Carlton (Oswego) Moraine (Muller, 1983b).

Surface-weathered zone.—

The one conspicuous and continuous surface unit is a weathering horizon 2 to 4 m deep at the bluff face. The zone is similarly developed in all reaches and is conspicuous in the drumlinized topography between Sodus Bay and West Ninemile Point. Conforming to drumlin profiles, the weathered zone cuts across and is developed on till and lacustrine units alike. It is distinguished by the yellow-brown colors of oxidation and hydration, the horizontal microfoliation (pseudobedding) in the silty tills, and particularly by the nearly vertical attitude of exposures compared to the sloping profiles of unweathered bluff materials below. Weathered and unweathered deposits are texturally similar, but the weathered units have partly disaggregated sandstone and crystalline clasts, weak vertical parting, lower standard-penetration resistance, and different colors (C. Bandoian, pers. commun., 1973).

REGIONAL CORRELATION AND GLACIAL CHRONOLOGY: SUMMARY

Ontario, Canada

Work of Feenstra (1972, 1981), Hegler (1974), and Barnett (1975) indicates that the Lake Ontario bluffs extending 65 km west of the Niagara River to Hamilton, Ontario, expose or are underlain locally by a glacial sequence similar to that in adjacent New York. The sequence includes, in ascending order above the Queenston bedrock, red sandy-silt till, gray to brownish-gray silty or clayey till, and proglacial Lake Iroquois sediments. Borings and canal-excavation exposures allow the tracing of the units south to the Niagara Escarpment, where the tills appear to be separated stratigraphically by fine-grained glaciolacustrine deposits. Analysis of the two till units (Feenstra, 1972) using various parameters, including traces of Ni and Cr (Barnett, 1975), suggests that both units north of the Niagara Escarpment should be correlated with the Halton Till Sheet of Port Huron age (Barnett, 1979). However, the lower till south of the escarpment has been correlated with the older Wentworth Till. The textural and carbonate compositions of units below the escarpment in Ontario correlate closely with the comparable till units in adjacent New York (Smith, 1990).

The southern limit of the Halton Till, deposited during the early Port Huron advance, coincides with a position just beyond the Crystal Beach and Wainfleet Moraines across the Niagara Peninsula in Ontario (Barnett, 1979; Feenstra, 1981) and, in turn, with the Hamburg Moraine in New York (Fig. 1, Calkin, 1970; Muller, 1975). Whereas the lower till north of the Niagara Escarpment in Ontario has been associated with the main Port Huron advance, the upper gray till has been correlated with a short readvance following glaciolacustrine deposition, which reached south only to the area of the Niagara Escarpment (Barnett, 1975; Feenstra, 1981).

New York

The upper till on the Erie shore is correlated with the Hamburg Moraine on the basis of stratigraphic position and extent northeast of Sturgeon Point (Fig. 1) and is therefore assigned to the Port Huron stade (Calkin and Feenstra, 1985; Muller and Prest, 1985). The lower till is tentatively assigned to the preceding glacial advance that formed the Lake Escarpment and Gowanda Moraines (Fig. 1). These are tentatively correlated with the Paris and Galt Moraines formed of Wentworth Till in Ontario and assigned to the late Port Bruce stade or early Mackinaw interstade (Calkin and Barnett, 1990).

The two till units with intervening glaciolacustrine drift exposed in the Lake Ontario bluffs can be traced southward from the type sites near Somerset, Smokey Point, and MacIntyres Bluff for 1 km or more using borings. In addition, borings and exposures southwest of Somerset permit tracing of the till units beneath Lake Iroquois deposits southward from Lake Ontario bluffs 18 km to the Niagara Escarpment (Smith, 1990). A single red-brown clayey-silt diamicton is recognized south of the Niagara Escarpment as far as Buffalo.

On the basis of stratigraphic position and relations with the units in Ontario, the Furnaceville Till is associated with the Port Huron stade. This advance overrode proglacial deposits in the area of the Lake Ontario bluffs and reached 50 km south to the Hamburg Moraine in western New York. Following recession north from the Ontario shore bluffs area and deposition of proglacial lacustrine deposits, possibly during an interval of retreat following the early Lake Warren phase, the ice margin readvanced. Massive purplish-gray and gray till was deposited as ice reached the area of the Niagara Escarpment in western New York, possibly to build the Batavia or Barre Moraines (Fig. 1). Drumlins near the Lake Ontario shore are believed to have received their present configurations at this time. Proglacial, subaqueously deposited diamicton units that formed during subsequent retreat were buried by glaciolacustrine deposits that became finer upward with expansion of Lake Iroquois following retreat of the ice margin from the Carlton Moraine west of Rochester and the Oswego Moraine to the east.

ACKNOWLEDGMENTS

The authors acknowledge the substantial contributions of their students, particularly Nena L. Solomon, who worked with Muller, and Richard J. Geier and Sandra F. Brennan, who worked with Calkin. Studies by Calkin and his students were funded through the New York State Sea Grant Program. The authors are also indebted to Dames and Moore geologists Martha W. Pendleton and Charles A. Bandoian as well as to Gordon G. Connally for discussions in the field and for bringing to their attention unclassified reports relevant to the Ontario bluff study. Appreciation is expressed for careful reviews by Jane L. Forsyth and Peter J. Barnett.

REFERENCES

Anderson, T. W., and Lewis, C. F. M., 1985, Postglacial water-level history of the Lake Ontario Basin, *in* Karrow, P. F., and Calkin, P. E., eds., Quaternary Evolution of the Great Lakes: Geological Association of Canada Special Paper 30, p. 231–253.

Barnett, P. J., 1975, Till matrix characteristics of the upper and lower tills of the Niagara Peninsula, Ontario: Unpublished M.S. Thesis, University of Waterloo, Waterloo, Ontario, 76 p.

BARNETT, P. J., 1979, Glacial Lake Whittlesey: the probable ice frontal position in the eastern end of the Erie basin: Canadian Journal of Earth Sciences, v. 16, p. 568–574.

BRENNAN, S. F., AND CALKIN, P. E., 1984, Analysis of bluff erosion along the southern coastline of Lake Ontario, New York: New York Sea Grant Institute, 74 p.

BUEHLER, E. J., AND TESMER, I., 1963, Geology of Erie County: Buffalo Society of Natural Sciences Bulletin 21, 118 p.

CALKIN, P. E., 1970, Strandlines and chronology of the glacial Great Lakes in northwestern New York: Ohio Journal of Science, v. 70, p. 78–96.

CALKIN, P. E., 1982, Glacial geology of the Erie Lowland and adjoining Allegheny Plateau, western New York, *in* Buehler, E. J., and Calkin, P. E., eds., Geology of the Northern Appalachian Basin, Western New York: Field Trip Guidebook for New York State Geological Association, 54th Annual Meeting, Amherst, New York, p. 121–148.

CALKIN, P. E., AND BARNETT, P. J., 1990, Glacial geology of the eastern Lake Erie basin, *in* McKenzie, D. I., ed., Quaternary Environs of Lakes Erie and Ontario, prepared for the first joint meeting of the American Quaternary Association and the Canadian Quaternary Association at University of Waterloo, June 4–6: Escart Press, University of Waterloo, Waterloo, Ontario, p. 1–86.

CALKIN, P. E., AND FEENSTRA, B. H., 1985, Evolution of the Erie-Basin Great Lakes, *in* Karrow, P. F., and Calkin, P. E., eds., Quaternary Evolution of the Great Lakes: Geological Association of Canada Special Paper 30, p. 149–170.

CARTER, C. H., 1977, Sediment-load measurements along the United States shore of Lake Erie: Ohio Geological Survey, Report of Investigations 102, 24 p.

CLARK, R. H., AND PERSOAGE, N. P., 1970, Some implications of crustal movement for engineering planning: Canadian Journal of Earth Sciences, v. 7, p. 628–633.

DAMES AND MOORE, INC., 1965, Site evaluation study, proposed Brookwood nuclear power plant, Ontario, New York: Rochester Gas and Electric Corporation, Rochester, New York (unpaginated).

DAMES AND MOORE, INC., 1973, The Sterling Power Project, New York, part S76, geology and seismology: Rochester Gas and Electric Corporation, Rochester, New York, p. S76.1–S76.4-2-3.

DAMES AND MOORE, INC., 1974, Cayuga Station application to the New York Board on Electrical Generation Siting and Environment, Somerset alternate site, part 76-S, geology and seismology: New York State Electric and Gas Corporation, Binghamton, New York, p. 76.1–76.2–91S.

DREIMANIS, A., 1982, Two origins of the stratified Catfish Creek Till at Plum Point, Ontario: Boreas, v. 11, p. 173–180.

DREIMANIS, A., AND KARROW, P. F., 1972, Glacial history of the Great Lakes–St. Lawrence region, the classification of the Wisconsin(an) Stage and its correlatives: 24th International Geological Congress, Montreal, Section 12, p. 5–15.

DREXHAGE, T. F., AND CALKIN, P. E., 1981, Historic bluff recession along the Lake Ontario coast, New York: New York Sea Grant Institute, 123 p.

EVENSON, E. B., DREIMANIS, A., AND NEWSOME, J. W., 1977, Subaquatic flow tills: a new interpretation for the genesis of some laminated till deposits: Boreas, v. 6, p. 115–133.

FAIRCHILD, H. L., 1907, Drumlins of central western New York: New York State Museum Bulletin 3, p. 391–443.

FAIRCHILD, H. L., 1929, New York drumlins: Rochester Academy of Science Proceedings, v. 7, p. 1–37.

FAIRCHILD, H. L., 1932, New York moraines: Geological Society of America Bulletin, v. 43, p. 627–662.

FEENSTRA, B. H., 1972, Quaternary geology of the Niagara area, southern Ontario: Ontario Division of Mines, Preliminary Map p. 764.

FEENSTRA, B. H., 1981, Quaternary geology and industrial minerals of the Niagara-Welland area, southern Ontario: Ontario Geological Survey Open-File Report 5361, 260 p.

FISHER, D. D., ISACHSEN, Y. W., AND RICKARD, L. V., 1970, Geologic map of New York: New York State Geological Survey, Map and Chart Series No. 15.

FRANCEK, MARK, 1990, Proglacial lake modification of the central New York drumlin field: a morphometric analysis: Northeastern Geology, v. 5, p. 211–217.

GEIER, R. J., 1980, Glacial stratigraphy and bluff recession along the Lake Erie coast in New York State: Unpublished M.A. Thesis, State University of New York, Buffalo, 91 p.

GEIER, R. J., AND CALKIN, P. E., 1983, Stratigraphy and bluff recession along the Lake Erie coast, New York: New York Sea Grant Institute, 58 p.

GILBERT, G. K., 1898, Boulder pavement at Wilson: New York Journal of Geology, v. 6, p. 771–775.

GILBERT, G. K., 1899, Dislocation at Thirtymile Point, New York: Geological Society of America Bulletin, v. 10, p. 131–134.

GILLETTE, T., 1940, Geology of the Clyde and Sodus Bay quadrangles, New York: New York State Museum Bulletin 320, 179 p.

HARTSHORN, J. H., 1958, Flow till in southeastern Massachusetts: Geological Society of America Bulletin, v. 69, p. 477–482.

HEGLER, D. P., 1974, Lake Ontario shoreline erosion in the regional municipality of Niagara: Unpublished Manuscript, Applied Science Department, Civil Engineering College, University of Waterloo, Waterloo, Ontario, 346 p.

HOUGH, J. L., 1959, Prehistoric Great Lakes of North America: American Scientist, v. 51, p. 84–109.

HUTCHINSON, D. R., POMEROY, P. W., WOLD, R. J., AND HILLS, H. C., 1979, A geophysical investigation concerning the continuation of the Clarendon-Linden fault across Lake Ontario: Geology, v. 7, p. 206–210.

KINDLE, E. M., AND TAYLOR, F. B., 1913, Niagara Folio, New York: U. S. Geological Survey Geological Atlas, Folio No. 190, 33 p.

LEVERETT, F., 1902, Glacial formations and drainage features of the Erie and Ohio basin: U. S. Geological Survey Monograph 41, 802 p.

MULLER, E. H., 1974, Origins of drumlins, *in* Coates, D. R., ed., Glacial Geomorphology: 5th Annual Geomorphology Symposium, State University of New York at Binghamton, p. 187–204.

MULLER, E. H., 1975, Quaternary geology of New York, Niagara Sheet 1:250,000: New York State Geological Survey, Map and Chart Series No. 28.

MULLER, E. H., 1977, Late glacial and early postglacial environments in western New York: New York Academy of Sciences, Annals v. 288, p. 223–233.

MULLER, E. H., 1983a, Dewatering during lodgement of till, *in* Evenson, E. B., Schluechter, C., and Rabassa, J., eds., Tills and Related Deposits: A. A. Balkema, Rotterdam, p. 13–18.

MULLER, E. H., 1983b, Glacial geology in west central New York: progress since the pioneering investigations of H. L. Fairchild: Rochester Academy of Science Proceedings, v. 15, p. 78–84.

MULLER, E. H., 1988, Introduction and overview, *in* Friends of the Pleistocene, 51st annual meeting—Guidebook—Late Wisconsinan Deglaciation of the Genesee Valley: Department of Geological Sciences, State University College, Genesee, New York, p. 6–7.

MULLER, E. H., AND PREST, V. K., 1985, Glacial lakes in the Ontario Basin: *in* Karrow, P. F., and Calkin, P. E., eds., Quaternary Evolution of the Great Lakes: Geological Association of Canada Special Paper 30, p. 213–229.

SALOMON, N. L., 1974, Stratigraphy of glacial deposits along the south shore of Lake Ontario, New York: Unpublished M.S. Thesis, Syracuse University, Syracuse, New York, 77 p.

SHEPARD, F. P., 1954, Nomenclature based on sand-silt-clay ratios: Journal of Sedimentary Petrology, v. 24, p. 151–158.

SLATER, G., 1929, Drumlins of New York State: New York State Museum Bulletin 281, p. 3–19.

SMITH, A., 1990, Glacial stratigraphy of Niagara County, New York: Unpublished M.A. Thesis, State University of New York, Buffalo, 159 p.

TESMER, I. H., 1963, Geology of Chautauqua County, New York, Part I stratigraphy and paleontology (Upper Devonian): New York State Museum Bulletin 391, 65 p.

YOUNG, R. A., 1983, The geologic evolution of the Genesee Valley region and early Lake Ontario: a review of recent progress: Rochester Academy of Science Proceedings, v. 15, p. 85–98.

LATE WISCONSINAN AND HOLOCENE COASTAL EVOLUTION OF THE SOUTHERN SHORE OF LAKE MICHIGAN

MICHAEL J. CHRZASTOWSKI

Illinois State Geological Survey, 615 E. Peabody Drive, Champaign, Illinois 61820–6964

AND

TODD A. THOMPSON

Indiana Geological Survey, 611 Walnut Grove, Bloomington, Indiana 47405

ABSTRACT: The late Wisconsinan and Holocene coastal evolution of southern Lake Michigan contrasts with the coeval history of ocean-coast settings. Multiple transgressive and regressive events occurred, and rates of lake-level change were often greater than the most rapid eustatic sea-level changes. A succession of lower high-lake maxima is recorded in mainland beaches, spits, and beach-ridge/dune complexes across the Chicago/Calumet lacustrine plain. The plain, which extends approximately 120 km from north of Chicago to the Indiana-Michigan border, was the sink for net-southerly littoral transport. During the high-lake phases between 14.5 ka and about 3.5 ka, littoral transport from the eastern and western lake shores terminated in separate spits on opposite ends of the lacustrine plain. Since about 3.5 ka, littoral transport converged along the southern shore. Gradual changes in coastal geomorphology, brought about by littoral processes acting within an overall trend of lake-level decline over the past 2,500 years, formed the modern coastal geography. The Chicago River was transformed from a westward- to an eastward-flowing drainage; littoral-sediment accretion resulted in an extensive beach-ridge/dune complex and a 35-km stream-mouth deflection forming the Grand Calumet River. A model for the coastal sedimentary evolution during the transgressive phases indicates minimal-sediment supply until rate of lake-level change declined and a peak lake level was reached. Wave erosion along the glacial-bluff lake margins could then supply the littoral-transport system. The overall depositional history of the south Lake Michigan coast is that of a regressive and progradational system.

INTRODUCTION

The late Wisconsinan and Holocene coastal evolution of the southern Lake Michigan coast contrasts markedly with the evolution of ocean coasts. Whereas most ocean coasts have experienced a single transgressive event, during the past 14,500 years the southern Lake Michigan coast has experienced multiple transgressive and regressive events, with lake level fluctuating as much as 18 m higher and 61 m or more lower than present (Hansel and others, 1985). Lake levels have changed rapidly, often at rates several times greater than the most rapid eustatic sea-level changes experienced by ocean coasts during the same time period.

The southern Lake Michigan coast preserves a record of this dynamic shoreline evolution across the Chicago/Calumet lacustrine plain (Fig. 1). The plain was transgressed repeatedly during a series of successively lower high-lake phases. Despite the urban development of the Chicago metropolitan region, the relict coastal depositional features remain geomorphically distinct. The coastal sedimentary record is preserved for study in the region's parks, cemeteries, and forest preserves.

This paper presents an overview of the evolution of the southern lake Michigan coast over the past 14,500 years. The coastal setting includes the type localities of coastal features used to define the late Wisconsinan and Holocene lake-level history of Lake Michigan and is described in the geologic literature dating from the late 1800s to the mid-1900s (e.g., Leverett, 1897; Alden, 1902; Goldthwait, 1908; Leverett and Taylor, 1915; Wright, 1918; Baker, 1920; Bretz, 1939, 1943, 1951, 1955). More recent studies have examined the coastal geology and processes of select segments of this coast (e.g., Larsen, 1985; Thompson, 1989, 1990; Fraser and others, 1990). This is the first study to examine the southern Lake Michigan shoreline from the perspective of a regional system and to focus on the late Wisconsinan and Holocene coastal sedimentary processes, evolution, and paleogeographic change.

GEOLOGIC SETTING

Lake Michigan occupies a depression formed by the erosion of multiple Pleistocene continental glaciations and most recently occupied in Wisconsinan time by the Lake Michigan lobe of the Laurentide ice sheet (Hough, 1958). Mean lake level is about 177 m above mean sea level (MSL). Lakes Michigan and Huron are connected through the Straits of Mackinac and function hydraulically as one lake with a single outlet at Port Huron (Fig. 1).

Lacustrine and glaciolacustrine silt and clay up to 18 m thick blanket the lake bottom at depths below 30 m (Lineback and others, 1972; Wickham and others, 1978). Shallower depths have a veneer of sand and mixed sand and gravel. Locally, the veneer is absent, exposing the underlying glacial till. The till has a maximum thickness in the range of 15 to 30 m and overlies bedrock of Devonian and Mississippian shales, siltstones, and dolomite (Hough, 1958). No lake-bottom end moraines have been identified across southern Lake Michigan (Lineback and others, 1974; Wickham and others, 1978).

Adjacent to or landward of the lacustrine plain is a series of juxtaposed end moraines consisting of gray, clayey glacial till with pebbles and cobbles (Willman, 1971). Crest elevations are generally in the range of 200 to 250 m, or about 25 to 75 m above present mean lake level. The moraines are breached by a Y-shaped cut formed by the junction of two channels, the Des Plaines channel on the north and the Sag channel on the south (Figs. 1, 2). The breach, known as the Chicago outlet, was an outlet for ancestral Lake Michigan when lake levels where at least 3 m above present. Lake drainage through the Chicago outlet led to the Illinois River and through it to the Mississippi River. The Chicago outlet is floored by a bedrock sill at an elevation of 180 m, which determined the minimum-lake elevation to activate the outlet.

Between the end moraines and the present lakeshore is the Chicago/Calumet lacustrine plain. The glacially formed,

Quaternary Coasts of the United States: Marine and Lacustrine Systems, SEPM Special Publication No. 48

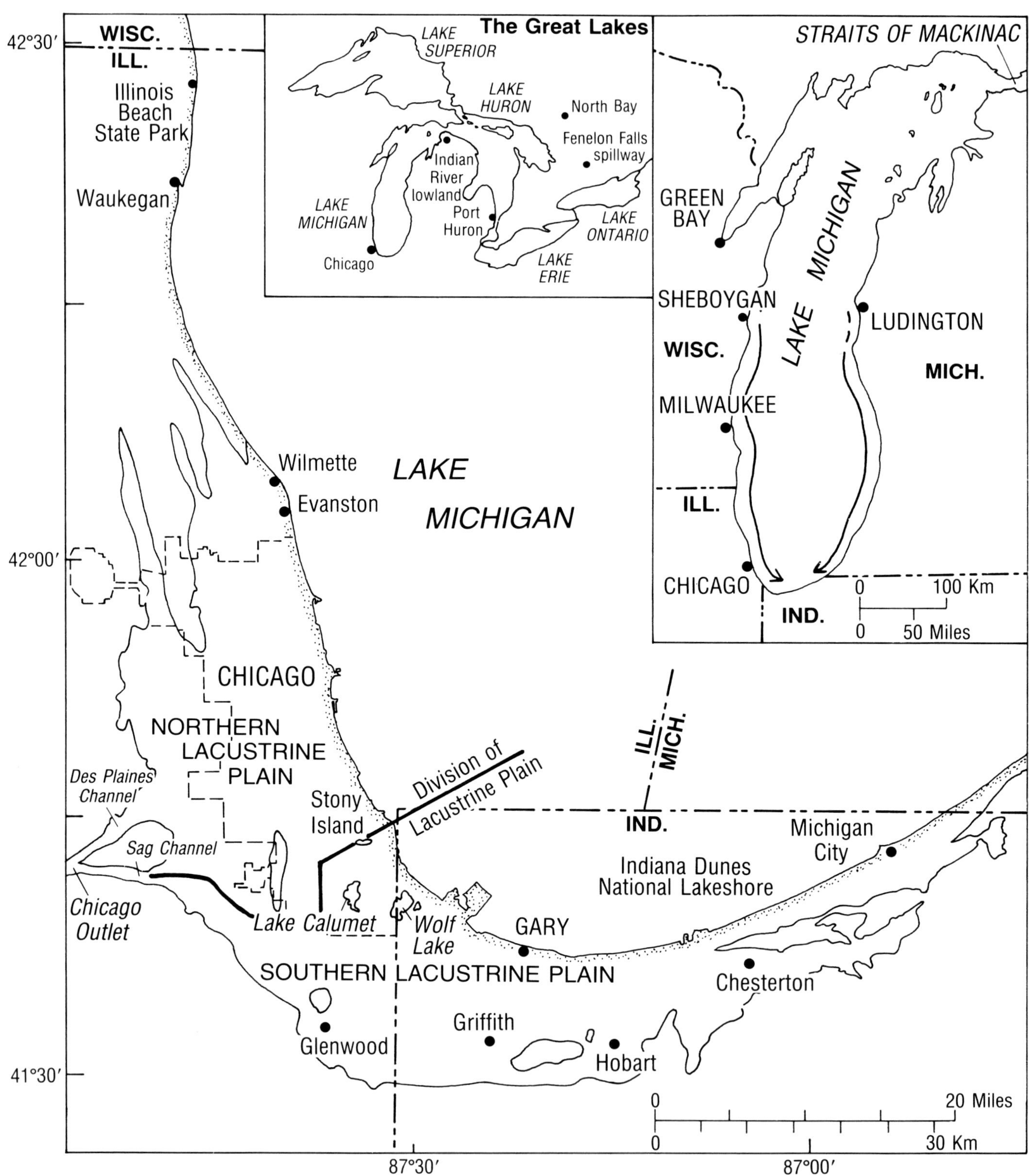

FIG. 1.—Location map of the Illinois/Indiana shore of southern Lake Michigan showing extent of the Chicago/Calumet lacustrine plain and its division into northern and southern parts. Lake Michigan inset shows extent of natural-state, net-southerly littoral transport along the western and eastern lakeshores.

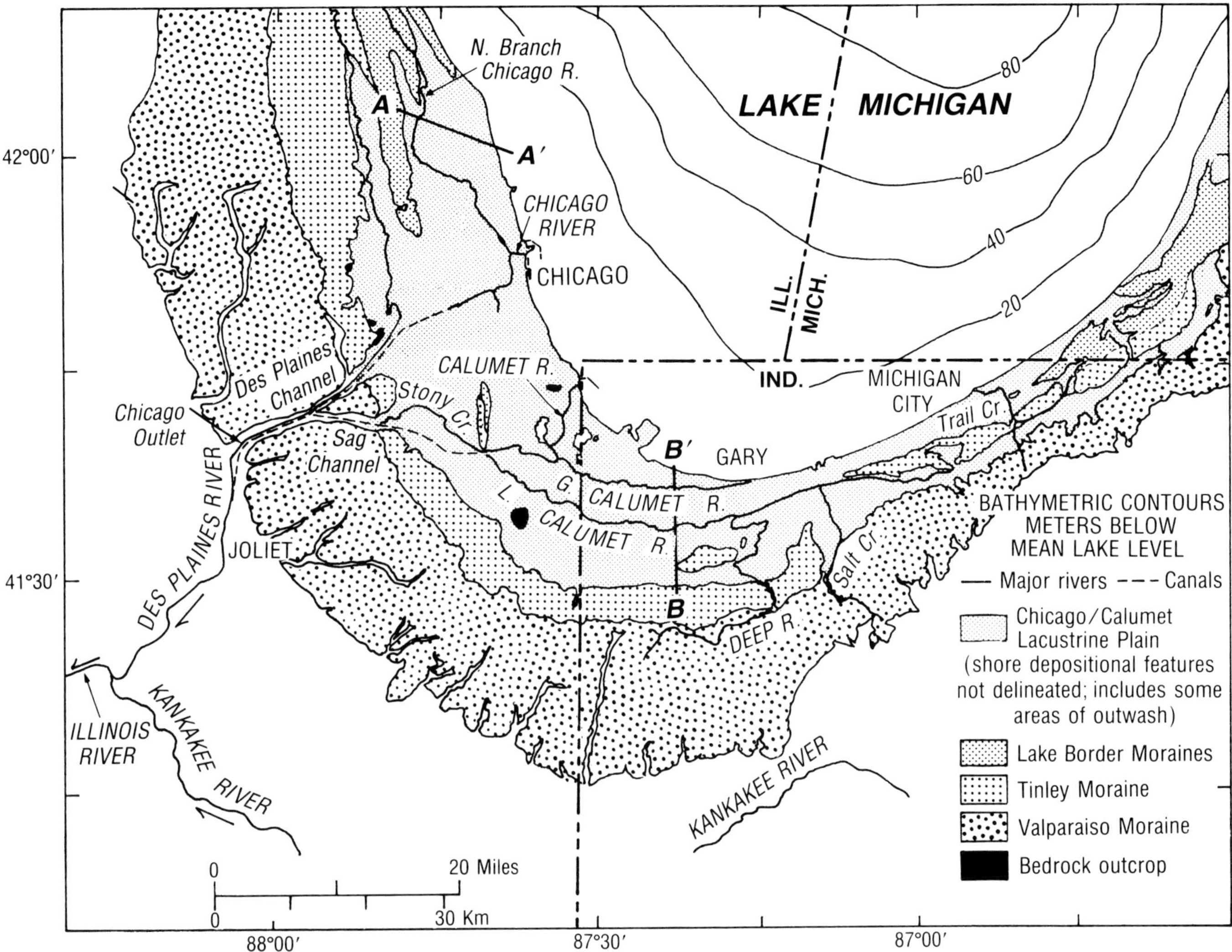

FIG. 2.—Geologic setting of the southern Lake Michigan coast showing end moraines, major streams, and generalized lake bathymetry (after Schneider and Keller, 1970; Willman and Lineback, 1970; and Wickham and others, 1978).

lakeward-sloping surface is a crescent-shaped area of approximately 2,300 km^2 extending about 120 km from near Wilmette, Illinois, to the Indiana-Michigan border and continuing northward in a narrow band along the Michigan shore. Approximate slope across the basal ground moraine ranges from a minimum of 1:1,100 near the Illinois-Indiana border to a maximum of 1:200 fronting the end moraines on the eastern part of the plain.

The lacustrine-plain surface is primarily a wave-scoured ground moraine. Lacustrine silts and clays blanket the surface in former back-barrier settings of the Indiana part of the plain (Schneider and Keller, 1970) and lower elevations of the Illinois part of the plain (Bretz, 1939, 1943; Willman and Lineback, 1970). The prominent depositional features on the plain are sand and gravelly sand relict spits, mainland beaches, and beach-ridge/dune complexes (Fig. 3). Insular, topographic highs of ground and end moraine rise 6 to 13 m above the adjacent plain and form Glenwood Island and Hobart Island near the cities of those names and Blue Island to the west of Lake Calumet (Fig. 1). Silurian dolomite reefs form mound-like bedrock outcrops (i.e., klintar) at several localities.

Modern Coastal Processes

The Illinois-Indiana coast faces a fetch of approximately 480 km to the north along the axis of Lake Michigan (Fig. 1). Northerly waves have the greatest height and net influence along these shores (Fox and Davis, 1976). For any given year, average maximum wave height along the Illinois shore is 2 m and individual waves rarely exceed 3 to 3.6 m (U.S. Army Corps of Engineers, 1953). The source of beach sediment is almost exclusively wave erosion of lake-border glacial bluffs, updrift beaches, and dune complexes. Stream contributions are insignificant. According to stream-mouth deflections shown on historical maps and other geomorphic indicators of littoral-transport directions (e.g., Hands, 1970; Hunter and others, 1979; Jacobson and Schwartz, 1981), the natural-state net-southerly littoral transport originated near Sheboygan, Wisconsin, on the

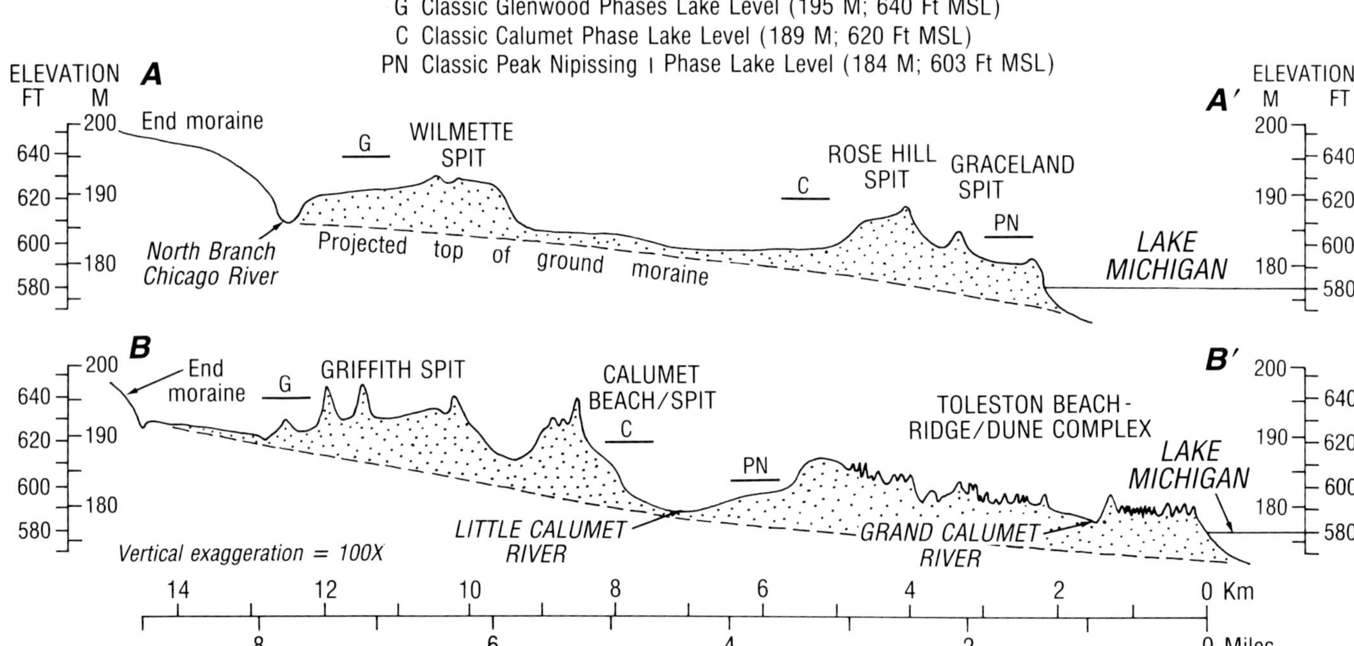

FIG. 3.—Cross sections showing the successively lower elevations of spits and beach-ridge/dune systems of the Chicago/Calumet lacustrine plain. See Figure 2 for cross-section locations.

western shore and Ludington, Michigan, on the eastern shore (Fig. 1). Chrzastowski (1990) estimated the natural-state littoral transport along the Illinois lakeshore to be at least 78,000 m^3/yr. An additional transport process in winter is ice rafting of sediments incorporated in nearshore ice (Miner 1989a, b: Barnes and others, 1990).

Division of Northern and Southern Lacustrine Plains

The history of littoral-sediment supply to the Chicago/Calumet lacustrine plain and the relief on the plain allow division of the plain into northern and southern areas (Fig. 1). The "divide" runs along the approach to the Sag channel, extends for about 5 km along a 3-m-high north-south scarp on the ground moraine, and crosses the 6- to 7.5-m-high bedrock outcrop of Stony Island. North of the line the littoral-sediment supply was primarily from glacial-bluff erosion of the western lakeshore, whereas south of the line the littoral-sediment supply prior to about 2.5 ka was from erosion of the eastern lakeshore. After about 2.5 ka, littoral-sediment supply to the southern lacustrine plain was from both the western and eastern shores.

LAKE-LEVEL HISTORY

During late Wisconsinan and early to mid-Holocene time, the level of ancestral Lake Michigan fluctuated as a result of several factors: (1) recession and advance of the ice margin, which opened and closed different outlets at different elevations; (2) differential isostatic changes in the elevation of outlets; (3) fluvial downcutting of outlets; and (4) major increases or decreases in the volume of water entering the lake basin. The history of lake-level change in southern Lake Michigan has been discussed by Leverett (1897), Alden (1902), Goldthwait (1908), Leverett and Taylor (1915), Wright (1918), Bretz (1955), and Hough (1955, 1958). Hansel and others (1985) and Hansel and Mickelson (1988) provide recent reviews with radiocarbon-age control.

Figure 4 shows a local relative lake-level curve for southern Lake Michigan for the past 14,500 years. The inclusion of an Algonquin Phase of higher-than-present lake level is a departure from previous studies and is based on our preliminary reevaluation of stratigraphic and geomorphic evidence along the eastern Indiana shore. The lake phases are defined for times of lake levels above or below present mean lake level of 177 m. The causes of lake-level fluctuations during the different phases are discussed in subsequent sections of this paper.

Elevations of the various lake levels are still debated. Early investigators defined the high-lake levels based on the elevation and geomorphology of abandoned-shore features. Recent stratigraphic studies by Thompson (1989) and Thompson and others (1991) of contacts between foreshore and shoreface deposits have shown that the "classic" lake levels are typically 1 to 4.5 m too high. For simplicity, we refer to the "classic" elevations for the late Wisconsinan and early Holocene deposits since the actual elevations are not well known. Actual elevations are used for the late Holocene (approximately 4.7 ka to present) based on studies by Thompson (1989), Thompson and others (1991), and supporting studies by Larsen (1985).

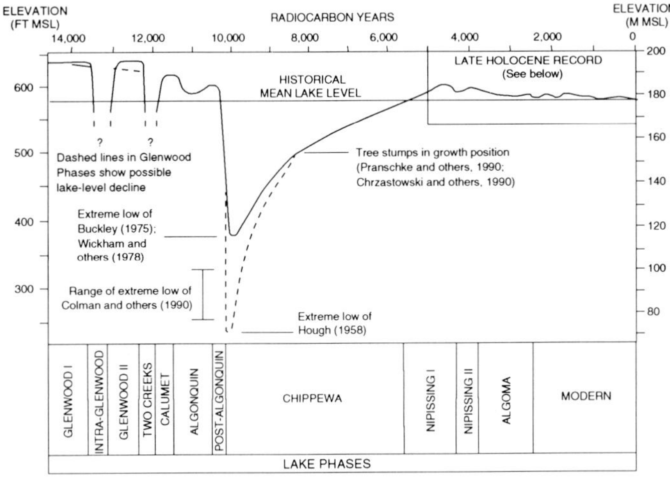

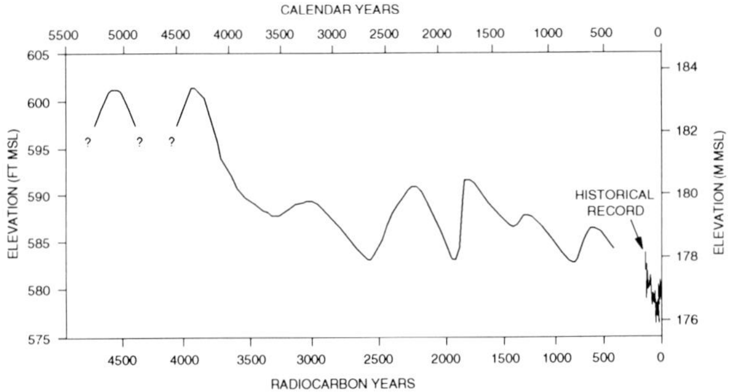

FIG. 4.—Southern Lake Michigan local relative lake-level curve for the past 14,500 radiocarbon years (after Hansel and others, 1985). Inset shows the upper limits of lake levels over the past 4,700 years based on beach-ridge litho- and chronostratigraphy (after Larsen, 1985; Thompson, 1989; Thompson and other, 1991).

Isostatic Rebound

The history of differential isostasy across the Chicago/ Calumet lacustrine plain is not well understood. Raised beaches on the Wisconsin and Michigan sides of the lake prompted Evenson (1973) to suggest that no tilting of Glenwood Phase to modern shorelines has occurred south of a line from near Sheboygan, Wisconsin, to the Indiana-Michigan border (Fig. 1). Recent modeling by Clark and others (1990) suggests that isobases run nearly north-south across the southern lake-shore and that the Illinois part of the lacustrine plain has subsided relative to most of the Indiana plain. Figure 3 illustrates how the equivalent-phase beaches of the northern lacustrine plain (cross section A-A′) are lower than those of the southern lacustrine plain (cross section B-B′) relative to the classic lake levels. This discrepancy could be due to the isostatic factor or could relate to differences in the timing of spit development during a range of falling-lake levels.

GLENWOOD PHASE COASTAL EVOLUTION

The Glenwood Phases began about 14.5 ka when recession of the Lake Michigan ice lobe formed a narrow, crescent-shaped lake (glacial Lake Chicago) between the ice margin and the morainal dam. Lake discharge was through the Chicago outlet. A distinction of Glenwood I and Glenwood II Phases (Fig. 4) is based on stratigraphic evidence for a low-lake level between 13.5 ka and 13.0 ka caused by ice recession allowing drainage through channels in the Straits of Mackinac vicinity (Farrand and Escham, 1974). How low the lake fell is unknown. Because of the scarcity of radiocarbon ages, the distinction of phases cannot be made with certainty for all Glenwood shore features, and in this discussion all such features are referred to as has having developed during the Glenwood Phases (plural). The Glenwood lake level of 195 m (640 ft) was defined by Leverett (1897) based on the geomorphology of relict beaches at Glenwood, Illinois (Fig. 1). Elsewhere, shore features identified as Glenwood may be below the 195-m elevation, in which case their identification is based on their being the highest elevation shore features and/or their occurrence above 189 m, which is the elevation of the next lower (i.e., Calumet) lake level.

The Glenwood Phases incorporate the only time when shoreline extended along all of the morainal highland that forms the landward margin of the plain (Fig. 5). As the ice margin receded northward, the earliest coastal development likely occurred along the southwestern and southern margin of the lake where fetch was greatest. Long periods with shore ice likely "locked up" wave-induced sediment erosion and transport. The two largest spits, Wilmette spit and Griffith spit, may represent transport rates that were high because of unstable and highly erodible morainal margins of the proglacial lake.

Northern Lacustrine Plain

Shoreline features at 195 m are poorly defined along the northern lacustrine plain. In places a shoreline may be suggested by a subtle notch or break in slope on the morainal highlands, but as noted on maps by Bretz (1939, 1943), designation of a shoreline at this elevation is based primarily on an extrapolation of the type-locality elevation. If lake level across the northern lacustrine plain reached 195 m, the lack of shore features is consistent with the weakness of wave and littoral-transport processes because of restricted fetch and prevalence of coastal ice.

A total of five individual spits developed on the northern lacustrine plain: the Wilmette, Oak Park, and La Grange spits, which were mainland-attached, and the St. Maria and Beverly Hills spits on Blue Island (Fig. 5). Differences in the morphology of the Glenwood spits relate to differences in littoral-sediment supply, local slope of the lacustrine plain, and probably timing and duration of spit development.

The largest of the northern-lacustrine-plain spits is the 8.5-km-long Wilmette spit, which built southward from the most lakeward of the Lake Border Moraines (i.e., Highland Park Moraine). This spit, a product of western lakeshore glacial-bluff erosion, partially enclosed Skokie Bay, which extended up the drowned valley of Skokie River. Extending 2.5 km southwesterly from the southern extent of spit ridges is a broad, fan-shaped spit platform comprising the distal

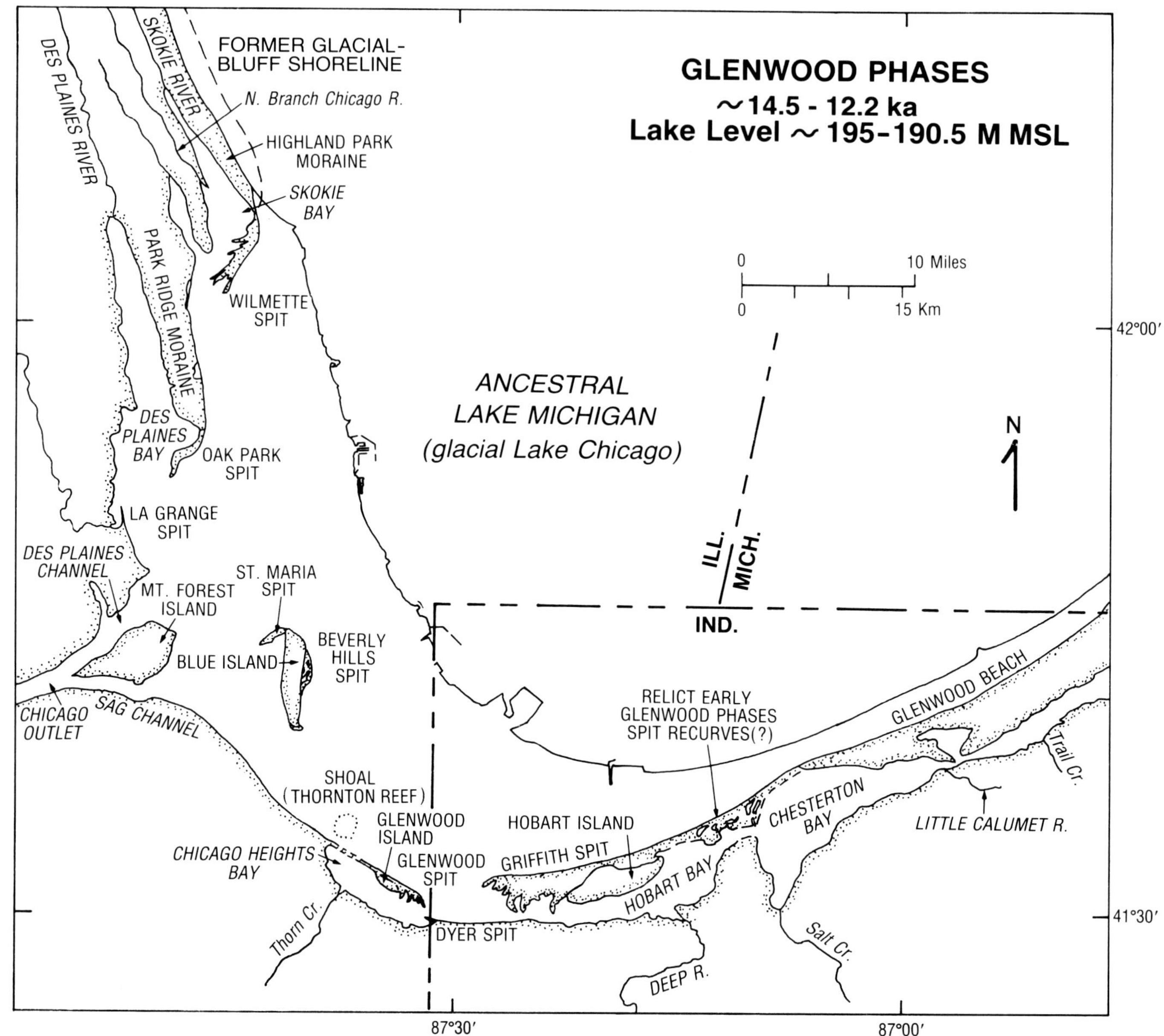

FIG. 5.—Coastal paleogeographic reconstruction for the Chicago/Calumet lacustrine plain during the Glenwood Phases (after Bretz, 1939, 1943; Schneider and Keller, 1970).

spit accretion. The spit platform is outlined by the arcuate course of the North Branch Chicago River (Fig. 2).

Southern Lacustrine Plain

The Glenwood beaches of the southern lacustrine plain are aligned along the lakeward side of the Tinley/Lake Border moraines and bridge between morainal islands and peninsulas (Fig. 5). Near the Illinois-Indiana state line, a pair of spits projects on opposite sides of a reentrant that corresponds to the point of maximum curvature in the arcuate morainal front. The reentrant became the terminus for net-littoral transport from the neighboring shores to the west and east. Within the reentrant, the short (2.8 km) Dyer spit likely predates maximum development of the neighboring spit pair.

Between the approach to the Sag channel and the embayment near Dyer, littoral transport bridged to the ground moraine that formed Glenwood Island and ultimately terminated in the 6-km-long Glenwood spit. Wave processes along this shore would have been influenced by the bedrock outcrop of Thornton Reef, which has crest elevations in the range 189 to 193 m. The outcrop was a shoal in the early Glenwood Phases and became an island and later a headland as lake level declined.

In the early Glenwood Phases of coastal evolution in Indiana, the discontinuities and crenulations in the morainal uplands prevented a continuous littoral-transport stream. Spits and bars developed off the downdrift (southwest) ends of the morainal islands and peninsulas. Southwest of the arrow-shaped peninsula of the Tinley/Lake Border moraines are sand bodies previously mapped as dune deposits (Schneider and Keller, 1970); however, their arcuate patterns indicate that these dune sands sit atop recurves of a relict spit. The proximal spit was apparently eroded by the crossing of the Little Calumet River during the late or post-Glenwood Phases.

Once a continuous beach extended from the arrow-shaped peninsula of the Tinley/Lake Border moraines to Hobart Island, Hobart Bay became partially enclosed. To the east, Chesterton Bay occupied the present valley of the Little Calumet River. The terminus for littoral transport was the Griffith spit, the largest of all Glenwood spits and built off the western end of Hobart Island. None of the spit ridges connect with the mainland shoreline to the south, and thus during early spit development Hobart Bay was connected with the lake westward around the distal end of the spit. The sediment source was eroding glacial deposits along the eastern lakeshore, but how far northward along this shore the littoral stream extended is unknown. During the Glenwood Phases numerous coastal embayments existed north of the Indiana-Michigan border that required bridging before continuous transport extended to the Griffith spit.

Spit Morphology and Lake-Level History

The existing model for lake level during the Glenwood Phases indicates a static level at 195 m (640 ft), but this may only be applicable to the oldest mainland beaches along the southern lacustrine plain. Figure 3 shows that the 195-m elevation is above the dunes of the Wilmette spit and near the tops of the dunes of the Griffith spit. Moreover, Thompson (1989) has shown that foreshore deposits along select Glenwood beaches of the Indiana shore ranged from 192 to 190.5 m (630 to 625 ft).

Common to the Wilmette and Griffith spits are successive spit ridges that have axes rotated progressively lakeward (Fig. 6). The smaller Oak Park and Glenwood spits also have lakeward-rotated axes. Since these four spits developed in different directions, the rotation is not related to changes in direction of maximum fetch or predominant wave approach. The lakeward rotation could be due to shoreface accretion and independent of lake-level change. However, because of the potential for lake-level change at that time, the rotation is considered consistent with a falling lake level. For each of the spits, the rather uniform width and areal extent of the lakeward recurves suggest that the lake-level decline was not a short-lived decline during the end of a high-lake phase, but rather a gradual trend. The La Grange, Dyer, and Blue Island spits do not record the lakeward rotation, probably because they were abandoned during different stages of the lake-level decline.

Present topography suggests the lakeward-rotated spits began forming at a lake level of about 193.5 m (635 ft) and continued until lake level fell to about 190.5 or possibly 189 m (625 to 620 ft). Development during a lake-level decline makes these depositional-regressive spits. The mechanism for a gradual lake-level decline during the Glenwood Phases was likely the progressive downcutting of the Chicago outlet to its bedrock sill. In its early history, the outlet would have had an accumulation of valley-train fill. Previous investigators recognized the importance of removing this material as a means to lower lake levels (Leverett, 1897; Alden, 1902; Goldthwait, 1908; Bretz, 1951, 1955; Hough, 1958). However, they did not recognize a long-term trend in falling lake levels through the Glenwood Phases that possibly ended with a lake level near or at the level of the next high-lake phase (i.e., Calumet Phase). Waves likely reworked and modified the lakeward parts of these spits to varying degrees during the initial peak lake level of the Calumet Phase.

TWO CREEKS PHASE LACUSTRINE-PLAIN DRAINAGE PATTERNS

The Two Creeks Phase began shortly before 12 ka (Fig. 4) when the receding Lake Michigan lobe opened a northern outlet across the Indian River lowlands near the Straits of Mackinac and later through the straits. The Two Creeks Phase lasted about 200 to 300 years. It is unknown how low lake level fell. During the Two Creeks Phase, the significant geomorphic developments across the lacustrine plain are related to drainage patterns.

At least two arterial streams would have drained the northern lacustrine plain. The ancestral North Branch Chicago River would have flowed southward on the west (bay) side of the relict Wilmette spit, followed a broad arc around the distal end of the spit platform, and likely headed lakeward beneath what is now the distal Rose Hill spit. The ancestral Stony Creek likely originated in the approach to the Sag channel and flowed eastward (Fig. 2). It is uncertain what the drainage pattern may have been for an ancestral South Branch Chicago River, but an eastward drainage is likely.

The arterial streams of the southern lacustrine plain did not have the area to bypass the Glenwood Phases beaches and therefore breached these strandlines. During the Two Creeks Phase, Thorn Creek, Deep River, Salt Creek, and the Little Calumet River would have each flowed from its Glenwood beach crossing northward across the northward-sloping lacustrine plain.

CALUMET PHASE COASTAL EVOLUTION

Readvance of the ice margin into the Lake Michigan basin at about 11.8 ka closed northern outlets, raised lake level to transgress the Chicago/Calumet lacustrine plain, and reactivated the Chicago outlet. This phase lasted about 600 to 800 years (Fig. 4). The ''classic'' Calumet lake level is 189 m (620 ft).

Northern Lacustrine Plain

The major change in coastal geography in the Calumet Phase compared to those of the Glenwood Phases is the reduced extent of drowned valleys, increased areal extent of Mt. Forest and Blue Islands, and the occurrence of Worth Island in the approach to the Sag channel (Fig. 7). The principal depositional feature was Rose Hill spit, built by net-southerly littoral transport from the glacial-bluff shore-

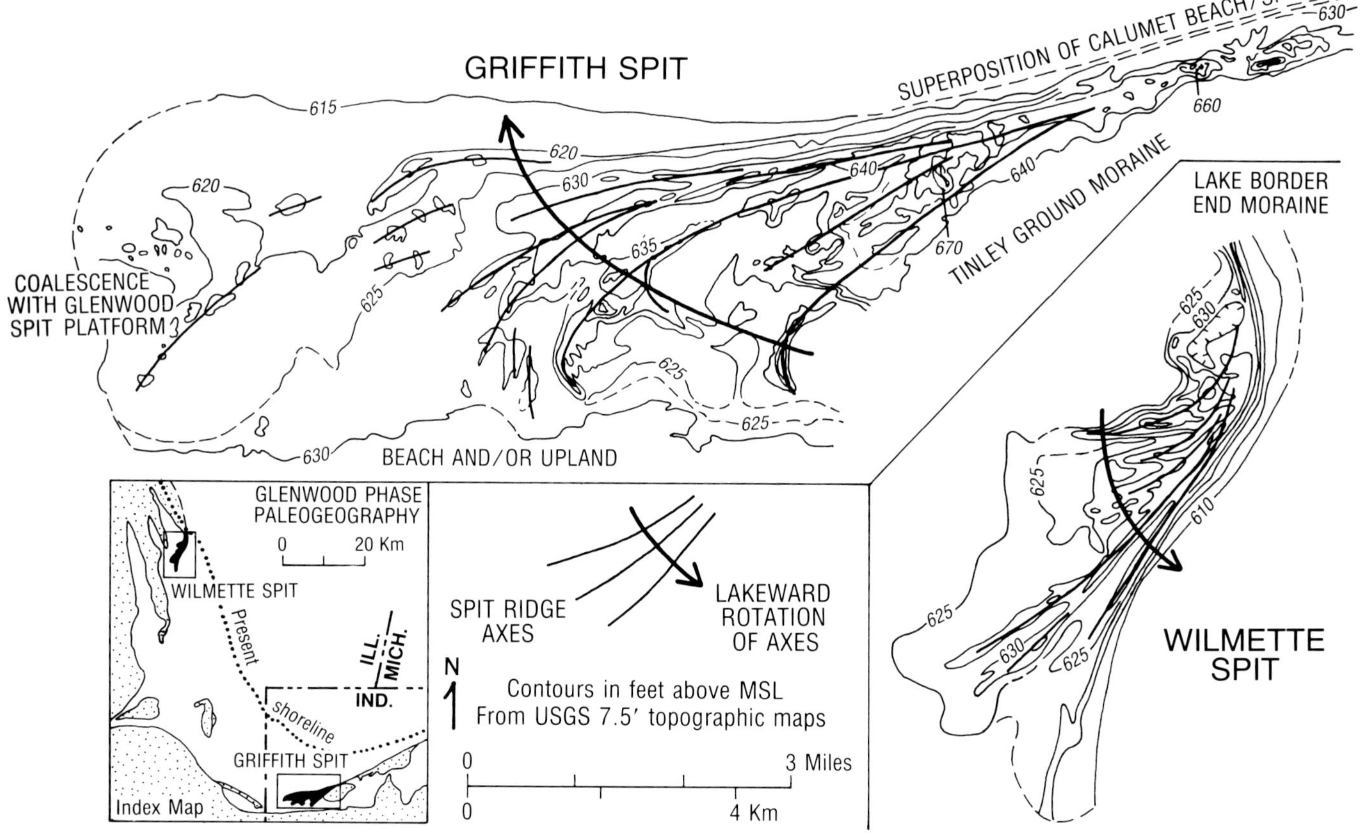

FIG. 6.—Contour maps of the Glenwood Phases Wilmette spit and Griffith spit showing lakeward rotation of the axes of the successively younger spit ridges.

lines north of the lacustrine plain (Schneider and Hansel, 1990). Not only has the bluff line eroded, but probably also the original proximal spit (Fig. 7).

Stratigraphic studies are needed to evaluate accurately the depositional history of Rose Hill spit, but geomorphic characteristics suggest a two-phase history of first a submerged spit and then emergence. As shown in Figure 3, the crest elevation of the spit is consistent with the spit being a submerged feature at a 189-m Calumet level. Differential isostatic adjustments could be responsible for part, but likely not all, of this low elevation. The linearity of the northern part of the spit is suggestive of deposition below water. If the northern part of the spit was emergent during its initial development, wave refraction would have likely produced an overall westward recurve comparable to that of the proximal Wilmette and Oak Park spits (Fig. 5).

The recurve of the distal end of the Rose Hill spit could have formed in the second phase of spit development when lake level fell from the Calumet level and temporarily held a few (4 to 5) meters lower. Baker (1920) first noted that the morphology of the distal Rose Hill spit suggested development below Calumet level but prior to the building of the adjacent and younger Graceland spit. An Algonquin Phase lake level fits the required timing and elevation (Fig. 4) and is a plausible hypothesis in light of newly identified Algonquin Phase coastal features on the Indiana shore discussed later.

Southern Lacustrine Plain

The Calumet Phase coastal geography of the southern lacustrine plain contrasts with those of the Glenwood Phases in that a more regular shoreline developed. In addition, the littoral-sediment transport from the eastern lakeshore extended farther westward, at least as far as Thornton Reef (Fig. 7). The Calumet spit and beach in Indiana was used by Leverett (1897) in defining the Calumet lake level.

The steep slope of the eastern lacustrine plain resulted in the Calumet beach being juxtaposed to the Glenwood beaches. Farther west, a more gradual lacustrine-plain slope caused the Calumet shoreline to diverge from the relict Griffith spit and build the 25-km-long Calumet spit. This spit was aligned to the headland formed by Thornton Reef. Recurved spit ridges suggest that at least for some time open water existed between the spit and headland. Beaches that curve around the headland and continue toward the approach to Sag channel are mapped as Calumet beaches based on elevation (Bretz, 1939, 1943), but lack of age control prevents ruling out possible late–Glenwood Phases development. Whether Calumet Phase littoral transport continued westward of Calumet spit is uncertain. Drainage from

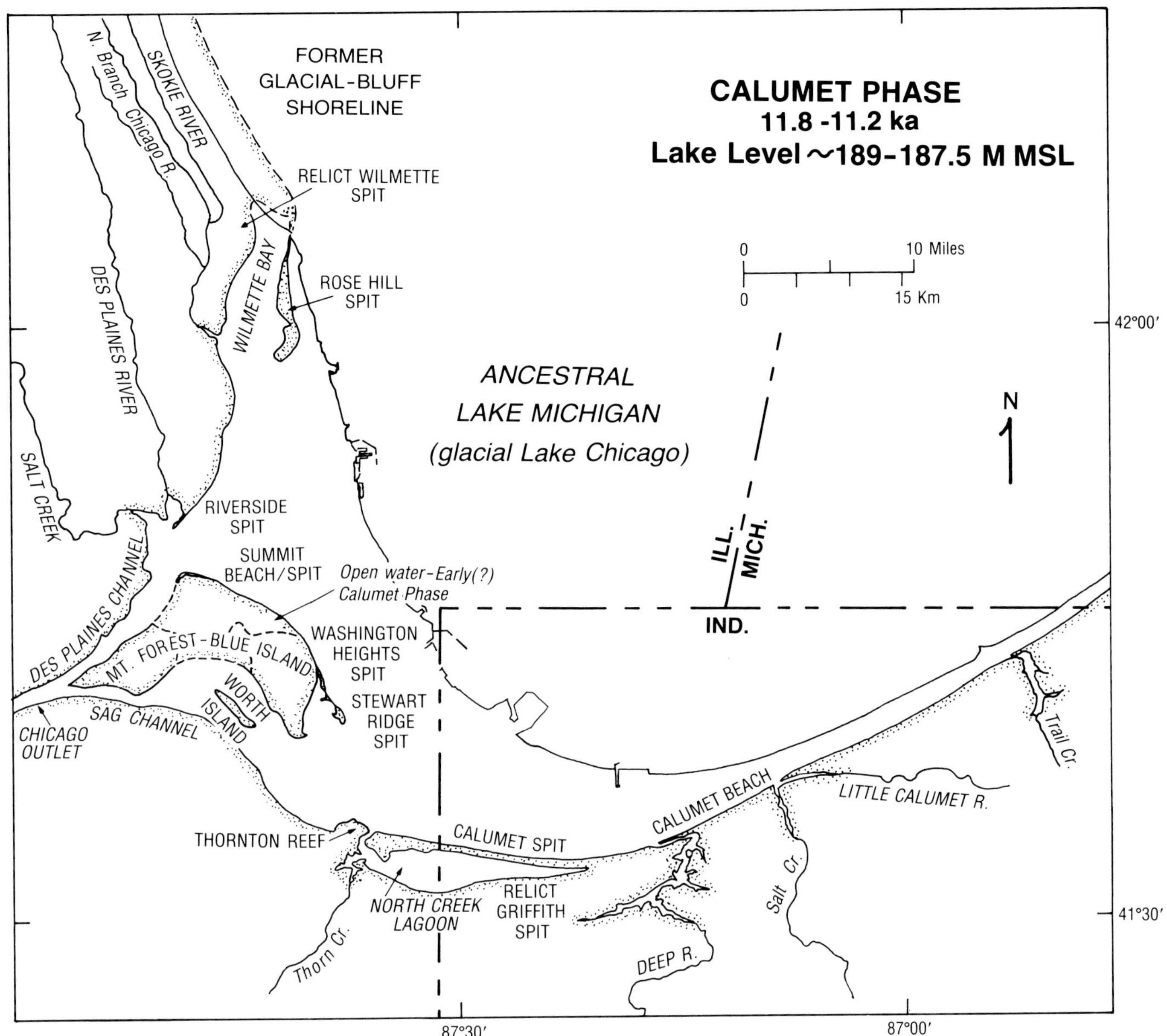

FIG. 7.—Coastal paleogeographic reconstruction for the Chicago/Calumet lacustrine plain during the Calumet Phase (after Bretz, 1939, 1943; Schneider and Keller, 1970).

Thorn Creek and North Creek lagoon would likely have caused at least an intermittent breach between the spit and the headland.

Development of the Calumet beach interfered with stream discharge to the lake and to varying degrees caused downdrift deflection of the stream mouths. The best example is the approximate 5-km westward deflection of Deep River (Fig. 7).

ALGONQUIN PHASE COASTAL EVOLUTION

The Calumet Phase ended about 11.2 ka when deglaciation of the Indian River lowland and later the nearby Straits of Mackinac joined Lakes Michigan and Huron and established the Algonquin Phase of ancestral Lake Michigan (Larsen, 1987). Ancestral Lakes Michigan-Huron then had the Chicago outlet as well as the additional connection to the Kirkfield outlet (i.e., Fenelon Falls spillway) in Ontario (Fig. 1). Along the southern shore of Lake Michigan, lake level rapidly fell when the two lakes were joined, possibly falling to about 180 m (592 ft; Fig. 4). Lake level subsequently rose slightly during the early part of this phase in response to isostatic rebound of the Kirkfield outlet (Larsen, 1987), and lake level reached a maximum altitude of 184 m (605 ft) about 10.6 ka (Futyma, 1981). The phase ended about 10.3 ka as ice recession opened successively lower northern outlets.

Although the Algonquin Phase is recognized in Lake Huron and in the northern part of Lake Michigan, Algonquin coastal features have not been previously recognized along the southern shore of Lake Michigan. Larsen (1987) suggested that the Main (or late) Algonquin Phase of Lake Huron is equivalent to the later part of the Calumet Phase in Lake Michigan. We propose that the two phases are distinguishable within the Chicago/Calumet lacustrine plain.

Thompson (1989, 1990) studied the eastern part of what had traditionally been identified as Calumet beach. However, the basal foreshore elevations (184 m; 605 ft) and the radiocarbon dates from beneath and within the nearshore sediments (10.4 to 10.6 ka) are consistent with this being an Algonquin beach. Rather than being related to a single-lake phase, this coastal deposition apparently records erosion into the Calumet beach and subsequent accretion and juxtaposition of an Algonquin beach. The mainland beach can be traced westward to Deep River where it terminates in a poorly defined ridge and platform that deflects Deep River to the west and is topographically below a more prominent ridge at a Calumet level. From Deep River westward to near the terminus of the Calumet spit, the lakeward margin of the Calumet beach is a 2.7-m-high scarp. The base of the scarp is at an altitude of 184 to 183 m and is equivalent in elevation to the basal foreshore deposits in the Algonquin beach to the east. This scarp apparently supplied sediment to the west, forming a sediment lobe lakeward of the distal end of the Calumet spit.

The poor development of Algonquin coastal features along southern Lake Michigan is likely due to low littoral-sediment supply before the lake level stabilized about 10.6 ka (possibly as late at 10.4 ka) and the short duration (approximately 300 years) of the peak lake level. Across most of the lacustrine plain, the Algonquin Phase was apparently a nondepositional or erosional event.

CHIPPEWA PHASE LACUSTRINE-PLAIN DRAINAGE PATTERNS

The Chippewa Phase lasted about 4,000 to 4,500 radiocarbon years and incorporates the extreme low-lake levels of ancestral Lake Michigan (Fig. 4). Shortly before 10 ka, the receding ice margin opened successively lower, isostatically depressed outlets across southern Ontario, and a series of short-lived, successively lower lakes occupied the Lake Michigan basin (post-Algonquin Phase). About 10 ka, the receding ice margin opened the isostatically depressed outlet at North Bay (Fig. 1), allowing eastward lake drainage through the ancestral Ottawa River. Ancestral Lakes Michigan and Huron rapidly fell to extreme low levels. For the southern Lake Michigan basin, previous studies have suggested extreme low elevations of 74 m (243 ft) and 116 m (381 ft; Hough, 1955, 1958; Buckley, 1975; Wickham and others 1978). Recent coring and geophysical studies by Colman and others (1990) and Foster and Colman (1991) suggest the extreme low elevation was between 74 and 116 m, possibly in the range 80 to 100 m. From the extreme low, subsequent lake-level rise was primarily a response to the isostatic uplift of the North Bay outlet (Hough, 1955, 1958; Harrison, 1972; Hansel and others, 1985).

Recent discovery of in-place tree stumps at 25-m depth (elev. 152 m) on the floor of southwestern Lake Michigan provides an unprecedented benchmark for the Chippewa Phase lake-level rise. Stumps of oak (Quercus sp.) and ash (Fraxinus sp.) provide radiocarbon dates averaging 8.3 ka (Chrzastowski and others, 1990; Pranschke and others, 1990). From the Chippewa Phase extreme low lake level at 10 ka to the stump elevation gives a rate of lake-level rise of 3.6 m/100 yrs using an assumed extreme low elevation estimate of 90 m. Seismic-reflection surveys within 1 km of the site document buried paleochannels incised up to 6 m into the lake-bottom glacial till. Lacustrine-plain streams that are candidates for possible association with these buried paleochannels are the ancestral streams of the Illinois-Indiana state-line vicinity, such as Stony Creek, Thorn Creek, or Deep River.

The effect of Chippewa Phase lake lowering on lacustrine-plain drainage was likely more significant than during the Two Creeks Phase because of the longer duration and extreme lowering of base level. The ancestral North Branch Chicago River likely rounded the distal spit platform of the Rose Hill spit and headed lakeward in a channel now buried by Nipissing Phases and younger shore deposits. Eastward drainage presumably occurred along the ancestral Stony Creek. An ancestral South Branch Chicago River likely originated near the approach to the Des Plaines channel and flowed eastward.

The arterial streams of the southern lacustrine plain had deflected stream mouths caused by Calumet and Algonquin Phases beach accretion, and during these phases the streams may have only intermittently breached bars across the stream mouths. The Chippewa Phase would have resulted in incision and continual occupancy of the breaches, from which stream flow would have been northward across the northward-sloping lacustrine plain.

NIPISSING AND ALGOMA PHASES COASTAL EVOLUTION

The Nipissing Phases began when continued isostatic uplift of the North Bay outlet caused southern Lake Michigan to exceed its present level about 6.0 to 5.5 ka (Gutschick and Gonsiewski, 1976; Hansel and others, 1985; Larsen, 1985). As lake level rose to 180 m, the Chicago outlet was reactivated, and for a short time ancestral Lakes Michigan and Huron had three outlets at North Bay, Chicago, and Port Huron (Fig. 1). Continued isostatic uplift abandoned the North Bay outlet. The peak Nipissing lake level of 184 m (603 ft) occurred at about 4.7 ka (Hansel and others, 1985), at which time about half the area of the Chicago/Calumet lacustrine plain was submerged or receiving littoral-sediment accretion. From the Chippewa Phase submerged-tree stumps (elev. 152 m; 8.3 ka) to the peak Nipissing level (elev. 184 m; 4.7 ka), a straight-line transgression rate is 88 cm/100 yrs. Following the peak Nipissing lake levels, there was a long-term lowering trend, largely due to downcutting of the Port Huron outlet. Short-term fluctuations in this downward trend are attributed to climatic influence (Fraser and others, 1990).

Northern Lacustrine Plain

As lake level rose to 184 m, the lacustrine plain was again submerged along embayments leading to the Des Plaines and Sag channels (Fig. 8). For the first time in a high-lake phase, the crest of Stony Island was emergent. An embayment again occupied the low-lying area on the landward side of the Rose Hill spit. In the early peak lake level, a shoreline likely reoccupied most or all of the spit's lakeward side, and waves may have modified the relict spit shoreline. The major depositional feature of the early Nipissing Phases was the Graceland spit, which extends 19 km from the point of shoreline reorientation at Wilmette to a terminus just north of the Chicago River (Fig. 8). A distinctive feature of the southern Graceland spit is an orientation change along a broad arc pointing the spit in a southeasterly direction toward the historical (and present) shoreline.

The earliest topographic and geologic mapping along the central Chicago lake shore gives no indication of a continuation of the Graceland spit on the south side of the Chicago River (Leverett, 1897; Alden, 1902; Bretz, 1939, 1943). Artist renderings of the earliest settlement near the Chicago River mouth suggest dunes of the then active beach (Andreas, 1884), but not a high ridge that could have been a Nipissing beach. Beginning about 6 km south of the Chicago River is a series of spits and beach ridges trending from the historical shoreline toward the southwest and south (Fig. 9). The orientation of the relict beaches north and south of the Chicago River and continuity of littoral transport require an ancient strandline lakeward of the historical shoreline, as shown in Figure 9 and as previously suggested by Baker (1920, plate 44).

Subsequent to the building of the Graceland spit, gradual lake-level decline during the late Nipissing II Phase and the Algoma Phase shifted strandlines eastward and a series of shore-parallel beach ridges formed lakeward of the Graceland spit (Fig. 9). The corresponding spits and beach ridges south of the central Chicago lakeshore have a progressive easterly rotation recording the falling lake level and beach deposition along the low-gradient plain.

Evolution of the Chicago River

The Chicago River, with its two branches and forks, is the single drainage for the northern lacustrine plain. The two branches of the river differ significantly. The North Branch is about 58 km long and originates within the Lake Border morainic system (Fig. 2). The South Branch is about 17 km long and originates on the lacustrine plain. In the natural setting a portage between the West Fork South Branch and a tributary to the Des Plaines River was a subtle drainage divide, readily facilitating stream flow and canoe passage during slightly elevated river stages (Andreas, 1884). Engineering has significantly altered the river, first with removal of a stream-mouth deflection, then construction of canals to reverse flow away from the lake to the Illinois River (Fig. 2).

Figure 10 shows a model for the evolution of the Chicago River based on present topography, the relict shoreline features shown in Figure 9, and the late Holocene lake-level curve (Fig. 4). About 4 ka the ancestral North Branch Chicago River entered an embayment landward of the early Graceland spit (Fig. 10A). By about 3.8 ka, continued southward development of spits and beach ridges formed a lagoon (here named Chicago lagoon) occupying the still-drowned lacustrine plain (Fig. 10B). Evidence of a lagoon within this area is lacustrine silt and clay recording quiet-water deposition (Bretz, 1939, 1943; Willman and Lineback, 1970).

Lake-level decline and emergence of a bar across the lagoon's inlet isolated the lagoon from the lake probably shortly after 3.8 ka. The lagoon then responded to water level in the Chicago outlet, which was influenced by continued lake drainage through the Sag channel. As declining water level in the lake and Chicago outlet reduced areal extent of the lagoon, the ancestral North Branch Chicago River flowed farther across the emerging lagoon bottom, ultimately becoming a tributary to the Des Plaines River (Fig. 10C). This drainage pattern may have persisted for some time. Erosion along the arcuate protrusion of the lakeshore would have aided the river in breaching the narrow divide between the river and lake, possibly during a river-flood stage. Such a breach would form the Chicago River main branch, reverse flow in what became the South Branch, and form the Chicago Portage between the head of the South Branch and a westward-flowing tributary to the Des Plaines River (Fig. 10D).

Southern Lacustrine Plain

All coastal deposition along the Indiana coast from the Nipissing Phases to present is part of a 61-km-long and up to 10-km-wide geomorphic feature called the Toleston Beach. The early Toleston Beach, built in the early Nipissing Phases (about 4.7 to 4.5 ka), was a barrier beach approximately 3 to 4.5 km lakeward of the relict Calumet and Algonquin beaches, with the separation increasing from east to west corresponding to a more gradual lacustrine-plain slope (Fig. 8). During the early Nipissing Phases, lake level was high enough to drown the back-barrier area and form a lagoon (here named Calumet lagoon). Marl, organic-rich sand, and peat more than 3 m thick occur along the lagoonal area. After about 3.8 ka, lake-level decline drained the lagoon, which then became the westward drainage of the Little Calumet River.

The Toleston spit extended to the approaches of the Sag channel, possibly by about 4 ka. Subsequent lake-level decline and continued littoral-sediment supply resulted in progradation along the western half of the spit. The Nipissing and Algoma Phases are recorded by a series of more than 30 beach ridges in the western, landward part of the Toleston Beach (Fig. 11). To the east, where the lacustrine-plain slope is greater, the ridges merge and are superimposed by parabolic dune fields.

Continued lake-level decline and reduced lake drainage through the Sag channel facilitated beach accretion to restrict the lake/channel connection. By about 2.5 ka or slightly before, there was no longer any significant lake drainage through the channel, and littoral transport from the western lake shore was continuous past Stony Island and southward to the Toleston Beach. This was the first time during the late Wisconsinan and Holocene coastal evolution that the western-lakeshore littoral transport reached the southern la-

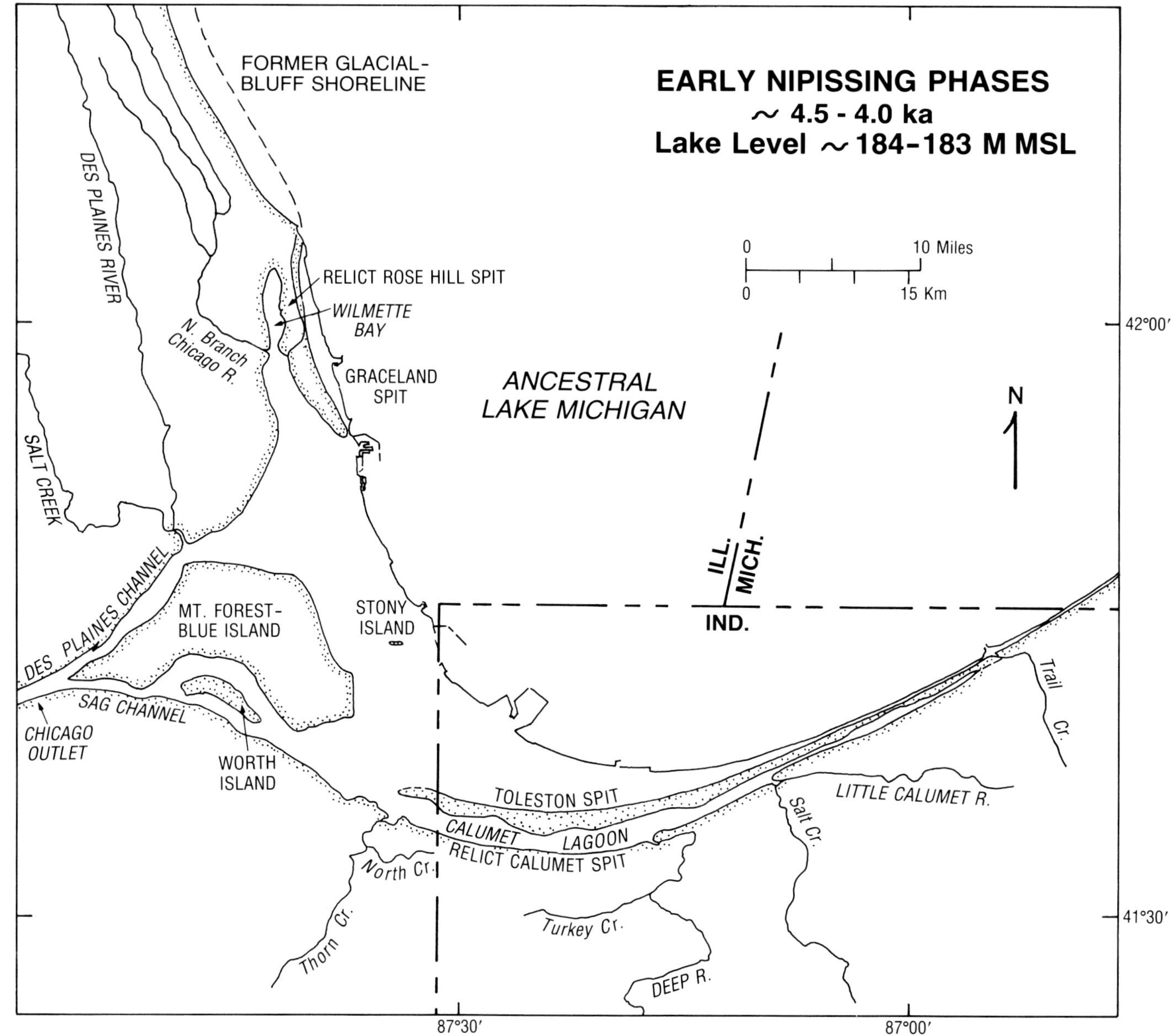

FIG. 8.—Coastal paleogeographic reconstruction for the Chicago/Calumet lacustrine plain during the early Nipissing Phases (after Bretz, 1939, 1943; Schneider and Keller, 1970).

custrine plain. The concave-lakeward configuration of the early and subsequent spits south of Stony Island strikingly records the wave-front divergence into the former lake embayment near the Illinois-Indiana border (Fig. 11).

MODERN PHASE COASTAL EVOLUTION

The Modern Phase encompasses the lake history from about 2.5 ka to the present (Fig. 4). During this time, an overall lake-level decline occurred, but climatically influenced highstands occurred at about 2.4 ka, 1.8 ka, 0.6 ka, and possibly 1.2 ka (Fig. 4).

Modern Phase coastal deposition along the Chicago/Calumet lacustrine plain has essentially been limited to the southern lacustrine plain, particularly along the western half of the Indiana coastal area where the Toleston Beach continued to prograde (Fig. 12). Here was the last shoreline reentrant across the lacustrine plain and the site of initial littoral-sediment convergence from the western and eastern lakeshores. Radiocarbon dating of organic materials in the inter-ridge swales provides the opportunity to place time lines on the beach-ridge development and to reconstruct the coastal evolution with some detail. By about 1.5 ka, a spit building southward from the Stony Island vicinity reached the southern lake shore and enclosed an open-water area, forming ancestral Lake Calumet (Fig. 12). A similar spit advance about 1.1 ka formed the ancestral Wolf Lake. In both cases the spit growth was apparently rapid, following

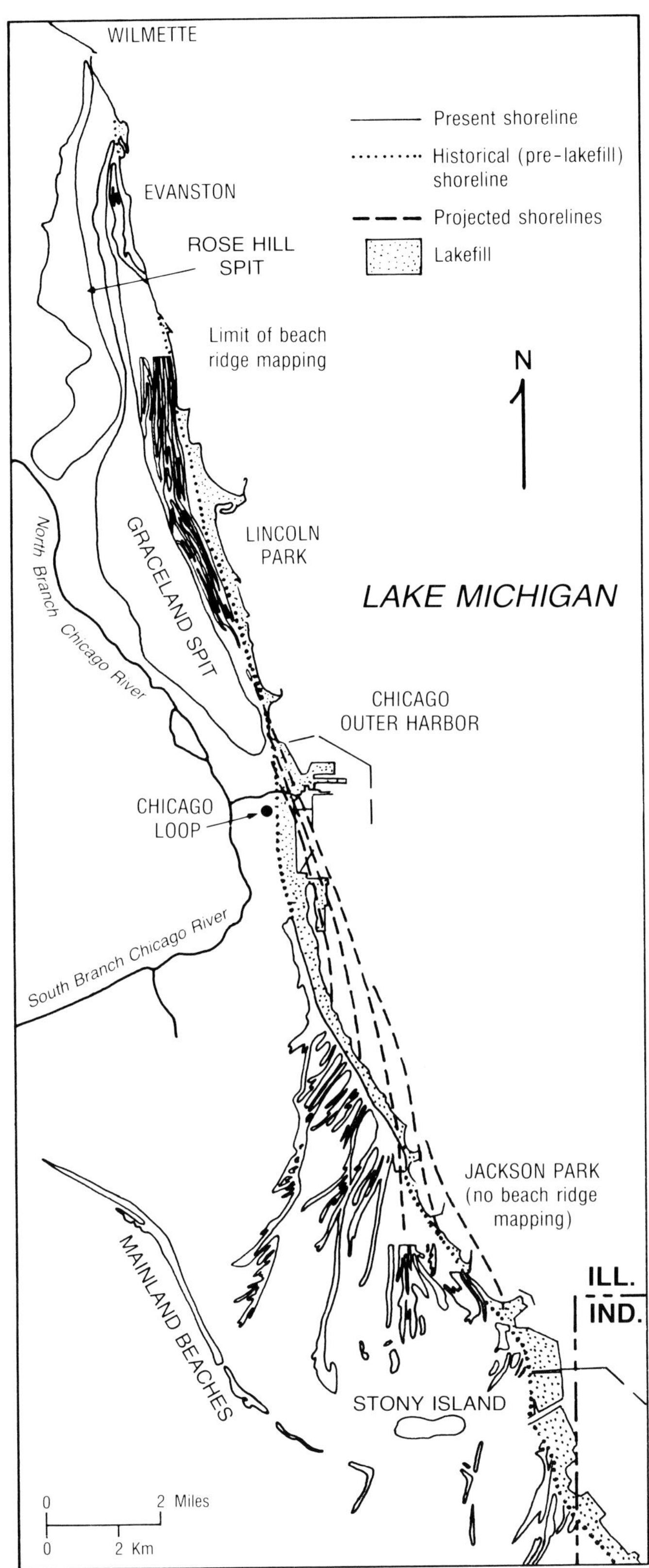

FIG. 9.—Nipissing to Modern Phases spits, beach ridges, and mainland beaches to the north and south of the Chicago central-lakeshore area as shown on maps of surficial geology by Alden (1902) and Bretz (1939; 1943). Dashed lines show projected arcuate shorelines linking the relict coastal features.

high-lake levels that possibly contributed to increased littoral-sediment supply. To within the past few hundred radiocarbon years, a headland was formed first by Stony Island and then by emergent or shoal areas of the lacustrine plain to the east. The preserved beach ridges in the state-line vicinity and the western Indiana shore intercept the historical shoreline and project lakeward, indicating erosion of former beach ridges in the present open-water area (Fig. 11).

The Little Calumet River was significantly influenced by declining lake levels and beach-ridge progradation. Its drainage was directed westward to the Sag channel when lake level was above the 180-m sill elevation of the Chicago outlet, and was directed eastward toward the lake when lake level fell below that threshold. Subsequent to the high-lake level of 1.8 ka, drainage was eastward, forming the Grand Calumet River, which was initially a tributary to Lake Calumet and later likely occupied the outlet of that lake (Fig. 12). With continued littoral-sediment supply from the western lake-shore, the stream mouth was continually deflected eastward, as recorded by recurved beach ridges along the north (lakeward) side of the river. By historical time the deflected stream mouth was 35 km eastward from the point where flow direction changed from westward to eastward in the approach to the Sag channel.

Since the initial convergence of western-and eastern-lakeshore littoral transport along the southern lake margin, the extent and limits of the zone of net convergence have likely migrated with time. Initially, the convergence and primary sediment sink was the beach-ridge complex of the western Indiana shore. The additional sink was the 0.5- to 1-km-wide domal and parabolic-dune fields that dominate the eastern half of the Indiana shore. The dunes of the Indiana Dunes National Lakeshore and State Park rise up to 56 m above present lake level and are the largest dunes of the lacustrine plain. During historical time, prior to lakeshore development, the zone of littoral-transport net convergence was along the dune shoreline based on the eastward-deflected mouth of the Grand Calumet River and the westward-deflected mouth of Trail Creek at Michigan City. Nearly all of the littoral-sediment supply is fine to medium sand capable of aeolian transport and dune building, which has apparently precluded coastal progradational features.

MODEL FOR LITTORAL-SEDIMENT SUPPLY

Models for the late Wisconsinan and Holocene evolution of ocean coasts involve the development of barrier-lagoon systems on the continental shelf and the landward and upward migration of these systems with glacio-eustatic sea-level rise. Shoreface erosion and overwash allow the sand body to migrate landward, keeping pace with sea-level rise (Swift, 1975). At extremely rapid rates of sea-level rise, beaches may be drowned in place (Sanders and Kumar, 1975). Thompson (1987, 1989) demonstrated that the core of the Calumet and Toleston beaches along the eastern Indiana lake shore preserve transgressive stratigraphic sequences similar to those of ocean-coast transgressive systems. Despite the similarities, we suggest that ocean-coast models for transgressive coastal evolution are probably not

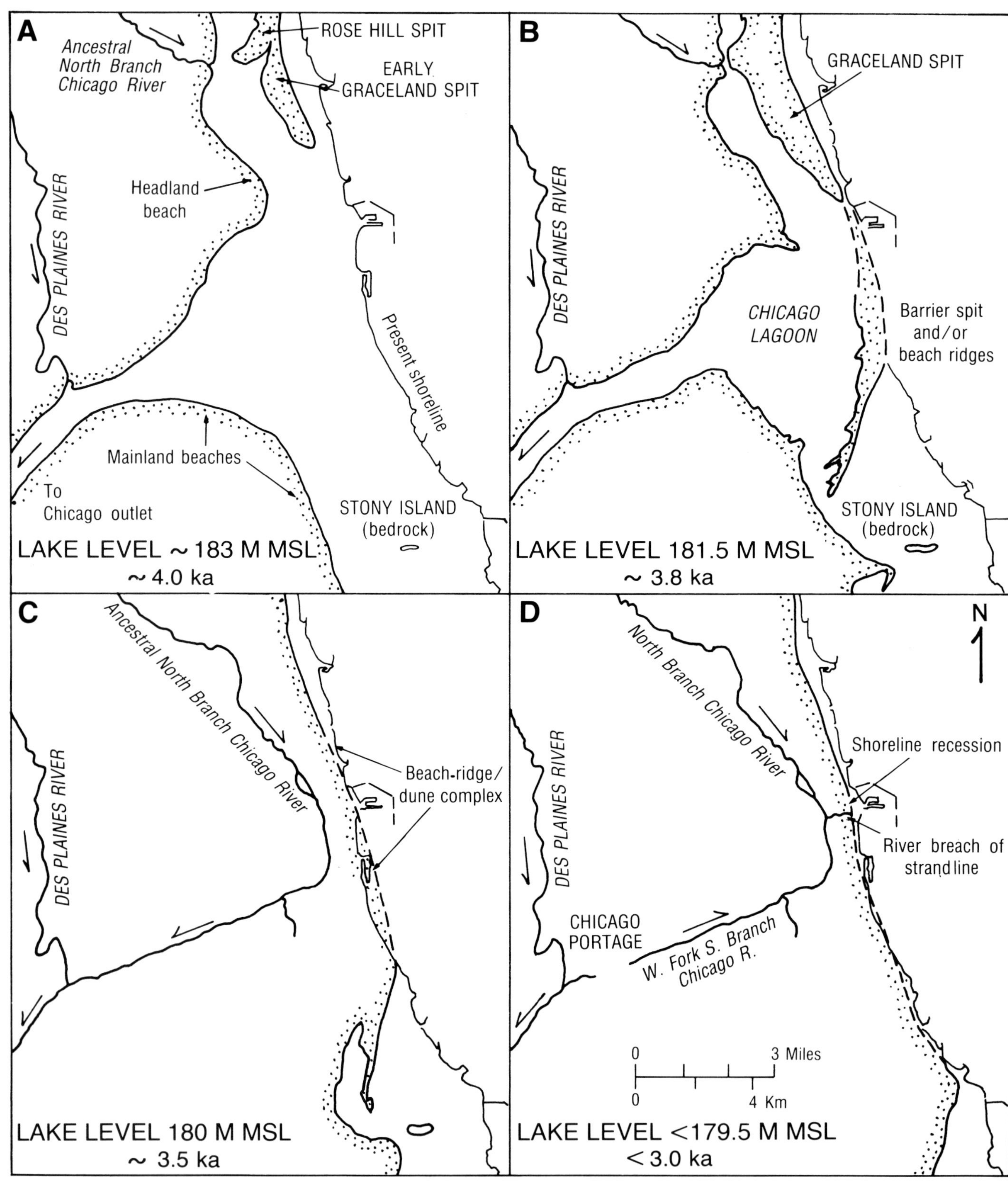

FIG. 10.—Model for the Nipissing to possible Modern Phases coastal evolution leading to the historical drainage pattern of the Chicago River.

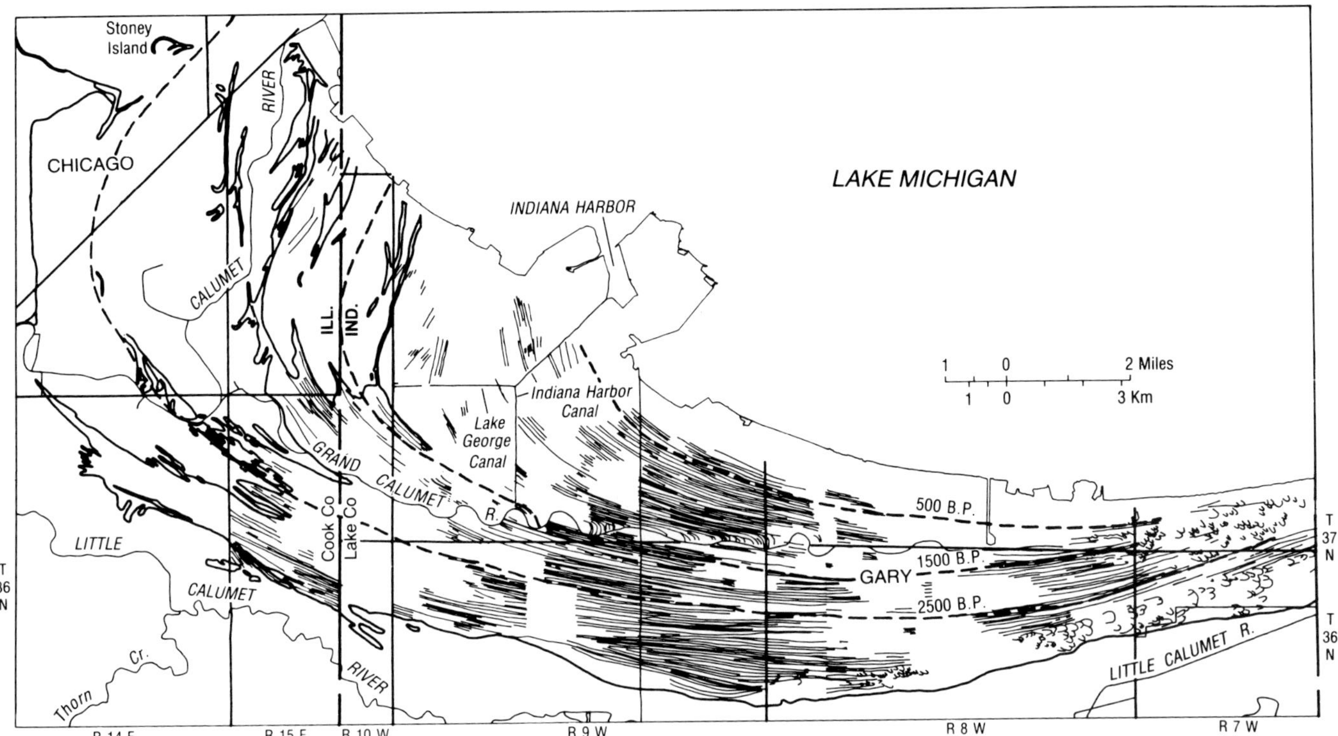

FIG. 11.—Beach-ridge and spit lineaments across the southern lacustrine plain. Time lines are based on the radiocarbon dating of basal marsh deposits from the inter-ridge swales (after Thompson and others, 1991).

applicable to the transgressive phases of the southern Lake Michigan coast, or are applicable only in the later stages of the transgression.

Three transgressive episodes have occurred along the southern Lake Michigan coast: the lake-level rise following the maximum lowstands of the Intra-Glenwood Phase (Intra-Glenwood transgression), the Two Creeks Phase (Calumet transgression), and the Chippewa Phase (Nipissing transgression). At the low-lake levels, waves were eroding the gently sloping ground moraine of the lake basin. Beaches would have developed, but the extremely rapid rates of lake-level rise would have precluded any significant landward and upward migration of the deposits. The general lack of lake-bottom features identified as strandlines (Lineback and others, 1972, 1974; Wickham and others, 1978; Foster and Colman, 1991) suggests that only in rare cases were beaches drowned in place and raises the question whether anything more than rudimentary beaches ever formed. For the Intra-Glenwood and Calumet transgressions, ice damming resulted in extremely rapid lake-level rise, possibly meters per year, since the proglacial lake temporarily had no outlet. The Nipissing transgression was slower, controlled by isostatic uplift of an outlet, but was at least as rapid as the most rapid glacio-eustatic sea-level rise.

Figure 13 shows a model summarizing the relation of rate of lake-level rise and rate of littoral-sediment supply for the southern Lake Michigan coast. The model shows topography and bathymetry depicting generalized relations in relief between the lake basin and border. During rapid lake-level rise, wave action was along the low-gradient lake basin, the rate of sediment supply was minimal, and beach systems were likely poorly developed. Once lake level reached a new controlling outlet, the rate of rise decreased and wave erosion was concentrated along the more steeply sloping lake-border moraines. As lake level then fluctuated at an elevation controlled by an outlet or climatic influence, the rate of littoral-sediment supply and transport reached a maximum. High rates of littoral-sediment supply would have persisted even during the gradual lake-level declines following a peak level because wave erosion still focused along the high-relief lake border.

SUMMARY

The Chicago/Calumet lacustrine plain was a sink for coastal deposition during a series of successively lower high-lake phases of ancestral Lake Michigan for the past 14,500 years. The plain was the terminus of net-southerly littoral transport along both margins of the lake. Differences in lake level between the high-lake phases altered the major loci of coastal deposition by changing the coastal geography. During the high-lake phases between about 14.5 ka and 2.5 ka, the termini of western- and eastern-lakeshore littoral transport were separate spit systems building into opposite ends of the lacustrine plain. After about 2.5 ka, western-lakeshore littoral transport could bypass the remnants of a shoreline reentrant in the present Lake Calumet vicinity, and for the first time littoral transport from both western and eastern lakeshores converged along the southern shore. Lake-level decline and coastal sedimentary processes since

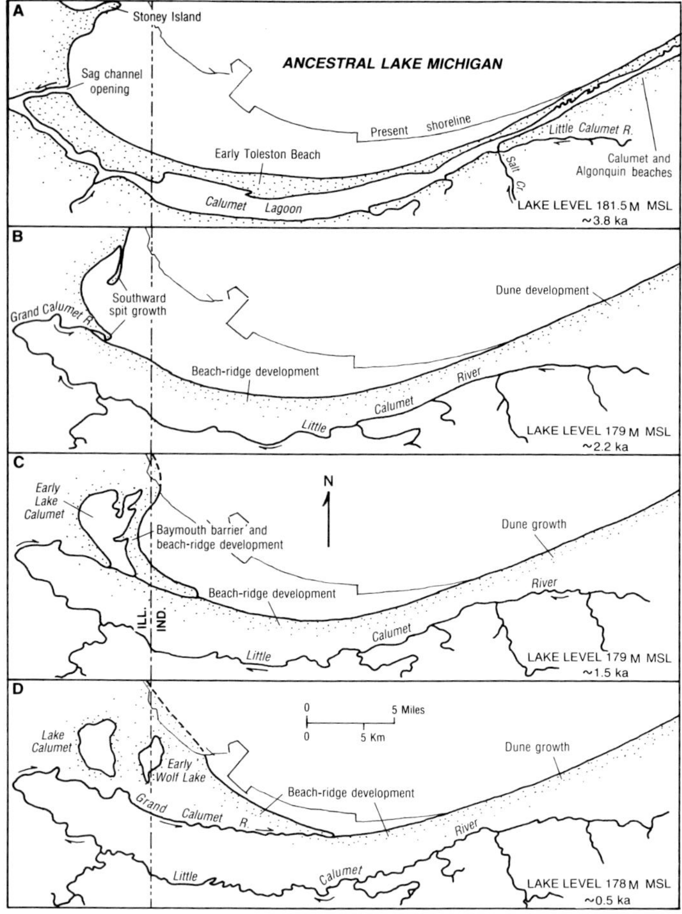

FIG. 12.—Paleogeographic reconstruction of the shoreline progradation of the Toleston Beach since 3.8 ka and the corresponding changes in drainage patterns along the Little and Grand Calumet Rivers.

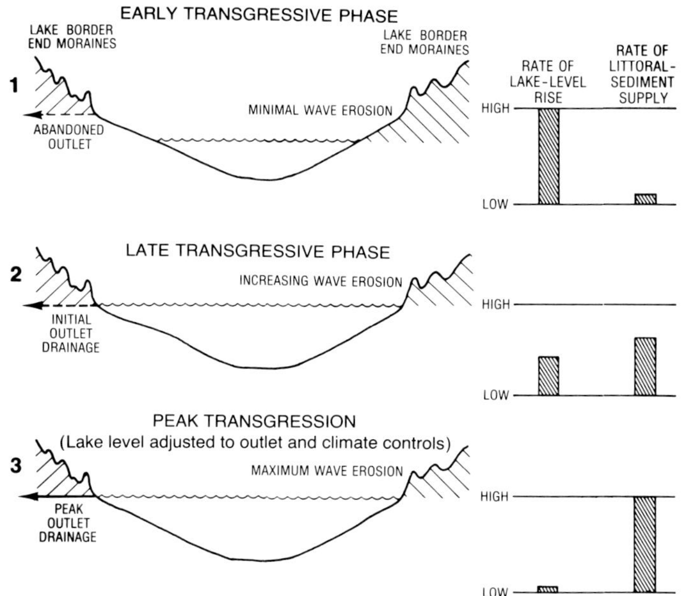

FIG. 13.—Model comparing the rate of lake-level rise and rate of littoral-sediment supply during a transgression of the southern Lake Michigan coast.

about 3.5 ka were critical factors in determining the present drainage patterns of the Chicago, Little Calumet, and Grand Calumet Rivers.

A model for the coastal sedimentary evolution of the south Lake Michigan coast emphasizes shore-parallel (longshore) rather than shore-normal transport processes. During the multiple transgressive phases caused by ice damming or isostatic rebound of outlets, rapid lake-level rise would have likely precluded any significant landward and upward migration of coastal sedimentary deposits. Not until a highstand was reached could wave erosion concentrate along the lake-border highlands and supply the littoral-sediment stream.

ACKNOWLEDGMENTS

This paper results from a study of southern Lake Michigan coastal geology by the Illinois State Geological Survey and the Indiana Geological Survey in cooperation with the U.S. Geological Survey Branches of Atlantic and Pacific Marine Geology. We thank Ralph Hunter, Paul Terpstra, and Richard Berg for their reviews and suggestions for improvement of the manuscript. All illustrations were prepared by the drafting and photography sections of the Indiana Geological Survey.

REFERENCES

ALDEN, W. C., 1902, Description of the Chicago District, Illinois-Indiana: U.S. Geological Survey Atlas, Folio 81, 14 p.

ANDREAS, A. T., 1884, History of Chicago from the Earliest Period to the Present Time, v. 1: A. T. Andreas, Publisher, Chicago, 648 p.

BAKER, F. C., 1920, The life of the Pleistocene or glacial period as recorded in deposits laid down by the great ice sheet: University of Illinois Bulletin, v. 17, 476 p.

BARNES, P. W., REIMNITZ, E., WEBER, W. S., AND KEMPEMA, E. W., 1990, Sediment content and beach profile modification by ice along the coast of southern Lake Michigan, *in* Barnes, P. W., ed., Coastal Sedimentary Processes in Southern Lake Michigan: Their Influences on Coastal Erosion: U.S. Geological Survey Open-File Report 90–295, p. 1–8.

BRETZ, J. H., 1939, Geology of the Chicago region: Part I—general: Illinois State Geological Survey Bulletin 65, 118 p.

BRETZ, J. H., 1943, Chicago areal geologic maps: Illinois State Geological Survey, 24-sheet folio.

BRETZ, J. H., 1951, The stages of Lake Chicago—their causes and correlations: American Journal of Science, v. 249, p. 401–429.

BRETZ, J. H., 1955, Geology of the Chicago region: Part II—the Pleistocene: Illinois State Geological Survey Bulletin 65, 132 p.

BUCKLEY, S. B., 1975, Study of post-Pleistocene ostracod distribution in the soft sediments of southern Lake Michigan: Unpublished Ph.D. Dissertation, University of Illinois, Urbana, Illinois, 189 p.

CHRZASTOWSKI, M. J., 1990, Estimate of natural-state littoral transport along the Chicago lakeshore, *in* Barnes, P. W., ed., Coastal Sedimentary Processes in Southern Lake Michigan: Their Influences on Coastal Erosion: U.S. Geological Survey Open-File Report 90–295, p. 19–26.

CHRZASTOWSKI, M. J., FOLGER, D. W., POLLONI, C. F., PRANSCHKE, F. A., AND SHABICA, C. W., 1990, Geologic setting of the Olson tree site, southern Lake Michigan: Geological Society of America Abstracts with Programs, v. 22, p. A85.

CLARK, J. A., PRANGER, H. S., II, WALSH, J. K., AND PRIMUS, J. A., 1990, A numerical model of glacial isostasy in the Lake Michigan basin, *in* Schneider, A. F., and Faser, G. S., eds., Late Quarternary

Quaternary History of the Lake Michigan Basin: Geological Society of America Special Paper 251, p. 111–123.

Colman, S. M., Jones, G. A., Forester, R. M., and Foster, D. S., 1990, Holocene paleoclimatic evidence and sedimentation rates from a core in southwestern Lake Michigan: Journal of Paleolimnology, v. 4, p. 269–284.

Evenson, E. B., 1973, Late Pleistocene shorelines and stratigraphic relations in the Lake Michigan basin: Geological Society of America Bulletin, v. 84, p. 2281–2298.

Farrand, W. R., and Esham, D. F., 1974, Glaciation of the southern peninsula of Michigan—a review: Academician, v. 7, p. 31–56.

Foster, D. S., and Colman, S. M., 1991, Preliminary interpretation of the high-resolution seismic stratigraphy beneath Lake Michigan: U.S. Geological Survey Open-File Report 91–21, 42 p.

Fox, W. T., and Davis, R. A., Jr., 1976, Weather and coastal processes, *in* Davis, R. A., Jr., and Ethington, R. L., eds., Beach and Nearshore Sedimentation: SEPM Special Publications 24, p. 1–23.

Fraser, G. S., Larson, C. E., and Hester, N. C., 1990, Climatic control of lake levels in the Lake Michigan and Lake Huron basins, *in* Schneider, A. F., and Fraser, G. S., eds., Late Quaternary History of the Lake Michigan Basin: Geological Society of America Special Paper 251, p. 75–89.

Futyma, R. P., 1981, The northern limits of glacial Lake Algonquin in upper Michigan: Quarternary Research, v. 15, p. 291–310.

Goldthwait, J. W., 1908, A reconstruction of water planes of the extinct glacial lakes in the Lake Michigan basin: Journal of Geology, v. 16, p. 459–476.

Gutschick, R. C., and Gonsiewski, J., 1976, Coastal geology of Mt. Baldy Indiana Dunes National Lakeshore south end of Lake Michigan, *in* Guidebook for Fieldtrips, North-Central Section: Geological Society of America Meeting, Western Michigan University, Kalamazoo, Michigan, p. 40–90.

Hands, E. B., 1970, A geomorphic map of Lake Michigan shoreline: International Association of Great Lakes Research, Proceedings, 13th Conference on Great Lakes Research, Buffalo, New York, Part 1, p. 250–265.

Hansel, A. K., and Mickelson, D. M., 1988, A reevaluation of the timing and causes of high lake phases in the Lake Michigan basin: Quaternary Research, v. 29, p. 113–128.

Hansel, A. K., Mickelson, D. M., Schneider, A. F., and Larsen, C. E., 1985, Late Wisconsinan and Holocene history of the Lake Michigan basin, *in* Karrow, P. F., and Calkin, P. E., eds., Quaternary Evolution of the Great Lakes: Geological Association of Canada Special Paper 30, Ottawa, Ontario, p. 39–53.

Harrison, J. E., 1972, Quaternary geology of the North Bay–Mattawa region: Geological Survey of Canada Paper 71–26, 36 p.

Hough, J. L., 1955, Lake Chippewa, a low stage of Lake Michigan indicated by bottom sediments: Geological Society of America Bulletin, v. 73, p. 613–620.

Hough, J. L., 1958, Geology of the Great Lakes: University of Illinois Press, Urbana, Illinois, 313 p.

Hunter, R. E., Sallenger, A. H., and Dupre, W. R., 1979, Maps showing directions of longshore sediment transport along the Alaskan Bering Sea coast: U.S. Geological Survey Miscellaneous Field Studies, Map MF 1049.

Jacobson, E. E., and Schwartz, M. L., 1981, The use of geomorphic indicators to determine the direction of net shore drift: Shore and Beach, v. 49, p. 38–43.

Larsen, C. E., 1985, A stratigraphic study of beach features on the southeastern shore of Lake Michigan: new evidence for Holocene lake level fluctuations: Illinois State Geological Survey, Environmental Geology Notes 112, 31 p.

Larsen, C. E., 1987, Geological history of Glacial Lake Algonquin and the Upper Great Lakes: U.S. Geological Survey Bulletin 1801, 36 p.

Leverett, F., 1897, The Pleistocene features and deposits of the Chicago area: Chicago Academy of Sciences, Geology and Natural History Survey Bulletin 2, 85 p.

Leverett, F., and Taylor, F. B., 1915, The Pleistocene of Indiana and Michigan and the history of the Great Lakes: U.S. Geological Survey Monograph 53, 529 p.

Lineback, J. A., Gross, D. L., and Meyer, R. P., 1972, Geologic cross sections derived from seismic profiles and sediment cores from southern Lake Michigan: Illinois State Geological Survey, Environmental Geology Notes 54, 43 p.

Lineback, J. A., Gross, D. L., and Meyer, R. P., 1974, Glacial tills under Lake Michigan: Illinois State Geological Survey, Environmental Geology Notes C9, 48 p.

Miner, J. J., 1989a, The Lake Michigan ice-foot complex: conditions of formation and destruction and an evaluation of winter erosion at Wilmette, Illinois: Unpublished M.S. Thesis, Northern Illinois University, Dekalb, Illinois, 89 p.

Miner, J. J., 1989b, Formation, destruction, and erosive potential of a Lake Michigan ice-foot complex: Geological Society of America Abstracts with Programs, v. 21, p. A153.

Pranschke, F. A., Shabica, C. W., and Chrzastowski, M. J., 1990, Discovery of an early Holocene relief forest on the floor of southern Lake Michigan: Geological Society of America Abstracts with Programs, v. 22, p. 42.

Sanders, J. E., and Kumar, N., 1975, Evidence of shoreface retreat and in-place "drowning" during Holocene submergence of barriers, shelf off Fire Island, New York: Geological Society of America Bulletin, v. 86, p. 65–76.

Schneider, A. F., and Hansel, A. K., 1990, Evidence for post–Two Creeks age of the type Calumet shoreline of glacial Lake Chicago, *in* Schneider, A. F., and Fraser, G. S., eds., Late Quaternary History of the Lake Michigan Basin: Geological Society of America Special Paper 251, p. 1–8.

Schneider, A. F., and Keller, S. J., 1970, Geologic map of the 1° × 2° Chicago Quadrangle, Indiana, Illinois and Michigan, showing bedrock and unconsolidated deposits: Indiana Geological Survey Regional Geologic Map 4, Part *B*, 1 sheet.

Swift, D. J. P., 1975, Barrier island genesis: evidence from the Middle Atlantic shelf of North America: Sedimentary Geology, v. 14, p. 1–43.

Thompson, T. A., 1987, Sedimentology, internal architecture and depositional history of the Indiana Dunes National Lakeshore and State Park: Unpublished Ph.D. Dissertation, Indiana University, Bloomington Indiana, 295 p.

Thompson, T. A., 1989, Anatomy of a transgression along the southeastern shore of Lake Michigan: Journal of Coastal Research, v. 5, p. 711–724.

Thompson, T. A., 1990, Dune and beach complex and back-barrier sediments along the southeastern shore of Lake Michigan; Cowles Bog area of the Indiana Dunes National Lakeshore, *in* Schneider, A. F., and Fraser, G. S., eds. Late Quaternary History of the Lake Michigan Basin: Geological Society of America Special Paper 251, p. 9–19.

Thompson, T. A., Fraser, G. S., and Hester, N. C., 1991, Lake-level variation in southern Lake Michigan: magnitude and timing of fluctuations over the past 4,000 years: Illinois-Indiana Sea Grant Special Report IL-IN-SR-91-2, Champaign, Illinois, 21 p.

U.S. Army Corps of Engineers, 1953, Illinois shore of Lake Michigan beach erosion control study: 83rd Congress, 1st Session, House Document 28, 137 p.

Wickham, J. T., Gross, D. L., Lineback, J. A., and Thomas, R. L., 1978, Late Quaternary sediments of Lake Michigan: Illinois State Geological Survey, Environmental Geology Notes 84, 26 p.

Willman, H. B., 1971, Summary of the geology of the Chicago area: Illinois State Geological Survey Circular 460, 77 p.

Willman, H. B., and Lineback, J. A., 1970, Surficial geology of the Chicago region (map), 1 sheet, *in* Willman, H. B., 1971, Summary of the Geology of the Chicago Area: Illinois State Geological Survey Circular 460, 77 p.

Wright, G. F., 1918, Explanation of the abandoned beaches about the south end of Lake Michigan: Geological Society of America Bulletin, v. 29, p. 235–244.

HOLOCENE TRANSGRESSION AND COASTAL-LANDFORM EVOLUTION IN NORTHEASTERN LAKE ERIE, CANADA

JOHN P. COAKLEY

Lakes Research Branch, National Water Research Institute, Canada Centre for Inland Waters, 867 Lakeshore Road, Burlington, Ontario, Canada L8S 3E9.

ABSTRACT: Holocene sediments from three deep boreholes from Long Point, a 35-km-long sandy foreland on the Canadian side of eastern Lake Erie, display a consistent coarsening-upward trend from sheltered-water clays and silts to well-sorted shoreface sands. This trend persists in spite of the documented record of lake-level rise, due primarily to postglacial isostatic rebound of the Niagara River outlet. To resolve this contradiction within the context of shoreline evolutionary trends in eastern Lake Erie, sediment data from the boreholes are analyzed in combination with data from other sources. The analysis suggests that the sedimentary sequence is linked to the history of the Long Point foreland and its precursors. The resulting evolutionary model, supported by relict shore features, postulates that Long Point began as a north-south-trending cuspate peninsula, once linked to the ancestral Presque Isle spit on the United States side. The location of the feature was controlled by the Norfolk moraine, a cross-lake ridge of glacial origin. As lake levels rose, the foreland retreated northward with the shoreline as a whole, becoming smaller and more rounded but with well-developed recurves extending northeast. Around 5 to 4 ka, an abrupt rise in lake levels, related to resumption of drainage from the Nipissing Phase in the Upper Great Lakes into the Erie basin, resulted in inundation of most of the ancestral foreland. When the lake later fell to previous levels, presumably due to widening of the Niagara River outlet, the foreland commenced its present east-west orientation. At present, the distal portions of the spit are prograding southward, while the proximal areas are still retreating northward; in other words, the feature is slowly rotating clockwise. The pattern of landward retreat of parts of the foreland over sheltered (lagoonal) deposits and the rotation of the distal portions are also noted in the two other north-shore forelands, Point Pelee and Pointe-aux-Pins, and comparison is made with the transgressive marine barrier islands of the United States east coast.

INTRODUCTION

The Canadian shoreline of Lake Erie extends more than 600 km from just south of Windsor to Fort Erie across the Niagara River from Buffalo, New York (Fig. 1). Composed almost entirely of unconsolidated glacial sediments, this shoreline shows considerable morphological variety, due in large measure to its glacial and postglacial history. The most striking coastal features are the three large, sandy cuspate forelands, Point Pelee, Pointe-aux-Pins, and Long Point, located at regular intervals along the shoreline. Associated with these features are large coastal wetlands and partially enclosed marshes. The remaining shoreline types range from 30-m-high till and glaciolacustrine sediment bluffs to low bedrock plain. Shoreline response to coastal processes ranges from rapid erosion (up to 5 m/yr in the high-bluff shoreline west of Long Point) to deposition and foreland evolution in the cuspate forelands to minimal changes in the bedrock-controlled areas.

The objective of this paper is to demonstrate the utility of sediment stratigraphy and relict geomorphologic features in reconstructing details of postglacial shoreline evolution in eastern Lake Erie. The area discussed comprises the northeastern section of the Lake Erie coastline, extending from Port Burwell to Fort Erie (Fig. 1), with the major focus being placed on the Long Point foreland.

POSTGLACIAL HISTORY OF LAKE ERIE

Deglaciation of the Erie basin began around 14.5 ka and was complete at approximately 12.5 ka with the retreat of the glacier north of the Niagara escarpment, opening up the Niagara River outlet. Rapid draining of the last major high-level lake, Lake Warren (Fig. 1), led to the initiation of Early Lake Erie (Leverett and Taylor, 1915; Lewis, 1969; Calkin and Feenstra, 1985). From lows of more than 35 m below present lake level at the onset of Early Lake Erie (Coakley and Lewis, 1985), lake levels rose at an exponentially declining rate to the present, under the influence of elastic and viscous isostatic rebound of the glacially depressed outlet and other neotectonic forces. A postglacial lake-level curve, reproduced from Coakley and Lewis (1985) and based on a large number of radiocarbon dates throughout the basin (Fig. 2), shows the initially rapid rise in lake levels followed by a slow stabilization to around 10 cm/century over the past 3,000 years. A noteworthy characteristic of the curve is the relatively abrupt rise to above present levels at about 3 ka. This peak is interpreted from data presented in Barnett (1985) and Barnett and others (1985) and appears to correspond to the resumption of drainage from the upper Great Lakes during the Nipissing Phase (Lewis 1969). After the peak, levels are interpreted as having fallen to near their present position, based on the evidence of ^{14}C dates from basal marsh peats from Point Pelee (Terasmae, 1970). This interpretation is argued in Coakley and Lewis (1985). Since then, the continuing rise in levels long after glacial rebound should have ceased has been linked to intracratonic crustal adjustments (Coakley and Lewis, 1985). This model of post-Nipissing lake levels differs somewhat from that presented in Barnett (1985) in that Barnett argues that peak lake levels occurred around 2 ka and that they have since fallen to their present position.

As the levels in the basin rose, the area of the water surface expanded through the gradual inundation of the low-lying areas and the recession of the high shore bluffs composed of unconsolidated glacial sediments. Longshore-sediment transport led to the accumulation of sandy forelands at intervals along the shoreline, with the locations of these forelands being controlled primarily by pre-existing glacial features, such as end moraines transverse to the lake axis (Fig. 1, Coakley, 1976, 1989). As a result of widespread erosion and lacustrine transgression since early Lake Erie, the nearshore slopes of the basin are virtually devoid of postglacial-sediment accumulation, and the only long-term records of postglacial sedimentation are found in associa-

Quaternary Coasts of the United States: Marine and Lacustrine Systems, SEPM Special Publication No. 48

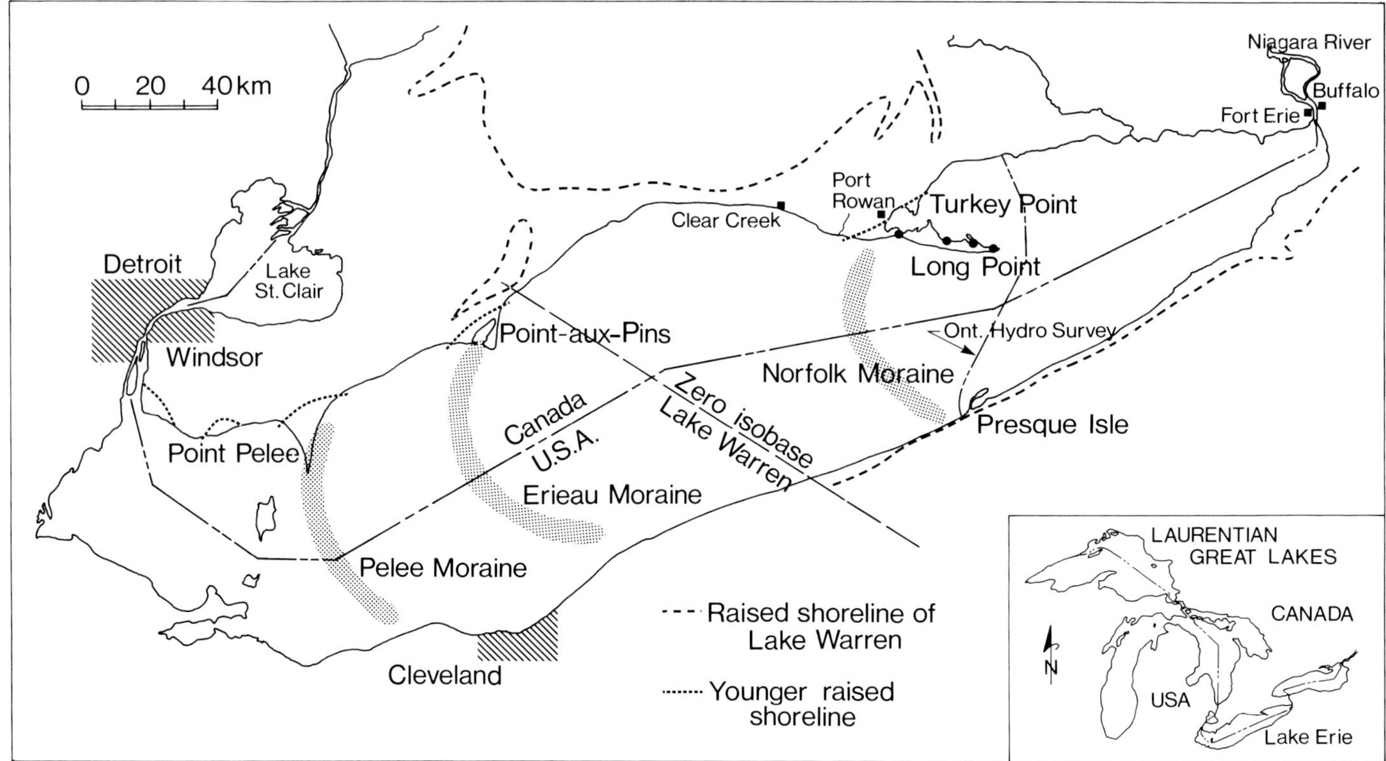

FIG. 1.—Location map of Lake Erie, showing study area. Geographic locations and glacial geologic features mentioned in the text are also shown.

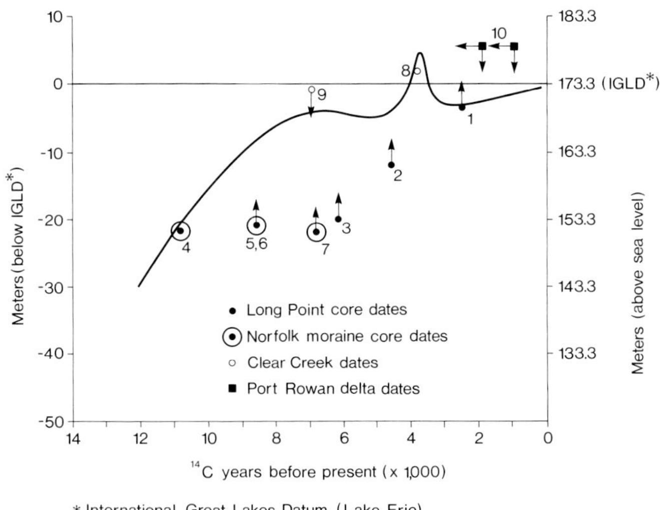

FIG. 2.—Summary of postglacial history of lake levels in the Erie basin (from Coakley and Lewis, 1985). Elevation and ^{14}C dates in the Long Point area also plotted. Arrows indicate interpreted direction to lake surface.

tion with the above cuspate forelands and in the less-accessible deep offshore deposits. The only other clues of previous lake history are found in erosional features preserved below or above the present lake level. Therefore, this study relies heavily on borehole and nearshore data from the Long Point area, supplemented by relict geomorphologic features.

DATA BASE

Methodology

Two boreholes (BH1 and BH2 in Fig. 3) were drilled on Long Point during the summer of 1980 using a wash-boring drilling technique with split-spoon sampling at 1- to 1.5-m intervals. The boreholes were sampled to depths of 40 and 59 m below lake level (173.3 m above sea level), respectively. Another borehole (BH3) was drilled to 38 m below lake level in 1981, using a hollow-stem auger rig that allowed continuous Shelby-tube and split-spoon sampling.

Other land-based boreholes drilled on Long Point for other purposes are also shown in Figure 3. The one near the tip, labeled LBH, was drilled earlier for geological investigative reasons, and a comprehensive log and grain-size profile are available (Lewis, 1966). Boreholes in the western part of the foreland were drilled for water-supply purposes and thus were logged only in a rudimentary fashion.

In addition to the land-based borehole data, unpublished data from short cores and water-jetting probes taken in the adjacent nearshore zone (Fig. 3; N. A. Rukavina, National Water Research Institute, Environment Canada, pers. commun., 1985) provided useful insight into the nature and thickness of the subaqueous-sediment cover around Long Point. These data complemented existing published reports on sediments in the vicinity of Long Point (St. Jacques and

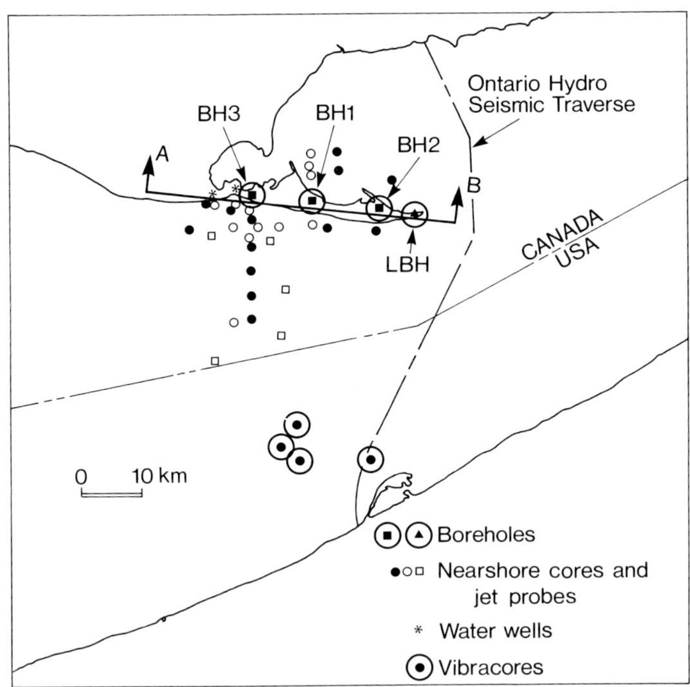

FIG. 3.—Locations of boreholes and other sources of information on subsurface sediments in the Long Point area. Location of cross section in Figure 6 is shown as A—B, Approximate location of Ontario Hydro seismic traverse across Lake Erie is also shown.

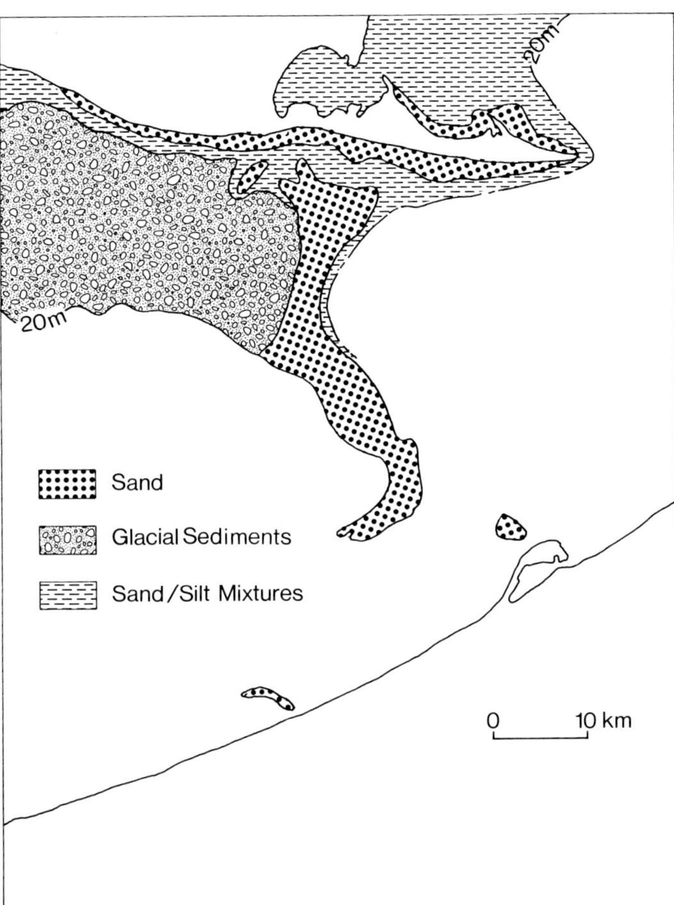

FIG. 4.—Surficial lake-bottom sediments in water less than 20 m deep, compiled from St. Jacques and Rukavina (1973) and Williams and Meisburger (1982).

Rukavina, 1973) and on the United States side (Williams and Meisburger, 1982). Figure 4 presents a compilation of nearshore-sediment types in the Long Point area based on the above sources. The author was also granted access to unpublished data on surface and subsurface features along a cross-lake seismic traverse conducted by Ontario Hydro (1981) near the east end of the lake (Figs. 1,3); these data were useful in the identification of relict shoreline indicators.

Particle-size analysis (0.5 phi [ϕ] intervals) was carried out on all samples, using a combined sieve–settling tube–pipette–Sedigraph procedure. As an aid in differentiating glacial and postglacial lacustrine fine sediments, profiles of undrained shear strength (fall-cone tests) and consolidation indices were also determined on Shelby-tube samples, mainly from BH3.

Sedimentary Record Preserved Below Long Point

The surficial-sediment distribution in the vicinity of Long Point is presented in Figure 4. Subsurface sediments encountered in the three boreholes from Long Point are described in Figure 5, together with profiles of parameters related to sediment texture and radiocarbon date locations. A less-detailed log of the LBH borehole (Lewis, 1966) is shown in Figure 6. Despite minor differences between boreholes, the overall sedimentary sequence is consistent enough to permit classification into distinct lithological units. Preliminary boundaries between the units are based on visible discontinuities, but when these are ill-defined, the units may be separated on the basis of breaks in the sediment-texture profile, especially sand-silt-clay ratios, mean grain size, and grain-size-curve types.

Unit 1: Compacted uniform clay.—

A compacted uniform clay unit forms the basal 18 m in BH3 and appears to comprise the oldest sediments encountered, with the possible exception of the till-like unit overlying bedrock in LBH. Unit 1 consists of a uniform fine clay, grayish-brown (2.5Y 5/2) in color. Structure consists of faint, contorted laminations, irregularly spaced silt patches, and isolated dropstone (?) pebbles. The grain-size distribution curves of subsamples from this unit are very uniform and consistently unimodal and negatively skewed (Fig. 5), with mean size ranging from 8.7 to 9.5 ϕ. In BH3, the Unit 1 sediments are discordantly overlain by a thin (8 cm) zone of polymict muddy sand material containing well-preserved mollusc shells. The contact with the overlying unit is abrupt and irregular, indicating an interval of erosion. In addition, geotechnical tests of the clay below reveal considerable overconsolidation, indicating that more than 20 m of overlying sediments had been eroded (Coakley, 1985). Mean size of the muddy sand layer is around 4 ϕ, and the sediment is strongly bimodal, with a dominant sand mode at 3.25 ϕ showing good to moderate sorting. This layer is

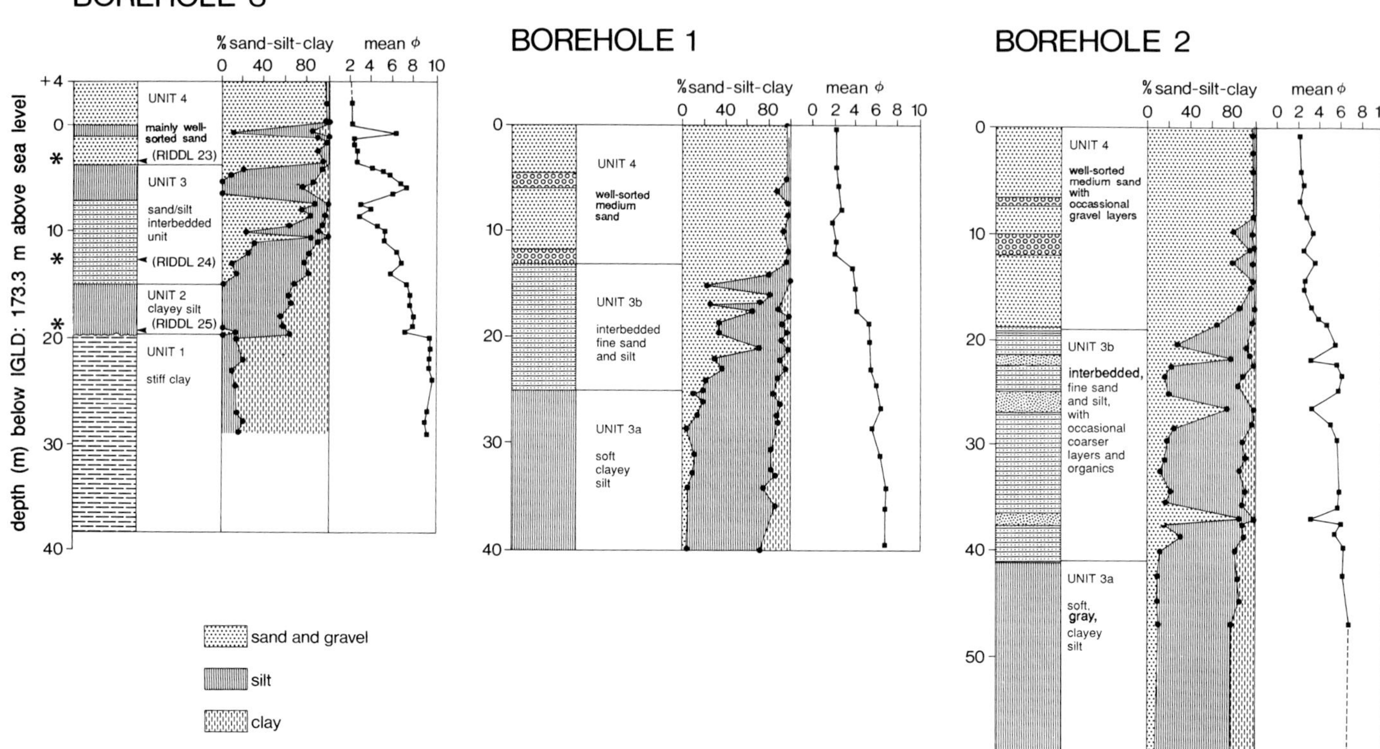

FIG. 5.—Stratigraphic description of three boreholes from Long Point; locations are given in Figure 3. Also shown (asterisks in BH3) are locations of ^{14}C-dated dating samples. Legend symbols refer to sand-slit-clay percentage profile only; symbols in sediment-type profile are explained in adjacent column.

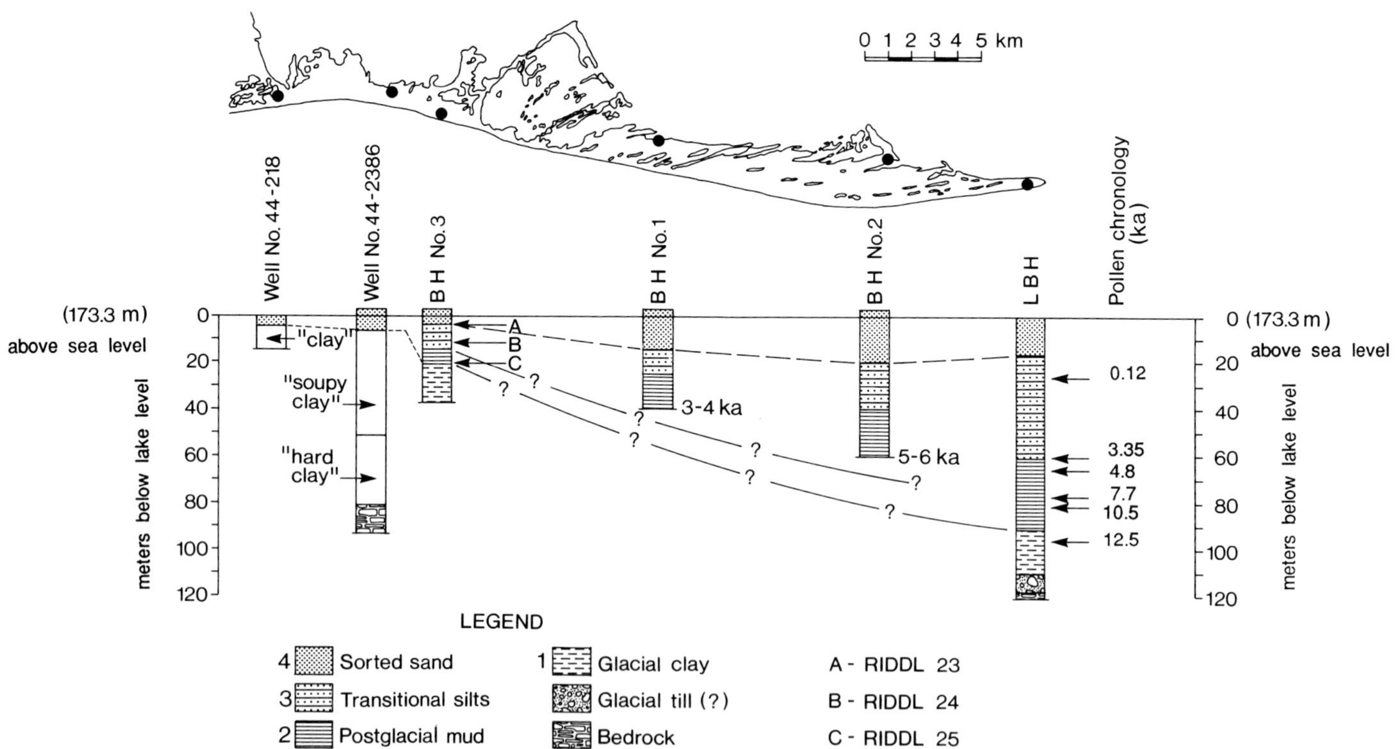

FIG. 6.—Longitudinal profile through boreholes and water wells on Long Point. Profile location shown in Figure 3. Stratigraphic control points from ^{14}C dating and pollen analysis are noted (A, B, and C refer to ^{14}C-dated samples shown in Fig. 5). Age estimations at base of BH1 and BH2 were inferred from pollen-profile comparisons. Descriptive terms from drillers-logs are also included.

TABLE 1.—RADIOCARBON DATES FROM THE LONG POINT AREA

Sample No.	Depth (m)*	Description of organic matter	C-14 Age (yrs)** (Lab. I.D.)
1) BH3-4A-1	4.1	Three shell slivers (0.102 g)	2,550±170* (RIDDL 23)
2) BH3-14	12.3	Assorted wood and bark flakes (0.052 g)	4,690±190* (RIDDL 24)
3) BH3-24-4B	20.0	Bivalve shells and shell fragments (0.147 g)	6,100±80* (RIDDL 25)
4) Core 28	22.0	Wood in firm, silty clay (*in situ*?)	10,800±190 (UM-1706)
5) Core 23	21.0	Wood at base of clean medium sand with clay layers (transported?)	8,545±150 (UM-1705)
6) Core 4	21.0	Wood in interbedded sand/clay complex (transported)	8,240±210 (UM-1703)
7) Core 18	22.0	Wood at base of massive fine-medium sand (transported)	6,870±150 (UM-1704)
8) C.Crk. 1	−1.8	Large wood fragment in wood bed (*in situ*)	3,900±100 (BGS-898)
9) C.Crk. 2	0.7	Peat layers in silt	6,980±120 (BGS-901)
10) Deltas	−5.0	Organic detritus and wood in alluvium (transported)	960±80 (BGS-980) 1,900±80 (BGS-928)

*Below present lake level (173.3 m above sea level)
**Quoted age is in radiocarbon years using the Libby half-life of 5,568 years.

TABLE 2.—DATED POLLEN SUCCESSION IN NORTHEASTERN NORTH AMERICA

Event	Years Before Present (Approx.)
Transition of spruce to pine	10.6 ka
Transition of pine to oak	7.6 ka
Hemlock rise	6.7 ka
Hemlock decline	4.8 ka
Hemlock recovery	3.5 ka
Ambrosia (ragweed) rise	0.12 ka

clearly a lag deposit built up on the eroded surface. Unit 1 is interpreted as a typical glaciolacustrine sediment, deposited in one of the earlier high-level phases (Barnett, 1985).

Unit 2: Soft clayey silt.—

Unit 2 is characteristically a uniform, dark gray (10Y 4/1), clayey silt (silt reaches 70 percent) and is much less consolidated than Unit 1. Mean size is around 6 to 8 ϕ, and, In contrast to Unit 1, skewness is always positive. Although organic matter and shells are rare, dark laminations and almost-black sulfide streaks occur throughout, indicating deposition in a restricted, chemically reducing environment. Unit 2 grades downward into the previously mentioned transitional muddy sand in BH3 but does not occur in BH1 and BH2. This unit is texturally comparable to the offshore muds of modern Lake Erie (Fig. 4), typical of deposition in deep water below wave base.

Unit 3: Interlaminated sand and silt.—

Unit 3 forms the basal 30 and 40 m, respectively, of BH1 and BH2. The unit can be divided into two sub-units on the basis of texture. The lower subunit (3a) is predominantly a sandy, clayey silt, becoming gradually more sandy upward; the upper subunit (3b) consists of distinct alternating silt and sand laminations. This classification is more difficult to apply to sediments in BH3, where the bottom part of the unit is more sandy than the top and the contact with Unit 2 is gradational. In BH3, an anomalous 2.5-m-thick fine-grained upper section, ranging in texture from sandy silt at the top to clayey silt at the base (Fig. 5), occurs at the top of Unit 3. Organic matter, consisting of plant debris and leached shell fragments, is common in this finer section and in the top portion of the unit in the other boreholes. Skewness is generally positive. Excluding the fine-grained upper section in BH3, unit 3 is characterized by a coarsening-upward and better-sorting-upward trend. Unit 3 is interpreted as a transitional or lower shoreface deposit (Relneck and Singh, 1975), very similar in texture to the sand/silt sediment found slightly offshore of Long Point (Fig. 4).

Unit 4: Sorted sand.—

The contact between Unit 3 and the top unit (Unit 4) is sharply defined in all cases, but no sign of erosion was noted, suggesting a transgressive succession. Unit 4 is best described as a well-sorted, medium sand (mean phi size between 2.0 and 2.5) gray to brownish-gray in color. Structures include laminations composed mainly of heavy-mineral concentrations and coarser grains or, in some cases, dark oxidized wood and charcoal. Occurrences of organic matter were also noted at the contact with Unit 3 in BH3. Unit 4 ranges in overall thickness from 6.5 m (BH3) to 20 m (BH2). In BH3, midway within the unit, a 40-cm-thick deposit of interbedded sand and olive-gray silty clay occurs abruptly (Fig. 5). This fine-grained layer contains much organic debris (leaves, roots, twigs). The sandy facies of Unit 4 is clearly an upper shoreface or beach deposit; whereas the high organic content of the fine sublayer suggests a back-beach lagoon or interdune-pond environment.

Sedimentary sequences from other sources.—

The sedimentary sequence encountered in the LBH borehole is described in Lewis (1966) and is included in the longitudinal section shown in Figure 6. The upper portion of this borehole shows sedimentary units closely resembling those from the other boreholes on Long Point. The drillers' logs for the borings carried out in the western section of Long Point are of limited stratigraphic value and serve mainly to define the contact between the sorted sands of Unit 4 and the undifferentiated "clay" below, and the total depth to bedrock.

Chronological Controls

C-14 dating.—

Except for those near the top, the borehole sediments are characterized by organic matter in quantities too low for conventional ^{14}C dating. However, in some cases, there is a sufficient amount of small organic debris for accelerator mass spectrometry dating. The procedure used to establish these dates (designated RIDDL in Table 1) is described in Nelson and others (1986). Other ^{14}C dates related to Long

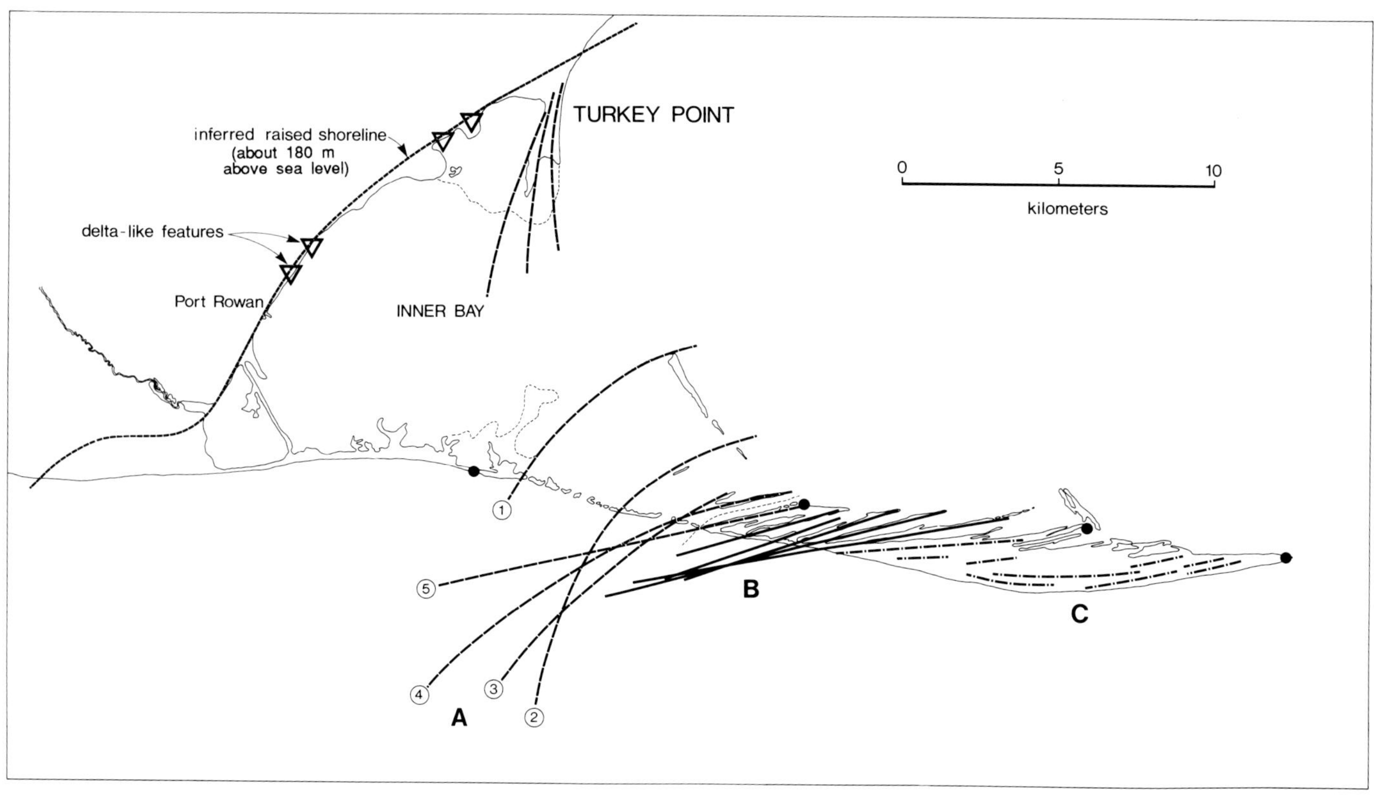

FIG. 7.—Summary of physiographic features in the Long Point and Turkey Point and Turkey Point area related to inferred paleogeography.

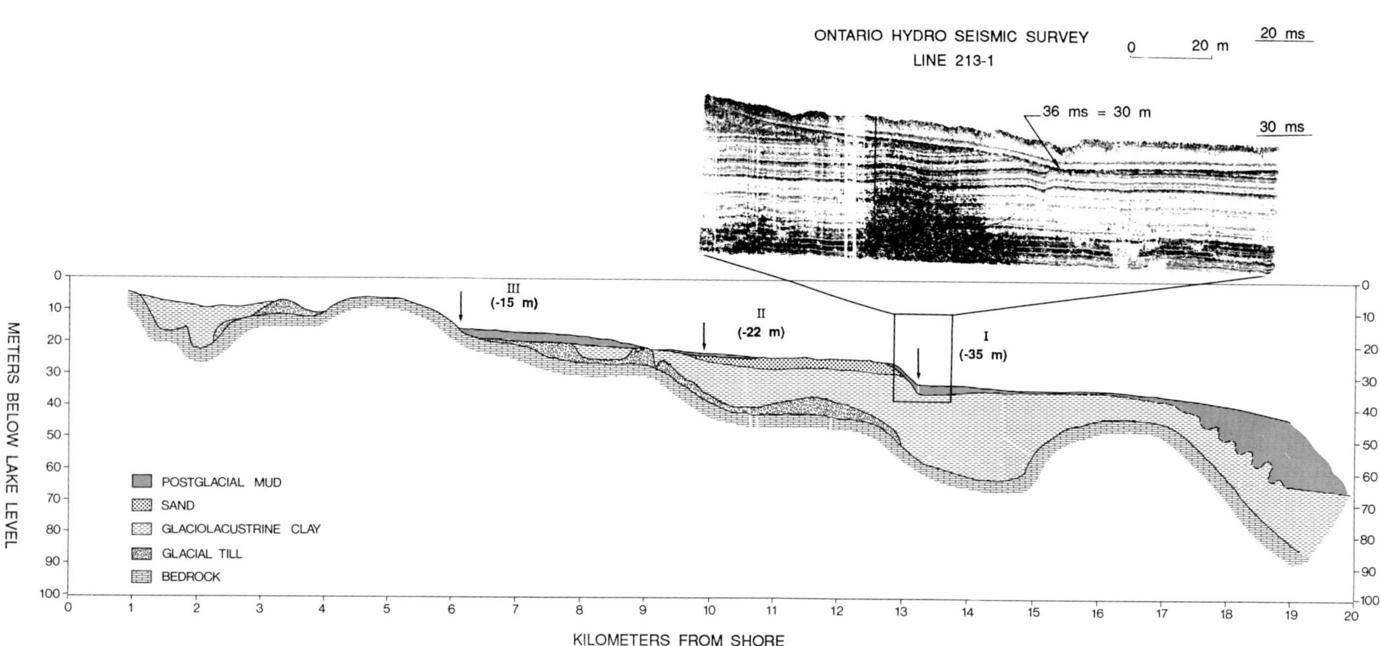

FIG. 8.—Interpreted geologic section through Quaternary deposits along the seismic-reflection survey line (location shown in Fig. 1 and 3). Reproduction of the original seismogram showing the truncated erosional feature (interpreted as relict shoreline 1) is shown in inset (vertical depth scale in milliseconds return travel time).

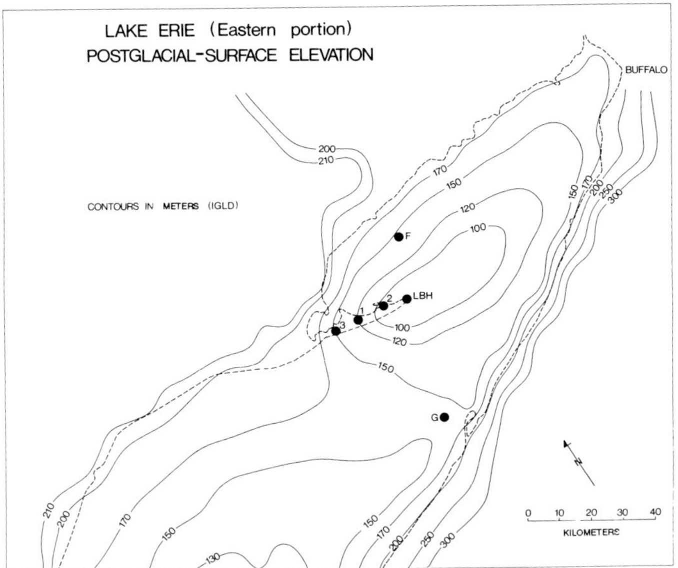

FIG. 9.—Reconstructed original surface of glacial sediments at the inception of Early Lake Erie (about 12.5 ka) based on raised Lake Warren shoreline data and models of postglacial crustal rebound. Long Point borehole positions are indicated; F and G refer to subsurface "notch" in Figure 8 and vibracore position shown in Figure 3.

Point history include four obtained using conventional techniques on organics in sedimentary sequences overlying the Norfolk Moraine south of Long Point (Fig. 1; Williams and Meisburger, 1982). Some of the dates are from wood that was either *in situ* (as roots in the moraine till) or deposited in very shallow-water sediments and thus, they are believed to be good indicators of shoreline position and lake level. All the ^{14}C dates are presented in Table 1, and those from BH3 are plotted at their respective stratigraphic position in Figures 5 and 6.

Pollen stratigraphy.—

The other source of chronological control for the Long Point borehole sediments is pollen stratigraphy. Pollen profiles obtained from LBH enabled relative datelines to be drawn on the basis of recognized time-markers (Table 2) for northeastern North America (Ogden, 1967; Anderson, 1971; Mott and Farley-Gill, 1978; T. W. Anderson, Geological Survey of Canada, pers. commun., 1991). Only the interpreted datelines for LBH are presented in Figure 6; the profiles are given in full in Coakley (1985).

Relict Shorelines and Coastal Features

Subaerial indications.—

Ongoing shore recession has removed almost all vestiges of previous shoreline positions in the study area. However, intersecting the shoreline and passing through the base of Long Point before going offshore east of Turkey Point is a readily identifiable abandoned shore cliff, whose base was estimated from local topographic maps at between 175 and 180 m above sea level, i.e., 2 to 6 m above present lake level (Fig. 7). Associated with this cliff are features identified as small deltas by Barnett (1985), indicating a still-stand of the lake some 5 m above present level. Organic detritus and wood from older alluvium graded to two of these deltas yielded dates of 960±80 yrs (BGS-980) and 1,900±80 yrs (BGS-928). These dates appear to be too young, since there is no doubt that this shoreline preceded the formation of the Long Point foreland, where much older dates were obtained for apparent spit sediments in BH3. Also, such a high lake level is more readily associated with the resumption of Nipissing Phase drainage into Lake Erie mentioned earlier, generally placed at 5 to 4 ka. Coakley and Lewis (1985) suggest that these dates might reflect contamination by modern organic materials.

Examination of the trend of beach ridges preserved on Long Point (A, B, and C, Fig. 7) permits discrimination of orientational changes believed to reflect evolutionary trends in shoreline (beach) position. Although barely visible on aerial photographs, the older (westernmost) ridges are concave lakeward, whereas those in the more distal section are convex lakeward. These trends indicate that unlike at present, when dominant wave energy is from the south-to-west quadrant, wave approach in the earlier shoreline stages was mainly from the east and southeast, reflected in the orientation of recurves to face these directions.

Subaqueous indications.—

Figure 8 presents an interpreted profile of surface and subsurface features detected during a seismic traverse of the east end of Lake Erie. In spite of the large vertical-scale exaggeration, several clear breaks in the nearshore slope are evident (labeled I, II, and III). The breaks in slope are accompanied by abraded and truncated glacial-sediment sections, which suggest that the breaks represent relict shoreline positions. However, their elevation falls below all the reconstructed postglacial raised-shoreline positions relative to the present lake outlet, suggesting that the breaks are not related to a lake-phase draining via the Niagara River (Coakley and Lewis, 1985). The most likely explanation is that they relate to a brief period when the lake basin was without an outlet, prior to its infilling by waters from the Algonquin Great Lakes.

Relict Offshore Sand Deposits

Another clue to the evolution of the eastern Lake Erie shoreline is found in the large offshore deposit of well-sorted sand located to the south of Long Point and extending almost to Presque Isle, Pennsylvania (Fig. 4). This deposit, now in more than 20 m of water, contains several-meter-thick sections of medium sand, totaling some 37×10^8 m^3 on the United States side of the border alone (Williams and Meisburger, 1982). Williams and Meisburger also found widespread fossilized wood in vibracores taken from the southern part of the deposit (Fig. 3). The orientation and thickness of the deposit, as well as the lake-level history of the basin, suggest strongly that the sands constitute a relict shoreface deposit once associated with a large, subaerial coastal landform, such as a spit. Dates for organics from the cores (Table 1) ranged from 10,800±190 yrs for wood embedded in the basal clay to 6,870±150 yrs for wood

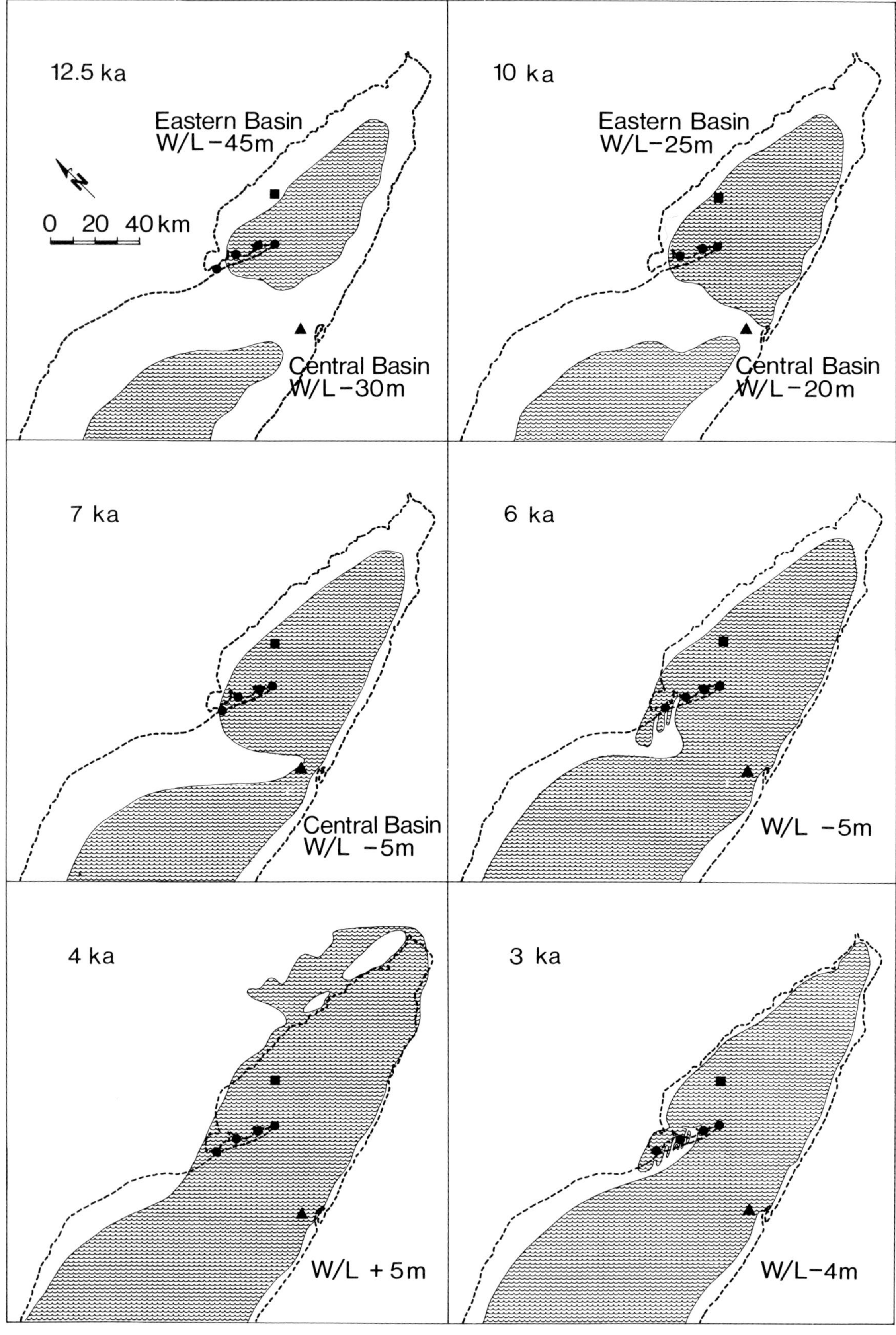
12.5 ka
Eastern Basin
W/L −45m
0 20 40km
Central Basin
W/L −30m
10 ka
Eastern Basin
W/L−25m
Central Basin
W/L −20m
7 ka
Central Basin
W/L −5m
6 ka
W/L −5m
4 ka
W/L +5m
3 ka
W/L−4m

from the upper sand units of the core. These dates place the existence of a north-south-trending foreland at this location between 11 and 7 ka. As lake levels rose, such a foreland would have been gradually reduced in size as it retreated upslope toward the north shore. Like its modern analogs, the sandy foreland would likely show attached recurves extending downdrift.

DATA SYNTHESIS

The foregoing lines of evidence are used as the basis for a model of the evolution of the eastern Lake Erie shoreline, including the morphologic and sedimentary sequences associated with the Long Point foreland.

Postglacial Shoreline Reconstruction

The reconstruction of the changing shoreline position in eastern Lake Erie is based on two factors: the interpreted lake-level curve (Fig. 2) and the response of the pre–Lake Erie basin topography to nonlinear glacio-isostatic rebound of the eastern half of the basin. The computer-based technique used to make the time-dependent adjustments is explained in detail in Coakley (1985) and so will only be summarized here.

The original paleosurface topography of Early Lake Erie was obtained from Lewis (1966). A square grid (10 km) was superimposed on the portion of the basin east of the Lake Warren zero isobase (Fig. 1), and present elevations at the grid intersects were entered into a computer. Time since unloading and distance from the zero-isobase line were entered into theoretical nonlinear (combination exponential and linear) models of crustal rebound adapted from the literature (Washburn and Stuiver, 1962; Andrews, 1970; Coakley and Lewis, 1985) to obtain the uplift-corrected elevation at each grid point. The original uplift corrected paleosurface for eastern Lake Erie at about 12.5 ka is shown in Figure 9. Applying the time-dependent crustal-rebound model to this surface and taking the corresponding lake level from Figure 2 resulted in the reconstructed shoreline as shown in Figure 10. Vertical control on model shoreline position was exercised by the shoreline indicators described earlier.

Evolution of Depositional Environments at Long Point

The dominant characteristic of the sedimentary sequences observed in the boreholes from Long Point is a consistent coarsening-upward trend in texture, from fine silty clays at the base, through interbedded sand/silts, to well-sorted sand evident in all boreholes. This pattern has been noted in sequences from the other Lake Erie cuspate forelands (Coakley, 1976; Coakley, 1989) and thus is directly related to basin-wide trends in depositional conditions, most likely water depth. However, the lake-level model presented here hypothesizes a steady rise in lake levels, a situation expected to result in a fining-upward sedimentary sequence as progressively deeper-water, and thus finer, sediments are deposited. The most likely explanation for the coarsening-upward sequence in progressively deepening water depths is that of transgression of a shoreface feature, such as a spit, over deeper or more sheltered depositional environments (Kraft and others, 1978; Moslow and Heron, 1981).

The development of Unit 1 at BH3 (and presumably at the other three borehole locations) most likely occurred on an existing paleoslope of till. The sediment textural properties suggest that deposition took place in low-energy conditions related to deep or ice-covered water (proglacial Lake Warren). After Niagara drainage resumed, ending these conditions soon after 12.5 ka, water levels dropped considerably, as shown by the early part of the lake-level curve (Fig. 2) and by the shallow-water or subaerial erosion of a large portion of Unit 1 at BH3. Fossilized wood apparently *in situ*, found in vibracores south of Long Point (Williams and Meisburger, 1982), is reasonable evidence of subaerial conditions on the moraine ridge between Long Point and Presque Isle up until 10.8 ka. The presence of wood in sand deposits dated as young as 6.87 ka in this area suggests that subaerial or shallow-water conditions existed in the topographic high south of Long Point until then. The sedimentary units deposited at the core sites are clearly time transgressive, with deeper-water or more restricted depositional environments apparently existing in the vicinity of LBH while Unit 2 silty clays were being deposited there. In the meantime, rapidly rising lake levels were causing the migration of the north-south-oriented early Long Point foreland northward up the local nearshore slope, thus reducing the area of the foreland. This change allowed relatively shallow-water but sheltered conditions to continue at BH3 leading to initiation of Unit 2 sedimentation there around 6.1 ka. Even while the transitional silts and interbedded sands of Unit 3 had begun to be deposited at the more shallow-water sites (BH1, 2, and 3), uniform fine sediments (resembling Unit 2) continued to be deposited at LBH up until around 3.5 ka. After about 4.69 ka (RIDDL 25), the situation changed and a progressive deposition of shoreface sands (Units 3 and 4) extended from the site of BH3 toward the more easterly (offshore) sites. This progradation is believed to be linked to the lateral migration and evolution of the Long Point foreland itself, described later. The occurrence of anomalous fine sequences in Units 3 and 4 is believed to be due to the occasional sheltering of previous shoreface depositional environments by the formation of baymouth bars across the ends of recurves.

Reconstruction of the Origin and Evolution of the Long Point Foreland

Isthmus and peninsula stages (12.5–ca. 8.0 ka).— The paleogeographic reconstruction of the eastern Lake Erie shoreline (Fig. 10) suggests that the initial phase was

←

FIG. 10.—Computer-assisted reconstruction of the eastern Lake Erie paleoshoreline about 12.5 to 3.0 ka. Interpreted water levels are taken from Figure 2. Positions of shoreline indicators and borehole and vibracore positions are shown as solid symbols.

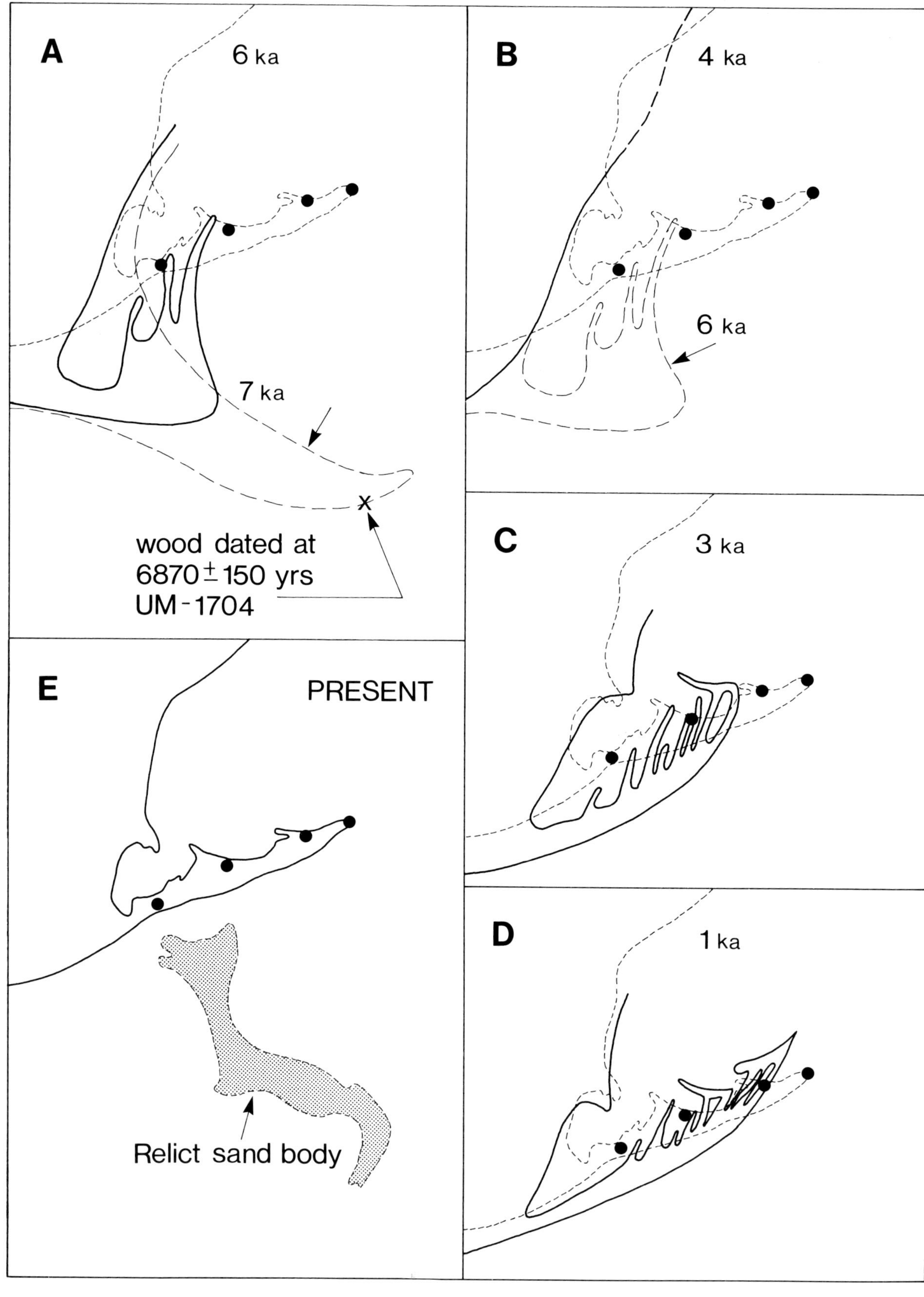
A
6 ka
7 ka
wood dated at
6870 ± 150 yrs
UM - 1704
B
4 ka
6 ka
C
3 ka
D
1 ka
E
PRESENT
Relict sand body

a broad morainal ridge or isthmus covered with glaciolacustrine sediments and spanning the lake. Rising levels in the central subbasin eventually led to the opening of the interbasin channel near the Presque Isle end of the structure, creating an elongated cuspate foreland oriented approximately north-south (Fig. 10 and 11). The separation interrupted supplies of sand to the southern end of the cross-lake structure and left the early Presque Isle foreland to develop on its own. The northern peninsula was likely the site of massive sand accumulation from littoral drift from sand-rich sources immediately to the west (Barnett, 1985).

Early Long Point (8–5 ka).—

Due to its position at the end of a long-fetch distance for the prevailing westerly winds and waves, the foreland migrated eastward as well as northward, up the glacial-sediment surface. Longshore-drifted sand would have been carried around the southern end to be eventually deposited as recurves on the eastern side (Zenkovich, 1967). By 6 ka, the foreland had assumed a more rounded form, with recurves extending northeastward and concave southeastward, i.e., toward the dominant wave direction in that part of the eastern subbasin. Modern examples of such rounded cuspate forelands with prominent recurves include the Toronto Islands in Lake Ontario; Presque Isle just across Lake Erie in Pennsylvania; and the La Coubre spit on the French Atlantic coast, near the Gironde estuary. These early Long Point recurve spits are believed to have been preserved as the low, concave-lakeward beach ridges still discernible on the present Long Point (labeled A in Fig. 7). This stage is believed to be associated first with the fine-grained, partially reduced (sulfide streaks) sediments of Unit 2 at the BH3 site, most likely a sheltered, moderate-depth area at the time (Fig. 11). Later, the interlaminated sand-silt Unit 3 could have been deposited as the recurve spits approached the area.

Modern Long Point (3.5 ka–present).—

One of the puzzling features in the reconstruction of Long Point history is the lack of any sign of a transition between the previous stage and that of the present. No relict spit deposits or structures are visible on the lake bottom, which suggests that the change from one stage to the next occurred fairly abruptly. The most reasonable hypothesis is that the transfer of Upper Lakes (Nipissing Phase) drainage waters to the Erie basin around 5 ka dramatically raised Lake Erie levels by 5 to 10 m (Fig. 2). This probably led to the submergence of the north-south-trending cuspate foreland then existing in the Long Point area (Figs. 10 and 11) and the creation of the relict shore bluff visible at the base of Long Point. Massive amounts of sand were probably dispersed to the northeast by wave action to create a broad sandy shoal, which later could serve as the spit platform (Meistrell, 1966) for the radically changed Long Point foreland that was to follow. The only recognizable trace of the previous foreland is the relict sand body south of Long Point and the indistinct beach ridges in the western part of the foreland (Fig. 7).

When lake levels again fell some 8 m to approximately 3 m below present level about 3.5 ka, the sediments making up the spit platform were reworked by shallow-water wave action to initiate the formation of the present eastward-trending "flying" spit (Fig. 11). The change to a more conventional evolutionary pattern (Zenkovich, 1967) is reflected in the abrupt change in the preserved beach-ridge pattern from concave- to convex-lakeward (labeled B in Fig. 7). The orientation of the more distal (younger) beach ridges also shows a gradual rotation from northeasterly to easterly, indicating the future evolutionary trend for the foreland. Furthermore, as lake levels rose, the area of the eastern subbasin increased and long-fetch waves could be generated more from the northeast and east than from the southeast. This would favor the development of bay-mouth bars with extensions across the tips of the spit recurves (Fig. 11). The obvious truncation of the beach ridges and relict bay-mouth bars on the north-facing side of the foreland indicates that such elements at one time extended as far north as the north shore, forming an asymmetrical cuspate foreland. Subsequent flooding and breaching of this connection could be responsible for the preservation of the Turkey Point feature on the north shore (Fig. 7).

Over the last 4,000 years or so, Long Point has been steadily migrating northward and eastward under the effect of slowly rising lake levels and the dominant southwesterly wave climate. The major processes now are the transfer inland of materials from the southern shoreface either by wind (Davidson-Arnott and Law, 1990) or washover through low-lying areas during storms and high-lake-level periods. Washover processes have at times resulted in major inlets across narrow parts of the foreland; some of these inlet breaches have been large enough to permit the passage of ships. The northern (landward) transgression of the foreland is clearly shown in the truncation of the beach ridges at their southern ends and in the overall narrowing of the point in its proximal (western) sections. However, in the distal (eastern) portions, shoreline monitoring over the past 40 years has indicated a pattern of accretion on the south side and erosion on the north side. In other words, the distal portions of the point are slowly rotating in a clockwise direction.

SUMMARY

The Long Point coastal structure began around 12.0 ka, when levels in Lake Erie were at their minimum, as a sandy, north-south-trending feature associated with the cross-lake Norfolk moraine. Extending initially as far as the Presque Isle area, it was later cut by a drainage channel from the central basin, and thereafter, retreated up the basin slope to its present position. Around 4–5 ka, an abrupt rise in

←

FIG. 11.—Schematic reconstruction of the evolution of the Long Point foreland from about 6 ka to present.

levels, related to resumption of drainage from the Nipissing Phase in the Upper Great Lakes into the Erie basin, resulted in inundation of most of the ancestral foreland. When the lake fell later to previous levels, presumably due to widening of the Niagara River outlet, the foreland commenced its present east-west orientation.

The pattern of landward retreat of parts of the foreland over sheltered (lagoonal) deposits and the rotation of the distal portions are also noted in the two other north-shore forelands, Point Pelee and Pointe-aux-Pins, and show similarities to the transgressive marine barrier islands of the United States east coast. In many respects, the evolutionary processes acting on these forelands (lateral migration by washover, deflation, and inlet opening) are comparable to those described in relation to barrier islands (Hoyt, 1967; Kraft and others, 1978; Moslow and Heron, 1981). Apart from scale, major differences that should temper direct comparison are the lack of both a tidal effect and a clearly defined lagoonal environment in Lake Erie.

ACKNOWLEDGMENTS

Drilling of BH3 was carried out by the University of Waterloo drill rig, and the material presented here represents part of a Ph.D. research project in the Department of Earth Sciences, University of Waterloo. The author thanks D. E. Nelson, J. S. Vogel, and J. R. Southon of Simon Fraser University for accelerator-spectrometry C-14 dates on microsamples from Long Point cores, and C. E. Winn and T. W. Anderson for providing access to unpublished pollen data. The Ontario Hydroelectric Corporation (J. Godawa) gave permission to use seismic-survey data. P. J. Barnett, B. H. Feenstra, and C. F. M. Lewis generously provided access to their unpublished data on Lake Erie Quaternary sediments, as well as many useful discussions and ideas on the postglacial history of Lake Erie. The author also thanks reviewers P. F. Karrow and R. A. Davidson-Arnott for their constructive comments and suggestions.

REFERENCES

ANDERSON, T. W., 1971, Postglacial vegetative changes in the Lake Huron–Lake Simcoe District, Ontario, with special reference to glacial Lake Algonquin: Unpublished Ph.D. Dissertation, University of Waterloo, Waterloo, Ontario, 245 p.

ANDREWS, J. T., 1970, A Geomorphological Study of Postglacial Uplift with Particular Reference to Arctic Canada: Institute of British Geographers, London, 156 p.

BARNETT, P. J., 1985, Glacial retreat and lake levels, north central Lake Erie basin, *in* Karrow, P. F., and Calkin, P., eds. Quaternary Evolution of the Great Lakes: Geological Association of Canada, Special Paper 30, p. 185–194.

BARNETT, P. J., COAKLEY, J. P., TERASMAE, AND J., WINN, C. E., 1985, Chronology and significance of a Holocene sedimentary profile from Clear Creek, Lake Erie shoreline: Canadian Journal of Earth Sciences, v. 22, p. 1133–1138.

CALKIN, P. E., AND FEENSTRA, B. H., 1985, Evolution of the Erie-basin Great Lakes, *in* Karrow, P. F., and Calkin, P., eds., Quaternary Evolution of the Great Lakes: Geological Association of Canada, Special Paper 30, p. 149–170.

COAKLEY, J. P., 1976, The formation and evolution of Point Pelee, western Lake Erie: Canadian Journal of Earth Sciences, v. 13, p. 136–144.

COAKLEY, J. P., 1985, Evolution of Lake Erie based on the postglacial sedimentary record below the Long Point, Point Pelee and Pointe-aux-Pins forelands: Unpublished Ph.D. Dissertation, University of Waterloo, Waterloo, Ontario, 362 p.

COAKLEY, J. P., 1989, The origin and evolution of a complex cuspate foreland: Pointe-aux-Pins, Lake Erie: Géographie Physique et Quaternaire, v. 43, p. 65–76.

COAKLEY, J. P., AND LEWIS, C. F. M., 1985, Postglacial lake levels in the Erie basin *in* Karrow, P. F., and Calkin, P., eds., Quaternary Evolution of the Great Lakes: Geological Association of Canada, Special Paper 30, p. 195–212.

DAVIDSON-ARNOTT, R. G. D., AND LAW, M. N., 1990, Seasonal patterns and controls on sediment supply to coastal foredunes, Long Point, Lake Erie, *in* Nordstrom K. F., Psuty, N. P., and Carter, R. W. G., eds. Coastal Dunes: Form and Process: John Wiley and Sons, New York, p. 177–200.

HOYT, H. J., 1967, Barrier island formation: Geological Society of America Bulletin, v. 78, p. 1125–1136.

KRAFT, J. C., ALLEN, E. A., AND MAURMEYER, E. M., 1978, The geological and paleo-geomorphological evolution of a spit system and its associated coastal environments: Journal of Sed Petr, v. 48, p. 211–226.

LEVERETT, F., AND TAYLOR, F. B., 1915, The Pleistocene of Indiana and Michigan and the history of the Great Lakes: U.S. Geological Survey Monograph 53, 523 p.

LEWIS, C. F. M., 1966, Sedimentation studies of unconsolidated deposits in the Erie basin: Unpublished Ph. D. Dissertation, University of Toronto, Toronto, Ontario, 135 p.

LEWIS, C. F. M., 1969, Late Quaternary history of lake levels in the Huron and Erie basins: Proceedings, 12th Conference on Great Lakes Research, International Association of Great Lakes Research, p. 250–270.

MEISTRELL, F. J., 1966, The spit-platform concept: laboratory observation of spit development, *in* Schwartz, M. L., ed., Spits and Bars: Dowden, Hutchinson, and Ros Inc Stroudsburg, Pennsylvania.

MOSLOW, T. F., AND HERON, D., 1981, Holocene depositional history of a microtidal cuspate foreland cape: Cape Lookout, North Carolina: Marine Geology, v. 41, p. 251–270.

MOTT, R. J., AND FARLEY-GILL, L. D., 1978, A Late Quaternary pollen profile from Woodstock, Ontario: Canadian Journal of Earth Sciences, v. 15, p. 1101–1111.

NELSON, D. E., VOGEL, J. S., SOUTHON, J. R., AND BROWN, T. A., 1986, Accelerator radiocarbon dating at SFU: Radiocarbon, v. 28, p. 215–222.

OGDEN, J. G., 1967, Radiocarbon and pollen evidence for sudden change in climate in the Great Lakes region approximately 10,000 years ago, *in* Cushing, E. J., and Wright, H. E., Jr., eds., Quaternary Paleoecology: Yale University Press, New Haven, p. 117–127.

ONTARIO HYDRO, 1981, Lake Erie marine cable crossing: Geotechnical Investigation—1980 program, unpublished Ontario Hydro Report 81169 (includes uninterpreted seismic-profiler records).

REINECK, H. E., AND SINGH, I. B., 1975, Depositional Sedimentary Environments: Springer-Verlag, Berlin, 439 p.

ST. JACQUES, D. A., AND RUKAVINA, N. A., 1973, Lake Erie nearshore sediments, Mohawk Point to Port Burwell, Ontario: Proceedings, 16th Conference on Great Lakes Research International Association of Great Lakes Research p. 454–467.

TERASMAE, J., 1970, Report on stratigraphic drilling in Point Pelee National Park, Ontario: Contract Report 66–200, Canada Department of Indian and Northern Affairs, National and Historic Parks Branch, Ottawa, 6 p.

WASHBURN, A. L., AND STUIVER, M., 1962, Radiocarbon-dated postglacial delevelling in northeast Greenland and its implications: Arctic, v. 15, p. 66–73.

WILLIAMS, S. J., AND MELSBURGER, E. P., 1982, Geological character and mineral resources of south central Lake Erie: U.S. Army Corps of Engineers, Coastal Engineering Research Centre, Miscellaneous Report 82–9, 62 p.

ZENKOVICH, V. P., 1967, Processes of Coastal Development: Interscience Publishers, New York, 738 p.

OBLITERATION OF SURFICIAL PALEOLAKE EVIDENCE IN THE TULE VALLEY SUBBASIN OF LAKE BONNEVILLE

DOROTHY SACK

Department of Geography, University of Wisconsin, Madison, Wisconsin 53706

ABSTRACT: This paper analyzes the extent of preserved and obliterated paleolake shoreline and sediment evidence in the Tule Valley subbasin of Lake Bonneville. It identifies the dominant processes responsible for effacing the paleolake evidence and estimates long-term average obliteration rates. Temporal control is provided by the well-constrained reconstructed chronology of Lake Bonneville.

Erosion associated with alluvial-fan processes is overwhelmingly responsible for the partial obliteration of the studied shorelines. Long-term average shoreline-obliteration rates vary from 3.0 to 5.7 percent per 1,000 yrs. Extent of preservation increases with the strength of a shoreline's geomorphic development.

Measurements of the areal extent of all lacustrine and nonlacustrine Quaternary deposits below the highest shoreline reveal that 57 percent of the paleolake sediments have been removed from surface exposure since the start of Lake Bonneville regression. This yields a long-term average obliteration rate of 3.9 percent (47 km^2) per 1,000 yrs. Alluvial-fan and eolian processes are responsible for 89 and 11 percent, respectively, of the removal of paleolake sediments from surficial exposure in Tule Valley.

INTRODUCTION

Lake Bonneville existed from approximately 28 to 13 ka (Currey, 1990) during the latest deep-lake cycle in the Bonneville basin. It was the largest of the western North American late Pleistocene paleolakes. Lake Bonneville attained a maximum depth of 372 m, at which level it covered over 51,000 km^2 in the northeastern Great Basin (Currey and others, 1984b) and was delineated by 6,800 km of shoreline (Fig. 1; D. R. Currey, pers. commun., 1990). Because of the lake's size, longevity, and recent existence, substantial sedimentologic and geomorphic evidence of it is found in the modern regional surficial landscape (Gilbert, 1890; Currey, 1982; Oviatt, 1989b; Sack, 1990). Lake Bonneville sediments and relict shorelines have been interpreted as paleolake features since Stansbury's survey of the region in 1849 (Stansbury, 1852; Sack, 1989c).

Although sedimentologic and coastal geomorphic evidence of Lake Bonneville is widespread, considerable portions of its major shorelines and deposits have been removed from surface exposure by postlacustrine subaerial processes (Oviatt, 1989b; Sack, 1990). Upon its formation, a lacustrine shoreline is laterally continuous around its basin. Lake Bonneville shoreline segments that are missing from the visible surficial record have been eroded, reworked, or buried since regression of the lake. Likewise, Lake Bonneville sediments at one time blanketed the basin below the elevation of the highest shoreline (Gilbert, 1890), which was maintained by intermittent external threshold control for approximately 850 years (Burr and Currey, 1988). The distribution of these sediments is also patchy now, due to subsequent subaerial erosion, reworking, and burial.

The purposes of this paper are to (1) describe the present extent of remnant and effaced paleolake shorelines and deposits in a selected portion of the Bonneville basin, (2) identify the dominant geomorphic processes leading to obliteration of the surficial paleolake evidence, and (3) estimate the long-term average rate of obliteration of that surficial evidence. Analyzing the extent of preserved and obliterated surficial evidence of late Pleistocene paleolakes requires detailed Quaternary mapping of the region of interest. Precise knowledge concerning the modern distribution of Bonneville basin lacustrine deposits and relict shorelines is just starting to become available as Quaternary mapping projects proceed in the region at a variety of map scales (e.g., Robison and McCalpin, 1987; Sack, 1989a, 1989b, 1990; Oviatt, 1989a, 1989b, 1991). The Tule Valley, Utah, subbasin of Lake Bonneville (Fig. 2) was chosen for study because, of the Bonneville basin's seven major subbasins, it has the most complete, large-scale (1:100,000 or larger) Quaternary geologic map coverage (Sack, 1989a, 1989b, 1990).

RATIONALE

No previous investigation has determined the extent of preserved and obliterated surficial evidence of late Pleistocene paleolakes in what is now a desert basin, although some general observations have been made. In an early work on lacustrine geomorphology, Gilbert (1885, p. 122) noted that "a system of shore topography, from which the parent lake has receded, is immediately exposed to the obliterating influence of land erosion and gradually, though very slowly, loses its character and definition. The strength of the record is directly proportioned to the duration of the lake and inversely to its antiquity." In his monumental monograph on Lake Bonneville, Gilbert (1890, p. 93) qualitatively compared the surficial lacustrine and subaerial records. He concluded that pre-Bonneville subaerial work greatly exceeded Bonneville lacustrine work. He interpreted the degree of preservation of the lacustrine features to mean that comparatively little alluvial accumulation had occurred in post-Bonneville time. Mifflin and Wheat (1979) suggested that shoreline features of late Pleistocene lakes in Nevada are better preserved in the cooler and moister northern portion of the state than in the warmer and drier southern portion. They estimated shoreline preservation to range from 10 to 80 percent. Currey (1990, p. 210) commented on the processes by which Great Basin paleolake records are obliterated, noting that "most basin-flank pelagial sediments were reworked long ago, by being washed onto basin floors or blown onto downwind terrain." In the report accompanying the Quaternary geologic map of Tule Valley, Sack (1990, Table 4) listed percentages of the map area covered by the various Quaternary and pre-Quaternary units, but the extent of lacustrine versus nonlacustrine Quaternary deposits lo-

Quaternary Coasts of the United States: Marine and Lacustrine Systems, SEPM Special Publication No. 48

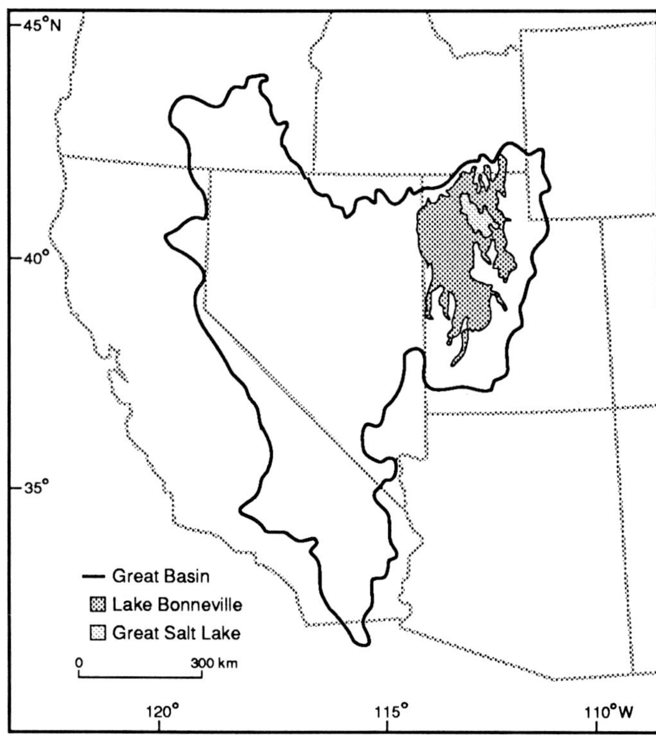

FIG. 1.—Map showing location of Lake Bonneville.

cated below the highest Lake Bonneville shoreline was not ascertained.

Measuring geologic rates is a basic objective of earth science because it improves the understanding of geologic processes and refines the ability to make geologic predictions. More than a century of research in the Bonneville basin has resulted in a geochronometrically well-constrained reconstruction of the lake's chronology (Fig. 3), which is considered highly reliable (e.g., Scott and others, 1983; Currey and Oviatt, 1985; Benson and others, 1990). Because of this reconstructed chronology and the measurability of effaced lacustrine evidence, the Bonneville basin provides an excellent opportunity to derive information on the long-term average rate of landscape evolution. Specifically, this analysis provides data on fundamental geomorphic questions concerning the amount of time required for landform assemblages to become adjusted to a new climatic regime (Chorley and others, 1984), and the extent of relict landforms in the present landscape.

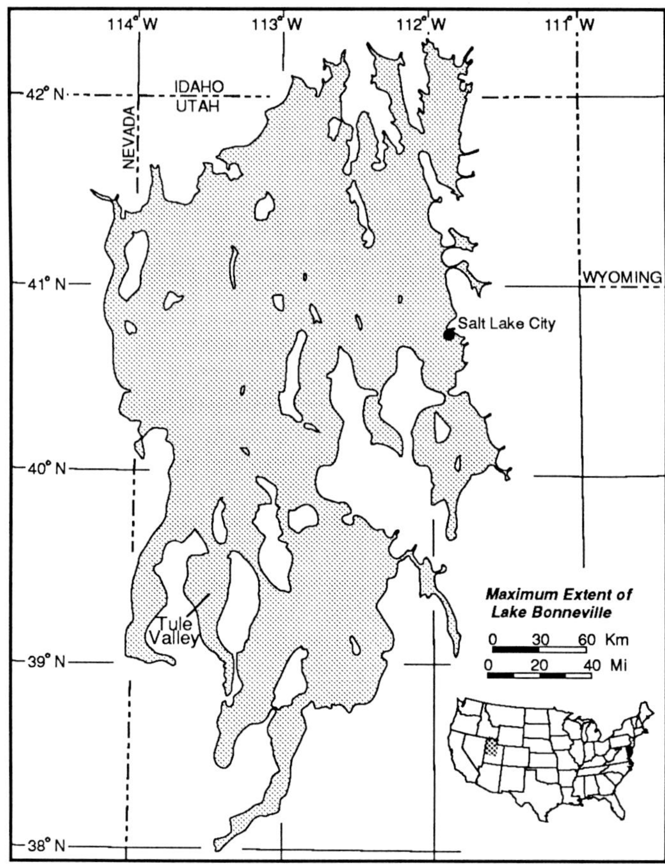

FIG. 2.—Map showing the Tule Valley subbasin with respect to Lake Bonneville at its maximum extent.

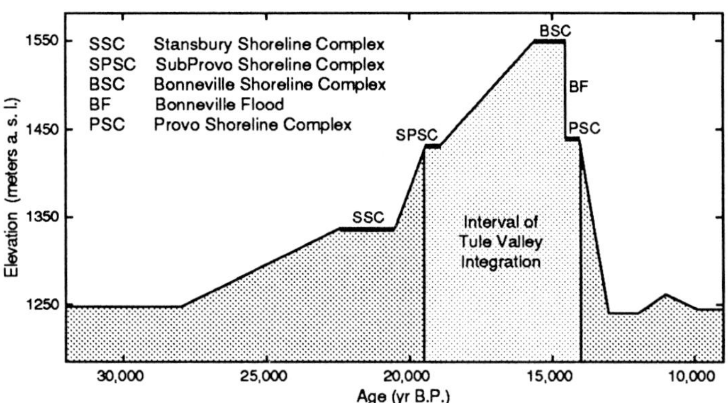

FIG. 3.—Generalized schematic hydrograph of Lake Bonneville (after Burr and Currey, 1988).

STUDY AREA

Tule Valley is located approximately 210 km southwest of Salt Lake City and 70 km west of Delta, Utah. It is a structural basin of interior drainage lying west of the House Range and east of the Confusion Range (Fig. 4). The valley contains a variety of surficial Quaternary sediments, including alluvial, eolian, lacustrine, mass-wasting, playa, and spring deposits. Tule Valley has a mid-latitude dry climate with an estimated mean annual temperature of 10°C, mean January temperature of −2.8°C, and mean July temperature of 24.4°C (Sack, 1990). It receives approximately 18 cm of precipitation annually (Sack, 1990). Vegetation consists primarily of desert shrubs and grasses.

During Lake Bonneville's higher stages, Tule Valley was an embayment in the southwest portion of the lake (Fig. 2; Gilbert, 1890; Currey and others, 1984a; Sack, 1990). Tule Valley is described as a subbasin of Lake Bonneville because it was integrated with the major lake for only part of

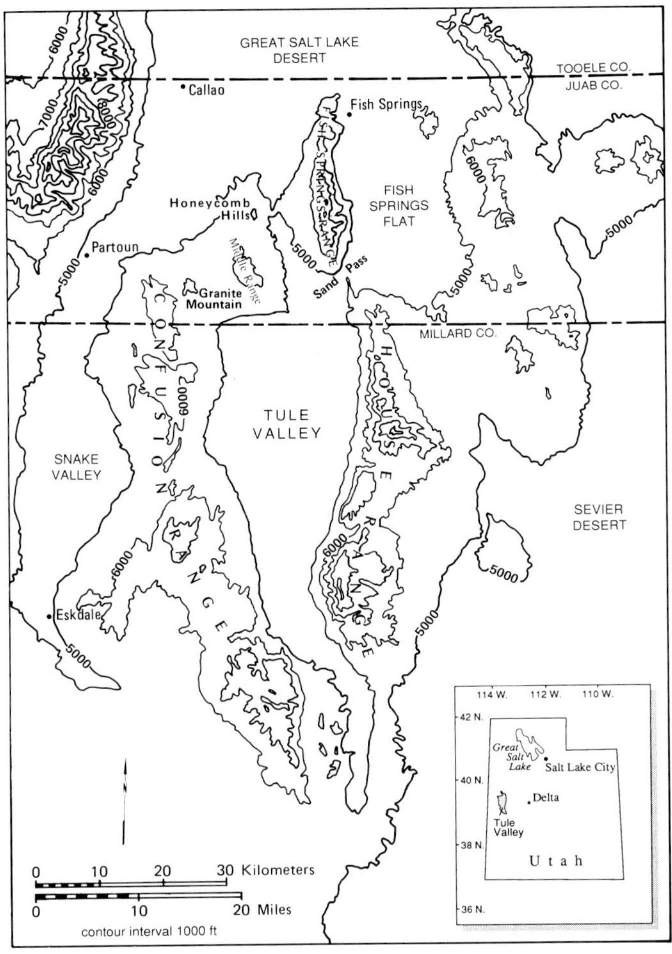

FIG. 4.—Regional map of the Tule Valley study area.

the Bonneville lacustral cycle (Eardley and others, 1957; Currey and Oviatt, 1985; Sack, 1990). The period of integration lasted from approximately 19.5 to shortly after 14 ka (Sack, 1990). Most of the major events in Lake Bonneville history occurred during this interval (Fig. 3). These include formation of the lake's highest and most conspicuous shoreline, named the Bonneville shoreline, occurrence of the Bonneville Flood (Gilbert, 1890; Malde, 1968; Jarrett and Malde, 1987), and formation of the Provo shoreline, which is the second most conspicuous Lake Bonneville level.

Prior and subsequent to the period of Lake Bonneville integration, Tule Valley contained an isolated paleolake, called Lake Tule (Sack, 1990). At its highest preintegration level, which was achieved shortly before inflow from Lake Bonneville, Lake Tule was approximately 76 m deep, covered 660 km^2, and was bounded by 188 km of coastline. Post-Bonneville Lake Tule regressed rapidly from the elevation of the internal Lake Bonneville threshold to very low levels (Sack, 1989a, 1990).

The study area of this investigation covers 1,193 km^2. It consists of that portion of Tule Valley lying at and below the Bonneville shoreline, and thus encompasses most of the Tule Valley piedmont and all of its basin-floor area. The few outcrops of pre-Quaternary bedrock that are located below the elevation of the Bonneville shoreline are not included in the study area.

METHODOLOGY

Three prominent Tule Valley shorelines are included in the analysis of shoreline preservation and obliteration (Fig. 5). Two of these, the Bonneville and Provo shorelines, are part of Lake Bonneville and were maintained by external threshold control for approximately 850 and 300 years, respectively (Burr and Currey, 1988). The third shoreline chosen for study is the highest preintegration Lake Tule shoreline, which was created under closed-basin conditions. For most of their lateral extent within the study area, all three of the shorelines were inscribed onto pre-Bonneville alluvial fans.

The preserved portions of these three shorelines were mapped by Sack (1990). The obliterated portions of the shorelines have been reconstructed from that compilation by delineating along proper contours between the preserved links. The lengths of mapped preserved and reconstructed obliterated shoreline segments were measured with a digitizing line meter.

An assumption of this investigation is that the study area was completely blanketed by lacustrine deposits just before the Bonneville Flood. This assumption conforms to observations concerning Lake Bonneville made by Gilbert (1890, p. 90), who maintained that "so much of the basin as lies below the highest shoreline has received lake sediments. . . ." Gilbert (1885) recognized early that lakes are sediment sinks and that the dominant geomorphic process acting beyond the littoral zone is deposition.

The sediments of Lake Bonneville and Lake Tule are not differentiated for this study. Deposits of the two lakes are lumped because (1) most of the lacustrine sediments in Tule Valley were deposited by the larger lake and (2) the lakes represent mutually exclusive subintervals of the same deep-lake cycle, the Bonneville lacustral cycle.

For three reasons the extent of preserved and obliterated lacustrine deposits in Tule Valley cannot be determined directly from previously published data regarding percentages of the valley covered by Quaternary geologic map units (Sack, 1990, Table 4). First, because Sack's (1990, Plate 1) map extends to the piedmont junction rather than just to the highest shoreline, those percentages include substantial areas above the elevation of the Bonneville shoreline. The percentages of the various Quaternary units below the Bonneville shoreline were determined for the present study by measuring their extent above the high lake level on the existing Quaternary geologic map and subtracting these values from the total area of each map unit (Sack, unpublished data). Second, Sack's (1990) data were calculated as percentages of total map area, which includes a pre-Quaternary bedrock unit. The area of bedrock below the elevation of the Bonneville shoreline is not considered in the present analysis. A third and more serious problem with the preexisting data stems from the fact that the published map employs a Quaternary unit defined as undifferentiated lacustrine and alluvial deposits (Qla). This unit includes pre-

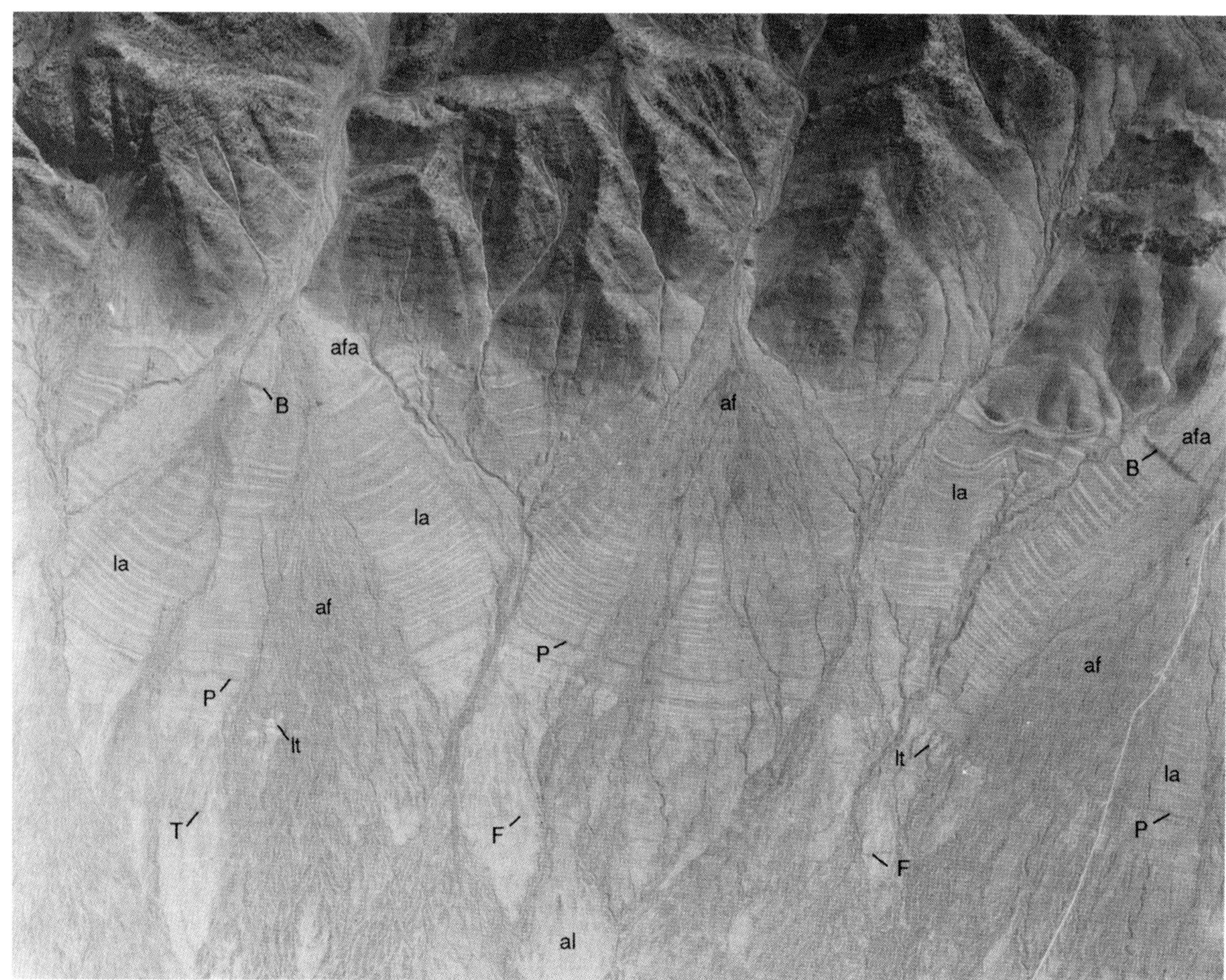

FIG. 5.—Vertical aerial photograph (CSR-F 17–159) of part of the House Range piedmont in northeastern Tule Valley. Scale is approximately 1:25,000. North is to left. Preserved segments of the Bonneville, Provo, and Lake Tule shorelines are indicated by B, P, and T, respectively. Note alluvial-fan obliteration of portions of all three major shorelines adjacent to their preserved segments. F = piedmont fault scarp, af = active alluvial-fan segments, afa = abandoned alluvial fans above the Bonneville shoreline, la = lake-reworked alluvial fans, al = fan-reworked lake deposits, lt = lacustrine tufa.

Bonneville alluvial fans that were slightly reworked by the lake, Lake Tule and Lake Bonneville deposits that have been only slightly reworked by postlacustrine alluvial-fan processes, and local areas where lake and fan deposits are not resolvable at the 1:100,000 scale (Sack, 1990, Plate 1). Therefore, additional mapping was required for the present study in order to divide the Qla unit into its lacustrine (lake-reworked fan deposits) and subaerial (fan-reworked lake deposits) components (Fig. 5). Once mapped, their respective areas were measured, like all areas in this research, with a digital electronic planimeter.

Depending on their elevation, Bonneville lake-cycle shorelines and sediments within Tule Valley have been exposed to postlacustrine subaerial processes for varying amounts of time. Temporal control on water level is derived from over 50 radiocarbon ages that provide the chronostratigraphic framework for the reconstructed chronology of Lake Bonneville (Fig. 3; Currey and Oviatt, 1985) and Lake Tule (Sack, 1990). Because of material failure at its external threshold, Lake Bonneville fell catastrophically 104 m from the Bonneville to the Provo shoreline at approximately 14.5 ka in the Bonneville Flood (Fig. 3; Gilbert, 1890; Malde, 1968; Burr and Currey, 1988). Therefore, all shorelines and sediments above the Provo shoreline have been subaerially exposed for an estimated 14,500 years. The lake regressed rapidly from the Provo shoreline due to climatic factors starting at about 14.2 ka (Fig. 3; Burr and Currey, 1988). Tule Valley became re-isolated from Lake Bonneville about 250 years later, and postintegration Lake Tule fell at least as fast as Lake Bonneville to very low levels (Sack, 1989a, 1990). Based on Lake Bonneville's estimated regression rate of approximately 0.09 m per yr (D. R. Currey, pers. commun., 1987), the highest preintegration Lake Tule shoreline would have been subaerially exposed at approximately 13.6 ka. Therefore, the ages of 14.5, 14.2, and 13.6 ka are used in determining the long-term average rate of obliteration of the Bonneville, Provo, and Lake Tule shorelines, respectively. The maximum-limiting age of 14.5 ka is used to determine the long-term average rate of obliteration of the lacustrine sediments in Tule Valley.

DATA

Reconstruction of the original perimeter of the three studied shorelines reveals that 53 percent of the Bonneville, 43 percent of the Provo, and 77 percent of the Lake Tule shoreline have been obliterated (Table 1). Using the ages of 14.5, 14.2, and 13.6 ka, respectively, for onset of Bonneville, Provo, and Lake Tule shoreline subaerial exposure, long-term average rates of obliteration for the three shorelines range from 3.0 to 5.7 percent per 1,000 yrs (Table 1). Aerial photographic interpretation reveals that almost all the shoreline obliteration occurred by erosion from alluvial-fan processes (Fig. 5), and many postlacustrine alluvial fans have developed with their apices located at one of these shorelines.

The Quaternary deposits covering the study area today consist of 51 percent alluvial, 6 percent eolian, and 43 percent relict lacustrine deposits, with minor amounts of playa, marsh, and mass-wasting sediments (Table 2). Since 14.5 ka, 57 percent of the surficial lacustrine sediments have been obliterated, primarily by alluvial-fan processes and secondarily by eolian processes. The long-term average rate of obliteration of the lacustrine deposits in the study area is 3.9 percent (47 km^2) per 1,000 yrs. Spatially, obliteration of the lacustrine deposits has occurred preferentially, although not entirely, in the piedmont zone. Alluvial processes account for most of the obliteration of lacustrine evidence on the piedmont, but eolian deposits have contributed to the obliteration within the lower piedmont sector, especially in the northeast (downwind) portion of the study area. Eolian sediments are also found on the Tule Valley floor, particularly in the center of the study area where they lie northeast of a small Holocene lake basin (Sack, 1989a).

DISCUSSION

Obliteration of paleolake shoreline evidence in Tule Valley has occurred predominantly by alluvial-fan channels and flows breaching shoreline segments. Shoreline sediments eroded by fan processes are reworked and redeposited lower on the piedmont as alluvial-fan sediments. It is possible that some shoreline segments obliterated from surface exposure are buried under alluvial-fan deposits and preserved in the stratigraphic record.

The Lake Tule shoreline is the most obliterated of the three shorelines for several possible reasons. The water body encompassed by the Lake Tule shoreline was much smaller, and therefore had a smaller fetch, than the water bodies encompassed by the Bonneville and Provo shorelines (Table 3). As a result, landforms of the Lake Tule level are not as large and well developed as those of the Bonneville and Provo levels, and are thus more easily obliterated (Fig. 5).

The Lake Tule shoreline was formed by the shallowest of the three water bodies (Table 3). It is located in the lower piedmont sector, where the size of available coastal detritus was relatively small. Streams of a given discharge can more easily erode the small-caliber clasts in the Lake Tule depositional features than the larger clasts comprising the higher Bonneville and Provo shorelines.

The nature of alluvial-fan processes may also help explain the degree of obliteration of the Lake Tule shoreline.

TABLE 2.–QUATERNARY DEPOSITS AS PERCENTAGE OF STUDY AREA

%	Deposit
	Lacustrine
23.80	Fine-grained lacustrine deposits
8.40	Lake-reworked fan deposits
5.98	Lacustrine gravel
2.99	Lacustrine marl
0.96	Lacustrine tufa
0.51	Lacustrine sand
0.14	Lacustrine lagoon deposits
42.79%	Total
	Alluvial
27.31	Fan-reworked lake deposits
23.95	Alluvial-fan deposits
0.10	Undifferentiated alluvium and colluvium
51.36%	Total
	Eolian
2.13	Eolian sand dunes
2.06	Clastic eolian sheet sand
1.37	Eolian gypsum sheet sand
0.03	Alluvially reworked eolian deposits
5.59%	Total
	Other
0.15	Playa mud
0.06	Spring-marsh deposits
0.05	Debris-flow deposits
0.21%	Total

TABLE 1.–PRESERVATION AND OBLITERATION OF TULE VALLEY SHORELINES

Shoreline	Total Reconstructed Shoreline (km)	Length of Shoreline Preserved (%)	Length of Shoreline Obliterated (%)	Estimated Age of Subaerial Exposure (ka)	Average Rate of Obliteration (% per 1,000 yrs)
Bonneville	251	47	53	14.5	3.7
Provo	217	57	43	14.2	3.0
Lake Tule	188	23	77	13.6	5.7

TABLE 3.—PHYSICAL AND TEMPORAL CHARACTERISTICS OF THE BONNEVILLE, PROVO, AND LAKE TULE SHORELINES IN TULE VALLEY

Shoreline	Area of Water Body (km^2)	Maximum Depth (m)	Minimum Age (ka)	Estimated Duration (yrs)
Bonneville	1258	243	14.5	850
Provo	839	129	14.2	300
Lake Tule	660	76	19.5	unknown

Because of their radial shape, alluvial fans impact a wider area in their distal sectors than in their proximal sectors. Postlacustrine alluvial fans would therefore interact with a greater percentage of the lower Lake Tule shoreline than the higher Bonneville and Provo shorelines.

The Lake Tule shoreline is the only one of the three shorelines that was submerged by deep water, and this may have had some effect on its present state of obliteration. Local erosion and burial of the Lake Tule shoreline occurred near the point of inflow as Lake Bonneville rapidly filled the water level in Tule Valley to equilibration, at approximately 19.5 ka (Sack, 1990). During its submerged period, the Lake Tule shoreline may have been somewhat muted by deposition of overlying pelagic Lake Bonneville sediments, although the Quaternary geologic map of Tule Valley shows that the shoreline was not much obliterated through burial by lacustrine fines (Sack, 1990). Erosion and reworking of the Lake Tule shoreline probably occurred to some extent as the postintegration Lake Tule water plane regressed rapidly over it. Furthermore, because of the long interval of submergence, the Lake Tule shoreline sediments would have been saturated at the start of the period of subaerial exposure. This may have resulted in some shoreline modification by mass wasting.

If the Lake Tule shoreline is the most obliterated for reasons related to the size of its water body and its location on the piedmont, it is reasonable to expect that the Bonneville shoreline would be the least obliterated of the three shorelines. This, however, is not the case. The Provo shoreline is better preserved than the Bonneville shoreline despite the facts that, compared to the Bonneville level, the Provo water body was smaller, shallower, and lower on the piedmont (Table 3). One explanation for the greater preservation of the Provo shoreline is that it is slightly younger than the Bonneville shoreline and has thus been subjected to subaerial reworking for a shorter time period. The time differential between the two, however, is only about 300 years (Fig. 3; Burr and Currey, 1988). Alternatively, although the Provo-level water body was smaller and of shorter duration than the Bonneville-level lake, Provo depositional features tend to be more massive, and therefore take more time to obliterate, than Bonneville depositional features (Gilbert, 1890). Morphostratigraphic studies reveal that the Provo-level depocenters are especially large because they overlie landforms created during a transgressive-phase Lake Bonneville stillstand or oscillation which occurred close to the elevation that the regressive-phase Provo shoreline would later occupy (Sack, 1990). Coastal deposits of that transgressive-phase stillstand or oscillation are referred to as the subProvo shoreline complex (Fig. 3).

Projecting the calculated long-term average obliteration rates of Tule Valley shorelines and deposits into the future yields estimates of when complete surface obliteration will be achieved. No evidence for the Lake Tule shoreline would remain by 4,100 radiocarbon years after present, Bonneville shoreline evidence would be effaced by 12,900 years after present, and surficial evidence of the Provo shoreline would be removed by 18,800 years after present. Using the 3.9 percent per 1,000 yrs obliteration rate of the deposits, it is calculated that they will be obliterated by 10,900 years after present. The fact that the obliteration rates of the Bonneville and Provo shorelines are slower than the rate for all deposits, which includes Bonneville and Provo shoreline deposits, emphasizes that (1) obliteration of geomorphic features of considerable relief may take substantially longer than obliteration of more blanket-like deposits, (2) the calculated rates are only long-term averages, and (3) the real relationship between amount of obliteration and time may not be a linear or even a smooth function (Kukal, 1990).

SUMMARY AND CONCLUSIONS

In the 14,500 years since the start of the regression of Lake Bonneville, 57 percent of the surficial sediment evidence of the Bonneville lacustral cycle has been obliterated from Tule Valley. Alluvial-fan and eolian processes have performed 89 and 11 percent of the effacement, respectively. Forty-three to 77 percent of the length of the studied shorelines has been obliterated from surface exposure in the postlacustrine interval. Size of shoreline features seems to be a major factor in determining the rate of shoreline obliteration. The Lake Tule shoreline, which is the least well developed, has been rapidly effaced. The Provo shoreline, which has the largest depositional features and is the youngest of the three shorelines, is the best preserved. The long-term average rates provided in this study give an indication of the speed with which landform and sediment assemblages in a paleolake basin have adjusted to the postlacustrine climate. These rates, however, should not be viewed as anything more than long-term averages of processes that are likely quite variable over time.

Tule Valley is representative of the smaller Lake Bonneville subbasins in being relatively long and narrow. Because of its shape, Tule Valley has a proportionally larger piedmont zone than the big, open Lake Bonneville subbasins, such as the Great Salt Lake, Great Salt Lake Desert, and the Sevier Desert basins. As a result of its geomorphic structure, the role of alluvial-fan processes in effacing relict paleolake evidence may be much greater in Tule Valley than in the big desert basins that have a larger percentage of their areas in basin floor. It is expected that eolian processes would obliterate a greater proportion of the lacustrine evidence in large, flat basins than they do in Tule Valley, given the same types of paleolake sediments. Fluvial processes may be significant in obliterating paleolake evidence in desert basins that have high or large drainage basins. Continued research in other basins will help determine whether principal obliterating processes vary with such factors as basin size, basin shape, piedmont-to-basin-floor ratio, basin relief, drainage area, and degree of aridity.

ACKNOWLEDGMENTS

I am grateful to R. H. Dott, Jr., and C. G. Oviatt for their careful reviews of this manuscript.

REFERENCES

Benson, L. V., Currey, D. R., Dorn, R. I., Lajoie, K. R., Oviatt, C. G., Robinson, S. W., Smith, G. I., Stine, S., and Thompson, R. W., 1990, Chronology of expansion and contraction of four Great Basin lake systems during the past 35,000 years: Palaeogeography, Palaeoclimatology, Palaeoecology, v. 78, p. 241–286.

BURR, T. N., AND CURREY, D. R., 1988, The Stockton Bar: Utah Geological and Mineral Survey Miscellaneous Publication 88–1, p. 66–73.

CHORLEY, R. J., SCHUMM, S. A., AND SUGDEN, D. E., 1984, Geomorphology: Methuen, New York, 605 p.

CURREY, D. R., 1982, Lake Bonneville: Selected features of relevance to neotectonic analysis: U.S. Geological Survey Open-File Report 82–1070, 31 p.

CURREY, D. R., 1990, Quaternary palaeolakes in the evolution of semidesert basins, with special emphasis on Lake Bonneville and the Great Basin, U.S.A.: Palaeogeography, Palaeoclimatology, Palaeoecology, v. 76, p. 189–214.

CURREY, D. R., ATWOOD, G., AND MABEY, D. R., 1984, Major levels of Great Salt Lake and Lake Bonneville: Utah Geological and Mineral Survey Map 73.

CURREY, D. R., AND OVIATT, C. G., 1985, Durations, average rates, and probable causes of Lake Bonneville expansions, stillstands, and contractions during the last deep-lake cycle, 32,000 to 10,000 years ago, *in* Kay, P. A., and Diaz, H. F., eds., Problems of and Prospects for Predicting Great Salt Lake Levels: University of Utah Center for Public Affairs and Administration, Salt Lake City, p. 9–24.

CURREY, D. R., OVIATT, C. G., AND CZARNOMSKI, J. E., 1984, Late Quaternary geology of Lake Bonneville and Lake Waring: Utah Geological Association Publication 13, p. 227–237.

EARDLEY, A. J., GVOSDETSKY, V. M., AND MARSELL, R. E., 1957, Hydrology of Lake Bonneville and sediments and soils of its basin: Geological Society of America Bulletin, v. 68, p. 1141–1201.

GILBERT, G. K., 1885, Topographic features of lake shores: U.S. Geological Survey Fifth Annual Report, p. 69–123.

GILBERT, G. K., 1890, Lake Bonneville: U.S. Geological Survey Monograph 1, 438 p.

JARRETT, R. D., AND MALDE, H. E., 1987, Paleodischarge of the late Pleistocene Bonneville Flood, Snake River, Idaho, computed from new evidence: Geological Society of America Bulletin, v. 99, p. 127–134.

KUKAL, Z., 1990, The rate of geological processes: Earth-Science Reviews, v. 28, 284 p.

MALDE, H. E., 1968, The catastrophic late Pleistocene Bonneville Flood in the Snake River Plain, Idaho: U.S. Geological Survey Professional Paper 595, 52 p.

MIFFLIN, M. D., AND WHEAT, M. M., 1979, Pluvial lakes and estimated pluvial climates of Nevada: Nevada Bureau of Mines and Geology Bulletin 94, 57 p.

OVIATT, C. G., 1989a, Quaternary geology of Fish Springs Flat, Juab County, Utah: Utah Geological and Mineral Survey Open-File Report 158, 53 p.

OVIATT, C. G., 1989b, Quaternary geology of part of the Sevier Desert, Millard County, Utah: Utah Geological and Mineral Survey Special Studies, 70, 41 p.

OVIATT, C. G., 1991, Quaternary geology of the Black Rock Desert, Millard County, Utah: Utah Geological and Mineral Survey Special Studies, 73, 23 p.

ROBISON, R. M., AND MCCALPIN, J. P., 1987, Surficial geology of Hansel Valley, Box Elder County, Utah: Utah Geological Association Publication 16, p. 335–349.

SACK, D., 1989a, Geologic map of the Coyote Knolls quadrangle, Millard County, Utah: Utah Geological and Mineral Survey Open-File Report 165, 20 p.

SACK, D., 1989b, Geologic map of the Swasey Peak NW quadrangle, Millard County, Utah: Utah Geological and Mineral Survey Open-File Report 164, 19 p.

SACK, D., 1989c, Reconstructing the chronology of Lake Bonneville: An historical review, *in* Tinkler, K. J., ed., History of Geomorphology: Unwin Hyman, Boston, p. 223–256.

SACK, D., 1990, Quaternary geology of Tule Valley, west-central Utah: Utah Geological and Mineral Survey Map 124, 26 p.

SCOTT, W. E., MCCOY, W. D., SHROBA, R. R., AND RUBIN, M., 1983, Reinterpretation of the exposed record of the last two cycles of Lake Bonneville, western United States: Quaternary Research, v. 20, p. 261–285.

STANSBURY, H., 1852, Exploration and survey of the valley of the Great Salt Lake of Utah: Congressional Document, 32nd Congress, U.S. Senate Special Session, March 1851, Executive Document 3, 487 p.

SAND PETROFACIES IN THE SHORE ZONE OF LAKE TAHOE, CALIFORNIA AND NEVADA: A MODEL FOR A LARGE LAKE

ROBERT H. OSBORNE

Department of Geological Sciences, University of Southern California, Los Angeles, California 900089–0740

ABSTRACT: Petrographic modal analysis of the medium sand fraction (0.35 to 0.50 mm in diameter) of 274 samples principally from the shore zone of Lake Tahoe indicates that source-rock composition and mechanical-energy level at the depositional site explain approximately 46 and 32 percent of the total sample variance, respectively. A strong mechanical-energy signal may persist within a restricted grain-size interval. Q-mode factor analysis and stepwise discriminant function analysis were used to define six petrofacies, which reflect source-rock composition and relative mechanical-energy levels. Mean values for total quartz, total feldspar, and total rock fragments for petrofacies 1, 2, 3, 4, and 6 fall in the lithic arkose field of McBride (1963), whereas the mean for petrofacies 5 falls in the feldspathic litharenite field. The areal distribution of these six petrofacies may be used to identify nine shore-zone divisions, one of which is due to a major beach-nourishment project completed near Tahoe City in the early 1900s. Given the 1-km sampling interval employed, the areal positions of seven of the nine divisional boundaries showed no change from September 1978 to October 1988. This compositional stability reflects the consistency of the upland drainage systems and that of the wind-derived current systems operative in Lake Tahoe. The importance of source-rock composition, relative mechanical-energy level, and the apparent temporal consistency of the areal distribution of the shore-zone petrofacies should prove useful in constraining interpretations of older siliciclastic shoreline deposits associated with large lakes.

INTRODUCTION

Petrographic modal analysis of modern sand from known source areas provides fundamental information concerning source control on sand composition in known tectonic settings and depositional environments. Resultant compositional data may be used to identify actualistic petrofacies, which, in turn, may better constrain provenance and tectonic interpretations applied to older sedimentary strata.

Ingersoll (1990) draws attention to problems associated with scale in petrofacies analysis. The model of Dickinson and Suczek (1979) and derivative models (e.g., Dickinson, 1985) represent third-order fields (e.g., magmatic arc, recycled orogen, and continental block) that are useful for continent-scale analyses of large data sets. Second-order fields (e.g., undissected, transitional, and dissected arcs of Dickinson, 1985) are useful at the scale of mountain ranges and basins. First-order fields (e.g., outcrops and individual mountains) reflect specific source rocks rather than tectonic setting, although the latter ultimately controls the former. The primary reason for compositional contrast among first-, second-, and third-order petrofacies is the sampling scale; namely, local drainages, tributary streams, and large rivers, respectively (Ingersoll, 1990).

This paper discusses the modern petrofacies present in the shore zone of Lake Tahoe a large, alpine lake that occupies much of its associated drainage basin. Inasmuch as all local drainages and backshore areas capable of contributing sand were sampled, this study produced a first-order, actualistic petrofacies model for the Tahoe basin. Although relatively few siliciclastic lacustrine shoreline sequences have been preserved in the stratigraphic record, where applicable, the results of this study should constrain associated interpretations. Furthermore, the absence of swell and the virtual absence of tides in Lake Tahoe contribute to the development of a petrofacies model that contrasts markedly with more familiar littoral marine models.

DESCRIPTION OF THE TAHOE BASIN

Geography

Lake Tahoe lies between two north-northwest-trending mountain ranges: the 2,744-m-high Sierra Nevada on the west and the 3,317-m-high Carson Range on the east (Fig. 1) Lake Tahoe is an irregularly shaped oval about 35.2 km long and 19.2 km wide with an area of 495 km^2. Hutchinson (1957) lists the maximum depth as 501 m, which ranks Lake Tahoe as the tenth deepest lake are in the world. However, a more recent study by Hyne (1969) records the maximum depth as 494 m, which moves its rank to eleventh place. With a volume of 1.51×10^9 m^3, Lake Tahoe is volumetrically the largest alpine lake on the North American continent.

The drainage basin is nearly symmetrical about the lake. The entering streams are all less than 5 km long, except for the Upper Truckee River, which is located at the southern margin of the basin. The Tahoe basin occupies 1,311 km^2, and the surface area of the lake is 495 km^2, thus the entire area available for drainage into the lake is only 2.6 times the area of the lake itself. Approximately 40 percent of the annual inflow enters the lake directly as rain or snow.

The subaqueous part of the basin is generally a flat floor with steep sides. The northern, northwestern, and southern margins have relatively wide shelves, with slopes commonly less than 1° (Hyne and others, 1972). Rubicon Point, Sugar Pine Point, Stateline Point, and Deadman Point are characterized by steep, sublacustrine slopes, in places exceeding 40° (Court and others, 1972). The broad, flat lake bottom is interrupted by numerous spurs and abundant slump structures along the margins, as demonstrated by seismic profiling (Hyne and others, 1972). Profiling also shows a 400-m-thick sequence of layered sediment overlying the bedrock that forms the lake floor (Hyne, 1969; Court and others, 1972). The lake is Plio-Pleistocene and a sedimentation rate of 7.5 cm per 1,000 years can be inferred from seismic evidence. The rate of sedimentation increased rapidly with accelerated urbanization of the Lake Tahoe basin in the 1960s (Glancy, 1971).

Climatology

The basin is influenced by Pacific air masses moving eastward and mixing with continental air masses. However, the effects of local topography can strongly influence bas-

Quaternary Coasts of the United States: Marine and Lacustrine Systems, SEPM Special Publication No. 48
 ISBN 0-918985-98-6

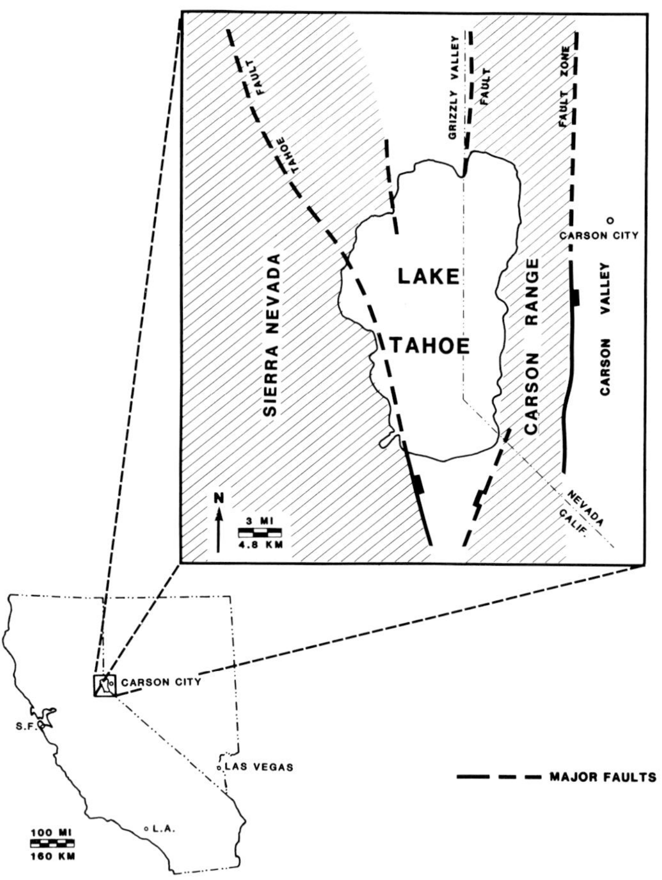

FIG. 1.—Map showing location of Lake Tahoe.

inal climatic variations. From 55 to 70 percent of the annual precipitation occurs from November through March, predominantly as snowfall. Summer precipitation is characterized by periodic afternoon thunderstorms. According to Crippen and Pavelka (1970); 85 percent of the sediment is delivered to Lake Tahoe during the April-May runoff period. The annual precipitation is greatest along the higher elevations of the western basinal margin and diminishes to a minimum near the east-central part of the lake. The western margin and the east-central lake receive approximately 127 cm and 38 cm of rainfall, respectively. The normal range of the snowpack is from 41 cm to 559 cm (Tahoe Regional Planning Agency, 1971).

Except for the gravity distribution of sediment, the principal mechanism for sediment transport is the force exerted by wave-induced currents. Wind speed and direction were measured continuously for 2.5 years at eight locations along the shore of Lake Tahoe (R. L. Moory, pers. commun., 1982). Data from these and other monitoring stations were used for the CALTECH 2-D Wind Model to develop wind-flow patterns across the lake (Mulberg, 1984). These data indicate that the principal wind direction is from the south and southwest. However, wind roses for each of the eight wind stations show wind from the northeast, east, and northwest contributing to bidirectional longshore sediment transport.

General Geology

Although traditionally included in the Sierra Nevada province, the Tahoe basin actually is part of the Basin and Range province. According to Birkeland (1963), the faulting that formed the Tahoe basin occurred less than 3 m.a. The major physiographic features around the Lake Tahoe basin formed during the late Pliocene to early Pleistocene (Birkeland, 1963). Tectonism and andesitic volcanism were the dominant geologic processes affecting the Tahoe basin during this period. Major deformation ceased just prior to the initiation of the younger of two pre-Wisconsin glaciations (Donner Lake; Birkeland, 1964). K-Ar age determinations of volcanics bounding the duration of active tectonism are consistent with the concept of Plio-Pleistocene deformation (Birkeland, 1963; Bateman and Wahrhaftig, 1966).

The lithologies in the Tahoe basin can be assigned to four major categories: metamorphic, acid plutonic, volcanic, and Quaternary sediment (Burnett, 1968). The subaerial distribution of these lithologies is shown in Figure 2. The metamorphic rocks generally occur as metasedimentary and

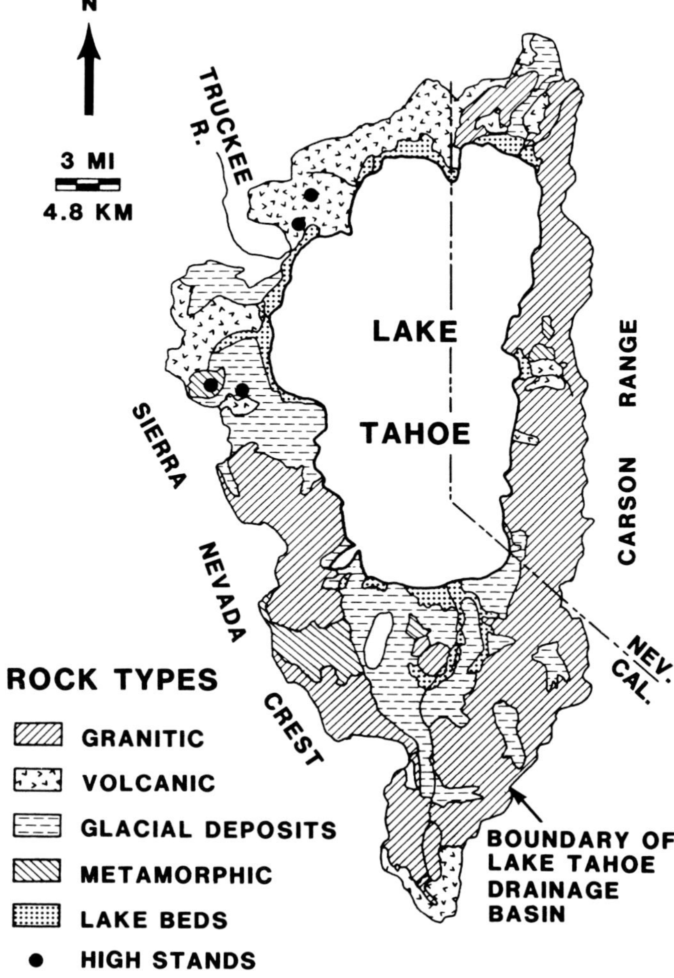

FIG. 2.—Generalized geologic map of the Lake Tahoe basin (after Hyne and others. 1972).

metavolcanic strata, which traditionally have been correlated with Paleozoic rocks of the eastern Sierra Nevada. The Paleozoic rocks are sparsely distributed throughout the range but have been described as a discontinuous belt extending from Bridgeport to Independence (Bateman and Wahrhaftig, 1966). Outcrops of metamorphosed Mesozoic strata occur in the southwestern part of the Tahoe basin. These rocks extend from Tahoe Mountain westward to Mt. Tallac (Blum, 1979) and have been described as roof pendants. These rocks have been metamorphosed to the hornblende-hornfels facies along their contact with younger acid plutonic intrusions. Granodiorite associated with the Sierra Nevada batholith is the most abundant bedrock exposed in the Tahoe basin. However, diorite, quartz diorite, and gabbro all occur as small bodies, usually adjacent to metamorphic rocks (Burnett, 1968). Bateman and Wahrhaftig (1966) explain the occurrence of the more mafic rock bodies as wall-rock contamination of the parent magma. K-Ar age determinations of biotites in the granodiorite yield dates of 106±my (Burnett, 1968). The northwestern part of the Tahoe basin is dominated by Pilocene basalt and latite volcanic lahars (Birkeland, 1963). These lahars presently dam the northern segment of Lake Tahoe and previously have blocked the Lower Truckee River, which resulted in elevated lake levels (Birkeland, 1963).

Pleistocene glaciation mostly affected the northern, western, and southern margins of the Tahoe basin. Only limited glaciation occurred along the Carson Range (Reid, 1911; Matthews, 1968), which may reflect a rain-shadow effect persisting since the Pleistocene (Reid, 1911; Hyne and others, 1972). Glaciation in the western part of the Tahoe basin is documented by thick and extensive morainal and outwash deposits. The northern and southern parts of the lake-basin proper show wide, gently sloping subaqueous shelves, which are the expression of large, outwash deltas formed from converging Pleistocene valley glaciers (Burnett, 1968; Hyne, 1969).

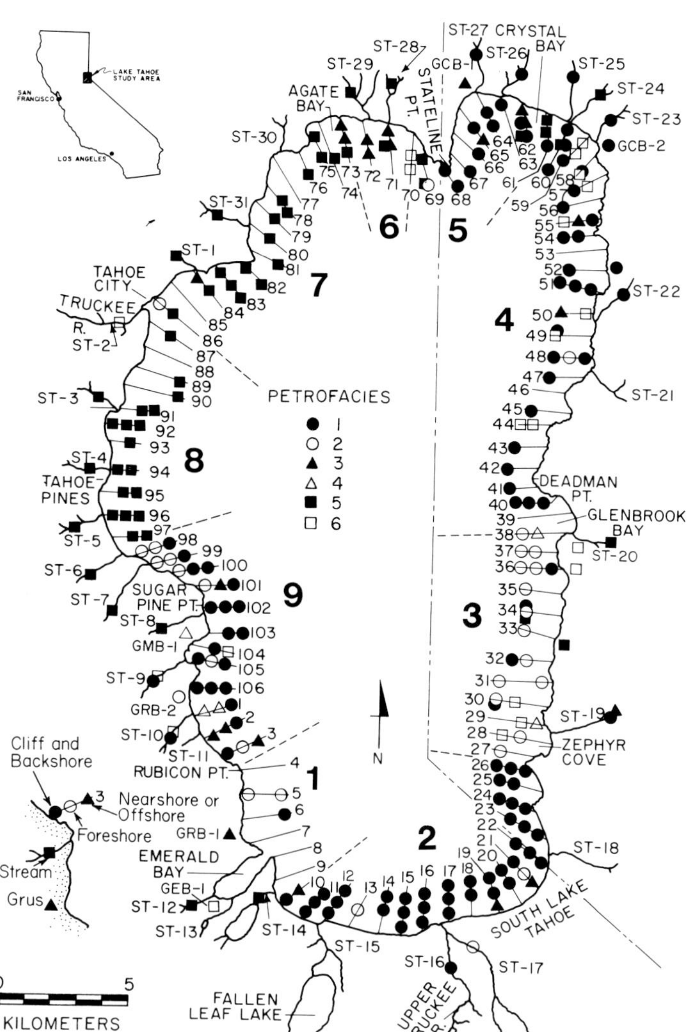

FIG. 3.—Map showing the location and petrofacies assignment of the sand samples used for modal analysis.

PROCEDURES AND RESULTS

Sampling

The analysis is based principally on a set of 318 lacustrine and 31 fluvial sediment samples collected during September 27–30, 1978. Backshore (above crest of associated berm), foreshore, and nearshore/offshore (from 6 [nearshore] to 9.2 [offshore] m below lake level) samples were collected at each of 106 stations spaced at approximately 1-km intervals around the perimeter of Lake Tahoe (Fig. 3). All stream samples were collected at least 3 m above lake level, which was at 1,897.72 m in September 1978 and at 1,897.24 m in October 1988. The stream, backshore, and foreshore samples were taken by inserting a tin can normal to stratification, thus obtaining a small core weighing about 700 g. The nearshore/offshore samples were obtained by dredging with a 7.6-*l* weighted bucket. An additional set of bluff and grus samples was collected in August 1982, and a final set of 25 foreshore samples was collected during October 21–22, 1988, to compare the areal consistency of the observed lacustrine petrofacies boundaries for a 10-year period.

Petrographic and Statistical Methods

The mineral composition of 0.35- to 0.50-mm (medium sand) fraction for 274 samples from the September 1978 and October 1988 sets was determined using the line method for grain thin sections (Galehouse, 1971). In petrofacies analysis, it is conventional to use a restricted grain-size interval to reduce the compositional variance arising from differences in grain size. Variation in grain size often reflects differences in mechanical-energy levels at the associated sampling stations. If mineralogic analyses were performed on the average grain-size fraction of each sample, resultant compositional differences would reflect both source-rock composition and mechanical-energy level at the sampling station. Such covariance would greatly complicate the interpretation of the resultant mineralogic data. The variables identified are listed in Table 1. The quartz-grain classification of Basu and others (1975) was employed. Inasmuch as a restricted grain size was used, the resultant values (Table 1) may be regarded as number percentages. At least

TABLE 1.—MEAN PERCENTAGE OF EACH LITHOLOGIC CONSTITUENT BY PETROFACIES. VALUES IN PARENTHESES INDICATE NUMBER OF SAMPLES ASSIGNED TO THAT PETROFACIES.

	Petrofacies					
Variable	1 (124)	2 (33)	3 (23)	4 (6)	5 (67)	6 (21)
Monocrystalline quartz						
Nonundulose quartz	1.4	1.4	2.5	0.3	1.2	0.5
Undulose quartz	32.2	25.1	28.1	4.8	12.8	19.9
Polycrystalline quartz						
2–3 crystals/grain	0.4	1.9	0.3	0.2	1.1	1.0
>3 crystals/grain	0.0	0.5	0.3	0.2	0.1	0.3
Unstable quartz	0.4	0.9	0.3	0.2	1.0	0.7
Plagioclase feldspar	30.0	29.4	26.0	8.6	17.8	29.1
Potassium feldspar	12.7	13.8	11.1	2.6	5.1	9.9
Opaque minerals	0.3	0.3	1.9	0.8	1.2	0.5
Pyroxene	0.4	0.0	0.5	0.2	3.8	0.1
Amphibole (non-hornblende)	0.1	0.1	0.2	0.0	0.2	0.0
Hornblende	2.3	2.6	2.0	0.8	0.6	1.8
Oxyhornblende	0.0	0.0	0.0	0.0	0.3	0.0
Biotite	3.2	3.0	9.9	65.4	1.9	13.5
Sphene	1.0	0.2	1.2	8.0	0.5	0.5
Volcanic-rock fragments	2. 5	1.7	6.6	2.2	47.9	7.3
Acid plutonic-rock fragments	12.1	17.0	8.4	1.5	3.6	12.6
Metamorphic-rock fragments	0.1	0.7	0.3	0.2	0.5	0.5
Sedimentary-rock fragments	0.0	0.2	0.1	3.6	0.3	1.1
Chert	0.0	0.0	0.0	0.3	0.3	0.1
Other accessory minerals	0.0	0.0	0.0	0.0	0.5	0.0
TOTAL	99.1	98.8	99.7	99.9	100.7	99.4

350 grains were identified per thin section; therefore, the true value of each constituent is within approximately ±5 percent of the obtained value at a 95 percent confidence level (Van der Plas and Tobi, 1965).

Q-mode factor analysis (Davis, 1986; Frane and Jennrich, 1977) was used to identify lithologic associations within an initial set of 140 randomly selected samples (Table 2 and Fig. 4). Factors 1 and 2 explain 45.6 and 31.8 percent of the total sample variance, respectively. The results shown in Figure 4 suggest the occurrence of six petrofacies. Stepwise discriminant function analysis (Davis, 1986; Jennrich and Sampson, 1977) was used for significance testing among the six potential petrofacies and to compute associated classificatory equations. Obtained values for the multivariate U-statistic or Wilks' Lamba, which tests for the equality of group means, range from 4.77 to 65.86 for each combination of group means. The critical value for the associated F-test (8 and 130 degrees of freedom) is approximately 2.66 for the 1 percent significance level. Therefore, the multivariate means for the lithologic associations shown in Figure 4 are statistically highly distinct, and the computed classificatory equations may be used to assign each of the

TABLE 2.—LOADINGS ≥±0.50 ORDERED ON FACTOR 1.

Variable	Factor 1	Factor 2
1. Volcanic-rock fragments	0.84	−0.15
2. Pyroxene	0.73	−0.18
3. Oxyhornblende	0.64	−0.11
4. Biotite	−0.03	0.81
5. Polycrystalline quartz (2–3)	−0.50	−0.11
6. Plagioclase feldspar	−0.62	−0.33
7. Undulose quartz	−0.62	0.08
8. Potassium feldspar	−0.65	−0.30
9. Acid plutonic-rock fragments	−0.72	−0.28

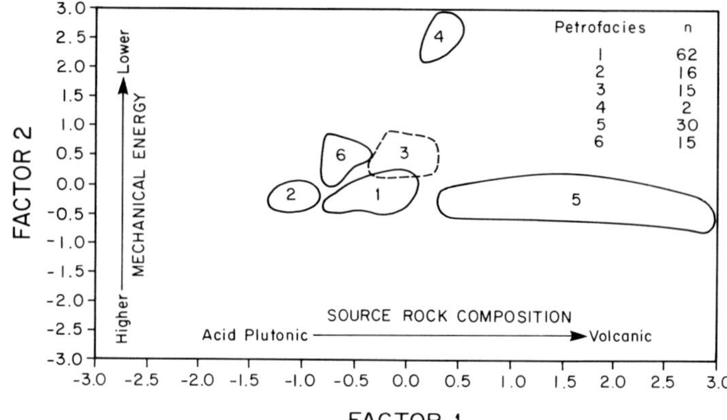

FIG. 4.—Plot of compositional fields derived from factor-score values, and interpretative assertions associated with factors 1 and 2.

134 samples processed during later stages of this study to its most likely petrofacies.

The mean values for each lithologic constituent in each of the six petrofacies are listed in Table 1, and the areal distribution of the samples assigned to each petrofacies is shown in Figure 3. The mean values for total quartz, total feldspar, and total rock fragments for petrofacies 1, 2, 3, 4, and 6 fall in the lithic arkose field of McBride (1963), whereas the mean of petrofacies 5 falls in the feldspathic litharenite field.

DISCUSSION OF RESULTS

Identification and Interpretation of Petrofacies

The loadings equal to or exceeding values of ±5.00 on factors 1 and 2 are listed in Table 2. For factor 1, the high positive values for volcanic-rock fragments, pyroxene, and oxyhornblende clearly reflect lithologic contributions from volcanic and metavolcanic rocks, whereas the high negative values for polycrystalline quartz (2–3 sub-units), plagioclase feldspar, undulose quartz, and acid plutonic-rock fragments reflect the contribution from granodioritic source rocks. Therefore, it is reasonable to interpret factor 1 as representing source-rock composition. Biotite is essentially independent of factor 1 but has a high positive loading on factor 2. The presence of abundant biotite in a sediment sample is suggestive of accumulation in relatively low mechanical-energy conditions (Neiheisel, 1965; Pomerancblum, 1966; Doyle and others, 1983); therefore, it is reasonable to interpret factor 2 as representing relative mechanical-energy levels at the depositional sites. Inasmuch as a small grain-size range (0.35 to 0.50 m) was used, it is surprising that differences in mechanical-energy level apparently account for nearly 32 percent of the total sample variance. Because a strong mechanical-energy signal may persist within a very restricted grain-size range, considerable caution must be exercised when interpreting mineralogic data for petrofacies identification.

Petrofacies 2, 5, and 4 are the compositional end members in Figure 4. As shown in Table 1, petrofacies 2 is enriched in acid plutonic-rock fragments, petrofacies 5 is highly-enriched in volcanic and/or metavolcanic-rock fragments, and petrofacies 4 is dominated by biotite. Petrofacies 1, 3, and 6 occupy intermediate field positions with respect to both source-rock composition and relative mechanical-energy level (Fig. 4 and Table 1). Although the lithologic composition of each of the defined petrofacies is stochastically distinct, petrofacies 1, 2, 3, 5, and 6 occupy a rather narrow band of factor score values (−0.74 to 0.97) with respect to factor 2. Therefore, the occurrence of samples assigned to petrofacies 3 or 6 may reflect only minor differences in mechanical-energy levels from those typical of petrofacies 1, 2, and 5 within or between different parts of the shore zone. Consequently, samples of petrofacies 3 and 6 may be closely associated with samples assigned to petrofacies 1 and 2, which also are derived predominantly from granodiorite.

Although the average wave-energy flux associated with a given shoreline segment may be computed, much of the shore zone at Lake Tahoe is characterized by variously sized pocket beaches wherein deviations from average mechanical-energy levels might be expected. Thus, the occurrence of numerous pocket beaches may contribute to the diversity of petrofacies within a given shoreline segment. Random processes generating sedimentologic "noise" also may play a greater role in sediment production and transport than is generally recognized. This, in turn, would also contribute to the diversity of petrofacies within a given shoreline segment. These considerations are reflected in the remainder of this section, which addresses the areal occurrences and associations of samples assigned to the six newly identified petrofacies.

Although petrofacies 2 contains more acid plutonic-rock fragments, it is compositionally quite similar to petrofacies 1, and both represent ultimate derivation from predominantly granodioritic source rocks (Table 1; Figs. 2 and 3). Samples assigned to petrofacies 1 occur most frequently, and they are present along the entire shore zone where backed by Sierran granodioritic rocks or sediment derived therefrom.

The samples assigned to petrofacies 2 are concentrated in the area from Zephyr Cove to Glenbrook Bay (stations 27 to 38) along the east shore and immediately north of Sugar Pine Point (stations 98 to 101) along the west shore of Lake Tahoe. The shore from Zephyr Cove to Glenbrook Bay consists of pocket beaches backed mostly by granodioritic bluffs and cliffs, whereas the area north of Sugar Pine Point is backed by older lake beds containing pebbles, cobbles, and boulders of granodiorite resedimented from associated moraines. The minor occurrence of petrofacies 2 along Rubicon Point (stations 3 and 5) is found in a topographic and geologic setting similar to that of Zephyr Cove to Glenbrook Bay.

Studies by Suttner and others (1981) and Grantham and Velbel (1988), among others, have contributed greatly to our understanding of the effect of climate on the lithologic composition of Holocene sand samples. Although it is tempting to attribute the abundance of granodioritic-rock fragments in petrofacies 2 to differences in effective precipitation and relief within the subbasins (Crippen and Pavelka, 1970) of the Lake Tahoe drainage basin, examination of Figure 3 shows that the lower reaches of streams 6, 7, and 8 in the Sugar Pine Point area are dominated by sediment assigned to petrofacies 5. Similarly, stream 19 entering Zephyr Cove contains samples assigned to petrofacies 1 and 3, neither of which is enriched in acid plutonic-rock fragments. Therefore, it is very difficult to argue for an upland source for the "excess" granodioritic-rock fragments occurring from Zephyr Cove to Glenbrook Bay along the east shore as well as in the area north of Sugar Pine Point. In both of these areas, the samples assigned to petrofacies 2 occur where there are granodioritic-bedrock exposures or unusual concentrations of granodioritic boulders derived from first-cycle erosion or resedimentation from morainal deposits and/or older lake beds at water depths as much as 6 m below lake level. The same can be said for the occurrences of samples assigned to petrofacies 2 along Rubicon Bay and Rubicon Point (stations 105, 3, and 5). It seems very likely that *in situ* weathering of these materials is responsible for the production of acid plutonic-rock fragments resulting in the enrichment typical of petrofacies 2. At this time, it is not known whether chemical or mechanical weathering of these nearshore materials is dominant in producing sand-size rock fragments; however, winter storms from the northwest, northeast, east, and southeast probably play an important role in the production and transport of such fragments. For example, near Tahoe Pines (station 95), winter-storm waves from the east and southeast transport pebbles and cobbles from water depths as great as 6 m below lake level and hurl them like artillery against and over an adjacent sea wall at Fleur du Lac Estates (Osborne and others, 1985). There can be little doubt that such activity here and at other localities along the Lake Tahoe shoreline results in the comminution of associated grains, thus contributing to the production of finer grained rock fragments largely by fracture.

The seemingly anomalous occurrence of a petrofacies 2 sample at Tahoe City (station 86) represents the slightly northward-transported remnant of a major beach-nourishment project completed at Tahoe Tavern (station 87) in the early 1900s. At that time, sand ultimately derived from the Santa Lucia Quartz Diorite (Combellick and Osborne, 1977) was imported by train from Monterey Bay and placed on the natural gravel beach (petrofacies 5) to enhance the enjoyment of the hotel's guests. Much of this sand was eroded by winter storms within a few years after placement and normally occurs as a ripple-marked sheet sand mostly offshore of its placement site. The occurrence of this sand at station 86 indicates that a small amount has been transported northward about 1 km during at least the last 65 years. Part of this sand sheet is now subaerially exposed due to recent drought conditions that reduced lake level to 1,897.26 m during September 1990.

The samples assigned to petrofacies 5, which reflect derivation predominantly from volcanic and/or metavolcanic rocks, are confined largely to the northwest quadrant of Lake Tahoe (stations 74 to 97). A variety of volcanic rocks including Tertiary andesite, pyroclastic and undifferentiated volcanic rocks, as well as Quaternary latite and undifferentiated volcanic rocks are exposed in the highlands bordering this part of the lake (Fig. 2). Therefore, the areal occurrence of samples assigned to petrofacies 5 along this part of the shore zone accurately reflects the geographic position of the associated source rocks.

Stream samples 6, 7, and 8 in the Sugar Pine Point area as well as stream samples 13 and 14 in the Emerald Bay area are assigned to petrofacies 5, but their volcanic affinities are not strongly reflected in the composition of associated shore-zone samples. Even though both the Sugar Pine Point and Emerald Bay areas are dominated by granodioritic bedrock, streams 6, 7, and 8 contain volcanic-rock fragments derived from relatively small exposures of undifferentiated Tertiary volcanic rocks exposed near their respective headwaters, and streams 12 and 14 contain metavolcanic-rock fragments from a rather large exposure west of Fallen Leaf Lake (Fig. 2) where these streams head. Although the composition of these stream samples faithfully reflects the source-rock composition, their sand contributions to the shore zone are sufficiently small or sufficiently diluted by grains derived from nearby rocks and sediment of acid plutonic affinities (Fig. 2) so that the strength of their volcanic signature is greatly reduced.

The occurrence of petrofacies 5 samples at stations 60, 61, and 62 reflects the presence of a sea wall containing andesitic and/or basaltic cobbles and small boulders. The fracturing of these volcanic clasts during winter storms has produced a sufficient number of sand-size volcanic-rock fragments to result in the backshore and foreshore samples at this site being assigned to petrofacies 5.

Petrofacies 4, which is dominated by biotite, is represented by only five samples, one of which is grus (GMB-1) at Sugar Pine Point. The sample at the southern end of Glenbrook Bay (station 38) was derived from granodiorite and undifferentiated Tertiary volcanic rocks, whereas the samples at Rubicon Bay (station 1) and Zephyr Point (station 29) were derived almost exclusively from granodiorite (Table 1; Fig. 2). Each of the lacustrine samples occurs in a sheltered portion of a pocket beach or re-entrant at or near the mouth of a stream. Thus, biotite supplied during one or more spring-runoff events remained at or near the mouths of the associated streams until at least late September when this sample set was collected. Biotite is highly and persistently concentrated at the southern end of Glenbrook Bay. This areal persistency may reflect not only the proximity to the mouth of stream-20 and relatively low mechanical energy but also the likely occurrence of a south-directed countereddy formed by the deflection of southerly winds by Deadman Point. All lacustrine samples assigned to petrofacies 4 have two factors in common: exclusive or predominant derivation from acid plutonic rocks and accumulation in environments characterized by relatively low mechanical energy.

Petrofacies 3 and 6 occupy intermediate positions in Figure 4 and thus represent samples enriched both in volcanic-rock fragments and biotite as compared to petrofacies 1 and 2. Although petrofacies 3 and 6 contain about 7 percent volcanic-rock fragments, examination of Table 1 and Figure 3 indicates that both petrofacies are derived predominantly from granodioritic source rocks. The ratio of mean percent acid plutonic/volcanic-rock fragments for petrofacies 6 and 3 is 1.7 and 0.8, respectively, which accounts for the position of these two petrofacies along factor 1 (Fig. 4). Although petrofacies 6 averages about 3.6 percent more biotite than petrofacies 3, both of these petrofacies occupy similar positions along factor 2.

The lacustrine samples assigned to petrofacies 3 are scattered from stations 2 to 84 (Fig. 3) and occur subequally in backshore, foreshore, and nearshore/offshore settings. The samples of petrofacies 3 from stations 71 to 73 are especially interesting because they are associated with the distal end of a longshore fining trend that extends from stations 70 to 73 (Osborne and others, 1985, their fig. 16). The occurrence of an eolian-dune field from stations 72 to 73 documents long-term, net longshore transport to the west in this area. Although bidirectional longshore transport does occur, synoptic winds from the south and southwest typically diverge at Stateline Point (station 68; Mulberg, 1984) and set up an associated westerly-directed longshore current system in this area. The average mechanical-energy levels at stations 71 to 73 are presumably slightly lower than those at stations 69 and 70, which would explain the enrichment of biotite in the samples assigned to petrofacies 3. On the other hand, lacustrine samples assigned to petrofacies 6 are concentrated somewhat along the eastern and northern shore zone from stations 28 to 59 but extend as far west as station 70. These samples occur in subequal numbers in foreshore and nearshore/offshore settings and occasionally occur in the backshore (stations 50 and 59). Samples assigned to both petrofacies 3 and 6 occur most frequently in small pocket beaches and shoreline re-entrants where sand ultimately derived from both acid plutonic and volcanic source rocks occurs and where the sediment is at least partially sheltered from associated wave energy, as reflected by the higher biotite content.

Areal Distribution of Petrofacies

The areal distribution of samples assigned to the six petrofacies may be used to identify nine shore-zone divisions (Figs. 2 and 3; Table 3). The position of the boundary between adjacent pairs of divisions permits monitoring of the direction and net rate of longshore sand transport. Divisions 1 and 3 are dominated by petrofacies 2 with large admixtures of petrofacies 1, whereas division 2 consists almost exclusively of petrofacies 1 (Table 3). Divisions 4 and 5 are dominated by petrofacies 1 with large admixtures of petrofacies 6 and 5, respectively. Division 6 consists of pe-

TABLE 3.—COMPOSITION OF SHORE-ZONE DIVISIONS BY PETROFACIES.

Shore-zone Division	Stations	Number of Samples	Frequency of Petrofacies (%)					
			1	2	3	4	5	6
1	04–10	03	33.3	66.7	0	0	0	0
2	11–26	44	88.6	4.6	6.8	0	0	0
3	27–38	23	21.7	56.5	0	8.7	0	13.1
4	39–60	38	65.8	2.6	5.3	0	2.6	23.7
5	61–70	22	50.0	4.6	13.6	0	22.7	9.1
6	71–73	08	0	0	75.0	0	25.0	0
7	74–85	16	0	0	6.2	0	93.8	0
8	86–97	20	0	5.0	0	0	95.0	0
9	98–03	34	55.9	23.5	11.8	5.9	0	2.9

trofacies 3 with an admixture of petrofacies 5. The dominance of petrofacies 3 between stations 71 and 73 reflects a predominantly granodioritic-source terrain and accumulation in partially sheltered beaches. Divisions 7 and 8 consist almost exclusively of samples assigned to petrofacies 5. The division indicated at station 86 (Fig. 3; Table 3) is based on the preserved remnant of the Monterey Bay sand body placed at station 87 and partially transported to station 86. The areal distribution of this sand body is very important in understanding the long-term, net shore-parallel and shore-normal transport vectors in the area (Osborne and others, 1985) and for this reason as well as its mineralogic composition serves to define a shore-zone boundary. Division 9 is dominated by petrofacies 1 with subordinate admixtures of mostly petrofacies 2 and 3. This composition contrasts sharply with that of division 1, which is dominated by petrofacies 2.

A set of 25 foreshore samples was collected during October 1988 to test the areal consistency of the petrofacies boundaries based on the September 1978 sample set. For this 10-year period, the boundaries for seven of the nine petrofacies remained unchanged, given the 1 km sampling interval employed. At station 73, the foreshore sample changed from petrofacies 3 in 1978 to 5 in 1988. The change may reflect longshore transport from the west and/or storm-related onshore transport of sediment assigned to petrofacies 5 at station 73. The foreshore sample at station 97 changed from petrofacies 5 in 1978 to a volcanic rock fragment-enriched petrofacies 1 in 1988, and the foreshore sample at station 98 changed from petrofacies 2 to 5. The change at station 97 must reflect the northwest-directed longshore transport of granodioritic sediment derived from Sugar Pine Point. The change at station 98 may reflect the downcoast transport of sediment assigned to petrofacies 5 during winter storms and/or the addition of petrofacies 5 sediment from streams 6 and 7 with minor longshore transport to this station.

Although the foreshore samples associated with two of the nine boundaries showed mineralogic changes from 1978 to 1988, the composition of the shorezone system as a whole appears to have been remarkably stable during this ten-year period. This condition is due to the stability of the upland drainage systems and of the wind-driven current systems operative in Lake Tahoe. Of particular importance is the fact that these data indicate little net longshore transport during this period.

CONCLUSIONS

1. Petrographic modal analysis demonstrates that source-rock lithology primarily controls the composition of the shore-zone samples at Lake Tahoe.
2. Although subordinate to source-rock lithology, relative mechanical energy at the depositional site is also an important factor in determining the composition of shore-zone samples: Therefore, caution must be exercised when using mineralogic data even for a restricted grain-size interval for petrofacies identification. Small differences in mechanical energy within the numerous pocket beaches at Lake Tahoe result in the close areal association of lower energy petrofacies 3 and 6 with slightly higher energy petrofacies 1 and 2.
3. Six shore-zone petrofacies reflecting both source-rock composition and mechanical-energy level were identified using Q-mode factor analysis and stepwise discriminant function analysis. Mean values for petrofacies 1, 2, 3, 4, and 6 are lithic arkose, whereas the mean value for petrofacies 5 is feldspathic litharenite.
4. Nine shore-zone divisions may be identified at Lake Tahoe based on the areal distribution of the defined petrofacies.
5. The foreshore samples associated with only two of the nine divisional boundaries showed compositional change from September 1978 to October 1988. Therefore, the shore-zone sedimentologic system as a whole appears to have experienced little change, particularly as a result of longshore transport during this 10-year period.

ACKNOWLEDGMENTS

The author thanks Randall L. Moory for his thoughtful administrative assistance and scientific input throughout the performance period of California State Lands Commission Contract No. C-8063, USC 53-4831-9023. The following individuals associated with the Sedimentary Petrology Laboratory at the University of Southern California contributed technical expertise essential to the completion of this study: Michael Edelman, James Waldron, Jane Mason, Sonja Saruba, and Pamela Vande Water. The manuscript was typed by Susan Turnbow and Desser Motion, and all line drawings were drafted by Janet Dodds. The author thanks Jinyou Liu and Chia-Chen Yeh for reviewing an early draft of this manuscript. Gary B. Griggs and Randall L. Moory reviewed the final draft.

REFERENCES

BASU, A., YOUNG, S. W., SUTTNER, L. J., JAMES, W. C., AND MACK, G. H., 1975, Re-evaluation of the use of undulatory extinction and polycrystallinity in detrital quartz for provenance interpretation: Journal of Sedimentary Petrology, v. 45, p. 875–882.

BATEMAN, P. E., AND WAHRHAFTIG, C., 1986, Geology of the Sierra Nevada, *in* Bailey, E. H., ed., Geology of Northern California: California Division of Mines and Geology Bulletin 190, p. 107–172.

BIRKELAND, P. W., 1963, Pleistocene volcanics and deformation of the Truckee area, north of Lake Tahoe, California: Geological Society of America Bulletin, v. 74, p. 1453–1464.

BIRKELAND, P. W., 1964, Pleistocene glaciation of the Northern Sierra Nevada, north of Lake Tahoe, California: Journal of Geology, v. 72, p. 810–823.

BLUM, J. L., 1979, Geologic and gravimetric investigations of the South Lake Tahoe groundwater basin, California: Unpublished M.S. Thesis, University of California, Davis, 96 p.

BURNETT, J. L., 1968, Geology of the Lake Tahoe basin, *in* Evans, J. R., and Matthews, R. A., eds., Geological Studies in the Lake Tahoe Area, California and Nevada: Sacramento, Annual Field Trip Guidebook, Geological Society of Sacramento, p. 1–13.

COMBELLICK, R. A., AND OSBORNE, R. H., 1977, Sources and petrology of beach sand from southern Monterey Bay, California: Journal of Sedimentary Petrology, v. 47, p. 891–907.

COURT, J. E., GOLDMAN, C. R., AND HYNE, N. J., 1972, Surface sediments in Lake Tahoe, California–Nevada: Journal of Sedimentary Petrology, v. 42, p. 359–377.

CRIPPEN, J. R., AND PAVELKA, B. R., 1970, The Lake Tahoe basin, California–Nevada: U.S. Geological Survey Water-Supply Paper 1972, 56 p.

DAVIS, J. C., 1986, Statistics and data analysis in geology: Wiley, New York, 646 p.

DICKINSON, W. R., 1985, Interpreting provenance relations from detrital modes of sandstones, *in* Zuffa, G. G., ed., Provenance of Arenites: D. Reidel, Boston, p. 333–361.

DICKINSON, W. R., AND SUCZEK, C. A., 1979, Plate tectonics and sandstone compositions: American Association of Petroleum Geologists Bulletin, v. 63, p. 2164–2182.

DOYLE, L. J., GARDNER, K. L., AND STEWARD, R. G., 1983, The hydraulic equivalence of mica: Journal of Sedimentary Petrology, v. 53, p. 643–648.

FRANE, J., AND JENNRICH, R., 1977, P4M Factor analysis, *in* Dixon, W. J., and Brown, M. B., eds., BMDP-77 Biomedical Computer Programs P-Series: University of California Press, Berkeley, p. 656–684.

GALEHOUSE, J. S., 1971, Point counting, *in* Carver, R. E., ed., Procedures in Sedimentary Petrology: Wiley-Interscience, New York, p. 385–407.

GLANCEY, P. A., 1971, A reconnaissance of streamflow and fluvial sediment transport, Incline Village area, Lake Tahoe, Nevada: State of Nevada, Department of Conservation and Natural Resources, Division of Water Resources, Water Resources Information Series Report Number 8, 28 p.

GRANTHAM, J. H., AND VELBEL, M. A., 1988, The influence of climate and topography on rock-fragment abundance in modern fluvial sands of the southern Blue Ridge Mountains, North Carolina: Journal of Sedimentary Petrology, v. 58, p. 219–227.

HUTCHINSON, G. E., 1957, A treatise of limnology, *in* Geography, Physics and Chemistry, v. 1: Wiley, New York, 1,015 p.

HYNE, N. J., 1969, Sedimentology and Pleistocene history of Lake Tahoe, California–Nevada: Unpublished Ph.D. Dissertation, University of Southern California, Los Angeles, 121 p.

HYNE, N. J., CHELMINSKI, P., GORSLINE, D. S., AND GOLDMAN, C. R., 1972, Quaternary history of Lake Tahoe, California–Nevada: Geological Society of America Bulletin, v. 83, p. 1435–1448.

INGERSOLL, R. V., 1990, Actualistic sandstone petrofacies: Discriminating modern and ancient source rocks: Geology, v. 18, p. 733–736.

JENNRICH, R., AND SAMPSON, P., 1977, P7M Stepwise discriminant analysis, *in* Dixon, W. J., and Brown, M. B., eds., BMPD-77 Biomedical Computer Programs P-Series: University of California Press, Berkeley, p. 711–753.

MATTHEWS, R. S., 1968, Geological hazards of Lake Tahoe basin area, *in* Evans, J. R., and Matthews, R. A., eds., Geological Studies in the Lake Tahoe Area, California and Nevada: Sacramento, Annual Field Trip Guidebook, Geological Society of Sacramento, p. 14–26.

MCBRIDE, E. F., 1963, A classification of common sandstones: Journal of Sedimentary Petrology, v. 33, p. 664–669.

MULBERG, E. J., 1984, Lake Tahoe seasonal wind study: California Air Resources Board, Sacramento, 19 p.

NEIHEISEL, J., 1965, Source and distribution of sediments at Brunswick Harbor and vicinity, Georgia: U.S. Army Coastal Engineering Research Center Technical Memorandum 12, 21 p.

OSBORNE, R. H., EDELMAN, M. C., GAYNOR, J. M., AND WALDRON, J. M., 1985, Structure of the littoral zone in Lake Tahoe, California–Nevada: California State Lands Commission, Sacramento, 85 p.

POMERANCBLUM, M., 1966, The distribution of heavy minerals and their hydraulic equivalents in sediments of the Mediterranean continental shelf off Israel: Journal of Sedimentary Petrology, v. 36, p. 162–174.

REID, J. A., 1911, The geomorphology of the Sierra Nevada northeast of Lake Tahoe: University of California Publications in Geology, Berkeley, v. 6, p. 89–161.

SUTTNER, L. J., BASU, A., AND MACK, G. H., 1981, Climate and the origin of quartz arenites: Journal of Sedimentary Petrology, v. 51, p. 1235–1246.

TAHOE REGIONAL PLANNING AGENCY, AND U.S. DEPT. OF AGRICULTURE, FOREST SERVICE, 1971, A guide for planning: South Lake Tahoe, California, 21 p.

VAN DER PLAS, L., AND TOBI, A. C., 1965, A chart for judging the reliability of point counting results: American Journal of Science, v. 263, p. 87–90.

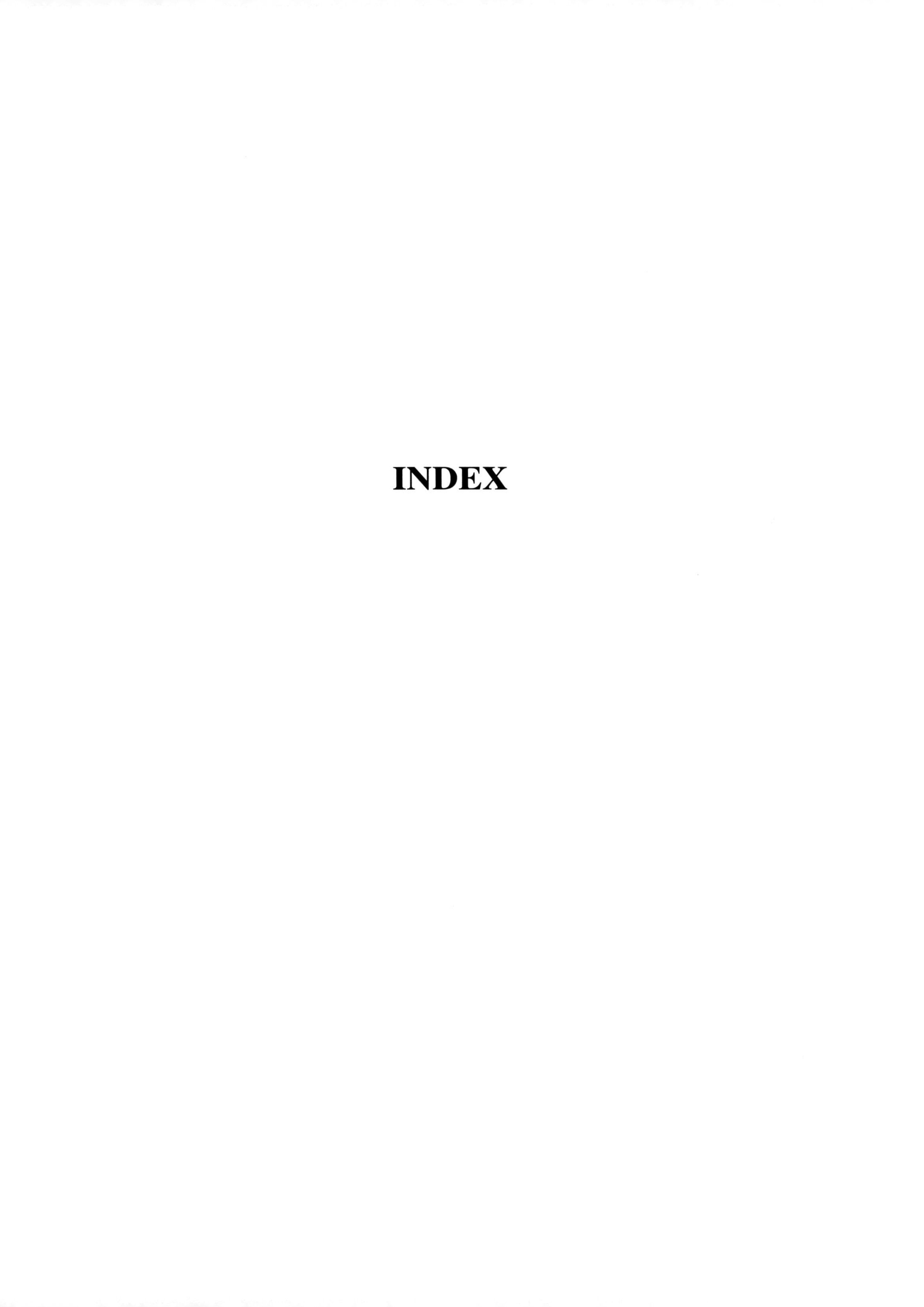

INDEX

INDEX

(citation page is first page of chapter in which citation occurs)